British Steam Locomotive Builders

Dedicated to my wife

First published in Great Britain in 1975 by TEE Publishing
Reprinted in this format in 2014 by
Pen & Sword Transport
An imprint of
Pen & Sword Books Ltd
47 Church Street
Barnsley, South Yorkshire
S70 2AS

ISBN 978 1 47382 289 4

A CIP catalogue record for this book is
available from the British Library

Printed and bound in England
By CPI Group (UK) Ltd, Croydon, CR0 4YY

Pen & Sword Books Ltd incorporates the Imprints of Aviation, Atlas,
Family History, Fiction, Maritime, Military, Discovery, Politics, History,
Archaeology, Select, Wharncliffe Local History, Wharncliffe True Crime,
Military Classics, Wharncliffe Transport, Leo Cooper, The Praetorian Press,
Remember When, Seaforth Publishing and Frontline Publishing

For a complete list of Pen & Sword titles please contact
PEN & SWORD BOOKS LIMITED
47 Church Street, Barnsley, South Yorkshire, S70 2AS, England
E-mail: enquiries@pen-and-sword.co.uk
Website: www.pen-and-sword.co.uk

British Steam Locomotive Builders

James W. Lowe

CONTENTS

PREFACE

James W. Lowe

It will be appreciated that there has been a multitude of articles, notes and a few books on locomotive builders ever since the first steam locomotive made its appearance. The information is so scattered that it is extremely difficult to know exactly where to find it. Some builders have been well served, but mainly the well known establishments of Swindon, Crewe, Derby, Doncaster, etc., building for their own lines, and also the big Glasgow firms, Hunslet and a few others.

The search for information of the lesser known firms has been both fascinating and frustrating but the most difficult part has been the investigation of firms which were supposed to have built one or more steam locomotives, the majority of which turned out to be agents for the more well known producers.

Conversely, in at least two cases it has only been by persistent correspondence that firms, having initially denied all knowledge of building locomotives, have had to admit that they were wrong and have even produced photographs of their creations in the end.

The information on the early builders already written could not be checked or amplified and therefore the reader will appreciate that repetition of this information was necessary and had to be accepted as correct.

To make the book as complete as possible, every known firm that produced one or more steam locomotives of 2 ft gauge and over has been included with a few exceptions.

No claim is made that the list of builders is exhaustive and there is no doubt that fresh evidence and claims will be forthcoming of other manufacturers. I think a lot more interesting information and data could be forthcoming, but either the owners of such knowledge are unaware of these efforts, or do not wish to part with such details for some peculiar reason best known to themselves.

The assistance I have received has been most generously given both by firms and individuals which has made the task pleasant, interesting and the means of making many new friends.

Strangely enough, the most interested and fervent followers of the locomotive are and were doctors, lawyers and clergymen, particularly the latter who have produced some excellent photographers, authors of railway literature, modellers and at least one with a 2 ft gauge railway complete with locomotives and rolling stock from nearby ironstone mines in his rectory garden.

For myself, railways were always in my blood, and what should be more natural than to become a premium apprentice on the Great Western Railway, first in 1928 at the Stafford Road Works, Wolverhampton, and then in 1931 to the Mecca of locomotive building, Swindon – that was when it was a name to be reckoned with, even if you were not partial to their products. Now, alas, due to nationalisation, control by men from other regions, centralisation, and what you will, the glory has departed never to return but, having been privileged to assist in building locomotives in those halcyon days, the memories crowd back and they at least cannot be scrapped.

The reader will find that in some sections an attempt has been made to compile a list of locomotives built; many are alas incomplete but it is hoped these may be augmented in the future.

In the 'customer' column, the *first* customer has been shown although in some cases this must be read as first *known* customer, and no attempt has been made to indicate their subsequent whereabouts and fortunes. As far as industrial locomotives are concerned, this aspect is excellently covered by the Industrial Locomotive Society's bulletins and pocket books. Cylinders and driving wheel sizes are given where known and, although very inadequate to give the full impression of the size of each locomotive, they are important. Other dimensions are as important or more so and it has been said by more than one pundit that the only dimension of any interest was the diameter of the blast pipe. This dimension up to recently was not widely publicised, particularly by the railway workshops.

Regarding the tables of dimensions of various classes of locomotives, the original published dimensions varied from journal to journal and even 'official' dimensions could not always be taken as gospel. In early days, variations in tube heating surfaces were due to the external tube surface being taken in some instances, and internal tube surfaces in other cases, in other words the water and fire surfaces respectively.

Many locomotive lists are available from the various societies, particularly the Stephenson Locomotive Society, and the well known makers' lists have been omitted for this reason and for the amount of space required.

Works numbers and lists have been much maligned, especially where they are unofficial, but where certain works numbers have been verified they can be of great value to the historian in attempting to trace the sources of missing locomotives.

Finally it is hoped that my efforts in assembling this multitude of builders will be of interest and of some use to those who have associated themselves with the steam locomotive either professionally or otherwise.

FOREWORD

Mr. James W. Lowe, has produced for all of us, professional or amateur, historian or railway enthusiast, a truly magnificent reference book on the steam locomotive from its inception to its demise. One hundred and fifty years of endeavour and with more information and detail in a single volume, covering the whole of that exciting period, than anyone could have thought possible.

When Mr. Lowe asked me to write a foreword for his book I was delighted to accept and felt particularly honoured as being perhaps one of the last of the steam locomotive manufacturers. I well remember as a young man the Liverpool and Manchester Railway Centenary in 1930 and on behalf of The Hunslet Engine Company, I was showing at Liverpool a 2-6-2 locomotive, Hunslet No. 1671 prior to its despatch for Tanganyika. Now in 1975 with the 150th Anniversary of Passenger Railways with us, Mr. Lowe presents us with a very full and detailed record of those momentous years.

The amount of information which Mr. Lowe has gathered together in a single volume setting out the abilities and capabilities of over 400 steam locomotive designers and manufacturers in the United Kingdom is unbelievable and I cannot imagine that any of us, no matter how steeped in the history of steam locomotives, can have had any idea that so many companies could ever have been involved in this specialised work, and this of course as main contractors. When I entered the industry over 50 years ago, the total list of manufacturers at that time was under 20, and indeed, very similar to the list shown on page 9 and during the last 50 years I have virtually seen them practically all disappear, but this of course does not only apply to the United Kingdom. The steam locomotive has had its day and what finer epitaph could it have than the publication of this magnificent volume.

Another recollection which comes to mind and particularly as Mr. Lowe refers to Baldwins in Philadelphia who built 1,000 locomotives within a year around the turn of the century, is a visit which I paid to Baldwins in 1932 which was of course during the big slump and their fantastic works were completely closed down and the only staff retained was seven draughtsmen in the drawing office. Undoubtedly the industry has had its ups and downs but happily Mr. Lowe does not allow us to forget its achievements. No wonder that Glasgow was proud of the North British Locomotive Company when in 1902 those three great locomotive building firms of Neilson Reid, Dübs and Sharp Stewart became a single entity.

Finally, we must not forget another achievement which Mr. Lowe has placed before us in this great volume, that is the bringing together of industry and the railways so that we have before us a complete work of steam locomotive achievement by all of Britain's locomotive engineers which includes those great railway men of which we are all so proud who worked exclusively for superiority of railways in Britain from Stephenson and Brunel through to Gresley and Stanier, as well as those other great designers and manufacturers producing over the years all types and sizes of steam locomotives for railways in other parts of the world, nor must we forget still others who have produced the more humble shunting locomotive built for both home and export with the results of all listed by Mr. Lowe and brought together in a most harmonious way which from every point of view, is most right and proper.

One must not assume however, that feelings between the railways and the manufacturers were always so harmonious. In 1875 the manufacturers took the view that the railways were paying more attention to the building of locomotives than to the running of their railways, so they joined together to form the Locomotive Manufacturers' Association and took out an injunction against the railways to prevent them extending their building interests beyond their own requirements, and this veto has lasted very nearly 100 years because it is only within the last year or two that the Government have rescinded the veto and encouraged railway workshops to expand their interests.

However, in this year of 1975, British Railways and the locomotive manufacturers together with all the Institutions come together to celebrate the 150th Anniversary of Passenger Railways and Mr. Lowe's very comprehensive book could not have arrived on the scene at a more propitious moment. I am sure that we all wish him the success which he most assuredly deserves.

June, 1975

John Alcock, O.B.E., M.A., C.Eng., F.I.Mech.E.
Chairman of Hunslet Holdings Ltd. and the Hunslet Engine Co. Ltd.

INTRODUCTION

Some notes on the development of locomotive building

The story of the steam railway locomotive goes back to the beginning of the 19th century. Great Britain was at war with France and Spain, Napoleon Bonaparte wreaking havoc in Europe, and Nelson gaining immortality at Trafalgar. The event at home was to have a greater impact on society, although not realised at the time, and more influence on promoting and enlarging industry than any other invention for the good of mankind. Its development was continuous and uninterrupted. The locomotive had an 'aura' of its own, it excited more interest than any other mechanical device because it was alive, it breathed fire and water and was admired by the technical and non-technical. Consider the difficulties and handicaps of the early locomotive builders. At first their workshops were ill-equipped, consisting of a smithy, forge, drilling machine and in most cases treadle lathes. The smith was the key figure, he bent the plates, rivetted and forged the iron work. They had no milling, slotting, planing, grinding or shaping machines. Machine tools were slowly developing and it was not until the 1850s that the locomotive could be said to have been machined to any degree of accuracy or finish. Yet when one examines the early products which have been preserved, one marvels at the standard of workmanship of those early days. It will be seen that the first locomotives were built by or for colliery owners but it was as early as 1759 when the power of steam was first utilised for propelling road carriages, the idea was taken up by William Murdoch and the 1784 trials of a three wheeled carriage are well known. The logical step forward was taken by Richard Trevithick, and it is he that we must honour more than anyone else for putting the steam engine on rails.

The main stumbling block to the development of the locomotive in the initial stages was the weakness of the track and although wrought iron rails were introduced in the Newcastle area during the time of the birth of the locomotive it was not until 1820 that wrought iron rails of lengths up to 15 feet were introduced with 'fish-belly' form which in turn was gradually superseded by parallel rail sections.

The weakness of the early track, with frequent fractures and distortions became so bad in some instances, that for a time the haulier reverted to horses.

An important and interesting facet of the contracting builders was the export market which from the start built up to one of the most fruitful sources of foreign trade.

For many years the makers built to their own designs with their attendant peculiarities and characteristics.

In the late 19th century and onwards the designs, specifications and inspection of locomotives, particularly those ordered through Crown Agents and other engineering consultants, was taken out of the hands of the locomotive manufacturers. This resulted in objections by some firms that they were tied down by unreasonable inspection of materials and finished products, whereas American builders were not subjected to the same stringent demands.

There was a great demand from India and South Africa, in excess of the capacity of British firms especially in the 1890s and 1900s. This of course let in foreign builders particularly America and Germany. Besides lack of capacity and its relationship with delivery, prices of locomotives produced in this country were high. An example of competition in the Indian market shows what home manufacturers were up against: a quotation for 10 locomotives for the Assam-Bengal railway and 40 for the East Indian railway resulted in the order being placed with a Hanover firm who quoted 25% less than the lowest price in Great Britain and a delivery in half the time. So a vital contract was lost and a trade connection established by the railways with a foreign firm who of course got further orders to the detriment of our own locomotive contractors.

Another sore point with British contractors was the decline of the home market and the equipping of railway workshops to deal with their own requirements. The London and North Western and Great Western railways in particular, built nearly all their own locomotives – whether they built them at less cost is a debatable point but as both of these railways

VALUE OF LOCOMOTIVES EXPORTED 1899–1901

	1899 £	1900 £	1901 £
South America	182,689	228,787	270,126
South Africa	35,554	80,064	281,158
Australasia	98,888	206,298	350,328
North America	3,017	909	3,630
Spain	30,285	37,972	53,721
Other European Countries	175,297	205,348	219,209
Total Value of Locomotives Exported 1899–1901			
	1,467,389	1,496,849	1,949,910
Value of Locomotives Exported During December			
South America	7,711	8,756	31,303
Brit. South Africa	14,457	4,318	64,039
Brit. India	88,356	38,714	42,756
Australasia	792	20,826	12,097

went in for effective standardisation, they probably did.

The lack of capacity around the end of the nineteenth century compelled some of the home railways, whose workshops had never been capable of building all the locomotives required, to order locomotives from America. The Midland Railway at this time had some two hundred locomotives on order with British contractors, one order for twenty placed in December 1898 could not be fulfilled for at least fifteen months, but an order placed in America for twenty 2-6-0s was completed in four months from date of order. The Great Northern, Great Central, Barry and Port Talbot railways all obtained American built locomotives in the same period.

This was pleasing to the American builders, their Norris 4-2-0s had been sent over in 1840 to show us how to climb the Lickey bank – and now it looked as if the market was re-opening, but the door did not open very wide for British imports but our export markets were permanently affected from that date.

The table below is based on the larger locomotive manufacturing firms' activities at the turn of the century and their capacities:–

	Men	Locos annum
Neilson Reid & Co	3500	300
Dubs & Co	2500	160
Sharp Stewart & Co	2000	150
Beyer Peacock	2000	140
Kitson & Co	1500	120
Vulcan Foundry	1200	110
Robert Stephenson & Co	700	60
Manning Wardle	500	40
Nasmyth Wilson	700	40
Hunslet Engine Co	300	20

So with a total of 14,700 workmen a total of 1140 locomotives were turned out per annum and taking the smaller firms into consideration the total rises to 2000 including the 400 to 500 produced in the railway shops. At this time Denmark and Holland bought nearly all their requirements in this country and markets were well established in most corners of the globe.

Comparing the above outputs with the Baldwin Locomotive Works gives an insight into the rapid growth of the latter.

In September 1899 the 17,000th locomotive was completed and eleven months later the 18,000th.

One should be cautious with statistics and whilst undoubtedly our own builders built locomotives to last with first class workmanship and materials, the Americans had a different outlook and technique in turning out machines which were not intended to last fifty years or more.

Exports up to 1920 were handicapped by the fact that locomotives had to be dismantled on the dock side and re-erected on arrival at their destinations which increased the cost. This state of affairs dramatically changed when Christian Smith introduced their Belship fleet specially equipped for taking complete units and the docks were equipped to deal with the increased weights.

History repeated itself in the 1950s and such traditional markets as India were gradually lost by intense competition from Japanese and Continental builders. In 1955 the British locomotive industry received no share in orders placed for over 400 locomotives by the Indian Railway Board, Japan, Germany, Hungary and Czechoslavakia receiving the bulk of the contracts. Other unsuccessful tenders came from

Italy, France, Belgium and the United States. As in many other contracting firms, the price governed the placing of orders, irrespective of previous quality, performance and service and although British prices were as keen as economics permitted and delivery satisfactory, the successful tenders comparatively speaking were unrealistic and for some foreign firms it was a question of getting a foot in the door with further orders in view and a chance to recoup their loss of profits. As a result of the escalation of wages and material prices, competition became impossible for firms which had to have a large order book to keep the wheels turning, and the inevitable closures took place.

Other industries are being affected in the same way, the car industry is a classic example, and unless productivity can be increased in proportion to wage demands the result will be the same. There is very little sentiment left in business today.

The basic steam locomotive has not changed at all. Various refinements have been added from time to time the more important being the change from feed pumps to injectors, the design of valve gears, superheating, front end design and draughting improvements. From the building point of view problems occurred as the size and weight increased in moving material, wheeling the locomotives – changing from the old shear legs to the powerful travelling cranes in the erecting shops, and the necessity for improved and larger capacity machines and more special purpose machine tools.

The total number of steam locomotives built by all builders cannot be accurately assessed but the approximate number would be 138,150.

The cost of locomotives varied considerably; an order for fifty would be cheaper per locomotive than an order for one or two. Price varied also according to the amount of ferrous and non-ferrous materials used, and an accurate comparison becomes very involved.

A 2-2-2 express locomotive built in the 1840s averaged £1750 with some builders quoting as low as £1000–£1200. Undercutting was and is a fatal pastime and many early builders fell by the wayside due to this fault.

Prices per ton for early locomotives were Robert Stephenson's 0-4-0, £78, G.N.R. Stirling 'Single' of 1877 £31, L.B.S.C. *Gladstone* £51.

Compare these figures with British Railways' *Britannia* 4-6-2 £160 to £210, *Class* 9 2-10-0 £200 to £280. More examples would be required to make a satisfactory review of prices. However to compare a *Britannia's* average cost of approximately £20,000 with the average cost of a *Deltic* diesel locomotive at £225,000 (including maker's profit) makes one think, even when availability and maintenance

OUTPUT OF STEAM LOCOMOTIVES BY MAIN CONTRACTORS

Builder	Period of Building	Number of Years	Output		
			Total	Max Annual Output	Average Annual Output
North British Locomotive Co.	1903–1958	56	11,318	577	202·1
Dubs & Co.	1865–1902	38	4 485	174	118·0
Neilson Reid & Co	1838–1902	65	5 864	235	90·2
Armstrong Whitworth & Co Ltd	1919–1937	19	1 455	150	76·5
Beyer Peacock & Co Ltd	1855–1958	104	7 761	134	74·6
Sharp Stewart & Co	1833–1902	70	5 088	126	72·6
Kitson & Co	1838–1938	101	5 405	112	53·5
Vulcan Foundry Co Ltd	1833–1956	124	6 210	186	50·0
R Stephenson & Hawthorn	1937–1959	23	*c.* 1 000	47	43·4
Kerr Stuart & Co Ltd	1892–1930	39	*c.* 1 500	78	38·4
R. Stephenson & Co Ltd	1825–1936	112	*c.* 4 190		37·4
W. Beardmore & Co Ltd	1920–1931	12	393	74	32·7
Manning Wardle & Co	1859–1926	68	2 004	67	29·4
Hawthorn Leslie & Co Ltd	1831–1938	108	2 611	68	24·1
W.G. Bagnall Ltd	1876–1957	81	*c.* 1 660	40	20·4
Hunslet Engine Co Ltd (a)	1865–1964	100	2 235	58	22·3
Yorkshire Engine Co	1866–1955	90	*c.* 800	36	22·2
Avonside Engine Co Ltd	1841–1935	95	1 960	85	20·6
A. Barclay Sons & Co Ltd	1859–1962	104	2 053	48	19·7
Hudswell Clarke & Co Ltd	1861–1961	101	1 807	73	18·0
Nasmyth Wilson & Co	1839–1938	100	1 531	125	15·3

(a) One more built in 1971

OUTPUT OF STEAM LOCOMOTIVES BY MAIN RAILWAY WORKSHOPS

Workshops	Period of Building	Number of Years	Output			Principal Railways
			Total	Max Annual Output	Average Annual Output	
Crewe	1845–1958	114	7331	149	64·3	London & North Western
Swindon	1846–1960	115	5964	155	51·8	Great Western
Derby	1851–1957	107	2995	82	28·0	Midland
Horwich	1889–1957	69	1840	77	26·6	Lancashire & Yorkshire
Doncaster	1867–1957	91	2224	60	24·4	Great Northern
Darlington	1864–1957	94	2269	43	24·1	North Eastern
Stratford	1851–1924	74	1702	84	23·0	Great Eastern
Wolverhampton	1859–1906	48	794	20	16·5	Great Western
Miles Platting	1847–1881	35	522	24	14·9	Lancashire & Yorkshire
Gateshead	1849–1910	72	1023	60	14·2	North Eastern
St. Rollox	1854–1928	75	1000	38	13·3	Caledonian
Nine Elms	1843–1908	66	815	38	12·3	London & South Western
Brighton	1852–1957	106	1211	61	11·4	London, Brighton & South Coast
Cowlairs	1844–1924	81	900	35	11·1	North British
Gorton	1858–1949	92	1006	35	10·9	Great Central
Greenock	1846–1855	10	97	20	9.7	Caledonian
Wolverton	1845–1863	19	178	45	9·3	London & Birmingham
Ashford	1853–1944	92	790	31	8·5	South Eastern
Eastleigh	1910–1950	40	320	21	8·0	London & South Western
Kilmarnock	1857–1921	65	392	14	6·0	Glasgow & South Western
Inchicore	(a) 1852–1940	89	403	11	4·5	Great Southern & Western
Bow	1863–1906	44	163	10	3·7	North London
Stoke	1868–1923	56	197	16	3·5	North Staffordshire
St. Margarets	1856–1869	14	39	5	2·7	North British
Broadstone	1879–1927	49	132	9	2·7	Midland Great Western
Cardiff	1856–1897	42	84	8	2·0	Taff Vale
Longhedge	1869–1904	36	54	5	1·5	London, Chatham & Dover
Great Victoria St.	1867–1882	16	19	3	1·2	Ulster
Lochgorm	1869–1906	38	41	4	1·0	Highland
Dundalk	1887–1937	51	47	5	0·9	Great Northern of Ireland
Grand Canal St.	1851–1911	61	44	2	0·7	Dublin & South Eastern
Belfast-York Rd.	1870–1942	73	37	4	0·5	Northern Counties Committee

(a) One more built in 1957

costs are taken into consideration – were not British Railways rather precipitate in ending steam and relying now on imported fuel, resulting in the run-down of the mining industry. With the advent of the Diesel engine and the decision of British Railways to rapidly phase out the steam locomotive, many of this form of transport were built by private firms especially by the English Electric Group who supplied 873 main line units over 1600 HP which was far more than the railway workshops turned out themselves.

There is now a pause and firms must of necessity concentrate more on the export market but competition is fierce and unrelenting. In a report on the Railway Workshops of British Railways in 1967, the idea was put forward that to promote fuller employment and efficiency, BR should tender for work suitable for employing the existing plant of the workshops in competition with private manufacturers both for the home market and overseas. Besides the obvious goods such as Diesel locomotives, carriages, wagons, and containers – metal fabrication, castings and general engineering work could be carried out.

The 1968 Transport Act indicated that the spare capacity of railway workshops could be utilised for providing manufactured material to customers, outside the railway framework. Since then work has been carried out for foreign and home customers including Diesel Locomotives for Coras Iompair Eirann, coaches for London Transport and Vans, Restaurant Cars and other work for various railways and firms.

Work other than that of a railway nature has been done utilising still further spare capacity. Apart from the building and maintenance of BR locomotives and rolling stock the workshops which can do this work are at Glasgow, Doncaster, Derby, Crewe, Swindon, Eastleigh and Horwich.

Fortunately many steam locomotives have been saved from the scrap heap, some to continue on active service on private railways run by Preservation Societies and others as static exhibits in Museums mostly on view to the public.

This is as it should be, in the land that gave birth to man's most useful servant.

ACKNOWLEDGEMENTS

It will be appreciated that over the years the amassing of information has come from many sources, some unknown and some unrecorded in the earlier years. There follows a list of firms, libraries and individuals who have been particularly helpful in supplying photographs, drawings and information. Space precludes a list of everyone who has helped and I hope that those who have been omitted will accept my apologies and also my repeated thanks for their co-operation.

Individuals

R. Abbott, Oxford; The late G. Alliez; G. E. Baddeley, Croydon; M. Billington, Nuneaton; Rev. E. Boston, Cadeby; R. E. Bowen, Dinas Powis; V. Bradley, Wirral; R. Brown, Kilmarnock; L. G. Charlton, Newcastle; R. H. Clark, Norwich; R. N. Clements, Co. Kildare; E. Craven, London W1; W. J. F. Davenport, Tollerton; S. R. Devlin, London SE1; G. Dow, Audlem; A. G. Dunbar, Glasgow; J. M. Fleming, Gosforth; S. H. P. Higgins, Wolverhampton; Dr. R. L. Hills, Manchester; K. Hoole, Scarborough; G. Horsman, Headingley; J. Houston, Co. Antrim; F. Jones, Loughton; O. Jones, Tredegar; S. A. Leleux, Keighley; W. E. Loveridge, Whitby; J. J. W. Lowe, Offenham; A. Mackenzie, Edinburgh; J. F. McEwan, Glasgow; D. Martin, Kirkintilloch; G. Moore, Beckenham; J. Morley, Rayleigh; J. M. Mowat, Glasgow; W. G. Pearson, Liverpool; J. D. Petty, Middlesboro'; A. R. Phillips, Leics; K. Plant, Sheffield; B. Radford, Derby; R. N. Redman, Horsforth; D. L. Smith, Ayr; J. J. Smith, Airdrie; W. A. Smyth, Birmingham; D. H. Stuart, Glasgow; C. H. Warrington, Dilhorne; C. R. Weaver, Kenilworth; R. Wear, Wellingborough; B. Webb, Scarborough; E. M. S. Wood, Ormskirk.

Libraries, Museums etc.

Belfast Transport Museum
Birkenhead — Williamson Museum
Birmingham — City Museum and Art Gallery
Bradford — City Art Gallery and Museum
Cardiff — County Record Office
Dublin — National Library of Ireland
Dundee Museum and Art Gallery
Eccles Central Library
Edinburgh — Scottish Record Office
Glasgow — Mitchell Library, Museum of Transport
Leicester — City and University Libraries
Liverpool Museum
London — The Science Museum
Manchester — Public Libraries, Local History Library
Poole Central Library

Companies

Allied Ironfounders Limited, Ketley
Atkinson Vehicles Limited, Preston
Aveling–Barford Limited, Grantham
British Railways Board — Swindon, York and London
British Steel Corporation — Cardiff and Dowlais
Clarke Chapman & Company Limited, Gateshead
Colvilles Limited, Glengarnock
De Beers Consolidated Mines Limited, Kimberley, SA
English Electric Company Limited, Newton-le-Willows
Hick Hargreaves & Company Limited, Bolton
Hunslet (Holdings) Limited, Leeds
Lillieshall Company Limited, St Georges
London Transport Executive
Markham & Company Limited, Chesterfield
Merryweather and Sons Limited, Greenwich
Redpath Dorman Long Limited, Warrington
Richardsons Westgarth (Hartlepool) Limited
Robey & Company Limited, Lincoln
Round Oak Steel Works Limited, Brierley Hill
Ruston and Hornsby Limited, Lincoln
A. Shanks and Son Limited, Arbroath
Yorkshire Patent Steam Wagon Company, Leeds

Society and Institution Journals

Industrial Railway Society
Institution of Mechanical Engineers
Irish Railway Record Society
Locomotive Engineers and Firemans Monthly Journal
Narrow Gauge Railway Society
Railway Correspondence and Travel Society
Stephenson Locomotive Society

Periodicals

Engineering The Engineer
The Locomotive, Carriage and Wagon Review
Practical Mechanics
Railway Magazine Railway News
Railway World

ABBREVIATIONS

BAGSR	Buenos Ayres Great Southern Rly
BAPR	Buenos Ayres Pacific Rly
BAWR	Buenos Ayres Western Rly
B&ER	Bristol & Exeter Rly
B&GR	Birmingham & Gloucester Rly
B&TR	Blythe & Tyne Rly
BNCR	Belfast & Northern Counties Rly
C&CR	Carmarthen & Cardigan Rly
CIE	Coras Iompair Eireann
CR	Caledonian Rly
D&DR	Dublin & Drogheda Rly
D&ER	Dundalk & Enniskillen Rly
D&KR	Dublin & Kingstown Rly
DP&AJR	Dundee Perth & Aberdeen Junc. Rly
E&GR	Edinburgh & Glasgow Rly
ECR	Eastern Counties Rly
ELR	East Lancs Rly
EP&DR	Edinburgh Perth & Dundee Rly
GCR	Great Central Rly
GD&CR	Glasgow Dumfries & Carlisle Rly
GER	Great Eastern Rly
GG&SJR	Great Grimsby & Sheffield Junc. Rly
GJR	Grand Junction Rly
GNR	Great Northern Rly
GNR(1)	Great Northern Rly (Ireland)
GP&GR	Glasgow Paisley & Greenock Rly
GPK&AR	Glasgow Paisley Kilmarnock & Ayr Rly
G&SWR	Glasgow & Southern Western Rly
GS&WR	Great Southern & Western Rly
GSR	Great Southern Rly
GWR	Great Western Rly
H&BR	Hull & Barnsley Rly
INWR	Indian North Western Rly
K&ESR	Kent & East Sussex Rly
LB&SCR	London Brighton & South Coast Rly
L&CR	Lancaster & Carlisle Rly
LC&DR	London Chatham & Dover Rly
LD&ECR	Lancashire Derbyshire & East Coast Rly
L&GR	London & Greenwich Rly
L&MR	Liverpool & Manchester Rly also London & Manchester Rly
LM&SR	London Midland & Scottish Rly
LNR	Leeds Northern Rly
L&NER	London & North Eastern Rly
L&NWR	London & North Western Rly
LR	Londonderry Rly
LRLTD	Light Railways Ltd
L&SWR	London & South Western Rly
LT	London Transport
L&YR	Lancashire & Yorkshire Rly
M&CR	Maryport & Carlisle Rly
MGNJR	Midland & Great Northern Joint Rly
MR	Midland Rly
M&GWR	Midland & Great Western Rly
M&KR	Monkland & Kirkintilloch Rly
M&LR	Manchester & Leeds Rly
MS&LR	Manchester Sheffield & Lincolnshire Rly
M&SMR	Madras & Southern Mahratta Rly
N&AR	Newry & Armagh Rly
NA&HR	Newport Abergavenny & Hereford Rly
N&BR	Neath & Brecon Rly
NBR	North British Rly
N&CJR	Newquay & Cornwall Junc. Rly
NCC	Northern Counties Committee
NER	North Eastern Rly
NLR	North London Rly
NMR	North Midland Rly
NWRR	Newry Warrenpoint & Rostrevor Rly
OW&WR	Oxford Worcester & Wolverhampton Rly
SAR	South African Railways
SA&MR	Sheffield Ashton-under-Lyne & Manchester Rly
S&BR	Shrewsbury & Birmingham Rly
S&CR	Shrewsbury & Chester Rly
SCR	Scottish Central Rly
S&DR	Stockton & Darlington Rly
S&DJR	Somerset & Dorset Joint Rly
SE&CR	Southern Eastern & Chatham Rly
SER	South Eastern Rly
SMJR	Scottish Midland Junc. Rly
SNER	Scottish North Eastern Rly
SR	Southern Rly
SWMR	South Wales Mineral Rly
SYR	South Yorkshire Rly
TVR	Taff Vale Rly
WHH&R	West Hartlepool Harbour & Rly Co.
YN&BR	York Newcastle & Berwick Rly
Y&NMR	York & North Midland Rly

CT	crane tank
DF	double frames
DW	driving wheels
F/Box	firebox
G	geared
HS	heating surface
IC	inside cylinders
IF	inside frames
OC	outside cylinders
OM	outside motion
PT	pannier tank
S	scrapped
S'heat	superheater elements heating surface
S/S	scrapped or sold
ST	saddle tank
T	side tank
VB	vertical boiler
VC	vertical cylinders
W'dr	withdrawn
WT	well tank

Cylinder dimensions, "diameter" x "stroke". Wheel arrangements throughout: Whyte system. Where, in the captions, a number appears before the building date this is the works number of the locomotive.

THE BUILDERS

The following list of builders names is compiled in the order in which they appear in the book. Page numbers are not given for each builder since they appear alphabetically and the figures printed after some names refer to pages where there is a further mention of that builder.

ABBOT, JOHN & CO. LTD.

Park Iron Works, Gateshead

This firm of steam and hydraulic engineers were reputed to have built a locomotive for their own use, but no confirmation is available or any details.

ABBOTT & COMPANY

This firm cannot be located, but at least one locomotive was built for the York and North Midland Railway in April 1845; this was No. 29 PLANET.

ADAMS, W. BRIDGES

Fairfield Works, Fairfield Road, Bow

William Bridges Adams, born in 1797, established his works at Bow in 1843. His father was a partner in a coachbuilding firm and his son William in 1837 wrote an authoritative book on 'English Pleasure Carriages' which was to be the first of many books on technical subjects which emanated from his fertile and inventive brain. His principal claims to fame were his patent fish-plate joint for the permanent way and a radial axlebox. His main interest however was in light railways and locomotives. The chief concepts upon which his designs were worked out were based on his rule that the power output must be proportionate to the load. The engine units all had a low centre of gravity and a long wheelbase to promote steadiness and adhesion even though in most cases he made the frames to take four wheels only. The locomotive was to be self contained, carrying its own fuel and water without the addition of a tender. The boiler pressure was to be relatively high, up to 120lb/sq.in. with a consequent

W. B. ADAMS

Built	Type	Cylinders inches	D.W. ft in	Customer	Name
4/1847	0-2-2RM	IC 3½ x 6	3 4	Eastern Counties Rly	Express
1848	0-2-0RM	IC 8 x 12	4 6	Bristol & Exeter Rly	Fairfield (a)(d)
1/1849	2-2-0RM	OC 8 x 12	5 0	Eastern Counties Rly	Enfield
8/1849	2-2-0WT			Eastern Counties Rly	Cambridge
1849	0-2-2WT	IC 8 x 12	5 0	Cork & Bandon Rly	Rith Teineadh (b)
1849	0-2-2WT	IC 8 x 12	5 0	Cork & Bandon Rly Fig. 1	Sighe Gaoithe (c)
1850	2-2-0WT	OC	5 0	Londonderry & Enniskillen Rly	No. 4

(a) Built to standard gauge converted to 7′ gauge for B & ER. (b) Translated 'Running Fire' sold to Cork Gas Works 1868. (c) 'Whirlwind' sold to Cork Gas Works 1868. (d) Vertical boiler.

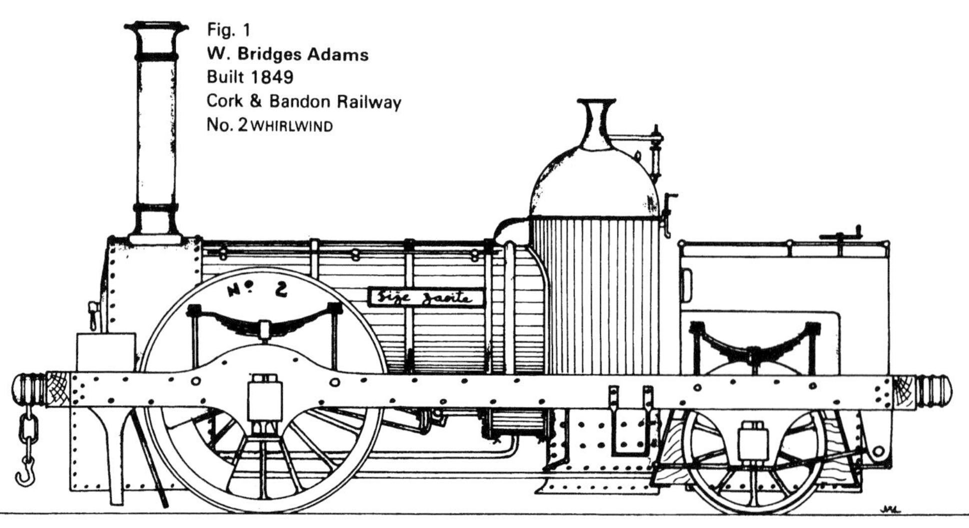

Fig. 1
W. Bridges Adams
Built 1849
Cork & Bandon Railway
No. 2 WHIRLWIND

reduction in the size of the firebox and boiler diameter and reduction in the quantity of water in the boiler and firebox. He regarded the boiler tubes as so many horizontal chimneys and the diameter of the tubes should be in proportion to their length to obtain all possible heat transfer without detracting the draught through them. Where the load would be excessive for one locomotive he recommended the coupling of two together to produce a four cylinder unit with double adhesion.

His first production was a manually driven unit at the request of Mr James Samuel, engineer to the Eastern Counties Railway. The Stratford works were close by and Samuel wanted a light inspection carriage. A logical step from this effort was a steam powered carriage and in collaboration with James Samuel a small unit was produced in April 1847. Named EXPRESS, with accommodation for seven passengers, it was a 0-2-2T with two inside cylinders $3\frac{1}{2}$" x 6", driving wheels 3′ 4″ diameter, total weight $1\frac{1}{4}$ ton. It was obviously a tiny engine, with a vertical boiler 1′ 7′ diameter x 4′ 3″ high with a total heating surface of $43\frac{1}{2}$ sq. ft. A 40 gallon tank provided the feed water. It was successful, however, and in six months ran a total of 5526 miles with an average coal consumption of 3lbs per mile.

A larger unit, FAIRFIELD, was built by Adams and Samuel in 1848. It is probable that this combined engine and coach was built to standard gauge and after trials converted to 7′ gauge as it became No. 29 on the Bristol and Exeter Railway, working the Tiverton Branch. The engine unit was mounted on an open platform, with a vertical boiler 3′ diameter x 6′ high. The inside cylinder, 8″ x 12″, drove a dummy crankshaft coupled to 4′ 6″ driving wheels. The carriage portion, which must have been articulated, due to the overall length of the frames being no less than 31′ 6″, was mounted on four 3′ 6″ diameter wheels and accommodated 16 first class and 32 second class passengers. It later worked the Clevedon and Weston branches. In 1851 it was rebuilt as a separate engine divorced from the carriage section in the form of a 0-2-2T. After a further alteration to a four coupled tank locomotive it was sold to Messrs Hutchinson and Ritson of Bridgwater in 1856.

ENFIELD was built in January 1849 for the Eastern Counties Railway. James Samuel was again associated with the design which took the form of a combined engine and coach similar to FAIRFIELD, but in this case the wheel arrangement was 2-2-0T for the engine, with loco type horizontal boiler 2′ 6″ diameter, outside cylinders 8″ x 12″ driving a pair of 5′ wheels. The coach section had four compartments carrying a total of 42 passengers and was carried on four wheels. Its total weight was 15 tons 7 cwt. A big increase in heating surface gave a total of 255 sq.ft. It was put to work between Angel Road and Enfield stations and performed satisfactorily. Later, as in the case of FAIRFIELD, the engine was separated from the coach and rebuilt as a 2-2-2T, the long frame proving too rigid in service.

CAMBRIDGE, for the Eastern Counties Railway, took the form of a separate locomotive and four wheeled carriage and was built in August 1849. Evidently the original idea of combined engine and coach had been discarded due to trouble encountered with the rigid wheelbase on curves. This locomotive was a small 2-2-0WT with loco type boiler, outside cylinders, but no dimensions have survived.

RUNNING FIRE and WHIRLWIND were two 0-2-2WTs built for the Cork & Bandon Railway in 1849. The names were inscribed in Gaelic (Fig.1). Each had sandwich frames, inside cylinders fixed to the frames behind the firebox and driving the forward wheels. The boiler was domeless with a semi-circular firebox. Water tanks were fixed beneath the smokebox and under the footplate. The weight in working order was $8\frac{1}{2}$ tons. The gauge was 5′ 3″. It is doubtful whether passengers were carried on the footplate.

No. 4, for the Londonderry & Enniskillen Railway, was built in 1850 as a 2-2-0WT with outside cylinders, conventional domed boiler and inside frames. A carriage unit was also built, mounted on four wheels for carrying passengers, the feed water being carried in a tank beneath the carriage.

It is not known when the factory ceased operations but no further locomotives can be traced. The works were later occupied by Messrs Bryant & May. Adams died in 1872.

ADAMSON, DANIEL & COMPANY

Newton Moor Iron Works, Hyde, Nr. Manchester

Established in 1842, boiler making was the main line of business and Mr Adamson was one of the earliest advocates of the introduction of steel for boiler construction. Many machines for their manufacture were designed and built in the works. Other products were stationary engines of all types, portable engines, bridges, cranes and general engineering.

Industrial locomotives were built from 1866 to 1890 for two firms only; Platt Bros & Co. Ltd. of Oldham had twelve between 1866 and 1896, and Oakeley Slate Quarry in North Wales were supplied with four—between 1885, and 1890.

The illustration of OLDHAM (Fig. 2) shows the steeply inclined 12″ x 24″ outside cylinders driving the rear pair of wheels. Access to the crosshead was effected by the flap on the tank housing, the tanks being cut back for this purpose. The safety valve was a standard type used for Lancashire and other stationary boilers. The regulator rod passed above the boiler through the dome to a valve between the chimney and dome. The buffers were rectangular and rubber cushioned. A feed pump and injector were fitted. The coupled wheels of 3′ 10″ diameter were steel tyred with cast iron centres.

These were novel locomotives with unusual features and the majority gave good service.

THE AIRDRIE IRON COMPANY

Standard Works, Airdrie, Lanarkshire

This general engineering firm built their first locomotive in 1869, to the order of the Drumgray Coal Co., and it was put on the rails on 5 April of that year. Little information is available on the locomotives built but they were probably all four coupled saddle tanks with outside cylinders. Like other Scottish builders the saddle tanks were 'ogee' in section, but later ones were more rounded, and the example illustrated (Fig. 4) had quite a Caledonian Railway appearance. Evidently built to a specified loading gauge, the chimney is short and the dome cover squat. Note also the single slide bar and round strap type big-end on the connecting rod.

Between twenty and thirty locomotives were built before the works closed down in 1913.

D. ADAMSON

Date	Type	Cylinders inches	D.W. ft in	Name	Gauge ft in	Customer	
1866	0-4-0T	IC 12 x 24	3 10	Oldham Fig. 2	Std	Platt Bros & Co Ltd Oldham	(a)
1873	0-4-0T	OC		Greenacres	Std	Platt Bros & Co Ltd Oldham	
1876	0-4-0T	IC		Chadderton	Std	Platt Bros & Co Ltd Oldham	(b)
1876	0-4-0T	OC		Crompton	Std	Platt Bros & Co Ltd Oldham	
1877	0-6-0T			Medlock	Std	Platt Bros & Co Ltd Oldham	
1878	0-4-0T			Hartford	Std	Platt Bros & Co Ltd Oldham	
1883	0-4-0T	OC		Werneth	Std	Platt Bros & Co Ltd Oldham	
1885	0-4-0T	OC		William	1 11½	Oakeley Slate Quarry	
1888	0-4-0T	OC		Mary Caroline	1 11½	Oakeley Slate Quarry	
1888	0-4-0T	OC		Edward	1 11½	Oakeley Slate Quarry	
1890	0-4-0T	OC		Charles	1 11½	Oakeley Slate Quarry	
1896	0-4-0ST	OC		Hollinwood	Std	Platt Bros & Co Ltd Oldham	
1896	0-6-0T			Milnrow	Std	Platt Bros & Co Ltd Oldham	

(a) Boiler pressure 130 lb/sq in. (b) Rebuilt HL.1900. N.B. There may have been two further locos built *c.* 1870 for Sutton Bridge Colly. Named LEES and COLDHURST.

Fig. 2
D. Adamson & Co
Built 1866
Platt Bros Oldham
OLDHAM

Fig. 3
Airdrie Iron Co
7/1875 0-4-0ST
Gartness Coal &
Iron Co No. 1
'Ogee' tank

Fig. 4
Airdrie Iron Co
1900 Glasgow
Corporation Gas
Works No. 5

AIRDRIE IRON CO

Works No	Date	Type	Cylinders inches	D.W. ft in	Customer	No.	Gauge
1	5/1869	0-4-0ST			Drumgray Coal Co		
2	10/1869	0-4-0ST			Lumphinnans Coal & Iron Co		
6	1875	0-4-0ST	OC		Gartness Coal & Iron Co		
7	1875	0-4-0ST	OC		Gartness Coal & Iron Co	1	
	1900	0-4-0ST	OC 12 x 20	3 5	Glasgow Corpn Gas Co	5	Std
	1900	0-4-0ST	OC 12 x 20	3 5	Glasgow Corpn Gas Co	6	Std

Others supplied to Robert Addie—Rosehall, Merry & Cunninghame—Carnbroe

Fig. 5
Allan Andrews
16/1878
Carron Co Falkirk.
No. 8

Fig. 6
Allan Andrews
0-4-2ST
Built before 1882
with firm's
later name

Fig. 7
Anderson
Probably built
by
Gibb & Hogg
HARE

ALLAN ANDREWS & CO.

Britannia Engineering Works, Kilmarnock

Commenced building locomotives in the early 1870s. Most of those built were 0-4-0ST with no particular characteristics. The saddle tanks were of all shapes including the 'ogee' pattern, straight sided and semi-circular. 'T' section cast iron spoked wheels were used and the step up to the footplate was lengthened to help the shunter. Brakes engaged at the rear of the trailing wheels only. One peculiarity was the design of the connecting rod little end which was almost as large as the big end and had two horizontal holes drilled through, for what purpose it is difficult to tell.

From about 1878 the firm's title was changed to ANDREWS, BARR & CO. and by 1882 had become BARR MORRISON & CO.

At least one 0-4-2ST was built for overseas and it is estimated that 35 to 40 locomotives were built altogether. In about 1883 the works were taken over by Dick, Kerr & Co., who continued with locomotive building.

Fig. 6 shows a 0-4-0ST with the name Barr Morrison & Co. imprinted by the photographer during the time of the firm's change of title.

ANDERSON, ALEX & SONS LTD.

Carfin Boiler Works, Motherwell, Lanarkshire

This firm of general engineers built locomotive boilers for firms such as Gibb & Hogg.

It is possible that locomotive repairs also were carried out and although an outside cylinder 0-4-0ST named HARE is attributed to this firm, it was probably a rebuild or heavy repair of a G & H loco as it bears all its characteristics. It has cylinders 12″ x 19″ and was sold to Newlandside Quarry Co., Durham, in 1908. (Fig. 7)

ANDREWS, BARR & CO.

See ALLAN ANDREWS & CO.

ALLAN ANDREWS & CO

Works No	Date	Type	Cylinders	Customer	No/Name
5	1874	0-4-0ST	OC	Carron Co Falkirk	No 7
	1875	0-4-0ST	OC	Moss Bay Iron & Steel Co	No 6
16	1878	0-4-0ST	OC	Carron Co Falkirk	No 8 (Fig 5)

ANDREWS BARR & CO

Works No	Date	Type	Cylinders	Customer	No/Name
24	1880	0-4-0ST	OC	Londonderry Colls.	
29	1881	0-4-0ST	OC	Glasgow & South Western Rly	218
30	1881	0-4-0ST	OC	Glasgow & South Western Rly	219
		0-4-0WT	OC	Threlkeld Granite Co Ltd	George V (a)
		0-4-0ST	OC	Cammell Laird & Co Ltd	Grimesthorpe (b)
32	1882	0-4-0ST	OC	Wm. Dixon Ltd, Calder	No 7
	1882	0-4-0ST		Wm. Dixon Ltd, Old Govan Colly	
		0-4-2ST	OC	Overseas	(Fig. 6) (b)

(a) 2′ 4″ gauge. (b) Not identified

APPLEBY BROS.
Southwark, London

Charles James Appleby was the principal of the works engaged in the manufacture of steam cranes, dredgers, brick making machinery, steam crabs, pile drivers, pumps, portable and stationary engines.

One 2′ 8″ gauge locomotive was supplied to Robert Campbell of Faringdon about 1871. It was a four coupled side tank, with marine type boiler, outside cylinders and motion of peculiar design as will be seen from the photograph (Fig. 8). It was named EDITH and whether the firm did actually build it is debatable, and it is probable that orders received were passed on to firms such as Fox Walker & Company, Avonside Engine Company and others.

In 1866 due to cramped conditions the works were transferred to East Greenwich.

There was a branch at Leicester trading as J. Jessop & Son but it is fairly certain no locomotives were built there.

ARMSTRONG, W. G. & CO.
Scotswood, Newcastle-on-Tyne

Later became ARMSTRONG WHITWORTH, SIR W. G. & CO. LTD.

Locomotive building was in three periods. The first locomotive was built in 1847, the second phase was between 1860 and 1864 and up to the latter date about fifty were built. The only details of early building is a 2-2-2 built in 1848 with vertical boiler, 7 ft diameter driving wheels, condensing apparatus, outside motion and 65lb/sq. in. boiler pressure.

Between 1861 and 1868 some 285 mixed traffic 2-4-0s were built for the East Indian railways by seven British and two Continental builders. W. G. Armstrong & Co. built twenty in this period and one was shown at the 1862 Exhibition.

The third and last phase was after the First World War. Immediately after the Armistice in November 1918 the conversion of Armstrong's shell factories at Scotswood to the construction of locomotives was put in hand and by 12 November 1919 the changeover had been completed and the first locomotive, a 0-8-0 superheater type for the North Eastern Railway, was steamed. The anticipated output of 300 to 400 main line locomotives per annum together with industrial tank locomotives for standard

Fig. 8
Appleby Bros 0-4-0T EDITH

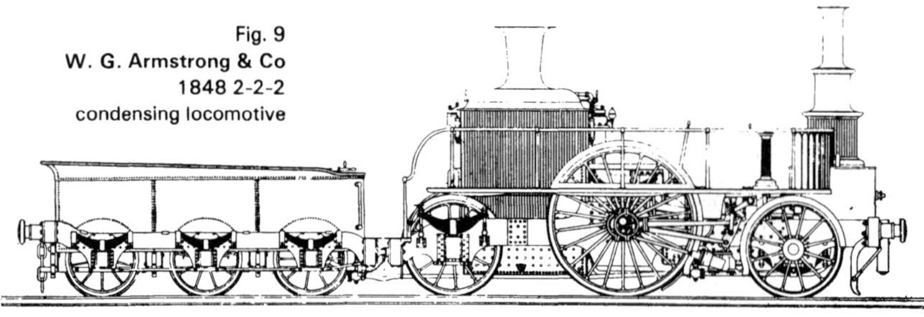
Fig. 9
W. G. Armstrong & Co
1848 2-2-2
condensing locomotive

and narrow gauge was never realised although the works and machinery had been laid out for this quantity. Many standard locomotives were supplied to various railway companies at home and abroad including two hundred 2-8-0s for Belgian State Railways and 327 mixed traffic 4-6-0s for the LMS. Alas the early promise did not continue and the final steam locomotive was turned out in 1937.

At the Openshaw works in Manchester where armour plate had been produced the shops were converted into a locomotive repair department, equipped with up to date machinery capable of repairing 400 locomotives a year. Many were repaired for the North Eastern, R.O.D. and the London & North Western Railways.

At their Elswick plant repairs were carried out to a number of North British Railway locomotives.

Scotswood works built a total of 1464 steam locomotives from 1919 to 1937 and building ceased in this year and the department closed in 1938. The most interesting locomotive built was the Ramsay Turbo-Electric locomotive (built in 1922, works No.160, Order No. E. 14) designed to overcome the heavy capital outlay required for the conventional type of electric traction. The locomotive comprised two sections, the front being in the form of a 2-6-0 with boiler, generator, turbines and two electric motors, the rear portion, equivalent to the tender, containing coal and water, condenser, air fan, hot well and electric motors to drive the 6-coupled wheels on this section.

The main turbine was placed under the boiler, and coupled to it, the electric generator; between the frames, the two electric motors coupled to a driving shaft. The shaft had flywheels attached and the coupled wheels were driven by means of this jack shaft which was situated behind the leading pair of coupled wheels. In the centre of the rear section, the condenser was fitted with a hot well beneath, so that the feed water was used repeatedly. The drive for the rear portion was similar to the fore part, so that in effect the locomotive combined the power of two 6-coupled engines.

The motor used was the slip ring type, 600 volts alternating current, each motor developing 275 bhp.

The condenser had a ring-shaped nest of tubes built into two headers. The nest of tubes formed into the shape of a drum rotated around

Fig. 10 A. W. 393/1921 LNER 2315 0-6-0T *Class J72*

ARMSTRONG, W. G. & CO.

Order No	Works No	Date	Type	Cylinders (inches)	D.W. ft in	Customer	Numbers	Gauge ft in
E4	1-50	1919-21	0-8-0	OC 20 x 26	4 7¼	North Eastern Rly	2253-2302	Std
E3	51-60	1920	4-6-0	OC 16 x 22	4 4	Leopoldina Rly	Various	Metre
E5	61-68	1920	4-6-0	OC 16 x 22	4 4	Leopoldina Rly	270-277	Metre
E1	69-93	1920	2-8-0	OC 22 x 26	4 8½	BB & CI Rly	122-146	5 6
E2	94-110	1920	2-8-0	OC 22 x 26	4 8½	M & SM Rly	483-499	5 6
E7	111-120	1921	4-4-0	IC 20½ x 26	6 6	Caledonian Rly	82-91	Std
E12	121-136	1920	4-8-0	OC 20 x 24	3 9	Nigerian Gov Rly	317-332	3 6
E6	137-159	1920	0-6-0	IC 20 x 26	5 1½	INW Rly	2484-2506	5 6
E14	160	1922	2-6-6-2	Ramsey Turbine		Ramsey Co Newcastle	—	Std
E10	161-170	1922	2-6-2T	OC 20 x 24	4 0	BAW Rly	824-833	5 6
E11	171-174	Not Built					—	
E13	175-179	1922	0-6-0	IC 19 x 26	5 8	MGW Rly	44-48	5 3
E15	180-184	1921	4-6-4T	OC 17¾ x 21$\frac{5}{8}$	4 5$\frac{1}{8}$	Java State Rly	1135-1139	3 6
E18	185-190	1921	4-6-0	4/14 x 26	6 7	GS & W Rly	407-409, 403-405	5 3
E19	191-390	1922	2-8-0	OC 24 x 28	5 0	Belgian State Rlys	5001-5200	Std
E21	391-415	1921	0-6-0T	IC 17 x 24	4 1¼	North Eastern Rly. Fig. 10	2313-2337	Std
E22	416-465	1921	0-6-0	IC 20 x 26	5 3	London Midland & Scottish Rly	3937-3986	Std
	466-467	Not Built					—	
E24	468-472	1922	0-6-0	IC 20 x 26	5 3	Somerset & Dorset Joint Rly	57-61	Std
	473-478	Not Built					—	
E29	479-487	1923	0-6-0	IC 20 x 26	5 1½	INW Rly	2536-2544	5 6
							2989-96	
E30	488-499	1923	4-4-0	IC 20 x 26	6 2	INW Rly	3006-09	5 6
E31	500-515	1923	2-6-4T	IC 20 x 26	5 1½	BB & CI Rly	265-280	5 6
E32	516-535	1923	0-6-0	IC 20 x 26	5 1½	Oudh & Rohilkund Rly	505—524	5 6
E33	536-552	1923	0-6-0	IC 20 x 26	5 1½	INW Rly	1390-1406	5 6
E34	553-564	1923	4-8-0	OC 19 x 22	3 3$\frac{3}{8}$	Loanda State Rly	201-212	Metre
E36	565-566	1924	4-6-6-4	4/16 x 22	3 4	FC Pacifico-Colombia	29-30	3 0
E37	567-591	1924	4-8-0	3/17½ x 26	4 7½	BAGS Rly	4201-4225	5 6
E35	592-595	1924	4-8-2	OC 20 x 24	3 9	Nigerian Govt Rly	701-704	3 6
	596	Not Built					—	
E39	597-602	1924	4-6-0	OC 15 x 22	3 3$\frac{3}{8}$	Loanda State Rly	101-106	Metre
E40	603-604	1924	0-4-4-0	Electric	2 4	New Modderfontein Gold Mine		
E41	605-616	1924	4-4-0	IC 20 x 26	6 9	London & North Eastern Rly	6390-6401	Std
E42	617-622	1924	4-6-0	OC 15 x 22	3 3$\frac{3}{8}$	Loanda State Rlv	107-112	Metre
E43	623-632	1925	4-6-2	OC 24 x 28	6 3	S. Australian Govt Rlys	600-609	5 3
E44	633-642	1925	4-8-2	OC 26 x 28	5 3	S. Australian Govt Rlys	500-509	5 3
E45	643-652	1925	2-8-2	OC 22 x 28	4 9	S. Australian Govt Rlys	700-709	5 3
	653-654	Not Built					—	
E47	655-701	1924	2-8-0	OC 21½ x 26	4 8	Bengal Nagpur Rly	700-729, 744-760	5 6
E48	702-707	1924	2-6-4T	OC 20 x 28	5 6	Metropolitan Rly (a) Fig. 11	111-116	Std
	708-709	Not Built					—	

E50	710-713	1926	4-8-2	OC 20 x 24	3 9	Nigerian Govt. Rly	705-708	3 6
E51	714-725	1925	4-6-2	3/18 x 26	5 11	BAP Rly	2101-2112	5 6
E52	726-760	1925	2-8-0	3/17½ x 26	4 7½	BAGS Rly	4301-4335	5 6
E53	761-769	1925	2-6-4T	OC 19 x 28	6 0	Southern Rly (a)	A791-A799	Std
E54	770	1925	4-6-0	16 x 22	4 4	Therezopolis Rly	31	
E55	771-801	1925	2-8-0	OC 21½ x 26	4 8	Bengal Nagpur Rly	761-791	5 6
E56	802-807	1926	4-6-0	OC 19 x 26	5 2	Palestine Rly (b)	876,884,891/2/6,901	Std
E56	808-813	1926	4-6-2T	OC 19 x 26	5 2	Palestine Rly (c)	7-12	Std
E57	814-821	1926	4-8-2	OC 20 x 24	3 9	Nigerian Govt Rly	709-716	3 6
	822-841	Not Locomotives					–	
E59	842-849	1926	4-8-2	OC 20 x 24	3 9	Nigerian Govt Rly	717-724	3 6
	850-874	1927	4-8-0	OC 17 x 22	3 9	Queensland Govt Rlys	802-26	3 6
	875-884	1927	4-8-4T	22½/31½ x 26	5 2	Central Argentine Rly (d)	501-510	5 6
	885-904	1928	2-6-0	OC 21 x 26	5 6¾	Egyptian State Rly	585-604	Std
	905-934	1928	4-8-0	3/17½ x 26	4 7½	BAGS Rly	4226-4255	5 6
	935-937	Not Locomotives						
	938-987	1928	0-6-2T	IC 18 x 26	4 7½	Great Western Rly	6650-6699	Std
	988-992	1928	4-8-2	OC 19 x 24	3 9	Nigerian Govt Rly	725-729	3 6
	993-1002	1928	4-6-2T	OC 16 x 22	4 4	Leopoldina Rly	350-359	Metre
	1003-1004	1928	4-6-0	OC 16 x 22	4 3¼	Trinidad Govt Rly	31, 32	Std
	1005-1015	1929	2-8-2	OC 22½ x 28	5 1½	M & SM Rly	853-863	5 6
	1016-1019	1929	4-6-0	OC 16½ x 22	4 3	GWR of Brazil	230-233	5 6
	1020-1023	1929	4-8-0	OC 18 x 22	3 6	GWR of Brazil	234-237	5 6
	1024-1025	1929	2-6-22-62	OC 14¼ x 20	3 6	GWR of Brazil	238 & 239	5 6
	1026-1037	1929	4-6-0	OC 18½ x 26	4 5½	Ceylon Govt Rly	279-290	5 6
	1038-1057	1930	4-8-4T	22½/31½ x 26	5 2	Central Argentine Rly (d)	511-530	5 6
	1058-1068	1930	4-6-2	OC 21½ x 28	6 2	East Bengal Rly	443-453	5 6
	1069-1080	1930	4-6-2	OC 21½ x 28	6 2	M & SM Rly	200-211	5 6
	1081-1100	1930	4-6-2	3/19½ x 26	6 2½	Central Argentine Rly (e)	1101-1120	5 6
	1101-1104	1930	4-6-2T	OC 16 x 22	4 4	Leopoldina Rly	360-363	Metre
	1105-1110	1930	4-8-0	OC 20½ x 28	5 8	Buenos Aires Western Rly	1500-1505	5 6
	1111-1130	1931	2-6-0	3/18½ x 26	5 8	London & North Eastern Rly	Various	Std
	1131-1155	1930	0-6-0PT	IC 17½ x 24	4 7½	Great Western Rly	7775-7799	Std
	1156-1165	1934	2-6-0	3/18½ x 26	5 8	London & North Eastern Rly	Various	Std
	1166-1265	1935	4-6-0	OC 18½ x 28	6 0	London Midland & Scottish Rly	5125-5224	Std
	1266-1269	1935	0-8-0	OC 16½ x 23½	3 11¼	Yueh Han Rly China	501-504	Std
	1270-1279	1936	2-6-0	3/18½ x 26	5 8	London & North Eastern Rly	Various	Std
	1280-1506	1936-7	4-6-0	OC 18½ x 28	6 0	London Midland & Scottish Rly	5225-5451 (Fig. 12)	Std

(a) Erected from parts supplied ex government. (b) Reconditioning order of Baldwin Locomotives. (c) Conversion of Baldwin 4-6-0s to 4-6-2Ts. (d) Two cylinder cross compounds. (e) Caprotti poppet valve gear.

a shaft supported at each end by bearings. The drum was fitted into a tank containing water through which the tubes passed. As the tubes left the water, air was passed over them at high velocity and the water left on the outside of the tubes evaporated rapidly with an equally rapid condensation of the steam inside the tubes. The condensate passed then into the hot well.

The locomotive ran trials on the L&Y section of the LMS, working between Horwich (where it was kept) Bolton and Southport.

From all accounts the locomotive acquitted itself quite well and it was reported that designs were being prepared for an improved version, but this did not materialise and the existing locomotive returned to the works, presumably to be scrapped.

Interesting news items in the *Engineer* dated 4 February, 11 March and 1 April 1921, were as follows:

4 February 1921

'Much satisfaction has been occasioned by the announcement that an important contract for the repair of locomotives in Russia has been secured. Although no exact figures are available it is known that more than 50% of locomotives owned by Russian railways are in need of repair and that many thousands of engines are therefore included in the contract. The negotiations which have led to this business coming to a British firm in face of persistent German attempts to secure it, have been in progress for a long time, and even now nothing will be done until the trade agreement between Great Britain and Russia has been signed. It is rumoured that it was necessary to quote a very low price to obtain the contract. The repair work will be carried out at the Company's Tyneside establishment.'

11 March 1921

'It is reported from Berlin that M. Krassin has entered into a contract with the Krupp Co. and other German interests for the delivery of 800 locomotives at a price which is said to be half that quoted by British firms.'

1 April 1921

'In addition to the contract for the repair of the large number of locomotives for Russian railways secured by the Armstrong company which, it is estimated, will provide employment for 2000 men for some years, an order has been obtained by the same company for 200 locomotives for the Belgian State railways.'

No records are available indicating whether this Russian contract was ever carried out.

Fig. 11 A. W. 704/1924 Metropolitan Railway No. 113

ASHMAN, WILLIAM

Radstock

Quoting from the *Somerset County Herald* dated 28 August, 1925: 'It is interesting to record that in 1825 – a significant date – Mr William Ashman of Clandown Colliery made a locomotive to haul coal tubs from Radstock to Midford, but the project failed owing to the rails (cast iron) breaking under its weight.' Here we have the news of the first locomotive in the county though, unfortunately, there is none concerning what became of it.

Described by an old inhabitant as a 'single arm' locomotive and subsequently used as a stationary engine.

ATKINSON–WALKER WAGGONS LIMITED

Frenchwood Works, Preston

Locomotives were built between 1926 and 1931 but detailed and specific information on them is extremely difficult to find, as all relevant records have long since been destroyed. The locomotives were of several different sizes and comprised wheel arrangements of 0-4-0 and 0-6-0 driven by geared vertical cylindered Uni-flow steam engines. The drive to the axles was via a Hans Renolds roller chain driving either directly from the engine crankshaft or through a simple reduction gear. In the smaller type of locomotive, steam was supplied by a vertical water tube boiler similar to that employed in the well known Atkinson steam wagon and the engine was in many respects similar to that of the wagon also. For the larger sizes of locomotives a separate design of 3 or 4 cylinder Uni-flow engine was used, producing some 250 bhp at 150 rpm.

In the early days of locomotive building when Atkinson had bought out the Leyland Steam Wagon Company of Chorley, quite a number of locomotives were fitted with the Poppett valve Leyland wagon engine and exported to such places as Singapore and South Africa.

Atkinson Waggons Ltd. were associated with Walker Bros (Wigan) Ltd. in producing these geared locomotives but the design was entirely Atkinson's.

Fig. 12 A. W. 1290/1936 LMS *Class 5* 4-6-0 5235 one of 327 built bv AW.

ATKINSON–WALKER WAGGONS LTD.

Various types were built:

Class A 2 cylinder vertical engine with 3ft diameter wheels and outside bearings.
Class B 3 cylinder vertical engine with 3ft diameter wheels and outside bearings.
Class C 2 cylinder horizontal engine with 2′ 3″ diameter wheels and inside bearings.
Class D 4 cylinder vertical engine with six wheeled chassis with outside bearings.

All cylinders were 7″ x 10″ with bevel gear driven shaft carrying sprockets with the roller chains to drive each side. The locomotives built for Singapore had the two axles coupled by outside chain drive.

No more than twenty-five units were built between 1927 and 1931; some were exported and quite a number of colliery and quarry owners used them in this country as far apart as Devon and Aberdeen.

Fig. 13 **Atkinson-Walker** 104/1928 Type A R. Briggs & Sons Ltd

Wks No	Date	Type	Gauge ft in	No/Name	Customer
101					
102	1927	4WTGVB	Std		Leatham Flour Mills, York
103		4WTGVB	Std		Hutchinson Hollingsworth & Co Ltd
104	1928	4WTGVB	Std	Lazarus	R. Briggs & Sons Ltd (Fig 13)
105	1928	4WTGVB		No 1	Singapore Municipal Council
106	1928	4WTGVB		No 2	Singapore Municipal Council
107	1928	4WTGVB		No 3	Singapore Municipal Council
108	1928	4WTGVB		No 4	Singapore Municipal Council
109	1928	4WTGVB	Std	No 2	Oxford & Shipton Cement Ltd
110	1928	6WTGVB (a)	Std	No 1	Oxford & Shipton Cement Ltd
111	1928	4WTGVB (c)	3 0	Lady Mallaby-Deeley	Ivybridge China Clay Co Ltd
112	1928	4WTGVB	Std	Felspar 2	Shap Granite Co
113	1928	4WTGVB (b)(d)	Std		Blaxters Ltd Northumberland
114	1928	4WTGVB	3 0	8	Clogher Valley Rly
115	1928	4WTGVB			
116	1928	4WTGVB			
117	1930	4WTGVB	Std	No 1	Harold Arnold & Sons Ltd
118	1930	4WTGVB	Std	No 2	Harold Arnold & Sons Ltd
119	1930	4WTGVB	Std	No 3	Harold Arnold & Sons Ltd

4WTGVB = 4 Wheeled tank, gear drive, vertical boiler. 6WTGVB = 6 Wheeled tank, gear drive, vertical boiler. (a)Class D. (b)Class C rail tractor. (c)cylinders 7 x 10 D.W. 2′ 4″. (d)cylinders 7x10 D.W. 2′ 6″

AVELING & PORTER LTD.
Rochester

In 1864 this firm built their first traction engine designed to run on rails and continued production of various types until 1926. They were made in sizes ranging from 4 to 20 nominal horse power. They were all four wheel types, some with both axles coupled by means of a chain drive, and others with a drive to one axle. On early engines they were simply traction engines with flanged wheels for rails, the wheels being driven by a chain from a countershaft and passing around gear wheels attached to both axles. Later machines were driven by direct gearing from the crankshaft. All boilers were of the locomotive type with cylinders placed on top of the barrel. The engines could work on a gradient of 1 in 20 and negotiate curves of 35ft radius. All early locomotives had single cylinders, the first built being 10 nhp. Later ones had the alternative of a single or compound cylinder. They worked at about half the cost of using horses and burned about half the fuel consumed by a direct coupled locomotive performing equal work. Another advantage was that the turning effort on the driving wheels, which needed no balance weights, was more even with the result that slipping was largely eliminated. The main disadvantage was they worked at slower speed than the direct coupled locomotive. In 1872 two of the chain driven engines of 6 nhp each were purchased by the Duke of Buckingham for use on the Wotton Tramway. Fortunately one of these (No. 807) has been restored at London Transport's works at Neasden and at a ceremony on 19 January 1957 it was handed over to them.

Another engine which has been preserved is No. 1607, which spent its entire working life at the Scout Moor Quarry near Ramsbottom, Lancs. and a third example was presented on 10 May, 1964 to the Bluebell Railway in Sussex. The latter was built in 1926 (No. 9449) and worked at the Associated Portland Cement Manufacturers at Holborough, Kent. It was 10 nhp, type TJ.

A departure from the above types was the Aveling-Greig Street Tram Engine which was built by Aveling & Porter in co-operation with Mr Alfred Greig. It was described as follows, in R. H. Clark's book *The Development of the English Traction Engine* (Goose and Son, rev. ed. 1974).

'Most of the working parts are enclosed to comply with legal requirements. Duplex cylinders $5\frac{1}{2}$" x 10" are in this case placed on the firebox end of the boiler with the valve chest outside. Roof type air condenser and cooling was assisted by a fan (Blackman air propeller) horizontally placed within, and partly overlapping the short chimney, thus aiding the draught as well. The drive is practically the same as the ordinary gear driven Aveling tramway locomotive, the crankshaft pinion (9" pcd) engage in the large gearwheel (40" pcd) on the stud shaft

Fig. 14 **Aveling & Porter** 1681/1881 Lees, Halling No. 3. Note feed pump and tank

Works No	Date Invoiced	NHP	Customer	Location	No/Name	Gauge ft in
129	21/2/1865	8	Admiralty	Chatham		Std
132	14/12/1864	8	Tilden & Co			
143	15/5/1865	10	W. Jackson			
151	27/7/1865	10	Grays Chalk Quarries Co	Essex		Std
167	27/11/1865	10	Grays Chalk Quarries Co	Essex		Std
182	1866	8	Admiralty	Chatham		Std
211	24/7/1866	8	A. Redfern & Co.			
212	11/6/1866	10	Grays Chalk Quarries Co	Essex		Std
218	3/8/1866	8	Admiralty	Chatham		Std
220	12/10/1866	12	W. Brassey & Co			
221	17/9/1866	6	W. Brassey & Lucas			
235	23/10/1866	10	W. Brassey & Lucas			
242	20/2/1866	10	Geo. Wythes			
252	28/3/1867	10	H. Lee & Sons			
259	21/6/1867	8	A. Gabrielli & Co			
299	27/8/1867	8	A. Gabrielli & Co			
313	16/12/1867	4	A. Gabrielli & Co			
314	26/12/1867	10	B. O. Harrison			
317	27/2/1868	10	A. Gabrielli & Co			
320	20/3/1868	6	I. Ball & Co			
341	20/4/1868	8	A. Gabrielli & Co			
385	30/6/1868	8	J. & A. Aird			
390	2/9/1868	8	J. & A. Aird	For Berlin		
447	16/2/1868	10	I. H. Ball & Co			
450	13/3/1869	10	Admiralty	Devonport		Std
454	23/4/1868	10	Admiralty	Portsmouth		Std
508	31/8/1869	10	Grays Chalk Quarries Co	Essex		Std
524	13/1/1870	6	U. lay			
611	12/12/1870	12	I. L. Reid			
629	26/4/1872	10	Jas. Oakes & Co Ltd	Alfreton	Jotto	Std
632	5/5/1872	10	Lucas Bros	Teynham		
696	5/1/1872	8	Colonial Co	Trinidad		
718	27/10/1871	9	Admiralty	Chatham		Std
719	27/10/1871	9	Admiralty	Chatham		Std
807	24/1/1872	6	Duke of Buckingham	Wotton		Std
822	1872		Admiralty		Steam Sapper No 3	Std
829	1872		Admiralty (j)		Steam Sapper No 4	Std
830	1872	6	Admiralty		Steam Sapper No 5	Std
831	1872		Admiralty (j)		Steam Sapper No 6	Std
832	1872		Admiralty (j)		Steam Sapper No 7	Std
846	22/6/1872	6	Duke of Buckingham	Aylesbury		

880	25/10/1872	10	Lucas Bros	Teynham		
915	6/3/1873	6	Francis & Co Cliffe	Cliffe, Kent		3 8½
925	26/4/1873	10	Ashton & Co	To Hong Kong		
939	1873		Admiralty (j)		Steam Sapper No. 8	Std
952	26/7/1873	4	Knight Bevan & Sturge	Northfleet		2 8½
1023	30/7/1874	9	Admiralty	Chatham		
1105	12/7/1875	6	Admiralty	Haulbowline		
1109	7/7/1875	6	W. Bird & Co	Vancouver		
1110	26/7/1875	6	W. Bird & Co	Vancouver		
1121	30/8/1875	10	Tunnel Portland Cement Co	W. Thurrock		
1199	23/5/1876	6	Peters Bros.	Wouldham		4 3½
1285	18/12/1876	6	G. & I. Wilson	Hoo		
1286	22/12/1876	8	Russian Govt	Via Paris Office		
1342	18/6/1877	6	Peters Bros	Wouldham	Victor	4 3½
1432	13/5/1878	6	Admiralty	Haulbowline		
1440	30/5/1878	10	Wm. Lee, Son & Co	Halling	2	4 3
1524	21/4/1879	5	Booth Bros	Borstal, Kent		Std
1574	21/11/1879	6	I. Butcher	Rainham		
1594	15/5/1880	6	Francis & Co Ltd	Cliffe, Kent		
1601	20/4/1880	5	G MacPherson & Co	Cliffe, Kent		
1602	27/4/1880	8	Brookes, Shoobridge & Co	Grays		
1604	12/5/1880	6	Peters Bros	Wouldham	Vivid	4 3½
1607	21/5/1880	5	J. Whittaker & Sons	Scout Moor		
1681	21/6/1881	10	Wm. Lee, Son & Co	Halling Fig. 14	3	4 3
1688		6				
1708	5/10/1881	6	Holman & Collard	For S. Gladden Adisham		
1740	16/3/1882	8	John M. Hooker	Sevenoaks		
1780	21/8/1882	8	Brookes, Shoobridge & Co	Grays		
1857	30/4/1883	12	Wm. Lee, Son & Co	Halling	4	4 3
1879	31/7/1883	6	W.O. School. Military Eng.	Chatham	(a)	
1888	1883	5	H. Brock & Co	For W. Perry		
2078	9/6/85	12	Formby's Cement Works Co Ltd	Halling	Formby	3 9
2229	15/3/90	8	Francis & Co Ltd (k)	Cliffe, Kent		3 8½
2428	30/10/88	6	Francis & Co Ltd	Cliffe, Kent		3 8½
2640	28/4/90	8	Peters Bros	Wouldham	Venture	4 3½
2774	26/1/91	6	Wouldham Cement Works Co.	Wouldham		n.g.
3187	17/4/93	12	Wm. Lee, Son & Co.	Halling	5	4 3
3277	23/10/93	C	William Ball	Frindsbury	(b) Alderman	Std
3567	26/7/95	10C	Beadle Bros	Erith	Sydenham	Std
3592	29/8/95	4	John Anderson			
3680	12/3/96	6	John Bazley, White & Bros	Swanscombe	Goliath	3 6
3730	22/6/96	10C	Fraser & Chalmers Ltd	Erith		Std
3766	10/9/96	10C	Glenlossie, Glenlivet Distillery Co	Elgin		
3787	10/10/96	10C	Henry Chambers	Faversham		
3888	15/4/97	S-C	Beadle Bros	Erith	Belvedere	Std

AVELING & PORTER LTD.

Works No	Date Invoiced	NHP	Customer	Location	No/Name	Gauge ft in
3978	1/10/97	S-10C	John Bazley, White & Bros	Swanscombe	Jubilee	3 6
3990	30/10/97	S-C	Manchester Corpn			
4006	25/11/97	S-C	Balmenach Glenlivet Distillery Co	Elgin	Haugh of Cromdale	
4141	21/6/98	6C	George Geddes	Elgin		
4155	27/7/98	S-C	Beadle Bros	Erith	Devon	
4176	31/8/98	S-C	John Bazley, White & Bros	Swanscombe	Samson	3 6
4248	21/12/98	S-C	William Ball	Frindsbury	Sabina	3 6
4347	15/6/99	4	Fischer & Son	Christiana		
4360	13/7/99	S-C	Francis & Co Ltd	Cliffe, Kent	3	3 8½
4399	1899	C	Aveling & Porter Ltd	Rochester	(Fig. 15)	Std
4371	26/7/99	S-C	Pepper & Son Ltd	Amberley	No. 1	Std
4402	22/9/99	S-C	Peters Bros	Wouldham	Ninety Nine	4 3½
4414	27/10/99	S-C	Wm. Lee, Son & Co	Halling		4 3
4430	21/12/99		Thos. Pascall & Sons Ltd	S. Norwood		Std
4444	29/12/99	S-C	Burham Brick, Lime & Cement Co Ltd		Meaway	Std
4445	2/1/1900	S-C	Crystal Palace District Gas Co		2	Std
4469	6/2/00	S-C	John Bazley, White & Bros	Swanscombe	Galley Hill	3 6
4497	23/3/00	S-C	Vickers, Sons & Maxim Ltd	Erith	Vickers	Std
4501	29/3/00	C	John Bazley, White & Bros	Swanscombe	Barnfield	3 6
4537	7/5/00	C	Francis & Co Ltd	Cliffe, Kent		3 8½
4612	3/10/00	14C	E. Kent Chalk Quarries Co	Gillingham	Venture	Std
4638	19/10/00	C	Birmingham Corpn		Forward No 7	Std
4754	4/5/01	C	Sussex Portland Cement Co Ltd	S. Heighton		Std
4780	5/12/01	C	Croydon Gas & Coke Co		Allen Lambert	Std
4909	16/11/01	C	Bath Gas, Light & Coke Co			Std
4941	6/1/02	C	Aldershot Gas & Water Co		Louise	Std
5039	29/5/02	S-C	Kynochs Ltd	Birmingham	1	Std
5082	31/7/02	S	E. Kent Chalk Quarries Co	Gillingham	Success (c)	Std
5441	2/2/04		Peters Bros	Wouldham	Vulcan (d)	4 3½
5568	18/8/04		Metropolitan Amal C & W Co Ltd	Oldbury		Std
5603	29/9/04		Peters Bros.	Wouldham	Victor	4 3½
5832	29/8/05	C	M. Fantauzzietcie	Paris	(e)	Metre
5935	25/1/06	C	Itter Ltd.	Hendon		Std
6040	3/7/06	C	APCM (1900) Ltd	Swanscombe	Progress	3 6
6158	20/12/06		Gypsum Mines Ltd	Mountfield	Sirapite	Std
6369	19/9/07	C	Mitchells & Butlers Ltd		John Barleycorn	Std
6419	6/12/07	C	APCM (1900) Ltd	Swanscombe	Enterprise	3 6
6828	23/6/09	C	APCM (1900) Ltd	Swanscombe	Reliance	3 6
7975	21/4/13	C	Queenborough Wharf Co Ltd		No 2	Std
8372	1/9/14	C	Sheffield United Gas Light Co	Grimesthorpe	No 2 Eric (f)	Std
8598	21/9/15	C	Cammell Laird & Co Ltd	Sheffield	(g)	Std
8647	3/9/15	C	Kynochs Ltd	Birmingham	3 (h)	Std

8695	19/1/16	C	Kynochs Ltd	Birmingham	2	Std
8800	12/1/17	C	Vickers Ltd	Sheffield	Sir Vincent	Std
8874	17/12/17	C	Cammell Laird & Co.Ltd	Sheffield	No 9	Std
8893	10/4/18		Aldershot Gas Co		Kathleen	Std
8944	18/11/18	C	Cammell Laird & Co Ltd	Sheffield	No 3	Std
9449	4/2/26	10	Holborough Cement Co Ltd	Snodland	(i)	Std
11087	24		Totternhoe Lime & Stone Co Ltd	Stanbridgeford		Std

NHP Nominal horse power — Suffix C denotes compound — Prefix 'S' denotes special
(a) Built as 'Sapper No 22' but with tram road gear. (b) First compound. (c) Brakes. (d) Additional tank. (e) Two injectors. (f) 217 gallon tank.
(g) Fitted with warping gear. (h) Fitted with spark arrester and sanding gear. (i) Single cylinder. (j) may have been built as road vehicles only.
(k) bought 1890 ex Brighton Tramways.

on the side of the barrel. Each driving axle was 4″ diameter and carried a large pinion (27″ pcd) receiving the drive from the stud shaft wheel to drive the track wheel (36″ diameter). Steam was used at 180 lb/sq. in., the small boiler having no less than 148 square feet of heating surface, the grate area being 5 square feet and when fully made up the coke fire would last for a run of four to six miles. The main dimensions were: wheelbase 4′ 6″, overall width 5′ 8″, overall length 11′ 6″, overall height 9′ 6″, weight in working order 10 tons. Gauge 3′ 6″. It is not known how many of these Street Tram engines were built, but one worked for several months on the Southwark Tramway in 1887 and was passed by the Board of Trade.'

In all about 130 locomotives of all types were built. The first locomotive supplied was works No.132 which was despatched on 14 December, 1864 to Tilden & Company, and works No. 129 was despatched to Chatham Dockyard on 2 January, 1865. The last one to be built was No. 9449, despatched on 4 February, 1926 to Holborough Cement Company near Rochester.

The company was founded in 1850 by Thomas Aveling as a small agricultural business. He made the first road traction engine in 1861 and it appeared at the Leeds Show in the same year. He was joined by Richard T. Porter in 1862 and the firm became Aveling & Porter Ltd. Locomotives were also exported to India and Australia.

Descriptions of Locomotives from various Catalogues

1874 The locomotive is simply a traction engine on wheels, the wheels being coupled by the pitch chain which is employed to communicate the motion to them. From the crankshaft the motion is taken by spur gearing to a counter shaft carrying at its end the pitch-chain wheel, the bearings of this shaft being placed in curved slots formed in the supporting brackets, so that the shaft can be raised and the chain tightened when necessary. The crankshaft carries a flywheel whereby it is practicable to employ the engine for driving fixed machinery when required.

The boiler is of the ordinary locomotive type and, with the large firebox invariably given in Aveling & Porter's engines, presents a greater heating surface than is customary in engines of the same type. The boiler, tender and underneath water tank are each independently connected to two wrought iron longitudinal frames joined at both ends by wooden buffer beams, and on the frames are rivetted wrought iron horn plates which act as guides to the axle bearing. By this arrangement the engine when

Fig. 15 **Aveling & Porter** 4399/1899 Compound. Works locomotive

required for other purposes and for convenience of repairs can be removed from the framework without interference with the driving wheels. The driving wheels are fitted with Vickers, Sons & Co's crucible steel tyres. The locomotives are made of any size from 4-horse to 20-horse power nominal, they will ascend inclines of 1 in 20 and can be made to work round curves of 35 feet radius. Type 0-4-0WT-CD. The Admiralty at their dockyards of Chatham and Portsmouth have twelve in regular work, they are also employed on some of the railway contracts of Messrs Brassey, Wythes, Lucas, Ogilvie, etc., and are becoming extensively known among chalk-quarry owners, brick manufacturers etc.

1879 The locomotive illustrated for this year is gear driven, the chain drive being discarded. The leading wheels are independent, the gearing driving the rear wheels which are larger than the leading wheels. Brake blocks are also fitted to the driving wheels together with sanding boxes and gear. Type 2-2-0WT-G.

1883 Similar to the 1879 locomotive, but not fitted with brakes or sanding gear.

1894 Up till now the wheels were solid, but this illustration shows spoked wheels, together with brakes and sanding gear. Type 2-2-0WT-G.

The description of the locomotive is similar and a sentence added: 'They are capable of hauling twice the load taken by ordinary direct acting tramway engines of equal nominal horsepower, with half their consumption of fuel'. In the same catalogue appears a testimonial from the Halling Lime and Cement Works of Wm. Lee, Son & Co. dated 25 February, 1893, part of which is of interest:

'We now have three in full employment (the first was bought in 1878) and you are making us a fourth. . . .

The track on which they run is about 1½ miles long from our pit to the works, with a gauge of 4′ 3″ and a gradient for the greater part of the distance averaging about 1 in 30, but in places where the engines have to draw the empty waggons, the gradient is as much as 1 in 20. The locomotives climb these inclines drawing 4 full waggons or 12 to 14 empty ones, with perfect ease at a good speed. . . .

The fuel consumption is about 7 cwt of coal per diem and the wages, paid for cleaning and attendance at the present time (and this has not varied much in the last ten years) are about 45/- per week of 56 hours.'

1900 This catalogue shows a more sophisticated locomotive with commodious cab, conventional buffers, cover over gearing, sanding fore and aft, brakes on all wheels, and gear drive to both axles. Locomotives made are of both the compound and high pressure types 0-4-0WT-G.

The two axles were connected with a stiffening bar with bearings outside the wheels on the centre line of the axles. The illustration of No. 4399 which was built in 1899 is on similar lines to that shown in the 1900 catalogue except that no cab was fitted and its characteristics can be clearly seen (Fig.15).

The note in the 1894 catalogue regarding the cost of cleaning and attendance is an interesting side light on wage levels and the stable conditions for a period of ten years from 1883!

AVONSIDE ENGINE CO. LTD.
Bristol

See STOTHERT, SLAUGHTER & COMPANY

BAGNALL, W. G.
Castle Engine Works, Stafford

Founded in 1875 by William Gordon Bagnall, who took over the small millwrighting business of Massey and Hill. For the first few years with a small work force he manufactured agricultural machinery, colliery plant and general engineers' requirements. The first locomotive was built in 1876 a standard gauge 0-4-0ST, No. 16 in Bagnall's list, and it was not until 1893 that locomotives bore consecutive works numbers. Previous to this date all products were included in the list.

No. 16 was sold to the Duke of Buckingham for his Wotton tramway followed by a second 0-4-0ST in 1877. A peculiarity of the early locomotives built was the tank which in shape was an inverted saddle tank passing under the boiler instead of over the top.

About ten years after W. G. Bagnall had started the firm Mr Samuel T. Price joined him, and having had considerable experience with

Fig. 16
W. G. Bagnall
2058/1917
typical narrow gauge
0-4-0ST
War Office

Fig. 17
W. G. Bagnall
2467/1932
metre gauge
0-6-0T
for China

steam engines at the Lilleshall Co. and on the Midland Railway at Derby, he was to prove the mainstay of the company for design and production for many years.

About 140 locomotives were built up to 1892 but a reputation for high class workmanship developed and orders were increased year by year.

In 1887 the firm became W. G. BAGNALL LTD.

All types of railway equipment was supplied, in fact the firm could supply complete railways, manufacturing turntables, carriages and wagons etc.

A considerable export trade developed for Europe, India, Ceylon, Java, Africa, Australia, South America and many other countries.

In 1891, E. E. Baguley joined the company and both he and Price influenced the subsequent success of the firm.

The former left in 1903 and started his own engineering business in Burton-on-Trent where he built a number of steam locomotives.

Specialising in light and narrow gauge locomotives it was not until 1926 that quantity orders were received for standard gauge stock including twenty-five LMS 0-6-0Ts and seven similar tanks for the Somerset and Dorset Railway. In 1930 fifty 0-6-0PTs were supplied to the GWR.

The first fireless locomotive appeared in 1924 for Edward Lloyd of Sittingbourne 2′ 6″ gauge. It was a 2-4-0T of conventional steam storage type, with a charging pressure of

Fig. 18
W. G. Bagnall
2508/1934
Glascote Colliery
AMINGTON No. 3

Fig. 19
W. G. Bagnall
2578/1937
Karabuk
Iron & Steel Co
(Turkey) No. 2

220lb/sq.in. An innovation was a reducing valve, giving a constant pressure of 80lb/sq.in. to the cylinders. Fifteen fireless locomotives were built with four and six coupled wheels.

During the second world war besides building locomotives for Turkey, India, Africa and Malaya many 0-4-0ST and 0-6-0ST type were built for use at home for the Ministry of Supply, War Office and for industry. Fifty-two 0-6-0ST *Austerity* class were built for the MOS during the war years.

After the war the works were busy catching up on their orders and in 1949 work commenced on an order for fifty 0-6-0PT for the GWR, the taper boilers being supplied from Swindon, and the last of this order was not completed until June 1954 well after nationalisation.

After the death of Mr W. S. Edwards who had acquired all the shares of the company, it was decided to place the firm on the market and in 1951 the firm became BRUSH–BAGNALL TRACTION LTD as an associate of Brush Electrical Engineering Co. Ltd., Loughborough.

By this time Bagnalls had entered the diesel locomotive market and orders for steam declined and the last one, delivered in 1957, was Works No. 3126, a 2-8-2 tender narrow gauge locomotive for Mysore Iron & Steel Works, India.

The firm changed ownership a number of times, eventually becoming part of the English Electric group and like many other works ceased building locomotives.

In all, approximately 1660 steam locomotives were built from 1876 to 1957.

The 'Castle' works was comparatively small, employing under 500 workmen. At the finish there were two erecting shops (a new one built for Diesels and the original shop) with the usual complementary machine shop, forge, brass and iron foundries, frame shop, wheel and press shop, pattern shop, boiler shop and other smaller departments.

Apart from locomotives there was a considerable business in spare boilers, wheel sets, boiler plates and spares.

The casting of steam locomotive cylinders was a speciality of the firm and many were made for Crown Agents, home railways, India and other overseas customers.

Fig. 20 W. G. **Bagnall** 2996/1951 Austin Motor Co Ltd VICTOR

BAGULEY, E. E. LTD.

Shobnall Road, Burton-on-Trent.

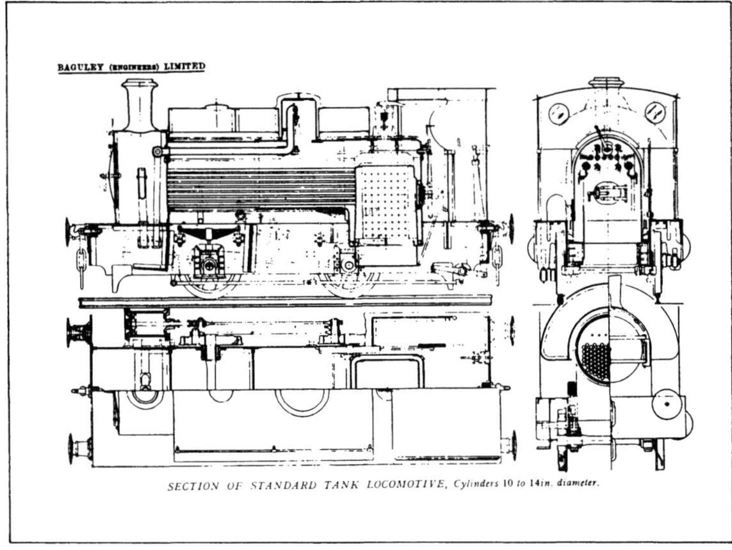

Fig. 21
E. E. Baguley
2001/1920
Section of the only standard gauge locomotive built

The firm was founded by Major Ernest E. Baguley in 1911 as BAGULEY (CARS) LTD. in the old Rykneild lorry works at Shobnall. Major Baguley had been at W. G. Bagnall Ltd., Stafford, and ultimately became their chief draughtsman and it was whilst he was with this firm that he designed and patented a locomotive valve gear in 1893. He left Bagnalls to become works manager of the Rykneild Co. in 1903 when the latter was established, and left them in 1907 for a similar position with the B.S.A. Co's motor division. His rank of Major was achieved in the Volunteers, serving in France with the BEF in 1914–15 and receiving the Territorial Decoration.

Baguley (Cars) Ltd. commenced building internal combustion road vehicles and continued with railcars, inspection cars, locomotives and rolling stock.

Usually locomotive building firms commenced with steam traction and introduced petrol and diesel at a later date. Not so at Baguleys – in this case we find steam locomotive construction starting some six years after the I.C. types had begun to be made.

In 1920 the firm's title was changed to BAGULEY (ENGINEERS) LTD. In that same year a standard gauge petrol-hydraulic locomotive, which was laid down in 1914 and held up due to the war – the 'campagne' transmission not having arrived – was converted into a steam locomotive. Its original works number (621) was altered to No. 2001 and all subsequent steam locomotives had their works numbers allocated in this series (2001 to 2032).

This prototype was a 0-4-0ST with outside cylinders 13″ x 18″ and 3ft(?) diameter wheels. It was the only one built to standard gauge, all the others being for narrow gauge railways; it was completed on 23 February, 1920 and was sold to Thomas Salt & Co. Ltd. of Burton (Fig. 21).

Baguley's valve gear was fitted to all the locomotives built here, either in the form of the original design or in a 'modified' version.

In the original valve gear, the movement is obtained from a third rod on the crankpin serving the connecting and coupling rod bearings. At the end of the rod is a forked rod attached to and swinging on a fixed pin on a curved link, providing the necessary lap and lead travel. On the same third rod between the forked rod and crankpin is a connecting rod and linkage to the upper die block in the curved link which gives the necessary valve spindle travel. The 'modified' version of this valve gear eliminated the curved link. The valve spindle was extended to the motion bracket and the spindle ran on an oscillating eccentric, the motion of which was derived from the third rod and its linkage. At least twenty-six of the locomotives produced had this valve gear and of course it was fitted to a number of Bagnall built narrow gauge locomotives.

The boilers fitted were either the conven-

Works No	Order Date	Del'd Date	Type	Cylinders inches	Gauge ft in		Name	Customer
2001		23/2/20	0-4-0ST	13 x 18	Std	i.f.(n)(p)(Fig. 21)		Thomas Salt & Co Ltd (a)
2002	6/2/19	8/11/20	0-4-0T	6 x 9	2 0	i.f.(o)(q)		L.R. Ltd (b)
2003	6/2/19	8/11/20	0-4-0T	6 x 9	2 0	i.f.(o)(q)		L.R. Ltd (b)
2004	6/2/19	17/11/20	0-4-0T	7 x 12	60 cm	i.f.(o)(q)	Flanders	L.R. Ltd (b)
2005				8 x 12		i.f.(o)(q)		For Stock
2006	6/2/19	18/6/21	0-4-0T	4 x 8	60 cm	i.f.(o)(q)	Winnie	Derwent Valley WB (c)
2007	7/7/19	28/11/21	0-4-0T	4 x 8	2 0	i.f.(o)(q)		L.R. Ltd India (d) (r)
2008	7/7/19	7/9/20	0-4-0T	5 x 8	2 0	i.f.(o)(q)	Winnie	Via Gellia Colour Co Ltd
2009	19/12/19	8/2/21	0-4-2T	6 x 9	75 cm	i.f.(n)(p)	Gallipoli No 1	L.R. Ltd Bangkok
2010	19/12/19	24/3/21	0-4-2T	6 x 9	75 cm	i.f.(n)(p)	Queen Mary	L.R. Ltd Bangkok
2011	5/5/20	10/1/21	0-4-0T	7 x 12	60 cm	o.f.(n)(q)	Flanders	L.R. Ltd
2012	5/5/20	18/3/21	0-4-0T	7 x 12	60 cm	o.f.(n)(q)		L.R. Ltd
2013	5/5/20	29/4/21	0-4-0T	7 x 12	60 cm	o.f.(n)(q)		L.R. Ltd
2014	16/10/20	23/5/21	0-4-0T	6 x 9	60 cm	i.f.(o)(q)	Eaglehurst	Henry Boot & Sons Ltd (e)
2015	16/10/20	31/5/21	0-4-0T	6 x 9	60 cm	i.f.(o)(q)	Trent	(f)
2016	11/11/21		0-4-0T	7 x 12	60 cm	i.f.(o)(q)(Fig. 22)	Violet	L.R. Ltd (g)
2017			0-4-0T	7 x 12	80 cm	i.f.(o)(q)		For stock
2018		9/3/22	0-4-0T	7 x 12	2 6	o.f.(n)(q)		L.R. Ltd India
2019		30/3/22	0-4-0T	7 x 12	2 6	o.f.(n)(q)		L.R. Ltd India
2020	23/3/23	31/3/25	0-4-0T	4/6 x 9	2 6	o.f.(n)(q)(Fig. 23)		L.R. Ltd (h)
2021	16/5/22		0-4-0T	6 x 9	60 cm	i.f.(m)(o)(q)		L.R. Ltd Singapore (i)
2022	16/5/22		0-4-0T	6 x 9	60 cm	i.f.(m)(o)(q)		L.R. Ltd
2023	16/5/22	31/3/25	0-4-0T	7 x 12	60 cm	i.f.(m)(o)(q)		L.R. Ltd India
2024	16/5/22		0-4-0T	7 x 12	60 cm	i.f.(m)(o)(q)		L.R. Ltd India
2025	2/7/22	30/11/22	0-4-0ST	6½ x 9	3 6	i.f.(m)(n)(q)		L.R. Ltd Australia
2026	2/7/22	19/12/22	0-4-0ST	6½ x 9	3 6	i.f.(m)(n)(q)	Kangaroo	L.R. Ltd Australia
2027	2/7/22	2/1/23	0-4-0ST	6½ x 9	3 6	i.f.(m)(n)(q)		L.R. Ltd Australia
2028	12/23	8/9/24	4WTG		75 cm			Egyptian Delta Light Rly (j)
(2029)	7/25		0-4-0F	6 x 9				High Pressure Boiler (k)
2030	3/26	15/9/26	0-4-2T	6 x 9	75 cm	i.f.(n)(p)		L.R. Ltd Bangkok
2031	9/27	14/12/27	0-4-0T	5 x 8	2 0	i.f.(m)(n)(q)	Enterprise	Jones Benton & Co Penang
2032	1/28		0-4-0T	6 x 9	60 cm	i.f.(m)(o)(q)		For stock
2033 to 2041	*c*.1930		Nine Power Units for Baguley-Devlin Steam Railcars					H. Cegielski Warsaw, Poland (l)

(a) Originally Works No 621. (b) Light Railways Ltd. (c) Built for the Yorkshire Engine Co. Their Works No 1878. (d) Restored by Messrs ICI Ltd Rishra Works, Calcutta. (e) Built for the Yorkshire Engine Co. Their Works No.1887. (f) Built for the Yorkshire Engine Co. Their Works No 1888. (g) Was also named 'FLANDERS'. (h) 'Rail-Grip' Locomotive. (i) 'Aden' Class 2021-2024. (j) Baguley-Clarkson. (k) High pressure superheated boiler (not built). (l) Piston Valves. (m) Wood burners. (n) Loco type boilers. (o) Circular firebox. (p) Baguley valve gear. (q) Modified Baguley valve gear. (r) Now preserved in England.

tional type of loco boiler and firebox, or with a circular firebox which had been developed by Mr Baguley whilst at Bagnalls, and were similar to those designed by Sir Arthur Heywood at Duffield Bank but of an improved design.

The circular type produced sufficient steam provided a good class fuel was used, but it was difficult to maintain pressure when using wood or other low grade fuels.

All the narrow gauge locomotives had single slide bars, sandbox astride the boiler and safety valves mounted over the round top fireboxes. The majority of them were built to the order of Light Railways Ltd. and were shipped abroad.

In November 1931 the Shobnall works were closed and as a temporary arrangement a small works in Clarence Street was occupied until a new works in Uxbridge Street was opened in the summer of 1934.

On comparing the date of order with date of delivery, in many cases the time taken varied considerably from three months to twenty four months and it is probable that priority was given to other work in many cases.

As will be seen from the works list, a total of thirty-one steam locomotives was built at the Shobnall Road works.

An experimental high pressure design was put forward with a Clarkson Thimble boiler, high speed two cylinder superheated engine with chain drive. It was proposed to build this basic design in a range of sizes from 50 to 150 hp and as an alternative to chain drive they could be adapted to geared drives with coupling rods for use on heavy incline work fitted with a patent two speed gear (No. 2029). The locomotive was never built.

Perhaps the most interesting locomotive built was the RAILGRIP (Works No. 2020) (Fig. 23).

This was a specially designed locomotive built to 2′6″ gauge to the order of the Light

Fig. 22 E. E. Baguley 2016/1921 0-4-0T VIOLET

Railways Ltd. in conjunction with the Drewry Car Co. Ltd. dated 23 March, 1923. It consisted of a standard 0-4-0T with two cylinders 6″ x 9″. Two further cylinders were mounted at the base of the smokebox and an additional pair of wheels located between the two normal axles were driven by these cylinders through a jackshaft geared through two sets of reduction gears to this axle. These wheels were known as 'Railgrip' and were raised or lowered on to the track. The wheels were grooved to prevent slip and the whole conception was to assist the locomotive when climbing heavy gradients but the rails on the heavy gradient sections would have to have corresponding notches into which the middle pair of wheels engaged. This was a form of rack and pinion principle but the idea was not developed and this locomotive remained the sole representative.

In 1914 a working arrangement was made with the Drewry Car Co. Ltd., a company established in 1903, for the manufacture of internal combustion Railcars and in 1915 Baguley purchased from the liquidator the old established internal combustion locomotive business of McEwan Pratt & Co., but this only affected the I.C. side of the business. In 1931 the general specifications for steam locomotives (although none were ordered or built after 1928) included for a boiler with a working pressure of 150lb/sq.in. The longitudinal seams of the barrel were double rivetted lap joints and the circular seams single rivetted lap joints. The firebox was of copper and the tubes were brass, 12 to 14 SWG expanded at the smoke box end and ferruled at the firebox end. Ramsbottom type safety valves were fitted and the boiler was insulated with wood battens clothed with planished steel sheets. The boiler was fed by two Gresham & Craven self acting restarting injectors. The axleboxes and guides were cast iron fitted with heavy phosphor bronze bearings lined with white metal with syphon lubrication. Hand brakes only were fitted. The valve motion was forged steel and narrow gauge locomotives were fitted with Baguley outside radial gear, and the standard gauge shunting engines with Stephenson's Link Motion.

The firm continues to build Diesel locomotives under the name of BAGULEY–DREWRY LTD.

Fig. 23 E. E. **Baguley** 2020/1923 'Rail-Grip' patent locomotive

BALLOCHNEY RAILWAY

Greenside, Nr. Coatbridge

Incorporated in May, 1826, this 4′ 6″ gauge railway was originally worked by horses and opened in part in 1828.

When locomotives were introduced they were supplied by the neighbouring Monkland and Kirkintilloch Railway, and the Greenside shops were used by both railways, the M&KR paying rent for this privilege. The shops were situated roughly on the site of the later Kipps locomotive shed, beside the M&KR a few hundred yards from its junction with the Ballochney Railway.

George Dodds, the M&KR's engineer, designed and built M&K No. 3, completed about May 1834, having vertical cylinders 10⅜″ x 24″ and 4′ 0″ driving wheels and generally similar to the *Killingworth* type, two of which were supplied to the railway, by an outside builder. George Dodds died in 1835 and was succeeded by George Lish who came from the Dundee & Newtyle Railway. However, William Dodds, the late George's son, supervised the engineering work of both companies from 1835 to 1838 and was responsible for the construction of the second locomotive built at Greenside. This was M&K No. 4 named VICTORIA with 10½″ x 24″ cylinders and 4′ 0″ driving wheels, completed late 1837 or early 1838. It was probably similar to No. 3 in design.

According to Mr Jas. F. McEwan, at least one locomotive was built by the Ballochney Railway, named BALLOCHNEY, and was completed in 1836. No further details are available.

In February 1838 the M&KR announced that the shops at Greenside were inadequate and that new shops were being built at Moss-side, across the railway from Greenside. They were completed in that year and all major work was transferred from Greenside.

In 1848 the Ballochney Railway became part of the Monkland & Kirkintilloch Railway. (See MONKLAND& KIRKINTILLOCH RAILWAY.)

BALMFORTH BROS

Peel Ings Foundry, Rodley, Yorks.

This firm were manufacturers of steam cranes, hoists and various type of lifting gear.

At least three locomotives can be traced, two worked at the Piel and Walney Gravel Co. Ltd., both were built about 1876 and were bought second hand from unknown sources, the third worked at Annan.

All had vertical cross tube boilers when new, outside inclined cylinders at the driving end, four coupled wheels with outside cranks. The connecting rods had marine type big ends. The Stephenson's link motion operated through rocking shafts.

At least one of the Piel & Walney locomotives was rebuilt with a conventional loco type boiler and both were scrapped in 1959.

BANKS, THOMAS & CO.

Date	Type	Cylinders inches	D.W. ft in	Customer	No/Name
7/1835	0-4-0			London & Southampton Rly	Alpha
1839	0-4-2	12 x 18	5 0	Liverpool & Manchester Rly	61 Mammoth
1839	0-4-2	12 x 18	5 0	Liverpool & Manchester Rly	63 Mastodon
1840	(a)			Birmingham & Gloucester Rly	16 Moseley
1840	(a)			Birmingham & Gloucester Rly	17 Pivot

(a) Possibly 4-2-0 'Norris' type

BANKS, THOMAS & COMPANY
Manchester

This firm of general engineers in 1833 tendered for the supply of a stationary engine for the Swannington incline on the Leicester and Swannington Railway but were unsuccessful, but from this it is probable that they were builders of stationary engines.

In 1835 they sold a 0-4-0 ballast locomotive to the London & Southampton Railway and it was used on the construction of the railway, but by 1839 it had been laid aside as unserviceable.

In 1839 two 0-4-2s were supplied to the Liverpool and Manchester Railway, Nos. 61 and 63 with 12″ x 18″ cylinders and 5ft diameter coupled wheels. No. 63 survived long enough to become Grand Junction Railway No. 119 in 1845, probably due to that fact that it had been rebuilt in 1841, whereas No. 61 had not been similarly dealt with.

The following year two locomotives were supplied to the Birmingham and Gloucester Railway, Nos. 16 and 17, but what type they were is not known. One was supplied with two sets of wheels 4′ and 5′ diameter.

It is possible that this firm were only agents for the supply of locomotives, but this is conjecture only.

BARCLAY, ANDREW SONS & CO. LIMITED
Caledonia Works, Kilmarnock

The firm was established in 1840 and locomotive building commenced in 1859 when the first of a long line of four coupled saddle tanks appeared. These were supplied principally to Scottish ironworks, collieries and contractors. The early tank locomotives were rather primitive in appearance with no cabs, square sided tanks, cast iron wheels and few had splashers. The design for both four and six coupled tank locomotives was gradually developed until very large and powerful machines were turned out and they became for many a colliery and steelworks, maids of all work. Until the last decade they could be found in abundance in South Wales, northern England and the Midlands. Whereas the Peckett tank – Barclays principal English competitor – rarely penetrated over the border, the Barclay tanks could be seen in most English counties.

The first departure from standard locomotives was a unique 'single' 2-2-2 for the Caledonian Railway in 1871. It was fitted with John Miller Ure's patent which provided for the driving wheels to be lifted off the rails when coasting. The apparatus was subsequently removed.

Fig. 24 **Balmforth Bros** Piel & Walney Gravel Co.

BARCLAY, ANDREW SONS & CO. LIMITED

Fig. 25
Andrew Barclay
109/1871
St. Bees Quarry,
Whitehaven
CHARLTON

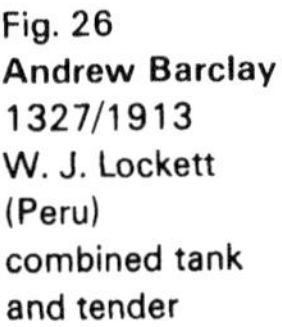

Fig. 26
Andrew Barclay
1327/1913
W. J. Lockett
(Peru)
combined tank
and tender

In 1881 a four coupled Crane tank was built for John Orr Ewing & Co. and was sent to Alexandria, and a total of 38 of this type was built, the last appearing in 1947. Another speciality was the fireless locomotive particularly useful in Munitions works and other factories with a high fire risk. The boiler became a receiver which was charged with steam from stationary boilers. Many standard locomotives were sent abroad and others of special designs went to Ireland, South America, Norway, Jersey, Spain and Canada. Some articulated locomotives went to New Zealand and in 1905 two railmotor units were built for the Great North of Scotland Railway. Nevertheless the main production was four and six coupled saddle tank locomotives. The largest batch orders were for twenty-five 0-6-0WT for the War Office in 1917 to 60 cm. gauge and for a similar quantity of London Midland & Scottish Railway standard Class 4 0-6-0s in 1927. Another comparatively large order was for ten 2′ 6″ gauge 4-8-0s for the Sierra Leone Government Railway in 1943. Fifteen *Austerity* 0-6-0Ts were built for the War Department from 1944 to 1946.

The works were busy for the first few years

after the War, but the demand for steam locomotives declined and few were built after 1955. The last one was works No. 2377 – a 0-6-2 for Indonesia the order having been transferred from W. G. Bagnall & Co. who had ceased building. It was steamed in September 1962. However the firm had conmmenced building diesel locomotives about 1935 and these are still being built.

BARCLAYS & CO. who had set up a separate establishment at the Riverbank Works were merged with the parent company in 1888 (See BARCLAYS & CO.).

Andrew Barclay, Sons & Co. are now the only locomotive manufacturers in Scotland having outlived the many firms which were established, including the largest of them all – the North British Locomotive Co. Ltd., whose goodwill, drawings, patterns etc. were acquired by Andrew Barclay. In 1972 the firm merged with the Hunslet Group which purchased the equity, and the Kilmarnock factory is being fully utilised.

The estimated number of steam locomotives built is 2052.

Fig. 27 **Andrew Barclay** 1654/1920 Sheffield Coal Co. VICTORY

Fig. 28 **Andrew Barclay** 1934/1927 Fireless locomotive Preston Corporation DUKE

BARCLAYS & CO.

River Bank Engine Works, Kilmarnock

This company was promoted by Andrew Barclay of ANDREW BARCLAY SONS & CO. LTD. in 1872 for his four sons and brother John who were made partners and these works were almost exclusively confined to the construction of locomotives. It was the probable intention of Andrew Barclay to concentrate the locomotive side of the business to Barclays & Co. and to concentrate on winding engines, pumps and other colliery work at Caledonia Works.

In a locally published *Life of Andrew Barclay* dated about 1879 it was stated that in seven years Barclays & Co. had built 76 locomotives from the commencing date of 1872.

Riverbank Engine Works lay adjacent to a mineral line running from Lanarkshire to the Kilmarnock & Troon line but by-passing Kilmarnock, providing easy access to the railway works, whereas the Caledonia works was separated by a roadway which was not so convenient, however, at the new works there was less room to expand being cheek by jowl with the firm of Glenfield & Kennedy.

In 1888 it was decided to concentrate all the work at the Caledonia works and the Riverbank works were sold. The two sets of works numbers were added together and after Andrew Barclay's No. 311 of 1888 the next one became No. 636 of 1888. One would assume that Barclay's & Co. had produced 324 locomotives during the period 1872–1888 but evidence indicates that the works numbers commenced at 201 and probably 124 is a more realistic figure for their output.

The majority of the output consisted of 0-4-0STs with some crane tanks and six coupled tanks.

BARNINGHAM, WILLIAM

Pendleton Ironworks, Manchester

About seven locomotives were built for their own use to standard gauge; both four and six coupled types. The firm was established about 1863, and covered the period 1867–1878. No photographs or details are available.

Fig. 29
Barclays & Co.
282/1881 Bairds & Scottish Steel — Clippens Limeworks

BARR & McNAB
Paisley

Three locomotives were built in 1840, two were for the Ardrossan Railway and were 0-4-0s of the *Bury* type with bar frames; cylinders were 12" x 18". According to Wishaw they cost £1150 each, complete with tenders. They were named FIREFLY and KING COLE, and were probably renamed EGLINTON and BLAIR respectively. They were not taken over by the Glasgow & South Western Railway when the Ardrossan Railway became part of the G&SWR in 1854.

In the same year a single was built for the Glasgow, Paisley & Greenock Railway (their No. 8 named HAWK) but no further particulars are known.

Date	Type	Cylinders inches	Name
1840	2-2-2		Hawk (b)(c)
7/1840	0-4-0	12 x 18	Firefly (a)(d)
8/1840	0-4-0	12 x 18	King Coil (a)(d)

(a) Cost £1150 each 'Bury' type. (b) No. 8
(c) Glasgow, Paisley & Greenock Railway
(d) Ardrossan

BARR MORRISON & CO.
See ALLAN ANDREWS & CO.

BARROW & COMPANY

This firm built for the York & North Midland Railway. Nothing else is known.

BARROW HAEMATITE STEEL CO. LIMITED
Barrow-in-Furness, Lancs

Built three 0-4-0ST locomotives with outside cylinders Nos. 4, 6 and 9. No. 9 was rebuilt from narrow gauge in 1914. No. 4 and No. 6 were built for the standard gauge system. The firm was orginally SCHNEIDER & HANNAY.

BEALE, J. T.
Whitechapel & Greenwich, London

The London & Croydon Railway's Engineer Mr Gibbs suggested to the Board on 6 November, 1837 that it should obtain prices for locomotives and this firm was one of those suggested.

Nevertheless no engines can be traced as being built by them.

WILLIAM BEARDMORE & CO. LIMITED

Dalmuir & Parkhead, Glasgow

A forge at Parkhead was purchased by Robert Napier in 1835 who ultimately became the 'father' of the shipbuilding industry on the Clyde. One of his first contracts of importance was the manufacture of the engines for the *British Queen*, the first steamship to cross the Atlantic in 1839. He acquired a yard at Govan and commenced ship building in the early 1840s.

Robert Napier retired in 1868 and William Rigby, his son-in-law, took over the Parkhead Forge, but it was during the 1860s whilst Napier was still in office that the original William Beardmore came into partnership with Rigby at Parkhead.

On the retirement of Rigby his place was taken by Isaac Beardmore, a brother of William and the firm became W. & I. BEARDMORE.

In 1886 young William Beardmore, son of the original William, took sole charge of the business.

The story of Beardmore's is a story of widely different fields of engineering—ships, armour plate, ordnance including guns, tanks, warships, aeroplanes, airships, and motor cars, outside the range of the present subject.

The Dalmuir works had been purchased in 1900 and were mainly concerned in the building of warships and ordnance. At the cessation of hostilities in 1919 all departments swung over to peace time production and at Dalmuir it was decided as soon as the gun contracts had been completed, to convert the shops for locomotive production. The heaviest task was the structural alterations to the girders and even the walls of the buildings for the installation of new and much heavier travelling cranes. In the erecting shop three pits were made running the entire length of the shop. The centre track incorporated five gauges: Metre 3′ 6″; Standard 5′ 3″ and 5′ 6″.

Special valve setting arrangements were provided on the centre track so that locomotives could pass on this track straight through to the steaming shed. The steaming shed was provided with friction rollers by means of which the finished locomotive could be subjected to a running test while still under cover.

A special advantage this firm had, was that every part was manufactured by the firm itself or its associated concerns. Another advantage was that the locomotive works adjoined the shipyard fitting out basin where vessels could be loaded up without any transit charges, either as complete locomotives and tenders, or in parts in crates, for the overseas markets.

At the time of conversion of the shops both at Dalmuir and Parkhead, they claimed to have one of the largest locomotive works in the world.

All new locomotives were constructed at Dalmuir and the majority of repairs and rebuilding at Parkhead; some repair work was done at Dalmuir when new orders diminished.

Locomotive building commenced in 1920 and most of those for abroad were built to the requirements of the Crown Agents for the Colonies, and Messrs. Rendel Palmer & Tritton,

Works Nos.	Date	Type	Cylinders inches	D.W. ft in	Customer	Running Numbers	Gauge ft in	Remarks
100-134	1920	2-8-0	OC 22 x 26	4 8½	East Indian Rly	1520-1554	5 6	B S
135-154	1920	4-6-0	IC 20 x 28	6 6	Great Eastern Rly	(a)	Std	B S
155-173	1920	4-8-0	OC 18 x 23	3 6¾	Nigerian Rlys	282-300	3 6	B S
174-234	1921	4-6-0	IC 20½ x 26	6 3	L & NW Rly	Various	Std	S
235-263	1922	4-6-0	IC 20½ x 26	6 3	L & NW Rly Fig. 32	Various	Std	S
264-268	1923	0-6-6-0	16/24¼ x 20	3 3	Burma Rlys Fig. 30+31	16-20	Metre	B S Mallet
269-293	1924	4-6-0	OC 21½ x 26	6 2	BB & CIR	540-564	5 6	B S
294-295	1924	4-6-0	OC 21½ x 26	6 2	M & SMR	784-785	5 6	B S
296-298	1924	4-6-0	OC 21½ x 26	6 2	E. Bengal State Rly	31-33	5 6	B S
299-303	1924	4-6-0	OC 21½ x 26	6 2	Indian States (NW) Rly	2875-2877 2879-2880	5 6	B S
304	1924	4-6-0	IC 20½ x 26	6 3	LM & S Rly	5845	Std	S (X)
305-324	1927	0-6-2T	IC 18 x 24	4 10	L & NE Rly	2642-2661	Std	B Class N.7
325-349	1928	0-6-0T	IC 18 x 26	4 7	LM & S Rly	16600-16624	Std	B
350-385	1928	0-6-0T	IC 18 x 26	4 7	LM & S Rly	16685-16720	Std	B
386-414	1929	0-6-0T	IC 18 x 26	4 7	LM & S Rly	16721-16749	Std	B
415-416	1928	Electric			BB & CI Rly		5 6	
417-460		2-8-2			East Indian Rly	(b)	5 6	
461-486	1930	4-6-2	OC 23 x 28	6 2	I. NW Rly	1842-1867	5 6	
487-490	1931	4-6-2	OC 23 x 28	6 2	BB & CI Rly	613-616	5 6	
491-492	1931	4-6-2	OC 23 x 28	6 2	East Indian Rly	1948-1949	5 6	
493-494		Diesel			I. NW Rly	30-31	5 6	

(a) Running numbers: 1541-1547, 1550, 1548, 1549, 1551-1560
(b) Running numbers: 2016-2031, 1981-2003, 2032-2036

WILLIAM BEARDMORE & CO. LIMITED

none were built to the firm's own designs and specifications.

Works numbers, for some unknown reason, commenced at number 100 and nearly 400 were built from 1920 to 1931, out of these 220 went to home railways and the remainder to India, Burma, and Nigeria.

The most interesting were six *Mallet* 0-6-6-0 locomotives for the Burma Railways. They were compound having high pressure cylinders 16″ diameter and low pressure 24½″ diameter and a stroke of 20″. The wheels were 3′ 3″ diameter, with two rigid wheelbases of 8′ 3″. The boiler had a total heating surface of 1403 sq.ft. including 224 sq.ft. of superheater tubes. The working pressure was 180 lb/sq.in. Engines weighed 60½ tons in working order and the eight wheeled tender 36¼ tons.

Another interesting locomotive was the one built for the LMS in May 1924 for the Wembley Exhibition of 1924, which was of the *Prince of Wales* type and specially named PRINCE OF WALES for the occasion. This engine had outside Walschaert valve gear similar to five others of the same class so fitted in 1923–4 due to previous Joy's valve gear which was fitted inside the frames and required a hole in the connecting rod to which the valve drive was connected. This was a weak point and resulted in bent rods in some circumstances. This exhibition engine was painted LMS red and was probably the only LNW type locomotive to carry a chimney with a brass rim.

During 1925 and 1926 no new locomotives were built and the shops were fully engaged in repairing and rebuilding a large number of locomotives from the home railways brought about by the immense backlog accumulated during the war.

Among the locomotive types which received heavy repairs after the 1914–1918 war were:

At Dalmuir		
Glasgow & South Western Rly	4-4-0	several including No. 355
Glasgow & South Western Rly	0-6-0	No. 367
London & North Western Rly	0-8-0	No. 1881
North British Rly	0-6-0T	No. 821
North British Rly	0-6-2T	No. 251
Caledonian Rly	0-6-0T	No. 787
Highland Rly	4-6-0	No. 141 (rebuilt)
Highland Rly	4-4-0	No. 94
Highland Rly	0-6-0	No. 19
At Parkhead		
Great Northern Rly	4-4-2	
Great Western Rly	0-6-0	No. 2447
London & North Western Rly	0-8-0	
Great Eastern Rly	4-4-0	No. 1809
North British Rly	4-4-0T	No. 1461
North British Rly	0-6-0	No. 58
Caledonian Rly	0-6-0	No. 583
Great North of Scotland Rly	4-4-0	

Fig. 30 **W. Beardmore & Co.** 264/1923 Burma Railway 0-6-6-0 superheated compound *Mallet*

WILLIAM BEARDMORE & CO. LIMITED

Fig. 31 **W. Beardmore & Co.**
Unit for *Mallet* locomotive

Fig. 32
W. Beardmore & Co.
263/1922 L&NWR
Prince of Wales
class No. 2043

Fig. 33
W. Beardmore & Co.
Erecting shop April 1920
Dalmuir Works

BEDFORD ENGINEERING COMPANY
Bedford

This firm whose principal product was the steam crane, built a number of vertical boiler locomotives adapted from the travelling steam crane design. Two horizontal cylinders fitted on the framework drove on to a crankshaft with flywheel cranks, and a pinion keyed on, driving a set of gears on to one or both axles.

At least one narrow gauge locomotive of this type went to India, and probably more were sent abroad.

No record of such machines working in this country can be found. Dates of building would be around 1910.

From the specifications it would seem that they were only intended for light railway operation turning the scale at 4 tons to 6 tons approximately.

BELFAST & COUNTY DOWN RAILWAY
Queens Quay, Belfast

Minutes dated February 1882 stated 'a new engine and tender built in the Company's workshops has been put on the line'. This was a 0-4-2 running No. 8 and according to one source was a rebuild using parts of an old Bury 2-2-2 of 1848 running No. 2.

Another version is that although two 0-4-2 locomotives were claimed to have been built at Queens Quay, it is fairly sure that these were Fairbairn/Sharp Stewart engines built for the Newry & Armagh Railway in 1864 (N&A No. 6 and No. 7) and which were sold to the Belfast and County Down Railway, after passing through other hands, in 1880.

BELFAST & NORTHERN COUNTIES RAILWAY
York Road, Belfast

Locomotive building commenced at York Road in 1870 although the works had carried out all repair work and rebuilding since its inception. Building took place in two periods 1870 to 1873 and 1901 to 1903. Six locomotives were built consisting of two 2-4-0, one 0-4-2 and three 4-4-0.

The most interesting of these were the 4-4-0s which were compound having cylinders 18″ and 26″ diameter x 24″ stroke. These two cylinder compounds were very successful and more were built. The compounding was on the Worsdell-von-Borries principle. Before this happened the B&NCR was taken over by the Midland Railway of England and became known in 1903 as the NORTHERN COUNTIES COMMITTEE (MR).

The locomotive superintendent up to this time was Bowman Malcolm and after building four more 4-4-0 compounds, and a 0-4-0ST for shunting work, York Road produced four 2-4-2Ts for service on the Ballymena and Larne section which was a narrow gauge line of 3′ 0″. They were designed by Bowman Malcolm and were compound locomotives with two cylinders, the high pressure 14¾″ diameter was on the left hand side outside the frames and the low pressure 21″ diameter on the right hand side. The last two, Nos.103 and 104 had their boilers supplied from the Derby works of the Midland Railway. They were handsome engines with side tanks the whole length of the boiler and smokebox and the two pony trucks had outside bearings. At the grouping of the railways in 1923 the railway became NORTHERN

Fig. 34 **B&NCR** 1873 *Class M*
No. 26 in NCC livery, Derby pattern chimney

BELFAST & NORTHERN COUNTIES RAILWAY

COUNTIES COMMITTEE (LM&SR) and building at York Road continued with a batch of eleven 4-4-0 simple engines and were built over a period from 1924 to 1936. Some had 18" cylinders (Class U.1.) and the others had 19" cylinders (Class U.2) all had 6ft coupled wheels.

In 1934 the building of the well known 2-6-0s was commenced and eleven of these were built. They were a tender version of the LMS standard 2-6-4Ts of that period, but with 6ft driving wheels.

After 1942 building of new locomotives ceased at York Road.

P No	Date	Type	Cylinders inches	D.W. ft in	No.	Name	Class	Notes	W'drawn
BELFAST & NORTHERN COUNTIES									
1	1870	2-4-0	15 x 20	5 6	5		-		1906
2	1871	2-4-0	15 x 20	5 6	4		-		1905
3	1873	0-4-2	16 x 24	5 1½	26		M	(Fig. 34)	1925
4	1901	4-4-0	18/26 x 24	6 0	34	Queen Alexandra	A	(a)	1950
5	1902	4-4-0	18/26 x 24	6 0	3	King Edward VII	A		1949
6	1903	4-4-0	18/26 x 24	6 0	4		A		1949
MIDLAND RLY – NCC									
7	1904	4-4-0	18/26 x 24	6 0	9		A	Re-No. 69	1954
8	1905	4-4-0	18/26 x 24	6 0	20		A	Re-No. 84	1929
9	1906	4-4-0	18/26 x 24	6 0	5		A	Re-No. 59 (b)	1934
10	1907	4-4-0	18/26 x 24	6 0	17		A	Re-No. 58	1954
11	11/1908	2-4-2T	14¾/21 x 20	3 9	112		S	Re-No. 102 then 42	1954
12	3/1909	2-4-2T	14¾/21 x 20	3 9	113		S	Re-No. 101 then 41	1954
13	1914	0-4-0ST	16 x 22	4 0	16		N		1951
14	9/1919	2-4-2T	14¾/21 x 20	3 9	103		S	(Fig. 35)	1938
15	3/1920	2-4-2T	14¾/21 x 20	3 9	104		S	Re-No. 43	1954
LM & S RLY – NCC									
16	1924	4-4-0	18 x 24	6 0	1	Glenshesk	U1		1947
17	1924	4-4-0	18 x 24	6 0	2	Glendun	U1		1947
18	1925	4-4-0	19 x 24	6 0	79	Kenbaan Castle	U2		1956
19	1925	4-4-0	19 x 24	6 0	80	Dunseverick Castle	U2		1961
20	1925	4-4-0	19 x 24	6 0	81	Carrickfergus Castle	U2		1957
21	1926	4-4-0	18 x 24	6 0	3	Glenaan	U1	(c) (Fig. 36)	1946
22	1929	4-4-0	19 x 24	6 0	84	Lissanoure Castle	U2		1961
23	1931	4-4-0	18 x 24	6 0	4	Glenariff	U1	Re-No. 4A	1949
24	1934	4-4-0	19 x 24	6 0	85		U2		1960
25	1934	4-4-0	19 x 24	6 0	86		U2		1960
26	1934	2-6-0	19 x 26	6 0	94	The Maine	W		1965
27	1934	2-6-0	19 x 26	6 0	95	The Braid	W		1964
28	1935	2-6-0	19 x 26	6 0	96	Silver Jubilee	W		1961
29	1935	2-6-0	19 x 26	6 0	97	Earl of Ulster	W		1965
30	1936	4-4-0	19 x 24	6 0	87	Queen Alexandra	U2		1957
31	1937	2-6-0	19 x 26	6 0	98	King Edward VIII	W		1964
32	1938	2-6-0	19 x 26	6 0	99	King George VI	W		1965
33	1938	2-6-0	19 x 26	6 0	100	Queen Elizabeth	W	(Fig. 37)	1959
34	1939	2-6-0	19 x 26	6 0	101	Lord Massereene	W		1956
35	1940	2-6-0	19 x 26	6 0	102		W		1956
36	1942	2-6-0	19 x 26	6 0	103	Thomas Somerset	W		1959
37	1942	2-6-0	19 x 26	6 0	104		W		1965

(a) Renamed KNOCKLAYD class A.1. (b) Rebuilt to class U.2. (c) Was GALGORM CASTLE until 1931.

Fig. 35 **B&NCR** 1919 narrow gauge 2-4-2T No. 103 compound

Fig. 36 **B&NCR** 1926 *Class U1* No. 3 GLENAAN Staff catching apparatus

Fig. 37 **B&NCR** 1938 *Class W* No. 100 QUEEN ELIZABETH

BELLIS & SEEKINGS
Broad Street, Birmingham

This firm took over the engineering business of R. Bach & Company in 1849. Although it has been stated that a number of small tank locomotives were built for contractors and industrial firms none can be traced and the drawing of the 0-4-0ST may have been used for advertising only.

In 1866 PRIMUS, a 0-4-2WT was built for Pike Bros. of Fayle in Dorset.

In 1872 the firm moved into premises in Ledsam Street, and it was there that the second locomotive for Pike Bros. was built in May 1874. This was a 0-6-0WT fitted with a marine type boiler outside cylinders and frames. Stephenson's valve gear was also fitted outside the frames (Fig. 38). Note the distinctive combined dome and safety valve. The cylinders were 6″ x 12″, boiler pressure 120lb/sq. in. and weight in working order 12 tons. The motion was enclosed in similar fashion to tram engines, with inspection covers. The boiler lagging created a step between boiler and smokebox. After a heavy repair at Stephen Lewin's factory at Poole in 1880, it received a new boiler in 1936 from Peckett & Son and about 1948 the cylinders were rebored by the Dorset Foundry. It was active until 1955 when the line was closed. The whole railway was bought as scrap by Messrs. Abelson of Sheldon, who offered SECUNDUS to the Birmingham City Museum and it was handed over at a ceremony at Abelson's on 25 July, 1955 and now reposes in the museum. This was a fine gesture, the locomotive being the only remaining locomotive built in Birmingham, and they were very few. It must be recorded that it was due to the initial efforts of the Birmingham Locomotive Club that the preservation became a reality.

Dimensions: cylinder 6″ x 12″; boiler pressure 120lb/sq.in.; weight W.O. 12 tons; wheelbase 6′ 0″; wheels 2′3″ diameter. Although the locomotive was known as 'Secundus' it did not actually bear that name.

In the maker's catalogue of the period is shown a four coupled saddle tank quite unlike the locomotives actually built. The inside cylinders drove on to a jack shaft located between the two pairs of wheels and coupled to them. The chimney and dome are the only similarities.

Fig. 38 **Bellis & Seekings** 1874 SECUNDUS Guards removed, note outside motion

BEYER PEACOCK & COMPANY
Gorton Foundry, Manchester

Fig. 39 **Beyer Peacock & Co** 328/1862 SER of Portugal, exhibited Paris Exhibition 1862

Founded in May 1854 by Charles F. Beyer and Richard Peacock. Mr Beyer had relinquished his post as manager of Sharp Bros. the well known locomotive builders. He came from Saxony to England at the age of 21, and soon had a number of Germans on the technical staff at Gorton Foundry. Richard Peacock had been apprenticed by Fenton Murray & Jackson, after which he became Locomotive Superintendent of the Leeds & Selby Railway and in 1840 he left and spent a short time on the Great Western Railway under Daniel Gooch, and within twelve months joined the Manchester, Sheffield and Lincolnshire Railway, quickly becoming Locomotive Superintendent. Both were therefore young men in partnership, one already fully conversant with competitive tendering and contracting and the other of high technical and managing ability.

The first locomotive was turned out and delivered on 31 July, 1855, being one of an order for eight 2-2-2 standard gauge engines for the Great Western.

From the outset their locomotives gained an enviable reputation for well built machines with clean and elegant lines. Many well known names in railway history were at one time or another associated with the company. H. A. Hoy from the Lancashire and Yorkshire Railway, John Ramsbottom from the LNWR and Robert H. Whitelegg from the Glasgow and South Western Railway. The close association with the Great Central Railway brought in Sir Sam Fay and J. G. Robinson.

Two of the many men who served their apprenticeship here were Gilbert Claughton and R. H. Burnett who became Locomotive Superintendent on the Metropolitan Railway and later returned to Gorton Foundry.

Orders for new engines and repair work came from all quarters, and besides the initial order from the GWR the following locomotives were put in hand:

10 – 2-2-2 for East Indian Railway
2 – 0-4-0WT for Great North of Scotland Railway
6 – 2-2-2 for Edinburgh & Glasgow Railway
2 – 0-4-2 for South Staffordshire Railway
6 – 2-4-0 for Swedish Govt Railway

Outstanding designs which followed were the 4-4-0 Side tanks with outside cylinders. Initially eight were supplied to the Tudela and Bilbao Railway in 1862, followed by eighteen for the Metropolitan Railway in 1864, five in 1866, five in 1867, five in 1868, six in 1869 and five in 1870. Five were supplied to the Rhenish Railway in 1871. The Metropolitan District Railway obtained twenty four in 1871, and finally the L&NWR had twenty two, seven (1871), nine (1872) and six (1876). Locomotives of similar design but with tenders were also built for the New South Wales Government Railway.

Another interesting tank engine of the same period was the 2-4-0WT to Beattie's design of which eighty two were supplied between 1863 and 1875, to the London and South Western Railway. The early history of the Dutch railways is the early history of Beyer Peacock. Hundreds of practically every type used by this company were supplied, and their reputation was such that orders were repeated to the extent no other builder enjoyed. Most of the railways in Britain bought locomotives from Gorton Foundry and some had no other builder's product on their lines.

Fig. 40 **Beyer Peacock & Co** 6119/1922 Semerang–Cheribon tramway, Java 4-6-0

Over 200 tram engines were built between 1881 and 1910 of which 71 were built to Wilkinson's patent with vertical boilers and geared drive, 97 were built for Java tramways, these being non-condensing type with loco-type boilers.

The first 'tram engine' was a combined street tramcar for the North Staffordshire tramways, but the remainder were separate units for towing cars and they were known as 'dummies'. The biggest fleet was supplied to the Manchester, Bury, Rochdale and Oldham tramways, amongst which was a compound with 9" and 14" x 14" stroke cylinders. Other tramways supplied were South Staffordshire, Coventry, Birmingham Central and North Staffordshire. All these of course were fitted with air condensers consisting of copper tubes connected to box chambers on the roof of the vehicle.

One example may be seen at the Crich Museum; it was originally built for service in New South Wales in 1885, but after trials was returned in 1889 and used by Beyer Peacock in their own yards as works shunter.

Notable features of design introduced by Beyer Peacock were the use of deep frames and welded horns at an early date. The Isle of Man 2-4-0Ts of 1873 were the first to have 'pony' trucks, the first square topped Belpaire boilers were produced in 1872 and put on two 2-4-0s ordered by the Malines and Terneuzen Railway.

When the Mersey Railway 0-6-4Ts were built in 1885 they were the most powerful tank engines in Britain and the same railway had the distinction of running the first 2-6-2T in Britain built in 1887. The first 4-4-4T was also built to the order of the Wirral Railway in 1896.

In 1887 occurred their first incursion into the design of rack railway locomotives, when some tank locomotives were supplied to the Puerto Cabello and Valencia Railway in Venezuela, built under the patent of Mr H. Livesay who was the son of J. Livesay, later of Livesay and Henderson the consultants, and Mr H. L. Lange who was the principal assistant to Mr Beyer.

In 1902 the private company was converted into a public company and by this change more capital was injected into the firm.

A demand for greater power within the normal loading gauges led to the evolution of the *Garratt* locomotive and represented an important development in steam locomotive design, and from a financial aspect, this type of locomotive could be constructed at less cost than any other type of locomotive of similar power output.

The chief characteristics of these locomotives were that the boiler was carried on a girder frame, being supported at both ends by pivotted bogies or power units.

The water tank was usually above one unit and the bunker over the other, with the boiler in between. Thus there was plenty of adhesion at both ends and the main fault of other articulated types, which was a 'shouldering' movement, was eliminated and the addition of check and side springs rendered unnecessary. The boiler could be of larger dimensions and if need be could lie low in the girder frames, which as they terminated at the pivots, there were no overhanging portions when the locomotive was on a curve. The boiler being in a separate frame a wide firebox could be used giving a large grate area. The steam regulator in most cases was placed in the dome and two steam pipes led from this point, one to the smokebox tubeplate and one to the firebox end, both internally and from these points suitable connections carried the steam to each set of cylinders. The joints were located centrally with the turning point of the bogies, the exhaust being conveyed to the chimney or each section could be exhausted to atmosphere independently. The first two *Garratts* were built in 1908 for Tasmania, and were four cylinder compounds 2′ gauge 0-4-0-0-4-0 since that time they have been built in all gauges and types, and resulted in the only form of articulated steam locomotive design to survive. The largest built were for the USSR in 1932 and were the 4-8-2-2-8-4 type weighing 262 tons in working order, even on the metre gauge they weighed up to 255 tons.

Principal customers were South Africa, South America, Burma and Australia, and apart from the L&NER 2-8-0-0-8-2 & 33 2-6-0-0-6-2s supplied to the LM&SR for the heavy coal traffic between Toton and Brent, the only other representatives in this country were four 0-4-0-0-4-0s as follows:

1924 (No.6172) Vivian & Sons Swansea.

1931 (No.6729) Sneyd Collieries Ltd. Burslem.

1934 (No.6779) Guest Keen Baldwins Iron & Steel Co.

1937 (No.6841) Baddesley Coll. Atherstone.

In 1926 the Ljungstrom turbo-electric locomotive of 2000 HP was built and ran prolonged trials on the LM&SR.

A strange order was for one Webb three cylinder compound locomotive for the Pennsylvania R.R. in 1888. It was a 2-2-2-0 with 14″ and 24″ x 30″ stroke cylinders and 6′ 3″ driving wheels. In 1889 three twin tank engines (Lange & Livesey patent) were built for the Inter-Oceanic R.R. of Mexico. They were 2-6-0-0-6-2s with 14½ x 20 cylinders and 3′ 3″ driving wheels. They had two boilers with fireboxes back to back rather similar to the *Fairlie* locomotives.

Fortunately one of the tram engines has been preserved and is at the Crich Tramway Museum.

Beyer Peacock were a major influence in good craftsmanship, simplicity and excellence in design and were paramount in elegance.

General Notes Beyer Peacock had started after the relatively crude experimental designs of the 1830s and 1840s had been replaced by more established designs. Even so, in the 1850s it was impossible to obtain iron plates long enough and wide enough for frames to be made in one piece and two or even three plates had to be welded together to make one frame. The most important man had always been the Smith and until mechanical forging and welding were employed, many difficult and intricate operations had to be carried out in the smithy and forge.

The introduction of mild steel in the 1880s made the manufacture of boilers and frames far easier and larger size plates could be obtained.

Beyer Peacock had their own Steel Foundry in 1899 and wheel centres and other parts were made there.

Their 5000th locomotive appeared in 1907 and up to then the approximate figures for the various customers were: 1600 for home railways, 1200 for European railways, 700 for Australian railways, 650 for S. American railways and 850 for various other countries.

Locomotives were built on pits, either side of a central track – a common form of erecting shop.

In 1927 the works occupied 23 acres and employed 3000 workmen.

In 1961 the works were re-organised to build diesel locomotives with hydraulic transmission, but anticipated orders fell short, principally because of British Railways decision to standardise on diesel-electric traction.

Dismantling of the works took place in 1966.

One of the many outside cylindered 4-4-0s supplied to Australia is preserved on a plinth at Canberra. Built in 1878 it hauled the first train to the site of the new capital. The locomotive was New South Wales Government Railway No.1210 and was withdrawn after 84 years of service.

Captions to illustrations on following pages:

Fig. 41 **Beyer Peacock & Co** 6735/1932
Designed by G. T. Glover for Belfast–Dublin Expresses Great Northern Railway (Ireland) No. 87 KESTREL *Class V* compound

Fig. 42 **Beyer Peacock & Co**
6744/1933 Arcos Ltd for Russia
5′ 0″ gauge 0-6-0T

Fig. 43 **Beyer Peacock & Co**
6833/1937 FCS Peru 2-8-0 No. 80

Fig. 44 **Beyer Peacock & Co**
6858/1937 Leopoldina Rly (Brazil) metre gauge 4-6-2T No. 364

Fig. 41

Fig. 42

Fig. 44

Fig. 45 **Beyer Peacock & Co.** 6905/1939 First 4-8-4 + 4-8-4 in the world, weight 186 tons wo. Kenya Uganda Railways & Harbours No. 77 MENGO

BICKLE & CO.
Plymouth

Fig. 46 is reproduced from an advertisement in the *Engineer* showing a narrow gauge locomotive, with two 4" x 7" inside cylinders under the footplate, and appears to be a 0-2-2ST and was intended for underground working in the Levant Mine in Cornwall.

It had a steel boiler and gunmetal tubes and worked at a pressure of 100lb/sq.in.

Built in 1892 it did not fulfil requirements, probably due to smoke emission.

BINGLEY & COMPANY
Harper Street, Leeds

Situated near Kirkgate, they were ironfounders and engineers. It is probable that they were subcontractors of Fenton Murray & Wood.

A 'Single' locomotive was built for the Sheffield & Rotherham Railway early in 1840 called ROTHERHAM probably a 2-2-2.

BIRMINGHAM & GLOUCESTER RAILWAY
Bromsgrove

A 2-2-2 single was built in 1844 and one other of the same type but no details are known. At the time of construction J. E. McConnell was the railway's engineer.

In 1845 he obtained the authority of the directors to build a powerful tank locomotive for the 1 in 37 Lickey incline. This was due to the performance of the American 4-2-0s which had been built by Norris & Co. of Philadelphia. They were pulling 33 tons up the bank at twelve to fifteen miles per hour, or a maximum load of 53¼ tons at eight and a half miles per hour. Edward Bury himself had driven one of his own locomotives, loaned by the London and Birmingham Railway, in an attempt to compete with the American type, but it only got half way up the bank. Further orders were placed in America for eight more 4-2-0s and the American press made the most of reports that America was showing England how to design and build locomotives.

McConnell's 1845 locomotive was a six coupled tank weighing 30 tons, with outside cylinders 18" x 26" and 3' 9' diameter wheels. It had an oval boiler 3' 10" high and 3' 9" wide. It was named GREAT BRITAIN and numbered 38, later renumbered 276, and after being rebuilt as a Well Tank given the number 300 in 1853.

Fig. 46 **Bickle** 1892 cylinders under firebox PIONEER Levant Mine

BIRMINGHAM & GLOUCESTER RAILWAY

It apparently was successful but in 1846 the railway was absorbed into the Midland Railway and its 37 locomotives taken over. In the same year McConnell took up the position of Locomotive Superintendent at the London & North Western Railway's Southern Division headquarters at Wolverton.

BLACK HAWTHORN & COMPANY

Gateshead-on-Tyne

After the failure of R. Coulthard & Co. in 1865, the works were taken over by William Black and Thomas Hawthorn in the same year. Building was concentrated more on industrial locomotives for collieries and ironworks but a number of tender locomotives were also built and went to many parts of the world.

The industrial saddle tanks were four and six coupled and these were supplied to many local industries and soon were being supplied to the North Eastern, Llynvi & Ogmore, Lynn and Fakenham and other railways. They were solidly built with no special features except the taper chimney, some had square windows on the spectacle plate and on some the dome was surmounted by Ramsbottom safety valves. The outside cylinders were slightly inclined.

Fig. 47
Black Hawthorn
695/1833 Metre gauge
4-4-0ST
Cartagente-Denia, Spain

Fig. 48
Black Hawthorn
903/1887 2' 6" gauge
SANDHLAN
Cape of Good Hope, Port Elizabeth

A number of Crane tanks were built with vertical boilers, others with conventional locomotive type boilers. An interesting example of the latter was one built in 1876 for John Spencer and Sons of Newcastle. It was basically a standard 0-4-0ST with frames extended to the rear with a four wheeled bogie with outside frames on which was mounted the crane post and crane, having two vertical cylinders and the lifting capacity of 4 tons on a fixed jib with counter balance weight fitted. It is interesting to compare this locomotive with the three 0-6-4CTs built by the Great Western Railway on the same principle except that the jib radius was adjustable.

Fig. 48 shows a 0-4-2ST with spark arrester rails above tank for wood storage, single slide bars and Ramsbottom safety valves with single exhaust.

Fig. 47 has a large saddle tank with safety valves and dome in characteristic position and typical chimney.

Fig. 49 is an interesting 2-4-2T for Japan, a novel feature being the feed pump driven off the crosshead.

The firm had developed into one of the premier builders in this area and up to 1896 when the firm ceased trading, over 1100 locomotives had been built besides a number built by other firms being completely rebuilt. The firm was bought by CHAPMAN & FURNEAUX who carried on in the same tradition and produced a further seventy locomotives the majority of which went to northern collieries and industry.

The last locomotive built was a 0-4-0ST in 1902 (Works No.1215) for Knight, Bevan and Sturge to their 2′ 8½″ gauge.

The same year the firm ceased trading and the goodwill, drawings, patterns and templates were purchased by R. & W. Hawthorn Leslie & Co. Ltd. of Newcastle.

Fig. 49
Black Hawthorn
1103/1894
3′ 6″ gauge 2-4-2T
Japan Miike Coal Mines
No. 4

Fig. 50
Chapman & Furneaux
1203/1901
FELLING No. 1

BLACKIE & COMPANY

Aberdeen

Two locomotives are known to have been built for the Aberdeen Railway in 1848. They were 0-4-2s with 13″ x 18″ cylinders and 5′ diameter coupled wheels.

The Aberdeen Railway became part of the Scottish North Eastern Railway in 1856.

BLACKIE & CO.

Date	Type	Cylinders inches	D.W. ft in	Railway	No	SNE No
1848	0-4-2	13 x 18	5 0	Aberdeen	22	69
1848	0-4-2	13 x 18	5 0	Aberdeen	23	70

SNE: Scottish North Eastern Rly.

BLAENAVON CO.

According to the Birmingham Locomotive Club's records this firm built a 3 ft gauge locomotive for the Pwlldu Limestone Quarries. It was named LLANFOIST and was scrapped in 1933. Type and date of building is unknown.

In addition the following were built, dates and customers not known.

0-6-0ST ic HARRY
0-6-0ST ic BLAENAVON
0-4-0ST LLANOVER (narrow gauge)

BLAINA IRON WORKS

According to the *Monmouthshire Merlin* dated 12 June, 1830, a locomotive was very nearly completed, to be used in their own works, but no other information has been discovered. The engineer was John Brown.

BLAIR & COMPANY

Stockton

See FOSSICK & HACKWORTH.

BLAYLOCK & PRATCHITT

Denton Iron Works, Denton Holme, Carlisle

This firm was formed in 1859 by William Pratchitt and John Blaylock, and was originally located at Long Island Ironworks in Carlisle. They specialised in vertical fixed steam engines and portable engines varying from 4 to 20 horse power. The firm expanded rapidly and a new site was purchased in 1863 at Denton Holme.

Just before their removal, they were requested to tender for the supply of locomotives for the narrow gauge Festiniog Railway in 1862.

Whether any locomotives were actually built by the firm is extremely doubtful. They were not successful in their tender (if they tendered) for the Festiniog locomotives.

However William Pratchitt was apprenticed at the age of sixteen in 1849 to the firm of Benjamin Hick & Son of Bolton and he remained with them for ten years. This firm of course, built locomotives from 1837 to the early 1850s and he would have had some experience in this field during his stay with the firm.

Blaylock and Pratchitt were building portable steam engines with horizontal boilers, cylinders mounted over the firebox and crankshaft and flywheel on the boiler barrel. Not much design work or additional plant would have been required for locomotive building. If such a policy was considered, it is fairly certain it never materialised.

It is interesting to note in passing, although irrelevant to the present subject, that this firm were suppliers of all railway clocks on the L&NWR main line and at all stations from Carlisle to Wigan; they also provided and erected the ironwork for the stations on the Cockermouth, Keswick & Penrith Railway and on the North British Railway from Carlisle to Langholm (Ref: Pratchitt Bros. Ltd. *A Hundred Years of Engineering*).

BLUNDELL, JOHNATHAN & SON
Pemberton, Lancs

This firm was reputed to have built a 0-4-0ST with 14″ x 20″ cylinders named KING. If this was so it would have been built before 1900 when the firm became the PEMBERTON COLLIERY COMPANY. No further information is available.

BLYTHE & TYNE RAILWAY
Percy Main

Although twenty locomotives were built in the Percy Main shops very little information can be obtained regarding them and very few photographs.

Between 1862 and 1876 fifteen 0-6-0s and five 2-4-0s were built. All were not completely new – parts from earlier locomotives were frequently used – the new locomotives carrying the same running number as the older replaced types.

The most interesting locomotive built was No. 35 in 1868 – a compound 0-6-0 with two outside cylinders 10″ x 24″ and one low pressure inside cylinder 20″ x 24″. The outside connecting rods had crank pins which coincided with each other so that both high pressure pistons moved simultaneously in the same direction. The object of this arrangement was to prevent rolling and lateral movement. It was intended for mineral traffic, the coupled wheels being 4′ 6″ diameter but the boiler was incapable of providing all the cylinders with sufficient steam and it was not long before a conversion was carried out to a two cylinder 'Simple'. On the other hand Mr J. Fleming an expert on railways in this area states that it is almost certain that No. 35 was never a compound. (Fig. 51)

At one period Joseph Cabry was the resident engineer. Besides locomotives, carriages and wagons were built.

The railway became part of the North Eastern Railway in 1874 and the last three built were NER locomotives, although the two built in 1874 carried B&T running numbers for a short time.

See table on next page.

Fig. 51
B&TR 1868
'Long Boiler' type
B&T No. 35
as NER 1335.
Early cabside
windows

BODMER, J. G.
Manchester

According to the Gauge Commissioners Minutes of Evidence, Mr J. G. Bodmer appeared before them in October 1845 and described himself as an engine-builder having shops at Manchester and that he had built some locomotives in which the reciprocating parts were balanced.

In 1834 he hired a workroom at Rothwell & Company's works at Bolton.

Some of Bodmer's locomotives were built by Sharp Roberts & Company and others by himself. They ran on the Sheffield and Manchester, London and Brighton and South Eastern railways.

The first design of 1834 was for a 2-2-0 which included many novel features. Two pistons were fitted in each cylinder, the rear piston rods were connected to the cranks in the usual manner but the front piston rods were connected to crossheads coupled rearwards by rods to arms rocking on a fixed cross shaft. The arms were attached to connecting rods attached to cranks fixed at 180° in relation to the cranks driven by the rear pistons. Each inside cylinder had three ports and balanced slide valves were fitted. The laminated springs had transverse equalising beams. The horn guides were curved, crank pins concave and the big end brasses convex, in order to equalise the lateral stresses. A rotary regulator valve was fitted in front of the dome, the regulator rod passing through a tube passing through the dome, outside steam pipes feeding the valve chests. A steam blower could also be operated through the same valve and a variable blast pipe was operated by a rack and pinion. Bodmer was a brilliant engineer and many of these features were patented by him but strangely enough it is doubtful whether this locomotive was built – if it was Rothwell & Company may have had a hand in the construction.

During 1837–8 he associated himself with cotton spinning, machine tools and foundry work.

The next known locomotive design was in 1842 and was a 2-2-2 with the characteristic curved frames of Sharp Roberts & Company's singles and may have been built by them, but the boiler design differed considerably with two pillar type safety valves on the barrel and a dome on the raised firebox. Cylinders were 16" x 24" with cylindrical valves and expansion could be varied by a cut off valve the shaft of which passed through the main valve. Two further locomotives were built in 1845 one becoming South Eastern Railway No.123 and the other London & Brighton Railway No. 20.

BLYTHE & TYNE RLY

Date	Type	Cylinders inches	D.W. ft in	B & T No	NER No (a)	Replaced
4/1862	0-6-0	16 x 24	4 6	3	1303	1891
4/1862	0-6-0	16 x 24	4 6	14	1314	1892
3/1863	0-6-0	16 x 24	4 6	7	1307	1891
3/1865	0-6-0	16 x 24	4 9	26	1326	1889
8/1865	0-6-0	16 x 24	4 6	27	1327	1889
8/1866	2-4-0	16 x 24	5 6	30	1330	1888
2/1867	2-4-0	16 x 24	5 6	31	1331	1884
6/1867	0-6-0	16 x 24	4 6	8	1308	1891
9/1867	0-6-0	16 x 24	4 9	9	1309	1891
6/1868	2-4-0	16 x 24	5 6	10	1310	1891
6/1868	0-6-0	16 x 24	4 6	34	1334	
8/1868	0-6-0	10/20 x 24(b)	4 6	35	1335	
6/1869	0-6-0	16 x 24	4 9	11	1311	
9/1871	2-4-0	16 x 24	5 6	36	1336	1887
12/1871	2-4-0	16 x 24	5 6	37	1337	1887
12/1872	0-6-0	16 x 24	4 6	16	1316	1896
8/1873	0-6-0	16 x 24	4 6	17	1317	
3/1874	0-6-0	16 x 24	4 6	20	1320	
6/1874	0-6-0	16 x 24	4 9	21	1321	
1/1876	0-6-0	16 x 24	4 6	–	1319	

(a) Railway acquired by North Eastern Railway in 1874. (b) See text

This pair had cylinders 16″ x 30″, 5′ 6″ driving wheels and domeless boilers with a total heating surface of 842 sq.ft. All three had Bodmer's opposed pistons.

Six wheeled tenders were fitted and a geared feed pump was operated off the centre axle. Sledge brakes were fitted. Both 1845 locomotives were reasonably successful and the Brighton one was subsequently rebuilt with conventional cylinders and lasted until 1892.

One of Bodmer's patent locomotives was supposed to have worked on the Sheffield, Ashton-under-Lyne and Manchester Railway, thought to be SA & MR No.9 BELLONA a 2-2-2 built by Sharp Bros. in 1844 (Works No. 269). The stroke of the cylinders is given as 20″, so that double pistons would not have been fitted.

The Manchester shops were closed at the end of 1845 and Bodmer left England in 1848.

BOLCKOW VAUGHAN & CO, LTD.

Cleveland Iron & Steel Works, South Bank, Middlesbrough

This old established iron and steel works, originally Bolckow & Vaughan until 1864, became Bolckow Vaughan & Co. Ltd. on 19 November, 1864, and became part of DORMAN, LONG & CO. LTD. on 1 November, 1929.

The rebuilding of old locomotives and the different categories of work carried out are detailed in notes on the Dorman Long Locomotives, which apply equally well to the following locomotives:

No.16 0-4-0T (1888) 18″ gauge locomotive with 9″ diameter outside cylinders, possibly of similar design to earlier *Black Hawthorn.* It was used to haul the trains conveying ingots from the soaking pits to the rolling mills.

No.152 0-6-0ST (1920) JAMES EVANS later renamed ALEXANDER supposedly a modified rebuild from a 15″ *Peckett* loco. The rebuilt engine had 16″ diameter inside cylinders and was withdrawn in 1958 and scrapped in that same year.

No.153 0-6-0T (1920) PATRIOT 15″ diameter inside cylinders and classed as a modified rebuild. Stated to have been locomotive No. 22 altered. (Fig. 52)

Fig. 52 **Bolckow Vaughan** 1920 0-6-0T PATRIOT

BOLCKOW VAUGHAN & CO. LTD.

Original Locomotive				Type	Name
BV No	Built	No	Date		
16	Black Hawthorn & Co			0-4-0T	
152	Peckett & Son			0-6-0ST	Alexander
153	Manning Wardle?			0-6-0T	Patriot
154	–	–	–	0-4-0ST	William Hewitt
161	Black Hawthorn & Co	609	1881	0-4-0ST	Marton
29	Peckett & Son	668	1897	0-4-0ST	Eston
162	Black Hawthorn & Co	519	1879	0-4-0ST	Whitworth
114	Kitson & Co	4624	1908	0-6-0ST	Airedale
140	Hawthorn Leslie	3209	1916	0-4-0ST	Haig

	New Locomotive	
BV No	Built, Rebuilt or replaced	
16	1888	(a)
152	1920	Virtually new (b)
153	1920	(c) Fig. 52
154	1921	new Fig. 53
161	1924	
29	1924	
162	1929	
114	1928	
140	1955	

(a) 18" gauge. (b) Originally JAMES EVANS. (c) No 22 altered. (d) Rebuilt from 0-6-0T.

No.154 (1921) WILLIAM HEWITT was a 0-4-0ST designed and built by Bolckow Vaughan and had outside cylinders 14" x 22" and 3' 4" wheels. The boiler had a Belpaire firebox, the dome housed the regulator valve and also two pop safety valves (Fig. 53). It was scrapped in 1961.

No.161 (1924) MARTON was a 0-4-0ST with 14" diameter outside cylinders, recorded as rebuilt from *Black Hawthorn* No. 609 of 1881.

No. 29 (1924) ESTON was a 0-4-0ST with 14" diameter outside cylinders, recorded as rebuilt from *Peckett* No. 668 of 1897.

No. 114 (probably 1928) AIREDALE was a 0-6-0ST with 18" diameter inside cylinders rebuilt from a side tank originally built by Kitson & Company No. 4624 of 1908. It was scrapped in 1963.

No.162 (1929) WHITWORTH was a 0-4-0ST with 14" diameter outside cylinders, recorded as rebuilt from *Black Hawthorn* No. 519 of 1879.

No.140 (1955) HAIG a 0-4-0ST with 14" diameter outside cylinders, recorded as rebuilt from *Hawthorn Leslie* No. 3209 of 1916. This locomotive was rebuilt after the firm became part of the Dorman Long Group. It was scrapped in 1963.

Fig. 53 **Bolckow Vaughan** 1921 0-4-0ST WILLIAM HEWITT

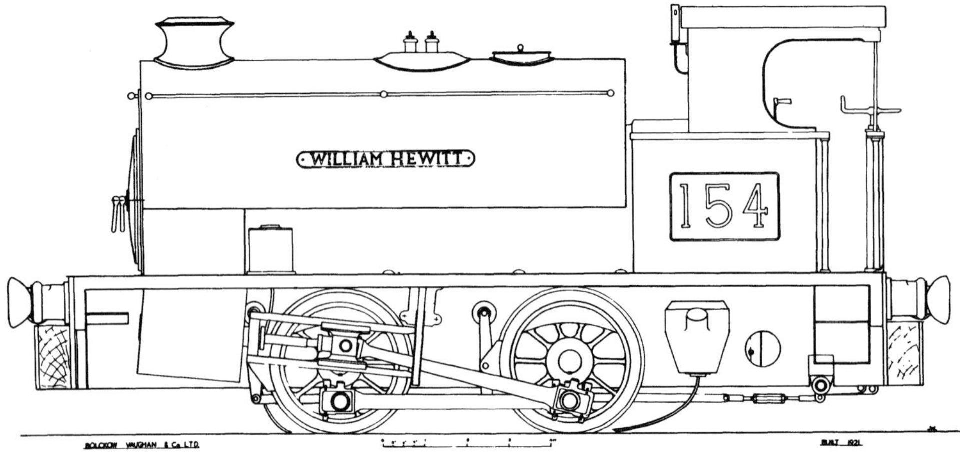

BORRIE, PETER & CO.
Tay Foundry, Dundee

Six locomotives were probably built. The firm offered five new engines and tenders to the Edinburgh & Glasgow Railway at a considerably reduced price in February 1841. Only one was sold to the Edinburgh & Glasgow Railway who named it EUCLID. Having the remaining four on their hands probably led to the company's bankruptcy in 1842.

One locomotive, known as CORYNDON, was built for John Chanter. It had Chanter's patent firebox which contained three vertical brick partitions between which the gases passed in order to ensure complete combustion before the smoke given off by the coal could be emitted. It was a 2-2-2 with 13″ x 18″ cylinders and 5′ 6″ driving wheels.

BORROWS, E. & SONS
Providence Works, St Helens

The firm was founded in 1865 by Edward Borrows who had been James Cross' assistant on the St. Helens Railway. In 1872, Borrows commenced building four coupled well-tanks of the same basic design created by James Cross in 1866. Up to 1913 about forty locomotives were built, the principal customers being Pilkington Bros. Ltd., United Alkali Co. Ltd., Brunner Mond & Co. Ltd. and a few collieries. All the locomotives were of the same design with outside cylinders and steam chests, the well tank was made integral with the main frames, the plates forming the sides of tank, thus providing an extremely strong and rigid frame. There was no room for the motion between the frames, so the eccentrics were fixed between the wheels and the frames, and Stephenson's link motion was used and coupled directly to the valve spindles, the port

Fig. 54
E. Borrows
33/1892 ROBY
Plate Works,
Pilkington Bros

Fig. 55
E. Borrows
53/1909 WINDLE
Presented to
Middleton Railway
Preservation Society,
October 1961

faces being vertical. Advantage was taken of the absence of valve gearing inside the frames to position the driving axle under the forward part of the firebox so that with the increased wheelbase obtainable, a corresponding shortening of the trailing end was accomplished, which was a decided advantage in the limited clearances and sharp curves of factories. The chimneys had a strong Drummond likeness.

In all, these engines were very compact and of extremely pleasing appearance – a well thought out design with a good overall length to wheelbase ratio. The smokebox was curved outwards at the base, containing the steam and exhaust pipework. The outside valve-gear was easily accessible to the running staff. It is interesting to note that the Pilkington engines, at least, had left hand driving positions.

During rebuilding, modifications were effected. New chimneys on those maintained by Pilkington's Plate Works staff spoiled their appearance, as did some of the modified cabs, but when these same engines had their well tanks modified by divorcing them from the frames, the rigidity which had been such a good feature of the original locomotives, was lost and severe racking took place with its attendant troubles.

The dimensions of these engines as built were as follows:

Cylinders 14½" x 20"; wheels 3' 4" diameter (2' 11$\frac{1}{16}$" without tyres); wheelbase 6' 6"; tubes 160" x 1¾" outside diameter; heating surface tubes 597·80; firebox 54·11; total 651·91 sq. ft.; tank capacity 321 galls; coal 8½ cwt; weight in W.O. 25 tons; length over buffers 20' 1½"; height to chimney top 11' 1¼"; working pressure 160 lb/sq.in.; tractive effort 13,000 lb.

Some of the earlier locomotives had 12" to 14" diameter cylinders. Pilkington engines rebuilt; tubes 138" X 1¾" outside diameter; heating surface tubes 454 sq. ft; firebox 48 sq. ft; total 502 sq. ft. Besides building locomotives Borrows built a number of steam rollers and some of the missing works numbers may have been allocated to these or other machinery, as they were general engineers.

Between 1912 and 1920 Kerr Stuart & Co. Ltd. built a number of very similar locomotives differing only in minor details. Seventeen were supplied to Brunner Mond & Co. Ltd.

In 1910 the locomotive business was taken over by H. W. JOHNSON & COMPANY.

Three *Borrows* type were built from 1913 to 1921. Building took place at Providence Works, later moving to Pocket Nook and then to Rainford.

Fig. 56 E. **Burrows 58/1921** KELVIN Plate Works, built by H. W. Johnson

Works No	Date	Name	Customer
1	1875		London & Manchester Plate Glass Co Sutton
2	1875		London & Manchester Plate Glass Co Sutton
3			
4	1875	Sutton	Pilkington Bros Ltd St Helens
5	1876	Windle	Pilkington Bros Ltd St Helens
6	1876	Victoria	Victoria Colliery Co Rainford
7			
8			
9	1879	St Helens	Pilkington Bros Ltd St Helens
10	1881	Solvay	Brunner Mond & Co Ltd Winnington
11	1882	Agnes	
12			
13	1882	Davy	Brunner Mond & Co Ltd Winnington (a)
14			
15	1883	Rainford	Pilkington Bros Ltd St Helens
16	1883	Agnes	United Alkali Co Ltd St Helens
17	1883	No 3	London & Manchester Plate Glass Co Sutton
18	1884	Emily	United Alkali Co Ltd St Helens
19	1884	Knowsley	Pilkington Bros Ltd St Helens
20	1885	Margery	United Alkali Co Ltd St Helens
21	1886	James Watt	Brunner Mond & Co Ltd Winnington
22	1887	AGK	United Alkali Co Ltd St Helens
23		Economy	
24	1888	Newton	Brunner Mond & Co Ltd Winnington
25			
26			
27			
28			
29			
30			
31	1891	Faraday	Brunner Mond & Co Ltd Winnington
32	1891	Syndicate	ICI
33	1892	Roby	Pilkington Bros Ltd St Helens (Fig. 54)
34			
35			
36		Eccleston	(b)
37	1898	No. 4	The Smelting Corporation Ellesmere Port
38			
39			
40			
41			
42			
43			
44	1902	Edenhurst	Pilkington Bros Ltd St Helens
45			
46	1904	Preston	Sutton Heath & Lea Green Colliery
47	1905	Wheathill	Pilkington Bros Ltd St Helens
48			
49	1905	Sutton	Pilkington Bros Ltd St Helens
50	1907	The Queen	Nuttall & Co (St Helens) Ltd
51	1908	Kelvin	Brunner Mond & Co Ltd Winnington
52	1908	Briars Hey	Pilkington Bros Ltd St Helens
53	1909	Windle	Pilkington Bros Ltd St Helens (c) (Fig. 55)
54	1909	St Helens	Pilkington Bros Ltd St Helens
55	1910	Hollies	Pilkington Bros Ltd St Helens
56	1913	Raven	Blackwell Colliery Co Ltd (e)
57	1921	Patience	Pilkington Bros Ltd St Helens (e)
58	1921	Kelvin	Pilkington Bros Ltd St Helens (e) (Fig. 56)
		Lucy	United Alkali Co Ltd St Helens
		Alice	United Alkali Co Ltd St Helens
		King	United Glass Bottle Ltd St Helens (d)
	1880	Ravenhead	Pilkington Bros Ltd St Helens

All were 0-4-0WT. (a) Works No could have been 14. (b) Works No could have been 35 built between 1893 and 1898. (c) presented by Pilkington Bros Ltd in 1961 to Middleton Railway Preservation Society. (d) Works No probably 48. (e) Built by H W Johnson & Co.

BOULTON, I. W.

Portland Street, Ashton-Under-Lyne

The activities of Mr Isaac Watt Boulton is so thoroughly dealt with by A. R. Bennett in his *Chronicles of Boulton's Siding* that nothing could possibly be added.

He bought, sold, repaired, rebuilt, hired, and built locomotives from 1858 to 1894 and what a unique and fascinating collection they were! His apprenticeship was served with the Sheffield, Ashton-under-Lyne and Manchester Railway at Newton, thence to Beyer Peacock & Company after which he set up on his own and built a road steam carriage and a steam boat. After this attempt at independence he rejoined the railway in 1854, which had now become the Manchester, Sheffield and Lincolnshire Railway and the locomotive department was now at Gorton. In 1856 he again set up on his own at the Portland Street works and commenced turning out a great variety of engineering products including machine tools, portable and stationary engines, road steamers and locomotives.

The locomotive department was instituted and a new locomotive was built in 1856, being a small four coupled geared locomotive. Many locomotives were bought secondhand with the intention of hiring them out to contractors and railways and this side of the business expanded rapidly and was very successful.

It was not until 1864 that a siding was laid down connecting the works with the MS&LR and the works were enlarged to cope with the increasing amount of locomotive work, and further additions were made in 1872.

There is no doubt that quite a number of new locomotives were built by Boulton, but many 'new' ones were really a hybrid erection of various parts of various locomotives which passed through the works in the course of repairs and rebuilding ready for sale or hire.

Unfortunately his son Thomas, who had been his father's manager and mainstay, died in 1880 and his death affected his father so much, that his interest lost its savour and the business gradually deteriorated until no work was carried out after 1894 and all was sold up by 1898. Contributary causes were that fewer contractors were wanting to buy or hire his locomotives which mainly were rebuilds, and many other larger manufacturers were now hiring out new engines and the demand for Boulton's products became negligible by 1890.

Boulton's first locomotive was a tiny creation with the cylinders driving a dummy shaft connected to the rear wheels with a 2:1 reduction gear. A similar one was built in 1860. In all about twelve were constructed with a geared drive. Also in the 1860s, a steam roller type engine was built with chain drive probably preceding Aveling & Porter's first railway locomotive of this pattern which was turned out in 1864.

Fig. 57 **I. W. Boulton** 0-4-0ST sold to T. Walmsley Sons & Co. Ltd. Bolton in 1872

Another interesting feature of Boulton's appeared in 1873 in the form of a cross water tube boiler. The basic design was for a vertical circular boiler with a large single flue without the customary firebox. Through this flue were fitted a number of inclined copper water tubes. The chief difficulty with water tubes was their accessibility in case of failure. This was overcome theoretically by attaching each end of the flue to the boiler end plates with bolts so that the whole flue with its tubes could be withdrawn.

In practice the boilers when new proved very efficient with considerable fuel saving, but after being in service, they gave a lot of trouble due to the variety of water used resulting in scaling and leaks, and the cost of maintenance was high due to the time taken in removing the flue and re-making the main joints when the flue was replaced.

BOURNE BARTLEY & COMPANY

Manchester

Reported to have built locomotive No.13 for the North Union Railway but the same engine is also credited to the Haigh Foundry and Summers, Grove & Day. It was a 2-2-2, with 12" x 18" cylinders and named ST. GEORGE. They also built another 2-2-2 named ST. DAVID for the Bolton and Leigh Railway.

The firm were machinists in Mayor Street between 1838 and 1840. It is probable that they were agents for various locomotive builders.

BOWES, JOHN ESQ., & PARTNERS

Marley Hill Colliery

One 0-4-0ST was built at the Colliery workshops in 1854 and was named DANIEL O'ROURKE after John Bowes' horse who won the Derby in 1852. It is possible such a locomotive was built here, particularly as two brothers (Thompson by name) had in all probability built a *Grasshopper* type much earlier in the same workshops, although no positive information can confirm this. (Ref: *The Bowes Railway* C. E. Mountford)

I. W. BOULTON

Date	Type	Cylinders inches	D. W. ft in	Name	Notes
1856	0-4-0G	OC 5¼ x 12	2 6		2:1 Gearing
1860	0-4-0G	OC 5½ x 12	2 0	Little Grimsby	2:1 Gearing
1860s	0-4-0WT	8 x 12		Rattlesnake	2:1 Gearing Chain Drive
1865	0-4-0ST	IC 6 x 12	2 3	Ashtonian	2:1 Gearing 3'6" Gauge
1867	0-4-0ST	IC 9 x 12	3 0		2½:1 Gearing
1867	0-6-0ST	IC 15 x 24	3 6	Briton	(c)
1868	0-6-0ST	IC 15 x 24	3 6	Cyclops	(c)
1871	0-6-0ST	IC 12 x 18	3 0	Stamford	(a) (b)
1874	0-6-0ST	IC 12 x 18	3 0	Ariadne	(a) (b)
1874	0-4-0ST	OC 12 x 20	3 3	Active	(b)
1874	0-4-0ST	IC 13¼ x 20	4 0	Brymbo	(c)
1875	0-6-0ST	IC 12 x 18	3 0	Black Knight	(a) (b)
1875	0-6-0ST	IC			(a)
1876	0-6-0ST	12 x 18	3 0	Raven	(a) (b)
	0-4-0ST	IC			
	0-4-0ST	OC 6 x 12	2 0	Lilliputian	2:1 Gearing 2' Gauge

(a) Rebuilt from ex GNR steam tenders. (b) Water tube boilers. (c) Made up from parts of various locos.

BRAITHWAITE & ERICSSON
New Road, London

John Braithwaite's father died in 1818 and left his engineering business to John and his brother Francis and the firm became J. BRAITHWAITE & COMPANY. In 1827 John Ericsson, a Swedish engineer became associated with John Braithwaite. Whether the firm was actually altered to BRAITHWAITE & ERICSSON seems to be in doubt.

Between them they designed NOVELTY which was intended to compete in the Rainhill trials. It was completed in seven weeks in 1829 and this haste probably led to its failure at the trials. Its general outline was similar to a fire engine built about the same time, having a four wheeled chassis with elliptical springs. A vertical boiler was mounted at one end and had a horizontal cylindrical portion 12′ long and 15″ diameter running above the wheel axles. The horizontal portion contained a return tube into which heated air was blown by means of bellows operated off a crank extension. The disadvantage of this method of draughting was that it only functioned when the locomotive was in motion. Vertical 6″ x 12″ cylinders were mounted over the other pair of wheels and drove the other axle adjacent to the boiler. The weight of the engine was between 3 and 4 tons. At the trials everyone was amazed at its speed, the engine running light reached 28 mph, but trouble was soon experienced and even after hasty repairs the locomotive failed and was withdrawn from the contest. The second and third locomotives were built in 1830 for the Liverpool and Manchester Railway and were named WILLIAM and QUEEN ADELAIDE. They were in general outline, similar to the NOVELTY except that the water was carried in a four wheeled tender and the forced draught equipment was replaced by an induced draught fan. They were not accepted by the L&MR and were eventually bought by the St. Helens Railway together with NOVELTY.

Date	Type	Cylinders inches	D.W. ft in	No/Name	Gauge	Customer
1829	0-2-2	6 x 12		Novelty	Std	Rainhill Trials VB
1830	2-2-0			William IV		Liverpool & Manchester Rly VB
1830	2-2-0			Queen Adelaide		Liverpool & Manchester Rly VB
1833	0-4-0			McNeill		Paterson & Hudson River RR
1833	0-4-0			Delaware		Allegheny Portage RR
1834	0-4-0			Allegheny		Allegheny Portage RR
1834	0-4-0		4 6	Mississippi		Natchez & Hamburg RR (Fig. 58)
1835	0-4-0		4 6	Natchez		Natchez & Hamburg RR
1835	0-4-0			Weldon		Petersburg RR

BRAITHWAITE, MILNER & CO

Date	Type	Cylinders inches	D.W. ft in	No/Name	Gauge	Customer
1837	0-4-0			Comet		Allegheny Portage RR
1837	0-4-0			Jefferson		R.F. & P.R.R.
5/1838	0-4-0	IC 10 x 16	4 6	Rocket		Philadelphia & Reading RR
1838	0-4-0			Firefly		Philadelphia & Reading RR
6/1838	0-4-0	9½ x 16	4 3	Spitfire		Philadelphia & Reading RR
1838	0-4-0			Dragon		Philadelphia & Reading RR
6/1838	0-4-0			Comet		Philadelphia & Reading RR
8/1838	0-4-0			Planet		Philadelphia & Reading RR
1838	0-4-0	14 x 18	5 0	Essex	5 0	Eastern Counties Rly
1838	0-4-0	14 x 18	5 0	Middlesex	5 0	Eastern Counties Rly
1838	0-4-0	14 x 18	5 0	Norfolk	5 0	Eastern Counties Rly
1838	0-4-0	14 x 18	5 0	Suffolk	5 0	Eastern Counties Rly
1839	2-2-0	12 x 18	6 0	1	5 0	Eastern Counties Rly
1839	2-2-0	12 x 18	6 0	2	5 0	Eastern Counties Rly
1839	2-2-0	12 x 18	6 0	3	5 0	Eastern Counties Rly
1839	2-2-0	12 x 18	6 0	4	5 0	Eastern Counties Rly
1839	2-2-0	12 x 18	6 0	5	5 0	Eastern Counties Rly
1839	2-2-0	12 x 18	6 0	6	5 0	Eastern Counties Rly
1839	0-4-0	14 x 18	5 0		5 0	Eastern Counties Rly
1839	0-4-0	14 x 18	5 0		5 0	Eastern Counties Rly
7/1840	0-4-0			Hecla		Philadelphia & Reading RR
3/1841	0-4-0			Gem		Philadelphia & Reading RR

VB Vertical Boiler R.F. & P.R.R: Richmond, Fredericksburg & Potomac RR

Between 1833 and 1838 at least fourteen locomotives were built for American railroads, they were all four coupled.

In 1834 John Braithwaite left the firm and joined the Eastern Counties Railway, and another brother Fred stepped into John's shoes and in 1836 the firm became BRAITHWAITE, MILNER & CO.

The firm failed in 1836 or 1837 but carried on manufacturing until 1841 presumably in the hands of the official receiver.

In 1838 four 0-4-0s were built for the Eastern Counties Railway, another two in 1839 and in the same year six 2-2-0s for the same railway, which at this time was 5′ gauge. These twelve engines were probably of the well known *Bury* type.

Ericsson emigrated to America in 1839.

In 1840–1 two further 0-4-0s were sent to the Philadelphia and Reading Railroad, six having been supplied in 1838.

BRAMAH & FOX
Birmingham

John Joseph Bramah first established at St. Barnabas' Church, Pimlico, making railway plant, then transferred to Smethwick as LONDON WORKS joining with Charles Fox and John Henderson and known later as FOX HENDERSON & CO. and were responsible for the construction of the Crystal Palace, London.

According to an article in the *Locomotive* on the locomotives of the Northern and Eastern Railway, one was supplied by this firm in October 1840 but it is improbable that it was actually built by them.

Fig. 58 **Braithwaite & Ericsson** 1834 Natchez & Hamburg RR America MISSISSIPPI

BRASSEY & COMPANY

Canada Works, Birkenhead

These works were established as a subsidiary to the contracting firm of Brassey Jackson Betts & Company, to build the greater part of the rolling stock for the Grand Trunk Railway of Canada. Before beginning production, two mechanics were sent in 1853 to visit various works in the United States, with a view to ensuring that the best and most suitable machinery would be used at the Birkenhead Factory.

A total of approximately fifty-five locomotives were built between 1854 and 1858 for Canada. The gauge was 5′ 6″, four 'singles' were built the remainder being 2-4-0s very similar to Alexander Allan's 'Crewe' type with outside cylinders 16″ (later ones 17″) x 20″ and 5ft driving wheels. The 'singles' had 15″ x 20″ cylinders and 6ft driving wheels. The gauge was altered in 1873 to standard 4′ 8½″

Fig. 59 **Brassey & Co** 1865 London Chatham & Dover Railway *Reindeer* class TEMPLAR

Qty	Customer	Nos	Type	Works Nos	Built
50	Grand Trunk Rly Canada	Misc	2-2-2 2-4-0	(a)	1854-58
1	Contract–North Devon Rly	—	2-4-0		1855
1	Contract–North Devon Rly		2-2-2		1855
3	Great Western Rly Canada	59-61	2-4-0	14-16	1855
6	Eastern Counties Rly	274-279	2-2-2	42-47	1856
6	Eastern Counties Rly	214-219	2-4-0		1856
6	Contract–Shrewsbury & Hereford	19-20 24-27	2-4-0	48-53	1857-9
3	Contract–Shrewsbury & Hereford	28-30	0-4-2	54-56	1861
4	Contract–Shrewsbury & Hereford	31-32	2-2-2		1857-8
1	Contract–North Devon Rly		2-2-2	66?	1857
4	Forth & Clyde Junc Rly		2-4-0		1859-61
3	Scottish North Eastern Rly	40-42	0-4-2		1861-2
3	Scottish North Eastern Rly	46-48	2-4-0		1855-7
13	S Slesvig Rly Denmark		2-2-2 2-4-0		1858-73
1 or 3	Contract – Australia NSW		2-2-2		1859
6	Finnish State Rlys	1-6	4-4-0	80-84, 142	1860-63
10	London, Chatham & Dover Rly	3-12	4-4-0		1861
16	East Indian Rly	Misc	2-4-0	120-122, 128-131 (part)	1862-4
20	Jutland-Fünen Rly	1-20	2-4-0	108-112, 150-152, 194-199	1862-6
1	Mt Cenis	—	Fell 0-4-OT		1863
6	London, Chatham & Dover Rly	44-49	2-4-0	182-187	1865
6	Central Argentine Rly	1-6	4-6-0?	188-193	1865
6	Central Argentine Rly	7-12	4-6-0	216-221	?
7	South Eastern Rly	235-241	0-4-4WT	200-206?	1866
2	Finnish State Rlys	7-8	4-4-0	212-213	1866
2	Buenos Ayres GtS Rly	9-10	4-4-0	222-223	1867
1	Egyptian State Rlys	196	2-4-0	224	1868
1	Egyptian State Rlys	197	0-6-0	225	1868
4	Roumanian State Rlys		0-6-0		1868
1	Roumanian State Rlys		0-6-OT		1869
11	Roumanian State Rlys		2-4-0		1869-70
6	Bolivar Rly Co	1-6	0-6-0	254-257, 262-263	1873-5
?	Italian Mediterranean Rly		?		?

(a) Wks Nos 1, 5-13, 17-41, 57-64, 67-79 (74-78 subcontracted to Fairbairn & Sons) and at least 30 were still running at this time and were converted.

Besides locomotives the works produced the whole of the ironwork for the tubes for the Victoria Bridge across the St. Lawrence river, consisting of 25 spans each approximately 245ft long.

The manager of the works was Mr George Harrison and from 1854 to 1875 over 260 locomotives were built for home and overseas railways.

From the table it will be seen that more locomotives were exported than delivered for home use.

According to local directories the firm's name was as follows:

1857 – Brassey Jackson Betts & Company, Canada Works, Wallasey Pool.

1864 – Peto, Brassey & Betts, Canada Works, Beaufort Road, Birkenhead.

1870–80 – Thomas Brassey & Co., Canada Works, Birkenhead.

Whether 'Brassey & Company' was used only for locomotive building is doubtful.

BRISTOL & EXETER RAILWAY
Bath Road, Bristol

Original headquarters of this railway was at Exeter, but to facilitate repairing locomotives and later for building them, new locomotive shops at Bath Road, Bristol were completed in September 1854 and for the first five years, repairs only were carried out and some extensive rebuilding. The first new locomotive did not appear until September 1859. A total of 35 were built, of which 23 were broad gauge, 10 standard gauge and 2 for 3ft. gauge. James Pearson was the locomotive superintendent and was responsible for the design of all these locomotives. The most spectacular of these were the 4-2-4Ts which were renewals of similar type supplied by Rothwell & Company. They had inside plate frames, flangeless driving wheels, brakes to all wheels. The water supply was from well and back tanks. The boiler was domeless and the total weight just under 50 tons. They were gigantic machines and any traveller familiar only with say the products of Nine Elms or Crewe, must have rubbed their eyes and stared in disbelief! Conversion of three of these 4-2-4Ts to 4-2-2 tender engines was carried out at Swindon. The broad gauge 2-4-0s, the largest class were also useful engines, albeit rough riders and most of them lasted until the gauge conversion in May 1892. The two 0-6-0Ts were built for the Culm Valley line and having been both extensively rebuilt lasted a lot longer than the other B&E locomotives and wandered far afield. No. 1377 ended up at Weymouth and after leaving there, No. 1376 served the Tanat Valley branch on the Cambrian section.

Fig. 60 **B&ER** 1871 7′ gauge B&E No. 5 running as GWR No. 2017

BRISTOL & EXETER RAILWAY

The 3′ gauge engines were built specially for the Burlescombe quarry line and when the B&ER's lease expired in 1898, they became redundant and were both sold later to the Bute Works Supply Co.

The GREAT WESTERN RAILWAY took over the B&E in January 1876 and at that time two 2-4-0 standard gauge tender locomotives were being built, and were completed under the auspices of the new owners.

Date	Type	Cylinders inches	D.W. ft in	B & E No	GWR No	Gauge		W'drawn
9/1859	4-2-4T	17 x 24	7 6	29	2006	7 0		9/1880
4/1862	4-2-4T	17 x 24	7 6	12	2005	7 0		12/1885
3/1866	0-6-OST	17 x 24	3 6	75	2092	7 0		6/1888
8/1867	0-6-OST	17 x 24	3 6	76	2093	7 0		12/1890
2/1868	4-2-4T	18 x 24	8 10	39	2001	7 0	(a)	12/1877
6/1868	4-2-4T	18 x 24	8 10	41	2003	7 0	(b)	6/1884
12/1868	4-2-4T	18 x 24	8 10	42	2004	7 0	(b)	12/1889
2/1870	2-4-0	17 x 24	6 7½	14	2020	7 0		5/1892
4/1870	2-4-0	17 x 24	6 7½	6	2018	7 0		6/1890
6/1870	2-4-0	17 x 24	6 7½	45	2023	7 0		6/1888
6/1870	2-4-0	17 x 24	6 7½	46	2024	7 0		12/1889
12/1870	2-4-0	17 x 24	6 7½	44	2022	7 0		12/1888
6/1871	2-4-0	17 x 24	6 7½	43	2021	7 0		5/1892
10/1871	2-4-0	17 x 24	6 7½	4	2016	7 0		5/1892
12/1871	2-4-0	17 x 24	6 7½	5	2017	7 0	(Fig. 60)	5/1892
6/1872	2-4-0	17 x 24	6 7½	8	2019	7 0		6/1889
8/1872	0-4-OWT	14 x 18	3 6	91	2094	7 0		5/1880
12/1872	2-4-0	17 x 24	6 7½	2	2015	7 0		6/1888
6/1873	4-2-4T	18 x 24	8 10	40	2002	7 0	(b)	12/1890
12/1873	0-4-OWT	8 x 12	2 0	112	1381	3 0	(c)	3/1899
5/1874	2-4-0	17 x 24	6 4	11	2025	7 0		6/1886
6/1874	2-4-0	17 x 24	6 4	3	1355	Std		10/1883
9/1874	0-6-OT	12 x 18	3 6	114	1376	Std		1/1934
10/1874	0-4-OWT	14 x 18	3 6	92	2095	7 0		6/1881
12/1874	2-4-0	17 x 24	6 4	20	2026	7 0		6/1886
1/1875	2-4-0	17 x 24	6 4	34	2027	7 0		1/1884
6/1875	2-4-0	17 x 24	6 4	1	1356	Std		10/1881
6/1875	2-4-0	17 x 24	6 4	16	1357	Std		12/1883
7/1875	0-4-OT	14 x 18	3 6	93	1378	Std		6/1880
8/1875	0-4-OWT	8 x 12	2 0	113	1382	3 0	(c)	3/1899
12/1875	0-6-OT	12 x 18	3 6	115	1377	Std		1/1927
12/1875	0-4-OT	14 x 18	3 6	94	1379	Std		6/1880
12/1875	0-4-OT	14 x 18	3 6	95	1380	Std		5/1880
BRISTOL—GREAT WESTERN RLY								
1876	2-4-OT	16 x 24	5 0	(30)	1358	Std		7/1888
1876	2-4-OT	16 x 24	5 0	(33)	1359	Std		3/1890

(a) Destroyed, Long Ashton 27/7/76. (b) Converted to Tender Loco 1877.
(c) Sold Bute Works Supply.

BRITISH RAILWAYS

On 1 January, 1948 the separate identities of the LMSR, LNER, GWR and SR were lost and became absorbed into the British Transport Commission together with London Passenger Transport Board.

At this time the following railway works were building new locomotives: LMS – Crewe, Derby, Horwich; LNER – Doncaster, Darlington, Gorton, Stratford; GWR – Swindon and SR – Eastleigh, Brighton, Ashford.

In addition to these the following were fully equipped to deal with heavy repairs and rebuilding: LMS – Kilmarnock, St. Rollox, Bow, Barrow, Inverness; LNER – Cowlairs, Inverurie, Gateshead; GWR – Wolverhampton, Caerphilly, Newton Abbot, Oswestry, Barry and LT – Neasden.

No new designs were independently carried out by the various regions, but contracts which had already been placed were allowed to stand, in addition to orders in the various railway works.

These regional locomotives built after nationalisation, between 1948 and 1956, came to a total of 1538 and comprised the following:

Built at Eastleigh 10 'Merchant Navy' 4-6-2 and 6 'West Country' 4-6-2.

Built at Brighton 34 'West Country' 4-6-2 and 41 LMS Class 4 2-6-4T

Built at Crewe 1 'Coronation' 4-6-2; 50 M.T. Class 5 4-6-0; 45 Class 2 2-6-0 and 110 Class 2 2-6-2T.

Built at Derby 106 Class 4 2-6-4T and 10 Class 2 2-6-2T.

Built at Horwich 50 M.T. Class 5 4-6-0; 72 Class 4 2-6-0 and 5 Class 0 0-4-0ST.

Built at Darlington 23 Class A1 4-6-2; 20 Class B1 4-6-0; 29 Class L1 2-6-4T; 28 Class J72 0-6-0T; 37 LMS Class 4 2-6-0 and 38 LMS Class 2 2-6-0.

Built at Doncaster 16 Class A1 4-6-2; 24 Class A2 4-6-2 and 50 LMS Class 4 2-6-0.

Built at Gorton 10 Class B1 4-6-0.

Built at Swindon 30 Castle 4-6-0; 49 Modified 'Hall' 4-6-0; 10 Manor 4-6-0; 2 2251 Class 0-6-0; 20 51xx 2-6-2T; 10 15xx 0-6-0PT; 70 16xx 0-6-0PT; 21 57xx 0-6-0PT; 20 67xx 0-6-0PT; 20 74xx 0-6-0PT and 25 LMS Class 2 2-6-0.

Outside Contractors: Built by North British Locomotive Co 106 LNE Class B1 4-6-0; 70 LNE Class K1 2-6-0 and 35 LNE Class L1 2-6-4T.

Built by Robert Stephenson & Hawthorn 35 LNE Class L1 2-6-4T and 100 GW 94xx 0-6-0PT.

Built by W. G. Bagnall 50 GW 94xx 0-6-0PT.

Built by Yorkshire Engine Co. 50 GW 94xx 0-6-0PT.

The most interesting event after nationalisation was the locomotive exchanges of 1948 in which specific types from all four regions were exchanged, to run on test routes, and all using the same grade of fuel. The fuel being a hard Yorkshire Coal had an unfortunate effect on ex GWR locomotives but other region's locomotives seemed to digest this coal quite satisfactorily. The GWR had always used the South Wales steam coal and had designed the blast pipes and draughting to suit this fuel. It was not surprising that the Great Western entrants did not cover themselves with glory, even on later tests burning Welsh coal. All the facts of these interesting exchanges were published but how they were interpreted by the technical staff is hard to follow when examining the BR standard designs.

During the continuance of steam locomotive building, rebuilding and repairs, a number of interesting modifications were carried out in the various shops. Work was continued in rebuilding the *Royal Scots* and *Patriots* with taper boilers. The Turbomotive was rebuilt as a Class 8 *Princess Pacific* which within a very short time met its end in the terrible Harrow accident.

Oil burning equipment was abandoned as a government policy and the apparatus was removed from *Castle, Hall, King Arthur,* Southern 2-6-0, and GWR 2-8-0 classes and quite a number more.

On the Southern the *Merchant Navy* and other *Pacifics* were rebuilt with Walschaerts valve gear as a substitute for Bulleid's chain driven gear in its oil-bath.

Double chimneys emerged on *Kings, Castles* and *Counties* and on A3 and A4 *Pacifics,* a few BR standard Class 4 4-6-0s, some V.2 2-6-2s and one 'Jubilee'.

High temperature superheat was introduced on *Kings* and *Castles* and higher pressure boilers on the LNE 2-6-2Ts.

Various modifications were carried out on the LNE 0.4 2-8-0 classes.

These are just some of the alterations which took place on the pre-nationalisation locomotive designs.

However in 1955 British Transport Commission announced their modernisation programme of electric and diesel traction, and to the effect that no further passenger hauling locomotives would be built after 1956 and the rapid run-down of steam locomotives was accelerated still further.

Selected workshops were re-equipped to deal with the new forms of traction and others were closed, as and when their capacity for dealing with steam locomotives became unnecessary.

Steps were taken in 1948 to plan a programme of new steam locomotive designs and this was effected by allocating various chosen types to drawing office staffs at the following works:

Derby – Class 6, 7, and 8 4-6-2s; Class 2 2-6-2T and Class 2 2-6-0.

Doncaster – Class 5 4-6-0 and Class 4 2-6-0.

Brighton – Class 4 4-6-0; Class 4 2-6-4T and Class 9 2-10-0

Swindon – Class 3 2-6-0 and Class 3 2-6-2T.

No less than twelve standard tenders were designed and built with water capacities ranging from 5625 to 3000 gallons, and coal from 10 tons to 4 tons to suit operating conditions for the twelve standard designs produced.

In addition the 2-8-0WD class of which a total of 732 were purchased from the Ministry of Supply was an additional 'standard' class and also the 25 WD 2-10-0s. Both these types were designed by Mr R. A. Riddles for war time 'austerity' conditions.

Many parts were standardised, for instance the LMS contributed: bogie and pony trucks; 2-bar slide bars and crosshead; water pick up gear and hornblocks and liners
The LNE contributed: 3-bar slide bars and crosshead; nameplates and brake blocks.
The GWR contributed: smokebox door; live steam injector and gauges and valves.
The Southern contributed: drawgear; top feed valve and tyre fastening.
This is just a brief list of standardised parts.

To compare the Railways Act of 1921 with the establishment of the British Transport Commission with the relevant executive bodies – the Railway Executive is the only one we are concerned with – is to compare a summer breeze to a hurricane.

The chief mechanical engineers of the big four were now answerable to a small group of officers with Mr R. A. Riddles at the head, assisted by four other LMS men, one from the Southern and one from the Great Western.

From this set up it will be seen that a very unbalanced group was formed with strong LMS bias. The results of this one sided faction can be quickly seen by referring to the locomotive designs – where the principal express designs were built – the number allocated to each works and the gradual running down of those which were not on the 'approved' list. This became more in evidence when the diesel programme got into its swing, but fortunately this has no place in this book.

However it speaks well for the team, that in January 1951 the first Class 7 *Pacific* appeared out of Crewe Shops, as No.70000 and in April the same year the first Class 5 4-6-0 No.73000 was finished at Derby. In May appeared 75000 Class 4 4-6-0 from Swindon and in December 1951 the first of the *Clan* Class 6 *Pacifics,* again from Crewe.

The *Britannia* class were eventually sent to all six regions, and were received, naturally, with mixed feelings. The most successful were those on the Eastern Region on the old Great Eastern section, where they revolutionised the running on the East Anglian expresses. On other regions apart from the London Midland, they did not show any conspicuous advantage over existing types.

The Class 5 4-6-0 was a dressed up version of the Stanier LMS mixed traffic, and therefore their performance was similar.

The *Clans* were a poor lot and it is puzzling to try and justify the building of these ten locomotives which all went to the Scottish region. Other superfluous designs must have been the Class 3 2-6-0 (77000) and the Class 3 2-6-2T (82000).

The Class 4 series of 2-6-0 and 2-6-4T were basically LMS designs and performed accordingly.

Perhaps the best all-rounder was the impressive looking Class 9 2-10-0 of which 251 were built and although mainly used on freight duties, they were occasionally pressed into passenger service and it was soon realised that they could travel at 80 mph with smoothness and ease and a speed of 90 mph was recorded down Stoke Bank on the LNE main line. Unfortunately this performance received such publicity that the authorities thought it prudent to prohibit such performances.

The only Class 8 locomotive to be built was No.71000 DUKE OF GLOUCESTER, a *Pacific.* It had three cylinders 18" x 28", Caprotti valve gear and double chimney, and was completed in

May 1954. Here again one wonders at this costly machine being built at this time, when a programme of 174 diesel-electric locomotives had been approved. Its basic design was good, and the steam consumption per IHP hour was the lowest achieved on any simple expansion locomotive in the world (*Locomotive Panorama Vol. 2* by E. S. Cox).

Ten of the Class 9 2-10-0s were fitted with Franco-Crosti boilers when new, but after about five years in service the heat exchanger drums were removed primarily because the expected saving of fuel due to the pre-heating of the feed water was nowhere near attained.

Mechanical stokers were fitted to three other Class 9s but these were not successful. It had been established many years before, that where large grate areas required firing such as the large American types, mechanical stokers were necessary and were successful, but for smaller grates – the largest grate area of the BR types was under 50 sq. ft. – hand firing was far more efficient and reliable for answering steam demands.

One other experiment was carried out on No. 92250 of the same class in the form of a Giesl ejector, but after exhaustive tests by Dr. Giesl himself with various qualities of coal, it was found that the objective of burning low grade coal could not be attained and no more were fitted.

Many BR locomotives were fitted with Roller Bearing axleboxes, but over their short life span no appreciable advantage was found; but based on continental practice, the maintenance costs in most cases were decreased considerably.

Caprotti valve gear was fitted to thirty of the Class 5 4-6-0s (Nos. 73125 to 73154). This was a type of poppet valve gear with less number of moving parts than the conventional Walschaert's gear, and a finer adjustment of inlet and exhaust valves. No long term results could be assessed but the locomotives' performance showed that this type of valve gear had a number of advantages.

Alas, all the test reports, the modifications, valve diagrams, testing station dossiers are now so much waste paper except to the historian.

The first BR locomotives to be scrapped were a Class 2 2-6-0, No. 78015 and Class 2 2-6-2T No. 84012 in 1963 at the age of nearly ten and just over ten years old respectively.

By October 1967 out of the 999 locomotives built to BR designs only 148 remained of which nearly half were Class 9 2-10-0s.

Many pre-nationalisation locomotives outlived their younger brethren – at the same date there were 249 LMS Class 5, 4-6-0s and 215 Stanier 8F 2-8-0s still on the books.

Were so many new or modified designs necessary, with all the expense of design, tooling, manufacture and teething troubles? The answer must be no to many of them. The only one that really justified itself was the 2-10-0 locomotive.

Of the works which built these last steam locomotives, Swindon had the melancholy honour of building the last one – a Class 9 2-10-0 No. 92220 appropriately named EVENING STAR. (*Sic transit gloria*).

See: Grand Junction Railway, Crewe; Great Northern Railway, Doncaster; Great Western Railway, Swindon; Lancashire & Yorkshire Railway, Horwich; London, Brighton & South Coast Railway, Brighton; London & South Western Railway, Eastleigh; Manchester, Sheffield & Lincolnshire Railway, Gorton; Midland Railway, Derby and North Eastern Railway, Darlington.)

BRITISH RAILWAYS–STANDARD CLASSES

Type	Class	Nos	1st Built	No Built	Cylinders inches	DW ft in	Tubes sq ft	Super sq ft	F'box sq ft
4-6-2	8	71000	1954	1	3/18 x 28	6 2	2264	691	226
4-6-2	7	70000-54	1951	55	20 x 28	6 2	2264	718	210
4-6-2	6	72000-09	1951	10	19½ x 28	6 2	1878	628	195
4-6-0	5	73000-171	1951	172	19 x 28	6 2	1479	369	171
4-6-0	4	75000-79	1951	80	18 x 28	5 8	1301	265	143
2-6-0	4	76000-114	1952	115	17½ x 26	5 3	1061	254	131
2-6-0	3	77000-19	1954	20	17½ x 26	5 3	932·9	190	118·4
2-6-0	2	78000-64	1952	65	16½ x 24	5 0	924	134	101
2-6-4T	4	80000-154	1951	155	18 x 28	5 8	1223	246	143
2-6-2T	3	82000-44	1952	45	17½ x 26	5 3	932·9	190	118·4
2-6-2T	2	84000-29	1953	30	16½ x 24	5 0	924	134	101
2-8-0T	* 8	90000-732	1943	733	19 x 28	4 8½	1512	310	168
2-10-0	9	92000-250	1954	251	20 x 28	5 0	1836	535	179
2-10-0	* 8	90750-74	1943	25	19 x 28	4 8½	1759	423	192

* WD Types

STEAM LOCOMOTIVES BUILT TO PREVIOUS COMPANIES DESIGNS

Works	1948	1949	1950	1951	1952	1953	1954	1955	1956	Totals
Brighton	19	8	30	18	–	–	–	–	–	75
Crewe	56	60	60	10	20	–	–	–	–	206
Darlington	41	36	42	48	8	–	–	–	–	175
Derby	46	30	30	–	10	–	–	–	–	116
Doncaster	23	17	20	24	6	–	–	–	–	90
Eastleigh	8	5	3	–	–	–	–	–	–	16
Gorton	2	8	–	–	–	–	–	–	–	10
Horwich	40	29	26	26	1	4	1	–	–	127
Swindon	64	75	73	20	12	13	5	15	–	277
Contractors	96	89	101	62	57	11	18	3	9	446
										1538

STEAM LOCOMOTIVES BUILT TO BRITISH RAILWAYS DESIGNS

Works	1951	1952	1953	1954	1955	1956	1957	1958	1959	1960	Totals
Brighton	17	27	18	22	22	20	4	–	–	–	130
Crewe	27	21	27	43	38	39	46	43	–	–	284
Darlington	–	4	8	33	10	10	10	–	–	–	75
Derby	29	11	20	29	26	20	10	–	–	–	145
Doncaster	–	4	11	20	34	25	28	–	–	–	122
Horwich	–	6	14	–	–	4	21	–	–	–	45
Swindon	16	24	25	37	26	11	22	19	15	3	198
											999

BOILER								
Total	WP	Tube Length	Grate	Boiler Diameter	Std No	TE	Weight WO ton cwt	Max Axle Load ton cwt
sq. ft	lb/sq ins	ft in	sq ft	ft in × ft in		lbs		
2955	250	17 0	48·6	5 9 x 6 5½	13	39080	101 5	22 0
3192	250	17 0	42	5 9 x 6 5½	1	32150	94 0	20 5
2701	225	17 0	36	5 4 x 6 1	2	27520	86 19	18 15
2019	225	13 2⅞	28·65	4 11$\frac{11}{16}$ x 5 8½	3	26120	76 4	19 14
1709	225	13 0	26·7	4 9 x 5 3	4	25100	69 0	17 5
1446	225	10 10½	23	4 9⅛ x 5 3	7	24170	59 2	16 15
1241·3	200	10 10½	20·35	4 5 x 5 0½	6	21490	57 9	16 5
1159	200	10 10½	17·5	4 3 x 4 8	8	18515	49 5	13 15
1612	225	12 3	26·7	4 9⅛ x 5 3	5	25100	86 13	17 19
1241·3	200	10 10½	20·35	4 5 x 5 0½	6	21490	74 1	16 6
1159	200	10 10½	17·5	4 3 x 4 8	8	18515	63 5	13 5
1990	225	12 0	28·6	4 7⅛		34215	72 0	
2374	250	15 3	40·2	5 9 x 6 1	9	39667	86 14	15 10
2550	225	15 8	40	5 7½ x 5 9⅞		34215	78 12	13 9

TOTAL STEAM LOCOMOTIVES BUILT AT BRITISH RAILWAYS WORKS

Works	1948	1949	1950	1951	1952	1953	1954	1955	1956	1957	1958	1959	1960	Totals
Brighton	19	8	30	35	27	18	22	22	20	4	–	–	–	205
Crewe	56	60	60	37	41	27	43	38	39	46	43	–	–	490
Darlington	41	36	42	48	12	8	33	10	10	10	–	–	–	250
Derby	46	30	30	29	21	20	29	26	20	10	–	–	–	261
Doncaster	23	17	20	24	10	11	20	34	25	13	–	–	–	212
Eastleigh	8	5	3	–	–	–	–	–	–	–	–	–	–	16
Gorton	2	8	–	–	–	–	–	–	–	–	–	–	–	10
Horwich	40	29	26	26	7	18	1	–	4	36	–	–	–	172
Swindon	64	75	73	36	36	38	42	41	11	22	19	15	3	475
														2091

SUMMARY

Steam Locomotives Built to Previous Designs	1092
Steam Locomotives Built by Contractors	446
Steam Locomotives Built to BR Designs	999
Grand Total	2537

BRORA WORKS
Brora, Sutherland

According to an article in the *Railway Magazine* for January 1960 by Mr Iain. D. O. Frew on 'The Brora Colliery Tramway', the Duke of Sutherland intended to make Brora the nerve centre of the railway which ran from Golspie and was built in 1871. He built a large engineering works at Brora in which he planned to manufacture everything from nuts and bolts to locomotives. This optimistic venture was not successful and the works were sold in 1890, after having built only one locomotive in 1871, being a 0-4-0ST named FLORENCE with a gauge of 1′ 8″. That the building of this locomotive was extremely doubtful, was shown by subsequent correspondence in the *Railway Magazine* and it is too much of a coincidence that according to the Manning Wardle works list No. 579 of 1875 is a 0-4-0ST 1′ 8″ gauge named FLORENCE for the Duke of Sutherland, Brora.

R. BROTHERHOOD
Railway Works, Chippenham

The works were established in 1842 and a great variety of work was carried out including railway fittings, points and crossings, wagons, bridges and signal equipment.

Locomotive building took place between 1857 and 1867. Information regarding the precise number, original customers and technical details has always been shrouded in mystery. To complicate matters the firm dealt in second hand locomotives which they no doubt repaired and rebuilt and some of these are no doubt among the locomotives built as new by the Brotherhoods.

Reference to Mr S. A. Leleux's book on this firm certainly clarifies many points which up to then had not been resolved, but it will never be certain as to how many new locomotives were built.

Rowland Brotherhood built a special workshop for his brother Peter to build these 'loco engines'.

The first one named MOLOCH was a six coupled broad gauge tank, with 17″ x 24″ cylinders and 4′ 8″ diameter wheels, which was eventually bought by the Bristol and Exeter Railway in 1874 and given the number 111. Whether it did any work between 1857 and 1874 is not known.

Fig. 61 **R. Brotherhood**
*c*1866 broad gauge locomotive

PHOENIX was also a six wheeled broad gauge tank with four coupled wheels, the cylinders were 12″ x 22″, and this locomotive appeared on the Newquay & Cornwall Junction Railway and was used by W. West and Sons of St. Blazey in building the line. It is thought that West may have built it.

A number (probably four) of standard gauge 0-4-0STs were built, and one named BEE was used at the opening of the Bishop's Castle Railway in 1866.

Two standard gauge 0-6-0s are associated with Brotherhoods, one worked the Scole Railway in Norfolk and the other was bought by the London, Chatham and Dover Railway.

The latter was the first 0-6-0 goods locomotive to work on the LC&DR and had 12″ x 18″ inside cylinders and 4ft. diameter wheels. It was named SWALE and was eventually converted to a saddle tank.

Due to financial complications the contents of the works were sold by auction in 1869–1872 and the works itself subsequently occupied by Saxby & Farmer, the signal manufacturers who were absorbed by the Westinghouse Brake & Signal Company.

BROWN J. B. & COMPANY

Cannon Street & Upper Thames Street, London

This firm advertised as locomotive builders in the 1860s but no actual record of any locomotive having been built has been discovered. Almost certainly they were only agents for this type of work.

BROWN & MAY

North Wiltshire Foundry, Devizes

Engineers and iron founders, this firm was founded in 1854. They built traction engines, road locomotives and portable and fixed steam engines.

Two locomotives were supplied to Brotherhood of Chippenham, with vertical boilers and multiplying gear between the driving wheels and the engines.

They quoted for a 12 hp locomotive for the Severn & Wye Railway for £450 with 6½″ x 13″ cylinders, 2′ 6″ driving wheels and 4′ 4″ wheelbase.

The works closed down in 1912.

BROTHERHOOD

Date	Type	Cylinders inches	D.W. ft in	Location*	Name*
1857	0-6-OT	16¾ x 24	4 8½	Bristol & Exeter Rly III	Moloch
c 1860	0-6-OT			South Wales	
c 1860	0-6-0	12 x 18	4 0	London, Chatham & Dover Rly	Swale
1862	0-4-OT			Road-rail design	
1863	-4-T			Newquay & Cornwall Junc Rly	Phoenix
1863	-4-T				
1863	0-6-0			Contractor	
1865	0-4-OST			Bishop's Castle Rly	Bee
	0-4-OST			Duston Iron Co	
c 1866	0-4-OST			I. W. Boulton	
	0-4-OST			Contractor	
c 1866/7	2-4-OST				
c 1866/7	2-4-OST				
c 1866/7	2-4-OST				
c 1866/7	2-4-OST			I. W. Boulton	(Fig. 61)
c 1866/7	2-4-OST			J. T. Williams	
	0-6-0			Frenze Estate Rly Norfolk (Scole Rly)	

* Not necessarily initial customer or original name.

BROWNE, SAMUEL
Haymarket Iron Works, Liverpool

Offered to make a locomotive with 11″ cylinders and coupled wheels for £1000, or for £900 if the wheels were not coupled, for the London and Croydon Railway, in July, 1837.

No locomotives were ordered from this quotation and no trace of any locomotive building has been discovered.

(Ref: *Locos of the SER* by D. L. Bradley (RCTS))

BRUNTON, WILLIAM
Ripley

See BUTTERLEY COMPANY

BRUSH BAGNALL TRACTION LIMITED
Stafford

See BAGNALL, W. G. LTD.

BRUSH ELECTRICAL ENGINEERING COMPANY
Loughborough

See HUGHES, HENRY & COMPANY

BURRELL, CHARLES & SONS LTD.
St. Nicholas Works, Thetford

Famous for their road engines and showman's road engines they also built two steam tram engines, one for Bradford and one for Birmingham.

1119 (1885) Bradford & Shelf Tramways No. 6 3′ 6″ gauge

1190 (1886) Birmingham Central Tramways No. 71 3′ 6″ gauge

Both had 2′ 6″ wheels, i.c. compound cylinder 10″/$17\frac{1}{2}$″ x 14″, Joy's valve gear, four coupled and boiler pressed at 160lb/sq.in.

An improved type of condenser for tram engines was patented, which consisted of a series of double longitudinal tubes. These were in effect a twin tube one inside the other, the space between providing a passage for the exhaust steam so that the air passed through the inner tube and over the external surface of the outer tube giving a more rapid condensation rate.

This type of condenser was used by Thomas Green & Son on their later tram engines.

The Birmingham tram was returned to Burrell's despite the fact that its performance was extremely good and was ultimately used as a stationary engine by Lord Iveagh of Elvedon. It had run 1048 miles on the Birmingham track on the King's Heath, Moseley, & Small Heath Routes. A full account with drawings is given in *Chronicles of a Country Works* by R. H. Clark.

Fig. 62 **Chas. Burrell** Tram engine and Milne trailer

BURSTALL, TIMOTHY

Leith

Built the PERSEVERANCE in 1829 for the Rainhill trials. It weighed 2 tons 17 cwts well within the stipulated maximum of 6 tons. Unfortunately it met with an accident on its conveyance to Liverpool, and although making a few tentative runs did not run in the trials.

The boiler was mounted on a simple frame midway between the two pairs of wheels. It was a vertical boiler with apparently a straight through flue with no cross tubes or baffles to transfer the heat and this is probably the chief reason for its poor performance, and speed which only reached three to four miles per hour. The drive was said to be similar to the steam road coaches which had been built by Burstall and Hill.

Fig. 63 **T. Burstall**
1829 PERSEVERANCE geared locomotive

BURY, EDWARD & COMPANY

Clarence Foundry, Love Lane, Liverpool

Edward Bury established an engineering works in the early 1820s; he was a keen business man, and obtained the services of James Kennedy who had had his training at the Stephensons' works. Bury made him works foreman and subsequently became a partner in the firm. For many years they concentrated on producing four wheeled locomotives, as single drivers, and four coupled types. The designs were sound and rivalled Stephensons' products of the 1830s.

In addition to being a private locomotive builder, Bury was at the same time locomotive superintendent of the London & Birmingham Railway, from its inception until 1846, after which he succeeded Mr Cubitt, in an advisory capacity in the locomotive department of the Great Northern Railway from 1847 to 1850.

This dual role evidently had advantages as Bury supplied over one hundred locomotives to the London and Birmingham Railway, thirteen to the Great Northern Railway (six of which were sub-contracted to William Fairbairn & Sons). His locomotives were also delivered as standard classes to the Eastern Counties, Manchester, Bolton, and Bury; Midland Counties, Lancaster & Preston Junction, and North Union railways.

The most notable characteristics of Bury's locomotives were the bar frames of forged iron, raised spherical topped copper firebox shell and internal copper firebox of semi-circular section, cranked axles, and inside horizontal cylinders, very slightly inclined to allow the piston rods to clear the leading axle by passing underneath. The two classes, namely 2-2-0 and 0-4-0 types varied only in their wheel diameter and size of cylinders. The boiler pressure averaged 120lb/sq.in. which was high for the period. The designs were purely functional and the workmanship excellent, and the locomotives were generally free from troubles which were encountered by other manufacturers.

His first locomotive named DREADNOUGHT, of which very little information is known, commenced running on the Liverpool & Manchester Railway in March, 1830. Most descriptions signify that it was modelled on Stephenson's or Hackworth's practice with outside cylinders, but the cylinders drove a jack-shaft which in turn was coupled to the driving wheels. It proved much too heavy for the track and very unsteady, and even after modifications to lower the centre of gravity and distribute the weight more evenly it was a failure.

The second was the LIVERPOOL, a four coupled locomotive with 6′ wheels, which after minor alterations was quite successful. Instead of a tubular boiler, convoluted flues were used, the blast being provided by means of bellows under the tender. It had a cranked driving axle which gave a lot of trouble. It was built in July, 1830 and ran trials on the Liverpool & Manchester Railway who refused to buy it. It was rebuilt and sold to the contractor on the Bolton and Kenyon Railway.

Locomotive work then gradually superseded other types of engineering work being carried out. Between 1831 and 1837 about twenty were supplied to railroads in America, and orders rolled in from the home railways, so much so that the works were unable to cope with such quantities and many were sub-contracted to firms like R. & W. Hawthorn, Benjamin Hick, Maudslay Son & Field and many others.

Fig. 64 E. Bury Plan of 2-2-0 showing frames, motion and section through firebox

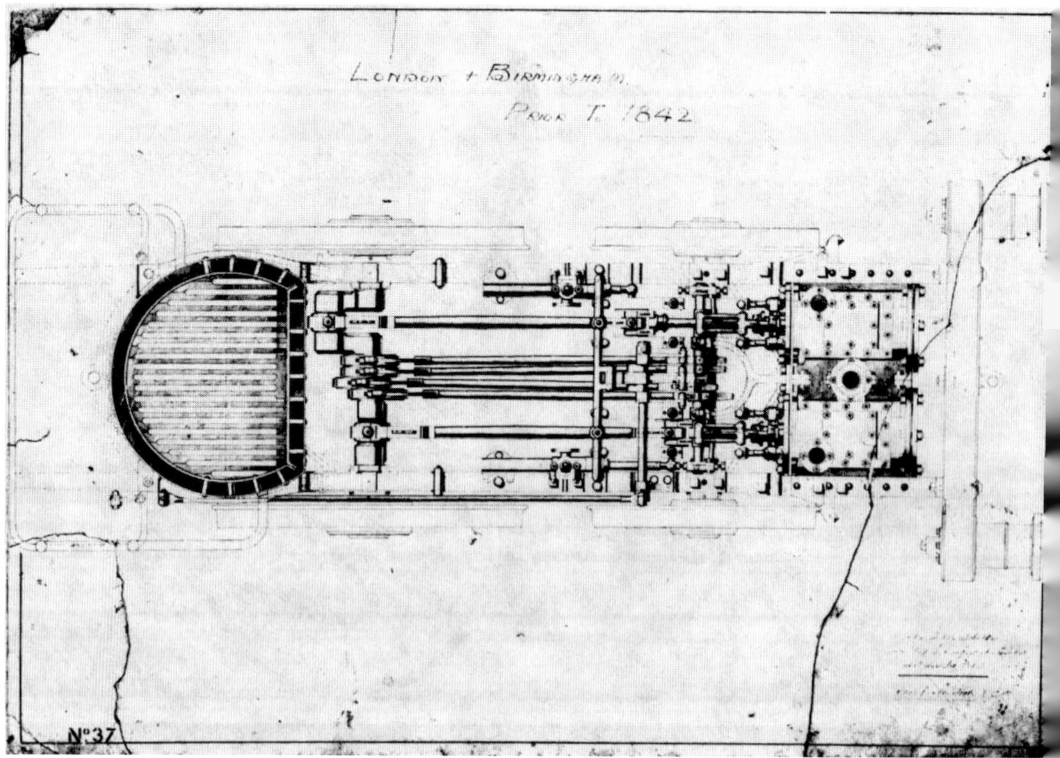

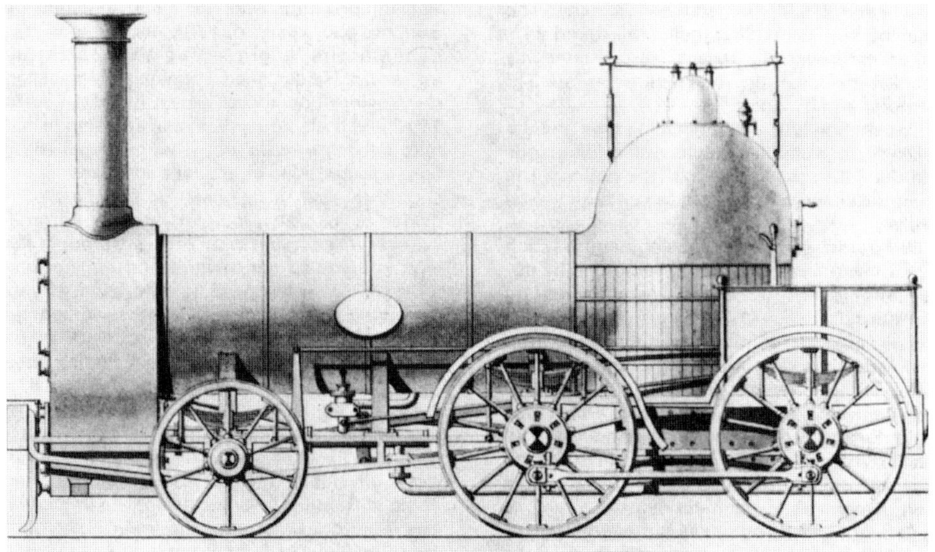

Fig. 65 E. **Bury** 1847 type 'F' 2-4-0

Edward Bury was asked by a Select Committee on Railways to state his reasons for his preference for four coupled locomotives and his main reasons given were (a) they were cheaper to build than six wheeled types, (b) took up less space, (c) were lighter and therefore required less power to move up banks, leaving more power available for hauling the trains, (d) safer as they adapted themselves better to the rails, being less likely to run off the rails at crossings and curves, (e) more economical in working with fewer parts in motion and consequently less friction, (f) parts more accessible, (g) buildings and turntables were not required to be on such a large scale, and (h) stoppages were not likely to take place on the journey.

Some of these reasons may seem very illogical now and overstatement of the case for his engines, but he managed to persuade many a locomotive superintendent to his own small locomotive policy.

180 2-2-0s were built for railways in this country and about twelve for abroad, eight of which went to America, but only about 110 were built at the Clarence Foundry. The last 2-2-0 was built in 1845.

Unfortunately he did not develop his designs quick enough to meet the increasing loads and speeds.

He had never favoured the six wheeled engine of any type but by 1846 he was forced to concede that the only future lay in these larger and more powerful locomotives. He built one six coupled engine for the Birmingham and Gloucester Railway but by the time he realised the large potential and demand for these larger types, he had lost the initiative and although a number of 2-2-2, 0-4-2 and 2-4-0s were built between 1847, it was too late and the firm was wound up in 1851. Edward Bury had retired in 1848 and died on 25 November, 1858.

The firm had become BURY CURTIS & KENNEDY in 1842, the works foreman becoming a partner.

350 locomotives were built by the summer of 1848. In this year the familiar D-shaped firebox was changed for the more conventional rectangular type with domeless boiler having the regulator valve in the smokebox. This applied to the new designs. Plate frames were also used from this date.

A description of two 0-4-0s built in 1846 for the Furness railway gives some interesting details. The inside cylinders were 14″ x 24″, the slide valves were between the cylinders. The width of steam ports were 1¼″, exhaust ports 3″ and 10½″ long and the valve lap ¾. Each cylinder was a separate casting and each had a separate exhaust pipe both of which converged into one.

The wheels were 4′ 9″ diameter with wrought iron centres and cast iron bosses and the iron tyres were secured to the rims with 1″

bolts. The axleboxes were gunmetal and the axle bearings $5\frac{1}{4}''$ diameter and $7''$ long. The coupling and connecting rods were round with forked ends and all brasses were secured by double gib and cotter. The crank pins were $2\frac{1}{4}''$ diameter and $2\frac{3}{4}''$ long.

The motion was the curved link type and the eccentric rods coupled direct, with crossed rods and the links suspended from the bottom. The eccentric sheaves were cast iron with gunmetal straps.

Bar frames were used rectangular in section $4'' \times 2''$ with a lower bar $2\frac{3}{8}''$ diameter. The upper and lower members were joined by pedestals.

The boiler barrel had three rings made from Low Moor iron, and each ring comprised two plates $\frac{1}{2}''$ thick, the rear ring was flanged back to join the firebox casing the crown of which was in the form of a dome made up of no less than twelve plates, eight of which went to make the crown. The seams throughout were single rivetted lap with $\frac{3}{4}''$ diameter rivets.

One of these locomotives is preserved.

A total of 415 locomotives were built in the twenty years the firm was in business. Other notable products were engines for steamships which were built up to 400 hp.

BUTTERLEY COMPANY
Ripley

This firm was established in 1790, and in 1808 William Brunton, who had been engaged by Boulton and Watt in 1796 as a mechanic, became the engineer at the Butterley Ironworks. He designed a 'mechanical traveller' or 'walking locomotive' which he patented in 1813 and built during that year. The engine had four smooth wheels which were propelled by two legs operating on the rails at the rear of the engine through a combination of levers connected to the piston rod of the single horizontal cylinder. The arrangement of levers brought the legs at each stroke towards the engine, one after the other, so pushing it forward. It was extremely ingenious but did not travel very far when it arrived at the Newbottle Colliery. When it did travel, its speed was about two to three miles per hour. Unfortunately on 31 July, 1815 the boiler exploded killing a number of people and causing a large amount of damage.

There were possibly two of these locomotives built (1813–14) as one is reported being at Rainton, besides the one at Newbottle. The Rainton locomotive however may have been the Newbottle one, repaired and put into service again about 1825.

From 1860 to 1907 Butterley built approximately twenty seven locos, originally for their own use, although as early as 1838 at least two singles were built for the Midland Counties Railway. (Fig. 66)

The earlier construction in the 1860s and 1870s included about six vertical boilered four wheeled locomotives with cylinders 4' diameter and a gear ratio of 3:1.

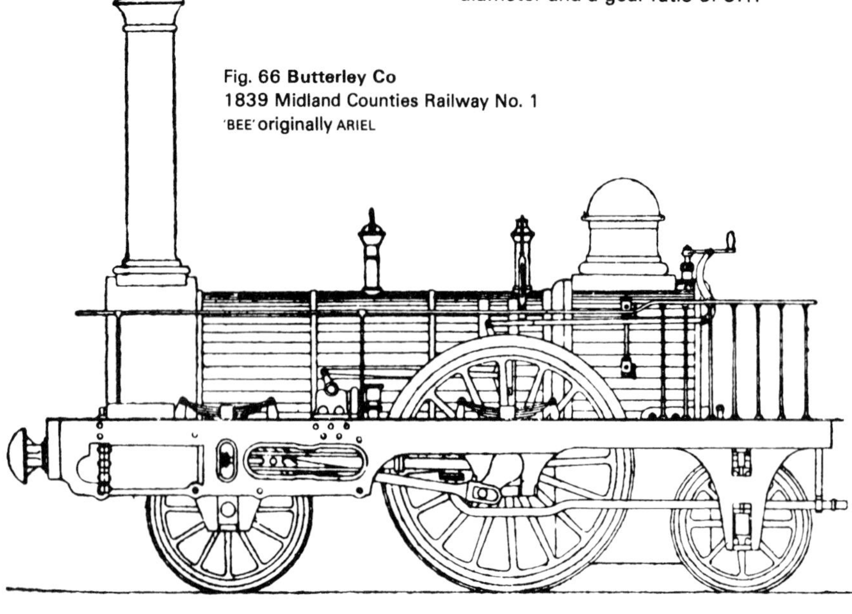

Fig. 66 **Butterley Co**
1839 Midland Counties Railway No. 1
'BEE' originally ARIEL

Date	Type	Customer	No	Name
1813	4W	Brunton's 'Walking' Locomotive		
1839	2-2-2	Midland Counties Rly	1	Ariel Fig. 66
1839	2-2-2	Midland Counties Rly	2	Hercules
1860-70	0-4-0VB-G	Butterley Co Ltd		
1860-70	0-4-0VB-G	Butterley Co Ltd		
1860-70	0-4-0VB-G	Butterley Co Ltd		
1860-70	0-4-0VB-G	Butterley Co Ltd		
1860-70	0-4-0VB-G	Butterley Co Ltd		
1860-70	0-4-0VB-G	Butterley Co Ltd		
1860-70	0-4-0T	Butterley Co Ltd	B 14 C	Gertrude
1887	0-4-0ST	Butterley Co Ltd	B 1 C	
1889	0-4-0ST	Butterley Co Ltd	B 2 C	
1892	0-6-0ST	Butterley Co Ltd	B 3 C	Fig. 68
1893	0-6-0ST	Butterley Co Ltd	B 4 C	
1896	0-4-0ST	Butterley Co Ltd	B 9 C	
1896	0-4-0ST	Butterley Co Ltd	B 10 C	
1900	0-6-0ST	Butterley Co Ltd	13	
1903	0-4-0ST	Butterley Co Ltd	B 14 C	
1904	0-6-0ST	Butterley Co Ltd	20	
1907	0-6-0ST	Butterley Co Ltd	21	
	0-4-0	Butterley Co Ltd		Adeline (?)
	0-4-0	Butterley Co Ltd		Ethel
	0-4-0	Butterley Co Ltd		Florence
	0-4-0	Butterley Co Ltd		Tiny
R1896	0-4-0ST	Butterley Co Ltd	B 5 C	
R1898	0-4-0ST	Butterley Co Ltd	B 6 C	
R1903	0-4-0ST	Butterley Co Ltd	B 7 C	
R1910	0-4-0ST	Butterley Co Ltd	B 8 C	
R1900 *c*	0-4-0ST	Butterley Co Ltd	B 11 C	Bessie
R1895	0-4-0ST	Butterley Co Ltd	B 12 C	Fig. 67

The rebuilt 0-4 OSTs may have been conversions of the early 0-4-0-VBGs or the 0-4-0s.

Fig. 67
Butterley Co
0-4-0ST as rebuilt
1895 B12C
at Ripley

Fig. 68
Butterley Co
1892 0-6-0ST
No. B3C
as NCB 25,
No. 5 (Shipley)
area

CAIRD & CO.
Greenock

This firm were well known shipbuilders and marine engineers and also built four single 2-2-2s for Scottish railways.

The two for the GP&GR had outside frames, inside cylinders and a boiler with a raised firebox with spring balance safety valves fitted above. The large dome was fitted on the barrel close to the chimney (Fig. 69). The valves were operated by Gab motion.

WASP had outside cylinders and single inside frames. The boiler had a dome placed close to the chimney with spring balance safety on top and a further safety valve on the raised top firebox.

No details are known concerning the Edinburgh & Glasgow railway single TREVITHICK.

These four locomotives are the only ones known to have been built by Caird & Co.

CALEDONIAN RAILWAY

See GLASGOW, PAISLEY & GREENOCK RAILWAY.

Works No	Date	Type	Cylinders inches	D.W. ft in	Railway	No	
E1	1840	2-2-2	IC 13 x 18	5 0	Glasgow, Paisley & Greenock Rly	11	Greenock (a)
E2	1840	2-2-2	IC 13 x 18	5 0	Glasgow, Paisley & Greenock Rly	12	Glasgow (b)
E3	1840	2-2-2	OC 13 x 18	5 6	Glasgow, Paisley, Kilmarnock & Ayr Rly	30	Wasp
E4	1841	2-2-2	13 x 18	5 0	Edinburgh & Glasgow Rly		Trevithick

(a) Became Caledonian Rly No 68. (b) Became Caledonian Rly No 69. Both withdrawn 1854.

Fig. 69 **Caird & Co.** 1840 2-2-2 for Glasgow, Paisley & Greenock Railway

CALEDONIAN RAILWAY

St. Rollox, Glasgow

Robert C. Sinclair (1849–1856); Benjamin Connor (1856–1876); George Brittain (1876–1882); Dugald Drummond (1882–1890); Hugh Smellie (1890); John Lambie (1890–1895); J. F. McIntosh (1895–1914); William Pickersgill (1914–1923).

Due to the cramped conditions of the Greenock Works of the Glasgow, Paisley & Greenock Railway which had become part of the Caledonian Railway in 1846 new works were built in 1853 and opened in 1854 at St. Rollox and the Greenock works were closed, all unfinished locomotives being taken to the new works for completion.

Robert Sinclair who had been responsible for the locomotive department at Greenock had produced a number of 2-2-2 and 2-4-0 *Crewe* type engines similar to Alexander Allan's designs and the first locomotive to be completed at St. Rollox in 1854 was a 2-4-0 of this pattern. Sinclair did not stay long at St. Rollox and in 1856 left for the south, becoming loco superintendent of the Eastern Counties Railway where he stayed for ten years. Sinclair had of course been influenced by Alexander Allan during his stay on the Grand Junction Railway and the *Crewe* type of locomotive had a strong bearing on the design of locomotives, not only the Caledonian but others too.

Benjamin Connor, Sinclair's successor had been Works Manager at Neilson & Mitchell. Connor carried on in the 'Allan' tradition and designed the 8ft 2-2-2 which had all the characteristics with outside cylinders, flared smokebox, raised firebox with dome on top, underslung leading springs and slotted splashers. Connor was the first to introduce crescent balance weights which were integral with the rims. He continued to rely on previous Allan experience and the 2-4-0 type was continued with 8ft (despite the banks), 7ft down to 5′ 2″ driving wheels.

When the Scottish Central and Scottish North Eastern railways were absorbed into the Caledonian system in 1866 one would have expected Alexander Allan who had been in charge of the locomotive department at Perth, to take charge at St. Rollox but this did not happen, Connor remaining in charge and Allan left Scotland for the post of manager of the Worcester Engine Co.

The 'Allan' influence continued however at St. Rollox and Connor carried on until 1876 when he died and his place was taken by George Brittain who had been an assistant of Allan's at Perth. Brittain is best remembered for his 4-4-0 *Oban* bogies, but other types built during his reign were not particularly successful.He produced a 2-4-2 radial tank for the Oban line which was not successful. The *Oban* bogies departed from the Allan influence somewhat; they were built in 1882, the last year of Brittain's office and although they had the customary outside cylinders they had inside frames which was a departure from preceding designs. They were the most successful of Brittain's locomotives and were used for the heavy holiday traffic on the Callander to Oban line. Ten of them were built by Dubs & Co.; the coupled wheels were 5′ 2″ and the cylinders were 18″ x 24″. Four wheeled tenders were fitted due to limited turning facilities.

In 1882 the dynamic Dugald Drummond took over. He came from the North British Railway's headquarters at Cowlairs and soon found that the passenger stock was quite inadequate for the increasing loads. He set about rebuilding a number of Connor's and Brittain's locomotive classes and the first new designs put on the drawing board were two 4-4-0s with 5′ 9″ and 6′ 6″ driving wheels, a 0-6-0 goods and a light tank with 0-4-4T wheels arrangement for suburban and branch line traffic. In addition Drummond paid great attention to the economic running of his department. Perth and Greenock ceased building new locomotives, all work of that nature being concentrated at St. Rollox. Perth remained a repair shop to the end of the Caledonian railway's separate existence and in LMS days too. Automatic continuous brakes were introduced, as well as steam heating for passenger stock and gas lighting.

The 6′ 6″ 4-4-0s were a success from the start. They had 18″ x 26″ cylinders with a modest boiler pressure of 150lb/sq.in., but subsequent batches with steel boilers had the pressure increased to 200lb/sq.in. The boilers were not large but Drummond's front end design was such that a more efficient use of the available steam was obtained, with full regulator and short cut-offs. During tests one of the higher pressure locomotives attained a power output approaching 1000 ihp at a speed of 51 mph and a percentage cut-off of 27 which was a wonderful demonstration and vindication of the design of these locomotives. The smaller wheeled version with the same size cylinders were put on the boat trains between Glasgow and Gourock.

A famous locomotive which appeared in 1886 was the 4-2-2 single built by Neilson & Co. for the Edinburgh exhibition. It was designed by Neilsons but had all the Drummond lines and after the exhibition it became CR No. 123 and was the only 4-2-2 to run on that railway. With 7ft driving wheels and 18″ x 26″ cylinders the weight was 42 tons with 16 tons available for adhesion and the locomotive was a

Fig. 70
Caledonian Railway
1863 2-2-2 (Connor)
as rebuilt by
D. Drummond 83A
8′ 2″ single

Fig. 71
Caledonian Railway
1872 2-4-0
Crewe type,
mixed traffic
No. 130

Fig. 72
Caledonian Railway
1896 4-4-0 No. 722
Dunalastair class
McIntosh

great success, taking part in the races of 1888 and 1895. In LMS days it was used on Directors' saloon specials and happily is now preserved. Although Drummond's stay at St. Rollox was comparatively short, he had transformed the locomotive stock into a reasonably efficient unit by the time he resigned in 1890 to make his brief impact in Australia.

He was succeeded by Hugh Smellie from the Glasgow and South Western Railway, who tragically was hardly in office before he died.

John Lambie was appointed in the same year and made no remarkable changes, but followed mainly on the lines of Drummond. The appearance of the locomotives was altered by Lambie taking the Ramsbottom safety valves off the top of the dome and mounting them over the firebox.

A new phase at St. Rollox commenced after John Lambie died in 1895, by the appointment of John Farquharson McIntosh. He had been in the locomotive department of the Scottish North Eastern Railway and became chief running inspector on the Caledonian Railway.

His first new passenger locomotive design was the 6′ 6″ DUNALASTAIR 4-4-0. It had larger cylinders and a larger boiler than any previous standard Caledonian design; the cylinders were 18¼″ x 26″, the total heating surface of boiler and firebox was 1403 sq. ft. The 1884 6′ 6″ 4-4-0 by Drummond had a heating surface of 1208·6 sq. ft. The diameter of the boiler was also increased which gave these engines an impressive appearance and their performance was in keeping with their massive outline (for that period).

In 1897 appeared the *Dunalastair II* with 19″ cylinders and a boiler with 1500 sq. ft. of heating surface, and the pressure increased to 175lb/sq.in. and in 1899 the *Dunalastair III* class appeared with a further increase of heating surface to 1540 sq. ft.

These new engines gained a reputation for fast running with heavy loads and boilers which steamed well, and were very popular with the running department. In fact their fame spread abroad, which led to Neilson & Co. building a batch of *Dunalastair II*'s for the Holland railway in 1898.

Charles Rous-Martin wrote at the time, that no living engineer had done more for the cause of railway progress than J. F. McIntosh.

For goods and mineral traffic McIntosh designed a six coupled locomotive with 5′ wheels and 18½″ x 26″ cylinders and using the *Dunalastair I*'s boiler.

For the suburban traffic three different classes of 0-4-4Ts were built, one for the Glasgow underground lines with condensing apparatus 18″ x 26″ cylinders and 5′ 9″ driving

Fig. 73 **Caledonian Railway** 1903 McIntosh 0-8-0 No. 606. Note wheel centres

Fig. 74 **Caledonian Railway** 1914 McIntosh superheated 4-6-0 No. 188

Built	Nos	Cylinders inches	D.W. ft in	H/S	BP
1902	55-59	19 x 26	5 0	1800	175
1903	49-50	21 x 26	6 6	2323	200
1905	51-54	21 x 26	6 6	2323	200
1906	903-907	20 x 26	6 6	2265.75	200
1906	908-917	19 x 26	5 9	2178	180
1906	918-922	19 x 26	5 0	2023	175
1913-14	179-182	19 x 26	5 9	1970	Superheated 170
1914	183-189	19 x 26	5 9	1970	Superheated 170

wheels (these locomotives were painted black); another type for the Cathcart circle with 17″ x 24″ cylinders and 4′ 6″ wheels; and for other services and branch lines with similar dimensions to the condensers but with a higher boiler pressure. The Cathcart locomotives had smaller wheels to enable them to get away quickly from the short distanced stations.

In 1902 McIntosh designed his first 4-6-0 and between this date and 1914 six different designs were built with a total of forty-two locomotives. The increasing practice of double heading emphasised the need for a larger and more powerful class but at the same time engineers had to bear in mind the limitations of axle loading. (See above table)

All the above 4-6-0s had inside cylinders driving the leading axles and round top flush fireboxes. They were equipped with large eight-wheeled tenders carrying 5000 gallons of water and 5 tons of coal. There were no water troughs.

The first Scottish 0-8-0 was also designed by McIntosh and six were turned out in 1903 together with a tank version. In 1912 five 2-6-0 Moguls were built and these also were the first of the type in Scotland. They had the same boiler and motion as the standard 0-6-0s and the addition of a pony truck did not improve their appearance or performance. McIntosh had also designed a projected 4-4-2 De Glehn compound and a four cylinder simple Pacific just before he retired in 1914 but neither was proceeded with by his successor.

William Pickersgill was appointed locomotive superintendent, and he had come from the Inverurie works of the Great North of Scotland Railway, where he had designed and built numerous 4-4-0s but nothing any larger. However, all the stock was well built and left in excellent condition. His first design at St. Rollox was a 4-4-0 on McIntosh lines, but some 4-6-0s were soon put in hand which were quite unlike the previous ones turned out at St. Rollox. These were outside cylinder locomotives 20″ x 26″ with 6′ 1″ diameter driving wheels, heating surface 1934·2 sq. ft and a large grate area of 25·5 sq. ft, the boiler pressure was 175lb/sq. in. They were massive machines, not notable in their performance. In fact, in their original state they were extremely sluggish, caused principally by the layout of the link motion.

Twelve 4-6-2T passenger tank locomotives were then designed by Pickersgill but they were not built at St. Rollox but by the North British Locomotive Co. in 1917. These were indeed giants, intended for the Clyde coast trains.

The last major design was for a three cylinder 4-6-0 and four were laid down in 1921, they were the largest of all the 4-6-0s built by McIntosh and Pickersgill with 18½″ x 26″ cylinders, 6′ 1″ coupled wheels and a boiler barrel diameter of 5′ 9″. Walschaerts valve gear was used with a complicated derived motion and this was the major contribution towards them being extremely disappointing locomotives. There is an engineers adage 'if it looks right – it is right' but these engines must have been the exception!

The works had been enlarged in 1870–1 and again in 1884 and by 1921 three thousand workmen were employed.

On 1 January, 1923 the Caledonian Railway became part of the LONDON MIDLAND & SCOTTISH RAILWAY.

William Pickersgill became divisional locomotive engineer. His '60' class, 4-6-0s of 1916 had been far more successful than the later three cylinder '956' built in 1921, and an order was issued to build a further twenty in 1925–6. They were numbered 14630 to 14649 and were the last Caledonian design to be built at St. Rollox. They had two outside cylinders 20″ x 26″ diameter coupled wheels. The boilers were parallel with a flush round topped firebox, the tube heating surface was 1529·5 sq. ft firebox 146·5 sq. ft and the superheater 258·2 sq. ft, giving a total of 1934·2 sq. ft. The grate area was 25·5 and working pressure 175lb/ sq. in. The weight in working order was 75 tons without tender.

Various reports on this class of engine indicate very erratic performances, but during trials with a Midland 4-4-0 compound, L&NWR *Prince of Wales* class and *Claughton* class and one of Hughes rebuilt L&YR 4-6-0s, the Caledonian locomotive certainly did not disgrace itself. The trials were between Carlisle and Preston. The overall figures were better than the Midland 4-4-0 but, like her sisters, No. 14630 was sluggish on the level and even downhill so that any advantage it showed on its hill climbing ability was cancelled out by this disadvantage. Their coal burning figures were high but, apart from the Midland Class 2 4-4-0s, had a lower maintenance cost figure than comparative locomotives of 4-4-0 and 4-6-0 types on the system.

The only other type built at St. Rollox was the standard Class 4 0-6-0, originating from Derby. Between 1924 and 1928 a total of sixty were built, the only variation from standard being the conforming to the Scottish loading gauge.

No further new building took place but the works were re-organised in 1968 to become the main Scottish repair works and absorbing the additional work previously carried out at Cowlairs.

An interesting footnote to the above is that the LM&SR acquired 1077 locomotives from the Caledonian railway and whereas the locomotive stock of the Highland, and Glasgow & South Western railways were rapidly withdrawn, the Caledonian had only lost 137 by 1932 and at the end of 1960 some 209 were still left. Officially, on 29 November, 1963 the last eight Caledonian Railway locomotives were withdrawn, all being 0-6-0s; the oldest was No.17296 built at St. Rollox in 1887.

St. Rollox built locomotives were certainly made to last, and their long record of service bears this out.

Details of LMS building at St. Rollox are as follows:

St Rollox Order No	Built	Type	LMS Nos
Y 132	1924	0-6-0(a)	4177-4178
Y 131	1925	4-6-0(b)	14630-14631
Y 132	1925	0-6-0(a)	4179-4191
Y 133	1925	0-6-0(a)	4192-4206
Y 131	1926	4-6-0(b)	14632-14649
Y 134	1927	0-6-0(a)	4312-4321
Y 135	1927	0-6-0(a)	4322
Y 135	1928	0-6-0(a)	4323-4331
Y 136	1928	0-6-0(a)	4467-4476

Classes: (a) Std 4F (b) CR '60'

Fig. 75 1925 *Class 4* standard goods 4182 for Scottish loading gauge

Date First Built	Type	Class	Cylinders inches	D.W. ft in	Heating Surface – sq ft				Grate Area sq ft	WP lb/sq in	Weight WO Ton Cwt
					Tubes	S' Heat	F'Box	Total			
B. CONNOR											
1859	2-2-2	115	OC 17½ x 24	8 2	1080	–	89	1169	13.9	120	30.13
D. DRUMMOND											
1883	0-6-0		18 x 26	5 0	1089·68	–	112·62	1202.3	19·5	150	40-6
1884	0-4-4T		16 x 22	5 0		–		672			37-15
1885	4-4-0		18 x 26	6 6	1088·7	–	122	1210.7	19·5	160	45-3
1888	4-4-0		18 x 26	5 9		–		936.5	16·75		42-7
1889	4-4-0		18 x 26	6 6	1088·7	–	122	1210.7	19·5	200	45-0
J. LAMBIE											
1894	4-4-0		18 x 26	6 6	1071·5	–	112·62	1184·12	19·5	160	45-5
J. F. McINTOSH											
1895	0-6-OT		18 x 26	4 6	975	–	110·9	1085·9	17·0	150	47-15
1896	4-4-0	(a)	18¼ x 26	6 6	1284·45	–	118·78	1403·28	20·6	160	46-19
1897	4-4-0	(b)	19 x 26	6 6	1381·22	–	118·78	1500	20·6	175	49-0
1899	4-4-0	900	19 x 26	6 6	1402	–	138	1540	22·0	180	51-14
1901	0-8-0		21 x 26	4 6	1970	–	138	2108	23·0	175	60-12
1903	4-6-0	918	19 x 26	5 0	1890	–	128	2018	21·0	175	60-8
1903	4-6-0	908	19 x 26	5 9	2050	–	128	2178	21·0	180	64-0
1903	4-6-0	49	21 x 26	6 6	2255	–	145	2400	26·0	200	73-0
1903	0-8-OT		19 x 26	4 6	1076·3	–	112·7	1189	19·5	175	60-12
1904	4-4-0	140	19 x 26	6 6	1470	–	145	1615	21·0	180	54-7
1906	4-6-0	(c)	20 x 26	6 6	2260	–	148·25	2408·25	26·0	200	73-0
1912	4-4-0	120	20¼ x 26	6 6	1220	295	145	1660	21·0	170	59-10
1912	0-6-0		19½ x 26	5 0	1071.3	266.92	118.78	1457	20·6	160	51-12
1913	4-6-0	179	19½ x 26	5 9	1439	403	128	1970	21·0	170	68-10
W. PICKERSGILL											
1916	4-4-0	113	20 x 26	6 6	1185	200	144	1529	20·7	175	61-5
1916	4-6-0	60	OC20 x 26	6 1	1529·5	258·2	146·5	1934·2	25·5	175	75-0
1918	0-6-0		18½ x 26	5 0	1332·92		118·78	1451·7	20·0	170	49-5
1921	4-6-0	956	3/18½ x 26	6 1	2200	270	170	2640	28·0	180	81-0

(a) Dunalastair (b) Dunalastair II (c) Cardean

CAMBRIAN RAILWAYS

Oswestry

The workshops were built in 1865 by Thomas Savin & Co. from designs provided by Messrs. Sharp Stewart & Co., Manchester, who in addition supplied most of the initial machinery. The erecting shop had eleven pits each side of a traverser. One set of pits was used for locomotive repairs and the other side for tenders and a boiler shop.

Only two locomotives were built at Oswestry, one in 1901 and the other in 1904. At this time Herbert Edward Jones was the locomotive superintendent. He had been a pupil under Matthew Kirtley at the Midland Railway's Derby works and continued working under Samuel W. Johnson there until 1899 when he obtained this post on the Cambrian Railways.

Both locomotives were of the '61' class, 4-4-0 passenger type; No.19 (July 1901) and No.11 (July 1904). There were twenty-two in the class, the majority being built by Sharp Stewart & Co. who designed them.

Two spare boilers supplied by Nasmyth Wilson in 1898 for this class were used and other parts were supplied by Robert Stephenson & Co. who had built four '61' class in 1897–8. Spare tenders were at first coupled to these two locomotives, but new ones were built at a later date.

No.19 received GWR No.1082 and was withdrawn in 1928, and No.11 became GWR No.1068 and was scrapped in 1924.

No.19 as No.1082 was optimistically put on the 'sales list', but not surprisingly there were no buyers.

CANNOCK & RUGELEY COLLIERY COMPANY LIMITED

Rawnsley

In 1888 one 2-4-0T No.7 BIRCH was built at the colliery's workshops at Rawnsley. It is unlikely that all the parts of this locomotive were made in the colliery workshops. Here again it is puzzling why a 2-4-0T should be chosen for the type to be built, especially as a 2-4-0T bought secondhand in 1905 was converted to a 0-6-0T in 1916 and lasted in this state until 1955. This locomotive was No. 8 HARRISON.

The cylinders of BIRCH were 15″ x 20″, with 3′ 6″ coupled wheels and the weight in working order was 31 tons 10 cwt.

CANNOCK CHASE COLLIERY CO. LIMITED

Chasetown

This firm built one 0-4-2ST FOGGO in 1946 from parts supplied by Messrs. Beyer Peacock. It may appear strange that a locomotive of this type and style was built in this year, but the colliery had been so satisfied with the performance and low maintenance costs of five locomotives of the same type which they had bought from Beyer Peacock between the years 1856 and 1872. One of these locomotives MCLEAN built in 1856 lasted for a century.

Fig. 76
Cambrian Railway 1904 4-4-0 *Class 61* No. 11

Fig. 77
Cannock & Rugeley Colliery
1888 2-4-0T
BIRCH No. 7

Fig. 78
J. & C. Carmichael
1833 0-2-2
EARL OF AIRLIE
Dundee & Newtyle Railway

Fig. 79
Carrett Marshall & Co
Colliery Guardian advertisement for 0-4-0ST

CARMICHAEL J. &. C.

Ward Foundry, Sessions Street, Dundee

James and Charles Carmichael founded the firm in 1810 and were engaged principally on marine work, weighbridges and turbines.

They produced the first locomotives in Scotland in 1833 for the Dundee and Newtyle railway. The first, No.1 named EARL OF AIRLIE was delivered on 20 September, and No. 2 LORD WHARNCLIFFE on 25 September. They cost £700 each and £30 for each water-butt tender.

They were both 0-2-4s and dimensionally the same, except that No.1 had 11″ x 18″ vertical cylinders whereas the diameter was increased to 11¼″ for No. 2. The cylinders were placed on each side of the boiler, the crossheads above the cylinders were connected to large bell cranks, the longer arms of which were coupled to the connecting rods fastened to outside cranks on the leading 5′ 4″ diameter wheels. All wheels had laminated springs above the frames. The steam admission was by means of a valve on the boiler barrel operated from the footplate by a handle and shaft. The feed check valve was on the side of the raised firebox.

The photograph (Fig. 78) was taken after they had both been withdrawn in 1854 and this one – said to be No.1 – was specially prepared for the occasion.

The boiler was apparently unlagged and worked at a pressure of 50 lb/sq. in. The locomotive's weight in working order was 9½ tons. Carmichael's own design of valve gear was used with one fixed eccentric for each cylinder.

Early in 1834 a third was built by Stirling of Dundee to the same general design and named TROTTER.

The gauge of the Dundee and Newtyle Railway was 4′6″, subsequently converted to standard gauge in 1849.

James Carmichael died in 1853 and Charles in 1843. The firm was carried on by their sons and became JAMES CARMICHAEL & CO. in 1853 and a limited liability company about 1894, closing down in 1929.

It appears that after their initial efforts in 1833 no further locomotives were built and marine engineering remained the principal activity.

CARRETT MARSHALL & COMPANY

Sun Foundry, Dewsbury Road, Leeds

In 1858 this firm took over from Charles Todd who, after leaving Shepherd & Todd had set up his own factory.

Carrett Marshall & Company may have built a few locomotives but little is known about this side of their activities. They were mainly known for their remarkable steam driven road vehicles.

Mr Carrett had left E. B. Wilson & Company to join Marshall as a partner in the new firm.

Many engineering articles were manufactured including water pumps and tanks.

An advertisement in the *Colliery Guardian* shows a drawing of a four-coupled saddle tank, with inclined outside cylinders fixed immediately behind the leading wheels. (Fig. 79)

One locomotive is reported to have been built in 1860 for the Natal Railway and named NATAL, and others for the Kendal & Windermere Railway.

CHANTER, JOHN

London

John Chanter did not, as far as is known, actually build any locomotives, but patented a coal burning firebox for more efficient combustion. It is thought that Thomas Vernon & Co. of Liverpool built the earliest ones, and Peter Borrie & Company of Dundee at least one. In 1839, in conjunction with Gray, a form of compound firebox was evolved to burn coal instead of coke. It consisted of a double firebox with a division between in the form of a water-way. The first was used for coking the coal, after which it was transferred to the second chamber. A locomotive called PRINCE GEORGE OF CAMBRIDGE was used for experiments with this patent firebox, and was built in 1838. In 1839 the DUKE OF SUSSEX appeared, and both these locomotives were run on the London & Croydon Railway, together with a third, named CORYNDON which was a 2-2-2 with 5′ 6″ driving wheels. It is known that CORYNDON at least, was built by Peter Borrie & Company.

Two more locomotives were supplied to the Hayle Railway which was worked by Mr Chanter. They were named CHANTER and CORYNDON. The latter was scrapped in 1850 and after CHANTER was repaired in 1851 in the Carn Brea Shops of the Hayle Railway it was renamed CORYNDON so that one time or another three locomotives had that name.

Summary: c 1837 CHANTER, renamed CORYNDON in 1851; 1837 CORYNDON, scrapped in 1850; 1838 PRINCE GEORGE, London & Croydon Railway; 1839 DUKE OF SUSSEX, London & Croydon Railway; *c* 1840 CORYNDON, London & Croydon Railway.

CHAPLIN, ALEXANDER & CO.

Cranstonhill Engine Works, Port Street, Anderson, Glasgow.

Alexander Chaplin built a number of patent vertical boilered locomotives from the 1860s to the early 1900s. The cylinders were bolted to the boiler, the cross head guides forming a casting bolted to the framework but the design layout varied considerably. The boiler was unusual having hanging tubes welded at their lower ends to the crown plate of an internal firebox and at right angles to it so that the gases passed around the tubes before entering the central chimney. The working pressure was 100lb/sq. in. and they were very efficient steam raisers.

The four wheels were coupled and in some instances the springing of the frame was by means of spiral springs. The drive was by gearing usually 2:1 ratio. The cylinders varied from 4½″ x 9″ to 9″ x 16″ and the wheels from 2′ 1½″ to 3ft diameter.

Many were sent abroad, four as shunters for the Danish State Railways between 1869 and 1872. Other products were cranes, road engines, boilers and stationary engines.

About 1890 the firm moved to Helen Street, Govan. The works numbers included all types of machinery. Manufacturing ceased in 1930.

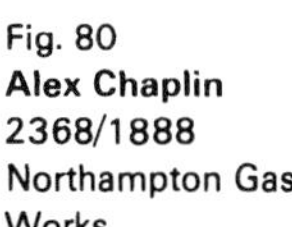

Fig. 80
Alex Chaplin
2368/1888
Northampton Gas Works

Works No	Built	Type	Vert. Cylinders inches	D.W. ft in	Customer	No/Name
140	1860	0-4-0TG			Grays Chalk Quarries Co Ltd	
153		0-4-0TG			Contractor	
188	1861	0-4-0TG			Grays Chalk Quarries Co Ltd	
232	1861	0-4-0TG			John Martin, Loughborough	
305	1862	0-4-0TG			Bolckow Vaughan & Co Ltd	
317	1863	0-4-0TG			Cochrane & Co Cargo Fleet	
358	1863	0-4-0TG			Cochrane & Co Cargo Fleet	
381	1863	0-4-0TG			Cochrane & Co Cargo Fleet	
382	1863	0-4-0TG			Cochrane & Co Cargo Fleet	
	11/1865	0-4-0TG			Southampton Docks	Chaplin
1056	1869	0-4-0TG			D. Murray, Blisworth	
1109	1869	0-4-0TG			Butterley Co	23
		0-4-0TG			Little Orme's Head Quarries	3 6 gauge
	1869	0-4-0TG			Danish State Rlys	
	1869	0-4-0TG			Danish State Rlys	
	1872	0-4-0TG			Danish State Rlys	
	1872	0-4-0TG			Danish State Rlys	
1290	1871	0-4-0TG			S. Cleveland Ironworks Co Ltd	
1458	1872	0-4-0TG			A. Hood Lasswade, E. Lothian	
1643	11/1874	0-4-0TG			Bridgewater Colls	
		0-4-0TG			Hälsingborgs Hamn, Sweden	
2368	1888	0-4-0TG	9 x 12	2 1½	Northampton Gas Works Fig. 80	

CHAPMAN & FURNEAUX

Gateshead

See BLACK HAWTHORN & COMPANY

CHAPMAN, WILLIAM & EDWARD

Murton House, Co. Durham

William and Edward Chapman took out a patent in 1812 for travelling engines which was a combination of locomotive and chain, the chain being fixed the full length of the railway and positioned between the rails. A gear on the locomotive engaged the chain, the gear being pressed into the chain by a friction roller. The gear was driven by a crankshaft and series of gears operated by steam cylinders, details are shown in C. F. Dendy Marshall's *History of Railway Locomotives, down to 1831.* Trials were carried out on the Heaton Colliery tramway in 1813 but were not successful. This apparatus may have been built by Butterley Company.

The second attempt at locomotive design took the form of an eight wheeled engine and was built by Phineas Crowther in 1814 and was tried out on the Lambton Colliery line in 1815. It had vertical cylinders. The designs of the Chapmans were important in producing the first bogie, as the eight wheels were in two groups of four – a load equalising device before springs were used – and for the application of two cylinders on a locomotive driving a single shaft with quartered cranks.

Two eight wheeled locomotives were reported to have been built, the other one for Wylam Colliery.

The Lambton locomotive hauled 18 wagons weighing 54 tons up a gradient of 1 in 115 at four miles per hour.

Fig. 81 **W. & E. Chapman** 1813 probably first locomotive built with a bogie

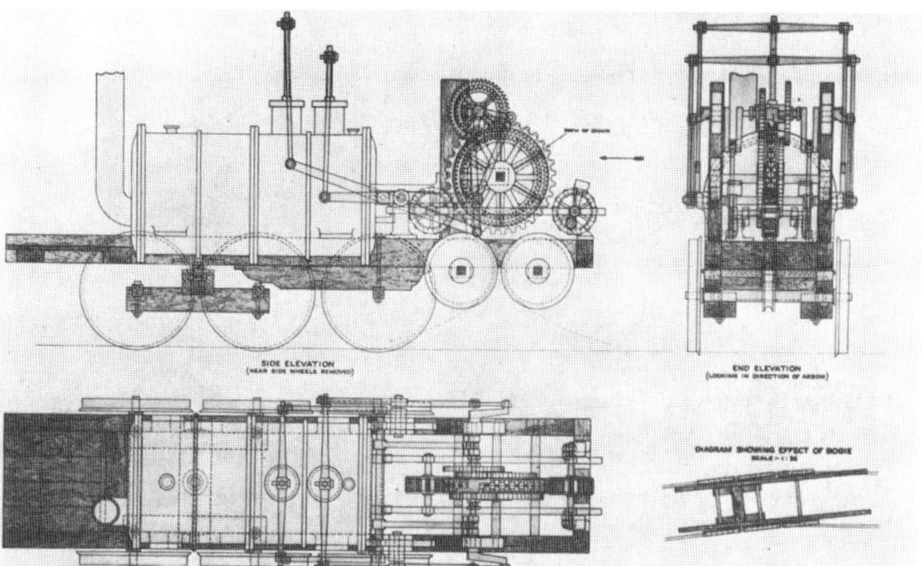

CHRISTIE ADAMS & HILL
Thames Bank Ironworks, Rotherhithe

Six 6′ 6″ singles were ordered from this firm by the London & South Western Railway in 1847. They had inclined outside cylinders with outside frames to the leading and trailing wheels which passed across the face of the driving wheels but did not support them. The boiler was domeless and the smokebox door was horizontally hinged. A raised round top firebox housed the combined dome and safety valves. The *Vesuvius* class built at Nine Elms were similar in outline. They lasted about twenty years and during that time they received new cylinders and new fireboxes.

CHRISTIE ADAMS & HILL

Built	Type	Cylinders inches	D.W. ft in	LSWR No	Name	W'drawn
10/1848	2-2-2	OC14¼ x 21	6 6	109	Rocklia	1869
11/1848	2-2-2	OC14¼ x 21	6 6	110	Avon	1868
12/1848	2-2-2	OC14¼ x 21	6 6	111	Test	1870
4/1849	2-2-2	OC14¼ x 22	6 6	112	Trent	1869
5/1849	2-2-2	OC14¼ x 22	6 6	113	Stour	1868
12/1849	2-2-2	OC14¼ x 22	6 6	114	Frome (Fig. 82)	1868

All built for London & South Western Railway

CHURCH, Dr W.
Birmingham

This locomotive was designed by Dr William Church for a Mr S. A. Goddard, presumably as a private venture, and was probably built by Mr Horton of Brierley Hill who at least supplied the boiler. It was a 0-2-2WT, with outside cylinders $11\frac{1}{4}$″ x 24″ positioned horizontally at the rear. The boiler had eighty-one 2″ diameter brass tubes, with circular firebox with a dome on top and safety valve, an additional dome was fitted to the boiler barrel. It was the first tank locomotive to be fitted with a multi-tube boiler. Other features were piston valves, light eccentric motion and plating covering the cylinders and wheels. The driving wheels were 6′ $2\frac{1}{2}$″ diameter and the total weight in working order 9 tons. In January 1838 it ran trials on the London and Birmingham Railway working as a ballast locomotive, then it was transferred to the Grand Junction railway where it carried the name VICTORIA. It was not a success on either railway.

Fig. 82 **Christie Adams & Hill** 1849 London & Southern Western Railway FROME

In 1840 it was agreed by the Birmingham and Gloucester Railway to try the SURPRISE as it was now called, on their metals and it was at Bromsgrove on 10 November, 1840 that the boiler exploded at Bromsgrove station and two of the B&GR employees were killed, and several injured. The two men, Thomas Scaife and John Rutherford, were buried in Bromsgrove churchyard and a memorial erected on which was carved one of the American Norris locomotives which has led to some misunderstanding, particularly as one of the Norris locomotives also exploded in the same location but not with such dire results.

Apparently the locomotive received another boiler after this accident and received yet another name ECLIPSE. In 1850 it was reported as standing at Camp Hill station by which time the Birmingham and Gloucester Railway had become part of the Midland Railway.

It then appeared on the Swansea Vale Railway in the late 1850s transformed into a six coupled tank locomotive and may have been purchased by them, but its ultimate fate is shrouded in mystery.

Dr Church was an amateur engineer and devised an expanding mandrel for fixing boiler tubes. The 'piston valves' were simply pistons working in an open pipe according to Zerah Colborn in his *Locomotive Engineering*.

CLARKE CHAPMAN & CO. LTD.

Victoria Works, Gateshead

Two Crane tanks were built in 1907 with vertical boilers for the Consett Iron Company, (Works Nos. 7519 & 7520) with Consett running numbers E.11 and E.12.

Although these two really come under the category of steam cranes, their appearance with direct drive to the coupled wheels are more like a locomotive than the usual geared drive used on steam cranes.

The gearing pillar and all principal castings were of cast steel. The jib was of the swan-neck type and was built up of mild steel plates and angles as was the carriage. The crane was fitted with worm gearing for varying the radius. There were three sets of engines – one for lifting the load and for varying the radius, and one for revolving, and one for travelling; each engine being equipped with double cylinders and link motion reversing gear. The boiler was a vertical multitubular type bolted to a heavy casting forming feed tank and back balance weight. The locomotive itself was fitted with spring and block buffers, spring draw hooks and brakes worked by hand wheel and screw, from the working position.

This firm's main business is marine and electrical engineering and boiler making.

See illustration on next page.

Fig. 83 **Dr Church** Diagram of locomotive from D. K. Clark's 'Railway Machinery'

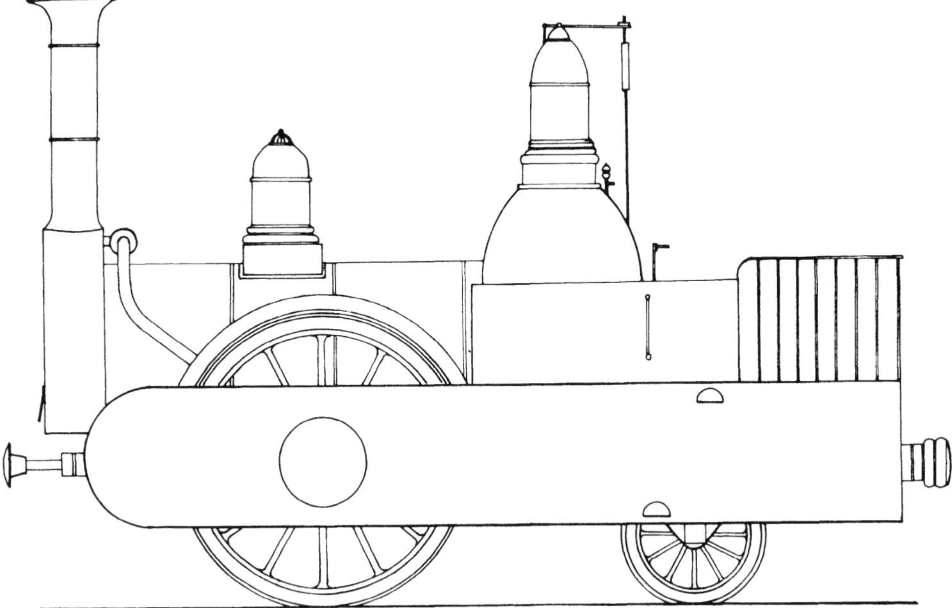

CLAYTON SHUTTLEWORTH & CO.
Stamp End Works, Lincoln

Established in 1842 by Joseph Shuttleworth and Nathaniel Clayton for the manufacture of agricultural machinery, locomotive and stationary boilers, oil engines, portable steam engines and steam traction engines.

A small number of steam trams were also built which were basically a traction engine placed on locomotive type frames carrying two pairs of wheels, both axles being driven by endless pitch chain and sprocket.

Small outside cylinder tank locomotives were also built for shunting purposes. Those which are known are:

PILOT 0-4-0WT built in May 1875 with outside cylinders 9″ x 14″ and 2′ 9″ coupled wheels. A peculiarity was the water tanks which were bolted to each frame on the outside and were 14′ 7½″ long, 1′ 6″ wide x 1′ 4″ deep, and held a total of 270 gallons. Stephenson's link motion was fitted inside the frames. The boiler was domeless and steam collected by a long perforated pipe in the steam space of the barrel. It was given Works No. 44701.

Two 0-4-0CTs were built, one in 1892 and the other in 1920 for the Consett Iron Co. (E No. 4 and E No.14). They had vertical boilers.

Two 4wh type were built in 1867 for Hall & Co. and were traction train engines of 10nhp cylinders 7½″ x 12″, wheels 3′ 0¾″.

The original works stood on 1½ acres and by 1918 this had increased to 100 acres with 4500 employees. Besides the erecting shop for railway rolling stock there was a steel foundry, stamping and forging shops, aeroplane shop and aerodrome.

The firm became CLAYTON & SHUTTLEWORTH LTD. at a later date and in 1929 were absorbed by MARSHALL SONS & CO. LTD.

Fig. 84 **Clarke Chapman**
7519/1907 Locomotive type steam crane.
Consett Iron Co Ltd E No. 11

CLAYTON WAGONS LTD.

Abbey Works, Lincoln

This firm was a subsidiary of Clayton & Shuttleworth Ltd. and the factory was built in 1917 for railway works. A number of vertical water tube boilered four wheel geared locomotives were built. The type 'A' had inside frames with a two cylinder engine fixed beneath the floor, the crankcase gearbox and reduction gearing were housed in one unit, which was derived from the road wagon engine. The cylinders were $6\frac{3}{4}''$ x 10", the piston valves were below the cylinders. Two eccentrics were mounted on a shaft driven from the crankshaft. The gear ratio was 3·04 to 1, the road wheels were 3' 6" diameter on a wheelbase of 7' 0". The boiler was placed at one end and had a heating surface of 85 sq. ft., with a superheater coil around the top of the firebox. The working pressure was 275lb/sq. in. The water tank and coal bunker were placed at the other end of the framing. The weight was approximately 20 tons.

A further type 'C' was in two sizes with wheelbases 3' 6" or 5' 0" with 2ft. diameter wheels. The drive was by a jackshaft. The frames were outside and the engine unit gave 100bhp.

In 1929 two large locomotives were built for the Indian N.W. Railways with four vertical cylinders 7" x 10" grouped in pairs and a White-Forster water tube boiler working at a pressure of 300lb/sq.in., consisting of a top drum and two smaller drums arranged as an inverted vee with two banks of curved tubes between them. They were 0-4-0s with a jackshaft between the two axles.

An earlier 'A' type was used as the works shunter at Lincoln and the second one, built in 1927 (Works No. L. 5001) was sent to G. & T. Earle Ltd., Hull for test but was found unsuitable and returned. In addition to the above locomotives which were built in the 1920s, Clayton Wagons Ltd. built approximately eighty steam rail cars for home and abroad between 1924 and 1928. They had light weight all steel bodies and were of advanced design. Special attention was paid to the weight of the units, and the seat ends were of aluminium-silica alloy. The web was about 3/16" thick and considerable savings were made in weight as compared with their wooden counterparts. The

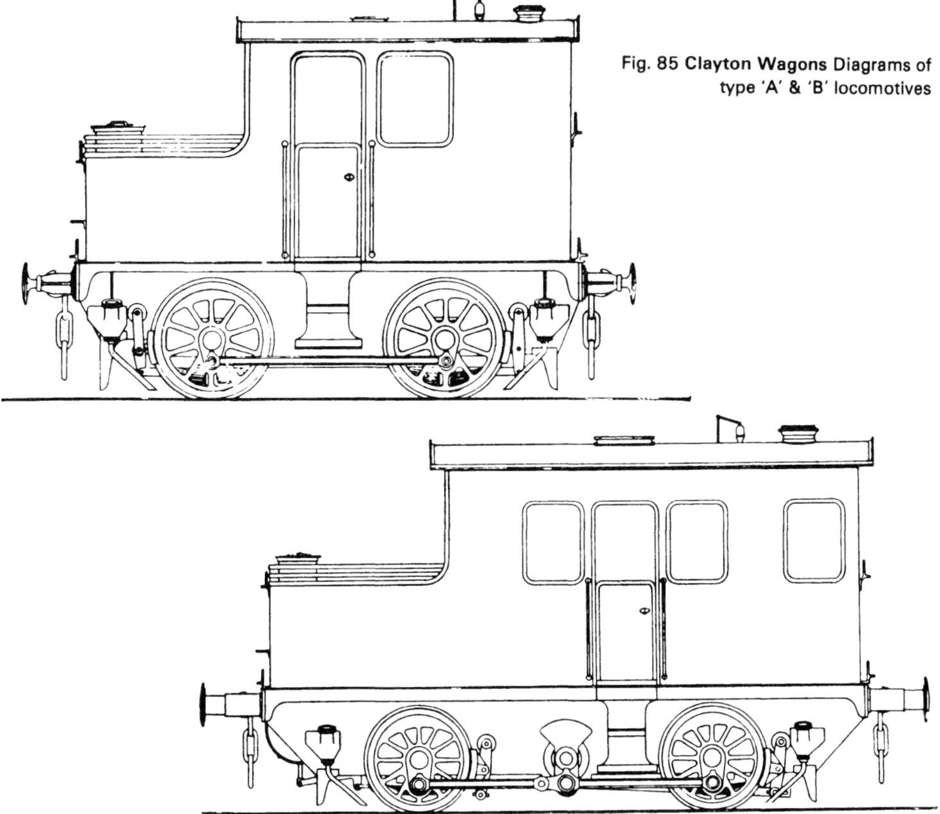

Fig. 85 **Clayton Wagons** Diagrams of type 'A' & 'B' locomotives

locomotive valve gear was of interest and gave a diagram similar to the Stephenson's link motion.

The valves were oscillated by eccentrics mounted on a cylindrical hollow cross shaft, driven by gearing from the crankshaft. Inside this cross shaft was a smaller shaft having four flats machined along the full length. By moving the inner shaft laterally in the outer shaft the throw of the eccentrics was varied by wedges sliding on flats on the inner shaft and passing through slots machined in the outer shaft. The simple and efficient mechanism was fully enclosed in the crankcase and was splash lubricated.

The eleven rail-cars built for the London and North Eastern Railway had water tube boilers with a working pressure of 275lb/sq. in., two cylinders 7" x 10" and coupled wheels 3ft. diameter. The engine unit was similar to the type 'A' locomotive with a carriage body added. Six of similar design were built for the Great Southern Railway (Ireland), (GSR Nos. 358–363), but the boiler pressure was 300lb/sq. in. As in most of the vertical boilers used on railcars and locomotives they were fired from the top through a stoking tube.

On both the L&NER and GSR the rail cars were not popular, mainly due to the performance of the boiler which was not a good steamer and as the tubes were pitched rather close together trouble was experienced when hard water was used, in scaling up, and as washing out was an awkward operation, the engine performance suffered. Spares also were difficult to obtain after 1930 and no further units were ordered.

The works were taken over by Smith-Clayton Forge Ltd. in 1930 and the drawings, patterns etc. for the locomotives were taken over by Richard Duckering Ltd. of Lincoln.

CLYDE LOCOMOTIVE CO. LTD.

Springburn, Glasgow

Founded in 1884 by Walter Montgomerie Neilson, the one-time owner of the Hyde Park Locomotive Works. They built eight locomotives for the Highland Railway in 1886, two for Girvan & Portpatrick Junction Railway in 1887, three 0-4-0STs and one 0-6-0ST.

SHARP STEWART & CO. of Manchester bought the works in 1888 and removed all their plant from Manchester. They called their new works Atlas Works. Clyde's order No. E6 was transferred to Sharp Stewart.

Fig. 86 **Clayton Wagons**
1928 L&NER
Railcar No. 287 ROYAL SAILOR

E No	Order No	Date	Type	Cylinders inches	D.W. ft in	Customer	No	Name etc
1	E1	1886	4-4-0	18 x 24	6 3	Highland Rly	76	Bruce (Fig. 87)
2	E1	1886	4-4-0	18 x 24	6 3	Highland Rly	77	Lovat
3	E1	1886	4-4-0	18 x 24	6 3	Highland Rly	78	Lochalsh
4	E1	1886	4-4-0	18 x 24	6 3	Highland Rly	79	Atholl
5	E1	1886	4-4-0	18 x 24	6 3	Highland Rly	80	Stafford
6	E1	1886	4-4-0	18 x 24	6 3	Highland Rly	81	Colville
7	E1	1886	4-4-0	18 x 24	6 3	Highland Rly	82	Fife
8	E1	1886	4-4-0	18 x 24	6 3	Highland Rly	83	Cadboll
9	E2	1887	0-4-0ST			John McAndrew Dalzell	2	
10	E2	1887	0-4-0ST			Watt & Wilson Greenock	4	
11	E3	1887	0-4-0ST			Eglinton Iron Co. Santander	2	La Pauline
12	E4	1887	0-6-0ST	IC 17½ x 26	4 3	Dowlais Ironworks		Clyde
13	E5	1887	0-6-0	17 x 26	5 0	Ayr & Wigtownshire Rly	6	Later G & SWR 304
14	E5	1887	0-6-0	17 x 26	5 0	Ayr & Wigtownshire Rly	7	Later G & SWR 305
15	E6	1888	4-4-0	OC 13 x 16	3 9	Brazilian Govt Rly	41	(a)
16	E6	1888	4-4-0	OC 13 x 16	3 9	Brazilian Govt Rly	43	
17	E6	1888	4-4-0	OC 13 x 16	3 9	Brazilian Govt Rly	42	
18	E6	1888	4-4-0	OC 13 x 16	3 9	Brazilian Govt Rly	44	
19	E6	1888	4-4-0	OC 13 x 16	3 9	Brazilian Govt Rly	46	
20	E6	1888	4-4-0	OC 13 x 16	3 9	Brazilian Govt Rly	45	
21	E6	1888	4-4-0	OC 13 x 16	3 9	Brazilian Govt Rly	47	
22	E6	1888	4-4-0	OC 13 x 16	3 9	Brazilian Govt Rly	48	
23	E7	1888	4-4-0	OC 14 x 20	4 2	Argentine Central Rly	19	(b) (d)
24	E7	1888	4-4-0	OC 14 x 20	4 2	Argentine Central Rly	20	
25	E7	1888	4-4-0	OC 14 x 20	4 2	Argentine Central Rly	21	
26	E7	1888	4-4-0	OC 14 x 20	4 2	Argentine Central Rly	22	
27	E7	1888	4-4-0	OC 14 x 20	4 2	Argentine Central Rly	23	
28	E7	1888	4-4-0	OC 14 x 20	4 2	Argentine Central Rly	24	
29	E8	1888	2-6-0	OC 16½ x 24	4 6	Uruguay Northern Rly	1	(c)
30	E8	1888	2-6-0	OC 16½ x 24	4 6	Uruguay Northern Rly	2	

(a) Built by Sharp Stewart & Co Works Nos 3291-3298. (b) Built by Sharp Stewart & Co Works Nos 3424-3429 (Metre Gauge). (c) Built by Sharp Stewart & Co Works Nos 3430-3431. (d) North extension.

Fig. 87 **Clyde Loco Co** 12/1887 Dowlais Iron Co No. 24 CLYDE

THE COALBROOKDALE COMPANY

Fig. 88 **Coalbrookdale Co** *c*1860 **ex Bardon Hill Quarries No. 5. Preserved at Coalbrookdale**

This very old established firm cast solid iron rails and plates as far back as 1767. The firm was founded in 1709. Their work attracted many famous engineers including Heslop, Hornblower, Sadler and Trevithick, and others had engines to their own designs built and tested.

In 1802 a locomotive was being constructed to Trevithick's design, but its later history is not reported and it is practically certain that it was abandoned before completion, but it was the first attempt ever made to produce a locomotive. On the other hand the Coalbrookdale Company claim that the first railway locomotive was built by them. Richard Trevithick was associated with the company in the early 1800s and they were building high pressure boilers for him and in 1802 carried out tests on both boiler and engine. The boiler worked at 50lb/sq. in. which was a high pressure at that time, and enabled the cumbersome ten or eleven foot long cylinder weighing five to six tons to be substituted for a cylinder much smaller acting directly on to a crankshaft without the heavy beams, housings and heavy parts previously used.

In Arthur Raistrick's book *Dynasty of Ironfounders,* which is a history of the Coalbrookdale Company he states, 'Trevithick had been at Coalbrookdale in 1796 and was there again in January 1802, presumably arranging for the making of one of his high pressure engines for use as a locomotive'. His biographer says 'In August 1802 the Coalbrookdale Co. were building for him a railway carriage or locomotive, though they feared to attempt its construction'.

The drawing of the Penydarren locomotive is thought in some quarters to be the Coalbrookdale locomotive which would confirm the local tradition that the company built the first steam locomotive in the world to run on rails.

There has been much speculation, debate, evidence for and against, but no definite decision can be brought to bear on this interesting problem of the rival claims of Penydarren and Coalbrookdale. A second phase of locomotive building was carried out in the 1860s and six 0-4-0STs were built, one of which is fortunately preserved at Coalbrookdale. This engine was purchased by the Netherseal Colliery Co. Ltd. who in turn sold it to Ellis & Everard Ltd. to work in their Bardon Hill quarries. After being laid aside for some time it was presented to The Coalbrookdale Company who have placed it on their unique museum site together with the remains of a second locomotive which had been rebuilt by Sentinel with geared drive and vertical boiler. Both are over a hundred years old and two locomotives from a batch of six for preservation is a wonderful tribute to the craftsmanship which must have gone into these locomotives. (Fig. 88)

Mr R. A. S. Abbott in his series of articles on Vertical Boiler Locomotives' in the *Engineer* for 1955 states that a number of such locomotives were built at Coalbrookdale for their own 2′ 4″ gauge system.

COCHRANE & CO.

North Ormsby Ironworks, Cargo Fleet, Middlesbrough

Approximately five vertical boilered locomotives were built for their own use between 1860 and 1880 for their standard gauge lines. The cylinders were vertical attached to the boiler and a water tank fixed on one end of the footplating. At the other end was a well giving access to the fire-hole. Unfortunately records regarding these interesting locomotives were destroyed.

The firm eventually became part of the Stanton and Staveley group.

COLTMAN H. & SONS LTD

Loughborough

This firm manufactured steam engines, boilers and for a short time motor cars, and were taken over by Herbert Morris Ltd., Crane Makers of Loughborough about 1914.

It is possible that they built a few locomotives of Hughes and Falcon design as sub-contractors to the local Falcon and Engine Car Works.

At one time Huram Coltman and Henry Hughes were partners at the Falcon Works.

No record of any locomotives built has been discovered or for what customers.

This firm should not be confused with another Loughborough firm Walter W. Coltman & Co. (Boilers) Ltd. who are still busily engaged in the manufacture of boilers. Walter was the son of Huram Coltman.

Fig. 89
Cork Bandon & SC
1901 4-4-0T No. 7

COED TALON COLLIERY CO.

Coed Talon, Flintshire

One locomotive was built in the colliery workshops in 1874 to the designs of the chief fitter Mr William Lea. It took the form of the frame from a coal wagon on which was mounted a portable horizontal steam engine. The drive was from a gear on the crankshaft, through reduction gearing to a gear fixed on the front axle. The engine was about 6 hp with one cylinder 9″ x 12″. The original horizontal boiler became unserviceable and a vertical boiler was fixed on the rear and the steam pipe coupled up to the cylinder.

One photograph of the locomotive appeared in the *Railway Magazine* for November 1904, page 403.

CORAS IOMPAIR EIREANN

See GREAT SOUTHERN & WESTERN RAILWAY.

CORK BANDON & SOUTH COAST RAILWAY

Rocksavage, Cork

A statement that one engine was built in 1901 in the Company's workshops, is not true. The locomotive in question was a 4-4-0T No. 7 which was probably a rebuild from the Fairbairn locomotive originally built as a 0-4-0ST in 1862, rebuilt in 1864 to a 0-4-2ST and again in 1889 to a 0-4-4T being withdrawn in 1899. Extensive rebuilding was carried out here and the rebuilt No. 7 no doubt had some parts from the old No. 7.

COULTHARD, JOHN & SON

Quarry Field Works, Gateshead

Commenced building locomotives in 1835 and produced about twenty up to 1865. Rebuilding was also carried out extensively. In April 1853 the partnership between John and Ralph Coulthard was dissolved and the firm became R. COULTHARD & CO. In 1865 Ralph retired and the work was taken over by Black Hawthorn & Co. Locomotives were built for northern England railways and collieries. The works numbers went up to 100.

COWLISHAW WALKER & CO. LTD.

Biddulph, Stoke-on-Trent

This firm was engaged in the manufacture of mining machinery and power presses. The engineering workshops of the Norton and Biddulph Collieries were taken over about 1930.

Although the building of new locomotives by this firm was possible none can be traced, but many locomotives of the Norton and Biddulph Collieries were rebuilt from 1930 to 1934.

J. COULTHARD & SON

Date	Type	Cylinders	D.W.	Customer	No	Later Nos
9/1847	2-4-0			York, Newcastle & Berwick Rly	156	(a)
1848	0-6-0			John Bowes	9	(b)
6/1849	2-4-0			North Eastern Rly	5	
1854	2-4-0			West Hartlepool Harbour & Rly	29	NE 604
1854	0-6-0			West Hartlepool Harbour & Rly	42	NE 614
1854	2-4-0			Blyth & Tyne Rly	12	
1854	2-4-0			Blyth & Tyne Rly	13	
1/1856	0-6-0			Blyth & Tyne Rly	23	NE 1323
4/1856	0-6-0				19	NE 1319
1862	0-6-0			Bowes Rly ?		
1864	0-6-0			Earl of Durham Colls Ltd	6	
1864	0-6-0			Earl of Durham Colls Ltd	7	
1864	0-6-0			Earl of Durham Colls Ltd	8	

(a) Named Jenny Lind Works No 42. (b) Rebuilt Saddle Tank 1900.

Fig. 90 J. **Coulthard** 1847 for York, Newcastle & Berwick Railway No. 156 JENNY LIND

CRAIG, A. F. & CO. LTD.
Caledonia Engineering Works, Paisley

Long established as ironfounders and engineers, this firm built a unique locomotive for use in their foundry in the 1870s. The gauge was 3′ 1½″ and no explanation has been found for this unusual figure.

The locomotive was designed and built by Robert Craig, a great uncle of the present Managing Director Mr A. F. Craig.

The original locomotive was fitted with a standard Cochran vertical cross-tube boiler placed on a girder frame and supplying steam to two 6″ x 8″ vertical cylinders driving a crankshaft on which was fitted a gear wheel driving the wheel axle through a train of gears. A flywheel was fitted to the crankshaft. The four wheels were 2′ 8″ diameter on a 3′ 6″ wheelbase. A water tank and fuel space were provided. An interesting feature was the 'steering' wheels fitted at the boiler end which brought the locomotive around extremely acute bends encountered in the foundry.

At some time early in the century the boiler was condemned and a second-hand cross tube boiler was fitted. It worked much better with this boiler as the exhaust did not pass through the tubes to induce a draught as it did in the Cochran boiler.

The locomotive apparently was rebuilt several times but no attempt was made to improve it or change its somewhat strange appearance which gave it the name of 'Paint Pot' in the works.

It proved very successful and was equal to the heavy loads required to be transported. The fuel used was gas char and it was given an overhaul every year during the Fair Holiday periods.

Its working life must have been nearly one hundred years, 1966 being the year of withdrawal, a modern petrol driven tractor taking its place.

Fig. 91 **A. F. Craig Works shunter as rebuilt**

CRAMPTON, THOMAS RUSSELL

Crampton was no locomotive builder, but he greatly influenced their design in the 1840s.

He took out a patent in 1843 for the disposition of the locomotive wheels by which the carrying wheels were in front of the firebox and the single drivers were behind the firebox, claiming the advantage that a low centre line of the boiler was made possible and consequently a low centre of gravity. In addition the size of driving wheels could be considerably increased, not being governed by boiler clearance. At this time locomotive engineers were apprehensive of the relatively high centre of gravity and the danger of overturning at speed. It was not until 1845 that an order was placed by the Namur and Liege Railway for two 4-2-0s. These were built by Tulk and Ley of Whitehaven. They had outside cylinders 16" x 20", double frames for the four leading wheels and inside frames for the 7' diameter driving wheels. The eccentrics were fixed on return cranks outside the connecting rods with all the valve gear outside. Completed in 1846 they were named NAMUR and LIEGE and after the former had been tried out on the Grand Junction Railway, one was put in hand at Crewe and completed in November 1847, named COURIER. In parallel with COURIER, the famous CORNWALL was built but in this case, to keep the centre of gravity at a minimum, the driving axle passed over the boiler!

McConnell at Wolverton also became interested in the *Crampton* type and ordered the LONDON (No. 200) with 8ft driving wheels, 18" x 20" cylinders with outside valve gear, and single frames. It was tried out on the York, Newcastle & Berwick Railway and running at 70 mph it fouled a platform but kept to the rails. Another one for the London & Birmingham Railway the LIVERPOOL was built in 1848 by Bury, Curtis & Kennedy. This had no less than six carrying wheels under the boiler. The driving wheels were 8ft diameter driven by 18" x 24" outside cylinders. The boiler and firebox had a large heating surface of 2290 sq. ft. and the locomotive weighed 35 tons in working order, 56 tons with the tender. It was found that the wheelbase was too rigid. It was shown at the 1851 Exhibition.

In 1849 Crampton designed an inside cylinder locomotive, the cylinders driving a jackshaft placed behind the carrying wheels, and connected to the driving wheels behind the firebox by coupling rods. This meant that the main cranks were inside the frame and the centre line of the boiler from rail level had to be increased accordingly to clear the boiler barrel, thus defeating one of the main objects of Crampton's patent.

One of the drawbacks to Crampton's locomotives was that the adhesive weight was considerably reduced by the position of the driving wheels and a lot of trouble was experienced in starting trains, but once on the way they were fast, steady but hard riders. The train loads were therefore comparatively light and in many cases the locomotives were rebuilt with four coupled wheels.

The Cramptons were not a success at home but those that were built for abroad, and those built abroad, were far more acceptable and well liked by the majority of foreign lines especially in France and Germany.

A total of 320 locomotives were built under his patents.

CROMFORD & HIGH PEAK RAILWAY

Cromford

Two locomotives were built in the company's workshops at Cromford but this probably was only the assembly of parts supplied by others. Some locomotive parts were supplied around 1840 by Falconer & Peach, Union Foundry, Derby. The locomotives were Nos. 4 and 5 (1859) 0-6-0ST with 10" x 12" outside cylinders and 3ft driving wheels.

John Leonard was the resident engineer and is purported to have designed these locomotives, and as early as 1841 was reported to have built an outside cylinder locomotive and intended to build two more. Whether these two were the ones built in 1859 is not known as no records exist of any previous building. The 0-6-0Ts must have been small, their weight being 12¾ tons.

After being taken over by the LNWR they became Loco Machinery 'C' and 'D' in 1876 and 1877 respectively and the former lasted until 1882. A photograph of one of these locomotives as 'D' can be seen in the *Railway Magazine* for December 1959. It has outside cylinders and steam pipes, saddle tank, no cab and no running plates over the wheels.

There is a strong probability that these two locomotives were 'rebuilds' using parts of withdrawn locomotives; this could be more readily assessed if a photograph of the other locomotive ('C') was available.

CROOK & DEANS

Phoenix Foundry, Little Bolton, Lancs

J. Crook and Wm. Deans were early builders of locomotives. Three were turned out for the Bolton & Leigh Railway. These were: *c.* 1831 IC IF 0-4-0 SALAMANDER which was renamed LIVERPOOL, *c.* 1831 IC IF 0-4-0 VETERAN May 1831 PHOENIX was tried on the Bolton & Leigh Railway but according to the *Bolton Chronicle* of January 1932 belonged to a W. Hulton Esq.

In June 1832 a locomotive named BOLTON was tried on the Bolton & Leigh Railway, intended for traffic between Bolton and Liverpool. This locomotive may have been VETERAN.

SALAMANDER may have been built between PHOENIX and BOLTON or after BOLTON. It is also a probability that it may have been one of them renamed.

By 1841 VETERAN had been rebuilt and may have had a pair of trailing wheels added. It became LNWR No. 218 and was replaced by July 1848.

No further information can be found.

CROSS, JAMES & COMPANY

Sutton Engine Works, St. Helens, Lancs

James Cross had been the locomotive engineer of the St. Helens and Runcorn Railway, and took over the old St. Helens workshops at Sutton shed, in 1864. He built locomotives from this date and remained in business until 1869.

The 'Borrows' type of shunting tank was initially designed and built by James Cross. Edward Borrows had been Cross's assistant on the St. Helens Railway.

Three 0-6-0Ts with outside cylinders and side tanks the length of the boiler and smokebox and as high as the top of the barrel were built for Portugal. The boiler was domed and Ramsbottom safety valves fitted over the firebox. Feed pumps were driven off the crossheads. Two additional 0-6-0Ts were probably similar in design and supplied to contractors. One eventually came into the hands of Isaac Boulton in 1872 who paid £500 for it and named it HERCULES No. 2. After hiring it up and down the country, Boulton sold it three years later to a Wigan colliery for £1000!

The Neath and Brecon locomotive PROGRESS was the first *Fairlie* patent locomotive having two boilers with one firebox between them. The two four coupled units each had two outside cylinders, and were pivotted on pins fixed to the underside of the boiler barrels.

Fig. 92 **James Cross** 1863 WHITE RAVEN as rebuilt. St. Helens Canal & Railway

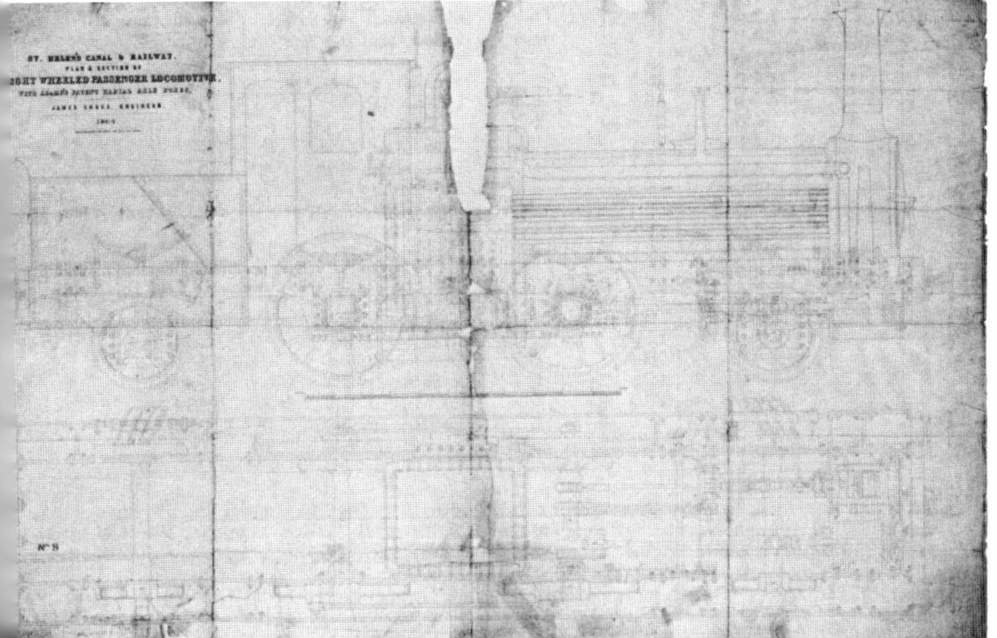

Works No.	Date	Type	Cylinders inches	D.W. ft in	Customer	No/Name
2	*c.* 1864				Mt. Cenis Rly.	
	c. 1864	0-4-0WT			For Own Use	
	c. 1864	0-4-0WT			For Own Use	
10	1865	0-6-0T	OC 16 x 24	4 6	Contractor (a)	
		0-6-0T	OC 16 x 24	4 6	(f)	
12	1865	0-6-0T	OC 16 x 24	4 6	S. Eastern Rly Portugal	No 1 Faro
13	1865	0-6-0T	OC 16 x 24	4 6	S. Eastern Rly Portugal	No 2 Cato
14	1865	0-6-0T	OC 16 x 24	4 6	S. Eastern Rly Portugal (g)	No 3 Extremes
	1865	0-4-4-0T	4/15 x 22	4 6	Neath & Brecon Rly (b)	Progress
17	1865	2-4-0T	OC 15 x 22	5 6	West Cork Rly (e)	Patience No 2
18	1865	2-4-0T	OC 15 x 22	5 6	West Cork Rly	Perseverance No 1
19	1866				FC C-C B Chile	7
20	1866				FC C-C B Chile	8
21	1866				FC C-C B Chile	9
	1866	0-4-40T	4/10 x 16	4 10	Anglesey Central Rly	Mountaineer
	1866	0-4-0WT			R. Evans & Co Ltd	Ant
	1868	0-4-0WT			R. Evans & Co Ltd	Bee
28	1866	0-6-6-0T	4/11 x 18	3 0	S & WR of Q (d)	
29	1866	0-6-6-0T	4/11 x 18	3 0	S & WR of Q (e)	
30	1866	0-6-6-0T	4/11 x 18	3 0	S & WR of Q (d)	
31	1867	2-4-0	16 x 22	5 7	East Indian Rly	336
32	1867	2-4-0	16 x 22	5 7	East Indian Rly	337
33	1867	2-4-0	16 x 22	5 7	East Indian Rly	338
34	1867	2-4-0	16 x 22	5 7	East Indian Rly	339
35	1867	2-4-0	16 x 22	5 7	East Indian Rly	340
36		2-4-0	16 x 22	5 7	East Indian Rly	341
37		2-4-0	16 x 22	5 7	East Indian Rly	342
38		2-4-0	16 x 22	5 7	East Indian Rly	343
39		2-4-0	16 x 22	5 7	East Indian Rly	344
40		2-4-0	16 x 22	5 7	East Indian Rly	345
41		2-4-0	16 x 22	5 7	East Indian Rly	346
42		2-4-0	16 x 22	5 7	East Indian Rly	347
43		2-4-0	16 x 22	5 7	East Indian Rly	348
44		2-4-0	16 x 22	5 7	East Indian Rly	349
45		2-4-0	16 x 22	5 7	East Indian Rly	350
46		2-4-0	16 x 22	5 7	East Indian Rly	351
47		2-4-0	16 x 22	5 7	East Indian Rly	352
48		2-4-0	16 x 22	5 7	East Indian Rly	353
49		2-4-0	16 x 22	5 7	East Indian Rly	354
50		2-4-0	16 x 22	5 7	East Indian Rly	355
51		2-4-0	16 x 22	5 7	East Indian Rly	356
52		2-4-0	16 x 22	5 7	East Indian Rly	357
53		2-4-0	16 x 22	5 7	East Indian Rly	358
54		2-4-0	16 x 22	5 7	East Indian Rly	359
55		2-4-0	16 x 22	5 7	East Indian Rly	360
56		2-4-0	16 x 22	5 7	East Indian Rly	361
57		2-4-0	16 x 22	5 7	East Indian Rly	362
58		2-4-0	16 x 22	5 7	East Indian Rly	363
59		2-4-0	16 x 22	5 7	East Indian Rly	364
60	1868	2-4-0	16 x 22	5 7	East Indian Rly	365

(a) Used at one time by J. Grainger, Contractor of Aberdeen. (b) The first 'Fairlie' Locomotive. (c) Order cancelled through shortage of funds, inspected by Ulster Rly but not bought. (d) Returned to England, rebuilt by Yorkshire Engine Co and sold to Uruguay Central Rly in 1874. (e) Returned to England, rebuilt by Yorkshire Engine Co and sold to Burry Port & Gwendraeth Valley Rly in 1873 and named VICTORIA. (f) Passed through Isaac Boulton's Hands (see text). (g) Diagram in Locomotive Magazine 15.8.1924 named GRATO. S & WR of Q: Southern & Western Rly of Queensland, FC C-C B Chile: FC Carrizal-Cerro Blanco Chile.

CROSS, JAMES & COMPANY

The steam pipes were coiled in the smokeboxes to compensate for the movement of the bogies. Trouble was experienced with the pipes fracturing and also draughting troubles due to no division in the common firebox, although a transverse water partition in the firebox was the designer's original requirement.

This was followed by a second *Fairlie* locomotive for the Anglesey Central Railway, but of smaller dimensions. It was transferred to the Neath & Brecon Railway but despite improved 'ball and socket' joints for the steam pipes and other modifications it was not a success and both *Fairlies* were soon withdrawn.

The largest order was for thirty 2-4-0s for the East Indian Railway which were built in 1867–8. These were part of the large order for 285 locomotives of this type which proved the undoing of more than one builder and it is significant that apart from one more 0-4-0WT this was the end of locomotive building by James Cross & Co.

This firm has been reported as building some *Fairlie* locomotives for the Venezuela Central Railway including notes in the *Engineer* and *Engineering* during 1866. None can however be traced, or the other railways identified to which they were supposed to have been sold at a later date.

One interesting locomotive (Fig. 92), which cannot definitely be identified as a new one, was 'rebuilt' in 1863 as a 2-4-2T for the St. Helens Canal and Railway and named WHITE RAVEN. It was the first 2-4-2T to have radial axleboxes, which were Adam's patent as were also the patent spring tyres fitted to the coupled wheels. It was subsequently rebuilt by the London & North Western Railway after the St. Helens line was absorbed, and did further work as a 2-4-0.

A portion of the works was taken over about 1872 by the London & North Western Railway and converted into a running shed known as Sutton Oak.

CROWLEY J. & COMPANY

Kelham Ironworks, Sheffield

No information available.

CROWTHER, PHINEAS

Ouseburn Foundry, Newcastle

Mr Crowther built for William & Edward Chapman and details will be found under that heading.

CUDWORTH, ARTHUR

St. Mark's Engineering Works, Wrexham

Two novel locomotives were built for W. H. Davis & Sons (Wagon) Ltd., Langwith Junction. They had vertical cross tube boilers and would appear to be adaptations of railway steam cranes.

They had horizontal cylinders with geared chain drive to the axle and also subsidiary gearing operating a winch. Both were used for winching wagons on to a traverser to which they were connected and then pulling or pushing the traversers to the appropriate set of rails. The wheels were double flanged to prevent them being pulled off the rails when winching.

One of these locomotives was transferred to the Neasden works of the same firm.

Date of building was about 1900.

DAGLISH, ROBERT

Wigan

Previously manager of the Haigh Foundry and Brock Mill Forge, Robert Daglish became manager of the Orrell Colliery near Wigan. Whilst there he constructed the railway connected with the colliery and claimed to have built the first railway locomotive in Lancashire in 1812. It was put on the colliery railway at the beginning of 1813, which incidentally was nearly two years before George Stephenson made a locomotive in Northumberland.

Daglish's locomotive was known as the YORKSHIRE HORSE which seems strange —having been built in Lancashire. The works had been established in 1798 and were managed by Lee Watson & Company becoming WATSON AND DAGLISH. Robert Daglish Junior was partner in this firm and then became sole proprietor.

At least two locomotives were built by arrangement with John Blenkinsop of Leeds on the lines of his patent, and would therefore have been rack engines. It has been suggested that these early engines were built for Daglish by the Haigh Foundry.

Further engines attributed to Daglish, who would of course be Robert Daglish Junior, as his father died in 1864, were two for St. Helens Canal & Railway Company.

These were No. 13 FORTH 4 coupled (1852) and No. 12 SARACEN 6 coupled, presumably built at Mill Forge in 1858.

Fig. 93 **A. Cudworth** Vertical boiler geared loco at Neasden

DARBISHIRES LTD.
Penmaenmawr

A scaled down version of a 3 ft gauge De Winton 0-4-0T was built by the firm's foreman fitter Mr Redstone in 1905 in the company's workshops. It was named REDSTONE after the builder and had two vertical cylinders 4″ x 6″ and 1′ 11″ diameter wheels. It was used on a short length of track in Mr Darbishire's garden and also tried out in the quarries but was not successful.

It is now preserved by Mr Hills of Llanberis.

DAVIES & METCALFE
Romiley, Manchester

This firm built two 2-6-2Ts for the Vale of Rheidol Railway, which is 1′ 11½″ gauge, and were the only locomotives built by this firm, who are perhaps better known for their injectors, exhaust steam ejectors and other boiler fittings.

The Vale of Rheidol locomotives were: No. 1 January 1902 EDWARD VII and No. 2 January 1902 PRINCE OF WALES.

The coupled wheels were 2′ 6″ and leading and trailing wheels 2′ 0″ diameter, the cylinders 11″ x 17″ and total weight 22 tons.

No. 1 became GWR 1212 and was withdrawn in December 1932 and put on the 'sales list' at Swindon and eventually cut up. No. 2 became GWR 1213 renumbered 9 in 1949 and is still running. The names originally bestowed were removed probably when taken over by the GWR but in 1956 No. 9 had its name restored.

When built it had outside frames and cylinders and outside Stephenson's link motion, but when overhauled at Swindon they were refitted with Walschaert's valve gear.

In their original form they were not unlike the three Lynton and Barnstaple Railway 2-6-2T locomotives built by Manning Wardle in 1897. The gauge was the same.

At the time the Rheidol engines were being built, the firm also extensively rebuilt two of the North Wales Narrow Gauge Railway's locomotives SNOWDON RANGER in 1902 and MOEL TRYFAN in 1903.

Fig. 94 Davies & Metcalfe 1/1902 Vale of Rheidol Railway EDWARD VII 1′ 11½″ gauge

DAVY BROS.

Park Ironworks, Brightside, Sheffield

This firm produced the first locomotive built in Sheffield, appropriately named SHEFFIELD. It began work on 1 April, 1840 on the Sheffield and Rotherham Railway.

A description of this locomotive can be found in the *Locomotive Magazine* for 15 June, 1943 in an article by 'P.C.D.' on Midland Railway locomotives—Sheffield and Rotherham section.

It was six wheeled with driving wheels approximately 5′ 6″ diameter. The other four wheels were connected to the driving wheels by pulleys fastened on extensions to the axles and outside the outside frames, on which 4″ belts were fitted. This method of driving was the invention of a Mr William Vickers, the belts and pulleys being used to enable wheels of different diameters to be coupled. Thus the whole weight of the locomotive was available for adhesion.

Little else is known about this novel design and its performance, although it is said that it was rebuilt as a conventional six-coupled locomotive.

Others may have been built but no confirmation has been obtained.

DEPTFORD IRON COMPANY

Built at least one locomotive for the Hartlepool Docks & Railway. This was a 0-4-0 goods engine built in July 1841 with 4′ wheels and 14½″ x 20″ cylinders. It became North Eastern Railway No. 130. No other locomotives can be traced.

DE WINTON

Union Works, Caernarvon

Founded in the 1840s by a Mr Owen Thomas. He was joined by J. P. de Winton in the early 1860s having been with the firm of Fawcett Preston & Company of Liverpool who had briefly entered the locomotive building business in 1848 but only built one.

The company manufactured a wide variety of machinery including marine engines, stationary engines, boilers, mining machinery and equipment and locomotives. The workshops were very well equipped and at their peak employed nearly 300 workmen.

All the locomotives built by De Winton had vertical boilers to which the two cylinders were bolted by plates and brackets. The flue or chimney was extended about 2 ft above the top of the boiler. The fuel was fed into the firebox, the door of which was below the footplating, by means of a trap door and chute. The water tank was placed at one end of the footplate and the coal bunker at the opposite end. The frames bellied out between the wheel centres, to accommodate the circular boiler which was 2′ 7½″ in diameter. The drive was not geared but direct on to one axle which was cranked and the connecting rods had marine type big ends. Joy's valve gear was used initially but on later locomotives an adaptation of Stephenson's motion was used.

They were designed specially for the narrow gauge quarry lines from 1′ 11½″ to 3′ 0″ gauges. Although most were built to standard dimensions many variations occurred. The average diameter of the four coupled wheels was 19″ to 20″, wheelbase 4′ to 4′ 6″ and cylinders varied from 5¾″ x 10″ to 6¾″ x 12″. Weight was between 4 and 5 tons.

The 1′ 11½″ and 2′ gauge locomotives had outside frames and the larger ones inside frames. No springs were used and in the earliest designs the axleboxes were bolted direct on to the underside of channel framing. They were very robust, well suited to the rough tracks and sharp curves.

One radical departure from standard was built in 1895 for the Pen-yr-Orsedd Slate quarry at Nantlle named ARTHUR. The cylinders were supported from the footplating and the water feed tank inserted between the boiler and the cylinder block. The drive was by means of a jack-shaft immediately under the cylinders. The steam pipe from the boiler was fitted with a large expansion loop.

It is estimated that approximately sixty were built but these locomotives were frequently transferred from quarry to quarry and in some instances renamed so that an accurate assessment is difficult.

Date	Type	Cyls	Name	Gauge ft in	Customer
c. 1869	0-4-0VBTC	VC		2 0	Dorothea Slate Co
c. 1870	0-4-0VBTC	VC		2 0	Dorothea Slate Co
1873	0-4-0T	VC		2 0	Welsh Granite Co. Trevor
c. 1874	0-4-0VBT	VC	Pert	2 0	Gorseddau Junc. & Portmadoc
1874	0-4-0VBT	VC	Glyn	2 0	Dorothea Slate Co
c. 1874	0-4-0VBT	VC		2 0	Croesor Quarry Co Ltd?
1875	0-4-0VBT	VC		2 0	Welsh Granite Co Trevor
c. 1875	0-4-0VBT	VC		2 0	Oakeley Slate Quarries Co
1875	0-4-0VBT	VC	Rhymney	2 0	Pen-y-Bryn Slate Co
c. 1876	0-4-0VBT	VC	Alice	2 0	Lord Penryhn's Slate Quarries
1876	0-4-0VBT	VC	Georgina	2 0	Lord Penryhn's Slate Quarries
1876	0-4-0VBT	VC	Ina	2 0	Lord Penryhn's Slate Quarries
1876	0-4-0VBT	VC	Lord Penryhn	2 0	Lord Penryhn's Slate Quarries
1876	0-4-0VBT	VC	Lady Penryhn	2 0	Lord Penryhn's Slate Quarries
c. 1876	0-4-0VBT	VC	Edward Sholto	2 0	Lord Penryhn's Slate Quarries
1876	0-4-0VBT	VC	Florinder	2 0	Alexandra Slate Co
1876	0-4-0VBT	VC	Lizzie	2 0	Cilgwyn Slate Co
1877	0-4-0VBT	VC	Gertrude	2 0	Cilgwyn Slate Co
1877	0-4-0VBT	VC	Efa	2 0	Alexandra Slate Co
1877	0-4-0VBT	VC	Kathleen	2 0	Lord Penryhn's Slate Quarries
1877	0-4-0VBT	VC	George Henry	2 0	Lord Penryhn's Slate Quarries
c. 1877	0-4-0T	IC	Hilda	2 0	Lord Penryhn's Slate Quarries
1877	0-4-0T	IC	Violet	2 0	Lord Penryhn's Slate Quarries
1877	0-4-0VBT	VC	Chaloner	2 0	Pen-y-Bryn Slate Co
1877	0-4-0VBT	VC	Inverlochy Fig. 95	2 0	Pen-yr-Orsedd Slate Quarry Co
c. 1878	0-4-0VBT	VC	Vron	2 0	Vron Slate Co
1878	0-4-0VBT	VC	Penmaen	3 0	Brundit & Co Penmaenmawr
1880	0-4-0VBT	VC	Moeleilia	2 0	Coedmadoc Slate Quarry
1880	0-4-0VBT	VC	Glynllifon	2 0	Pen-yr-Orsedd Slate Quarry Co
1891	0-4-0VBT	VC	Lilian	3 0	Darbishires Ltd Penmaenmawr
1892	0-4-0VBT	VC	Louisa	3 0	Darbishires Ltd Penmaenmawr
1892	0-4-0VBT	VC	Ada	3 0	Darbishires Ltd Penmaenmawr
1893	0-4-0VBT	VC	Watkin	3 0	Darbishires Ltd Penmaenmawr
1893	0-4-0VBT	VC	Puffin	3 0	Brundit & Co Penmaenmawr
1893	0-4-0VBT	VC	Gelli	2 0	Pen-yr-Orsedd Slate Quarry Co
1894	0-4-0VBT	VC	Pendyffryn	3 0	Pen-yr-Orsedd Slate Quarry Co
1894	0-4-0VBT	VC	Harold	3 0	Darbishires Ltd Penmaenmawr
1895	0-4-0VBT	VC	Llanfair	3 0	Brundit & Co Penmaenmawr
1895	0-4-0VBT	VC	Arthur Fig. 96	2 0	Pen-yr-Orsedd Slate Quarry Co
c. 1895	0-4-0VBT	VC	Emily	2 0	Llanberis Slate Co
1896	0-4-0VBT	VC	Carnarvon Castle	2 0	Llanberis Slate Co
1897	0-4-0VBT	VC	Victoria †	2 0	Pen-yr-Orsedd Slate Quarry Co
			† Works No. 201		

The following also built – dates unknown

	Type	Cyls	Name	Gauge	Customer
	0-4-0VBT	VC	Wellington	1 10¾	Dinorwic Slate Quarries Co Ltd
	0-4-0VBT	VC		2 0	Braich Slate Quarry
	0-4-0VBT	VC	Baladeulyn	2 0	Pen-yr-Orsedd Slate Quarry Co
	0-4-0VBT	VC	Starstone	2 0	Pen-yr-Orsedd Slate Quarry Co
	0-4-0VBT	VC		2 0	Talysarn Slate Co Ltd
	0-4-0VBT	VC	Freda	2 0	Llechwedd Slate Quarries
	0-4-0VBT	VC		2 0	Llechwedd Slate Quarries

The following geared locomotives said to have been built

Date	Type	Cyls	Name	Gauge	Customer
c. 1865?	0-4-0TG			2 0	Glynrhonwy Slate Co Ltd
c. 1869	0-4-0VBTG	VC		2 0	Dorothea Slate Co
c. 1870	0-4-0VBTG	VC		2 0	Dorothea Slate Co
a. 1870	0-4-0TG			2 0	Glandwr Slate Mill?

Another non-standard locomotive was built in 1897 named VICTORIA which was fitted with a launch type engine and may have been the last built. One further locomotive was tendered for at the turn of the century for a company on Christmas Island, but it is not known whether it was actually built.

The firm closed down in 1902.

Fig. 95 **De Winton** 1877 Pen-yr-Orsedd Quarry, Nantlle INVERLOCHY

Fig. 96 **De Winton** 1895 Jackshaft drive, Pen-yr-Orsedd Quarry, ARTHUR

DICK, W. B. & CO
Britannia Engineering Works, Kilmarnock

This firm was well known for its tramway equipment supplying all classes of cars, rolling stock, street tramways and narrow gauge railways.

The title changed to DICK, KERR & CO. LTD in 1883.

Steam locomotives were built from 1883 to 1918. Many steam trams were supplied at home and abroad and they were the sole makers of Morrison & Kerr's patent tramway engines.

The steam locomotives were quite conventional except that T-section spokes were used in the cast iron wheel centres.

The works were sold in 1919 to the Kilmarnock Engineering Co. and the works of Dick Kerr & Co. Ltd. concentrated at Preston.

It is difficult to assess the number built, but it would be over fifty.

When Hartley, Arnoux & Fanning were taken over by Kerr Stuart, the railway and tramway plant section was bought by Dick Kerr & Co. and Mr Hartley became manager at Kilmarnock.

Fig. 97
c 1886 Tram engine as built, Schull & Skibbereen Lt. Rly. 3′ gauge No. 3 ILEN

Fig. 98
1915 for Ministry of Munitions at Stanton Ironworks

DICK, W. B. & CO.

Date	Type	Cylinders inches	D.W. ft in	Customer	No/Name	Gauge ft in
1883		5 x 10	1 8		8	NG
1883		5 x 10	1 8		11	NG
1885	0-4-0TR	IC 7 x 12	2 4	Sutton & Alford Tramway	No 3	2 6
1885	0-4-0TR	IC 7 x 12	2 4	Penang Tramways		2 6
1886	0-4-0TR	9½ x 16	2 6	Schull & Skibbereen Rly	No. 1 Marion	3 0
1886	0-4-0TR	9½ x 16	2 6	Schull & Skibbereen Rly	No. 2 Ida	3 0
1886	0-4-0TR	9½ x 16	2 6	Schull & Skibbereen Rly	No. 3 Ilen	3 0
1886	0-4-0TR	IC 8 x 14	2 6	North London Tramways	16	Std
1886	0-4-0TR	IC 8 x 14	2 6	North London Tramways	17	Std
1886	0-4-0TR	IC 8 x 14	2 6	North London Tramways	18	Std
	0-4-0TR	IC 8 x 14	2 6	North London Tramways	19	Std
	0-4-0TR	IC 8 x 14	2 6	North London Tramways	20	Std
	0-4-0TR	IC 8 x 14	2 6	North London Tramways	21	Std
	0-4-0TR	IC 8 x 14	2 6	North London Tramways	22	Std
	0-4-0TR	IC 8 x 14	2 6	North London Tramways	23	Std
	0-4-0TR	IC 8 x 14	2 6	North London Tramways	24	Std
1887	0-4-0TR	IC 8 x 14	2 6	North London Tramways	25	Std
1889	0-4-0ST	OC		Thos. W. Ward Grays (b)	TW510	Std
1890	0-4-0TR	IC 8 x 14	2 6	Calcutta Tramways		
1890	0-4-0TR	IC 8 x 14	2 6	Calcutta Tramways		
1890	0-4-0TR	IC 8 x 14	2 6	Calcutta Tramways		
1893	0-4-2T	OC		Selangor Govt Rly	Sisyphus	Metre
1893	0-4-0ST	12X		Ministry of Munitions, Hayes		Std
1901	0-4-0ST	OC		Joseph Rank, Birkenhead (b)	Rainham	
1901	0-4-0ST	OC 10 x 18		S. A. Clarke, Barkingside (b)	No 3 Maybury	
1901	0-4-0ST	OC 12 x 22	3 6½	Cadbury Bros Ltd	No 3	Std
1901	0-4-0ST	OC 8 x 16	2 6	J. & J. White, Rutherglen (b)		Std
1903	0-4-0ST	OC		Manchester Corporation Electricity Works	No 1	Std
1904	0-4-0T			Cape Copper Co.	11	Std
1905	0-4-2T	OC 9½ x 15	2 9	B. North Borneo Ex Syn Co (c)	Marudu	Metre
1905	0-4-2T	OC 9½ x 15	2 9	B. North Borneo Ex Syn Co (c)	Enterprise	Metre
1905	0-4-0T			Cape Copper Co	13	
1905	0-4-0ST	OC		Marconi's Wireless T'grph Co Ltd (d)		2 0
1907	0-4-2T	OC 9 x 16	2 6	Burma Corpn Ltd	1	2 0
1907	0-4-2T	OC 9 x 16	2 6	Burma Corpn Ltd	2	2 0
1907	0-4-2T	OC 9 x 16	2 6	Burma Corpn Ltd	3	2 0
1907	0-4-2T	OC 9 x 16	2 6	Burma Corpn Ltd	4	2 0
1907	0-4-2T	OC 9 x 16	2 6	Burma Corpn Ltd	5	2 0
1912	0-4-0ST	OC		Admiralty - Chatham	Thistle No. 20	1 6
1913	0-6-0T	OC		?	Salsomaggiore	
1915	0-4-0ST	OC 6 x 10	1 8	Ipswich Dock Commissioners	No 70	Std
1915	0-6-2T	OC 13 x 16	2 6	War Dept – Chattenden	No 40 Fisher	2 6
1915	0-4-0ST?			War Dept – Longparish		Std
1915	0-6-0T	OC 14 x		Ministry of Munitions – Chilwell		Std
1915	0-6-0T	OC 14 x 20	3 2½	Ministry of Munitions – Hackney	51	Std
1915	0-4-0ST			Cashmore & Co (b)		NG
1915	0-4-0ST	OC 6 x		Marcus Bain, Mauchline (b)		
1916	0-4-0ST	OC 6 x		Harry Stephenson & Sons Ltd		Std
1916	0-4-0ST	OC 6 x		Jos. Place & Sons Ltd		Std
1918	0-4-0T	OC 8 x 14	2 5	Islip Iron Co	Slipton No 6	3 0
	0-4-0ST			Cadbury Bros Ltd	1	Std
	0-4-0ST			Cadbury Bros Ltd	2	Std
				Ipswich Dock Commissioners (e)		Std
				River Wear Commissioners		Std
				River Wear Commissioners		Std
				J. Williamson & Sons, Lancaster		NG
				J. Williamson & Sons, Lancaster		NG
	0-4-0ST	14 x		Wilsons & Clyde Coal Co		Std
	0-4-2T			Shap Granite Co Ltd	Popsy	3 0

(a) TR : Tram Loco. (b) Not original customer. (c) British North Borneo Exploration Syndicate Ltd later by British North Borneo Rly. (d) Clifden, Connemara. (e) Probably built before 1915.

DICK & STEVENSON

Airdrie Engine Works, Bell Street, Airdrie

Founded about 1790 this firm was of considerable importance, and built many types of machinery, stationary engines, and included a large size iron foundry. Many products were sent abroad.

In the early 1860s the firm was run by Alexander and John Dick, and Graham Stevenson under the title DICK, STEVENSON & DICK. After John Dick retired the title became Dick & Stevenson and under the latter title in 1864 they built their first locomotive. It is believed the first customer was the Langloan Iron Company, and over the years six were supplied to this company.

There was no direct rail link between the works and the Monkland railway and all locomotives leaving the works were either towed to the railhead or traversed the streets under their own steam.

Several were sent abroad to Holland, Spain and Singapore. In later years they built a standard 0-4-0ST with 14″ cylinders and wheels up to 3′ 6″ diameter. These locomotives could be identified by the unusual curved footplating, single slide bars and some had a convex bulge in front of the cab. Organ pipe whistles were fitted.

The later locomotives were numbered on even numbers only, the last two being 98 and 100 in 1890. Some came into the hands of Sharp Henderson & Co. for repairs who invariably put their own works numbers on the repaired locomotives, but did not actually build any. The works closed in 1890. See Fig. 99.

Works No	Date	Type	Cylinders inches	D.W. ft in	Customer	No/Name
	1864	0-4-0ST			Langloan Iron Co (a)	
	1865	0-4-0ST	OC 12 x 18	3 0	Monkland Rlŷ (b)	
11	?	0-4-0ST	OC		Holwell Iron Co (c)	Holwell No 7
16	1867	0-4-0ST	13 x 20	3 7	H. C. Patterson (d)	
20		0-4-0ST	14 x 22	3 7	Wm. Dixon & Co (e)	Govan No 3
	1868	?	12 x		Salter & Eskett Park Mining Co	Mountaineer
26	1868	0-4-0ST	OC		Glasgow I & S Co Broomside	No 6
32		0-4-0ST	OC 14 x	3 4	United Colls Ltd W. Lothian (f)	No 7
		0-4-0ST	OC 12 x		United Colls Ltd W. Lothian	No 8
	1870	0-4-0ST	OC 12 x 20	3 0½	Southampton Dock Co	Canute
41		0-4-0ST	OC 12 x 20		Whitehaven Colly Co Ltd	Dinah
43		0-4-0ST			Udston Coal Co Hamilton (f)	
45		0-4-0ST	OC		John Murray & Sons (f)(g)	
60	1875	0-4-0ST	OC		H. Briggs, Son & Co(h)	
98	1890	?			Jas. Waldie	
100	1890	?				
Building dates unknown;						
		0-4-0ST			Robt. Addie, Coatbridge	Marquis
		0-4-0ST			Champfleurie Oil Co Ltd	
		0-4-0ST			Jas. Nimmo & Co Ltd Redford Colly	
		?			Thompson	
		0-4-0ST			Coltness Iron Co	No 5
		0-6-0T			Wm. Dixon & Co Carfin (j)	
					Provenhall Colly	
					Provenhall Colly	

Also: Three locos said to have been built for Uphall Oil Co W. Lothian 0-4-0STs with OC Uphall Nos 2, 3 & 4 probably built before 1867. Two or three locos supplied to Spain (narrow gauge) & shunters to Singapore. One supplied to Berry Hill Colly (Second hand).

(a) First loco built. (b) Probably bought second hand. Became NBR No 240 in 1869. (c) 2 8½ gauge. (d) Wm. Baird 1873/4 No 9. (e) Carried Sharp Henderson plates dated 1879 (Rebuild date?). (f) Carried Sharp Henderson Plates. (g) Sold to Sheffield & District Gas Co PHYLLIS. (h) Sold to Whitwood Colly VICTORIA. (j) Returned to Sharp Henderson.

DIXON, WILLIAM LTD.
Calder and Govan, Glasgow

The Calder Iron Works were built in 1795, and a few years later William Dixon became the owner. About 1830 he formed the Govan Iron Works, and after his death in 1859 was succeeded by his son William Smith Dixon, who in 1872 formed a private limited company, and in 1906 it became a public limited company.

Such a large concern had extensive engineering workshops and a number of shunting locomotives were built in these shops for internal use. They were all 0-4-0STs, five were built in the Govan shops between 1860 and 1901, and one at Calder in 1892.

By the appearance of No. 2, illustrated in the *Locomotive Magazine* dated 15 May 1930, Vol XXXVI, it had many Neilson & Co's characteristics and parts may have been supplied by that company, and by Andrew Barclay Sons & Co. Ltd. for others that were built.

GOVAN WORKS

Built	Type	Cylinders inches	D.W. ft in	Dixon No.	
1860	0-4-0ST	OC 12 x 20	3 6	2	(a)
1866	0-4-0ST	OC 12 x 20	3 6	4	(a)
1874	0-4-0ST	OC 12 x 20	3 6	9	
1901	0-4-0ST	OC 12 x 20	3 5	1	

CALDER WORKS

Built	Type	Cylinders inches	D.W. ft in	Dixon No.	
1892	0-4-0ST	OC 13 x 22	3 6	8	(b)

(a) Fitted with Neilson indirect valve gear. (b) Fitted with Barclay's Motion.

No. 1 Worked at Govan, the others at Calder originally

Fig. 99 **Dick & Stevenson** *c*1870. Originally marked C. D. Phillips, Newport. Note characteristic continuous splashers

DODDS & SON

Holmes Engine & Railway Works, Rotherham

This firm was established by Isaac Dodds, who was well known for his patent wedge motion valve gear. He had been in charge of the Sheffield & Rotherham Railway locomotive department from 1839 to 1842 as 'superintendent and resident engineer' and had previously been with the Horsley Iron & Coal Co. of Tipton, Staffs. He took over part of Messrs. S. Walker & Company's works at Rotherham. This firm had made the ironworks for the Southwark bridge. He had started the works during his period of office with the S&RR, but it was not until 1849 that locomotive building commenced, which he continued to do until 1868.

It was not ideally situated, there being no railway access to the works, locomotives had to be manhandled to the wharfside and then barged on the River Don.

He received orders from the GIPR, Isabel II Railway, London, Brighton & South Coast, Deeside, and Newport, Abergavenny & Hereford Railways. In all about seventy locomotives were built in this period. YSABEL one of a series of 2-4-0s built for the Isabel II Railway was tried out on the Lickey Incline in 1853, where experiments were carried out under the supervision of Mr Stalvies the locomotive manager at the Bromsgrove workshops.

Various loads were pulled up the incline: with a weight of 45 tons 12 cwt 3q—time taken 12 min. 12 secs; with a weight of 29 tons 4 cwt 1q—time taken 7 min. 5 secs (speed 18 mph).

The general dimensions of this engine were: cylinders 14¼" x 20"; driving wheels 4' 6"; weight of engine and tender 24 tons 17 cwt; water capacity of tender 970 gallons.

In 1866 the firm quoted the London Brighton & South Coast Railway for a number of 2-2-2s and an order was obtained for two. The three month delivery period could not be maintained and the order was cancelled, but due to the cost of the work already carried out, the two locomotives were proceeded with and were eventually completed in January 1868. It was not until March 1871 that they appeared on the railway company's metals.

This pair were the last to be built by Dodds and from 1867 the firm were in the hands of the official receiver and the two locomotives were completed under these conditions. A total of seventy were built but many cannot be traced.

Date	Type	Cylinders inches	D.W. ft in	Customer	No	Name
c. 1849	0-4-2	16 x 24		South Yorkshire Rly	3	Fitzwilliam (Fig. 100)
c. 1850	0-4-2			Great Indian Peninsula Rly	10	
1850	0-4-2			Great Indian Peninsula Rly	11	
1853	2-4-0			Alar & Santander Rly		Ysabel
1854	0-4-2			Deeside Rly	3	
1854	0-4-2	14 x 20	4 6	NA & H Rly	20	(a)
1854	0-4-2	14 x 20	4 6	NA & H Rly	21	
1854	0-4-2	14 x 20	4 6	NA & H Rly	22	
1856	2-4-0			Great Indian Peninsula Rly		-
1856	2-4-0			Great Indian Peninsula Rly		
1856	2-4-0	14¼ x 20	4 6	Isabel II Rly	1	Ysabel II
1856	2-4-0	14¼ x 20	4 6	Isabel II Rly	2	Fran Asis
1856	2-4-0	14¼ x 20	4 6	Isabel II Rly	3	Perseveranc
1856	2-4-0	14¼ x 20	4 6	Isabel II Rly	4	Santander
1856	2-4-0	14¼ x 20	4 6	Isabel II Rly	5	Castilla
1856	2-4-0	14¼ x 20	4 6	Isabel II Rly	6	Habana
1856	2-4-0	14¼ x 20	4 6	Isabel II Rly	7	Velasco
1856	2-4-0	14¼ x 20	4 6	Isabel II Rly	8	B. De Garay
1856	2-4-0	14¼ x 20	4 6	Isabel II Rly	9	Alfonso
1856	2-4-0	14¼ x 20	4 6	Isabel II Rly	10	Princesa
1856	2-4-0	14¼ x 20	4 6	Isabel II Rly	11	Cantabria
1856	2-4-0	14¼ x 20	4 6	Isabel II Rly	12	Bonifaz
1868	2-2-2	17 x 22	6 6	LB & SC Rly	127	(b)
1868	2-2-2	17 x 22	6 6	LB & SC Rly	128	

(a) Supplied to Thomas Brassey who contracted for supplying locomotives to the NA & HR (Newport, Abergavenny & Hereford Rly). (b) These two locomotives were the last built, and not delivered until 1871 to the London, Brighton & South Coast Rly.
List not complete. The twelve 2-4-0s for Isabel II Rly were not paid for.

Fig. 100 **Dodds & Son** 1849 for South Yorkshire Railway No. 3 FITZWILLIAM
Note pump eccentric on trailing coupled wheels

Fig. 101 **Alfred Dodman & Co** 1893 2-2-2WT shown as rebuilt in 1911 to 0-4-2WT

DODMAN, ALFRED & CO. LTD.

Highgate Works, King's Lynn

This firm was established by Alfred Dodman in 1850 and engaged in agricultural machinery, including portable engines, and later many traction engines.

In 1896 a steam locomotive was built to the order of a Mr William Burkitt, a corn merchant of Lynn and Chesterfield who used it for travelling between these two towns. It was a 2-2-2 Well Tank named GAZELLE having inside cylinders 4″ x 9″ and driving wheels 3′ 9″ diameter, the driving axle was behind the firebox and the connecting rods were fitted between the frames and the narrow firebox.

The boiler had a raised firebox on which the dome was fitted, surmounted by a spring balance safety valve. The operating pressure was 140 lb/sq. in.

How often it was permitted to work on the main lines is not known but it was bought by Messrs. T. W. Ward who sold it to the Shropshire & Montgomeryshire Light Railway in 1911. The same year it was converted to a 0-4-2WT by Messrs. W. G. Bagnall Ltd. Its weight was just over 5 tons. It is now preserved at Longmoor.

A similar locomotive was built in the same year for the West Norfolk Farmers Manure & Chemical Co-Operative Co. Ltd., South Lynn and it was eventually sold and was shipped to Australia.

GAZELLE in its original form had no cab, and Bagnalls fitted one with a small 'saloon' over the trailing wheels. The coupled wheels were 2′ 3″, the original larger splashers being left as they were.

Further particulars of this interesting locomotive are given in R. H. Clark's *A Short History of the Midland & Great Northern Joint Railway* (Goose & Son 1967).

DORMAN, LONG (STEEL) LTD.

Britannia Works, Middlesbrough

This firm was known as Dorman, Long & Co. Ltd. up to 2 October 1954.

Locomotives were built between 1949 and 1959, and as far as is known seven in all were built.

Messrs. Dorman, Long (Steel) Ltd. have kindly furnished particulars of the procedures adopted. It is important to consider all the ways in which spare parts for their locomotives were prepared, and how in turn this affected repairs, rebuilding and the construction of new locomotives.

The provision of spare parts for maintenance purposes was solved by either: (a) measuring up parts requiring renewal and then preparing detailed drawings, which in turn enabled the spare parts to be made, or (b) making spare parts 'to sample'. In this method the sizes of the parts concerned were taken by the men in the shops and transferred to new material. For example a pattern maker would measure up a part and make a pattern to suit. When the casting was available it would be sent to the machine shop and machined to the sizes taken from the old part or 'sample', due allowance being made for any sizes on the sample, which were obviously too large or too small, due to wear. In the same way a pair of frames could be set out and cut from plates by taking the dimensions from the existing original frames. Other parts such as forgings would be prepared in a similar manner. Modifications to parts might be made, where service had proved this to be necessary, during all the above methods of providing spares.

Obviously as a locomotive increased in age and more parts were made for maintenance purposes there would come a time when by means of the drawings, templates, or recorded shop details from 'samples', it would be possible to renew any part of the locomotive as required. This in turn enabled three main types of work to be carried out on a locomotive: (a) Immediate breakdown or routine maintenance when only a single, or a few parts had to be replaced, (b) A rebuilding when a large proportion of parts required replacement. The term generally used to describe such a locomotive afterwards would be a 'Rebuild', (c) Scrapping the entire locomotive, if this was necessary, and replacing it by a new one built as a replica of the old. This type of new locomotive might be termed a 'Replacement'.

A 'Rebuild' or a 'Replacement' might be an exact replica of the original locomotive or embody modifications depending, as explained earlier, on how the spare parts had been prepared, or if details from other locomotives were built into it.

DORMAN, LONG (STEEL) LTD.

In most, if not all, of the above cases an arrangement drawing was not prepared and was not necessary. The skill and experience of the men and the years of shop practice enabled locomotives to be assembled and built satisfactorily without arrangement drawings.

As opposed to the above 'repair' work, however, a locomotive could be designed, drawings made, and the locomotive then built in the shops. This locomotive could be described as 'New Work'. In addition a new locomotive could be built by the shops as a replica of an existing locomotive by the methods already described under 'repair' work. In this case the new locomotive would be what might be termed a 'Duplicate' of an existing locomotive. The original and the 'Duplicate' would both continue to work and the original locomotive was not scrapped.

Therefore the terms 'Rebuild', 'Replacement', 'New Work' and 'Duplicate' will be used in connection with the remarks on the following locomotives.

The locomotives of Acklam, Britannia, Ayrton, and Newport, including a few from Port Clarence, were renumbered in 1943 into a new combined series of numbers. Locomotives numbered 1, 22, 27, 35, 36, 39, and 43 according to Messrs. Dorman Long (Steel) Ltd., were either very heavy rebuilds, replacements, or duplicates as shown in the table.

Unfortunately no drawings or photographs have been discovered of these locomotives; they would have been of considerable help and interest in assessing which category each locomotive should be placed.

DORSET IRON FOUNDRY

West Quay Road, Poole

At least one narrow gauge locomotive was built in 1878 after the former works manager of Stephen Lewin, Mr W. J. Tarrant had established the firm. This was a narrow gauge saddle tank with a flywheel at footplate level sent to New Zealand to Messrs. Auckland & Company. It had all the characteristics of the typical Lewin product.

BRITANNIA WORKS – DORMAN LONG

Previous Locomotive Details						Replacement Locomotive		
Loco	1943 No	Type	Built	No	Date	Built	Type	No
Old Britannia No 43	43	0-4-0ST	CF	1199	1900	1949	0-4-0ST	43
Old Ayrton No 3	39	0-4-0ST	BH	?	1874	1951	0-4-0ST	39
Old Newport No 6	27	0-4-0ST	K	2363	1881	1952	0-4-0ST	27
Old Acklam No 8	35	0-4-0ST	BH	1124	1896	1952	0-4-0ST	35
Old Britannia No 7	36	0-4-0ST	P	462	1887	1953	0-4-0ST	36
Old Britannia No 11	22	0-4-0ST	HL	2905	1911	1953	0-4-0ST	22 (a)
Old Acklam No 3	1	0-4-0ST	AB	1888	1926	1959	0-4-0ST	1 (b)

AB : Andrew Barclay Sons & Co Ltd. BH : Black Hawthorn & Co CF : Chapman & Furneaux.
K : Kitson & Co P : Peckett & Sons Ltd.

(a) Old Britannia No 11 was not scrapped until 1962 according to the BLC, and if this was so then the new and old locos shared a common existence for nine years and the new loco would come under the category of a 'duplicate'. The HL loco was transferred to Warrenby Works in 1938 and numbered 19.
(b) This loco built in 1959 after firm changed its title to Dorman, Long (Steel) Limited.

DRUMMOND, D. & SON
Helen Street, Govan

The firm was founded by Dugald Drummond in 1891 after he had worked in various capacities at locomotive establishments at Glasgow, Birkenhead, Cowlairs, Inverness, Brighton, Cowlairs again, St Rollox and Australia.

Drummond had left his mark on all the products he had dealt with, and surely few men had had so many changes of venue. Perhaps his peregrinations might not have been so varied, if his personality had been more subdued. So it was after his disappointing visit to Australia that he resolved to be independent and to have full scope for developing his own ideas and methods. He was to build locomotives to order and for any customer, assuming that the men and companies with whom he had associated would be only too eager to have locomotives built by such an experienced and eminent a person. Alas orders were hard to come by, and very few were built. In 1895 he obtained the post of locomotive superintendent on the London & South Western Railway at Nine Elms and left the works in charge of his sons who formed a limited company.

THE GLASGOW RAILWAY ENGINEERING CO. LTD., were entered as 'locomotive makers', and this designation continued until 1901. The firm remained in Helen Street until 1937 when the address became 139 Rigby Street, Glasgow E.1 and continued there until 1958. For the last forty or more years of its existence, the description of work carried out was for railway wheels and axles.

During the locomotive building period, seven narrow gauge 0-4-0WTs were built in 1894 for the Glasgow Corporation Gas Works, railcars for New South Wales and in 1897 two railmotors were supplied to the Alexandra Dock and Railway Co., Newport. They had 9″ x 14″ cylinders and 3 ft diameter coupled wheels. The first coach had a length of 59′ 0″ over buffers, and the other 64′ 10″. In the same year a 2 ft gauge 0-4-0ST named LITTLE TICH was built for the Premier Cement Co. Irthlingboro. Railway equipment of all kinds was manufactured and at one time there was a wheel and axle division at Parkhead, Glasgow operated in association with Wm. Beardmore & Co. Ltd.

The photograph (Fig. 102) cannot be identified but is obviously for an overseas narrow gauge railway, probably Brazil.

Fig. 102 **D. Drummond** 1898(?) 4-4-0 No. 38 customer unknown

DUBLIN & DROGHEDA RAILWAY

Amiens Street, Dublin

According to records two locomotives were built here, but due to limited equipment in their workshops it is doubtful whether these were entirely new locomotives. They were: D&DR No. 1 (1861) 0-4-2 14½" x 20" 5' diameter and D&DR No. 2 (1866) 0-4-2 14½" x 20" 5' diameter.

It has been stated also that old D&DR No. 15 which was Beyer Peacock 2-2-2 built in 1859 (Works No. 105) was heavily rebuilt in 1880, and though claimed in some quarters to be a new engine, it could not have been entirely new.

DUBLIN & KINGSTOWN RAILWAY

Grand Canal Street, Dublin

The locomotive works were established in 1839 at Grand Canal Street, the railway having bought an unused Distillery and adapted it to their requirements. Previously, repairs had been carried out at the Serpentine Avenue works which had also been used by the builders of the earlier D&KR locomotives.

John Melling was the locomotive superintendent until 1840 when he was replaced by Richard Pim who persuaded the Board to use the new works to build their own stock. In April 1841 the first locomotive built in

Fig. 103 **D&KR** Grand Canal Street Works, showing lifting gear

Fig. 104 **D&KR** 1847 running as DW&WR No. 27 rebuilt in 1887

Fig. 105 1890 No. 9 DALKEY 2-4-0T name removed

a railway workshop in the British Isles was completed, also having the honour of being the first built in Ireland. It was named PRINCESS, a 2-2-2WT with outside cylinders, inside frames for the leading and driving axles, a boiler with raised firebox on which was fitted the dome and two pillar-type safety valves on the boiler barrel. Further tank locomotives generally similar in design were built to the standard gauge, and were modelled on Forrester's VAUXHALL which had been rebuilt as a 2-2-2WT.

In 1843 Pim was succeeded by James Rawlins and BURGOYNE, built in 1845 had larger cylinders 14″ x 20″ and three more followed in 1847–8.

In 1851 COMET although not entirely new, was entirely rebuilt and was very similar to PRINCESS. This was done under the supervision of Samuel Wilfred Haughton who had succeeded Rawlins in 1849.

In 1860 the railway (after being converted to 5′ 3″ gauge in 1855) became part of the newly formed DUBLIN, WICKLOW & WEXFORD RAILWAY.

No further new building took place until 1871 when the first of seven more 2-2-2WTs was put in hand. According to Ahrons, one or two of this batch had stove pipe chimneys but the others were copper capped. The tank was located under the coal bunker.

In 1882 the first of a series of 2-4-0Ts was built and continued until 1891. These tanks had domed boilers and round topped fireboxes. The dome had a spring loaded lock-up safety valve fitted on top and the dome cover was open at the top, shaped like a small chimney. A further safety valve was located over the firebox. Volute springs were used on the leading wheel axleboxes and the splashers over the leading coupled wheels had a series of slots. Some were later converted to 2-4-2Ts.

A further series of 2-4-0Ts was built from 1893 to 1896 which had larger cylinders and were quite an improvement on the previous members of this wheel arrangement.

It was found that the converted 2-4-0Ts into 2-4-2Ts were far more satisfactory on the road and a number of new ones of this type were built together with four six-coupled goods locomotives.

In 1907 the railway's title was changed to DUBLIN & SOUTH EASTERN RAILWAY, and very little new building took place although the shops were busy rebuilding and repairing.

One 2-4-2T appeared in 1909, a 0-6-0 in 1910 and finally a 4-4-2T in 1911.

No further building took place after 1911 and in 1925 the works were closed down when under Great Southern Railway ownership.

The 2-2-2WTs, 2-4-0Ts and 2-4-2Ts were mainly used on the suburban traffic at the Dublin end of the line and a frequent service was run to Kingstown and Bray.

The locomotive superintendents after Mr Haughton were: 1864–1865 W. Meikle, 1865–1882 J. Wakefield; 1882–1894 W. Wakefield; 1894–1897 T. Grierson; 1897–1917 R. Cronin, and 1917–1925 G. H. Wild.

GRAND CANAL ST. DUBLIN

P No	Date	Type	Cylinders inches	D.W. ft in	DWW No	Name*	GSR Class	GSR No	W'dr
DUBLIN & KINGSTOWN RLY									
1	4/1841	2-2-2WT	12 x 20	5 6		Princess (a)			1884
2	8/1841	2-2-2WT	12 x 20	5 6		Belleisle (b)			(e)
3	11/1842	2-2-2WT	12 x 20	5 6		Shamrock			(e)
4	11/1843	2-2-2WT	12 x 20	5 6		Erin			(e)
5	7/1844	2-2-2WT	12 x 20	5 6		Albert			1874
6	9/1845	2-2-2WT	14 x 20	5 6		Burgoyne			(e)
7	3/1847	2-2-2WT	14 x 20	5 6	27	Cyclops			1887
8	*c.* 1847	2-2-2WT	14 x 20	5 6	28	Vulcan			1187
9	*c.* 1848	2-2-2WT	14 x 20	5 6		Jupiter			1876
10	1851	2-2-2WT	14 x 20	5 6		Comet (c)			(e)
DUBLIN, WICKLOW & WEXFORD RLY									
11	1871	2-2-2WT	15 x 20	5 6	29				1906
12	1873	2-2-2WT	15 x 20	5 6	30				1902
13	1873	2-2-2WT	15 x 20	5 6	31				1905
14	1874	2-2-2WT	15 x 20	5 6	36				1902
15	1874	2-2-2WT	15 x 20	5 6	40				1905
16	1879	2-2-2WT	15 x 20	5 6	4				1908
17	1882	2-4-0T	15 x 22	5 5	41	Delgany			1925
18	1885	2-4-0T	16 x 24	5 5	2	Glenageary			1925
19	1886	2-4-0T	16 x 24	5 5	45	St Kieran (d)	F.2	432	1957
20	1887	2-2-2WT	15 x 20	5 6	27	Fig. 104			1906
21	1887	2-4-0T	16 x 24	5 5	28	St Laurence (d)	F.2	431	1950
22	1888	2-4-0T	16 x 24	5 5	46	Princess Mary (d)	F.2	433	1957
23	1889	2-4-0T	16 x 24	5 5	47	Stillorgan	G.1	425	1953
24	1889	0-4-2	16 x 24	4 10½	48				1913
25	1890	2-4-0T	16 x 24	5 5	9	Dalkey Fig. 105	G.1	424	1952
26	1891	2-4-0	16 x 24	5 5	49	Carrickmines	G.1	423	1955
27	1893	2-4-0T	16 x 24	5 5	1				1925
28	1894	2-4-0T	16 x 24	5 5	6				1925
29	1895	2-4-0T	16 x 24	5 5	7		G.1	(426)	1926
30	1896	2-4-0T	17 x 24	5 5	10	St Senanus (d)	F.2	(429)	1925
31	1896	2-4-0T	17 x 24	5 5	11	St Kevin (d)	F.2	430	1952
32	1898	2-4-2T	17 x 24	5 6	3	St Patrick	F.2	428	1953
33	1899	0-6-0	17 x 24	5 0	17	Wicklow	J.20	440	1929
34	1900	0-6-0	17 x 24	5 0	36	Wexford	J.14	441	1935
35	1901	2-4-2T	17 x 24	5 6	12	St. Brigid	F.1	435	1950
36	1902	2-4-2T	17 x 24	5 6	40	St. Selskar	F.1	439	1952
37	1903	2-4-2T	17 x 24	5 6	8	St. Brendan	F.1	434	1950
38	1904	0-6-0	18 x 26	5 0	13	Waterford	J.8	442	1930
39	1905	0-6-0	18 x 26	5 1	14	Limerick	J.8	443	1955
40	1906	2-4-2T	17 x 24	5 6	29	St. Mantan	F.1	437	1951
41	1907	2-4-2T	17 x 24	5 6	27	St. Aidan	F.1	436	1953
DUBLIN & SOUTH EASTERN RLY									
42	1909	2-4-2T	17 x 24	5 6	30	St. Iberius	F.1	438	1952
43	1910	0-6-0	18 x 26	4 11½	18	Enniscorthy	J.8	444	1957
44	1911	4-4-2T	18 x 26	6 0	20	King George	C.2	455	1959

Dublin & Kingstown Rly. Built to 4'8½" gauge converted to 5'3" in 1855.

(a) First locomotive built in Ireland and the first in the world by a railway works, (b) Built from parts of older locomotives. (c) Re-constructed from tank loco. Built G. Forrester 1835. (d) Subsequently rebuilt as 2-4-2T. (e) Withdrawn between 1864 and 1872.

GSR: Great Southern Railways.

* The DW & WR started naming locomotives about 1898.

† DNW & D & SE Numbers

Fig. 106 **DW&WR** 1902 *No. 12 class* No. 40 ST SELSKAR

Fig. 107 **D&SER** 1911 No. 20 KING GEORGE as GSR No. 455

DUBS & CO.

Glasgow Locomotive Works, Polmadie, Glasgow

Henry Dubs severed his connection with Neilson & Co. where he was managing partner, in 1863 and established his own locomotive factory at Little Govan later to be re-named Polmadie. Building of the works started in 1864 and the excavated clay used for making the bricks required. The well known diamond shaped works plate was also used as a trademark even for the bricks. Within twelve months the first locomotive was built. It was one of an order for ten 0-4-2s for the Caledonian railway with outside cylinder 16½″ x 22″ and 5′ 2″ coupled wheels.

Dubs had persuaded Sampson George Goodall-Copestake to leave Neilson's with him, to be Chief Draughtsman and his services proved invaluable. The firm soon gained a reputation for excellent workmanship and from 1867 began building for India, Cuba, Spain, Finland and Russia, including a traction engine for the Ottoman railway. Their locomotives went to many parts of the world at the same time maintaining a steady supply to the home market. The locomotive crane tank was first produced by the firm and many were built for iron and steel works where they were in great demand.

Twelve 2′ 6″ gauge 0-4-0Ts were built for the New Zealand Government railways in 1873. They had outside cylinders 8″ x 15″, shallow side tanks the full length of the locomotive and 2′ 6″ diameter wheels. One named MABEL worked for 75 years and has been preserved.

Japan in 1870 received two 2-4-0Ts to which were attached four wheeled brake vans, also built by Dubs & Co.

At home the Midland Railway was a good customer and many 0-6-0s, 4-4-0s and 0-4-4Ts were supplied. An interesting 4-4-0 design was built for the Intercolonial Railway (Canadian Pacific) with outside cylinders 17″ x 24″ and coupled wheels 5′ 3″ diameter. Forty were built in 1882 with bar frames for the coupled wheels, the smokebox and cylinders were supported by the front bogie. A large double window cab was fitted and a large spark arrester on top of the chimney. The tender ran on two bogies.

Fig. 108 **Dubs & Co** 1560/1882 Intercolonial Railway, Canada 4-4-0 No. 4, bar frames

Fig. 109 **Dubs & Co** 2571/1890 Inter Oceanic Railway, Panama 4-6-0 No. 35

Fig. 110 **Dubs & Co** 2629/1890 New South Wales Govt Railway 2-6-0 No. 64

Fig. 111 **Dubs & Co** 2841/1891 Norwegian Govt Railway 4-4-0 No. 68

Fig. 112 **Dubs & Co** 3989/1901 Egyptian Govt Railway 4-4-2 No. 601

Two 0-4-0STs were built for the Imperial Chinese railway; they were named SPEEDY PEACE and FLYING VICTORY and delivered in 1886. Although not the first to arrive in China, their history was tragic. Chinese land-owners and priests were violently opposed to these devil's machines which would desecrate their ancestors graves, and to prevent their progress, these extremists halted the trains by throwing coolies in front of the trains and many were run over and killed. The line was forced to close but in later, enlightened times many were supplied by Dubs & Co. to China.

Copestake and Dubs, between them patented a special coupling to enable two tank locomotives to be coupled bunker to bunker providing flexibility as well as rigidity. The idea was intended to offset the double engine design of Robert Fairlie.

All the main Scottish railways ordered from Dubs & Co., and many English lines. In 1901 the firm exhibited a Class T.9 4-4-0, at the

Glasgow Exhibition, which was built for the London & South Western Railway. It was equipped with Drummond's water tube firebox, for which he claimed a fuel saving of 5 lb per mile.

In 1903 with Neilson Reid and Sharp Stewart the three firms amalgamated and the works became known as the 'Queens Park Works' and the works plates remained with the familiar diamond shape.

The Dubs list ran on to No. 4492 although previous to this North British Loco. Co. plates were used. For convenience the number of Dubs locomotives is taken to this total, and deducting cancellations, the total built was 4485. (See NORTH BRITISH LOCOMOTIVE CO.)

DUNCAN & WILSON

Liverpool

The *English Mechanic* for 1881 refers to experiments at Liverpool with steam locomotives and this firm took part but whether they actually built any locomotives is doubtful.

EARL OF DUDLEY

Round Oak Works, Castle Mill, Dudley

Two standard gauge locomotives were built: LADY HONOR (1912) and LORD EDNAM (1915). Both were 0-6-0ST with OC 16″ x 24″ and 4′ diameter wheels. Many earlier 0-4-0s built by R. Stephenson, E. B. Wilson and Manning Wardle were extensively rebuilt as 0-4-0STs including DUDLEY (1906), ALMA (1908), FREDERICK (1903), EDWARD VII (1902) and ALEXANDRA (1904). Also built was at least one small o.c. 0-4-0ST for the narrow gauge (2′ 3½″) railway that served the old Blast Furnaces in 1882 bearing a plate 'Castle Mill 1882' named GNAT and a 0-6-0ST of the same gauge at an unknown date named DOCTOR JIM.

The two narrow gauge locomotives were scrapped by 1933 but ceased work in April 1921 when the old blast furnaces were closed down.

See W. K. V. Gale's *A History of the Pensnett Railway* (Goose and Son, 1975).

Fig. 113 **Earl of Durham** 1894 No. 26 0-6-0. Cab was altered later for clearance

EARL OF DURHAM COLLIERIES LIMITED

Philadelphia Engine Works, Durham

The works were built originally to carry out repairs to the numerous Lambton Collieries' locomotives and other work associated with the Collieries. Despite limited facilities a six coupled locomotive was built in 1877. It had strong North Eastern Railway characteristics, as had the other two of the same wheel arrangement built in 1890 and 1894 respectively. The three 0-6-0s built were: No. 9 (1877), No. 25 (1890) and No. 26 (1894). To cope with the increasing amount of work that the shops had to deal with, a new erecting shop was built in 1882 with two roads. The first locomotive to be actually built in the new shop was a four coupled saddle tank with outside cylinders, completed in 1884 (No. 12).

After 1894 no new stock was built but the works were kept busy with repairs and rebuilding, and the Collieries became part of the National Coal Board in 1947. Locomotives were sent to Philadelphia from other collieries in the Durham area and to the early 1960s the works were extremely busy. In 1958 a hybrid 0-6-0PT was 'built' out of parts of three condemned locomotives, but could hardly be classed as a new locomotive.

A characteristic of the Lambton locomotives was the sloping top half of the cab sides made necessary by the Staithes tunnel. Side windows were also fitted.

The works were well equipped and could cope with all repairs, although it is probable that new boilers and wheels were bought out.

With the amalgamation of the Hetton Coal Co. Ltd., the title became LAMBTON & HETTON COLLIERIES LTD. in 1911 and in 1924 the James Joicey & Co. Ltd. Collieries were combined to form the LAMBTON, HETTON & JOICEY COLLIERIES LTD., and then in 1947 NCB No. 2 (Mid-East Durham) area Durham Division.

Philadelphia Locomotives

The NCB South Durham Area Spennymoor have kindly supplied the following dimensions:

No. 9: Cylinders 17″ x 24″; wheelbase 16′ 0″; weight/locomotive 26 ton 10 cwt; weight/tender 14 ton 14 cwt; total in w.o. 55 ton 14 cwt; number of tubes 159; diameter 2″; heating surface/tubes 954 sq. ft; heating surface/firebox 90 sq. ft; grate area 14 sq. ft; working pressure 150 lb/sq. in.

Nos. 25 and 26: Cylinders 17″ x 24″; wheelbase 16′ 0″; weight/locomotive 28 ton 10 cwt; weight/tender 14 ton 10 cwt; total in w.o. 56 ton 0 cwt; number of tubes 150; diameter 2″; heating surface/tubes 910 sq. ft; heating surface/firebox 90 sq. ft; grate area 14 sq. ft; working pressure 160 lb/sq. in.

These locomotives have now been cut up.

EASTERN COUNTIES RAILWAY

Stratford, London

The original locomotive repair shops of the ECR were at Romford and with no opportunity to expand on that site it was decided to build a new factory at Stratford which was completed in 1847 and all machinery transferred from Romford. In 1848 all repair work was being carried out at the new works. The erecting shop was 348 ft long and 142 ft wide comprising four bays, and 50 locomotives could be accommodated at one time, it had the customary ancillary shops including boiler shop, machine shop, iron and brass foundry, wheel shop and drop hammers.

At the time of its opening the Locomotive Superintendent was John Hunter (1846–1850) then John V. Gooch (1850–1856) followed by Robert Sinclair in September 1856.

It is interesting to note that the original gauge of the ECR was 5 ft, from 1839 to 1844 when it was decided to conform to the standard gauge.

John Viret Gooch who had been locomotive superintendent of the London & South Western Railway was the first to design a class of locomotive to be built in the new works. Six 2-2-2Ts were put in hand in 1851 with double frames, the 11″ x 22″ cylinders being fixed between them. The leading and trailing wheels had outside bearings and the driving wheels 6′ 6″ diameter inside bearings. The boiler pressure was 110 lb/sq. in. They were small locomotives and weighed 23 ton 15 cwt in working order. They were well tanks and known as Class A. The cylinders were subsequently enlarged to 12″ diameter. Six more followed from 1852 to 1854, originally built with 12″ diameter cylinders, they were known as Class B. Running numbers were 20 to 25 and 7 to 12 respectively. One was converted into an Inspection Saloon as a 4-2-4T and lasted until 1883. Ten more were built in 1854 and 1855 followed by five 0-6-0 tender locomotives.

All of Gooch's designs had domeless boilers. Gooch was the brother of the Great Western Railway's chairman. His place was taken by Mr Robert Sinclair in 1856 who had been locomotive superintendent on the Caledonian Railway to this date.

Faced with a multitude of nondescript locomotives which had accumulated from various minor railways taken over by the ECR he planned to modernise the stock and standardise on a few types to serve the various requirements of the company. He designed a mixed traffic 2-4-0 with 17″ x 24″ cylinders and 6′ 1″ diameter coupled wheels, and a passenger 2-2-2 with 16″ x 24″ cylinders and 7′ 1″ drivers.

One hundred and ten of the former were built from 1859 to 1866 and thirty-one 2-2-2s

(1862–1867) and all were built by outside contractors including the firm of Schneider et Cie of France.

In 1862 occurred the formation of the GREAT EASTERN RAILWAY by the passing of the Great Eastern Railway Act of 7 August 1862 which brought together the Eastern Counties, Eastern Union, Norfolk, East Anglian, and Newmarket railways. Only the Eastern Counties Railway had built new locomotives in its own workshops.

Robert Sinclair continued as locomotive superintendent of the new Great Eastern Railway, and his standard Y class 2-4-0s continued to be built and although mixed traffic locomotives, were used extensively on passenger services to supplement the W class 2-2-2s.

Up to this period the locomotives had been painted green, Johnson introducing a darker shade. W. Adams changed the livery to black with red lining and this was continued until 1882.

Some further 0-4-2Ts were built in 1877–8 and were the only locomotives designed by William Adams to be built at Stratford.

He did however produce a 0-4-4T design of which fifty were built outside and also twenty 4-4-0s with 18″ x 26″ cylinders and 6′ 1″ diameter driving wheels known as 'Ironclads' which proved too heavy for passenger services and were relegated to goods traffic.

Adams is well known for his introduction of a 2-6-0 goods locomotive and were the first of this type in this country. Fifteen were built by Neilson & Co. with 19″ x 26″ outside cylinders

Fig. 114
GER 1893 2-2-2
7′0″ drivers
designed by
J. Holden No. 1006

His other design was for a 2-4-0WT, five of which were built at Stratford in 1862.

In 1866 Samuel Waite Johnson was appointed in Sinclair's place. Various 2-4-0s were built, and thirteen built between 1868 and 1872.

His next designs were for an 0-4-2T for branch line working, and a 4-4-0 for express passenger work. Their claim to fame – two were built – was that they were the first inside cylinder, inside frame 4-4-0s to be built in England. They were actually built in 1874, the year after Mr Johnson had left the GER to take up a similar post on the Midland Railway. His sojourn was brief, but many of his locomotives had the characteristics of his later Midland designs – rimmed chimney, combined domes and spring balance safety valves, graceful safety valve covers, short cab.

With Johnson's departure to Derby, there came a William Adams who was later to go to the London & South Western Railway.

and 4′ 10″ diameter driving wheels. They weighed 46 ton 10 cwt., and did not go into traffic until after Adams had left. The name MOGUL was painted on the central box-like splashers of the first one (No. 527). The name was attributed to the Great Moguls of Delhi who were much in the news (1879–1880).

Once again Stratford became the jumping off ground for locomotive superintendents who aspired to greater things, which also conferred on Stratford the doubtful honour of having no less than twelve of these gentlemen from ECR days in 1839 to 1923 when the GER became part of the London & North Eastern Railway. Add to these appointments, three more who acted in a temporary capacity at various times.

William Adams introduced a stove-pipe chimney with its diameter larger at the top than the bottom with a small rim around the top. He also introduced larger cabs for tender locomotives with side windows and Ramsbottom safety valves were fitted.

Fig. 115 1894 2-4-0 5′8″ No. 485

Mr Massey Bromley took office in 1878 for an even shorter reign than his predecessor. 0-4-4Ts continued to be built for the intensive suburban traffic out of Liverpool Street. Known as Class E.10 sixty were built from 1878 to 1883. They had 16½″ x 22″ cylinders and 4′ 10″ diameter coupled wheels.

He designed a most successful 4-2-2 very similar in appearance to Patrick Stirling's famous 8 ft singles built in 1870. Bromley's singles had 7′ 6″ diameter driving wheels and 18″ x 24″ outside cylinders. They were not built at Stratford.

A small number of new locomotives had been built at Stratford up to 1878 but from this date a greater proportion were built by the company. From 1851 to 1880 just over a hundred had been built, and by 1890 the 500th had appeared.

Another type he introduced was the 0-6-0T for shunting purposes, ten of which were built in 1880 with 16″ x 22″ cylinders and 4′ 10″ diameter wheels. The boiler pressure was usually kept at 140 lb/sq. in. for all locomotives with few exceptions.

Fig. 116 1902 The 'Decapod' No. 20. 335 tons to 30 mph in 30 secs

Mr Massey Bromley resigned in August 1881 and Mr M. Gillies, the works manager acted as temporary superintendent until in February 1882 the post was filled by Thomas William Worsdell who had been trained at Crewe, went to America and returned to Crewe and served under F. W. Webb as works manager from 1871 to 1881. Here was another short period of superintendency, Worsdell leaving in 1885 for the North Eastern Railway.

In 1882 Worsdell produced a 2-4-0 passenger class with 18″ x 24″ cylinders and 7 ft diameter coupled wheels. Twenty were built 1882–1883, Joy's valve gear was used and radial axleboxes; this was the Crewe training becoming evident. They had continuous splashers over the coupled wheels, stove pipe chimney and large cabs.

He then built eleven 4-4-0 compound locomotives in appearance similar to the 2-4-0s except for the bogie. The HP cylinder was 18″ and LP cylinder 26″ both having a stroke of 24″, driving wheels were 7 ft diameter and boiler pressure increased to 160 lb/sq. in., otherwise the boiler was similar with a grate area of 17·3 sq. ft. and heating surface approx. 1200 sq. ft. They were not a success and were converted to simple expansion in 1892 with 18″ diameter cylinders.

A more successful type was a 2-4-2T for suburban work with inside cylinders 18″ x 24″ and 5′ 4″ diameter driving wheels. Joy's valve gear was fitted and forty were built 1884–1887. They were very heavy on coal and were nicknamed 'Gobblers'. The type increased by another 120 after Worsdell had left, with various modifications.

A six coupled locomotive was designed, the first appearing in 1883 as the Y14 class. Cylinders were 17½″ x 24″, 4′ 11″ diameter wheels, total heating surface 1199·5 sq. ft and a grate area of 18 sq. ft. This class was used all over the system with success. The boiler centre line was low, a stove pipe chimney was fitted as standard, the dome was central on the barrel.

In December 1891 one of this type No. 930 built during James Holden's term of office was 'built' in 9¾ working hours. This beat previous records set up in 1888 by Crewe assembling a 0-6-0, No. 2153 in 25½ hours and later in the same year the Pennsylvania RR in their Altoona shops built one in 16¼ hours. What these stunts achieved, is questionable, apart from an exercise in organising the various gangs to follow on in sequence and not fall over one another. No. 930 ran 1,127,750 miles in 43 years, which speaks well for the class of workmanship employed at Stratford.

Fig. 117 1910 S. D. Holden's 19″ 0-6-0 No. 1239 cylinders 19″ × 26″

Fig. 118 1913 A. J. Hill's 4′ 0-6-0T *Class C72* No. 22

The remaining design of Mr Worsdell was a 0-4-0 Tram engine with 11″ x 15″ inside cylinders, 3′ 1″ diameter wheels and weighing 21 ton 5 cwts. Ten were built over a period from 1883–1897. They were built for use on the Wisbech & Upwell Tramway. The motion was covered in and the upper part housed in coachwork.

Worsdell obtained the post of locomotive superintendent on the North Eastern Railway at Gateshead and left for the north in May 1885. He left behind him a locomotive stock well maintained and in a reasonable state of efficiency. He altered the colour of locomotives to Royal Blue with polished brasswork, cast number plates with vermilion background, and coupling rods of the same colour, resulting in a finish second to none.

His successor was James Holden who, strangely enough had been apprenticed at Gateshead under Edward Fletcher. He joined the Great Western Railway in 1865 as manager of the carriage and wagon department under Joseph Armstrong, and on Mr Armstrong's death became chief assistant to William Dean until August 1885.

When he had familiarised himself with Stratford works and the locomotives in service, he reorganised the workshops and brought them up to a pitch of efficiency that had never been previously reached or attempted.

Some degree of standardisation had been tried, but not taken to sufficient lengths and with his twenty years experience at Swindon, Holden began to study the possibilities of standard interchangeable parts, large and small including cylinders, boilers, wheels, etc. Nevertheless for some years he made no drastic alteration to styling for its own sake and made judicious use of existing material, templates and jigs.

The 2-4-2T 'Gobblers' designed by Worsdell continued to be built until 1909 reaching a total of 160 with slight modifications. Another Worsdell class was the 0-6-0 Y14 class which was built at intervals until 1913 making a total of 270 locomotives.

Two designs of 0-6-0 were introduced by Holden, the first in 1888 with 17½″ x 24″ cylinders and 4′ 11″ diameter wheels, the boiler had a heating surface of 1199·5 sq. ft. They had Stephenson's valve gear and the valve chests were below the cylinders. The boiler had a round top firebox, and the dome was placed on the first ring of the boiler, close to the chimney a characteristic of Holden's boilers for many years. Holden also fitted a patent variable blastpipe designed by Macallan and consisted of an annular ring on top of the blastpipe which could be raised or lowered to increase or decrease the cross sectional area of the blast pipe. It was operated from the cab through a crank and rod mechanism, and became a standard fitting on most of the locomotives on the GER.

Although they looked similar to Worsdell's Y14 class, their performance was very different. They were very sluggish and were prone to priming. Eighty-one were built from 1893 to 1898 and were class N31. They were not long lived, withdrawal commencing in 1908 but some were rebuilt and eighteen lasted long enough to become part of the LNER stock but all had gone by 1925.

The second design appeared in 1900 as

class F48. They had standard 4′ 11″ diameter wheels with 19″ x 26″ cylinders and a large boiler with 1624·38 sq. ft of heating surface. One of this class was built new with a boiler fitted with a Belpaire firebox, so that comparisons could be made, and the 1905 batch and later orders up to 1911. All had Belpaire fireboxes.

In 1898 Holden built ten 4-2-2s; he had already built twenty-one 2-2-2s between 1889 and 1893 which on light trains were moderately successful.

The 4-2-2s were handsome locomotives. The bogie had inside frames and the driving and trailing wheels double frames. They bore quite a resemblance to the Great Western singles, and had plenty of brass and copper embellishments including the chimney rim, axleboxes, beading etc. The cylinders were 18″ x 26″, and driving wheels 7 ft diameter. Boiler pressure was 170 lb/sq. in.; the total heating surface 1293 sq. ft and a grate area of 21·3 sq. ft. They were oil fired.

They were built during a period of increasing train loads and although creditable performers with medium loads something more powerful was required and only two years after their manufacture appeared the first of a famous class of 4-4-0s.

Oil burning was first tried by Holden in 1887, not because of any coal shortage but to endeavour to get rid of the residue of the oil-gas plant at Stratford. Complaints were constantly received concerning the fouling of the River Lea & Channelsea by the sanitary authorities. The oil, which was a mixture of coal tar and creosote oil with a specific gravity of 1·16 at 60°F and a flash point of 230°F, was passed through a tubular heater and heated by exhaust steam from the air brake pump being thus warmed to a temperature of 200°F. The burners were placed below the footplate and holes made in the firebox about 12″ above the firebars. The oil was sprayed into the firebox by an annular steam jet in the centre of each burner and then atomised and distributed by small steam jets issuing from ring blowers placed immediately behind the nozzles. Air was induced into the firebox.

Petroleum residues were also tried but commercial quantities were not sufficient although the fuel was more efficient.

Returning to the 4-4-0, this appeared on 17 March 1900. It was one of the few locomotives that bore a name on the GER and it was CLAUD HAMILTON with the appropriate running number 1900. Much larger than any previous Great Eastern 4-4-0 with boilers 4′ 9″ diameter, round top fireboxes, new style cab with side windows. The inside cylinders were 19″ x 26″ and the coupled wheels 7 ft diameter. Heating surface was 1630·5 sq. ft and grate area 21·3. Total weight in working order 50 tons 6 cwt. The running numbers 'descended' the next two batches being 1890–99, then 1880–89 and so on. From 1903 they were built with Belpaire fireboxes and altogether a total of 111 *Claud Hamilton*'s were built from 1900–1911, ten more were built in 1923 after the amalgamation.

Fig. 119 **LNE** (Stratford) 1923 'Super Claud' *Class D16* No. 1784E later 8784

Few classes can have had so many alterations, modifications and rebuildings, details of which can be found in other books. The principal alterations were the addition of Robinson type superheaters, and extended smokeboxes.

They were used on all express duties and performed them with vigour and dash and were the backbone of the passenger locomotives. No. 1900 was sent over to the Paris Exhibition of 1900 and won a gold medal for excellent workmanship.

His most startling design was for the well-known DECAPOD, a 0-10-0WT. Its inception was caused by the GER being threatened by two Bills to be presented to Parliament for the building of electrified railways from the City of London to its north east suburbs to provide faster transport for the ever growing passenger traffic to the dormitory towns. Holden took up the challenge by putting forward a locomotive which would accelerate a train to reach a speed of 30 mph in 30 secs. from a dead stand.

A description of this remarkable machine should be of interest. It had three cylinders 18½″ x 24″, the centre cylinder placed a short distance in front of the two outside ones. The outside connecting rods were to the third pair of wheels which were 4′ 6″ diameter and were flangeless. The inside connecting rod was to the second axle and to clear the leading axle the rod was triangular in shape and surrounded the axle which was cranked to give sufficient clearance, so arranged that the cranked portion was phased with the rise and fall of the connecting rod, which was virtually solid for half its length but had three lightening holes. The cranks were set at 120°.

The LH cylinder had its own steam chest but the middle and RH cylinders had a common steam chest with valves working back to back all operated by Stephenson's valve gear, with screw reverse, the three sets operating off the third axle.

The boiler was 5′ 3″ diameter x 15′ 10⅞″ between tubeplates with 395 – 1¾″ outside diameter tubes giving 2878·3 sq. ft of heating surface. The firebox was of the 'Wootten' type, shallow and wide and resting on the main frames. The firebox heating surface was 131·2 sq. ft giving a total of 3010 sq. ft and a grate area of 42 sq. ft. The boiler pressure of 200 lb/sq. in. was the highest that Holden had used. Four 3½″ Ramsbottom safety valves were fitted over the firebox, and four more on the first ring of the boiler.

The smokebox had two lower doors to assist in ash removal. The chimney was only 8″ high. The well tanks were under the bunker, and in front of the firebox between the frames with an equalising pipe between them and the tanks were filled through the bunker. The wheel base was 19′ 8″ and sanding was operated by compressed air. The cost of the locomotive was approximately £5000.

This locomotive did indeed do what it was designed to do, but unfortunately the track and bridges could not stand up to the weight and stresses. However, it was instrumental in defeating the two Bills.

In its original form it did little work and in 1906 was rebuilt as a 0-8-0 tender locomotive and was shedded at March, but not showing any startling improvement on the current 0-6-0s, it was cut up in 1913. So ended a remarkable experiment which on strengthened modern track might have achieved some startling results.

Regarding the 2-4-0s Holden built, the most important were the T.19s and the T.26s.

The T.19s had 7 ft driving wheels and 18″ x 24″ cylinders, and the T.26s 17½″ x 24″ cylinders and 5′ 8″ diameter driving wheels. Both types had the same boiler, outside frames to the leading wheels only.

As the diameter of the driving wheels indicate, the T.19s were for main line passenger traffic whereas the T.26 or 'Intermediates' as they were known were for secondary passenger traffic and fitted goods trains.

Holden had certainly made his mark at Stratford. The locomotives were well built, robust in design and usually well maintained and on his retiring in 1907 he left behind efficient workshops and an excellent spirit. Holden was succeeded by his son, Stephen Dewar Holden who took up office in 1908.

He continued to build his father's main designs, but in 1909 designed a new 2-4-2T, to replace the older 0-6-0Ts on the branch lines. They were necessarily light locomotives weighing 45 ton 14 cwt in working order with 15″ x 22″ cylinders and 4′ 10″ diameter coupled wheels. Twelve were built, but were not as popular as the 0-6-0Ts they replaced. They were given the name CRYSTAL PALACES due to the large overall cabs with side windows and large shaped spectacles front and rear.

Holden Junior's 4-6-0 brought out in December 1911 (No. 1500) was a logical development of the *Claud* 4-4-0s with two inside cylinders 20″ x 28″ and coupled wheels 6′ 6″ diameter. Fifty-one were built at Stratford up to 1920 and all were fitted with Belpaire boilers with superheaters with a total heating surface of 1919 sq. ft.

They were finished in the same manner as the *Clauds* and were excellent performers on the Hook of Holland boat trains and other main line trains. Most routes were bedevilled with

STRATFORD – GE RLY

Date	Type	Class	Cylinders inches	D.W. ft in	Heating Surface – sq ft				Grate Area sq ft	WP lb/sq in	Weight WO ton cwt	LNE Class	
					Tubes	S Heat	F'Box	Total					
S. W. JOHNSON													
1869	2-4-0	L.7	16 x 22	5 7		–				140	29 3	–	
MASSEY BROMLEY													
1878	0-4-4T	E.10	16½ x 22	4 10		–		983.21		140	44 16	–	
T. W. WORSDELL													
1882	2-4-0	G.14	18 x 24	7 0		–		1122.8		140	41 3	–	
1883	0-6-0	Y.14	17½ x 24	4 11	1055.1	–	105.5	1160.6	18.0	160	37 2	J.15	
1884	2-4-2T	M.15	17½ x 24	5 4	1018	–	98.4	1116.4	15.3	160	51 10	F.4	
1884	4-4-0	G.16	18/26 x 24	7 0		–		1200		160	44 10	–	
J. HOLDEN													
1886	0-6-0T	T.18	16½ x 22	4 0	889.2	–	78.0	967.2	12.4	140	40 5	J.66	
1886	2-4-0	T.19	18 x 24	7 0	1098.6	–	100.9	1199.5	18.0	160	42 0	–	(a)
1888	0-6-0T	E.22	14 x 20	4 0	844.6	–	77.5	922.1	12.4	140	36 10	J.65	
1890	0-6-0T	R.24	16½ x 22	4 0	881.2	–	78.0	959.2	12.4	140	40 0	J.67	
1891	2-4-0	T.26	17½ x 24	5 8	1098.6	–	100.9	1199.5	18.0	160	40 6	E.4	
1893	0.6.0	N.31	17½ x 24	4 11	1116.18	–	100.9	1217.08	18.0	160	38 18	J.14	
1893	2-4-2T	C.32	17½ x 24	5 8	1098.6	–	100.9	1199.5	18.0	160	58 12	F.3	
1898	4-2-2	P.43	18 x 26	7 0	1178.5	–	114.2	1292.73	21.3	160	50 0	–	
1898	0-4-4T	S.44	17 x 24	4 11	989.1	–	94.9	1084	15.3	160	53 7	G.4	
1900	4-4-0	S.46	19 x 26	7 0	1516.5	–	114	1630.5	21.3	180	50 8	D.14	
1900	0-6-0	F.48	19 x 26	4 11	1516.5	–	114	1630.5	21.3	180	44 11	J.16	
1902	0-10-0WT	A.55	3/18½ x 24	4 6	2878.3	–	131.7	3010	42.0	200	80 0	–	
1903	4-4-0	D.56	19 x 26	7 0	1588.9	–	117.7	1706.6	21.6	180	51 12	D.15	
1905	0-6-0	G.58	19 x 26	4 11	1157.4	226	117.7	1501.1	21.6	180	45 7	J.17	

S. D. HOLDEN

Date	Type	Class	Cylinders inches	D.W. ft in	Heating Surface – sq ft Tubes	S Heat	F'Box	Total	Grate Area sq ft	WP lb/sq in	Weight WO ton cwt	LNE Class
1909	2-4-2T	Y.65	15 x 22	4 10	797.2	—	75.7	872.9	12.2	160	45 13	F.7
1911	4-6-0	S.69	20 x 28	6 6	1489.1	286.4	143.5	1919	26.5	180	62 19	B12/1

A. J. HILL

Date	Type	Class	Cylinders inches	D.W. ft in	Heating Surface – sq ft Tubes	S Heat	F'Box	Total	Grate Area sq ft	WP lb/sq in	Weight WO ton cwt	LNE Class	
1912	0-6-0	E.72	20 x 28	4 11	1157.4	154.8	117.7	1429.9	21.6	160	47 7	J.18	
1913	0-4-0T	B.74	17 x 20	3 10		—		980.5	13.9	180	38 1	Y.4	
1914	0-6-2T	L.77	18 x 24	4 10	1076.3	102.4	113	1291.7	17.7	180	61 12	N.7	
1914	4-4-0		19 x 26	7 0	1157.4	154.8	117.7	1429.9	21.6	180	52 4	D.15	(b)
1920	0-6-0	D.81	20 x 28	4 11	1489.1	201.6	143.5	1834.2	26.5	180	54 15	J.20	

(a) Rebuilt as 4-4-0 LNE Class D13. (b) Robinson Superheater

numerous speed restrictions and some bad gradients and these locomotives acquitted themselves well under these arduous conditions.

In 1912, S. D. Holden was succeeded by A. J. Hill, first as acting locomotive superintendent in October and November 1912, when his appointment was confirmed and in 1915 his title was altered to chief mechanical engineer. A new repair shop was built at the beginning of the First World War 450′ long by 150′ wide in three bays – subsequently converted to diesel repairs in 1957.

Hill's first design was for a superheated 0-6-0 class E72 with the same boiler as the *Claud* 4-4-0s. Cylinders were 20″ x 28″ with 10″ piston valves and Stephenson's motion. Thirty-five were built followed by the ultimate, as far as GER 0-6-0s were concerned. The D81 class of which twenty-five were built from 1920 to 1922 had the same cylinders as the E72s, the same boiler as the 4-6-0s with a pressure of 180 lb/sq. in. The wheels were the standard 4′ 11″ diameter. They were the most powerful 0-6-0s in the country with a tractive effort of 29,044 lb and remained so, until the advent of Bulleid's Q1 class in 1942.

In 1915 he introduced a 0-6-2T locomotive for London suburban traffic. Their 4′ 10″ coupled wheels gave them rapid acceleration. The first was fitted with a saturated boiler and the second one was superheated. These two ran comparative trials for a long period, and the result was that later locomotives were fitted with superheated boilers. An unusual feature was the Walschaerts valve gear used instead of Stephenson's motion, inside the frames.

For shunting around the North London Docks, Hill designed a very powerful 0-4-0T, with a short wheelbase of 6′ 0″ to enable short radius curves to be traversed. Five were built from 1913 to 1921, they had outside cylinders 17″ x 20″, 3′ 10″ diameter wheels, outside Walschaert's valve gear and boiler with Belpaire firebox. The heating surface was 980·5 sq. ft, and weighed 38 tons in working order.

At the amalgamation into the LONDON AND NORTH EASTERN RAILWAY in 1923 A. J. Hill retired and with that the Great Eastern Railway lost its identity but, as later recorded, some of the designs were perpetuated and more were built under LNER ownership speaking well of their design and first class workmanship.

At the grouping in 1923 the last ten 4-4-0s were laid down. They were GER Nos. 1780–1789 and were known as *Super Clauds*. The main difference was in boiler dimensions, the boiler having the same diameter as the GE 4-6-0s and only 9″ shorter tubes. The cylinders remained the same as previous batches, the larger steam capacity giving the improved performance with heavier loads and higher speeds. They had Belpaire fireboxes and Ross pop safety valves. Mr A. J. Hill gave way to Nigel Gresley. Gresley commenced rebuilding the *Clauds*, reverting to a round top firebox, modified splashers to give better accessibility, standard chimneys, some were fitted with long travel piston valves fitted over the cylinders in place of the previous slide valves which were under the cylinders. Cylinders were the same size but the steam and exhaust passages were enlarged. A marked improvement in performance was the result, and gave them an extra lease of life. No. 8900 (old No. 1900) retained its name CLAUD HAMILTON and when withdrawn in 1947 the name was transferred to No. 2546 (old No. 1855). The LNE renumbering scheme had been implemented in 1946. The LNE classification was D16/3 for these rebuilds.

The 1500 class 4-6-0s were also rebuilt in a similar fashion and a few had already been fitted with ACFI feed water heaters and although those so fitted showed a small saving in fuel, the heaters were not fitted to the Gresley rebuilds.

Building continued at Stratford and after the completion of the 4-4-0s, ten 0-6-0Ts were built in the same year. These were Holden's design with 16½" x 22" cylinders and 4 ft diameter wheels. They had tall cabs with side windows, dome near the chimney. They were equipped with steam brakes.

Ten further 0-6-2Ts were built similar to those built in 1921 by Mr Hill. They were approved by the LNE and 112 were built to a slightly modified design but none at Stratford. They were acknowledged as the most efficient type of locomotive designed by Mr Hill. These 0-6-2Ts were the last steam locomotives built at Stratford making a grand total of 1702 from 1851 to 1924.

Miscellaneous notes – GER At the end of 1922, as far as can be assessed 1672 locomotives had been built at Stratford. The early days of the Eastern Counties Railway had a complicated locomotive history not fully documented and it is difficult to get true totals for this period.

Mr S. W. Johnson introduced works order numbers comprising a letter and number and this practice was continued by his successors. Each batch of locomotives put in hand was allocated one of these order numbers and the first works order number of a series of similar locomotives was used as the locomotive class number. An example was an order for twenty-five 0-6-0s: Works Order No. D81—Loco Nos. 1270–4 (built 1920); Works Order No. M87—Loco Nos. 1275–84 (built 1922); Works Order No. Y87 – Loco Nos. 1285–94 (built 1922). The locomotive class would be D81.

Most large orders were dealt with in batches of ten at a time, and since order numbers included tenders, boilers and other work, there were gaps in the locomotive order numbers.

The Westinghouse Brake was used probably from Johnson's time and the pump was fitted to the right hand side of the locomotive.

When locomotives were placed on the duplicate list the number had a nought placed in front.

Fig. 120 1924 A. J. Hill *Class N7* 0-6-2T condensing No. 995E later 7995

EBBW VALE STEEL IRON & COAL COMPANY

Eight locomotives built between 1905 and 1917. They were all 0-4-0STs with outside cylinders.

1905	HOLLEY	No. 20
1907	NASMYTH	No. 22
1909	SIEMENS	No. 25
1910	WHITWORTH	No. 17
1911*	TREFIL	No. 18
1912	CWM-TA-FECHAN	No. 26
1914	TYLLWYN	No. 27
1917	ECONOMY	No. 19

* rebuilt from Peckett 969/1904.

Some of the above locomotives including NASMYTH and SIEMENS were built from parts supplied by Peckett & Sons of Bristol.

EDINBURGH & GLASGOW RAILWAY

Cowlairs, Glasgow

The railway was opened for traffic in 1842 and locomotives were ordered from R. & W. Hawthorn and Bury. The workshops were built in 1841 in readiness for repair work, and was one of the earliest railway workshops established.

William Paton was appointed locomotive superintendent. The first two locomotives were built in 1844 for working the Cowlairs incline in an endeavour to eliminate rope haulage which had become extremely troublesome.

They were both six coupled well tanks with 4′ 3½″ diameter wheels. No. 21 HERCULES had 15½″ x 25″ inclined outside cylinders and No. 22 SAMSON had 16½″ x 25″ cylinders, but otherwise they were identical. The boilers had large ornamental domes, steam brakes were fitted and water jets for cleaning the greasy rails in the tunnel. They were very powerful but unfortunately played havoc with the track and caused concern when passing through the tunnel, the sharp exhaust having a bad effect on the brickwork. After various trials, with these two, and two 0-6-0s built by R. & W. Hawthorn, rope haulage had to be re-instated.

In 1848 Paton designed and built two 2-2-2 passenger locomotives, No. 52 ORION and No. 58 SIRIUS. They had 6 ft diameter driving wheels and 15″ x 20″ cylinders and had Gooch's valve gear fitted. They were used on the main line expresses.

Fig. 121 **E&GR** 1862 ex No. 101 as NBR 2-4-0 No. 354

EDINBURGH & GLASGOW RAILWAY

Date	Type	Cylinders inches	D.W. ft in	E & G No	Name	1st NBR No	W'drawn
1844	0-6-0WT	15½ x 25	4 3½	21	Hercules		
1844	0-6-0WT	16½ x 25	4 3½	22	Samson		
1848	2-2-2	15 x 20	6 0	58	Sirius	223	1868
1848	2-2-2	15 x 20	6 0	59	Orion	224	1871
1850	0-4-0						
1850	0-4-0T						
1861	2-4-0	16 x 22	6 0	101		351	1915
1861	2-4-0	16 x 22	6 0	102		352	1915
1862	2-4-0	16 x 22	6 0	103		353	1912
1862	2-4-0	16 x 22	6 0	104		354	1913
1864	2-4-0	16 x 22	6 0	105		355	1914
1864	2-4-0	16 x 22	6 0	106		356	1912
1864	0-4-2	16 x 22	5 0	83		329	1912
1864	0-4-2	16 x 22	5 0	84		330	1913
1864	0-4-2	16 x 22	5 0	85		331	1913
1864	0-4-2	16 x 22	5 0	86		332	1914
1864	0-4-2	16 x 22	5 0	87		333	1912
1864	0-4-2	16 x 22	5 0	88		334	1912
1865	2-4-0	16 x 22	6 0	35		236	1910
1865	2-4-0	16 x 22	6 0	39		235	1914
1865	2-4-0	16 x 22	6 0	12		239	1911
1866	0-4-2	15 x 22	5 0	—		262	1907

Absorbed by the North British Rly July 1865

In 1850 two more banking locomotives were built but this time they were four coupled, one a tank and the other a tender. Even these lighter types were not a success but their exploits are shrouded in mystery and it is probable that the tender version was not tried at all. Six 0-4-2 goods locomotives were built following the same pattern as those supplied by Beyer Peacock & Co.

For passenger traffic a series of double framed 2-4-0s were built from 1861, the year that William Steel Brown took over from William Paton. Paton incidentally in 1845 was found guilty of culpable homicide after standing trial for using one of his Bury locomotives in an unfit state. It was trying to haul a 'special', with intermittent stops due to loss of steam when it was struck in the rear by another train and the passenger of the 'special' was killed. Paton was sentenced to twelve months imprisonment, after which he resumed his post at Cowlairs.

The 2-4-0s had 16" x 22" cylinders and 6 ft coupled wheels and building went on after Samuel W. Johnson had replaced Mr Brown who had left at the end of 1863.

In 1865 the Edinburgh and Glasgow became part of the NORTH BRITISH RAILWAY.

Up to this time approximately 22 locomotives had been built at Cowlairs.

North British Railway The headquarters of the locomotive department of the NBR had been at St Margarets, Edinburgh. (See NORTH BRITISH RAILWAY—ST. MARGARETS). In 1866 Cowlairs was made the principal locomotive works, there being more room for expansion.

No more new locomotives were built at St. Margarets but repairs to locomotive stock continued.

S. W. Johnson was appointed locomotive superintendent at the amalgamation but his increased responsibility was short lived and in the same year he obtained a similar post on the Great Eastern Railway.

To take his place came Thomas Wheatley who had been trained on the Leeds and Selby railway, and gaining experience on the MS&LR and the LNWR where he had been locomotive superintendent (Southern Division). The following were the successive locomotive superintendents: 1875–1882 Dugald Drummond; 1882–1903 Matthew Holmes; 1903–1919 William Paton Reid; and 1920–1922 Walter Chalmers.

Thomas Wheatley had the job of sorting out the varied collection of inherited locomotives. The Edinburgh and Glasgow stock was the best, but the older NBR locomotives which had been built and maintained at St. Margarets were a motley group.

Cowlairs works were reorganised and he set about designing and building larger and more powerful machines to meet the traffic requirements.

At the time of the amalgamation Cowlairs had turned out over 20 new locomotives.

T. Wheatley's first locomotives were two 0-4-0 goods, with four wheeled tenders built in 1868 and numbered 357 and 358. They had domeless boilers, 15″ x 24″ cylinders and 5′ 1″ diameter wheels. Some parts were used off scrapped locomotives and No. 358 lasted until LNER days and was not condemned until 1925.

In 1870 two 2-4-0s were built with 6′ 6″ diameter coupled wheels and 16″ x 24″ cylinders. Domeless boilers were fitted. 1871 marked an important year in locomotive construction when the first two inside frames, inside cylindered 4-4-0s were built at Cowlairs. Cylinders were 17″ x 24″, coupled wheels 6′ 6″ diameter. The bogie had 2′ 9″ diameter wheels with solid centres and the driving wheel splashers were slotted. A domeless boiler was fitted and the combined dome and safety valve placed over the round top firebox. They were numbered 224 and 264. No. 224 was the locomotive involved in the Tay Bridge disaster and was later salvaged and ran in various forms until 1919. Four more were built in 1873 with various modifications including a domed boiler.

For secondary services eight more 6 ft 2-4-0s were built with 16″ x 22″ cylinders, and for goods traffic he designed a 5 ft 0-6-0 with 17″ x 24″ cylinders. Eighty six were built from 1869 to 1873 of which sixty two were built at Cowlairs.

For coal traffic a 4 ft version was built with 16″ x 24″ cylinders, thirty-seven were built and as was customary on many railways some were rebuilt as saddle tanks, as required, for shunting.

An unusual requirement made it necessary for six 0-6-0STs of small dimensions to be built in 1874 for moving wagons and other vehicles off train ferries. They had 13″ x 18″ cylinders and 3 ft diameter wheels, except one isolated example which had 4′ 3″ diameter wheels.

Thomas Wheatley left the service of the North British Railway at the end of 1874. He had done some fine work at Cowlairs and had endeavoured to create some sort of standardisation, but his chief claim to fame was his 4-4-0s.

William Stroudley had already left the NBR in 1865 for the Highland Railway and in 1869 was appointed locomotive superintendent of the London Brighton & South Coast Railway.

Dugald Drummond had served under Stroudley at Cowlairs, followed him to the Highland Railway and to Brighton where Drummond was appointed works manager. In 1875

Fig. 122 1877 originally No. 151 GUARDBRIDGE running as LNE 10330 *Class J82*

he left Stroudley and returned to Cowlairs as locomotive superintendent. Much has been written about Dugald Drummond, suffice it to say that although he had a somewhat ferocious temperament he was strictly fair and had the good of the department at heart.

Having settled in and made his presence felt he first designed a small passenger tank locomotive bearing most of the characteristics of Stroudley's 'Terriers' of Brighton. They were larger, with 15″ x 22″ cylinders and 4′ 6″ diameter wheels. The side tanks had rounded top edges and the same sort of feedwater heating and pump was installed. The cylinders had short direct steam ports and double exhaust ports one at each end of the cylinders. The dome mounted on the boiler had a pair of Ramsbottom's safety valves on top encased in a copper casing – with the easing lever omitted. The total heating surface was 664 sq. ft with a grate area of 13·3 sq. ft and a boiler pressure, standard at this time, of 140 lb/sq. in. The chimney was squatter than the graceful 'Terrier' chimney. Side wings to the front of the smokebox were used and in addition Drummond adopted the practice of naming his locomotives, with large gold transfer letters, after local places where the particular class might work. Twenty-five of these useful 0-6-0Ts were built from 1875 to 1878.

Drummond introduced the Westinghouse brake as standard due principally to the opening of the Settle – Carlisle section of the Midland Railway in 1876, the Midland using this type of brake on all its Scottish expresses. Trials were carried out with Smith's vacuum brake in 1876 but the Westinghouse brake was preferred.

In 1876–7 a batch of 0-6-0s on the lines of the LB&SCR's Class C was built, with 18″ x 26″ cylinders and 5 ft diameter wheels. Pressure was increased to 150 lb/sq. in. and Stephenson's link motion used.

His first 4-4-0 passenger design came out in 1876 and 1878 with a two ring boiler giving 1099·3 sq. ft of heating surface, and a grate area of 21·0 sq. ft. The inside cylinders were 18″ x 26″ and coupled wheels 6′ 6″ diameter. Cowlairs built four and Neilson's eight.

One of these locomotives No. 493 NETHERBY was sent to Newcastle in 1881 for the Stephenson centenary exhibition. These were the ancestors of the numerous 4-4-0s Drummond designed for the Caledonian Railway. These 1876, 4-4-0s were badly needed to replace the 2-4-0s on the heavier expresses, and in 1876 they were probably the largest and most powerful passenger locomotive in the British Isles. They worked over the 'Waverley' route, which was by no means an easy road, with great success.

Six 0-4-2Ts were built in 1877 similar to Stroudley's 'D' class but with 5′ 9″ diameter instead of 5′ 6″ diameter coupled wheels and in 1880 he introduced the 4-4-0Ts which were used for suburban traffic. Twenty-four were built at Cowlairs from 1880–1882. The cylinders were 16″ x 22″ and coupled wheels 5 ft diameter. A small domed boiler gave a heating surface of 647·46 sq. ft. Although the first was built at Cowlairs, Drummond had three larger locomotives of the same wheel arrangement built by Neilsons in 1879 which were used on the fast business trains on the Clyde Coast line to Glasgow.

His last design was for a smaller 0-6-0 Class D initially with 17″ x 24″ cylinders, but later ones had 17″ x 26″ fitted. The wheels were 5 ft diameter and the boiler gave 1061 sq. ft of heating surface with a grate area of 16·5 sq. ft as compared with his larger Class C locomotives with 1235 sq. ft and 17 sq. ft respectively.

One hundred and one were built, and all but five at Cowlairs from 1879 to 1883.

These were not so limited in route availability, and found their way to most parts of the system.

In June 1882 Drummond resigned on receiving an offer from the Caledonian Railway, which he accepted, to fill the post of locomotive superintendent vacated by George Brittain.

Although his early NBR locomotives were based on Stroudley's principles his 4-4-0s bore no resemblance to Brighton practice. His comparative short stay at Cowlairs did not enable him to really establish a full complement of designs, but what he did build were good and solid.

He was succeeded by Matthew Holmes who had been on the NBR since 1859 and was fully conversant with all the problems of the locomotive department.

Holmes carried on following the trends set by Drummond apart from superficial modifications. The 17″ 0-6-0s and the 4-4-0T classes were added to, but in 1884 Holmes built six small 4-4-0s and in 1886–1888 twelve 7 ft 4-4-0s with 18″ x 26″ cylinders (Nos. 592–603) in preparation for the opening of the second Tay Bridge. This was not completed until 1887 and through workings were introduced between Burntisland and Aberdeen via Dundee. No. 592 was shown at the Edinburgh Exhibition of 1886.

To the detriment of the footplate crew Holmes introduced a rounded cab with a foreshortened cab roof, very similar to Patrick Stirling's cabs on the Great Northern Railway.

Twelve 0-4-4Ts were built, six in 1886 and six in 1888, and were very similar to Drummond's class of 0-4-2Ts which were subsequently rebuilt as 0-4-4Ts.

The Forth Bridge having opened in 1890 accelerated services were expected from Edinburgh to the north and for this work a batch of twelve 4-4-0s were built in 1890 with 18″ x 26″ cylinders and 6′ 6″ diameter coupled wheels. The boiler gave a total heating surface of 1266 sq. ft, with a grate area of 20 sq. ft, the working pressure being the standard 140 lb/sq. in. Twelve more were added to the class in 1894–5 but with an increased boiler pressure of 150 lb/sq. in. In the famous 1895 'Race to the North' these 4-4-0s were used from Edinburgh to Aberdeen.

Another class of 4-4-0 built specially for the West Highland Railway came out in 1893–1896. They had 5′ 7″ diameter coupled wheels and a slightly smaller boiler was fitted.

To keep up with the increasing goods traffic, Holmes built a variety of 0-6-0s. The 17″ x 26″ with 5 ft wheels was a development of Drummond's smaller 0-6-0s and the 18″ x 26″ with the same diameter wheels became the most numerous class on the NBR. Like all Holmes's designs the grate was level and not sloping like Drummond's. This class was built from 1888 to 1900 to a total of 168 locomotives. Except for fifteen built by Neilson & Co. in 1891, and fifteen by Sharp Stewart & Co. in 1892 all were built at Cowlairs. As built they had the usual combined dome and lock up safety valves on the boiler and round cab. Twenty-five of this class were sent overseas during the 1914–1918 war and on their return were given appropriate names such as YPRES, OLE BILL and so on. The names were put on the centre splashers by transfer.

Holmes final 4-4-0 design came out in 1903. Twelve locomotives were built (Nos. 317–328) with inside cylinders 19″ x 26″, piston valves and Stephenson's link motion. The boiler had a total heating surface of 1577 sq. ft and working pressure 200 lb/sq. in. The coupled wheels were 6′ 6″ diameter and the weight in working order 52 tons. They were fitted with steam reversing gear and a new type of cab which incorporated a side window. Unfortunately before the class was completed Matthew Holmes died in July 1903.

Holmes had carried on the work of Drummond and although some of his earlier designs were not entirely successful, he had built up the stock of 4-4-0s and 0-6-0s of sound basic design. He discontinued the practice of naming locomotives, and when Drummond's classes

Fig. 123 1887 Holmes 7′ 4-4-0 No. 595 rebuilt by Reid in 1911 shown as LNE 9595

Fig. 124 1906 Reid 5′ 0-6-0 No. 330

were rebuilt – and considerable rebuilding was done, with standardised boilers – they lost their names.

William Paton Reid took his place in the same year. He was a NBR man, having been locomotive foreman at St. Margaret's works and then Holmes' assistant. 'Paton' was a family name, as William Paton previously mentioned was a relation.

He followed the designs of Holmes, but due to increased loadings his locomotives were necessarily larger.

The first locomotives built under his direction were of Holmes' design, consisting of twelve six coupled dock shunters with 3′ 9″ diameter wheels, outside cylinders 15″ x 22″ and weighing 36 tons 5 cwt in working order. The class eventually grew to a total of thirty-five and building was carried out from 1905 to 1919.

Although not built at Cowlairs, mention must be made of Reid's express passenger 4-4-2 design. These massive locomotives were of course unlike anything that had appeared on NBR metals or any other Scottish line. The boiler was 5′ 6″ diameter, and had a Belpaire firebox. The total heating surface was 2256·2 sq. ft and the grate area 28·5 sq. ft. Two outside cylinders 20″ x 28″ were fitted with piston valves and reversing was by steam. Three safety valves in triangular formation were fitted on the firebox. The cab was large with two windows each side. Fourteen were built by the North British Locomotive Co. in 1906 and Robert Stephenson & Co. supplied another six in 1910. The practice of naming locomotives was revived in the case of the Atlantics and all passenger locomotives subsequently built were given names, but what a difference in names – much more in keeping with the character of the locomotives and the railway they ran on – THANE OF FIFE, BORDERER and COCK O' THE NORTH stir the imagination far more than CLYDEBANK, BONNYBRIDGE or GRETNA!

They immediately took over the major expresses from Carlisle, Edinburgh and Aberdeen. The boiler size ensured plenty of steam in all conditions but were very heavy on fuel. This was amply demonstrated when No. 881 BORDERER competing with a LNWR *Experiment* class locomotive REDGAUNTLET consumed no less than 71 lb coal/mile as an average figure on the Preston to Carlisle section. After the *Atlantics* were superheated the coal consumption dropped – but only to about 60 lb/mile which was still excessive.

In 1906 a new class of 0-6-0 was developed from the Holmes 19″ goods. Cylinders 18½″ x 26″ were fitted and the wheels kept at the standard 5 ft diameter. A much larger boiler was fitted with a total heating surface of 1748 sq. ft and a grate area of 19·8 sq. ft. The pressure was 180 lb/sq. in. A few were fitted with piston valves, and all had Stephenson's valve gear. Seventy six were built from 1906 to 1913 of which thirty-six came from the Cowlair works.

In 1906–7 a new 6 ft class of 4-4-0s were built Nos. 882–893. They were intended for mixed traffic working, especially fitted goods trains carrying fish and other perishable goods to the markets. They had 19″ x 26″ cylinders with piston valves. The boiler had a total heating surface of 1760 sq. ft and a grate area of 22·5 sq. ft and a working pressure of 180 lb/sq. in. They weighed 53 tons without tender. They were very good locomotives on the services for which they were designed and were often pressed into service for excursion and semi-fast trains. No more were built and they were not named.

As loads increased, more and more 4-4-0s were needed and two classes were introduced both being developments of the 1906 mixed traffic class. Both came out in 1909, the first was a 6′ 6″ class, the 1909 batch of six being built by the North British Locomotive Co., and ten more at Cowlairs in 1911. Cylinders were 19″ x 26″ with piston valves and the boiler was smaller giving a total heating surface of 1618·12 sq. ft and a grate area 21·13 sq. ft. They were known as the *Scott* class being named after Sir Walter Scott's characters. A much larger tender was fitted, holding 4235 gallons of water as compared with 3525 gallons for the mixed traffic locomotives, to enable them to work non-stop from Edinburgh to Carlisle. In 1911 No. 897 REDGAUNTLET was fitted with a Phoenix smokebox superheater but it was removed after a short time.

The second class was a 6 ft version of the *Scott* class, with the same boiler and cylinders. The large capacity tender was fitted and the class known as 'Intermediates'. They weighed 15 cwt lighter than the *Scott's*, turning the scales at 54 tons 1 cwt. They were more widely dispersed over the system. This class was not named. Twelve were built 1909–10.

In 1912 the two last locomotives laid down on the *Scott* design – the original order was for twelve at Cowlairs and only ten were built as *Scott's*, the two remaining were fitted with 20″ x 26″ cylinders with piston valves, but above the cylinders in this case and operated by rocking shafts. The boilers were fitted with Schmidt superheaters but subsequent locomotives of this class were equipped with Robinson superheaters and twenty-five were built between 1914 and 1920. The coupled wheels were 6′ 6″ diameter. They took their part in the workings operated by the *Scott* class.

This class also bore names relating to Scott's novels.

The final development of the NBR 4-4-0s came in 1913 in the form of a superheated version of the 1909 *Intermediates* with increased

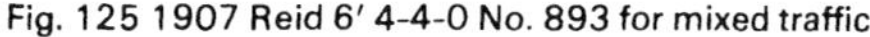
Fig. 125 1907 Reid 6′ 4-4-0 No. 893 for mixed traffic

size cylinders 20″ x 26″. The weight in working order was 57 tons 4 cwt and they were named after Scottish glens. Many worked over the West Highland section which was a difficult line and they stood up to this hard work extremely well.

A new class of 0-6-0 was designed by Reid, the building commencing in 1914. The cylinders were 19½″ x 26″ with piston valves and Stephenson's link motion. The boiler was superheated working at a pressure of 165 lb, and was pitched 8′ 6½″ from rail level which gave an impression of great hauling power, and they coped extremely well with the heavy war-time goods trains. One hundred and four were built, Cowlairs turning out thirty-five from 1914 to 1921.

Reid introduced various other classes which do not come within the scope of this section. Two classes of 4-4-2Ts the first appearing in 1911, thirty being built, all by the Yorkshire Engine Co., 1911–13, then a superheated version comprising twenty-one locomotives all by the North British Locomotive Co. 1915–21, who also built Reid's twelve 0-4-4Ts in 1909.

In 1919 William Reid retired and had been one of the chief architects in building a modern fleet of locomotives. He introduced superheating, the large locomotive with the large boiler and modernised all the older locomotives which were worth re-boilering and rebuilding.

He was succeeded by his chief draughtsman Walter Chalmers in January 1920, he was solely a NBR man having spent all his life at Cowlairs. His new title was Chief Mechanical Engineer and it was unfortunate that with the impending amalgamation he had no encourage-

NORTH BRITISH RAILWAY

DUGALD DRUMMOND

LNE Class	First Built	No Built	Type	Cylinders inches	D.W. ft in	H. Surface sq ft	Grate Area sq ft	WP lb/sq in	Weight WO ton cwt	Notes
J32	1876	12	0-6-0	18 x 26	5 0	1099.3	21.0	150	39 10	(a)
D27	1877	4	4-4-0	18 x 26	6 6	1099.3	21.0	150	44 5	(g)
D51	1880	30	4-4-0T	16 x 22	5 0	647.5	14.5	140	35 4	
G8	1877	6	0-4-4T	17 x 24	5 9	1075.0	15.9	150	47 17	

MATTHEW HOLMES

LNE Class	First Built	No Built	Type	Cylinders inches	D.W. ft in	H. Surface sq ft	Grate Area sq ft	WP lb/sq in	Weight WO ton cwt	Notes
J36	1888	168	0-6-0	18 x 26	5 0	1235.13	17.0	150	40 4	(b)
D25	1886	12	4-4-0	18 x 26	7 0	1102.0	21.0	150	45 5	
D26	1903	12	4-4-0	19 x 26	6 6	1577.0	22.5	200	52 0	
G7	1886	12	0-4-4T	17 x 24	5 9	1059.0	17.0	150	49 12	

WILLIAM P. REID

LNE Class	First Built	No Built	Type	Cylinders inches	D.W. ft in	H. Surface sq ft	Grate Area sq ft	WP lb/sq in	Weight WO ton cwt	Notes
J37	1914	104	0-6-0	19½ x 26	5 0	1732.22	19.8	180	54 14	(c) (S) PV
D29	1909	10	4-4-0	19 x 26	6 6	1618.12	21.13	190	54 16	(d) (S) PV
D34	1913	32	4-4-0	20 x 26	6 0	1346.06	21.13	165	57 4	(S) PV
C15	1911	30	4-4-2T	18 x 26	5 9	1309.0	16.6	175	68 16	(e)
C16	1915	21	4-4-2T	19 x 26	5 9	1245.0	16.6	165	72 10	(f) (S)
C9	1909	12	0-4-4T	18 x 26	5 9	1309.0	16.6	175	58 6	(f)

(a) 20 Built by Neilson & Co. in addition. (b) Including 15 each by Neilson & Co., and Sharp Stewart & Co. (c) Including 69 by NB Locomotive Co. (d) 6 Built by NB Locomotive Co. in addition. (e) All built by Yorkshire Engine Co. (f) All built by NB Locomotive Co. (g) Built by Neilson & Co.

(S) : Superheated. PV : Piston Valves,

The North British Railway had no loco. classification, but used a load Class A to G and S for goods, and H to R for passenger locomotives in descending order of power. The scheme was introduced in 1913 when a new system of train control was put into effect lasting until about 1928.

ment to proceed very far with locomotive development. Consequently he concerned himself more with the continuance of rebuilding previous classes built by Holmes, with standard boilers. Two more *Atlantics* were built in 1920 with superheated boilers but this order was entrusted to the North British Locomotive Co.

Few new locomotives were built at Cowlairs during this difficult period. Five superheated *Scotts* appeared in 1920, and twelve *Glens*, in the same year, and in 1921 ten 0-6-0s of Reid's 19½" class commenced in 1914.

After the war Cowlairs was seriously in arrears with locomotive repairs and many NBR Locos were sent to Beardmores, North British Loco Co., Hawthorn Leslie & Co., Armstrong Whitworth & Co., and Vickers Ltd., Barrow-in-Furness.

The Railway Act was passed in 1921 and the North British Railway became part of the LONDON & NORTH EASTERN RAILWAY, as and from 1 January 1923.

Up to this time approx. 850 locomotives had been built at Cowlairs from 1842 to the amalgamation.

H. N. Gresley became the Chief Mechanical Engineer of the LNER and Walter Chalmers continued at Cowlairs as an Area Mechanical Engineer until 1924 when he retired. The whole of the Scottish area was then taken over by Mr R. A. Thom.

Very little new locomotive building took place at Cowlairs after the amalgamation. In 1923 twelve of Reid's 4′ 6″, 0-6-2Ts were built followed by eight more in 1924. Many of the previously built locomotives of this class were, used for banking purposes on the Cowlairs incline. The new batch were intended for shunting and goods traffic. They were classified N.15 (LNE) and had 18″ x 26″ cylinders, 4′ 6″ diameter coupled wheels and weighed 60 tons 18 cwt in working order.

Cowlairs then became a repair centre only and after the re-organisation of St Rollox works in the 1960s it was closed down.

Approximately 900 steam locomotives had been built from 1844 to 1924.

EDINBURGH, PERTH & DUNDEE RAILWAY

Burntisland

The workshops were established in 1847 by the Edinburgh and Northern Railway which was renamed as above in 1849.

Robert Nicholson was locomotive superintendent and like the North British Railway had a fleet of Hawthorn built locomotives to look after.

Building commenced in 1861 with a 2-2-2 express passenger locomotive with inside frames, followed by four 0-6-0s with 16″ x 24″ cylinders and 5 ft diameter wheels.

It is almost certain, with the minimum of equipment that was available at Burntisland – the railway had an agreement with the Scottish Central Railway for carrying out overhauls and rebuilding of EP&DR locomotives at their Perth workshops – that the locomotives which were recorded as being built at Burntisland must have used boilers, cylinders etc. from some of their older stock.

In 1862 the company became part of the North British Railway and only repair work was carried out until closure in September 1923.

EDINBURGH, PERTH & DUNDEE RLY BURNTISLAND

Date	EPD No	Type	Cylinders inches	D.W. ft in	NBR No	EP & DR. No	W'drawn
1861	37	2-2-2	15 x 20	6 0	147	37	1882
1861	40	0-6-0	16 x 24	5 0	150	40	1883
1862	35	0-6-0	16 x 24	5 0	145	35	1889
1862	(49)	0-6-0	16 x 24	5 0	159	(a) 49	1896
1863	(50)	0-6-0	16 x 24	5 0	160	(a) 50	1885

(a) May have been laid down as EP & DR Nos 49 & 50, but were completed after the North British Rly took over.

EDINGTON, THOMAS & SONS

Phoenix Iron Works, Glasgow

This firm built four known locomotives in 1840–1 for the Glasgow, Paisley, Kilmarnock and Ayr Railway. They were all 2-2-2s built to the specifications of the railway company's engineer Mr J. Miller, and had inside cylinders and outside frames with driving wheels approx. 5′ 6″ diameter.

The boiler had a combined dome and safety valve fitted on the barrel and on the raised firebox, a second dome/safety valve assembly. Why two steam domes were fitted is not known. The boiler was supported by two plates fastened to the frames and to the boiler on each side and the smoke box was similarly supported, a common method of the period. Four wheeled tenders were supplied and were fitted with spring buffers whereas stuffed leather buffers were fixed to the locomotive buffer beam. They were named as follows: GPK&A No. 7 PHOENIX, GPK&A No. 8 PRINCE ALBERT, GPK&A No. 10 GARNOCK, and GPK&A No. 15. KYLE.

No. 8 was sold to the L&NWR in July 1847 and renamed DUNLOP.

ENGLAND GEO. & COMPANY

Hatcham Ironworks, Pomeroy Street, New Cross, London

The works were established in about 1839 and the main business at first was the production of screw jacks. Extensions were effected in 1853 and 1862 and the works were closed down in 1869 when they were taken over by the Fairlie Engine & Steam Carriage Co.

Test tracks were laid down in an adjacent field. About 250 locomotives were built including some to Fairlie's patent. George England had attended the trials of MOUNTAINEER a *Fairlie* built by James Cross of St. Helens in 1866, which resulted in a meeting with Robert Fairlie which developed into friendly relations and orders from Fairlie for his type of locomotive, and eventually resulted in the purchase of the business by Robert Fairlie so that he could build his own engines.

England exhibited at the International Exhibitions of 1851 and 1862.

Avonside Engine Company took over the Fairlie patent manufacturing rights in 1871.

England's first design was for a small 2-2-2WT with 9″ x 12″ outside cylinders and 4′ 6″ driving wheels, one of which was shown at the 1851 Exhibition resplendent in Prussian Blue – a colour which the makers painted most of the locomotives they built.

A large number of 2-4-0 tender engines and 2-4-0 tank engines were built. The early tender locomotives supplied to the Somerset Central

Fig. 126 **T. Edington** Built in the early 1840s for the GPK&AR PHOENIX

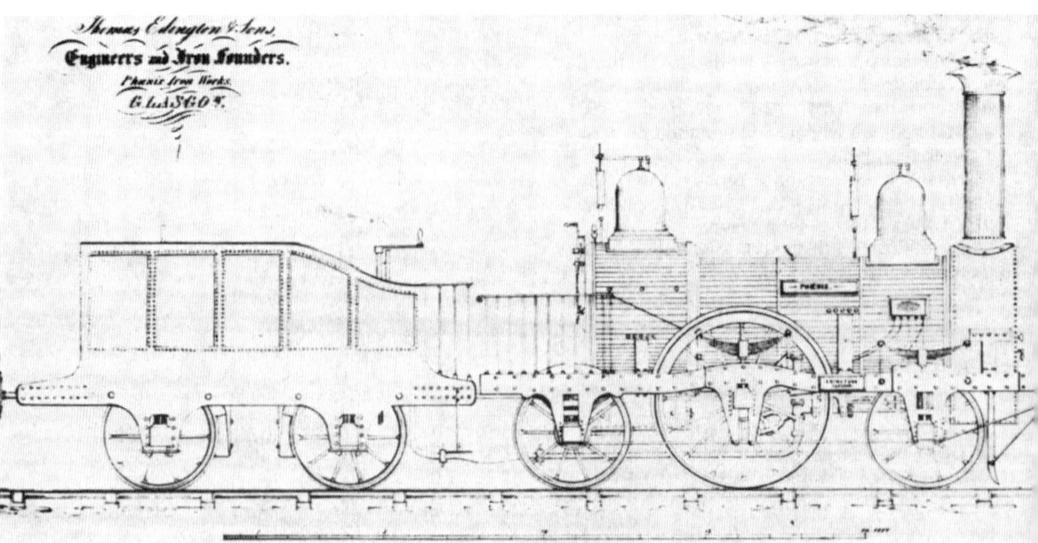

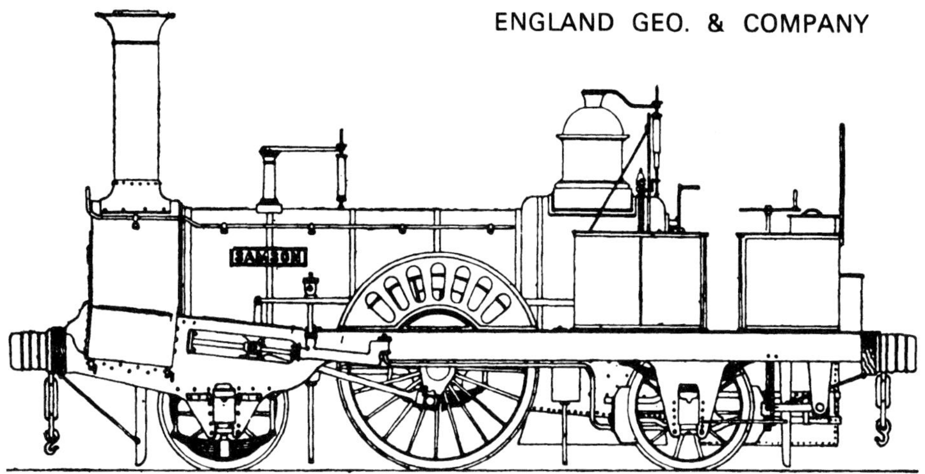

Fig. 127 **Geo England** 1852 2-2-2WT SAMSON London & Blackwall Rly

Railway were of this type with outside frames to the leading wheels and inside frames for the coupled wheels. They had domeless boilers, with the safety valves mounted over the round topped firebox. Others went to the Caledonian, London & South Western, Victorian Govt., and Belgian Railways. In 1862 eight 2-4-0s were built for the GWR, these had sandwich frames, inside cylinders 16″ x 24″ and 6′ 6″ coupled wheels, and equipped with Stephenson's link motion they were all subsequently rebuilt and most of them gave useful work for 57 to 58 years.

In 1863 appeared the first locomotives for the Festiniog Railway to the order of Charles Spooner who was general manager and engineer of the railway at that time. Nos. 1 and 2 were 0-4-0Ts and were followed by two more in 1864.

Charles Spooner was the champion of the narrow gauge system and despite severe criticism proved that 2 ft gauge working was quite practicable, decreasing the initial capital costs involved in laying the tracks, cost of rolling stock and maintenance.

Fig. 128 **Geo England** 1861 2-4-0T for Somerset & Dorset Rly No. 11 as L&SWR No. 21 SCOTT

ENGLAND, GEO. & COMPANY

Works No	Built	Type	Cylinders inches	D.W. ft in	Gauge	Customer	No/Name
	1849	2-2-2WT	IC 9 x 12	4 6	Std	D.P. & A.R	Eclipse (a)
	1849	2-2-2WT	IC 9 x 12	4 6	Std	London & Blackwall	Dwarf (a)
	1850	2-2-2WT	IC 9 x 12	4 6	Std	London & Blackwall	Pigmy Giant
	1850	2-2-2WT	IC 9 x 12	4 6	Std	Edinburgh & Glasgow	England
	1850	2-2-2WT	IC 9 x 12	4 6	Std	L.C. & S.R.	England (a)
	1850	2-2-2WT	IC 9 x 12	4 6	Std	Exhibition Loco	Little England
	1850	2-2-2WT	IC 9 x 12	4 6	Std	London & Blackwall	8. Samson
	1850	2-2-2WT	IC 9 x 12	4 6	Std	London & Blackwall	9. Hercules
	1852/3	2-4-0	OC 16 x 20	5 0	Std	Caledonian Rly	144 (b)
	1852/3	2-4-0	OC 16 x 20	5 0	Std	Caledonian Rly	145
	1852/3	2-4-0	OC 16 x 20	5 0	Std	Caledonian Rly	146
	1852/3	2-4-0	OC 16 x 20	5 0	Std	Caledonian Rly	147
	1852/3	2-4-0	OC 16 x 20	5 0	Std	Caledonian Rly	148
	1852/3	2-4-0	OC 16 x 20	5 0	Std	Caledonian Rly	149
	1852/3	2-4-0	OC 16 x 20	5 0	Std	Caledonian Rly	150
	1852/3	2-4-0	OC 16 x 20	5 0	Std	Caledonian Rly	182
	1852/3	2-4-0	OC 16 x 20	5 0	Std	Caledonian Rly	183
	1852/3	2-4-0	16 x 22	4 9	Std	D.P. & A.R.	13 Scorpion
	1852/3	2-4-0	16 x 22	4 9	Std	D.P. & A.R.	14 Spitfire
	1853/3	2-4-0	15 x 22	5 0	Std	D.P. & A.R.	15 Sprite (c)
	1854/5				Std	West Flanders Rly	
	1854/5				Std	West Flanders Rly	
	1855				Std	Sambre & Meuse Rly	
139	1856	2-2-2WT	IC 14 x 18	4 6	Std	MS & LR	123 Watkin
	1856	2-4-0	OC15½ x 20	5 6	Std	South Yorkshire Rly	13
	1857	2-4-0	OC 15 x 18	5 0	Std	L & SWR Eng Dept	Hawkshaw
	1857	2-2-2	14 x 22	6 6	5 3	Geelong & S. Suburban	1
	1857	0-6-0	16 x 22	5 0	5 3	Geelong & S. Suburban	2
	1857	0-6-0	16 x 22	5 0	5 3	Geelong & S. Suburban	3
	1857	0-6-0	16 x 22	5 0	5 3	Geelong & S. Suburban	4
	1857	0-6-0	16 x 22	5 0	5 3	Geelong & S. Suburban	5
	1857	0-4-0WT	9 x 12	3 0	Std	Sandy & Potton Rly	
	1858	2-4-0	OC 15 x 18	5 0	Std	L & SWR Eng Dept	Brunel
	1859	2-2-2	14 x 22	6 6	5 3	Victorian Govt Rlys	12
142	1859	0-6-0	16 x 22	5 0	5 3	Victorian Govt Rlys	13
	1859	0-6-0	16 x 22	5 0	5 3	Victorian Govt Rlys	15
145	1859	0-6-0	16 x 22	5 0	5 3	Victorian Govt Rlys	17
	1859	2-4-0WT	11 x 17	4 0	5 6	BB & Central India	3
	1859	2-4-0WT	11 x 17	4 0	5 6	BB & Central India	4
	1859	2-4-0WT	11 x 17	4 0	5 6		(d)
	1859	2-4-0WT	11 x 17	4 0	5 6		(d)
	1859	2-4-0WT	11 x 17	4 0	5 6		(d)
	1859	2-4-0WT	11 x 17	4 0	5 6		(d)
	1859	2-4-0	OC 15 x 18	5 0	Std	L & SWR Eng Dept	Hesketh
160	1860	2-4-0T	15 x 20		5 3	Melbourne & Suburban	Hawthorn
161	1860	2-4-0T	15 x 20		5 3	Melbourne & Suburban	Richmond
	1860	2-4-0T	15 x 22	4 0	Std	London & Blackwall	10
	1860	2-4-0T	15 x 22	4 0	Std	London & Blackwall	11
156	1861	2-4-0ST	16 x 22	5 0	5 3	Victorian Govt. Rly	14
157	1861	2-4-0ST	16 x 22	5 0	5 3	Victorian Govt. Rly	16
158	1861	2-4-0ST	16 x 22	5 0	5 3	Victorian Govt. Rly	18
159	1861	2-4-0ST	16 x 22	5 0	5 3	Victorian Govt. Rly	20
164	1861	2-4-0ST	16 x 22	5 0	5 3	Victorian Govt. Rly	22
165	1861	2-4-0ST	16 x 22	5 0	5 3	Victorian Govt. Rly	24
166	1861	2-4-0ST	16 x 22	5 0	5 3	Victorian Govt. Rly	26
	1861	2-4-0	OC 14 x 18	5 0	Std	L & SWR Eng. Dept	Locke
	1861	2-4-0	OC 14 x 18	5 0	Std	L & SWR Eng. Dept	Stephenson
	1861	2-4-0	OC 16 x 18	5 0	Std	L & SWR Eng. Dept	Smeaton
	1861	2-4-0	OC 16 x 18	5 0	Std	L & SWR Eng. Dept	Telford
	1861	2-4-0T	OC 11 x 16	3 10	Std	L & SWR Eng. Dept	Scott Fig. 128
	1861	2-4-0	IC 15 x 18	5 0	Std	Somerset Central Rly.	1
	1861	2-4-0	IC 15 x 18	5 0	Std	Somerset Central Rly	2
	1861	2-4-0	IC 15 x 18	5 0	Std	Somerset Central Rly	3

Works No	Date	Type	Cylinders inches	D.W. ft in	Gauge ft in	Customer	No/Name
	1861	2-4-0	IC 15 x 18	5 0	Std	Somerset Central Rly	4
	1861	2-4-0	IC 15 x 18	5 0	Std	Somerset Central Rly	5
	1861	2-4-0	IC 15 x 18	5 0	Std	Somerset Central Rly	6
	1861	2-4-0	IC 15 x 18	5 0	Std	Somerset Central Rly	7
	1861	2-4-0T	IC 15 x 18	5 0	Std	Somerset Central Rly	8
	1862	2-4-0T	OC 11 x 17	4 0	Std	Somerset & Dorset Rly	11 (e)
185	1862	2-4-0	IC 16 x 24	6 0	Std	Great Western Rly	149
186	1862	2-4-0	IC 16 x 24	6 0	Std	Great Western Rly	150
187	1862	2-4-0	IC 16 x 24	6 0	Std	Great Western Rly	151
188	1862	2-4-0	IC 16 x 24	6 0	Std	Great Western Rly	152
189	1862	2-4-0	IC 16 x 24	6 0	Std	Great Western Rly	153
190	1862	2-4-0	IC 16 x 24	6 0	Std	Great Western Rly	154
191	1862	2-4-0	IC 16 x 24	6 0	Std	Great Western Rly	155
192	1862	2-4-0	IC 16 x 24	6 0	Std	Great Western Rly	156
	1863	0-4-0T & T	OC 8 x 12	2 0	1 11½	Festiniog Rly	1 Princess (f)
	1863	0-4-0T & T	OC 8 x 12	2 0	1 11½	Festiniog Rly	2 The Prince (f)
	1863	2-4-0T	11 x 16		Std	Colne Val. & Halstead	Cam (g)
	1863	2-4-0T	11 x 16		Std	Colne Val. & Halstead	Colne (g)
	1863	2-4-0	IC 16 x 18	5 0	Std	Somerset & Dorset Rly	9
	1863	2-4-0	IC 16 x 18	5 0	Std	Somerset & Dorset Rly	10
	1864	0-4-0T & T	OC 8 x 12	2 0	1 11½	Festiniog Rly	3 Mountaineer (h)
	1864	0-4-0T & T	OC 8 x 12	2 0	1 11½	Festiniog Rly	4 Palmerston (h)
	1864	2-4-0	IC 16 x 18	5 0	Std	Somerset & Dorset Rly	12
	1864	2-4-0	IC 16 x 18	5 0	Std	Somerset & Dorset Rly	13
	1864	2-4-0	IC 16 x 18	5 0	Std	Somerset & Dorset Rly	14
	1864	2-4-0	IC 16 x 18	5 0	Std	Somerset & Dorset Rly	15
	1865	2-4-0	OC 16 x 18	5 0	Std	L & SW Rly	201
	1865	2-4-0	OC 16 x 18	5 0	Std	L & SW Rly	202
	1865	2-4-0	IC 16 x 24	6 0	Std	Somerset & Dorset Rly	17
	1865	2-4-0	IC 16 x 24	6 0	Std	Somerset & Dorset Rly	18
	1865	2-4-0	IC 16 x 24	6 0	Std	South Eastern Rly	215
	1865	2-4-0	IC 16 x 24	6 0	Std	South Eastern Rly	216
	1865	2-4-0	IC 16 x 24	6 0	Std	South Eastern Rly	217
	1865	2-4-0	IC 16 x 24	6 0	Std	South Eastern Rly	218
	1865	2-4-0	40 x 61	184	Std	Flandres Orientale	(i)
	1865	2-4-0	40 x 61	184	Std	Flandres Orientale	
	1865	2-4-0	40 x 61	184	Std	Flandres Orientale	
	1865	2-4-0	40 x 61	184	Std	Flandres Orientale	
	1865	2-4-0	40 x 61	184	Std	Flandres Orientale	
	1865	2-4-0	40 x 61	184	Std	Flandres Orientale	
	1865	2-4-0	40 x 61	184	Std	Flandres Orientale	
	1865	2-4-0	40 x 61	184	Std	Flandres Orientale	
		0-4-0ST				2nd Hand to I. W. Boulton	Phospho
234	1867	0-4-0ST & T	OC $8\frac{1}{8}$ x 12	2 2	1 11½	Festiniog Rly	5 Welsh Pony (h)
235	1867	0-4-0ST & T	OC $8\frac{1}{8}$ x 12	2 2	1 11½	Festiniog Rly	6 Little Giant (h)

FAIRLIE ENGINE & STEAM CARRIAGE CO.

Works No	Date	Type	Cylinders inches	D.W. ft in	Gauge ft in	Customer	No/Name
	1869	0-4-4-0T	4/8¼ x 13	2 4	1 11½	Festiniog Rly	7 Little Wonder (j)
	1869	0-4-4-0T	4/10 x 18	3 6	Std	Nassjo Oscarshamn Rly	(j) (k)
	1870*c*	0-4-4-0T	4/10 x 18	3 6	Std	Nassjo Oscarshamn Rly	(j)
	1870*c*	0-4-4-0T	4/10 x 18	3 6	Std	C. de F. de la Vendée	(j)
	1870	0-6-6-0T	4/15 x 20	3 6	Std	Iquique Rly	Tarapaca (j)
	1870	Steam Car				?	

(a) May have been same locomotive. (b) Ordered by C. Dunlop & Co. (c) Bought 1855. (d) Lost at Sea. (e) At 1862 exhibition, sold 1863. (f) Side tank and tender. (g) Built for contractor and returned. (h) Side and Saddle Tanks. (i) Later became Etat Belge Nos. 433-440. Dimensions in cms. (j) Built to Fairlie's Patent. (k) Delivered to Burry Port & Gwendraeth Valley Rly
MOUNTAINEER. Closed Down. 1872

After supplying two more four-coupled tanks, Spooner in co-operation with Mr Robert Fairlie decided to order one of the latter's patent double bogie locomotive and entrusted the construction of it to George England. Traffic was getting heavier and it was thought that an engine of this description would be ideal, spreading the axle load over the comparatively light track, the additional length of the machine being compensated for by the two flexible driving units which could cope with the severe curves. This two-in-one locomotive was the first narrow gauge *Fairlie* and had four 8½″ x 13″ cylinders, 2′ 4″ driving wheels and a total wheelbase of 19 ft – compared with 4′ 6″ for No. 1. The double boiler had a single firebox between the two boiler barrels, with a central mid-feather – the firebox being fed from one side only. This locomotive attracted an enormous amount of interest and attention and in 1869–70 trials were carried out with LITTLE WONDER as the Fairlie was called with the other conventional tank locomotives built for the line by George England. Engineers appeared from all over the world who acted as judges at the trials. The Fairlie came through with flying colours, emphasising the inferiority of the four coupled tank in comparative pulling power, fuel consumption and riding qualities.

Due to these demonstrations George England received orders through Robert Fairlie for patent locomotives from the Iquique Railway, Chemin de Fer de la Vendée, Nassio-Oscarsham Railway. It is interesting to note that Robert Fairlie was George England's son-in-law.

Late in 1869 George England leased the works to a company whose title was FAIRLIE ENGINE & STEAM CARRIAGE COMPANY, which consisted of Robert Fairlie, George England's son and Mr J. S. Frazer from the GWR.

This enabled Fairlie to construct his patent locomotives without hindrance, and this continued until 1872 when the works were closed and plant and machinery sold by auction.

The early failure of the company was precipitated by George England Junior's early death.

George England's career is interestingly described by Mr C. H. Dickson in the *Journal of the Stephenson Locomotive Society* (1961 volume). In addition Mr Dickson gives a list of locomotives built by George England, as far as information was available.

EVANS, J. CAMPBELL

Morden Iron Works, Greenwich

A tender for three locomotives was forwarded for the Festiniog Railway, dated 30 October, 1863, but there is no record of this firm having built any locomotives.

Fig. 129 **Geo England** 1867 Festiniog Rly LITTLE GIANT 0-4-0ST and tender

EVANS, RICHARD & COMPANY
Haydock Colliery

Six interesting locomotives were built for use in the Haydock Colliery, between 1869 and 1887. The design was on similar lines to the two 0-4-0WTs built for the colliery by James Cross in the 1860s.

They were among the first to be fitted with piston valves. The valve gear was outside and the first two had Stephenson's open links and the rest Gooch's box links and direct drive. Equalising beams were provided betweeen the leading and driving wheels. Cylinders were outside 16" x 22" and the wheels 4′ 0½". BELLEROPHON weighed 32 tons.

An illustration and description of GOLBOURNE may be found in the *Locomotive Magazine* for December 1901.

Locomotives built:

A	1869	No. 1	HERCULES
B	1871	No. 2	AMAZON
C	1874	No. 3	BELLEROPHON
D	1876	No. 4	MAKERFIELD
E	1886	No. 5	PARR
F	1887	No. 6	GOLBOURNE

The letters were in place of works numbers.

All were 0-6-0Ts.

FAIRBAIRN, WILLIAM & SONS
Canal Street Works, Manchester

Founded in 1816 as an iron foundry and in 1817 Fairbairn was joined by Mr Lillie and then known as Fairbairn & Lillie Engine Makers at an address in Mattier Street. They moved shortly afterwards to Canal Street and were then designated iron steam boat builders.

They became general engineers, millwrights, and bridge builders.

In 1839 they entered into the locomotive building business and in April of that year turned out four 0-4-0s for the Manchester & Leeds Railway of the *Bury* type. Lillie had now left and the title was as shown above.

The locomotives in general that were built by this firm were either to the customer's designs or were similar to those built by other Manchester firms such as Sharp Bros. and Edward Bury. In the case of *Bury* types the bar frames, haystack fireboxes and other details were copied, the only differences being superficial such as the splashers over the wheels. Many of this type were built and also singles similar to the *Sharpies*.

The main orders came from the home railways and only a small percentage of building went abroad, although at this time the market was wide open but the Fairbairns seemed content to obtain orders from most of the railways in the British Isles.

Fig. 130 **R. Evans & Co** 1874 Haydock Colliery BELLEROPHON. Note valve gear

Fig. 131 W. Fairbairn 1849 Shrewsbury & Birmingham Rly VULCAN

The best customer was the Manchester & Leeds Railway; at least sixty-nine locomotives were supplied of 0-4-0, 2-2-2, 2-4-0 and 0-4-2 types.

The claim that they were the first to build tank locomotives to run in service, is not true, that honour going to George Forrester & Co. who supplied two outside cylinder 2-2-0WTs to the Dublin and Kingstown Railway in 1836.

Some small 2-2-2WTs were built in 1850 for the 'Little' North Western Railway and others went to Ireland for the Dublin, Wicklow and Wexford Railway and the Midland Great Western Railway. These were notable in having solid plate welded frames, the boilers were domeless with raised fireboxes with safety valves on top.

Some large 2-2-2s were built for the Southern division of the London & North Western Railway to the designs of J. E. McConnell for working the London to Birmingham expresses. They had inside cylinders 18" x 24" and 7' 6" diameter driving wheels with double frames and outside bearings. The boiler was large with a total heating surface of 1230 sq. ft and included McConnell's patent combustion chamber, and firebox with longitudinal feather with two separate fire-doors. The cylinders were low set and attached to the smokebox and outside frames and this proved to be the weak spot in design.

The 2-2-2 singles continued to be built following Sharp's and Stephenson's designs, with outside sandwich frames and bearings. Others of the same type built in 1862 for the Great Eastern Railway were to Robert Sinclair's designs with 16" x 24" outside cylinders, and outside frames to leading and trailing wheels. The cylinders followed Allan's designs, Gooch's valve gear was used and the cab had side windows – quite unusual for this period.

Double framed 0-6-0s were supplied to the Midland Railway, and West Midland Railway in the early 1860s, and in 1861 Fairbairn supplied the Furness Railway with two 0-4-0s of the *Bury* type even though Bury, Curtis & Kennedy had long since ceased building locomotives.

A few locomotives were sent abroad to Australia, India, Brazil, France and Canada and the final approximate totals were: England 269; Scotland 29; Ireland 58; Europe 23; India and Colonies 16; a total of approximately 395 locomotives.

Building continued into 1863 when the locomotive side of the business was bought by Sharp Stewart & Co. who finished off two 0-4-2s which had been laid down for the Newry & Armagh Railway.

Fig. 132 **W. Fairbairn** 1851 Midland Great Western Rly No. 28 ELF at Broadstone

Fig. 133 **W. Fairbairn** 1855 Waterford & Tramore Rly No. 1

FAIRLIE ENGINE & STEAM CARRIAGE COMPANY

Hatcham Ironworks

See ENGLAND, GEORGE & COMPANY.

FALCON RAILWAY PLANT WORKS

Loughborough

See HUGHES, HENRY & COMPANY.

FAWCETT PRESTON & COMPANY

Phoenix Foundry, York Street, Liverpool

Established in 1758 and mainly concerned with marine engineering this firm prepared to enter the locomotive business in 1848 and turned out a six coupled tender engine and in December 1849, after unsuccessful attempts to sell it in August 1848 to the LNWR (Northern Division), managed to enter into a contract with the East Lancs Railway which empowered the firm to run this engine named PHOENIX on the railway at their own risk, as a banking engine at Accrington. After a fortnight of successful trials the ELR bought the engine for £2000 and gave it their number 43 and retained the name.

Cylinders were 18" x 24" and the wheels 5' 0" diameter. It was rebuilt in 1869 and withdrawn in 1881.

It is interesting to note that a Mr Jeffreys Parry de Winton worked for this firm, leaving it in the early 1860s to join Mr Owen Thomas at his works at Caernarvon to form the locomotive building firm of De Winton.

At some later period the firm became railway supply agents. Manning Wardle supplied a number of locomotives to this firm, which were shipped abroad including five (MW Nos. 874, 875, 878, and 879 of 1883 and No. 938 of 1885) to Brazil.

FENTON CRAVEN & COMPANY

Leeds

See SHEPHERD & TODD.

FENTON, MURRAY & WOOD

Round Foundry, Water Lane, Holbeck, Leeds

The firm was founded in 1795 by Matthew Murray and his partner David Wood. Two years later James Fenton joined the firm together with William Lister who subscribed considerable financial aid.

At first they built machine tools and stationary steam engines which became the envy of Boulton and Watt, and Murray was more interested in this field than locomotives.

The first four locomotives that were built were for the Middleton Colliery in 1812 and 1813, they were: PRINCE REGENT, SALAMANCA, LORD WILLINGTON and MARQUIS WELLINGTON. Two similar locomotives were built for the Kenton & Coxlodge Colliery.

John Blenkinsop was the agent to Mr C. J. Brandling at Middleton in 1808 and it was he who, doubtful of the adhesion of a smooth wheel on a smooth rail, introduced the rack rail for propulsion which he patented in 1811. Matthew Murray was entrusted with designing and building the locomotives incorporating this patent.

The locomotive designed had wooden beams as frames, the wheels running on unsprung plummer blocks. The boiler for the first locomotive was of cast iron and oval in section, with a single flue and was bolted together in two halves. There were two vertical cylinders sunk into the top of the boiler both on the longitudinal centre line. They were 8" diameter bore and 24" stroke each operating a separate crankshaft with cranks at right angles to one another. Plug steam valves were driven by fixed eccentrics and reversing was effected by turning the valves through 90°.

The rack wheel was carried on a shaft between the two crankshafts, and was driven by gears on each crankshaft engaging a gear fixed to the rack wheel shaft. The rack rail was laid on one side of the running track and there was only one rack wheel on the left hand side of the locomotive. The boiler and gearing were carried on four flanged wheels which were not driven. The steam was exhausted through a separate pipe and not into the chimney, and subsequently a silencer was fitted to the exhaust pipe. The boiler was lagged with wooden boards. Driving was carried out at one end and firing at the other. The whole weighed 5 tons and the fitting of two cylinders was unique at that time, thus obviating the use of a flywheel. At this time a locomotive was expected to haul four times its own weight by plain driving wheels on a smooth rail. It is true that Blenkinsop had no confidence in a pure friction drive but what he wanted was a higher ratio of engine weight to load hauled and this was actually achieved by the rack rail method, his 5 ton locomotives regularly hauled 90 tons.

It is interesting to note that as far as this country was concerned, a period of sixty years was to elapse before further locomotives were built using a rack system.

The first two locomotives went into service in August 1812, and two more followed in 1813 although there is some doubt whether there was one or two locomotives delivered in 1813. They were all scrapped in 1835.

No more were built until 1831. Before this, Murray died in 1826 and soon after the firm became FENTON, MURRAY & JACKSON and locomotive building was re-instituted from 1831 to 1842, the firm ceasing business in 1843.

Nearly eighty locomotives were built altogether. Most of their locomotives were of standard Stephenson's design with double frames and inside cylinders derived from the *Planet* type 2-2-0 and its successor the *Patentee* 2-2-2. The firm was among those who received a number of orders from Stephenson's who could not cope with all the orders they received. Fenton, Murray & Jackson's works had a circular erecting shop presumably in the form of a roundhouse. Their initial output was two or three locomotives per annum but by 1840 they were turning out a total of twenty-two. Unfortunately orders dwindled and the state of the trade in the early 'forties' forced them to close down.

James Fenton took over the Railway Foundry at Leeds previously owned by Shepherd and Todd in 1846 and continued locomotive building.

Many of their locomotives went abroad mostly to France where orders were received from: Roanes Andrézieux, Nr. Lyons; Lyons & St. Etienne; Paris & Versailles R. & L. Bank; Montpelier & Cette; and Paris – Orleans.

One went to the South Carolina RR, two *Patentee* 2-2-2s to the Munich & Augsburg and one to the Belgian State Railway.

Fig. 134 **Fenton Murray & Wood** 1812 Blenkinsop's rack locomotive for Middleton Colliery

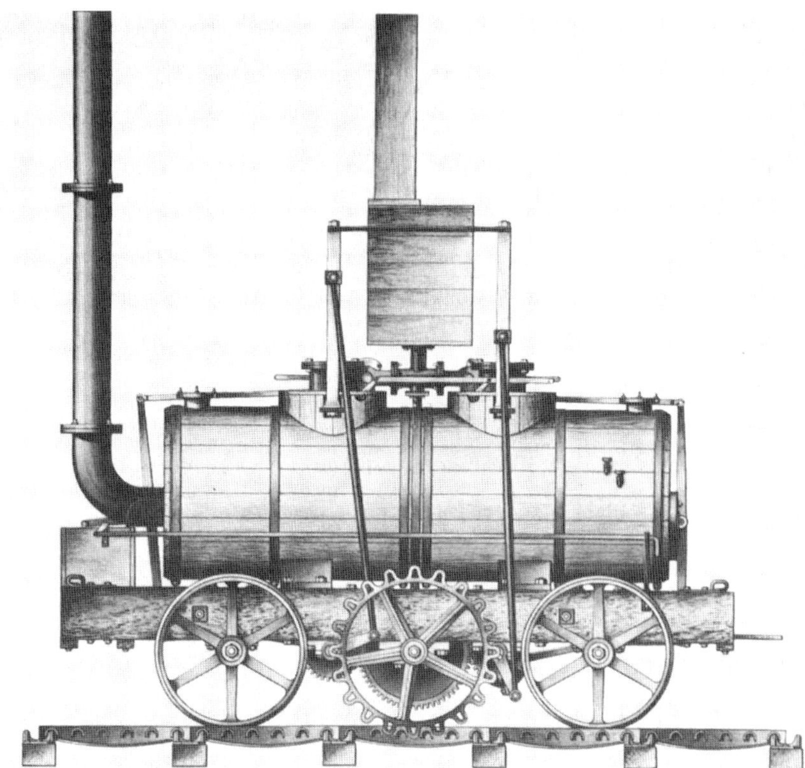

Date	Type	Cylinders inches	D.W. ft in	Customer	No/Name	Gauge ft in
6/1812	4W. Rack	9 x 22	3 0	Middleton Colliery	Prince Regent	4 1
8/1812	4W. Rack	8 x 24		Middleton Colliery	Salamanca	4 1
8/1813	4W. Rack	8 x 24		Middleton Colliery	Lord Willington	4 1
11/1813	4W. Rack	8 x 24		Middleton Colliery	Marquis Wellington	4 1
9/1813	4W. Rack			Kenton & Coxlodge Colly		4 7½
c 1813	4W. Rack			Kenton & Coxlodge Colly		4 7½
before 1824				For Russia		
5/1831	2-2-0	11 x 16	5 0	Liverpool & Manchester Rly	19. Vulcan (a)	Std
8/1831	2-2-0	11 x 16	5 0	Liverpool & Manchester Rly	21. Fury (a)	Std
1832	0-4-0			Roanes Andrézieux	2	
1832	0-4-0			Roanes Andrézieux	3	
1/1833	2-2-0	11 x 16	5 0	Liverpool & Manchester Rly	30. Leeds (a)	Std
1834				S. Carolina RR	Columbia	Std
9/1834	2-2-0	11 x 18	5 0	Leeds & Selby Rly	Nelson (a)	Std
1836	2-2-0	11 x 18	5 0	Leeds & Selby Rly	Exmouth	Std
1835	0-4-0			Lyons & St. Etienne	13	
1836	2-2-0			Paris & Versailles Rt. Bank	Jackson	
1836	2-2-0			Paris & Versailles Rt. Bank	Denis Papin	
1836	2-2-2			Leeds & Selby Rly	Hawke	Std
12/1836	0-4-0			London & Birmingham Rly		Std
1837	2-2-2			Leeds & Selby Rly	Anson	Std
1837	2-2-2			Munich & Augsburg Rly	Vulcan (b)	
1838	2-2-2			Munich & Augsburg Rly	Mars (b)	
1838	2-2-0			Paris & Versailles Left Bank	Matthew Murray	
1838	2-2-0			Montpelier & Cette Rly	1	
1838	2-2-0			Montpelier & Cette Rly	2	
1838	2-2-0			Montpelier & Cette Rly	3	
1838	2-2-0			Montpelier & Cette Rly	4	
1838	2-2-0			Montpelier & Cette Rly	5	
1839	2-2-0			Paris & Versailles Left Bank	Jackson	
4/1839	2-2-2		5 6	Sheffield & Rotherham Rly	Agilis	Std
7/1839	2-2-2			Leeds & Selby Rly	Dart	Std
10/1839	2-2-2			Leeds & Selby Rly	Express	Std
6/1839	2-2-2	13 x		London & Southampton Rly	31 Leeds	Std
8/1839	2-2-2	13 x		London & Southampton Rly	32 Eclipse	Std
1840	2-2-2			Paris & Versailles Rt. Bank	La Versailles	
3/1840	2-2-2	12 x 18	5 0	Belgian State Rly	Firefly	
4/1840	2-2-2	12 x 18	5 6	Hull & Selby Rly	Kingston	Std
5/1840	2-2-2	12 x 18	5 6	Hull & Selby Rly	Exley	Std
5/1840	2-2-2	12 x 18	5 6	Hull & Selby Rly	Selby	Std
6/1840	2-2-2	12 x 18	5 6	Hull & Selby Rly	Collingwood	Std
7/1840	2-2-2	12 x 18	5 6	Hull & Selby Rly	Andrew Marvel	Std
7/1840	2-2-2	12 x 18	5 6	Hull & Selby Rly	Wellington	Std
5/1840	2-2-2			North Midland Rly		Std
4/1840	2-4-0	15 x 18	5 0	Great Western Rly (c)	Hecla	7 0
4/1840	2-4-0	15 x 18	5 0	Great Western Rly (c)	Stromboli	7 0
6/1840	2-4-0	15 x 18	5 0	Great Western Rly (c)	Etna	7 0
5/1840	2-2-2			London & Southampton Rly	33. Phoenix	
8/1840	2-2-2			London & Southampton Rly	34. Crescent	
5/1840	2-2-2	15 x 18	7 0	Great Western Rly	Charon	7 0
6/1840	2-2-2	15 x 18	7 0	Great Western Rly	Cyclops	7 0
10/1840	2-2-2	15 x 18	7 0	Great Western Rly	Cerberus	7 0
9/1840	0-4-0	13 x 18	4 0	Great North of England (d)		Std
10/1840	2-2-2			North Midland Rly		Std
10/1840	2-2-2			North Midland Rly		Std
10/1840	2-2-2			North Midland Rly		Std
10/1840	2-2-2			North Midland Rly		Std
10/1840	2-2-2			North Midland Rly		Std
10/1840	2-2-2			North Midland Rly		Std
1840	2-2-2			Paris-Orleans Rly		Std
1840	2			Paris-Orleans Rly		Std
8/1841	2-2-2	15 x 18	7 0	Great Western Rly	Pluto	7 0
8/1841	2-2-2	15 x 18	7 0	Great Western Rly	Harpy	7 0
9/1841	2-2-2	15 x 18	7 0	Great Western Rly	Minos	7 0
10/1841	2-2-2	15 x 18	7 0	Great Western Rly	Ixion	7 0
10/1841	2-2-2	15 x 18	7 0	Great Western Rly	Gorgon	7 0

Date	Type	Cylinders inches	D.W. ft in	Customer	No/Name	Gauge ft in
11/1841	2-2-2	15 x 18	7 0	Great Western Rly	Hecate	7 0
12/1841	2-2-2	15 x 18	7 0	Great Western Rly	Vesta	7 0
1/1842	2-2-2	15 x 18	7 0	Great Western Rly	Acheron	7 0
2/1842	2-2-2	15 x 18	7 0	Great Western Rly	Erebus	7 0
3/1842	2-2-2	15 x 18	7 0	Great Western Rly	Medea	7 0
4/1842	2-2-2	15 x 18	7 0	Great Western Rly	Hydra	7 0
4/1842	2-2-2	15 x 18	7 0	Great Western Rly	Lethe	7 0
5/1842	2-2-2	15 x 18	7 0	Great Western Rly	Phlegethon	7 0
6/1842	2-2-2	15 x 18	7 0	Great Western Rly	Medusa	7 0
6/1842	2-2-2	15 x 18	7 0	Great Western Rly	Proserpine	7 0
7/1842	2-2-2	15 x 18	7 0	Great Western Rly	Ganxmede	7 0
8/1842	2-2-2	15 x 18	7 0	Great Western Rly	Argus	7 0

(a) Planet (b) Patentee (c) Leo Class (d) NER 65

At home the Great Western Railway's broad gauge was supplied with twenty 2-2-2s with 7′ diameter driving wheels of the *Firefly* class and three 2-4-0s of the *Leo* class with 5′ diameter driving wheels.

There were 62 *Firefly* class locomotives built altogether, the other 42 being built by other contractors but Gooch the locomotive engineer of the GWR stated in a report that the Fenton, Murray & Jackson batch were the best. They had the standard sandwich frames which consisted of oak or ash 3″ thick planking in between two $\frac{1}{2}$″ thick iron plates with bolts holding them together. To lighten these frames slots were cut out of the framing, and it is stated that this design of outside framing minimised the vibration in running over the longitudinal sleepers which were a feature of the 7′ broad gauge track.

John Chester Craven, David Joy, Benjamin

Fig. 135 1836 2-2-0 for Paris & Versailles Rly (Right Bank) JACKSON

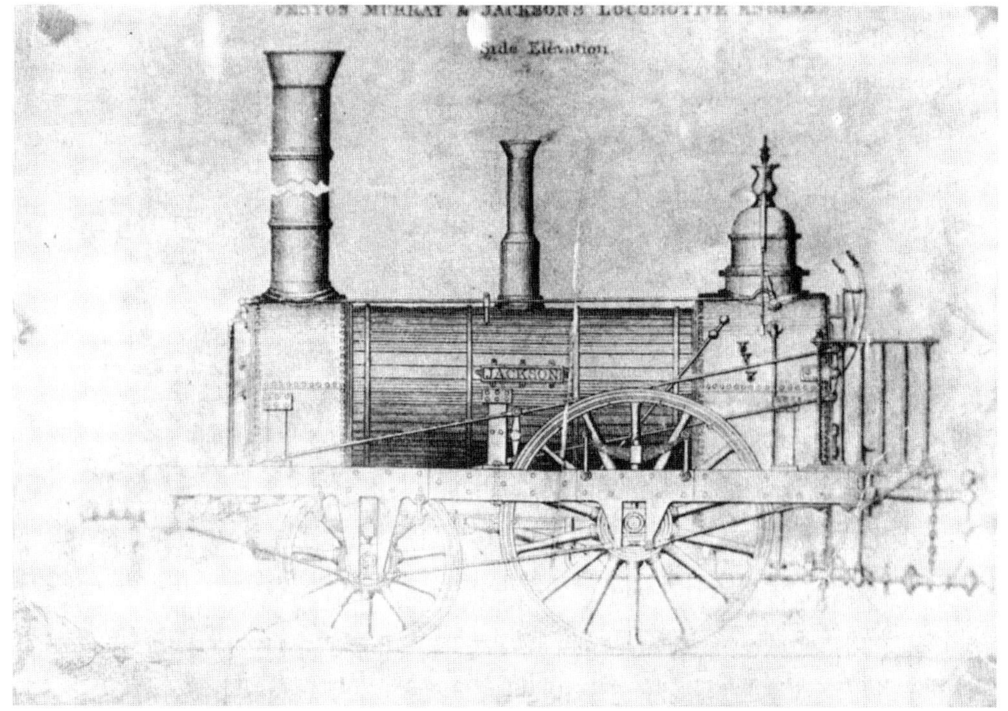

Hick and Richard Peacock were some of the men who received their training at the Round Foundry, all to achieve prominence in the locomotive field later on.

After Fenton, Murray and Jackson ceased building in 1843, for a short time the business was kept going by a consortium of workers, on a profit-sharing basis, but their methods were such, that they became known as the 'Forty Thieves'.

This co-operative effort did not last long, and the Foundry was taken over by Smith, Beacock & Tannett, renaming it the 'Victoria Foundry' and they continued in business as machine tool manufacturers until 1894.

FESTINIOG RAILWAY

Boston Lodge Works, Portmadoc

The railway built two locomotives of Fairlie's Patent design: in 1879 0-4-4-0 No. 10 MERDDIN EMRYS and in 1885 0-4-4-0 No. 11 LIVINGSTON THOMPSON.

Both were designed by G. P. Spooner, engineer to the Railway. Although the Boston Lodge works was well equipped for the heavy repair of its locomotives, it is improbable that the two locomotives mentioned would have been completely built without any outside help. The boilers for both would have been built by Avonside Engine Company, especially as these Fairlie locomotives were double ended with boilers at each end, a common firebox in the middle, with a central mid-feather and separate fire-doors.

The two engine units for each locomotive were mounted on central pivots to provide the flexibility for which this type of locomotive was noted. Each locomotive had four cylinders 9″ x 14″, 12′ 9¼″ diameter wheels, boilers pressed at 160 lb/sq. in.

These two *Fairlie*s were excellent machines, they rode the track very well with their pivotted units and were cheap to maintain.

Fig. 136 **Festiniog Railway** 1879 Fairlie patent MERDDIN EMRYS

FLETCHER BURROWS

Chanters & Howe Bridge Collieries, Atherton, Lancs.

One locomotive reported to have been built by this firm in 1888 at their Gibfield Works, a 0-4-0ST named ELECTRIC with outside cylinders. No further details available.

FLETCHER JENNINGS & COMPANY

Lowca Works, Whitehaven

See TULK & LEY.

FODENS LTD.

Elworth Works, Sandbach

Only one railway locomotive was built and was to the order of a local customer Palmer Mann and Co., Salt Producers, a works situated some 1½ miles from Fodens Ltd.

This particular locomotive was built over a period of eighteen months and delivered on 17 February, 1930 (Works No. 13292).

Temporary railway lines had to be laid by Fodens to connect up with the Manchester to Crewe main line, from whence it was able to be diverted into the customer's railway siding, where it was used for shunting wagons loaded with bulk salt for a period of over twenty years.

The engine used was Foden's type E2 with two vertical cylinders 7" x 10" with crankshafts and motion enclosed in a crankcase, the crankshaft running in roller bearings. The water tube boiler operated at 250 lb/sq. in. with a smokebox superheater as fitted to the Type E steam road wagons.

The drive was by double helical gears to cardan shafts with worm and worm wheels as the final drive to both axles. The wheels were 3 ft diameter with laminated springs and radius bars. One boiler feed pump and one injector were fitted. The design was neat incorporating some novel features, but only one was built.

Fig. 137 **Fodens Ltd** 1930 for Palmer Mann & Co. Sandbach

FORREST & BARR
Glasgow

No information available.

FORRESTER, GEO. & COMPANY
Vauxhall Foundry, Liverpool

Commenced building locomotives in 1834. They were entered in the Liverpool directory from 1827 to 1890 as iron founders, but locomotive building ceased *c*1847.

Forrester favoured outside cylinders, motion and frames for accessibility. They had neat lines.

Many were rough riders due to the outside cylinders and frames, with a severe racking motion on the track.

The idea of using four fixed eccentrics instead of two loose ones was first applied by Forrester in 1834.

Forrester designed an outside cylinder 2-2-2 which was sold to a number of railways. One of the largest single orders was from the South Eastern Railway for fifteen 2-4-0s in 1847.

They had Stephenson's long boilers with outside cylinders 15″ x 22″ and 5′ 6″ diameter coupled wheels. The large dome was on the first ring of the boiler immediately behind the chimney and the raised round top firebox had the safety valve fitted with a tall casing surrounding it. All the springs were above the frames. They were put on express passenger duties but were soon relegated to suburban services and their average life was fifteen years.

The Dublin and Kingstown Railway ordered three 2-2-0s with outside cylinders 11″ x 18″ and 5′ diameter driving wheels. They were named KINGSTOWN, DUBLIN and VAUXHALL, built in 1834 and were converted to tank locomotives in 1837 to 1840. They had three sets of frames, the cylinders being bolted between the outside and intermediate frames, and the horn plates for the driving wheels in the inner frames and the leading wheels had the hornplates in the intermediate frames – a complicated arrangement.

Other characteristics of Forrester's locomotives were the driving axles which had bosses on each end to which the wheels were bolted, the boss diameter being large enough to enable the wheels to pass over the outside cranks.

Fig. 138 **Geo Forrester** 1834 for Liverpool & Manchester Rly No. 36 SWIFTSURE

Alexander Allan was the works manager until February 1840 when he took charge of the Edge Hill Shops of the Grand Junction Railway; and later became foreman of locomotives at Crewe until he went to Scotland in 1853.

Allan's locomotives for the GJR and LNWR bore a strong resemblance to Forrester's practice, especially the front end and the deep heavy frames.

Forrester's received another order from the Dublin and Kingstown Railway in 1835, for two tank locomotives VICTORIA and COMET and these had the distinction of being the first tank locomotives to work on a public railway.

The earlier locomotives did not have slide bars, the piston rods were guided by a parallel motion similar to stationary engine practice, but these two tanks had slide bars at the request of the D&K's locomotive superintendent, so that it is probable that the parallel motion was not successful. Forrester was the first to employ outside horizontal cylinders.

FOSSICK & HACKWORTH
Stockton

The firm was founded about 1838–9. Thomas Hackworth brother of Timothy Hackworth was the co-partner, and his brother's influence is very evident in Fossick & Hackworth's designs. George Fossick probably financed the project and bought the factory in Stockton. Thomas had been manager at the Soho works which were then known as Hackworth and Downing.

They commenced building locomotives in 1839 together with railway wagons and at a later stage marine engines, the firm providing the engines for the first two iron ships built at Stockton in 1853. This side of the business increased as the locomotive side declined.

The first two 0-6-0s had horizontal outside cylinders according to E. L. Ahrons in *The British Steam Railway Locomotive 1825–1925.* If this was so they quickly reverted to the high-set inclined outside cylinders which was more in keeping with Timothy Hackworth's locomotives.

The 0-6-0 tender locomotive built for the Llanelly Railway in 1864 and named VICTOR was a well known example with the inclined outside cylinders bolted to the smoke box and driving the rear coupled wheels with very lengthy connecting rods. The valve gear was a radial type patented by John Hackworth who also built stationary engines and machinery in Darlington. The VICTOR was sent to Swindon after the amalgamation of the Llanelly Railway where it was an outdoor museum piece for many years.

Fig. 139 **Fossick & Hackworth** 1864 Llanelly Rly & Dock Co VICTOR Preserved at Swindon until 1889

Locomotives were also built for the West Hartlepool Harbour & Railway, Hartlepool Dock Company, Stockton and Darlington, and Irish railways.

They were peculiarly archaic in design and their performance was poor.

In 1865 Thomas Hackworth retired and the firm became FOSSICK & BLAIR.

The following year George Fossick died and Mr G. T. Blair, the new partner who had been works manager carried on as BLAIR & CO.

Very little locomotive building took place after Fossick's death.

Despite the fact that the Llanelly locomotives were not successful, at least eight were supplied from 1842 to 1866, probably explained by the fact that Messrs. I'Anson, Fossick and Hackworth had contracted to work the railway from 1850 to 1853, and from 1864 the U.K. Rolling Stock Co. obtained a similar contract and Fossick & Hackworth were again involved having a financial interest in this company.

The LEO supplied in 1858 was converted to a 0-8-0T to work on the severe gradients of the railway as a banking locomotive, its conversion being completed by the GWR but did very little work.

About 120 locomotives are reported to have been built by this firm, and although the list is by no means complete, and a number were ordered from abroad it is doubtful whether this figure was reached.

Date	Type	Cylinders inches	D.W. ft in	Customer	No/Name	Later Nos.
6/1839	0-6-0	14 x 18	4 0	Hartlepool Dock Co		NE 119
6/1841	0-6-0	15 x 24	4 0	Stockton & Darlington Rly	Stockton No. 4	
6/1842	0-6-0	14 x 18	4 0	Hartlepool Dock Co		NE 120
1842	0-6-0	14½ x 24	4 0	Hartlepool Dock Co		NE 123
1842	0-6-0	14 x 18	4 0	Llanelly Rly. & Dock Co	Princess Royal	
3/1845	2-2-2	14 x 20	5 6	Gravesend & Rochester Rly	Van Tromp	SE 116
7/1847	2-4-0	16 x 22	5 0	GG & SJR	50. Grecian	
9/1847	2-4-0	16 x 22	5 0	GG & SJR	51. Minotaur	
12/1847	2-4-0	16 x 22	5 0	GG & SJR	52. Proteus	
1/1848	2-4-0	16 x 22	5 0	GG & SJR	53. Polyphemus	
7/1848	2-4-0	16 x 22	5 0	GG & SJR	54. Terror	
11/1848	2-4-0	15 x 22	5 0	GG & SJR	63. Daedalion	
1855	0-6-0	16 x 21	4 6	WHH & Rly Co	45	NE 616
1856	0-6-0	16½ x 21	4 6	WHH & Rly Co	48	NE 618
1856	0-6-0	16 x 24	4 6	WHH & Rly Co	57	NE 625
1857	0-6-0	16½ x 21	4 6	WHH & Rly Co	53	NE 621
1857	0-6-0	16½ x 21	4 6	WHH & Rly Co	58	NE 626
1857	0-6-0	16½ x 21	4 6	WHH & Rly Co	59	NE 627
1857	0-6-0	16½ x 21	4 6	WHH & Rly Co	60	NE 628
8/1858	0-6-0	16 x 24	4 9	Llanelly Rly & Dock Co	Arthur	GW 903
1858	0-6-0	16 x 24	4 9	Llanelly Rly & Dock Co	Leo	GW 902
1859	0-6-0	16½ x 21	4 6	WHH & Rly Co	69	NE 637
1860	0-4-2	16 x 20	5 0	Llanelly Rly & Dock Co	Beatrice	GW 895
1860	0-6-0			Llanelly Rly & Dock Co	Louisa	GW 904
1861	0-6-0	16½ x 21	4 6	WHH & Rly Co	71	NE 639
1861	0-4-0T	15 x 22	4 0	WHH & Rly Co	72	NE 640
1861	0-4-0T	15 x 22	4 0	WHH & Rly Co	73	NE 641
1862	0-6-0T	13 x 24	4 0	Londonderry & Lough Swilly Rly	1	(a)
1862	0-6-0T	13 x 24	4 0	Londonderry & Lough Swilly Rly	2	(a)
1862	2-2-2			Dublin & Meath Rly		
1862	0-4-2	15 x 22	4 9	Waterford & Limerick Rly	22	
1862	0-4-2	15 x 22	4 9	Waterford & Limerick Rly	23	
1862	0-4-2	15 x 22	4 9	Waterford & Limerick Rly	24	
1864	2-4-0	16 x 22	5 1	Dublin & Meath Rly	7 Meath	
1864	0-6-0	16 x 24	4 9	Llanelly Rly & Dock Co	Victor	
4/1866*	0-6-0	16 x 24	4 9	Llanelly Rly & Dock Co	Ernest	GW 905
9/1866*	0-6-0	16 x 24	4 9	Llanelly Rly & Dock Co	Edinburgh	GW 906

(a) 5' 3" Gauge sub-contracted to Gilkes Wilson & Co GG & SJ Rly: Great Grimsby & Sheffield Junc. Rly WHH & RLY Co: West Hartlepool Harbour & Railway Co. NE: North Eastern Railway. GW: Great Western Railway. *Fossick & Blair

FOSTER, RASTRICK & CO.
Stourbridge

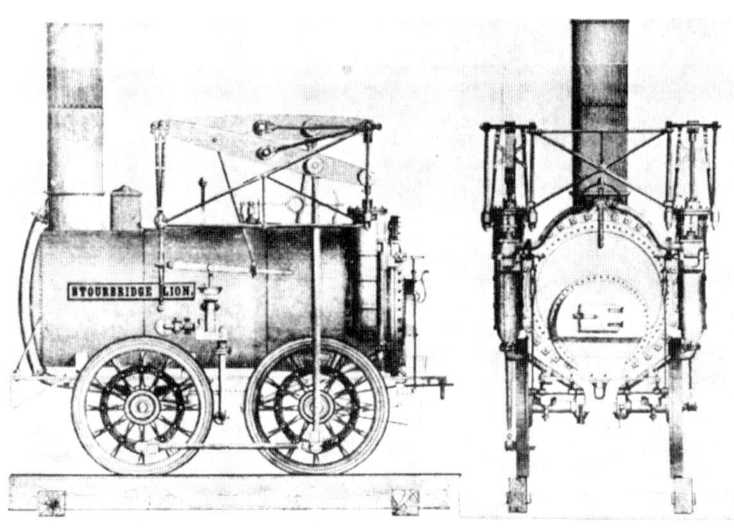

Fig. 140
Foster Rastrick
1829
Delaware
& Hudson RR
STOURBRIDGE LION

Date	Delivered	Type	Cylinders inches	D.W. ft in	Customer	Name
1828	5/1829	0-4-0	8¾ x 36	4 0	Delaware & Hudson RR	Stourbridge Lion
1829	8/1829	0-4-0	8¾ x 36	4 0	Delaware & Hudson RR	Delaware (?)
1829	9/1829	0-4-0	8¾ x 36	4 0	Delaware & Hudson RR	Hudson (?)
1829	6/1829	0-4-0	8½ x 36	4 0	Shutt End Railway	Agenoria (a)

(a) South Kensington Museum in 1885 then to York

Fig. 141
Foster Rastrick
1829 Shutt End Rly
AGENORIA

FOSTER, RASTRICK & CO.

William Foster and John Urpeth Rastrick became partners about 1816 and in 1829 began manufacturing steam locomotives. Three were supplied in 1829 to the Delaware and Hudson Railroad and were named STOURBRIDGE LION, DELAWARE and HUDSON. The first named was the first locomotive to be tried in America. Unfortunately the track was too weak to allow running and the other two were put aside.

A fourth was built in the same year for the Shutt End railway and named AGENORIA.

All four were similar and had four coupled wheels 4 ft diameter driven by two vertical cylinders 8½" x 36" at the rear end. The crossheads were guided by grasshopper parallel gear with the connecting rods attached to intermediate points on the beams fixed in front of the cylinders reducing the stroke to 27". Slide valves were fitted driven by loose eccentrics controlled by stops fixed to the axle. For reversing, hand gear was provided for working the valves. The boiler feed pump was driven off one of the beams. The boiler itself was approx. 4 ft diameter and 10 ft long and the double flues joined together at the firing end to provide a grate with an area of 8·5 sq. ft. The chimney was fixed on top of the boiler and the flues after forming one chamber had a vertical flue passing through the boiler barrel vertically into the chimney. Springs were fitted to the leading wheels. On AGENORIA probably the first mechanical lubricators on a locomotive were fitted to the axleboxes, comprising toothed rings on the axles driving leather 'throwers' at twice the speed of the axle. The wheels had cast iron centres and wrought iron tyres.

Unlike the three locomotives sent to America, AGENORIA had a long working life and in 1884 was presented to the South Kensington Museum.

J. U. Rastrick was chosen as one of the judges at the Rainhills trials of 1829.

Among other work carried out was the supply of cast iron rails.

JOHN FOWLER & CO.
Steam Plough & Locomotive Works, Leeds

This firm was established in 1850 and engaged in the manufacture of traction engines, steam rollers and agricultural machinery. The firm's founder John Fowler devoted his life to the design and production of steam ploughs and other farming equipment. Unfortunately he died as a result of a hunting accident at the early age of 38 in December 1864. He had joined the firm of Gilkes Wilson & Co. of Middlesbrough when he was 21.

It was not until after the death of John Fowler that the first steam locomotives were built in 1866, when six 0-6-0s for the London, Chatham & Dover Railway, eleven 2-4-0s for the Imperial Mexican Railway, two 2-4-0Ts for the Waterford & Kilkenny Railway and three 2-4-0s for the Great Northern Railway were built, all in the same year.

From 1867 to 1871 forty 0-6-0s were built for the Great Indian Peninsula Railway and others of the same type for the Somerset & Dorset Railway, Ceylon, and the Great Northern Railway.

After this encouraging start in the manufacture of standard gauge and main line locomotives the emphasis shifted more to industrial and narrow gauge locomotives, and it was in this field that Fowlers gained a reputation, particularly in the supply of equipment for overseas sugar plantations, quarries, forestry etc.

The last large order for standard gauge locomotives was fulfilled in 1885/6 being six double framed 0-6-0STs for the Brecon & Merthyr Tydfil Junction Railway the last of which was not condemned until 1934 and was subsequently sold to the Hazelrigg & Burradon Coal Co. being scrapped in 1944 which was a fair indication of the good workmanship put into them.

Narrow gauge locomotives of varied wheel arrangements were built for India, Australia, Chile, Assam, Egypt, South Africa and many other countries.

In 1886 the firm's title was changed to JOHN FOWLER & CO. (LEEDS) LTD.

Besides locomotives, rolling stock of all types were built. The locomotives were designed for the varied conditions abroad. Many were equipped for burning wood or bagasse. The majority had outside frames for the coupled wheels with outside Walschaert valve gear. Efforts were made towards a degree of standardisation (see tables) and various combinations of parts were made possible but small batches and single orders were more often than not the order of the day.

There was no separate works list for steam locomotives, all products being included.

Between 1857 and 1927, 16,000 items were turned out and up to 1935 the number had risen to 22,000.

Fig. 142 **J. Fowler** 1866 London Chatham & Dover Rly CONSTANTINE

Internal combustion engines were introduced in 1923 and in the 1930s the demand for steam propulsion slackened off.

It is probable that the last one built was No. 21496, a 0-6-0WT for a forestry railway in Pakistan in 1935. Two further orders obtained by Fowler were built by Kitson & Co. and Hudswell Clarke & Co. Ltd.

Diesel locomotives were turned out in large quantities for many years until the intense competition altered the firm's policy to continue manufacturing other more profitable equipment and in 1968 the goodwill of the locomotive side of the business was sold to Andrew Barclay Sons & Co. Ltd of Kilmarnock together with all drawings, spares etc.

It is difficult to estimate how many steam locomotives were built due to incomplete records but it would be in the region of between 200 and 300.

Typical specification for special fittings if required: Copper firebox and brass tubes; Enlarged water tanks or separate tenders; Radial driving axles for sharp curves; Enclosed cabs and double roofs for hot countries; Stephenson or other valve gear (Walschaerts – standard); Oil burning apparatus; Ash ejector in smokebox; Saddle instead of side or well tanks; Acetylene or electric lamps; Adjustable wedges in axle horn blocks; Motion covers; Pumps in addition to injectors; Steam sanding; Spark arresters; Rocking fire grates for use with inferior fuel; Steam water lifters.

Fig. 143 **J. Fowler** 1935 0-6-2T AIRDMILLAN for a Queensland sugar estate

Gauge inches	HP	Cylinders inches	Wheel Dia D ft in	Wheel Dia T ft in	Heating Surface sq ft	Grate Area sq ft	Wkg Press lb/sq in	Tank Gallons	Weight WO ton cwt	Rigid W'Base ft in	Tractive Effort LBS
0-4-2T											
24	60	8½ x 12	2 0	1 8	222	5.7	180	300	12 8	4 0	4870
30	60	8½ x 12	2 0	1 8	222	5.7	180	300	12 14	4 0	4870
24	90	9½ x 14	2 4	1 8	322	6.3	180	350	13 18	4 6	6070
30	90	9½ x 14	2 4	1 8	322	6.3	180	350	14 4	4 6	6070
36	150	10½ x 16	2 6	1 9	443	9.0	180	450	18 10	5 0	7938
0-4-0T											
24	18	5½ x 10	1 8	WT	83.75	3.0	180	85	5 8	3 0	2040
30	18	5½ x 10	1 8	WT	83.75	3.0	180	110	5 16	3 0	2040
24	40	7 x 12	1 10	WT	137	4.0	180	125	7 6	3 6	3600
30	40	7 x 12	1 10	WT	137	4.0	180	150	7 12	3 6	3600
24	60	8½ x 12	2 0	T	222	5.4	180	200	11 8	4 0	4870
30	60	8½ x 12	2 0	T	222	5.4	180	200	11 12	4 0	4870
24	90	9½ x 14	2 4	T	327	6.3	180	250	13 4	4 6	6070
30	90	9½ x 14	2 4	T	327	6.3	180	250	13 8	4 6	6070
42	90	9½ x 14	2 4	T	327	6.3	180	250	13 14	4 6	6070
36	150	10½ x 16	2 6	T	449	9.0	180	450	17 12	5 0	7930
0-6-0T											
24	40	7 x 12	1 10	WT	137	4.0	180	125	8 6	4 0	3600
30	40	7 x 12	1 10	WT	137	4.0	180	150	8 12	4 0	3600
24	60	8½ x 12	2 1	T	256.5	5.7	180	300	13 0	5 0	4680
30	60	8½ x 12	2 1	T	256.5	5.7	180	300	13 4	5 0	4680
24	90	9½ x 14	2 4	T	334	6.3	180	350	14 8	5 9	6070
30	90	9½ x 14	2 4	T	334	6.3	180	350	14 14	5 9	6070
42	90	9½ x 14	2 4	T	334	6.3	180	350	15 4	5 9	6070
36	190	12 x 16	2 6	T	510	10.0	180	500	20 0	6 2	10360
0-6-2T	FOR WOOD OR INFERIOR FUEL										
24	60	8½ x 12	2 1	1 8	256.5	5.7	180	300	13 14	5 6	4680
24	90	9½ x 14	2 4	1 8	334	6.3	180	350	15 6	6 0	6070
30	90	9½ x 14	2 4	1 8	334	6.3	180	350	15 16	6 0	6070
36	190	12 x 16	2 6	1 9	497	13.4	180	600	22 16	6 2	10360

Horse Power is 60% of IHP

FOX WALKER & CO.
Atlas Engine Works, St. George, Bristol

Fig. 144
Peckett built 1884 0-4-0ST
Class 436 of their own design

This firm commenced building steam locomotives in 1864 and catered for industrial requirements in the shape of four coupled and six coupled tank locomotives, mostly saddle tanks. The early records are very incomplete and little is known of the details of the first ten years of trading. They also built a number of stationary steam engines and were one of the pioneers in steam tramway propulsion. The first steam tram was tried out on the Bristol tramway in December 1877 and it is probable that the tram engine patented by James Matthews was built by Fox Walker & Co. In 1878 a number of 0-4-0 tram engines were sent to Rouen to the order of G. P. Harding, and classed TE and in the same year some 0-4-2 tram engines were built for the same customer, type SWTE. They all had 8″ x 9″ cylinders and 2′ 6″ diameter coupled wheels. Six TE type and twelve type SWTEs were built from 1877 to 1879.

A considerable number of saddle tank locomotives were sent abroad to Tasmania, Sweden, Holland and the Sudan.

Ten outside cylinder 2-4-0Ts with 15″ x 22″ cylinders and 4′ 6½″ diameter coupled wheels were built for the East Argentina Railway in 1873/4, and four dimensionally similar for the Berlin – Stettin railway who were also supplied with six tender versions of the 2-4-0T in 1873/4.

Four 2-4-0s, six 0-4-2s and ten 0-6-0s were built for the Cordoba and Tucuman railway in 1874. Their main output at this time were 0-6-0STs with inside cylinders and many with outside cylinders, the majority going to steelworks, mines and docks in this country. The standard six coupled tank had outside cylinders 13″ or 14″ x 20″ and 3′ 6″ diameter wheels.

In 1876 five 0-6-0Ts and seven 2-4-0Ts were built for the Java Government railway, all had outside cylinders 14″ x 18″.

An interesting order was from the Admiralty, Chatham for six 1′ 6″ gauge 2-4-2Ts. They were designed for trench engines with Handyside's steep gradient apparatus.

A number of home railways were supplied with six coupled tank locomotives for shunting and goods train duties including the Eastern & Midlands, Whitland & Cardigan, Gwendraeth Valley, and Midland Railways.

Nine powerful inside cylinder 0-6-0STs were supplied to the Somerset and Dorset Railway (Nos. 1–9) with cylinders 18″ x 24″ and 4 ft diameter wheels.

The last building that took place was the Class SWTE tram engines in 1879 followed by four semi-portable horizontal high pressure engines with 14″ x 20″ cylinders.

The last works number was 424 and from 1862 to 1880 approximately 410 steam locomotives were built.

They were taken over by Thomas Peckett in 1880 as PECKETT & SONS, ATLAS ENGINE WORKS, BRISTOL.

Peckett & Sons followed on with the well-established industrial saddle tank locomotives, the designs at first being changed but little. The characteristic chimney with copper top, the brass dome and safety valve cover and spring balance safety valves and the shapely cab.

In 1897 a 2-4-2T was built for the West Sussex Railway. It was named SELSEY and had outside cylinders 10″ x 15″ and 2′ 9″ diameter coupled wheels.

Many industrial saddle tank locomotives both four and six coupled, inside and outside cylinder varieties were built and these continued to be the bulk of Peckett's production.

The Schull and Skibbereen Tramway ordered a 4-4-0T which was built in 1905 and named GABRIEL. It had outside cylinders 12″ x 18″ and 3′ 0½″ coupled wheels and weighed 26½ tons in working order, and another slightly smaller 4-4-0T was delivered to the same tramway in 1914 named CONCILIATION which, for political reasons was afterwards renamed KENT.

In the same year two practically identical locomotives of the same wheel arrangement and gauge were built for the Sarawak Railway named BULAN and BINTANG. On all this type outside frames were used for the coupled wheels.

Some small 0-4-0Ts were built from time to time known as the ALUMINIUM class with outside cylinders 7″ x 10″, wheels 1′ 8″ diameter and a wheelbase of 3′ 3″ weighing only 7 tons. They were designed for gauge 2′ 6″ to 3′ 6″.

Perhaps the most typical Peckett saddle tank was the six coupled inside cylinder type. and a specification for Class 1094 should prove interesting:

The engine to be of the inside cylinder class, having six coupled wheels 3′ 10″ diameter and for a gauge of 4′ 8½″. The wheel base to be 11′ 9″ and the flanges of the tyres of the driving wheels to be turned thin to facilitate passage round sharp curves.

Boiler The boiler to be 14′ 3″ long. The barrel to be 3′ 11″ diameter, ½″ thick secured by rivets, 1⅞″ from centre to centre on the vertical seams. The longitudinal seams above water-line to be butt jointed with butt strips each side of the same thickness as the barrel plates, secured with four rows of rivets. The rivets to thoroughly fill the holes and all joints to be fullered on both sides. The smokebox tube plate to be ¾″ thick and flanged over to receive the smokebox. All flanged and boiler shell plates to be of the best Siemens-Martin's mild steel, and all flanged plates flanged by special hydraulic machinery at one operation. All plate edges planed, and rivet holes drilled and hydraulic rivetted. The boiler to be suitable for a working pressure of 160 lb/sq. in., and tested by hydraulic pressure to 240 lb/sq. in., and by steam pressure to 165 lb/sq. in. A certificate of the boiler test to be given. After being tested the boiler to be lagged with yellow pine battens and sheet steel, and fitted with ash-pan and damper. Suitable expansion brackets to be fitted at the firebox end. Cast steel seatings rivetted on for back pressure valves, also fitted with fire bar frame, set of fire bars, fire doors and slides, the whole to be so arranged that the boiler can be taken out in a few hours for repairs, without taking down the working machinery. The boiler to be fitted with sixteen washing-out plugs and four hand-hole doors at bottom of external firebox, also two large hand-hole doors on side plates above top of firebox.

External firebox The external firebox to be made of the best Siemens-Martin's mild steel plates, the casing to be made in one plate ½″ thick, the front plate ½″ thick, and the back plate $\frac{9}{16}$″ thick. Double thickness of plate at bottom, so as to ensure having a good thread for the mud plugs. Foundation and firehole rings to be solid forgings, machined where required for joints. The foundation ring to be double rivetted at the corners.

Internal firebox The internal firebox to be made of the best selected copper 4′ 6″ long, 3′ 3¾″ top, 3′ 6″ bottom, wide, 4′ 0¾″ high, the plates to be ½″ thick except the plate where the tubes pass through, where it is to be 1″ thick. To be rivetted together with copper rivets. Also fitted with a fire brick arch.

Fig. 145 **Peckett** 429/1883 GKN Dowlais MAGPIE one of the few i.c. 0-4-0STs.

Fig. 146 1362/1914 Sarawak Rly BULAN 3′ gauge

Steel Tubes The steel tubes to be placed in vertical rows and to be 164 in number, of the best solid drawn steel, 1¾″ diameter inside, and No. 10 BWG thick, and 9′ 8″ long. Steel ferrules to be fitted in at the firebox end. Special care to be taken that the tubes are readily accessible for cleaning and repairs.

Staying The shell to be stayed by strong gusset stays the top of the firebox by wrought iron bridge stays, attached by sling stays and cross-tee pieces to the roof of the firebox casing, and the flat surfaces of the inner and outer fireboxes by screwed copper stays, 15/16″ diameter, arranged as near 4″ pitch as possible, and rivetted over at both ends.

Smokebox Secured to the tube plate, and fitted with door, crossbar, bolt and handles, and also with chimney, fitted with base and polished copper top.

Feed Apparatus The boiler to be fed by two No. 8 gun-metal self-acting, re-starting injectors, with removal cones and fitted with all necessary steam, feed and overflow valves, each injector of ample capacity to supply the boiler with water, and of the best and simplest construction, all handles being worked from the footplate.

Safety Valves A pair of 3″ safety valves to be fitted on the steam dome cover, and held down by wrought iron levers and adjustable spring balances. Safety valves to be covered with a polished brass dome.

Regulator To be fixed in smokebox and of easy access for repairs, and fitted with Roscoe's lubricator for lubricating the cylinders. The steam to be taken from the top of the dome to ensure having it dry.

Pipes All the steam feed and delivery pipes to be of solid drawn copper.

Tank To be placed on the boiler and to contain 1175 gallons of water, and connected by a copper equalising pipe, fitted with drain cock. To be securely fastened to the frames by upright stays, and also to the smokebox.

Coal Bunker A coal bunker to be placed on the foot-plate and to have a capacity of 35 cwt.

Cab A weatherboard and awning to be fixed over the footplate, and fitted with four eye-glasses and hinged brass frames.

Frames The frames to be each made from one solid plate of steel of best boiler quality 2′ 11″ deep, 1⅛″ thick, to be bolted together while being machined to ensure perfect accuracy; when fixed to be strongly secured to one another by cross-stays. The buffer beams to be made of steel plates 20″ deep x 1½″ thick. The buffers to be fitted with best volute springs, and have cast steel plungers fitted with steel heads 18″ diameter.

Axle Boxes The axle boxes to be of strong cast iron with heavy phosphor-bronze bearings. The boxes to be arranged for lubrication by oil, and to be accurately fitted to the horn blocks.

Horn Blocks Horn blocks to be each carefully fitted to its own place on the frames, held in position by steel bolts, turned so as to be driving fit through both frames and horn blocks, all holes being reamered out with the blocks in position.

Wheels and Axles The wheel centres to be made of tough cast iron, accurately turned and bored, crank pins of polished hard steel to be fitted in; the tyres to be made of best hard Siemens steel 3″ deep on the tread and 5½″ wide, bored and shrunk on, and secured to the wheel centres by turned steel bolts. Wheels all turned to equal diameter and gauge. The axles to be made of the best Siemen-Martin's steel with journals 6½" diameter and 7½" wide. The wheels to be forced on the axles by hydraulic pressure. Crank pin holes bored on a special machine exactly at right angles and of equal radii from wheel centre after same have been forced on the axles.

Brakes A powerful screw brake to be fitted so as to lock all the wheels, the handle being on the right hand side of the engine. Brake blocks of cast iron fitted to all wheels, bearing on the flanges, as well as the tread of the tyre. Also fitted with steam brake, with improved regulator, copper pipes and relief valve.

Sand Boxes To be placed on each side of the engine at the leading end, coupled together and worked by levers from the foot-plate; also two boxes at trailing end, coupled together. All four boxes to be self-acting.

Springs The springs to be made of the best spring steel, with suitable links and pins made from the best hammered iron.

Draw Bars The draw hooks to be of a heavy pattern, and made from best hammered iron, to be fitted with links and pins, and the bars to have volute springs fitted on.

Cylinders The cylinders to be of the best hard, close grained cylinder metal, accurately bored out to 16" diameter and 22" stroke, to be faced and strongly bolted together with steel bolts, turned so as to be a driving fit into reamered holes through frames, and cylinder flanges, and then securely fastened to the frames. Waste water cocks to be fitted to each cylinder and worked by suitable levers from the foot-plate. Cylinders fitted with gun-metal glands and neck rings for the piston and valve spindle rods.

Pistons The pistons to be 3" in width, and fitted with spring metallic rings. The rods to be of hard steel 2½" diameter secured by gun-metal nuts to the pistons, and by cotters to the crossheads. Metallic packing for piston and valve spindle rods.

Crossheads and Slide Bars The crossheads to be of the best solid forged steel, properly turned and fitted with steel pins and phosphor-bronze shoes, with ample wearing surfaces. The slide bars to be of hard steel, accurately planed and polished, secured at one end to a strong motion plate and to the cylinder covers at the other end.

Valve Motion The whole of the valve motion to be thoroughly fitted together and then case-hardened, after which it to be accurately refitted. The gearing to be the ordinary link motion, reversing from the footplate, worked from

Fig. 147 1396/1915 Lysaght's Scunthorpe Works No. 13

four eccentric sheaves and straps, made in halves, with solid quadrants and dies. Slide valves to be of hard metal accurately planed and scraped to the cylinder faces.
Connecting and Coupling Rods The connecting and coupling rods to be of the best solid forged mild steel, polished and to have heavy bearings of the best phosphor-bonze, fitted in and secured by suitable steel cotters, with ample provision for taking up the wear. Oil syphons to be forged on the rods solid.
General Fittings The engine to be fitted with two sets of Asbestos-packed water gauge cocks, fitted with polished plate-glass guards, instantly removable, two back-pressure valves, jet valve, steam gauge and cock, blow-off cock, sixteen washing out plugs, two safety plugs, four cylinder cocks, two injector steam valves, two feed cocks, whistle and whistle valve, two tallow cocks, lubricators fixed on the slide bars, and regulator; all of the best gun-metal. The two injector steam valves, pressure gauge cock, whistle and jet valve all to take their steam from one column, which is arranged with a valve so that the steam can be shut off and all these fittings examined with the engine in full steam.
Tools The following tools to be supplied with the engine, viz: One 25 ton traversing screw jack, two lock-up tool boxes, containing a complete set of spanners, one hand hammer and chisel, one coal hammer and one lead hammer, two spare gauge glasses with India-rubber rings, hand brush, two oil cans, one oil bottle, head, tail, gauge and hand signal lamps, one crowbar and one complete set of firing irons and shovel, a few assorted split pins and some spare fire-bars.
Painting During construction all the plates to be thoroughly cleaned free from rust, and to receive two coats of priming, three coats of filling up, rubbed down to a smooth surface, and when completed to receive two coats of grey, two coats of green, picked out with black, and finally fine lined in a neat and artistic style, and varnished twice with best Engine Copal varnish. If for shipment, the last coat to be grey ready to receive its finishing coat at its destination.
General All working parts to be interchangeable with locomotives of the same class and Whitworth standards used throughout.
Testing Under Steam When completed, the Engine to be thoroughly tested under steam on a line of considerable length, and up and down an incline specially constructed for the purpose.
Weight etc. Weight: empty 32 tons 10 cwts; loaded 40 tons 0 cwts. Heating surface: Tubes 716 sq. ft; Firebox 81 sq. ft; Total 797 sq. ft. Grate Area 15·75 sq. ft.

The materials and workmanship throughout to be of the best description.

Peckett's were specialists in this field and endeavoured to standardise as many parts as possible to facilitate interchangeability and to reduce the variety of stock items. They built their standard range for stock and as the above specification shows, all the materials were 'of the best'. Many indeed had very long lives despite the lack of maintenance and arduous duties they had to perform. They were to be found all over the country, Wales was particularly partial to them and a few even penetrated into Scotland – the stronghold of Barclay and other Scottish made tank locomotives.

Class	Date	Type	Cylinders inches	D.W. ft in	W'Base ft in	Weight WO ton cwt	Gauge ft in	Tank Gallons	WP lb/sq in
Cranmore		0-4-0ST	OC 7 x 10	1 8	3 6	7 10	2 0	125	160
Jurassic	1903	0-6-0ST	OC 7 x 10	1 8	6 6	7 10	1 9	160	160
Yorktown	1920	0-4-0ST	OC 7 x 12	2 0	4 6	11 10	Std	320	160
'1287'	1912	0-4-0T	OC 8 x 12	2 3	4 6	12 10	Std	230	160
'1682'	1925	0-4-0ST	OC 9 x 14	2 3	4 6	14 10	Std	350	180
'1903'	1936	0-4-0ST	OC 10 x 15	2 9	5 0	18 0	Std	475	180
'1853'	1934	0-4-0ST	OC 12 x 20	3 0½	5 3	23 0	Std	690	180
'1691'	1925	0-6-0ST	IC 13 x 18	3 0½	10 0	28 10	Std	800	170
Greenhithe	1927	0-4-0ST	OC 14 x 22	3 2½	5 6	28 10	Std	830	180
'1960'	1939	0-4-0ST	OC 14 x 22	3 2½	5 6	30 0	Std	920	180
'1860'	1933	0-6-0ST	OC 14 x 22	3 7	10 0	35 0	Std	920	180
'1464'	1916	0-4-0ST	OC 15 x 21	3 7	6 6	34 10	Std	1000	180
'1895'	1935	0-4-0ST	IC 16 x 22	3 10	7 6	38 0	Std	1000	180
'1892'	1935	0-4-0ST	OC 16 x 24	3 10	7 0	37 0	Std	1100	180
'1859'	1932	0-6-0ST	OC 16 x 24	3 10	11 0	43 0	Std	1100	180
'1519'	1919	0-6-0ST	IC 16 x 22	3 10	11 9	41 0	Std	1175	160
Ebbw Vale	1904	0-6-0ST	IC 18 x 24	4 0½	15 0	47 10	Std	1350	180

Examples of typical tank locomotives. Many other classes were designated by letters.

FOX, WALKER & CO.

At an unknown date the firm became PECKETT & SONS LTD.

In 1931 a 0-8-0 tender locomotive was built for the Christmas Island Phosphate Co. Ltd with outside cylinders 18″ x 26″ and 4′ 6″ diameter wheels. This was the largest type built weighing with tender some 73 tons.

Both during the 1914 and 1939 wars, they were busily engaged in building tank locomotives for industry but after the latter war the going proved very hard and only five locomotives left the works in 1947. In the early 1950s orders did increase to some extent and in 1956 the first diesel-mechanical locomotive appeared, but the market had been entered at too late a stage to make any impact.

The last steam locomotive Works No. 2165 left the works in 1958, a 3′ gauge 0-6-0T, for the Sena Sugar Estates, Mozambique.

The total output of Fox Walker/Peckett steam locomotives was approximately 1900 but this may be a conservative figure as there were many blanks in the Works List, some of these were intentional blanks for reasons unknown and other blanks may be unidentified locomotives.

In addition some customers were supplied with the necessary parts to assemble at their own works. These included such firms as John Lysaght Ltd, Skinningove Iron Co. Ltd and the Ebbw Vale Steel, Iron and Coal Co.

Pecketts also built a few locomotives for other locomotive builders as a sub-contractor.

In 1958 the firm was taken over by the Reed Crane & Hoist Co. Ltd and Peckett spare parts were supplied for a time until this firm went into liquidation.

A firm trading as Peckett and Sons Ltd of Ongar, Essex are in existence and spares are still supplied.

FRAZER & CHALMERS ENGINEERING WORKS LTD
Erith

At least two locomotives are reported to have been built by this firm for abroad, but no confirmation can be obtained.

FURNESS RAILWAY
Barrow-in-Furness

One steam railcar was built at Barrow in 1905 to the designs of William Frank Pettigrew the locomotive superintendent at this time (1897–1918).

The cylinders were 11″ x 14″ with four coupled wheels 2′ 10″ diameter.

The boiler pressure was 160 lb/sq. in. and the overall length over buffers was 60′ 11″. It was intended for the Ulverston to Lake Side branch but it was not successful and soon withdrawn.

GALLOWAY, BOWMAN & GLASGOW
Caledonian Foundry, Great Bridgewater Street, Manchester

Founded in Manchester in 1790, by William Galloway and joined by James Bowman about 1811 this firm was known as Galloway, Bowman & Company. In 1820 William Glasgow joined the firm which became Galloway, Bowman & Glasgow; their work was described as 'millwrights and engineers' but in 1830 a decision was made to commence building locomotives. The first one emerged in August 1831, taking the form of a 2-2-0 named MANCHESTER, and was offered to the Liverpool and Manchester Railway. It had vertical cylinders in front, driving a dummy crankshaft which in turn drove the driving wheels at the rear. Extensive rebuilding took place shortly afterwards with the cylinders and dummy crankshaft in the middle, and the chimney which had previously been placed adjacent to the firebox, was moved to the front of the engine. According to C. F. Dendy Marshall this rebuilt engine was renamed CALEDONIAN and was bought by the Liverpool and Manchester Railway and became their No. 28. The cylinders were 12″ x 16″ and the driving wheels 5′ 0″ diameter. Other sources state that the CALEDONIAN was a new engine with 0-4-0 wheel arrangement, the second built by the firm in October 1832, and there is evidence that both engines were working on the Liverpool and Manchester Railway in 1833.

The claim that the MANCHESTER was the first to be made in Manchester was made by Galloway, Bowman and Glasgow, and also by Sharp Roberts with their 2-2-0 EXPERIMENT built in 1833, so it is reasonable to assume that the former had a justifiable claim for that honour.

GALLOWAY, BOWMAN & GLASGOW

Only two or possibly three locomotives were built, as the potential was not realised by the management.

The two Galloway brothers, John and William, severed their connections with the firm in 1835, and opened the Knott Mill Ironworks as W. & J. Galloway which concentrated on the manufacture of patent boilers and continued as Galloways' Ltd. until closure in the 1920s.

Galloway, Bowman & Glasgow closed down in 1838–9. A locomotive MERSEY, used by the contractors building part of the Midland Counties Railway in 1838, may have been CALEDONIAN renamed. The CALEDONIAN had been sold to the London & Birmingham Railway in May 1837.

An exhaustive article regarding these locomotives appears in the SLS for November 1953 (No. 342) by P. C. Dewhurst and S. H. P. Higgins, and attempts to resolve how many locomotives were built by this firm and what happened to them.

GARFORTH W. J. & J

Date	Type	Cylinders inches	D.W. ft in	SA & M Rly.	
				No	Name
1/1847	2-4-0	15 x 20	5 0	32	Fury
6/1847	2-4-0	15 x 20	5 0	33	Vulcan
4/1848	0-4-2	15½ x 22	5 0	37	Mentor
6/1848	0-4-2	15½ x 22	5 0	39	Mars
8/1848	0-4-2	15½ x 22	5 0	40	Triton
11/1848	0-4-2	15½ x 22	5 0	41	Neptune
9/1849	0-4-2	16 x 22	5 0	42	Taurus
11/1849	0-4-2	16 x 22	5 0	43	Plutus

All for Sheffield, Ashton-under-Lyne & Manchester Rly.

GARFORTH W. J. & J.
Manchester

This firm was a sub-contractor to Sharp Bros. & Co. and built two 2-4-0s and six 0-4-2s for the Sheffield, Ashton-under-Lyne and Manchester Railway, to Sharp's drawings. They were well built and some lasted twenty-five years. No further building can be traced.

GARRETT, RICHARD & SONS LTD
Leiston

This famous Suffolk firm were noted for their traction engines, agricultural machinery, steam wagons and stationary engines. Like many such firms they became interested in railway traction with a view to incorporating their steam wagon unit and boiler in a locomotive or steam railcar. In 1925 an enquiry was received from the Emu Bay Railway, Tasmania for a steam railcar and the Metropolitan-Cammell Carriage, Wagon & Finance Co. were approached regarding the provision of the car body and seating. A considerable amount of design work was put into this project using the steam wagon boiler but incorporating a new engine unit.

Although a few quotations and specifications were sent out in answer to various enquiries, no orders were forthcoming.

The same happened to a proposed shunting locomotive about the same time. The London & North Eastern Railway had proposals submitted but these were turned down on the grounds that the boiler was too small. Although brochures were produced for both types nothing came of the venture and none were built.

Fig. 148 **Galloway Bowman & Glasgow** 1831 First locomotive built in Manchester

GARSWOOD HALL COLLIERY CO. LTD.

Garswood Hall, Lancs.

Built a 0-6-0ST, with inside cylinders and named ARTHUR in 1884 and given No. 1 in the Garswood Hall list. It was scrapped in 1933. No further details.

GAS, LIGHT & COKE COMPANY

Beckton Gas Works

This extensive system with standard gauge track had a large stock of steam locomotives. Between about 1878 and 1900 three standard types of locomotives were adopted: 10″ x 18″ Side tank; 10″ x 18″ Saddle tank; and 12″ x 20″ Side tank.

The saddle tanks had cut down fittings for low head room working and were known as *Jumbos.* All were built by Neilson of Glasgow except one 10″ x 18″ side tank and one *Jumbo* by Black Hawthorn.

In 1900 Neilsons became part of the North British Locomotive Company, and when Beckton required two more 10″ x 18″ side tanks they built them at Beckton, obviously from Neilson drawings, and probably borrowed their patterns.

The link motion, side rods etc. were probably purchased from the North British Locomotive Company, who were able to supply these parts until they went out of business.

The two side tanks which were built at Beckton were Nos. 30 and 31 and both appeared in 1902. They both had 2′ 9″ diameter wheels. The casing on top of the boiler between the chimney and dome was to conceal a steamcock which was used to shut off steam to the ejector when it needed cleaning.

Fig. 149 **Gas Light & Coke** No. 31 one of two built at Beckton

GIBB & HOGG

Victoria Engine Works, Airdrie

Although locomotive building did not commence until the 1890s this firm was established in 1866 as founders and engineers.

When McCulloch Sons and Kennedy of Kilmarnock closed down in 1890, Gibb & Hogg acquired their patterns and drawings.

About twenty locomotives were built for local industries mostly four coupled saddle tanks with typical horizontal outside cylinders. Stove pipe chimneys were fitted similar to the previous maker's pattern. In fact they were typical Kilmarnock saddle tanks. Perhaps the chief characteristic was the cylinder cleading which surrounded half the cylindrical portion and continued vertically to meet the underside of the footplating. The wheel spokes were the usual I section.

The firm closed down in 1912.

GIBB & HOGG

Works No	Date	Type	Cylinders inches	D.W. ft in	Customer	No/Name
		0-4-0ST	14 x 22	3 6	Wm. Dixon	Calder No 11
		0-4-0ST	12 x 20	3 6	Wm. Dixon	Calder No 12
	1898	0-4-0ST			Korsnäs AB. Sweden	
5	1899	0-4-0ST	14 x 22		Dalmellington Iron Co	Dalmellington No 5
	1900	0-4-0ST	14 x 22		Niddrie & Benhar Coal Co	No 3
	1902	0-4-0ST			Oakbank Oil Co Ltd	No 2
	1902	0-4-0ST			Shotts Colly	
	1903	0-4-0ST			Meyer. Widnes	
54	1904	0-6-0ST			Eden Colliery	No 6
56	1905	0-4-0ST			Niddrie & Benhar Coal Co	No 4
58	1905	0-4-0ST			Babcock & Wilcox Dumbarton	

GILKES WILSON & CO.

Teesside Engine Works, Middlesbrough

Established 1843–4 locomotive building commenced in 1847 although locomotive repairs were carried out before this. Edgar Gilkes came from the Stockton and Darlington Railway and a considerable number of the total locomotives built by this firm were supplied to this railway and later to the North Eastern Railway. Other customers were the York, Newcastle & Berwick Railway, Leeds & Thirsk Railway, Liskeard & Caradon Railway, Llanelly Railway and the Newmarket & Great Chesterford Railway.

Over 100 2-4-0 and 0-6-0 types were built for the S & D Railway. The first batch of six 2-4-0s had outside cylinders 15″ x 22″ and 5′ 6″ coupled wheels. The 0-6-0s had inside cylinders varying from 15″ x 24″ to 17″ x 24″ and coupled wheels 4′ 3″ to 5′ diameter.

Fig. 150 **Gibb & Hogg** 1903 0-4-0ST Meyer (Northern) Ltd Widnes. Note cast iron wheels and cylinder position.

A number of four coupled and six coupled industrial tank locomotives were supplied to various collieries.

In 1865 a merger was carried out with the firm of Hopkins & Company with their Rolling Mills and the title of the firm became HOPKINS GILKES & CO. LTD.

Locomotive building continued until 1875 and just previous to this the title was again changed to TEES-SIDE IRON & ENGINE WORKS CO. LTD.

Complete closure took place in 1880 and altogether 351 locomotives were built from 1847 to 1875.

A prominent part of their business was bridge building and they took over the building of the Tay Bridge in 1873 from Messrs De Bergh and Company. Designed by Thomas Bouch, it met with a disastrous end on the night of 28 December, 1879 a contributory factor being the quality of the cast iron in the columns and poor workmanship.

Fig. 151 **Gilkes Wilson** 75/1856 Stockton & Darlington Rly No. 117 NUNTHORPE

Fig. 152 264/1869 (Hopkins Gilkes) Liskeard & Caradon Rly KILMAR

GLASGOW, PAISLEY & GREENOCK RAILWAY

Greenock

Robert Sinclair was the locomotive superintendent of the GP&GR. It was decided in 1845 to commence building the company's own locomotives and three 2-2-2s were put in hand and completed in 1846. They were based on Alexander Allan's *Crewe* design. The characteristics were double frames with the cylinders fixed between them. The driving wheels had inside bearings only and the leading and trailing wheels, outside bearings. The boiler had a raised top firebox on which the dome and spring balance safety valve were fitted with an additional safety valve in the middle of the boiler barrel.

They were numbered 17, 18 and 19 and were the last built for the company, for in 1847 the railway became part of the CALEDONIAN RAILWAY.

Sinclair was given the same post on the Caledonian Railway, and Greenock became the locomotive headquarters. Building continued with 2-2-2s similar to the original three, followed by larger *Crewe* type 2-2-2s with outside cylinders 15″ x 20″ and 6 ft diameter driving wheels.

The Greenock Works were originally under a lease until 1852 when the premises were bought.

In 1848–9 ten 0-4-2s were built, the first five having outside cylinders 16″ x 18″ and the remainder 17″ x 18″. All had 4′ 6″ diameter coupled wheels. The boiler had a raised firebox and similar arrangement of dome and safety valves to the 2-2-2s. They had Gooch's link motion. The boiler pressure was 90 lb/sq. in. and the weight in working order without tender was 24 tons 13 cwts.

The next class was a six coupled mineral locomotive, ten were built 1849–50. Outside cylinders were 16″ x 20″ and wheels 4′ 7″ diameter.

Two 2-2-2WTs were built in 1851, they were tiny locomotives with outside cylinders 9″ x 15″ a 2′ 8″ diameter boiler with a total heating surface of 386 sq. ft and 8·27 sq. ft of grate. They weighed just over 26½ tons in working order.

In 1853 eight tank locomotives were built and although designed for working in colliery sidings the wheelbase was 12 ft, the drive being off a dummy crankshaft mid-way between the two pairs of wheels, and driven by outside cylinders. Four tanks were fitted, a saddle tank and three well tanks between the frames. They were all interconnected and filled from the saddle tank. The smoke box was built flush with the top of the tank and the firebox was raised, with dome and safety valves fitted on top. They were soon altered to 0-4-2Ts, the dummy crankshaft being replaced by moving the rear pair of wheels to this position and adding the trailing wheels with new horn guides etc. They were numbered 136 to 143.

Further 2-2-2s were then built, similar in design to those previously built. A boiler with raised firebox on which the dome and safety valve gear was fitted. Twelve were built in of 891 sq. ft and a grate area of 12·12 sq. ft. The pressure was 110 lb/sq. in. Gooch's box link valve gear was fitted. Twelve were built in 1854–5 and numbered 65–76. The boiler of No. 76 exploded, less than twelve months after building, at Rockliffe station and the locomotive was written off.

The next class built was a 5′ 2″ version with a smaller boiler, a pressure of 120 lb/sq. in. and larger cylinders. The design still followed the Alexander Allan outline of framing and cylinders. Ten were built, the first four with 16″ x 20″ cylinders and the remainder 17″ x 20″, all in 1855.

In the same year a second batch of 15″ 2-4-0 were built, and the boilers of this series had flush topped fireboxes. Eleven were ordered but only three were completed at Greenock the remainder being the first class to be built at the St. Rollox works.

Considering the size of the workshops the output was no mean feat and from 1846 to 1855 a total of 97 locomotives were built.

The machinery was gradually transferred to the new factory at St. Rollox where all new locomotives were built from 1855.

GLASGOW, PAISLEY & GREENOCK RLY: GREENOCK

Date	Type	Cylinders inches	D.W. ft in	Running No	Type
1846	2-2-2	14 x 20	5 6	17	Reno Cal. Rly 1 Crewe
1846	2-2-2	14 x 20	5 6	18	Reno Cal. Rly 2 Crewe
1846	2-2-2	14 x 20	5 6	19	Reno Cal. Rly 3 Crewe

GLASGOW, PAISLEY & GREENOCK RAILWAY

CALEDONIAN RLY – GREENOCK

Date	Type	Cylinders inches	D.W. ft in	CR No	Type
1847	2-2-2	OC 15 x 20	6 0	4	Crewe
1847	2-2-2	OC 15 x 20	6 0	5	Crewe
1847	2-2-2	OC 15 x 20	6 0	6	Crewe
1847	2-2-2	OC 15 x 20	6 0	7	Crewe
1847	2-2-2	OC 15 x 20	6 0	8	Crewe
1847	2-2-2	OC 15 x 20	6 0	9	Crewe
1847	2-2-2	OC 15 x 20	6 0	42	Crewe
1847	2-2-2	OC 15 x 20	6 0	43	Crewe
1847	2-2-2	OC 15 x 20	6 0	44	Crewe
1848	2-2-2	OC 15 x 20	6 0	32	Crewe
1848	2-2-2	OC 15 x 20	6 0	45	Crewe
1848	2-2-2	OC 15 x 20	6 0	46	Crewe
1848	2-2-2	OC 15 x 20	6 0	47	Crewe
1848	2-2-2	OC 15 x 20	6 0	48	Crewe
1848	2-2-2	OC 15 x 20	6 0	49	Crewe
1848	2-2-2	OC 15 x 20	6 0	50	Crewe
1848	2-2-2	OC 15 x 20	6 0	51	Crewe
1848	2-2-2	OC 15 x 20	6 0	52	Crewe
1848	2-2-2	OC 15 x 20	6 0	53	Crewe
1848	2-2-2	OC 15 x 20	6 0	54	Crewe
1848	2-2-2	OC 15 x 20	6 0	55	Crewe
1848	2-2-2	OC 15 x 20	6 0	56	Crewe
1848	2-2-2	OC 15 x 20	6 0	57	Crewe
1848	2-2-2	OC 15 x 20	6 0	58	Crewe
1848	0-4-2	OC 16 x 18	4 6	101	
1848	0-4-2	OC 16 x 18	4 6	102	
1848	0-4-2	OC 16 x 18	4 6	103	
1848	0-4-2	OC 16 x 18	4 6	104	
1848	0-4-2	OC 16 x 18	4 6	105	
1849	2-2-2	OC 15 x 20	6 0	31	Crewe
1849	2-2-2	OC 15 x 20	6 0	33	Crewe
1849	2-2-2	OC 15 x 20	6 0	34	'Crewe'
1849	2-2-2	OC 15 x 20	6 0	35	'Crewe'
1849	0-4-2	OC 17 x 18	4 6	106	
1849	0-4-2	OC 17 x 18	4 6	107	
1849	0-4-2	OC 17 x 18	4 6	108	
1849	0-4-2	OC 17 x 18	4 6	109	
1849	0-4-2	OC 17 x 18	4 6	110	
1849	0-6-0	16 x 20	4 7	122	
1849	0-6-0	16 x 20	4 7	123	
1849	0-6-0	16 x 20	4 7	124	
1850	2-2-2	OC 15 x 20	6 0	36	'Crewe'
1850	2-2-2	OC 15 x 20	6 0	37	'Crewe'
1850	0-6-0	OC 16 x 20	4 7	125	
1850	0-6-0	OC 16 x 20	4 7	126	
1850	0-6-0	OC 16 x 20	4 7	127	
1850	0-6-0	OC 16 x 20	4 7	128	
1850	0-6-0	OC 16 x 20	4 7	129	
1850	0-6-0	OC 16 x 20	4 7	130	
1850	0-6-0	OC 16 x 20	4 7	131	
1851	2-2-2WT	OC 9 x 15	5 1	77	
1851	2-2-2WT	OC 9 x 15	5 1	78	
1853	0-4-0SWT	OC 12 x 18	4 7	136	Crampton, Intermediate Shaft
1853	0-4-0SWT	OC 12 x 18	4 7	137	Crampton, Intermediate Shaft
1853	0-4-0SWT	OC 12 x 18	4 7	138	Crampton, Intermediate Shaft
1853	0-4-0SWT	OC 12 x 18	4 7	139	Crampton, Intermediate Shaft
1853	0-4-0SWT	OC 12 x 18	4 7	140	Crampton, Intermediate Shaft
1853	0-4-0SWT	OC 12 x 18	4 7	141	Crampton, Intermediate Shaft
1853	0-4-0SWT	OC 12 x 18	4 7	142	Crampton, Intermediate Shaft
1853	0-4-0SWT	OC 12 x 18	4 7	143	Crampton, Intermediate Shaft
1854	2-2-2	OC 15½ x 20	7 2	65	
1854	2-2-2	OC 15½ x 20	7 2	66	
1854	2-2-2	OC 15½ x 20	7 2	67	

GLASGOW, PAISLEY & GREENOCK RAILWAY

CALEDONIAN RLY: GREENOCK

Date	Type	Cylinders inches	D.W. ft in	CR No.	Type
1854	2-2-2	OC 15½ x 20	7 2	68	
1854	2-2-2	OC 15½ x 20	7 2	69	
1854	2-4-0	OC 16 x 20	5 2	152	
1854	2-4-0	OC 16 x 20	5 2	153	
1854	2-4-0	OC 16 x 20	5 2	154	
1854	2-4-0	OC 16 x 20	5 2	155	
1854	2-4-0	OC 16 x 20	5 2	156	
1854	2-4-0	OC 16 x 20	5 2	157	
1854	2-4-0	OC 16 x 20	5 2	158	
1854	2-4-0	OC 16 x 20	5 2	159	
1854	2-4-0	OC 16 x 20	5 2	160	
1855	2-2-2	OC 15½ x 20	7 2	70	
1855	2-2-2	OC 15½ x 20	7 2	71	
1855	2-2-2	OC 15½ x 20	7 2	72	
1855	2-2-2	OC 15½ x 20	7 2	73	
1855	2-2-2	OC 15½ x 20	7 2	74	
1855	2-2-2	OC 15½ x 20	7 2	75	
1855	2-2-2	OC 15½ x 20	7 2	76	
1855	2-4-0	OC 16 x 20	5 2	161	
1855	2-4-0	OC 16 x 20	5 2	162	
1855	2-4-0	OC 16 x 20	5 2	163	
1855	2-2-2	OC 16 x 20	5 2	164	
1855	2-4-0	OC 17 x 20	5 2	165	
1855	2-4-0	OC 17 x 20	5 2	166	
1855	2-4-0	OC 17 x 20	5 2	167	
1855	2-4-0	OC 17 x 20	5 2	168	
1855	2-4-0	OC 17 x 20	5 2	169	
1855	2-4-0	OC 17 x 20	5 2	170	
1855	2-4-0	OC 15 x 22	5 2	171	
1855	2-4-0	OC 15 x 22	5 2	172	
1855	2-4-0	OC 15 x 22	5 2	173	

Fig. 153 **GPK&AR** 1846 2-2-2 LIGHTNING

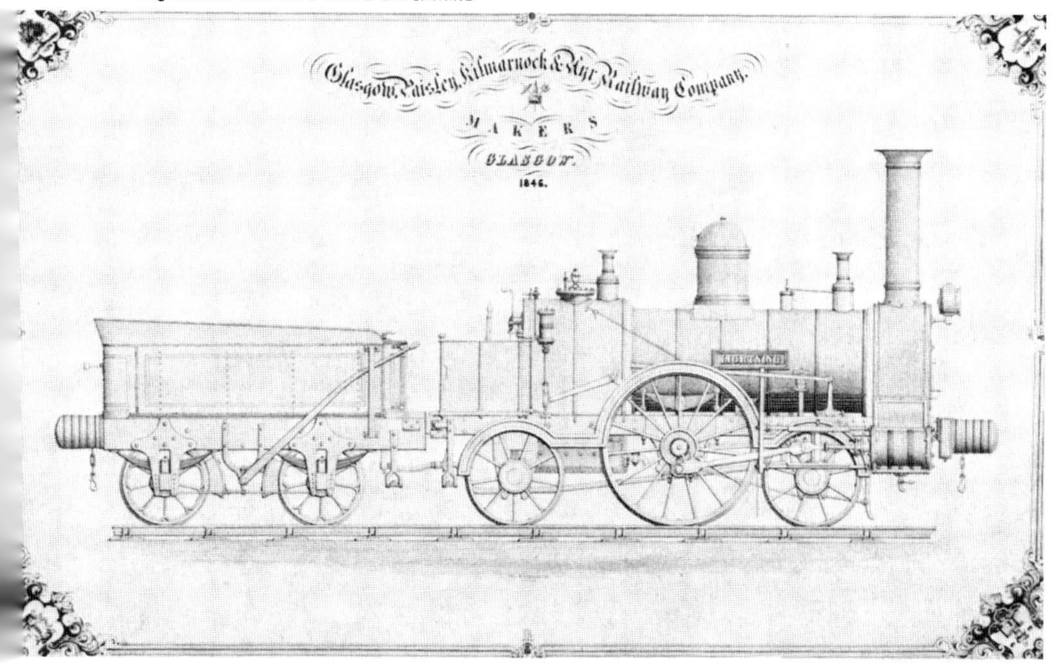

GLASGOW, PAISLEY, KILMARNOCK & AYR RAILWAY

Cook Street, Glasgow

Locomotive Superintendents D. B. Stark 1840; Peter Robertson 1840–1853; Patrick Stirling 1853–1866. Work transferred to new factory at Kilmarnock in 1856.

Repairs only were carried out from 1839 when the workshops were established. The locomotive stock was made up of *Bury* 2-2-0s, and 0-4-0s and 2-2-2 types. The company's engineer J. Miller provided the specifications for the 2-2-2s which were built by a number of locomotive building contractors in Scotland. In 1845 it was decided to use Cook Street works to build new and more powerful locomotives and six 2-2-2s were built up to 1848. They all had outside cylinders with domes on the front ring of the boilers. Similar ones were built by Caird & Co. (one) and Kinmond, Hutton & Steel (six). Boiler heating surfaces and cylinder bores varied considerably.

The superintendent of locomotives at this time was P. Robertson. In 1850 the railway, together with the Glasgow, Dumfries, & Carlisle Railway combined to form the GLASGOW & SOUTH WESTERN RAILWAY.

Two further locomotives were built in 1851–2, but with the increased number of locomotives to maintain, Cook Street works proved inadequate and expansion was impossible in the closely confined area.

New shops were built at Kilmarnock which were opened in 1856 and all work was transferred there. Construction and repairs were concentrated at the new works.

GLASGOW & SOUTH WESTERN RAILWAY

Kilmarnock

Locomotive Superintendents: Patrick Stirling 1853–1866; James Stirling 1866–1878; Hugh Smellie 1878–1890; James Manson 1890–1912; Peter Drummond 1912–1918; Robert H. Whitelegg 1918–1923.

The first locomotives built at Kilmarnock under Patrick Stirling's supervision were 2-2-2s with outside cylinders, domed boilers, with no cab and slotted driving wheel splashers. The cylinders were 16″ x 21″ and the driving wheels 6′ 6″ diameter. Stirling had been trained at his uncle's Dundee works, and he also spent a short time on the Dumbarton & Balloch Joint Railway, Neilson's, and Hawthorns at Newcastle. The first 2-2-2 appeared in 1857 and had strong Hawthorn characteristics but later locomotives of the same class appeared with domeless boilers, a feature of Stirling's that gave all his products that distinctive appearance which in later years on the GNR enhanced his famous 8 ft singles.

He then designed a domeless 0-4-2 which however were built by Sharp Stewart, and were the first GSW locomotives to be equipped with the Giffard injector. His first locomotives had no cabs but in the early 1860s he introduced his round top cab with the port hole type of cab side windows, the cab coming down flush with the opening to the tender.

GLASGOW, PAISLEY, KILMARNOCK & AYR RLY

Date	Type	Cylinders inches	D.W. ft in	No	Name
10/1845	2-2-2	OC 13½ x 18	5 6		Thunderbolt
1846	2-2-2	OC 13½ x 18	5 6		Lightning (Fig 153)
1846	2-2-2	OC 13½ x 18	5 6		Firebrand
1847	2-2-2	OC 14 x 18	5 6		Orion
1847	2-2-2	OC 15 x 18	5 6		Atlas
1848	2-2-2	OC 15 x 18	5 6		Vesuvius

GLASGOW & SOUTH WESTERN RLY

Date	Type	Cylinders inches	D.W. ft in	No	Name
1851	2-2-2	15 x 20	6 0	8	
1852	0-4-0	15 x 20	5 0	7	

Six 0-6-0s were built at Kilmarnock between 1862–3. They had inside cylinders, domeless boiler, 16″ x 22″ cylinders and 5 ft diameter wheels. A 0-4-2 version was built by R. & W. Hawthorn Ltd.

His last design was for a series of 2-2-2 main line locomotives and was a considerable improvement. The driving wheels were 7 ft diameter with 16″ x 24″ outside cylinders and domeless boilers. They were not very successful but in outline gave an indication of what was to come from Doncaster, where Patrick Stirling went in 1866 as locomotive superintendent.

His post on the GSWR was taken by his brother James who up till then had been works manager under Patrick.

He retained the domeless boiler but the tall brass safety valve cover gave place in 1874 to the popular Ramsbottom valves with no outer casings. The severe cab was also retained but slightly cut away as a concession, to enable the driver and fireman to lean out of the cab in more comfort.

All locomotives designed by James Stirling had inside cylinders and Stephenson's valve gear. The reversing gear received attention and at first the conventional lever reverse was fitted, and then a vertical screw reverser, followed by his steam reversing gear which attracted great interest and favourable comment. The cylinder for the reversing gear was placed inside the cab on the right hand side.

GLASGOW & SOUTH WESTERN RAILWAY

His first batch of locomotives were some very neat 2-4-0s with 17″ × 24″ cylinders and 6′ 7″ diameter driving wheels. Fifteen were built in 1868–9 followed by ten similar locomotives built in 1870–1, the only difference being in the driving wheels which were reduced to 6′ 1″ and were used mainly on the branch lines.

No fewer than 22 small 0-4-0 tender locomotives appeared between 1871–4; this type was popular on the Scottish railways and were useful for the sharp curved sidings. These were fitted with the vertical screw reversing gear.

Undoubtedly the twenty-two No. 6 class 4-4-0s were James Stirling's most spectacular design; No. 6 came out in 1873. It had 18″ x 26″ cylinders and large driving wheels 7′ 1½″ diameter. The main duties of this class were to haul the Anglo Scottish expresses from Carlisle to St. Enoch and in the reverse direction which they did with satisfactory results. Most of them were equipped with steam reversers, but by no means were they kept on these duties. They did a lot of work on the much heavier trains on the coast lines to Ayr and Ardrossan.

He also designed a 0-6-0 goods tender loco, a 0-4-4T, and 0-4-0T which were all built at Kilmarnock.

In 1878 James Stirling obtained the post of locomotive superintendent of the South Eastern Railway and left for Ashford.

Fig. 154 **G&SWR** 1871 Stirling 2-4-0 as 104A domeless boiler

GLASGOW & SOUTH WESTERN RAILWAY

Fig. 155 **G&SWR** 1887 Smellie's *153 class* 4-4-0 No. 56

His successor was Hugh Smellie who at one time had been works manager at Kilmarnock under James Stirling but had left in 1870 to take charge of the loco department of the Maryport & Carlisle Railway.

Again, the basic design laid down by Patrick Stirling, and reasonably perpetuated by his brother James was carried on by Smellie – the domeless boiler, steam reverser, Stephenson's link motion and the round topped cab.

Many 0-4-2 type locomotives were introduced by Stirling and from 1858 to 1878, 151 were built (but none at Kilmarnock) and mention must be made of this type as the last ninety were Stirling's first real mixed traffic locomotives on the G&SWR the latest having 18″ x 24″ cylinders and 5′ 7½″ coupled wheels.

Mr Smellie's first design was strangely enough a reversion to the 2-4-0 type. Whether he had a mistrust of the bogie or whether he wanted to be on safe ground is not known but these locomotives came out in 1879 and 1880. The cylinders were the same as Stirling's 4-4-0s but the wheels were reduced to 6′ 9½″ and the heating surface increased from 1112 sq. ft to 1206 sq. ft. They were used to supplement Stirling's 4-4-0s and on the Stranraer turns where the 4-4-0s could not penetrate because of the inadequate turntables. They were good engines and did their work with great economy. An interesting point of design was the leading axleboxes which had inclined planes having a lateral movement of half an inch on either side. Then followed a six coupled goods tender engine similar to the 2-4-0s except that the heating surface was reduced to 1061 sq. ft and the wheels were 5′ 1¼″ diameter. No less than sixty-four were built, forty-four from Kilmarnock. They were the first to be fitted with the steam brake on the GSWR. The Kilmarnock batch was turned out between 1881 and 1891.

More engine power was needed for the coast trains and a 4-4-0 was designed with 18¼″ x 26″ cylinders and 6′ 1¼″ wheels. Despite the increase in size of the cylinders the boiler was still of modest proportions 964 sq. ft of tube heating surface and 101 sq. ft for the firebox. All twenty four of them were built at Kilmarnock between 1882 and 1885 they were unofficially known as *Wee Bogies* or *Greenock Bogies*. A further series had 6′ 9½″ diameter driving wheels with a slightly larger boiler. These were known as *Big Bogies* and were chiefly used on the Anglo-Scottish expresses.

The Westinghouse brake had been standard for many years but from 1886 all new locomotives were fitted with vacuum brakes and this was standardised, primarily because the Midland Railway had adopted this brake as standard. The Midland were on excellent terms with the GSWR, and about this time amalgamation was a popular rumour.

In 1890 Hugh Smellie was appointed locomotive superintendent of the Caledonian Railway. Whilst at Kilmarnock, he had kept pace with traffic demands and had produced a sound series of locomotives with no startling features but constructed on sound principles and workmanship.

His successor was James Manson who had received his initial training at Kilmarnock and had gone on to gain experience in marine engineering and was appointed locomotive superintendent of the Great North of Scotland Railway in 1883. At Kittybrewster he had brought the locomotive department to a reasonable state of efficiency. At Kilmarnock the Stirling/Smellie traditions were put on one side and his own characteristic features appeared – a tall dome centrally placed on the boiler barrel, Ramsbottom safety valves with no casing around them, inside cylinders for most designs and Stephenson's motion driving direct to the valves on his goods locomotives, placed between the cylinders. Rocker arms were used on the passenger types, and balanced valves fitted on top of the cylinders and his 0-4-4Ts were similarly fitted. The Stirling round top cab vanished and a more orthodox and protective cab provided.

The 1892, 0-6-0 goods had been ordered from Dubs and were designed by Manson although the cabs and frames were similar to Smellie's designs, but this was probably to expedite delivery.

His first design for a 4-4-0 was very similar to the GNSR No. 12 class for which he had been responsible. In all he produced five classes of 4-4-0.

Class	Cylinders	H.S.	B.P.	1st Loco.	Total
8	18¼ x 26	1203	150	1892	57
336	18¼ x 26	1205	165	1895	25*
11	IC 14½ x 26 OC 12½ x 24	1173	165	1897	1
240	18¼ x 26	1434	170	1904	15
18	18¼ x 26	1407	170	1907	13

All had 6' 9½" DW except 336 with 6' 1¼" DW.
* Built by Dubs & Co.

The No. 8 class were sound locomotives and their performance was on a par with Smellie's 153 class, the 336 class were a smaller wheeled version of the No. 8 class with increased boiler pressure.

No. 11 was of course the most well known locomotive designed by Manson and was the first four cylinder 4-4-0 in the British Isles. It had two inside and two outside cylinders operated by two sets of Stephenson's valve gear both between the frames, one set driving the outside valves by means of a rocking shaft. In its original form it was not a brilliant performer, trouble was experienced in valve setting and its fuel consumption was heavier than the '8' class. The volume of the four cylinders was only slightly in excess of the normal two 18¼" x 26" fitted as standard on the other four classes. It was subsequently rebuilt twice.

The No. 18 class 4-4-0s differed in one respect to the other classes, the firebox was

Fig. 156 **G&SWR** 1905 Manson's *240 class* 4-4-0 No. 388. Note cab and capuchon

long and shallow instead of the usual deep firebox, this gave an increase of grate area from 18·2 sq. ft to 22 sq. ft. The first 4-6-0 design was Manson's next venture with 20" x 26" outside cylinders, 6' 6" diameter driving wheels and a heating surface of 1852 sq. ft. They were fine looking locomotives and did good work on the Anglo-Scottish expresses. Another departure from previous practice was the fitting of Belpaire fireboxes. The first ten were built by the North British Locomotive Co. in 1903 and Kilmarnock built seven more in 1910–11.

In 1912 Weir feedwater equipment was tried out by Manson on three locomotives: No. 389, non-superheated 4-6-0, No. 129 superheated 4-6-0 and No. 27, a 4-4-0 of the '18' class. The 4-6-0s had the drum fixed on top of the boiler whereas on the 4-4-0 it was put on the side footplating.

Although not built at Kilmarnock, Manson designed two superheated 4-6-0s which were built by the North British Locomotive Co. in 1911 numbered 128 and 129. They had outside cylinders 21" x 28", 6' 6" diameter coupled wheels and a total heating surface of 2005 sq. ft including 445 sq. ft for the Schmidt superheater. The grate area was 24·5 sq. ft. These two 4-6-0s differed greatly in detail from his former 4-6-0s, and were far more successful. They spent a lot of their time on the 2 pm Glasgow to Carlisle and gave some excellent performances and were fitting monuments to his last design.

Other types turned out by Manson were three rail motors for branch line working, fifteen 0-6-0Ts for shunting purposes, five of which were built by Drummond, and three series of 0-6-0 to cope with increased mineral traffic. The second series, the 361 class, were mostly built by outside contractors but two came from Kilmarnock shops in 1910.

Some 0-4-0Ts were built in 1907–9 for Greenock and Ardrossan harbour lines. A feature of these outside cylinder locomotives was that the valves were placed on top of the cylinders with inside Stephenson's valve gear operating the valves from the front end by means of rocking shafts.

James Manson retired in 1912 and lived to the ripe old age of 90.

He was succeeded by Peter Drummond, brother of Dugald Drummond, coming from the Highland Railway where he had been locomotive superintendent since 1896.

Up to this time, with a few exceptions the policy had been 'small locomotives' and an acceptance that, in the summer particularly, the heavy passenger trains would be assisted.

Peter Drummond on the other hand was a big engine man and this he proceeded to show

Fig. 157 **G&SWR** 1910 Manson 2-cylinder 4-6-0 No. 120

in his designs for the GSWR. Kilmarnock works were not very busy at this time and out of sixty-nine new locomotives built during the time Drummond was in office only six were built in the railway's own workshops, the rest being built by the North British Locomotive Co.

The six from Kilmarnock were a batch of 4-4-0s built in 1914–15. These had a strong L&SWR flavour and resembled their '463' class in appearance. They had inside cylinders with piston valves, Walschaerts valve gear and marine type big ends. This was obviously very much the influence of his brother Dugald who had unfortunately died in the same year as Peter's appointment at Kilmarnock.

One practice he did not copy was the water-tube firebox and his tenders were the more conventional six wheeled outside frame pattern. The marine type big ends gave constant trouble by over-heating and a return was made to the cottered type. On these 4-4-0s Schmidt superheaters were fitted, and Weir feedwater heaters.

The most radical change was the changeover to the left hand side of the footplate for the driver, an alteration which did not go down at all well. Manson's well known 4-4-0 No. 11 was given a larger boiler in 1915.

Although not built at Kilmarnock, Drummond designed a 2-6-0 with inside cylinders 19½" x 26" and 5 ft diameter driving wheels. The boiler was equipped with a Robinson superheater. These were very successful locomotives, very economical in fuel and water. The introduction of superheating to various classes improved their performance considerably.

Two tank designs were also built 'outside', a 0-6-0T and 0-6-2T, the latter being a shortened version of his 0-6-4Ts of the Highland Railway.

After being in office only six years Peter Drummond died in June 1918. Robert H. Whitelegg was selected to fill the post. He came from the London, Tilbury & Southend Railway where he had succeeded his father Thomas Whitelegg in 1911 as locomotive superintendent. The aftermath of war resulted in an endeavour to get the existing stock into reasonable running order and Whitelegg designed four boilers which he intended to be standard units for the types he intended to retain. In all ten standard boilers were planned but only four were built. These unfortunately showed themselves as poor steamers and heavy on coal.

He met with no more success when he redesigned Manson's valve gear and this had to be re-modified before it was at all successful.

He altered the driver's position back to the right hand side.

The four cylinder 4-4-0 which had been fitted with a larger boiler by Drummond, was entirely rebuilt in 1922. It was practically a new locomotive retaining only the wheels and part of the motion. The inside and outside cylinders were 14" x 26" and 14" x 24" respectively with two piston valves, one to each pair of cylinders. It was reboilered again and the heating surface increased from the original boiler's 1173 sq. ft to no less than 1803 sq. ft including 211 sq. ft of superheater tubes.

The name LORD GLENARTHUR was given to this rebuild and it performed in a satisfactory manner on the Glasgow–Ayr run.

Robert Whitelegg had already built a 4-6-4 Baltic tank for the LT&SR and inevitably the same type was introduced by him on the GSWR. They were not built at Kilmarnock but from the North British Locomotive Co. Six were built in 1922.

From 1 January, 1923 the Glasgow & South Western Railway became part of the LMS group. Whitelegg remained at Kilmarnock until 1 March, 1923 when he left to become general manager of Beyer Peacock & Co.

With the amalgamation came the standardisation plan and as the Caledonian Railway was the largest railway in the Scottish division, Kilmarnock lost its glory and tradition.

Kilmarnock had produced 392 new locomotives between 1857 and 1921 but only fourteen from 1914 to the end. A tremendous amount of rebuilding took place over the years, but the works were very limited and this was probably the reason why so many locomotives were built by outside contractors who could produce them cheaper and quicker. Repairs were carried out until 1959, the works closing on 4 July, 1959.

GLASGOW & SOUTH WESTERN RAILWAY

KILMARNOCK WORKS—OUTPUT

Date Built	Type	Class	No. Built	
1857-1860	2-2-2	2	13	
1861-1864	2-2-2	40	10	
1862-1863	0-6-0	46	6	
1864-1868	2-2-2	45	11	
1865-1866	0-4-0	52	6	
1866-1867	0-6-0	58	6	
1868-1870	2-4-0	8	15	
1870-1871	2-4-0	71	10	
1871-1874	0-4-0	65	22	
1873-1877	4-4-0	6	22	
1875-1876	0-4-0T	14	6	
1877-1878	0-6-0	13	12	
1879	0-4-4T	1	4	
1879-1880	2-4-0	157	12	
1881-1891	0-6-0	22	44	
1882-1885	4-4-0	119	24	
1886-1889	4-4-0	153	20	
1892-1904	4-4-0	8	57	
1896	0-6-0T	14	4	
1897	4-4-0	11	1	
1897-1898	0-6-0	160	18	
1902-1906	4-4-0	240	15	
1903	0-6-0T	14	6	
1904-1905	0-4-0RM	—	3	
1906	0-4-4T	266	6	
1907-1909	4-4-0	18	12	
1907-1909	0-4-0T	272	6	
1910-1911	4-6-0	381	7	
1910	0-6-0	361	2	
1912	4-4-0	18	1	
1914-1915	4-4-0	137	6	
1914	0-6-0T	14	5	
1921	4-4-0	8 Reb	1	
1921	0-6-0	101 Reb	2	

GLASGOW & SOUTH WESTERN RAILWAY

P. STIRLING

Date	Class	Type	No Built	Cylinders inches	D.W. ft in	Heating Surface Sq Ft Tubes	S'heat	F'box	Total	Grate Area sq ft	WP lb/sq in	Weight WO ton cwt	
1860-4	40	2-2-2	10	OC 16 x 21	6 6	860	-	67	927	14.0	120	24 10	
1862-3	46	0-6-0	6	16 x 22	5 0	851	-	78	929	13.3	125	30 6	
1865-6	52	0-4-0	6	16 x 22	5 0	878	-	67	945	14.0	120	26 4	
J. STIRLING													
1870-1	71	2-4-0	10	17 x 24	6 1	779	-	82	861	15.0	130	30 12	
1873-7	6	4-4-0	22	18 x 26	7 1	1028	-	84	1112	16.0	140	39 0	
1877-8	13	0-6-0	12	18 x 26	5 1	1059	-	98	1157	16.0	130	35 10	
H. SMELLIE													
1879-80	157	2-4-0	12	18 x 26	6 9½	1105	-	101	1206	16.0	140	38 10	
1882-85	119	4-4-0	22	18¼ x 26	6 1¼	964	-	101	1065	16.0	140	41 10	
J. MANSON													
1892-04	8	4-4-0	57	18¼ x 26	6 9½	1092	-	111	1203	18.2	150	45 0	
1896-14	14	0-6-0T	15	16 x 22	4 7½	820	-	70	890	12.0	140	40 2	
1897	11	4-4-0	1	IC 14½ x 26 OC 12½ x 24 IC 14 x 26	6 9½	1062	-	111	1173	18.0	165	48 14	
	11	4-4-0	1	OC 14 x 24	6 9½	1444	211	148	1803	27.6	180	61 8	
1910	361	0-6-0	2	18 x 26	5 1½	1097	-	111	1208	18.0	165	44 0	(a)
1910-11	381	4-6-0	7	OC 20 x 26	6 6	1721	-	131	1852	24.5	180	67 2	(b)
1904-5	RM	0-4-0RM	3	OC 9 x 15	3 6	400	-	40	440	8.0	180	40 0	(c)
1904-6	240	4-4-0	15	18¼ x 26	6 9½	1315	-	119	1434	18.2	170	50 8	
1906	266	0-4-4T	6	16 x 22	4 7½	820	-	70	890	12.0	140	42 8	
1907-9	272	0-4-0T	6	OC 16 x 24	4 7½	815	-	70	815	12.0	140	39 12	
1907-12	18	4-4-0	13	18¼ x 26	6 9½	1297	-	110	1407	22.0	170	51 14	
P. DRUMMOND													
1914-15	137	4-4-0	6	19½ x 26	6 0	1444	330	148	1922	27.6	180	64 0	

(a) 32 more by outside contractors. (b) 10 more by outside contractors. (c) Weight including coach. Loco-type boiler

GLASGOW RAILWAY ENGINEERING CO. LTD.

See D. DRUMMOND & SON.

GLENGARNOCK IRON & STEEL WORKS

Glengarnock, Ayrshire

One locomotive was built in 1913 and was a typical Scottish 0-4-0ST with outside cylinders $14\frac{1}{2}$" x 22" and 3' 6" diameter wheels. It had a circular window in each cab side and bore all the characteristics of a *Grant Ritchie* locomotive.

It is almost certain that parts were supplied from a locomotive builder and erected at Glengarnock.

The steel works were acquired by Colvilles Ltd before 1920.

GORDON, ADAM

Deptford Green, London

Primarily marine engineers, this firm built at least two locomotives and on 21 November, 1839 they sought permission 'to try a new engine of their construction on the line between New Cross and Croydon'. The railway company agreed and asked Gordon if they could hire this 'small engine' but there is no further record.

They were asked to quote for a locomotive for the London and Croydon Railway in 1837.

GORTON & CO.

This company built vertical boilered locomotives, and one 0-4-0T named BEE was supplied to Pease & Partners for Normanby Iron Works in 1887.

GOURLAY BROS. & CO. LTD

Dundee Foundry, Dundee

Later known as GOURLAY MUDIE & CO. See JAMES STIRLING & COMPANY.

Fig. 158 **Glengarnock I. & S. Co**

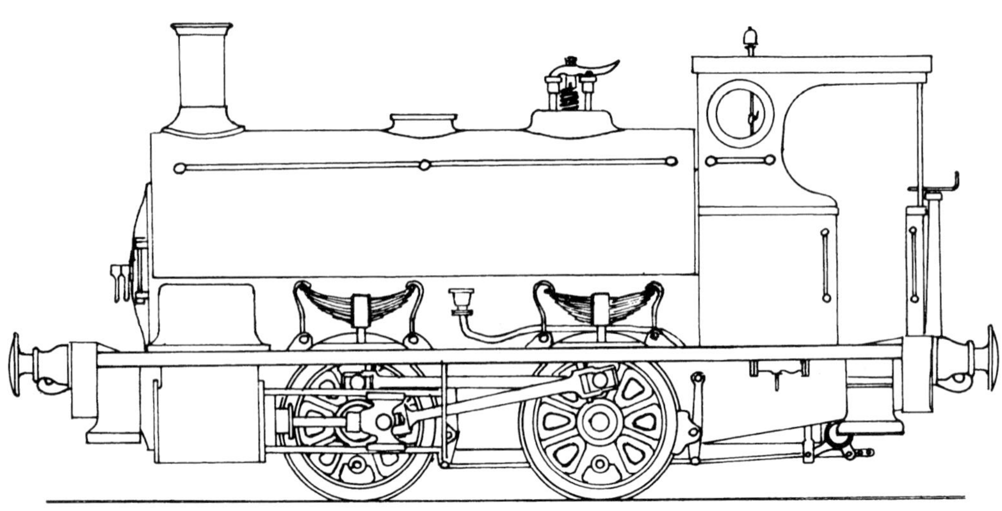

GRAND JUNCTION RAILWAY

Edge Hill, Liverpool

Locomotive Superintendents: Thomas Melling 1837–39; William B. Buddicom 1840–41; Francis Trevithick 1841–43.

These were the original locomotive repair shops of the Grand Junction Railway, established about 1839.

Extensive rebuilding was carried out. The original inside cylinder 2-2-2 passenger locomotives which were built by a multitude of builders during 1837–39 gave quite a lot of trouble mainly due to broken crank axles, and instructions were issued to design a replacement locomotive and to rebuild existing locomotives to this pattern. From this instruction there evolved the well known 2-2-2 with outside cylinders with distinctive front end and outside frames to leading and trailing wheels.

They were later to become known as the *Old Crewe* type. The design is credited to Alexander Allan who was foreman of locomotives, but other sources state that Buddicom laid down the basic design.

The first locomotive to be rebuilt was No. 26 AEOLUS built by Chas. Tayleur in 1838. It was fitted with outside cylinders 13″ x 20″ and 5′ 6″ diameter driving wheels. Two others followed in 1841: No. 39 TARTARUS and No. 42 SUNBEAM. They were officially listed as new locomotives, but it is doubtful whether they were completely new.

A new locomotive works was being built at Crewe and from 1843 all new work was concentrated there, involving the subsequent shut down of Longsight and Edge Hill as locomotive factories.

GRAND JUNCTION RAILWAY

Crewe

The original locomotive works of the Grand Junction Railway were at Edge Hill, Liverpool. All repairs and extensive rebuilding of the original stock was carried out, and two new locomotives were probably built there.

More extensive buildings were required and Crewe was the site decided upon and the new works were ready in 1843. The first entirely new locomotive was completed on 20 February, 1845.

Francis Trevithick, son of Richard Trevithick, was the locomotive superintendent and he organised the removal of plant etc., to the small village of Crewe. The new works covered some three acres of ground and employed about 160 men.

Alexander Allan had been the works manager at Edge Hill and at Crewe became 'foreman of locomotives'.

In all, twelve 2-2-2s and four 2-4-0s were built as Grand Junction Railway locomotives. The 2-2-2s subsequently became known as the *Old Crewe* type and the 2-4-0s *Crewe goods*.

COLUMBINE was *Crewe* No. 1* (GJR No. 49) and had outside cylinders, 15″ x 20″, 6 ft diameter driving wheels with 3′ 6″ leading and trailing wheels. The boiler pressure was 120 lb/sq. in. and the locomotive weighed 19 tons. The twelve of this type had small fireboxes with safety valve mounted above, and another one midway along the boiler barrel. The *Crewe Goods* had 5 ft diameter driving wheels.

On 16 July, 1846 the amalgamation took place of the Grand Junction (which had already amalgamated with the Liverpool and Manchester Railway in 1845), London and Birmingham, and the Manchester and Birmingham railways to form the LONDON & NORTH WESTERN RAILWAY.

The Grand Junction section became the Northern Division with headquarters at Crewe, but it was not until 1861, when John Ramsbottom was appointed locomotive superintendent of the whole railway, that Crewe became responsible for all new locomotive construction.

Between the 1846 amalgamation and 1861 the other divisions (i.e. North East Division at Longsight and Southern Division at Wolverton) had independent action regarding design and new construction. The North East Division was absorbed into the Northern Division in 1857.

* This is hypothetical. Seventeen locos were built at Crewe before No. 49 (see Table).

Built	Type	Cylinders Outside inches	D.W. ft in	GJR No	GJR Name
10/1843	2-2-2	13 x 20	5 6	32	Tamerlane
10/1843	2-2-2	13 x 20	5 6	27	Merlin
12/1843	2-2-2	13 x 20	5 6	13	Prospero
1/1844	2-2-2	13 x 20	5 6	6	Stentor
2/1844	2-2-2	13 x 20	5 6	18	Cerberus
3/1844	2-2-2	13 x 20	5 6	20	Eagle
4/1844	2-2-2	13 x 20	5 6	5	Falcon
4/1844	2-2-2	13 x 20	5 6	15	Phalaris
9/1844	2-2-2	13 x 20	5 6	30	Sirius
10/1844	2-4-0	13 x 20	5 0	2	Hecla
10/1844	2-2-2	13 x 20	5 6	8	Wildfire
10/1844	2-2-2	13 x 20	5 6	17	Caliban
1/1845	2-2-2	13 x 20	5 6	3	Shark
2/1845	2-2-2	13 x 20	5 6	10	Dragon
2/1845	2-2-2	13 x 20	5 6	21	Wizard
3/1845	2-2-2	13 x 20	5 6	11	Zamiel
4/1845	2-2-2	13 x 20	5 6	22	Basilisk
6/1845	2-2-2	14½ x 20	5 6	73	Prince (a)
7/1845	2-2-2	14½ x 20	5 6	19	Princess
7/1845	2-2-2	14½ x 20	6 0	49	Columbine
9/1845	2-2-2	14½ x 20	6 0	76	Albion
9/1845	2-2-2	14½ x 20	6 0	74	Deva
10/1845	2-2-2	14½ x 20	6 0	75	Apollo
11/1845	2-2-2	14½ x 20	6 0	77	Mersey
2/1845	2-4-0	15 x 20	5 0	78	Lonsdale
2/1845	2-2-2	15 x 20	6 0	79	Belted Will
3/1845	2-2-2	15 x 20	6 0	80	Dalemain
4/1845	2-4-0	15 x 20	5 0	81	Greystoke
4/1845	2-4-0	15 x 20	5 0	82	Wordsworth
4/1845	2-4-0	15 x 20	5 0	83	Windermere (b)
4/1845	2-2-2	15 x 20	6 0	84	Saddleback
5/1845	2-2-2	15 x 20	6 0	85	Ingestre
6/1845	2-2-2	15 x 20	6 0	86	Sandon

(a) Renamed PRINCE ALBERT in 1847.
(b) Renamed SKIDDAW in 1850

The locomotive superintendents at Crewe were as follows: Francis Trevithick 1846–57; John Ramsbottom 1857–71; Francis William Webb 1871–1903; George Whale 1903–09; Charles John Bowen-Cooke 1909–20; Capt. Hewitt Pearson Montague Beames 1920–21; George Hughes 1922.

Meanwhile at Crewe, building continued of the two classes built during the GJR period and altogether 398 of the *Old Crewe* type were built from 1845 to 1858, comprising 5 ft Goods with straight and crooked frames, and 6 ft, 7 ft and 8 ft Passenger types, also with the two types of frames. All had indirect motion.

The early locomotives of the *Crewe* pattern had small fireboxes but from 1853 larger fireboxes were fitted.

Two experimental *Crewe* type were built during this period. CORNWALL was a single, with 8′ 6″ diameter driving wheels, and to give it a low centre of gravity the boiler passed under the driving axle. Trevithick was responsible for the design and it was completed in November 1847. VELOCIPEDE, another single with 7′ diameter driving wheels and 15″ x 20″ outside cylinders, was designed by Alexander Allan, and completed in March 1847.

This was at the time of the 'Battle of the Gauges' and the broad gauge Great Western Railway, with its large 4-2-2 tender locomotives, were a challenge to their narrow gauge neighbours. With an average coal consumption of 27 lbs per mile, they were unrivalled and the directors of the L&NWR sanctioned the building of VELOCIPEDE and CORNWALL.

The VELOCIPEDE proved very successful on the Northern Division expresses. CORNWALL in its original condition was moderately successful but the design was not perpetuated and it was rebuilt 1863–4 by John Ramsbottom with a boiler above the driving axle.

In addition two *Cramptons* were ordered – LONDON built by Tulk & Ley and the COURIER built at Crewe.

LONDON had 8 ft driving wheels, cylinders 18″ x 20″ and a large boiler with 1529 sq. ft of heating surface. COURIER had 7′ drivers.

The *Cramptons* had their driving axles behind the firebox, which was the main characteristic of Crampton's patent, the object being to provide large diameter driving wheels with a low centre of gravity.

Many of the *Crewe Goods* were converted to side tanks.

Francis Trevithick retired in 1857 and Alexander Allan had already left to take charge of the locomotive department of the Scottish Central Railway in 1853.

John Ramsbottom took over the department from Trevithick and his first design was the famous 'DX' Class D express goods locomotive, a class which was to run to a total of 943 from 1858 to 1874. They were six coupled with inside cylinders 17″ x 24″, 5 ft diameter wheels and a boiler with 1102 sq. ft of heating surface, working at a pressure of 120 lb/sq. in. The smokebox front was sloping with a horizontally hinged door.

Included in the 943 locomotives built were some supplied new to other railways: two for Portpatrick & Wigtownshire Railway 1861–62 and 86 to Lancashire & Yorkshire Railway 1871–74.

Other types were also supplied new to other railways at this time, which incensed the private locomotive builders to such an extent that an injunction was secured against the L&NWR and the practice was stopped.

Ramsbottom's first passenger class was a

Fig. 159
222/1852
Allan 2-2-2
originally 291
PRINCE OF WALES

'single' with 7′ 7½″ diameter driving wheels with outside cylinders 16″ x 24″. The weight in working order was 27 tons, the same as the DX goods. They were good looking machines but were unsteady at speed due to the short wheelbase, light weight, and outside cylinders placed well forward.

Sixty were built from 1859 to 1865 and were known as the *Problem* class and later as the *Lady of the Lake* class.

He made all tenders with timber frames as, in his view, this vehicle should be the weakest part of the train. The timber would break up first in case of a collision and thus save the passenger carriages.

Up to 1859 all locomotives were named irrespective of class but with the 'DX' goods only fifty-four were named and after this the passenger locomotives bore names – and what a variety of names they were. Fascinating because the names did not follow a pattern; BUFFALO and THEOREM, QUICKSILVER and FALSTAFF.

PROBLEM was the first locomotive in England to be fitted with Giffard's injectors. Giffard had had great difficulty in 'selling' the idea of the injector, most engineers preferring to carry on with their boiler feed pumps.

Due to the rapid expansion of shops at Crewe the internal network of 18″ gauge track was such that locomotives were necessary and in 1862 a start was made in building suitable works locomotives for this purpose. The first was named TINY as indeed it was, and PET, NIPPER, TOPSY and MIDGE followed, the latter being completed in 1870.

Some small 0-4-0STs were built from 1863 with 4 ft coupled wheels, inside cylinders 14″ x 20″ and Stephenson's valve gear. They were an interesting type, being the first on the LNWR to have cast-iron wheels, the first with solid bushes to the connecting rods. As there was insufficient clearance between crank axle and firebox a marine type boiler was introduced with a circular firebox inside the barrel.

They proved very successful as light shunters. Ramsbottom built thirty-six and Webb added another twenty-two, the last batch appearing some 29 years after the prototype.

In the same year appeared the *Samson* class of 2-4-0s, departing from the previous designs in that inside cylinders and inside bearings to all wheels were introduced. The cylinders were 15¼″ ×20″, coupled wheels 6 ft diameter, total heating surface 1083 sq. ft with a pressure of 120 lb/sq. in. Later *Samson* class locomotives had 16″ cylinders and an increased boiler pressure of 140 lb/sq. in.

They were very small locomotives with inadequate cylinder and boiler capacity, but despite this a total of ninety were built from 1863 to 1879. An interesting experiment was carried out by Webb on one of this class No. 757 BANSHEE. The coupling rods were removed and friction wheels were fitted which could be brought into contact with both pairs of 6 ft diameter wheels by being forced against them when required.

In 1866 a further 2-4-0 design appeared with larger cylinders 17″ x 24″ and 6′ 7½″ diameter coupled wheels, but the boiler had about the same heating surface. They gave a moderate performance. They were mainly used between Crewe and Carlisle and ninety-six were built from 1866 to 1873.

Larger tank locomotives were required and Ramsbottom designed the *Special* 0-6-0STs with 17″ x 24″ cylinders and 4′ 3″ diameter wheels.

Two hundred and fifty-eight were built, the last one surviving until 1941.

John Ramsbottom had created a useful stud

of locomotives, with a nucleus of standard classes. He had enlarged the works and the capacity increased from fifteen new locomotives in 1845 to 129 in 1867. He insisted upon high standards of workmanship which were to be upheld as the Crewe tradition to the present day.

Ramsbottom, before he came to Crewe, had invented some important improvements on the locomotive. They were the split piston ring, the double beat regulator valve, displacement lubricator and, perhaps the best known, the 'Ramsbottom Safety Valve'. He improved on the screw reversing gear and used a type of Stephenson's link motion for most classes.

Whilst at Crewe he designed a system of water troughs and pick-up apparatus on the tenders. The first were installed between Llandudno and Chester in 1860 and others were added until about sixteen sets were fitted and were more closely spaced than average which explains the fact that there was no need for large tenders, and LNW tenders carried a maximum of 3000 gallons.

In 1871 John Ramsbottom retired.

Francis William Webb was appointed chief mechanical engineer in 1871. He had been chief draughtsman and later works manager under Ramsbottom but had left the railway in 1866 to become manager of the Bolton Iron & Steel Co.

He had been a pupil at Crewe Works under Francis Trevithick, becoming his assistant after the end of his pupilage. At the age of twenty-four he was appointed chief draughtsman and later works manager under Ramsbottom. Here were two powerful personalities and it may well have caused Webb to leave the railway in 1866. Whether he was recalled in 1871, or applied for the post that became vacant, is not known.

There has never been anyone quite like Francis Webb, but much has been written about him, so suffice it to say that he was an autocrat among CMEs, brooking no interference or criticism from anyone.

At first he continued building Ramsbottom locomotives. The shape of things to come started in 1878 when he converted an old Trevithick 6 ft single named MEDUSA to a two cylinder compound with a high pressure cylinder 9½″ diameter and a low pressure cylinder 15″ diameter.

Whilst tests were carried out he designed a 2-4-2T with 17″ x 24″ inside cylinders and 4′ 6″ diameter driving wheels. In addition he brought out an 18″ goods 0-6-0 with a stroke of 24″ and 5′ 1½″ diameter wheels. These were later called *Cauliflowers* as the emblem on the centre splasher closely resembled the vegetable! They had Joy's valve gear and were used on passenger trains as well as freight. Before this in 1873 the *Coal Class* 0-6-0 was brought out and 500 were built up to 1892. They were probably the simplest and cheapest type ever built in Britain and were a good example of standardisation. It was claimed that they were built for £400 each and one was erected in 25½ hours – a record at that time.

Fig. 160 1988/1876 2-4-0T No. 2238. Remained in service until 1925

Two new passenger types (2-4-0s) were introduced in 1874 both with inside cylinders 17″ x 24″ and frames. The first to appear was the *Precursor* in April 1874 with 5′ 6″ diameter coupled wheels, the second was the *Precedent* in October 1874 with 6′ 6″ diameter coupled wheels. The boilers had the same heating surface of 980 sq. ft but the *Precursor's* firebox was 94·6 sq. ft compared with a slightly larger one for *Precedents* of 103·6 sq. ft. Both had Allan's straight link motion.

Primarily it was intended that the *Precursor* class should work the Crewe/Carlisle main line and the *Precedents* the Crewe to London, Liverpool and Manchester lines. The *Precedents* were, in the main, enlargements of Ramsbottom's four coupled locomotives and although on the light side were quite successful. The *Precursors* however were not so successful and were eventually rebuilt as 2-4-2Ts.

The most noteworthy of the *Precedents* was CHARLES DICKENS which between February 1882 and September 1891 ran the round trip from London to Manchester and back daily except when stopped for maintenance purposes. In fact it ran 1,000,000 miles in 9 years 219 days, an average of 105,398 miles per annum. This was indeed a superb performance at that period and showed what excellent workmanship went into these locomotives. In twenty years it had run 2,000,000 miles, still on the same duties and when CHARLES DICKENS was withdrawn in 1912 the total mileage run was 2,330,000. Another *Precedent* HARDWICKE took part in the famous 'Race to Aberdeen' series and on the night of the record run, HARDWICKE ran from Crewe to Carlisle, 141¾ miles, in 126 minutes, at an average speed of 67·2 mph.

By the middle of 1876 over 2000 locomotives had been built at Crewe.

The 2000th was one of a series of 2-4-0Ts with 4′ 7½″ diameter driving wheels, the leading wheels were fitted with Mr Webb's patent radial axleboxes. The cylinders were 17″ x 24″. Most of them were rebuilt, with trailing wheels, into 2-4-2Ts with extended cabs and coal bunkers.

Further 2-4-2Ts were built with 5′ 8″ diameter driving wheels with radial axleboxes on the trailing wheels but standard axleboxes on the leading wheels.

Other tank types were 0-6-2T with 4′ 3″ diameter driving wheels, and a four coupled tank on which the driver stood on one end and the fireman at the other end. The idea was for either man to couple or uncouple vehicles during shunting operations without getting off the footplate. A light flexible coupling of rope or leather was used. Other novel features were the coupling rods, each made up of two steel circular rods with brass ends cast on. The frames formed the sides of two water tanks and the boiler had a combustion chamber and water tube firegrate. The smokebox was formed inside the boiler, the chimney being carried through the dome. The wheels were 2′ 6″ diameter, inside cylinders 9″ x 12″ and weight in working order 15 ton 3 cwt. As a result of the success of the experimental compound Mr Webb designed a system of compounding which he patented. Under the patent the use of three or four (but preferably three) cylinders were specified. In addition no coupling rods were used on the initial locomotives.

The high and low pressure cylinders were connected to separate pairs of wheels.

The first compound built as new was in 1882 and was 2-2-2-0 formation. Two independent pairs of driving wheels 6′ 6″ diameter were used, the leading wheels were placed well forward under the low pressure cylinder and fitted with radial axleboxes. The first pair of driving wheels were coupled to the inside single low pressure cylinder and the rear pair coupled to the two high pressure cylinders.

The two HP cylinders were 11½″ x 24″ and the LP cylinder 26″ x 24″. To enable comparisons to be made the boiler was the same as the *Precedent* class. The valves for the HP cylinders were on the underside and for the LP cylinder on top. The valve motion was David Joy's patent. The locomotive started work on 3 April 1882 and was named EXPERIMENT. As would be expected, slipping on starting was the main trouble and this was aggravated by the fact that the locomotive was designed to start as a compound and not as a simple engine. Nevertheless its performance satisfied Mr Webb and orders were placed for twenty-nine similar compounds except that the diameter of the high pressure cylinders was increased to 13″. This series was completed in July 1884. One of the Beyer Peacock 4-4-0Ts was then converted to a compound with a view to more economical running.

The next design was for a larger *Experiment* class. This was the DREADNOUGHT of the same wheel arrangement but 6′ 3″ diameter wheels, two HP cylinders 14″ x 24″ and one LP cylinder 30″ x 24″ and a much larger boiler with a total heating surface of 1379 sq. ft. These were heavy on coal and maintenance.

A 2-4-2T locomotive was next converted in 1885 with two HP cylinders 14″ x 18″ and one LP cylinder 26″ x 24″. This was the first locomotive to employ different strokes for HP and LP cylinders. Joy's valve gear was again used and an apparatus was fitted to discharge the exhaust steam into the tanks when working underground. This was No. 687.

A further tank No. 777 was converted in 1887 but this took the form of a 2-2-4-0T where the trailing coupled wheels were driven by the two HP cylinders 14″ x 24″ and the leading driving wheels were connected to the LP cylinder 30″ x 24″. The locomotive was also fitted with an adjustable blast pipe and Ramsbottom's water pick up gear. In the same year a new compound tank locomotive was built at Crewe. This was No. 600 and the 3000th built at Crewe.

An enlarged 'Dreadnought' was then built for express passenger work, the difference being in the driving wheels which were increased in diameter to 7′ 1″. Ten were built and were known as the *Teutonic* class.

Bigger but not better compounds followed with the *Greater Britain* and *John Hick* classes. These had a much larger boiler divided into two by a combustion chamber, which had an ash hopper fitted which could be operated from the footplate. The lower pressure cylinder slide valve was worked by a loose eccentric and the HP cylinders worked by ordinary curved link motion placed inside the frames. These locomotives were not as successful as the *Teutonics.*

The *John Hick* class was the least successful of all. They were intended for the Crewe to Carlisle section and had 6′ 3″ diameter driving wheels. One of the *Greater Britain* class No. 2054 QUEEN EMPRESS was exhibited at the Chicago Exhibition of 1893.

The first mineral locomotive to be turned out as a compound was No. 50 in 1893. It had three cylinders all in line and all driving the second axle. The connecting rods of the two outside HP cylinders drove on crank pins set at right angles to each other, whilst the centre crank was set at an angle of 135° to the HP cranks. The valve gear was the same as the *Greater Britain* class. The wheels were all coupled which was certainly an improvement.

In 1897 Webb built two 4-4-0s. One was a four cylinder simple (No. 1501 IRON DUKE) and the other a four cylinder compound (No. 1502 BLACK PRINCE built for comparison purposes. Apparently the compound was assessed as being superior to the simple and in 1899–1900, 38 compounds similar to No. 1502 were built, followed by 40 more, with larger boilers, known as the *Alfred the Great* class from 1901–1903.

The performance of all 79 locomotives followed the previous pattern, being erratic and sluggish, mainly due to the valve gear. Two sets of valve gear operated the four valves, and as the HP and LP cylinders were notched up

Fig. 161 2794/1887 Webb compound No. 777 2-2-4-0T

together, fast running could not be accomplished, the exhaust port openings being so restricted.

It was unfortunate that Webb would not listen to constructive criticism and his iron hand over Crewe brought a reaction gradual at first but culminating in Webb's request, at the end of 1902, to be relieved. There must have been a great wave of relief and joy when he finally retired – not that he was an unjust man, but his rule was so autocratic that very few must have mourned his retirement.

Some of his compounds, particularly the *Teutonics* acquitted themselves well, but taking a broad view of his activities, his 'simple' designs were far more successful.

George Whale took over the post of chief mechanical engineer in May 1903. He was a man who had spent most of his working life on the railway and for the past four years had been running superintendent for the whole railway and previous to this had occupied the same position on the Northern Division. With this experience, his knowledge of Webb's compounds must have been unique, and it was logical that he soon started to iron out some of the inherent faults. The modification of the valve gear on the four cylinder 4-4-0 compounds was one of his first actions and this was accomplished by separating the reversing gear to the HP and LP cylinders. By this means the two sets of cylinders could be linked up simultaneously. This improved their performance considerably. Nevertheless he was not long in getting rid of the three cylinder compounds.

His first design was for a two cylinder simple 4-4-0 which was to be the fore-runner of many similar designs. From bitter experience compounds were out, although the tail end of orders placed before George Whale took over were finished at the beginning of his taking up the reins.

The new 4-4-0 had inside cylinders 19″ x 26″, driving wheels 6′ 9″ diameter, a boiler with a modest pressure of 175 lb/sq. in and a total heating surface of 2009·7 sq. ft which was a big boiler for that period particularly for a 4-4-0. The grate area was 22·4 sq. ft. With these proportions, success was assured and 'PRECURSOR as it was named was a tremendous advance in design and performance, but in external lines Whale continued the LNWR tradition.

Tenders up to this time had wooden frames but Whale designed a steel frame tender, and instead of a handle to operate the water scoop, the new design incorporated a screw operated scoop. Other improvements were an extended cab roof supported by pillars. One hundred and thirty *Precursors* were built from 1904 to 1907.

For the more demanding Crewe to Carlisle section a 4-6-0 was put in hand, known as the *Experiment* class. It had the same cylinders as the *Precursors*, driving wheels 6′ 3″ diameter and a boiler not much larger, giving a total heating surface of 2041 sq. ft and a grate area 25 sq. ft. The driving wheels were put close together and therefore necessitated a shallow firebox creating draughting troubles and awkward firing. One hundred and five were built from 1905 to 1910.

In performance they were, in general, inferior to the *Precursors*, being sluggish and heavy on coal. They could run very fast down hill, but uphill they had to be thrashed.

In 1906 a 'goods' version of the *Experiment* came out with 5′ 2½″ diameter coupled wheels, of which 170 were built and were successful at mixed traffic work. In the same year a tank version of the *Precursors* a 4-4-2T appeared with 6′ 3″ diameter coupled wheels. They were excellent locomotives especially suited for the London suburban traffic.

During this period most of the 4 cylinder compound 4-4-0s were rebuilt as two cylinder simples, they were known as the *Renown* class.

In common with other railways Mr Whale introduced a number of steam rail cars in 1905 and 1906. They had inside cylinders 9½″ x 15″ driving 3′ 9″ diameter wheels.

George Whale retired in 1909 but during his short term of office he had laid down a progressive policy of locomotive building and design.

He was succeeded by C. J. Bowen Cooke in 1908 and his first design was the *George the Fifth* class of 4-4-0 of which ninety were built from 1910 to 1915.

These locomotives were the first fitted with superheaters (Schmidt) on the L&NW. The cylinders were 20½″ x 26″ and driving wheels 6′ 9″. The boiler had a total heating surface of 1849·6 sq. ft which included 302·5 sq. ft of superheater tubes.

Joy's valve gear was fitted and piston valves in place of slide valves.

For comparison purposes ten *Queen Mary* class were built with non-superheated boilers, and as built had 19″ x 26″ cylinders, but all were subsequently brought into line with the *George the Fifth* class.

An interesting feature of these first superheated locomotives was that each one was equipped with a pyrometer showing the superheat temperature to the driver, but this bit of instrumentation was not continued. The economical advantage of these superheated 4-4-0s in saving between 10% and 20% in coal burned and their superior running, encouraged Bowen Cooke to fit all subsequent express

Fig. 162
3354/1893
Waterloo or
Whitworth class
No. 634 ELLESMERE
Note brake gear

Fig. 163
4661/1907
Precursor class
No. 218 DAPHNE

Fig. 164
5057/1912
G1 class
0-8-0 No. 2001

Fig. 165
5089/1912
Fast suburban tank
4-6-2T No. 2004

passenger types with superheaters and to rebuild many earlier locomotives with this refinement.

A 4-6-2T was then built with 20″ x 26″ cylinders and driving wheels 5″ 8½″ diameter. Like the new 4-4-0s ten were built without superheaters and the remainder had boilers with 248·2 sq. ft of superheater and a total heating surface of 1333·5 sq. ft. A new feature was the Belpaire firebox. In all forty-seven were built from 1910 to 1916, the ten 'saturated' locomotives were subsequently rebuilt to conform to the other thirty-seven.

In October 1911 No. 819 PRINCE OF WALES appeared. This was a development of the *Experiment* class and had superheated boilers, 20½″ x 26″ cylinders and Joy's valve gear.

They were eminently successful locomotives, not particularly fast, but strong and like all L&NWR locomotives they were worked to their limit and they stood up to it.

Departing from their policy of building all locomotives required at Crewe, twenty *Prince of Wales* were built by the North British Locomotive Co. in 1916 and no less than ninety by William Beardmore & Co. Ltd in 1921 and 1922. Two hundred and forty-five were built from 1911 to 1922.

For freight working and banking duties Bowen Cooke designed the following: 0-8-0 Class G saturated; 0-8-0 Class G1 superheated; 0-8-2T tank version of G1 but saturated.

The 0-8-0s were very similar to Whale's 0-8-0s except for minor details and superheated boilers in the case of the G1s.

The tank version had similar cylinders 20½″ x 24″ and a useful adhesive weight of 62 tons.

Bowen Cooke's masterpiece appeared in January 1913 in the form of No. 2222 SIR GILBERT CLAUGHTON a four cylinder simple expansion 4-6-0. The cylinders were 16″ x 26″ and driving wheels 6′ 9″ diameter, a boiler with Belpaire firebox, a total heating surface of 2128 sq. ft of which 379·3 sq. ft was superheater tubes. The drive from all four cylinders was on to the leading coupled axle. The design was clean and handsome and having the characteristic L&NW lines except for the replacement of Joy's valve gear by outside Walschaerts gear with rocking levers to connect the inside valves.

The general arrangement of the *Claughtons* was theoretically good, but at first trouble was experienced with indifferent steaming, probably due to the unfamiliar shape of the grate which had its rear half horizontal and the front half sloping towards the front. A different firing technique was required. Other faults were poor front end design and leaking piston valves which gave trouble with all the superheated locomotives. Nevertheless some fine performances were recorded despite these faults and when they were rebuilt with large boilers their performance left nothing to be desired.

One of the *Experiments* was rebuilt in 1915 with Dendy-Marshall's patent valve gear. Four cylinders were fitted 14″ x 26″ and a Schmidt superheater. The engine was No. 1361 PROSPERO. The main idea of this particular valve gear was to provide one single set of valve gear for two cylinders. This was effected by a valve chest placed between the outside and the adjacent inside cylinders and by a series of ports in the common valve chest, the piston valve operated admission and exhaust to the two cylinders. No other locomotive was fitted with this valve gear and no results were ever published.

Unfortunately C. J. Bowen Cooke died on 19 October, 1920 but his locomotive building schedules and maintenance methods were carried on by his successor Capt. H. P. M. Beames who had been his deputy since 1919.

Crewe during this period continued building *Claughtons* 0-8-0s (G1) and receiving *Prince of Wales* class from Beardmore's.

Beames did in fact design a 0-8-4T but in January 1922 the amalgamation took place between the L&NWR and the Lancashire & Yorkshire Railway and George Hughes was appointed CME over the new company. George Hughes was Beames counterpart at Horwich and as the general manager Arthur Watson was an old L&YR man appointed in 1920 it was significant that George Hughes replaced Capt. Beames. It was a pointer of things to come, for on 1 January, 1923 the amalgamation took place of the railways which formed the LONDON MIDLAND & SCOTTISH RAILWAY of which George Hughes became the first chief mechanical engineer.

For details of locomotive building during this period 1923–47 please see under LONDON MIDLAND & SCOTTISH RAILWAY and also Tables showing locomotives built at Crewe.

LNWR CREWE - J. RAMSBOTTOM

First Built	Type	Class	No Built	First No	Cylinders inches	D.W. ft in	Boiler H/S Tubes	Boiler H/S F'box	Boiler H/S Total	Grate Area	WP	Weight WO ton cwt	
1858	0-6-0	DX	943	355	17 x 24	5 0			1102	15	120	27 0	(a)
11/1859	2-2-2	Problem	60	184	16 x 24	7 6			1098	14.9	120	27 0	
5/1863	2-4-0	Samson	90	633	16 x 20	6 0			1083		120	26 2	(b)
11/1864	0-4-0ST		58	1357	14 x 20	4 1	395	20	415	6	120	23 0	(c)
4/1866	2-4-0	Newton	96	1480	17 x 24	6 6	924	89	1013	14.9	120	28 13	(d)
4/1870	0-6-0ST	Special Tank	260	1750	17 x 24	4 5	980	95	1075	17	120	34 0	

(a) Includes 86 sold to other railways. (b) Includes 43 built by F. W. Webb. (c) Includes 22 built by F. W. Webb. (d) Includes 30 built by F. W. Webb.

SIMPLE LOCOS BY F. W. WEBB

Date	Type	Class	No Built	First No	Cylinders inches	D.W. ft in	Boiler Tubes	Boiler F'box	Boiler H/S Total	Boiler Grate	Boiler WP	Weight WO ton cwt	
2/1873	0-6-0	17" Coal	500	944	17 x 24	4 5	980	94.6	1074.6	17.1	140	29 10	(a)
4/1874	2-4-0	Precursor	40	2145	17 x 24	5 6	980	94.6	1074.6	17.1	140	31 8	
12/1874	2-4-0	Precedent	70	2175	17 x 24	6 7	980	103.5	1083.5	17.1	140	32 15	
1876	2-4-0T		50	2233	17 x 20	4 6	885	85	970	14	140	38 0	
1879	2-4-2T		180	1384	17 x 20	4 6	885	85	970	14	140	46 0	
1880	0-6-0	18" Coal	310	968	18 x 24	5 2	985.2	94.6	1079.8	17.1	140	33 7	
1881	0-6-2T	Coal	300	602	17 x 24	4 5	980	94.6	1074.6	17.1	140	44 0	
R.1881	0-6-0	Special DX	500R		17 x 24	5 2	980	94.6	1074.6	15	140	29 0	
R.1887	2-4-0	REB Newton	96		17 x 24	6 6	980	104	1084	17.1	150	35 12	
R.1890	2-4-2T		160R		17 x 24	5 6	980	94.6	1074.6	17.1	160	50 10	
1892	0-8-0		1	2524	19½ x 24	4 5	1130	115	1245	20.5	160	46 16	(b)
R.1893	2-4-0	REB Precedent	70R										(c)
11.1896	0-4-2ST		20	317	17 x 24	4 5½	882	85	967	14	150	34 17	
1898	0-6-2T	Radial Tank	80	1597	18 x 24	5 2	980	103.5	1083.5	17.1	160	52 0	

(a) One built in 25½ hours No 1140 (Crewe No 2153). (b) Fitted with combustion chamber. (c) Known as 'Jumbos'

F. W. WEBB COMPOUND LOCOMOTIVES L & NWR

Built	Type	Cyl Bores				Stroke in	Name	1st No	No Built	Dia Drivers ft in	H/S	GA	Weight WO ton cwt	Pressure lb/sq in
		HP		LP										
		No	Dia in	No	Dia in									
1878	2-2-2	1	9	1	15	20	(a)	1874	1	6 0	709.0	10.5	23 5	120
1/1882	2-2-2-0	2	11½	1	26	24	Experiment	66	1	6 6	1083.5	17.1	37 15	150
3/1883	2-2-2-0	2	13	1	26	24	Compound	300	29	6 6	1083.5	17.1	37 15	150
6/1884	4-2-2-0T	2	13	1	26	24	(b)	2063	1	5 9	1028.43	14.6	46 17	150
9/1884	2-2-2-0	2	14	1	30	24	Dreadnought	503	40	6 3	1401.5	20.5	42 10	175
9/1885	2-2-2-2T	2	14	1	26	H18 L24	(c)	687	1	4 8½	993.6	14.2	50 17	160
3/1887	2-2-4-0T	2	14	1	30	24	(c)	777	1	5 2½	1098.9	17.1	55 0	160
7/1887	2-2-2-2T	2	14	1	26	H20 L24	(c)	600	1	5 8½	993.6	14.2	52 0	160
3/1889	2-2-2-0	2	14	1	30	24	Teutonic	1301	10	7 1	1401.5	20.5	46 10	175
10/1891	2-2-2-2	2	15	1	30	24	Greater Britain	2535	10	7 1	1505.7	20.5	52 2	175
8/1893	0-8-0	2	15	1	30	24		50	111	4 5½	1489.0	20.5	49 5	175
2/1894	2-2-2-2	2	15	1	30	24	John Hick	20	10	6 3	1505.7	20.5	52 0	175
6/1897	4-4-0	2	15	2	19½	24	Black Prince	1502	1	7 1	1400.4	20.5	53 18	175
3/1899	4-4-0	2	15	2	20½	24	Iron Duke	1903	38	7 1	1379.6	20.5	54 8	200
5/1901	4-4-0	2	16	2	20½	24	Alfred The Great	1941	40	7 1	1557.5	20.5	57 12	200
8/1901	0-8-0	2	15	2	20½	24		1881	170	4 5½	1753.0	20.5	53 10	200
2/1903	4-6-0	2	15	2	20½	24		1400	30	5 0	1753.0	20.5	60 0	200

(a) Rebuilt from No 54 MEDUSA built Crewe 1846, altered in 1895 to triple expansion type named TRIPLEX. (b) Rebuilt from Standard Beyer Peacock 4-4-0T. (c) Rebuilt from Standard 2-4-2T.

Date	Type	Class	First No	No Built	Cylinders inches	D.W. ft in	Heating Surface				Grate Area	Boiler Press lb/sq in	Weight WO ton cwt
							Tubes	S'heat	F'box	Total			
GEORGE WHALE													
3/1904	4-4-0	Precursor	513	130	19 x 26	6 9	1848.4	—	161.3	2009.7	22.4	175	59 15
4/1905	4-6-0	Experiment	66	105	19 x 26	6 3	1908	—	133	2041	25.0	175	65 15
4/1905	RMO-4-0	Railmotor		6	9½ x 15	3 9		—		317.27	6.38	175	43 8
12/1906	4-6-0	Mixed Traffic	285	170	19 x 26	5 2½	1840.5	—	144.3	1984.8	25.0	175	63 0
5/1906	4-4-2T		528	50	19 x 26	6 3	1777.5	—	161.5	1939	22.4	175	74 15
1908	0-8-0	Class D		63R	19½ x 24	4 5	2033	—	147	2180	23.6	175	59 0
R/1908	4-4-0	Renown		70R	18½ x 24	7 1	1330	—	120	1450	20.5	175	56 0
C. J. BOWEN COOKE													
1/1910	0-8-0	Class G	2653	60	20½ x 24	4 5	2033	—	147	2180	23.6	160	59 0
7/1910	4-4-0	George The Fifth	2663	90	20½ x 26	6 9	1380	300	160	1840	22.4	175	60 0
12/1910	4-6-2T		2665	47	20 x 26	6 9	942	250	138	1330	24	175	78 0
10/1911	4-6-0	Prince of Wales	819	245	20½ x 26	6 3	1440	300	133	1873	25	175	66 0
12/1911	0-8-2T		1185	30	20½ x 24	4 5	1803	—	147	1950	23.6	170	72 0
2/1912	0-8-0	Class G1	1329	120	20½ x 24	4 5	1623	380	147	2150	23.6	160	61 0
1/1913	4-6-0	Sir Gilbert Claughton	2222	130	4/16 x 26	6 9	1645	414	171	2230	30.5	175	78 0
R/1915	4-6-0	Prospero	1361	1	4/14 x 26	6 3	1440	300	133	1873	25	175	69 0
CAPT. H. P. J. BEAMES													
6/1921	0-8-0	Class G2	485	60	20½ x 24	4 5	1545	360	147	2052	23.6	175	62 0
2/1923	0-8-4T		380	30	20½ x 24	4 5	1545	360	147	2052	23.6	185	88 0

LMSR – CREWE

	2-8-0	0-8-0	4-6-2	4-6-0	2-6-0	0-6-0	4-4-0	0-8-4T	2-6-2T
1923								7930-58	
1924						4107		7959	
1925						4108-58			
1926					2730-35	4159-76 4302-11			
1927					2736-99	4437-46			
1928					2800-07	4447-56 4507-56			
1929		9500-99			2808-09				
1930		9600-03			2850-2924				
1931		9604-39			2925-34		636-660		
1932		9640-74	6200	5502-19, 5523	2935-44		686-700		
1933			6201	5524/29-32/36-41	2945-60				
1934				5542-5551 5607-5646	2961-84				
1935	8000-11		6202-12	5000-19/70-74 5647-54/65-81					
1936	8012-14			5682-5742					
1937	8015-26		6220-24			4562-76			185-195
1938	8096-97		6225-34	5252-71					196-209
1939	8098-8125		6235-39						
1940			6240-44						
1941	8126-39								
1942	8140-57								
1943	8158-75 8301-16		6245-48						
1944	8317-30		6249-52	4826-60					
1945				4861-4920					
1946			6253-55	4921-31 4967-81	6400-09				1200-08
1947			6256	4758-82	6410-19				1209

All given LMSR numbers. Note: 2730-2809 were 13030-13109, and 2850-2984 were 13150-13284

Fig. 166
5553/1921
Claughton class
4-6-0 No. 150
ILLUSTRIOUS

Fig. 167
5614/1923
C. J. Bowen Cooke's
0-8-4T with
L&NWR No. 1189

Fig. 168
106/1932
Stanier's first
Pacific No. 6200
THE PRINCESS ROYAL

Fig. 169
113/1933
Stanier taper boiler
2-6-0 13245 with
GWR safety valve
bonnet,
later removed.

From 1 January, 1948 the LM&SR became part of the nationalised BRITISH RAILWAYS and at first work proceeded with the building of standard LMS classes.

The last LMS *Pacific* No. 46257 was completed in 1948 and to the end of 1950 only LMS types were built. Fifty Class 5 4-6-0, 2 cylinder locomotives were built. They were not all standard, and to compare with the single and double chimney types with Walschaerts valve gear, and the double chimney locomotive fitted with Stephenson's valve gear; the following were built with Caprotti valve gear: Nos. 44738–47 with plain bearings and single chimney; Nos. 44748–54 with Timken bearings and single chimney; Nos. 44755–57 with Timken bearings and double chimney. In addition twenty more (Nos. 44718–37) were built with steel fireboxes.

Other standard LMS types built were: forty-five Class 2 2-6-0s and eighty Class 2 2-6-2Ts.

1951 brought the eagerly awaited BR Class 6 and 7 *Pacifics.* Fifty-five Class 7 and ten Class 6 were completed in 1954 and 1952 respectively.

In 1954 came the solitary Class 8 *Pacific*, which had not originally come into the scheme of things. However as a result of the Harrow disaster the newly rebuilt *Pacific* No. 46202 PRINCESS ANNE had to be written off and permission was obtained for a replacement *Pacific* to be designed and built, and so the Class 8 *Pacific* No. 71000 DUKE OF GLOUCESTER came into being. The engine part was excellent and had a cylinder efficiency unmatched, but the boiler was reluctant to steam. No doubt given further opportunity this would have been rectified but the writing was already on the wall, and what could have been the 'ultimate' in British locomotive design had to be left in this unsatisfactory state until withdrawn.

Further Class 2 2-6-2Ts were built in 1953 which, of course, followed closely on Ivatt's Class 2 design.

There remained the 2-10-0 design put in hand in 1955 and completed in 1958. Ten were fitted with the Crosti type of boiler, but unfortunately the savings in fitting pre-heaters did not come up to expectations and maintenance costs were too high. All ten had their drums removed and became similar to the rest. Mechanical stokers and Giesl ejectors were fitted to selected locomotives but no outstanding results were achieved. Here again if time and money had been justified better results may have been achieved.

So the long saga of Crewe steam locomotives came to an end on 15 December, 1958 when No. 92250 was steamed – the last of 198 2-10-0s built at Crewe and the 7331st built there, spanning a period of 114 years.

A summary of locomotives built at Crewe during British Railways management is as follows:

SUMMARY OF LOCOMOTIVES BUILT AT CREWE – BRITISH RAILWAYS

1948	1	LMS	Class 8	4-6-2	46257.		
1948	20	LMS	Class 5	4-6-0	44738-57		
1948	15	LMS	Class 2	2-6-0	46420-34		
1948	20	LMS	Class 2	2-6-2T	41210-29		
1949	30	LMS	Class 5	4-6-0	44658-67,	44718-37	
1949	30	LMS	Class 2	2-6-2T	41230-59		
1950	30	LMS	Class 2	2-6-2T	41260-89		
1950	30	LMS	Class 2	2-6-0	46435-64		
1951	25	BR	Class 7	4-6-2	70000-24		
1951	2	BR	Class 6	4-6-2	72000-01		
1951	10	LMS	Class 2	2-6-2T	41290-99		
1952	20	LMS	Class 2	2-6-2T	41300-19		
1952	13	BR	Class 7	4-6-2	70025-37		
1952	8	BR	Class 6	4-6-2	72002-09		
1953	7	BR	Class 7	4-6-2	70038-44		
1953	20	BR	Class 2	2-6-2T	84000-19		
1954	1	BR	Class 8	4-6-2	71000		
1954	10	BR	Class 7	4-6-2	70045-54		
1954	32	BR	Class 9	2-10-0	92000-19,	92030-41	
1955	38	BR	Class 9	2-10-0	92020-29,	92042-69	
1956	39	BR	Class 9	2-10-0	92070-86,	92097-118	
1957	46	BR	Class 9	2-10-0	92119-62,	92168-69	
1958	43	BR	Class 9	2-10-0	92163-67,	92170-77,	92221-50
Total	490						

Crewe Works

The original works occupied about three acres and employed 161 workmen. It was extended in 1862, 1864 and 1870. Crewe always ranked as one of the leading locomotive building establishments and many leading mechanical and locomotive engineers received their initial training in these works, from many parts of the world. As an economic unit it was badly sited, bounded on the east by the main line to Liverpool and on the south by the Chester line forming a narrow strip squeezed in the middle, the northern boundary being formed by the town itself which had grown from a village to a railway town.

The point was reached when further expansion was impossible and reorganisation essential and a complicated scheme was put in hand just after the amalgamation and completed in 1927. Before this there had been no less than nine separate erecting shops dotted about from east to west the two extremes being about one and a half miles from each other, so that the time spent in moving materials from department to department was far from efficient or economical so that a definite flow pattern was evolved to minimise such wasteful methods employed in the past. Nevertheless the output of the Old Works was impressive, attaining 100 locomotives in 1864, 129 in 1867 and a maximum of 138 in 1874, albeit many were major rebuilds of older locomotives.

This reorganisation, as far as the erecting shops were concerned, employed the principle of moving the work to the men rather than the men to the work with a continuity of work carefully timed and co-ordinated. It was a large edition of the car assembly belt system in principle.

Only one new building was necessary and this was the south erecting shop built on the side of the existing large erecting shops. The nine old shops had a total of 256 locomotive pits, whereas the new scheme had only 72 all located in the same group of shops, but with the greater efficiency of operation the time spent in repairs and overhauls was drastically cut from 30 to 50 days for the old shops, to 8 to 12 days by the new methods according to size of locomotive.

The work was divided into twelve stages from the initial stripping to the final valve-setting after wheeling, fitting up the motion etc.

Each 'belt' or pit had a winch at the end outside the shop and at times indicated by clocks at each stage the locomotive under repair was winched to the next stage.

To obtain these results the other major shops were reorganised, machinery re-arranged and sectionalised, and the boiler and tender shops were likewise altered.

The whole scheme was extremely well thought out although many snags had to be ironed out, within a year of operating over 850 locomotives had been repaired. The output potential of new locomotives was 100 .

Works numbers were given to L&NWR locomotives built at Crewe and were amended twice, once in 1923 and again in 1931 when the works numbers recommenced at No. 1 with 0-8-0, 9620. This seemed rather pointless and only complicated the assessment of new building.

The last and 7331st new locomotive to be built was turned out on 15 December, 1958 and the last steam locomotive to be repaired, *Britannia* class No. 70013 on 2 February, 1967.

The works is at present engaged on diesel and electric locomotive repairs and electric locomotive building.

Early Crewe locomotives were painted a shade of green (some sources describe it as bluish green) until 1873 when black was specified as the standard colour.

Until 1878 the locomotive running number was painted on the cab sides, but in this year brass number plates were introduced and subsequently Webb standardised on a cast iron number plate with surrounding beading, the numbers being raised and painted yellow, the background being red.

The L&NWR tender had no sign of ownership except a small rectangular number plate fixed on the rear of the tender.

The name plates were brass strip with sunk letters filled in with black wax. The strip was either curved or straight according to the splasher on which it was fixed. The date of building was on one end of the name plate and 'Crewe Works' inscribed at the opposite end.

Numbering was not in any sequence and a new locomotive took the lowest number available, a few series of numbers were used such as the 1901 to 1980 series for the 4-4-0 compounds, and a few other examples.

The duplicate list of numbers for locomotives that had been condemned or replaced was at first from 1101 to 1300, and when these numbers were wanted for additional stock a new series appeared from 1801 to 2000, and finally the duplicate series was from 3000 to 3477, 3551 to 3585 and 3776.

To trace the history of some of these locomotives is extremely complicated and in some instances impossible. One typical example will suffice: Crewe Works No. 2289, 0-6-0ST built October 1879 and numbered 331;

Fig. 170 244/1935 2-8-0 No. 8003 Stainer *Class 7F* (later 8F)

Fig. 171 279/1935 Stanier 3-cylinder *Jubilee* class No. 5681 ABOUKIR

Fig. 172 1937 streamlined *Pacific* No. 6221 QUEEN ELIZABETH

renumbering: 1805 in October 1896, 3278 in April 1897, 1726 in November 1908, 3092 in July 1919, 1726 in December 1919, 3092 in July 1920 7420 in October 1928 (LMS No.). From this it seems the locomotive was placed on the duplicate list four times and then returned to running stock and this particular one survived until December 1932, having first been put on the duplicate list thirty-five years before being cut up. They were certainly built to last (in most cases).

The design of the LNW chimney was improved greatly after the passing of the Ramsbottom type which had a pronounced vee-shaped rim in which were cut-out sections. The standard chimney introduced by Webb had a cast iron rim and an unusual square flange type base, bolted to the smokebox.

The L&NW locomotive in motion could be picked out from any other, firstly by the exhaust beats, which were usually two strong, then two lighter beats attributed to the Joy's valve gear, and secondly by the metallic ringing of the air pump valves. There were other characteristic sounds, as on all designs.

Smokebox doors did not seem particularly air tight in many instances, indicated by a large patch of burnt metal around the lower half of the door.

The works themselves progressively increased in area from the humble three acres of ground in 1843 to some 160 acres including rolling mills which produced their own rails for many years.

Works numbers had passed 5700 but many locomotives carried new works numbers after heavy rebuilding and also the locomotives delivered from outside contractors in the 1920s were also given Crewe numbers. 'Motion' numbers were introduced in August 1857.

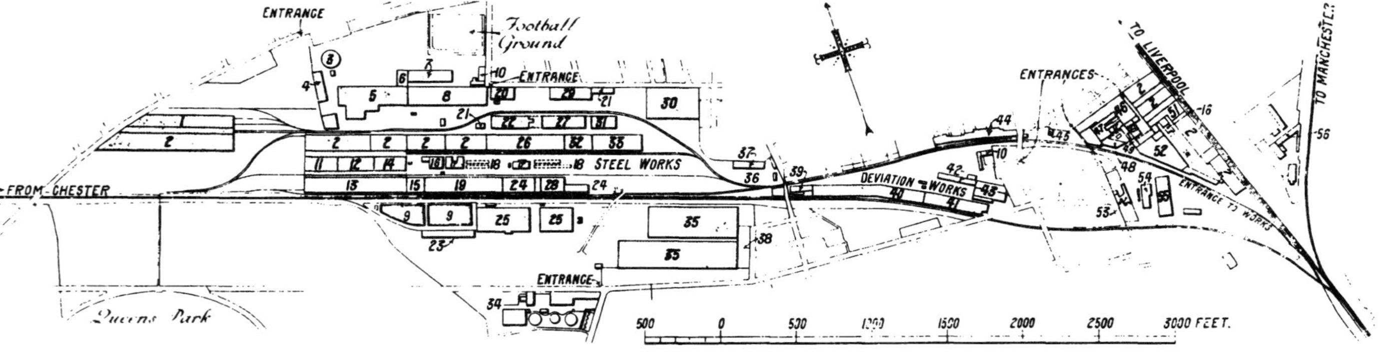

BEFORE REORGANISATION.

1. Fitting Shop. 2. Erecting Shops. 3. Brick Kiln. 4. Drying Shed. 5. Iron Foundry. 6. Pattern Shop. 7. Pattern Stores. 8. Tender Shop. 9. Cooling Ponds. 10. Dining Rooms. 11. Plate Stores. 12. Flanging Shop. 13. Boiler Shop. 14. Angle Iron Smithy. 15. Points and Crossings Shop. 16. Electric Power Station. 17. Boilers. 18. Gas Producers. 19. Rail Mill. 20. Brass Foundry. 21. Stores. 22. Coppersmiths' Shop. 23. Carriage Washing Shed. 24. Siemens Martin Furnaces. 25. Carriage Repairing Shops. 26. Axle Forge. 27. Wheel Shop. 28. Spring Mill and Ingot Stripping Shop. 29. Signal Shop. 30. Paint Shop. 31. Nut and Bolt Shop. 32. Iron Forge. 33. Steel Foundry. 34. Gas Works. 35. Carriage Store Sheds. 36. Store Yard. 37. Mortar Mill. 38. Clothing Factory. 39. Testing Shop. 40. Millwrights' Shop. 41. Joiners' Shop. 42. Timber Shed. 43. Sawmills. 44. Locomotive Offices. 45. Pay Office. 46. Smiths' Shop. 47. Drop Hammers. 48. Iron Stores. 49. Millshop. 50. Spring Shop. 51. Valve Motion Shop. 52. Fitting Shop. 53. Hospital. 54. Locomotive Stores. 55. Out Station Shop. 56. Grease Works.

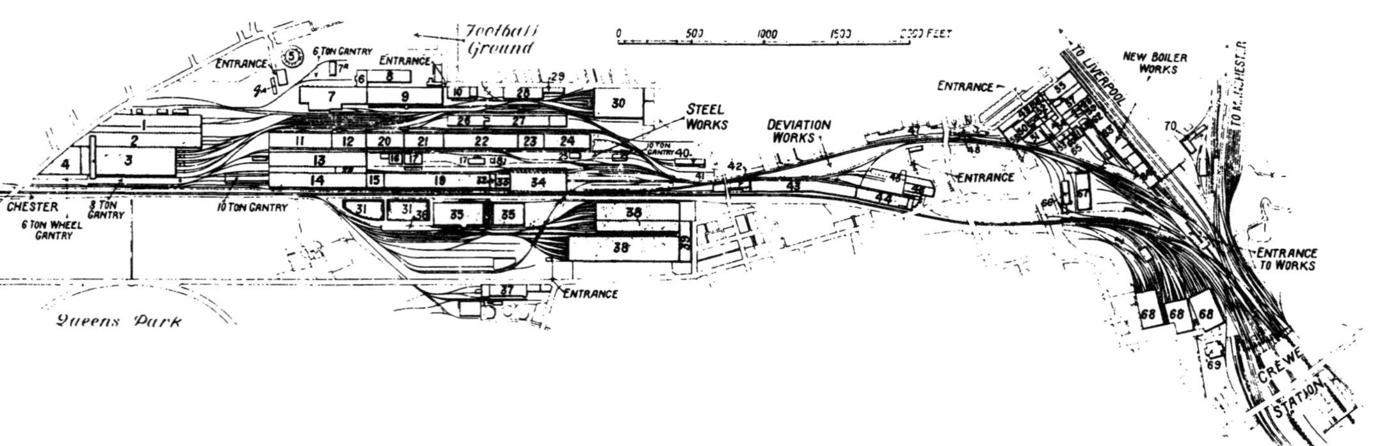

1927

AFTER REORGANISATION.

1. Fitting and Machine Shops. 2. No. 9 Erecting Shop. 3. New Erecting Shop. 4. Machine Shop. 4a. Progress Office. 5. Brick Kiln. 6. Pattern Shop. 7. Iron Foundry. 7a. Chair Foundry. 8. Pattern Stores. 9. Tender Shop. 10. Brass Foundry. 11. Finished Part Stores, and Welders Superheater Element Tubes. 12. Mounting Shop. 13. Smithy. 14. Boiler Shop Repairs. 15. Points and Crossings Shop. 16. Electric Power House. 17. Boilers. 18. Gas Machines. 19. Rail Mill. 20. Tube Shop. 21. Frame and Cylinder Shop. 22. Axle Forge. 23. Iron Forge. 24. Steel Foundry. 25. Laboratory. 26. Coppersmiths' Shop. 27. Wheel Shop. 28. Brass and Finishing Shop. 29. Stores. 30. Paint Shop. 31. Cooling Ponds. 32. Millwrights' Shop. 33. Spring Mill. 34. 45- and 65-ton Steel Furnaces. 35. Carriage Repairing Shops. 36. Carriage Washing Shed. 37. Gas Works. 38. Carriage Store Sheds. 39. Clothing Factory. 40. Mortar Mill. 41. Stone Yard. 42. Testing Shop. 43. Millwrights' Shop. 44. Joiners' Shop. 45. Timber Shed. 46. Saw Mills. 47. Locomotive Offices. 48. Pay Office. 49. Smiths' Shop. 50. Drop Hammer Shop. 51. Iron Stores. 52. Tube Shop. 53. Die Sinking [illegible] 57. Mounting and Assembly Shop. 58. Boilers. 59. Stay Shop. 60. Galvanising Shop. 61. Tin Shop.

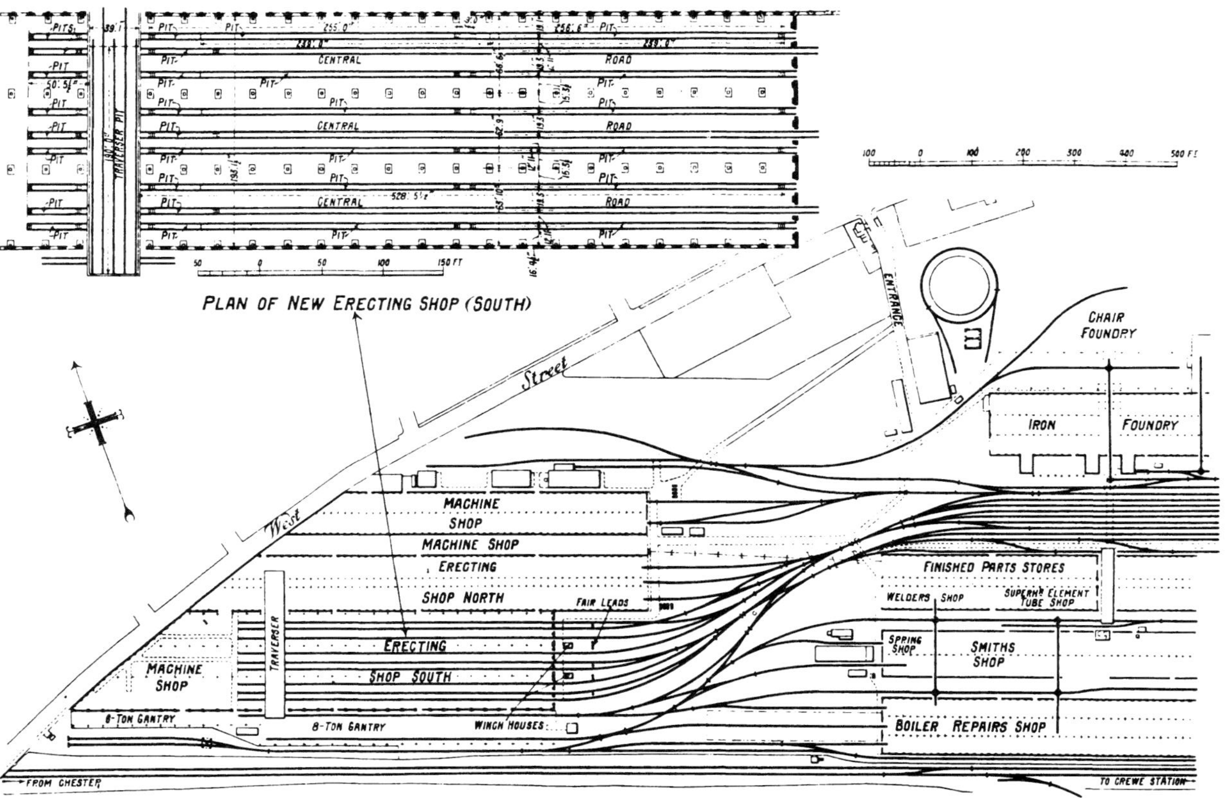

Layout of Western Portion of Crewe Works, with Plan of New Erecting Shop (South).

Fig. 173 1950 Ivatt *Class 2* 2-6-0 46460

Fig. 174 1951 *Britannia* class 70018 FLYING DUTCHMAN

Fig. 175 1953 *Class 2* 2-6-2T No. 84016

Fig. 176 1954 3-cylinder *Pacific* No. 71000 DUKE OF GLOUCESTER

F. W. Webb's Compounds

Mr F. W. Webb's first experimental compound used Mr Anatole Mallet's system. The *Medusa* had its left hand cylinder lined up to 9" diameter. Now, Mallet's system provided for the steam to be admitted to the low pressure cylinder when required for starting, and direct discharge into the chimney of exhaust from the high pressure cylinder when working as a simple engine. It was set to work on the Ashby and Nuneaton branch and with the experience gained, Webb was convinced that compound working should be proceeded with, and designed the *Experiment* to test its capabilities on main line passenger duties. Many trial runs were carried out with this locomotive and its behaviour over twelve months closely watched. It was very steady in running due to the position of the three cylinders and for a period worked the Irish Mail from Crewe to London and returned with the limited Scottish Mail – a daily run of 310 miles. Its average coal consumption was 26·6 lbs per train mile as compared with 34·6 lbs. for the four coupled simple locomotives used on the same duties.

Mr Webb decided that compounds should work a high proportion of the main line expresses and proceeded to build twenty-nine similar locomotives with some modifications. An alteration was made to the steam ports of the LP cylinder increasing their size from 1½" x 14" to 2" x 16" and the valve gear was modified.

The HP steam chests were placed below the cylinders and the valves were of the Trick or Allan type having a travel of 3⅛" in full gear; the lap was ¾" and the lead ⅛". The high pressure steam ports were 1⅛" x 9" and exhaust ports 2½" x 9". The Joy's valve gear to these cylinders had the motion discs carrying the quadrant bars fastened to the underside of the slide bars. The quadrant bars were each grooved to a radius equal to the length of the valve rod link. Working in the grooves were brass slide blocks, carried by the lifting links, to the lower end of which was attached the valve rod link, and to the upper end a compensating link on the connecting rod. A rod attached to a return crank on the HP crank pin controlled the upper end of the compensating link.

The reversing shaft was placed outside the trailing wheels, and attached to the quadrant bars, which were extended below the discs for this purpose. The reversing was carried out by a screw and lever arrangement, and the gear for HP and LP cylinders worked independently.

The steam chest was on top of the low pressure cylinder which was carried between two cross steel plates between the main frames. In full gear the valve travelled 4½" with 1" lap and 3/16" lead. The exhaust ports were 3¼" x 16" a relief valve set at 75 lb/sq. in. was fitted to the cover of the LP steam chest. A pressure gauge in the cab indicated the steam pressure at the low pressure cylinder inlet.

Other features of the locomotives were, the firebox had a water space under the grate with the ash-pan between the two. This idea had been used before by Webb on some of his goods engines, the object was to eliminate the rigid foundation ring and promote the circulation of water. A good feature was the size of the journal bearings. The leading axle having 6" diameter x 10" long, the LP drivers 7" diameter x 13½", and the HP drivers 7" diameter x 9" long bearings respectively.

The steam for the HP cylinders was taken

from the dome to a T pipe on the smokebox tube plate and thence by means of two 3″ copper pipes to the HP steam chests. The 4″ exhaust pipes followed back to the smokebox and thence to the LP cylinder. Passing back into the smokebox the pipes were in contact with the flue gases. Provision was also made to feed steam direct to the LP cylinder. The blast pipe was $4\frac{5}{8}$″ diameter. Later, *Experiment* was fitted with piston valves and a boiler with a pressure of 180 lb/sq. in. The weight of passenger trains continued to increase and in 1884 the *Dreadnought* class was turned out, being much more powerful than their predecessors. All the slide valves were balanced and driven by Joy's motion. When difficult starting conditions prevailed, the driver could admit steam into the LP cylinder by means of an independent valve.

The *Teutonics* had a slight increase in adhesive weight and larger driving wheels. All previous compounds had inside clearance to the valves, but for the *Teutonics* this was not adopted for the LP cylinder, as it was found that by delaying the exhaust somewhat, the large piston was cushioned and the surging that was present on previous compounds was eliminated. Later *Teutonics* had Webb's eccentric for actuating the LP valve.

JEANIE DEANS was exhibited at the 1890 Edinburgh Exhibition.

The *Greater Britain* class was the first 8 wheeled compound tender locomotive. The valves of the HP cylinders were placed inside the frames and were driven by Stephenson's link motion and the LP valve by Webb's single eccentric. The boiler had a combustion chamber complete with inspection cover, hopper for discharging the ashes, and steam blast apparatus for tube cleaning. They were originally fitted with piston valves but, because of leakage troubles, slide valves were substituted.

The QUEEN EMPRESS was sent to the Chicago World's Fair in 1893 and came back with a Gold Medal.

The 0-8-0 compound of 1893 had all three cylinders in line and bolted together, and the three connecting rods drove the same axle, the second one. The inside driving axle was built up, consisting of three pieces, and Stephenson's valve gear was fitted.

The *John Hick* class of 1894 had fireboxes fitted without the water space, similar to the *Greater Britain* class.

The four cylinder compounds had their cylinders all in line with the low pressure pair inside. The high pressure cylinders had piston valves fitted, but slide valves operated the low pressure side, working on top of the cylinders. Two sets of valve gear were fitted to work the four cylinders.

The valve spindle of the LP valve was extended to the front of the steam chest and by means of a lever, operated the adjacent valve of the HP cylinder. As mentioned before, Webb's compounds in the main were not an unqualified success. On the earlier designs without coupling rods, the starting was very critical as the two driven axles were independent, and surging was very evident in running. Yet there were patches of brilliant running – GREATER BRITAIN ran six days without a break, on test between Euston and Carlisle with a load of $160\frac{1}{2}$ tons tare, averaging nearly 48 mph with a coal consumption of just under 30 lb per mile.

JEANIE DEANS of the *Teutonic* class regularly worked the famous 2 pm Scottish express between Euston and Crewe for eight months with very little time lost. Coupling rods were omitted due to the bulk of the outside valve gear and it was only with the appearance of the 4-4-0 compounds that inside valve gear was used, which seems strange as there was no doubt that the lack of coupling rods and the retention of the original type of valve gear were the prime causes of the poor performances of most of the previous types. Even so the 4-4-0s were sluggish as the HP and LP cylinders were notched up together, preventing the proper expansion and expulsion of steam.

GRANGE IRON WORKS
Durham

This works built a few locomotives for collieries and ironworks. 1866 0-4-0ST CARBON; 1873 0-4-0ST West Hartlepool Steel & Iron Co. Ltd.

GRANT, WILLIAM
Belfast

A tram locomotive was suppied by this firm to the Cave Hill and Whitewell Tramway, which ran between Belfast and Cave Hill. It was supplied in 1886 and William Grant either acted as agent only, or the parts were supplied by Kitson & Co. and the locomotive erected at Belfast. It is thought to be Kitson Tram Works No. 51 and therefore the parts would have been made in 1882 and sent over later. The tram had cylinders $7\frac{1}{4}$″ x 12″ and wheels 2′ $4\frac{1}{4}$″ diameter and was numbered 3 by the tramway, which was to the unusual gauge of 4′ 8″.

GRANT RITCHIE & CO.

Townholme Engine Works, Kilmarnock.

This firm *formerly* GRANT BROS. built industrial tank locos, also colliery winding engines. Works numbers were allocated to all pieces of machinery. They built forty-five locomotives between 1879 and 1920.

The principal types built were 0-4-0ST, 0-4-0CT, 0-6-0ST, and 0-4-2ST. They were of typical Scottish design and in some cases bore a strong resemblance to its neighbour's products – Andrew Barclay Sons & Co.

The saddle tanks covered boiler and firebox and were flat sided with rounded lower edges. The wheels had cast iron centres with I section spokes. Both single and double slide bars were used, and cylinders were horizontal or slightly inclined.

The earlier locomotives had chimneys which were unusual in having two built up rings below the top flange giving a very heavy appearance and the cabs had circular side windows.

At least three crane locomotives were built, two going to the Glengarnock Iron & Steel Co. They had a fixed slewing jib positioned half way along the boiler.

The 0-4-2STs which were built, provided for a larger bunker without a six coupled arrangement.

Their products did not penetrate very far over the border – principally because locomotive building was not their main preoccupation, but they were good solid machines and gave good service.

Fig. 177
Grant Ritchie & Co
435/1903
Piel & Walney Gravel Co
Walney Island
3′1″ gauge
0-6-0ST

Fig. 178
Grant Ritchie & Co
538/1917
0-4-2ST
William Dixon, Govan
enlarged 0-4-0ST to provide more room for coal and better weight distribution

GRANT RITCHIE & CO.

Works No	Date	Type	Cylinders inches	D.W. ft in	Gauge ft in	Customer	No/Name
		0-4-0ST	OC			Eglinton Iron Co.	Lugar No 5
			OC 12 x			Callender Coal Co. Ltd	No 4
	1879					Scottish Iron & Steel Co.	
39	1879	0-4-0ST	OC		Std	Whitehill Colly	No 2
40	1880	0-4-0ST	OC 12 x 20	3 6		Steel Co. of Scotland	
60	1880	0-4-0ST	OC 14 x 21	3 5	4 4	Fordell Rly	Alice
112	1883	0-4-0ST	OC 12 x			J. & R. Howie	
119	1883	0-6-0ST				Tredegar Iron & Coal Co. Ltd	
146		0-4-0ST	OC		Std	Fife Coal Co.	14 (a)
164	1886	0-6-0ST	OC		3 0	Piel & Walney Gravel Co.	Express
174	1887	0-4-0ST	OC 14 x 22	3 6	Std	Methil Dock Co.	Jubilee (b)
179	1887	0-4-0ST	OC 12 x 20	3 8	Std	Dalmellington Iron Co. Ltd	12
209	1890	0-4-0CT	OC 12 x 22		Std	Glengarnock Iron & Steel Co.	
213	1893	0-4-0CT	OC 12 x 22		Std	Lanarkshire Steel Co.	
234		0-4-0ST	OC			Bent Colly. Co. Ltd	
272	1894	0-4-0ST	OC		Std	Fife Coal Co.	21
343	1898	0-4-0ST	OC 14 x		Std	Fife Coal Co.	19
344	1898	0-4-0ST	OC			Stewarts & Lloyds Ltd	
354	1899	0-4-0ST			Std	Colvilles. Glengarnock	15
355	1899	0-4-0ST	OC 14 x			Jas Dunlop & Co Ltd	Calderbank
362	1899	0-4-0ST	OC		Std	W. Roberts (Tipton) Ltd	R.4
	1899	0-4-0T	OC		Std	David Colville & Sons Ltd	13
387	1901	0-4-0ST			Std	Fife Coal Co.	25
408	1901	0-4-0ST	OC			Hargreaves Colls. Ltd	John
427	1903	0-4-0ST	OC		Std	Fife Coal Co.	15
435	1903	0-6-0ST	OC		3 0	Piel & Walney Gravel Co.	Wadham
480	1905	0-4-0ST	OC		Std	Fife Coal Co.	4
514	1907	0-4-0ST	OC 14 x		Std	W. Baird	Muirkirk No 6
521		0-4-0ST	OC		Std	Kinneil Colly.	
522	1907	0-4-0ST	OC 15 x 22	3 6	Std	Young's PL & MO Co. Ltd	Uphall (c)
527	1908	0-4-2ST	OC		Std	Lothian Coal Co.	5
531	1911	0-4-0ST	OC 14 x 22	3 8	Std	William Baird & Co.	Eglinton No 6
535	1913	0-4-0ST	OC		Std	W. Roberts (Tipton) Ltd	R.3
536	1914	0-4-2ST	OC		Std	Lothian Coal Co.	7
538	1917	0-4-0CT	OC 12 x 22		Std	Glengarnock I & S Co.	
539	1917	0-4-2ST	OC		Std	Glengarnock I & S Co.	No 7
540	1917	0-4-2ST	OC 16 x 22		Std	Glengarnock I & S Co.	No 6
	1917	0-4-0ST	OC 12 x 22		Std	Glengarnock I & S Co.	No 16
582	1918	0-4-0ST			Std	W. Roberts (Tipton) Ltd	R.6
698	1920	0-4-0ST	OC 16 x 22			Oakbank Oil Co.	No 2
727						Fordell Colly.	
769	1920	0-4-0ST	OC 16 x 24		Std	L.H. & J. Colls Ltd	43
805	1918	0-4-0ST	OC 16 x 24	3 9	Std	Baird & Scottish Steel Co.	No 10
865	1919	0-4-0ST	OC			Motherwell I & S Co.	No 1
	1920	0-4-0ST	OC 12½x20	3 8	Std	Dalmellington Iron Co. Ltd	21

(a) Date given as 1890 probably built previous to this for stock. (b) Became North British Rly No 612 in 1891. (c) Youngs Paraffin, Light & Mineral Oil Co. Ltd. Pumpherston, L.H. & J.C. Ltd. Lambton, Hetton & Joicey Collieries Ltd.

GREAT CENTRAL RAILWAY
Gorton

See MANCHESTER, SHEFFIELD & LINCOLNSHIRE RAILWAY.

GREAT EASTERN RAILWAY
Stratford

See EASTERN COUNTIES RAILWAY.

GREAT NORTHERN RAILWAY
Doncaster

Locomotive Superintendents: 1850–1866 Archibald Sturrock; 1866–1895 Patrick Stirling; 1896–1911 H. A. Ivatt; 1911–1922 H. N. Gresley.

The original locomotive headquarters of the GNR was at Boston, Lincolnshire. Established in 1847, it was not designed to cope with such a rapid increase of locomotive stock. By 1853 over 200 locomotives were working and it was obvious that Boston could not manage the large amount of repair work required. No new work was carried out at Boston under Mr Benjamin Cubitt, who was the first locomotive superintendent. Unfortunately he died before the first train ran and Edward Bury, the Liverpool locomotive builder, was invited to manage the line and be responsible for locomotive matters. Bury could not give his undivided attention to this post, he had a thriving locomotive business in Liverpool to attend to and so after a very short term of office he either decided to leave, or was asked to leave, and the vacant post was advertised in 1850.

Fig. 179 **GNR** 50/1870 Stirling's 8′ single No. 1 with large tender

Fig. 180 **GNR** 446/1887 Stirling 2-2-2 7′6″ inside cylinder No. 239

Fig. 181 GNR 874/1900 Small *Atlantic* No. 982

The successful candidate was Archibald Sturrock. He had been apprenticed to the firm of J. & C. Carmichael at Dundee, after which he had a short spell at William Fairbairn's. In 1840 he joined the Great Western Railway at the age of 24 as assistant locomotive superintendent under Daniel Gooch and in 1842 was made responsible for supervising the erection of the new locomotive shops at Swindon and became works manager when they became operative in January 1843.

Having joined the Great Northern Railway he again encountered the building of a new locomotive factory, this time at Doncaster which was ready for occupation in 1853.

Mr Sturrock adopted two fundamental principles in designing his locomotives; a high boiler pressure and a firebox of ample proportions. A pressure of 150 lb/sq. in. was introduced as standard. No locomotives were built at Doncaster whilst he was in office, but over 270 were ordered from various contractors many of which were 0-6-0s from W. Fairbairn & Sons, E. B. Wilson & Co. and many others. The most noteworthy locomotives supplied to his specifications were thirty 0-6-0s with steam tenders, the forerunners of boosters introduced much later. They were not popular with the footplate men and were heavy on coal. Two eight coupled tanks were supplied by the Avonside Engine Co. with condensing apparatus for working goods trains in the Metropolitan area. Sturrock favoured the domeless boiler and most of his passenger types had that feature. Finally his 4-2-2 of 1853 was a remarkable locomotive built by R. W. Hawthorn with two inside cylinders 17″ x 24″ and 7′ 6″ diameter driving wheels. A large domeless boiler with 1719 sq. ft of heating surface was fitted and the entire wheelbase was rigid. It has been said that it was rather premature and was withdrawn in 1860.

The works here expanded in 1865 with a further twelve pits in the erecting shop, and the boilershop was extended.

He resigned in 1866 at the early age of 50 deciding to live the life of a country gentleman which he did do until the ripe old age of 93.

Patrick Stirling took his place, coming from the Glasgow & South Western Railway at Kilmarnock.

Before he obtained the post of locomotive superintendent on the G&SWR he had wandered far and wide. Apprenticed by Stirlings of Dundee, he obtained marine engineering experience on the Clyde, became a foreman at Neilson & Co., locomotive superintendent on the Caledonian & Dumbartonshire Railway, back to marine engineering, then to R. & W. Hawthorn at Newcastle.

With this great and varied experience he settled down to make his mark at Doncaster. In 1867 appeared the first locomotives to be built in the company's works. They were mixed traffic 0-4-2s with 17″ x 24″ cylinders and 5′7″ diameter coupled wheels. Small and light they were able to cope with branch line requirements and semi-fast trains. Ninety-one were built at Doncaster between 1867 and 1879 and thirty by outside contractors.

In 1870 appeared the first of Stirling's eight foot 'singles' in the form of a 4-2-2 with outside cylinders 18″ x 28″ – they were the only outside cylinder design by Stirling, and took the form of one of the most handsome locomotives ever produced. The boiler was domeless with 1165 sq. ft of heating surface including a mid-feather of copper in the firebox from the tubeplate half way to the firedoor. The cab was the wrap-around type which looked, and was, very uncomfortable, thirty-seven were built from 1870 to 1882 and ten more followed in 1884 to 1887 with modifications. They were used on all the main line trains for twenty-five years. In 1895 they took part in the famous 'Race to the North' where average speeds 'start to stop' between King's Cross and York were well over 60 mph.

Stirling's designs emphasised simplicity of outline and indeed the 4-2-2s looked right and lived up to appearances. His final batch of six 8 ft singles were larger with 19½″ x 28″ cylinders, increased boiler pressure of 170 lb/sq. in. larger grate area, but oddly enough the heating surface was reduced to 910 sq. ft. Besides this class of 'singles' Stirling built some 2-2-2s the most successful being a class of 23 with 7′ 6″ diameter driving wheels with inside cylinders 18″ x 26″, inside bearings to the driving wheels and outside bearings to the leading and trailing wheels. Various classes of 2-4-0s were also built and many were used on the second class expresses and branch lines.

For goods traffic various 0-6-0s were built, the standard class having 17½″ x 26″ cylinders and 5′ 1″ wheels, but the most interesting were a series of six built in 1871 with 19″ x 28″ cylinders and 5′ 1″ wheels, the comparatively large boiler producing a total heating surface of 1352 sq. ft. The snag with these locomotives was that they could pull far many more wagons than the refuge sidings could accommodate, so that they could not be utilised to their full capacity.

To work the suburban services particularly in the London area Stirling brought out a 0-4-2WT in 1868 and a 0-4-4WT in 1872 and side tank development came in 1881 with the same wheel arrangement as the 1872 design culminating in the smart looking 0-4-4-Ts in 1889 still with domeless boiler, and condensing apparatus. Finally for shunting and trip working he designed various classes of 0-6-0STs between 1868 and 1893, the standard pattern having 17½″ x 26″ cylinders, 4′ 7″ wheels, 40 tons in weight and the saddle tank holding 1000 gallons of water.

Boilers were standardised to a great extent and being domeless had the regulator valve in the smokebox and a perforated collector pipe running the length of the boiler. The regulator was of the pull-out type. Cylinder sizes had also been kept to a minimum and the 17/17½″ x 24″ and 17½″ x 26″ sizes were fitted to over 700 of his locomotives.

Patrick Stirling died in office on 11 November 1895 and it was not until 1896 that his successor H. A. Ivatt was appointed. He had been apprenticed at Crewe and had been locomotive superintendent of the Great Southern and Western Railway at Inchicore for the past ten years.

Fig. 182 **GNR** 922/1901 Ivatt 6′7½″ 4-4-0 No. 1385

At this time traffic was getting heavier and engine mileage increasing and the demand for new locomotives was such that Doncaster could not cope with all the requirements – in fact since its inception the company had to rely on outside contractors to a considerable degree. Patrick Stirling had 190 locomotives built outside.

Ivatt put in hand some 2-4-0s and 4-4-0s for the passenger traffic; both types had 17½″ x 26″ cylinders and 6′ 7½″ driving wheels. The 4-4-0s were of two classes, the larger having a domed boiler with a heating surface of 1250 sq. ft and a grate area of 20·8 sq. ft and the smaller type 1124 sq. ft and 17·8 sq. ft of grate area. The boiler length was the same at 10′ 1″ but the diameters were 4′ 8″ and 4′ 4″ respectively. A continuous splasher was fitted over the driving wheels, with one or two exceptions, and straight framing.

Ivatt in 1898 reverted to the 4-2-2 type and built twelve, the first having 18″ and the others 19″ cylinders and 26″ stroke. The driving wheels were 7′ 7½″ diameter.

It is puzzling to try to understand the reason for building these 'singles' as they were now quite unsuitable for the heavy expresses to the north.

One hundred and thirty-three new 0-6-0s were built at Doncaster from 1897 to 1901 with 17½″ x 26″ cylinders and 5′ 1½″ diameter wheels and the same boiler as the smaller 4-4-0s.

A further class with 5′ 8″ wheels was built in 1908/9.

Perhaps Ivatt's best known locomotives were the *Atlantics*; No. 990 was the first 4-4-2 built in Britain and made its appearance in 1898. It had two outside cylinders 18¾″ x 24″, 6′ 7½″ diameter driving wheels and a 4′ 8″ diameter boiler with a total heating surface of 1439 sq. ft and a grate area of 24·5 sq. ft. The trailing wheels had outside frames. Twenty one were built altogether, the last batch in 1903.

In 1902 another *Atlantic* was built with four cylinders – all in line, 15″ diameter and 20″ stroke. It had a number of unusual features, including double headed piston valves with only two sets of motion, one set controlling the inside and outside cylinders on one side. It was not a success and went through numerous modifications, finally ending up as an inside two cylinder 4-4-2 with superheated boiler and became an excellent performer.

In the same year appeared No. 251 a much larger *Atlantic* weighing some 10 tons heavier than the previous class, but the most significant feature was the boiler which produced a total heating surface of 2500 sq. ft and a large grate area of 31 sq. ft, the firebox being the wide Wootten type which extended over the frames at each side. They were a success from the start and eighty were built between 1904 and 1908.

In addition two more of the same wheel arrangement were built, Nos. 292 (1905) and 1421 (1907). They were 4 cylinder compounds, the first having LP cylinders (inside) 16″ x 26″ and HP cylinders 13″ x 20″. Walschaerts valve gear controlled the HP cylinders and link motion the LP cylinders. It was possible to work the locomotive as a 4 cylinder simple.

On the second compound the LP cylinders were 18″ diameter. The boiler pressure in both cases was 200 lb/sq. in. as compared with the usual 175 lb/sq. in. at this time.

Regarding the large increase in goods traffic and loading, Ivatt designed a 0-8-0 tender locomotive in 1901 and built fifty-five between that date and 1906. They had inside cylinders 19¾″ x 26″ and 4′ 7½″ diameter wheels; the boiler was the same as fitted to the small *Atlantic*. A tank version was also built as a 0-8-2T with identical cylinders, motion and boiler but with large tanks originally holding 2000 gallons but subsequently shortened with 1500 gallon capacity to lighten the axle load. The weight was consequently reduced to 70 tons 5 cwt. They were originally intended for London but even when their weight was reduced, they played havoc with the permanent way. Their original condensing gear was removed and they were transferred to the Leeds and Nottingham areas.

Tank versions of his 4-4-0s were also introduced in 1898 in the form of 4-4-2Ts and were used extensively on London suburban services and in the provinces; sixty were built 1898–1901.

Other types introduced were six rail motors in 1905, four of which were built by outside contractors, and some 0-6-2T radial tanks with 18″ x 26″ inside cylinders and 5′ 8″ wheels; a type which was to prove popular in later years for suburban traffic.

In 1911 Henry Ivatt retired, having been responsible for building over 700 locomotives; in the majority of cases they were a big step forward in the development of Great Northern motive power. He had experimented with various types of superheaters, brought standardisation up to date with standard cylinders and boilers particularly, and had not been afraid to increase the boiler pressure at the same time.

His place was taken by Mr Herbert Nigel Gresley in the same year. Like his predecessor he had been trained at Crewe under Mr. F. W. Webb and after he had served on the Lancashire & Yorkshire Railway in various capacities he had obtained the post of carriage and wagon superintendent on the GNR in 1905.

Fig. 183 GNR 1102/1905 0-8-2T No. 132 Found too heavy for suburban work

Fig. 184 **GNR** 1486/1919 Gresley 5'2" 0-6-0 *Class J22* No. 605

Fig. 185 **GNR**1509/1920 3-cylinder 2-6-0 No. 1001 large 6' diameter boiler

Fig. 186 **GNR** 1536/1922 The first Gresley *A1 Pacific* No. 1470 GREAT NORTHERN

Fig. 187 1789/1934 *Class P2* 2-8-2 COCK O' THE NORTH Lentz poppet valve gear

Fig. 188 1837/1936 Gresley *Class V2* 2-6-2 No. 4771 GREEN ARROW

Fig. 189 1870/1938 Gresley *Class A4* 4-6-2 No. 4468 MALLARD World's record holder for steam at 126 mph

Fig. 190 1984/1945 Thompson *Class L1* 2-6-4T No. 9000

His first locomotives were mixed traffic 0-6-0s with 19″ x 26″ cylinders and 5′ 8″ wheels, with the same boiler as the last 0-6-0s built by Ivatt. In fact they were probably on the drawing board before Gresley came to office. The first actual design was for a 2-6-0 mixed traffic locomotive. Many Ivatt features were retained, but 10″ piston valves were employed, actuated by Walschaerts valve gear. The running plate was high above the wheels. The driving wheels were the established 'mixed traffic' size of 5′ 8″. The boiler was superheated and the total heating surface was 1421 sq. ft. Ten were built and were the forerunners of a large family of 2-6-0s which Gresley developed.

The next batch had larger boilers 5′ 6″ diameter with an increased heating surface totalling 2070 sq. ft.

Then in 1920 appeared the huge 3 cylinder 2-6-0 No. 1000 which had a boiler with the unheard of diameter of 6 ft. The cylinders were 18½″ x 26″ and this was the first locomotive to be fitted with a derived valve gear for three cylinder operation, which was to become standard.

The inside piston valve was operated by the external valve gear by two horizontal rocking levers pivotted on their length in the ratio 2:1 and were connected to the front tail rods of the piston valves.

These locomotives although primarily intended for fast freight traffic were painted in passenger colours and besides working fish and fitted freight trains were quite at home on express workings. They were very free running and the valve chest design contributed a great deal to their success.

For the slower heavier goods trains, he designed a 2-8-0 with two outside cylinders 21″ x 28″ and 4′ 8″ diameter coupled wheels. This type first appeared in 1913 and five were built at Doncaster. The boiler barrel was 15′ 5″ long and the total heating surface was 2514 sq. ft, boiler pressure 170 lb/sq. in. and piston valves 10″ diameter. The pony truck was pivotted on a double swing bolster which provided a good deal of flexibility and steadiness on curves.

In 1918 a three cylinder version appeared with the same boiler but cylinders reduced to 18″ x 26″ and fitted in line. The drive was off the second pair of coupled wheels instead of the third pair for the 2 cylinder version. The cylinders were inclined steeply to meet the short connecting rods. The crossheads moved on four-bar slide bars. The cranks were set at 120° to each other. The inner valve was driven from the outside valve gear with vertical levers forming the derived motion which was rather complicated as compared with the simpler gear on the 3 cylinder 2-6-0s.

Gresley also designed a 0-6-0 side tank departing from the usual saddle tank design. They were usually allocated in the Leeds area and became known as *Ardsley* tanks and after amalgamation were a standard tank class.

In 1920 a new 0-6-2T larger than Ivatt's class and fitted with Gresley's own design of twin-tube superheater and also condensing apparatus for working in the London area. Ten were built at Doncaster, and fifty by the North British Locomotive Co. Ltd. They had 19″ x 26″ cylinders, 5′ 8″ diameter driving wheels and a standard boiler with 1205 sq. ft of heating surface including 207 sq. ft of superheater surface. They were heavy tanks and weighed 70 tons 5 cwt in working order.

In 1922 appeared Gresley's masterpiece, as far as Great Northern days were concerned, a 4-6-2 *Pacific*, the forerunner of many more to be built over the years in many forms and with varying success.

No. 1470 later named GREAT NORTHERN had most of the GNR characteristics and showed the development from its forebears. It had three cylinders 20″ x 26″ and 6′ 8″ coupled wheels. The boiler was enormous, the diameter 5′ 9″ at the smokebox tapering up to 6′ 5″ at the rear tubeplate. The firebox extended into the boiler barrel to form a combustion chamber. The wide firebox similar in pattern to the *Atlantics* gave a grate area of 41·25 sq. ft and the total heating surface was 3455 sq. ft including a Robinson type superheater. The weight in working order was 92·5 tons and a new pattern eight wheeled tender added another 56·25 tons.

The two outside cylinders were placed forward of the inside one as in the 3 cylinder 2-6-0s and Gresley's derived motion was again employed with the 2:1 lever.. The maximum valve travel was $4\frac{9}{16}$″ with $1\frac{1}{4}$″ lap and maximum cut off of 65% – which gave no opportunity for short cut off working. Despite this short-coming, working at long cut-off percentages, their performance in their early days was such that an order was put through for ten more to be built.

The boiler pressure was surprisingly kept down to 180 lb/sq. in. and the firebox, although round-topped as for all previous Doncaster locomotives, tapered inwards on both sides towards the cab as did the top crown plate. The cab was a radical departure from the austere short roof variety in that it had a high arched roof and two windows on each side making it exceptionally roomy and with a good look-out. The regulator was the usual pull-out type with handles on either side of the cab.

No. 1470 appeared in April 1922 and the second, No. 1471, in July the same year and was subsequently named SIR FREDERICK BANBURY. It was the 1539th locomotive to be built at Doncaster Works or, as it was usually called 'The Plant'.

This was the last to be turned out under Great Northern Railway ownership but the GNR traditions continued and so did H. N. Gresley continue as chief mechanical engineer on the LONDON & NORTH EASTERN RAILWAY formed on 1 January, 1923.

Fig. 191 2029/1948 class *A2 Pacific* No. 60538 VELOCITY

P. STIRLING

Date First Built	Type	No Built	Cylinders inches	D.W. ft in	Heating Surface sq ft			Grate Area sq ft	WP lb/sq in	Weight WO ton cwt
					Tubes	F'box	Total			
1867	0-4-2	46	17 x 24	5 7	975	100	1075	16.25	140	31 18
1868	2-2-2	12	17 x 24	7 1	922.25	89.5	1011.75	16.4	130	33 0
1868	0-4-2WT	13	17½ x 24	5 7	817.5	100	917.5	14.0	140	41 13
1870	4-2-2	37	OC 18 x 28	8 1	1043	122	1165	17.6	140	38 9
1871	0-6-0	6	19 x 28	5 1	1240	112	1352	18.75		40 0
1872	0-4-4T	22	17½ x 24	5 7	806	81	887			40 14
1874	2-4-0	19	17½ x 26	6 7	897.8	95	992.8			38 12
1874	0-4-2	25	17½ x 24	5 7	743	94.5	837.5			32 3
1876	0-4-2ST	4	17½ x 26	5 1½	689	74	763	12.75		37 5
1881	0-6-0ST	43	17½ x 26	4 7	715	83	798	16.0		42 12
1884	0-4-4T	16	17½ x 24	5 1	830	92.4	922.4	16.25	160	53 10
1886	0-6-0	72	17½ x 26	5 1½	830	92.4	922.4	16.25	160	36 10
1888	2-4-0	56	17½ x 26	6 7½	836.9	92.4	929.3	16.25	160	39 0
1894	4-2-2	6	19½ x 28	8 1½	909.98	121.72	1031.7	20.0	170	49 11

IVATT

Class		Date	Type	Cylinders inches	D.W. ft in	Heating Surface - sq ft				Grate Area sq ft	WP lb/sq in	Weight WO ton cwt
GN	LNE					Tubes	S'heat	F'box	Total			
C1	C2	1898	4-4-2	18¾ x 24	6 7½	1302	–	137	1439	24.5	175	59 0
D1	D2	1898	4-4-0	17½ x 26	6 7½	1130	–	120	1250	20.8	170	
	–	1898	4-2-2	18 x 26	7 7½	1144	–	126	1270	23.2	170	
C2	C12	1899	4-4-2T	17½ x 26	5 7½	1023	–	103	1126	17.8	170	62 6
K1	Q1	1901	0-8-0	19¾ x 26	4 7½	1302	–	137	1439	24.5	175	54 12
C1	C1	1902	4-4-2	18¾ x 26	6 8	2359	–	141	2500	31.0	175	68 6
L1	R1	1903	0-8-2T	19¾ x 26	4 8	1302	–	137	1439	24.5	175	70 5
C1		1905	4-4-2	13 x 20 16 x 26	6 8	2359	–	141	2500	31.0	200	69 2
N1	N1	1906	0-6-2T	18 x 26	5 8	1130	–	120	1250	20.8	170	65 17
C1		1907	4-4-2	13 x 20 18 x 26	6 8	2359	–	141	2500	31.0	200	69 2
J21	J1	1908	0-6-0	18 x 26	5 8	1130	–	120	1250	19.0	170	46 14
J22	J5	1909	0-6-0	18 x 26	5 2	1130	–	120	1250	19.0	160	47 6
K1	Q2	1909	0-8-0	21 x 26	4 7½	1027	343	137	1507	24.5	160	
C2	C1	1910	4-4-2	20 x 24	6 8	2022	427	141	2449	31.0	170	69 12
D1	D1	1911	4-4-0	18½ x 26	6 8	852	258	120	1230	19.0	160	53 5

GRESLEY

Class GN	Class LNE	Date	Type	First No	Cylinders inches	D.W. ft in	Heating Surface - sq ft Tubes	S'heat	F'box	Total	Grate Area sq ft	WP lb/sq in	Weight WO ton cwt
J22	J6	1912	0-6-0	71	19 x 26	5 2	852	258	120	1230	19.0	170	51 0
H2	K1	1912	2-6-0	1630	20 x 26	5 8	981	303	137	1421	24.5	170	61 14
0.1	0.1	1913	2-8-0	456	21 x 28	4 8	1922	430	163	2514	27.0	170	76 4
J23	J51	1913	0-6-0T	157	18 x 26	4 8	869	—	111	980	17.8	175	56 10
H3	K2	1914	2-6-0	1640	20 x 26	5 8	1523	403	144	2070	24.0	170	63 14
02	02	1918	2-8-0	461	3/18 x 26	4 8	1922	570	163	2655	27.0	170	
N2	N2	1920	0-6-2T	1606	19 x 26	5 8	880	207	118	1205	19.0	170	70 5
H4	K3	1920	2-6-0	1000	3/18½ x 26	5 8	1719	407	123	2249	28.0	180	
A1	A1	1922	4-6-2	1470	3/20 x 26	6 8	2715	525	215	3455	41.25	180	92 9
J23	J50	1922	0-6-0T	221	18½ x 26	4 8	1016	—	103	1119	16.25	170	57 0

For details of locomotive building during the period 1923–47 please see under LONDON & NORTH EASTERN RAILWAY and also table showing locomotives built at Doncaster.

From 1 January, 1948 the L&NER became part of the nationalised BRITISH RAILWAYS and in common with other works, from 1948 to 1952 standard classes of L&NER and LM&SR designs were proceeded with. The last A1 and A2 *Pacifics* were built and then a quantity of LMS Class 4 2-6-0s which were very similar to the later BR 2-6-0 design.

The following steam locomotives were built for BR.

LNE	9	4-6-2	Class A1	60114-60122	1948
LNE	14	4-6-2	Class A2	60526-60539	1948
LNE	17	4-6-2	Class A1	60123-60129, 60153-60162	1949
LMS	20	2-6-0	Class 4	43050-43069	1950
LMS	24	2-6-0	Class 4	43107-43111, 43137-43155	1951
LMS	6	2-6-0	Class 4	43156-43161	1952
BR	4	2-6-0	Class 4	76020-76023	1952
BR	11	2-6-0	Class 4	76024-76034	1953
BR	10	2-6-0	Class 4	76035-76044	1954
BR	10	2-6-4T	Class 4	80106-80115	1954
BR	19	4-6-0	Class 5	73100-73108, 73110-73119	1955
BR	15	2-6-0	Class 4	76045-76049, 76053-76062	1955
BR	10	4-6-0	Class 5	73109-73120-73124, 73155-73158	1956
BR	15	2-6-0	Class 4	76050-76052, 76063-76074	1956
BR	13	4-6-0	Class 5	73159-73171	1957
BR	15	2-6-0	Class 4	76100-76114	1957
Total	212				

The last steam locomotive to be built at Doncaster 'Plant' was No. 76114 steamed on 16 October, 1957, the 2223rd steam locomotive ending a period from 1867 to 1957 noted for three cylinder locomotives, large parallel boilers with round top fireboxes, derived motion and a consistent 'large' locomotive policy.

For many years new building was carried out in a separate shop known as the new erecting shop; the main repair shop was known as the 'Crimpsall', standing on reclaimed land of that name.

See BRITISH RAILWAYS.

GREAT NORTHERN OF IRELAND RAILWAY

Great Victoria Street, Belfast

See ULSTER RAILWAY.

GREAT NORTHERN OF IRELAND RAILWAY

Dundalk

Locomotive Superintendents: 1881–1895 J. C. Park; 1895–1912 Charles Clifford; 1912–1933 George Glover; 1933–1939 G. B. Howden; 1939–1950 H. R. McIntosh.

The company was formed in 1876 by the amalgamation of a number of companies, and the locomotives thus absorbed were of infinite variety, most interesting to look at, but a real headache for the man made responsible for running and maintaining them.

The new works at Dundalk were not established until 1882 and were built on the site of the DUBLIN AND BELFAST JUNCTION'S old railway works, which became one of the GNR's constituents. It was not until 1887 that the first completely new locomotive was built in the new works, and was a 4-4-0T numbered 100 which was soon after changed to No. 1 for obvious reasons. Altogether ten of this type were built from 1887 to 1893, the first having inside cylinders 14″ x 18″. The other nine had the cylinder bore increased to 15″. The coupled wheels were 4′ 7″ diameter and the small leading bogie 2′ 7½″ diameter. The total heating surface was 628 sq. ft and the tank capacity 570 gallons. They were used on the short branch lines of the system, particularly in the Belfast area.

During his term of office Park set about the task of introducing standard types of locomotives to replace the heterogeneous groups of locomotives he had acquired by the amalgamation. In 1885 he introduced two 4-2-2 passenger tender locomotives, built by Beyer Peacock, which were the only bogie singles to work in Ireland. They were subsequently rebuilt as 4-4-0s.

The only other type built in the company's works during Park's time was six coupled goods, of which five appeared.

Due to the urgency for replacements many others were built by Beyer Peacock including 2-4-0, 4-4-0 and 0-6-0 classes.

In 1895 J. C. Park retired and was succeeded by Charles Clifford who, until the amalgamation, had been with the Irish North Western Railway. He carried on the standardisation plans but the train loadings were increasing and it was apparent that more powerful locomotives would be required.

An enlarged 4-4-0 was designed and three were built in 1896 and three more in 1898, all by Beyer Peacock. They had 6′ 7″ diameter coupled wheels and inside cylinders 18″ x 24″ and a total boiler heating surface of 1131 sq. ft. These were followed by larger 4-4-0s in 1899–1900 built by Neilson Reid, four in 1899 and three in 1900 with 18½″ x 26″ cylinders and a large boiler with 1361·5 sq. ft of heating surface.

At Dundalk the two 4-2-2s built in 1885 were rebuilt as 4-4-0s in 1904; they were practically new locomotives with new frames, wheels etc, and in the following year two similar 4-4-0s were built, Nos. 104 and 105, and in 1911, two more, 25 and 43.

Additional power was needed to haul the Dublin local traffic and between 1895 and 1902 six 2-4-2Ts were built at Dundalk with 5′ 7″ coupled wheels. The first two had 16″ x 22″ cylinders with a boiler having 960·5 sq. ft of heating surface, but the other four had 16½″ x 22″ cylinders and a larger boiler with 1066·5 sq. ft of heating surface.

Fig. 192 **GNR(I)** 13/1894 *Class AL* 0-6-0 No. 32 formerly DROGHEDA

GREAT NORTHERN OF IRELAND RAILWAY

Charles Clifford started the practise of naming the locomotives in 1895 and these were carried on brass plates until the beginning of the first World War and it was not until 1932 that new locomotives were given names (the Class V 4-4-0).

In 1912 Charles Clifford was succeeded by George T. Glover and during the latter's term of office no new locomotives were built in the company's own works at Dundalk. Whether it was a clear cut policy or not, all were built by Beyer Peacock and Nasmyth Wilson as follows:

Beyer Peacock			
5' 9"	4-4-2T	1913 to 1929	15 locos.
5' 1"	0-6-0	1912 to 1921	25 locos.
5' 9"	4-4-0	1915	5 locos.
6' 7"	4-4-0	1913 to 1932	13 locos.
Nasmyth Wilson			
5' 9"	4-4-2T	1924	10 locos.
5' 1"	0-6-0	1924	5 locos.

One non-standard locomotive was supplied in 1928 by Hawthorn Leslie in the form of a 0-6-0 Crane tank.

Nevertheless the works during this period were very busy on rebuilding and reboilering existing classes and in 1919 Glover commenced a superheating programme for the older existing types.

In 1932 Mr Glover introduced a new design of 4-4-0. It was a three cylinder compound and five were built (Nos. 83–87). The Boyne Viaduct was previously restricted to a 17 ton axleload and was rebuilt in the same year to take the increased axle loadings. Accelerated services between Belfast and Dublin were then made possible. They were built by Beyer Peacock and were extremely modern and efficient. The boiler working pressure was 250 lb/sq. in. but when, for economic reasons, the timings of many trains were eased the boiler pressure was reduced to 215 lb/sq. in.

Clifford's 4-4-0s were first dealt with and an interesting point is that in some classes the piston valves were placed between the inside cylinders, made possible by the 5′ 3″ gauge, which obviated the necessity of raising the boiler using rocking arms.

In 1933 Mr Glover relinquished the post and his place was taken by Mr George B. Howden, who inherited a relatively modern stud of locomotives. He carried out a certain amount of rebuilding and the only new building which took place at Dundalk were five 0-6-0s in 1937 Nos. 78 to 82 and eight 4-4-0s in 1938 and 1939. The 0-6-0s were the first to have side window cabs and were intended for light goods and excursion traffic. They had a wide route availability with an axle load of 15 tons. The boiler, cylinders and motion were interchangeable with the Class T2 4-4-2T and Class U 4-4-0 types.

A Robinson type 16 element superheater was fitted, with a heating surface of 168·6 sq. ft, the total heating surface of the boiler being 1031·6 sq. ft. The firebox was the usual round top flush type fitted with Ross pop safety valves mounted on the top. The tender's capacity was 3500 gallons of water and six tons of coal.

The 4-4-0 renewals of 1938–39 were of two classes. The S class were renewals of the 1913 Beyer Peacock 4-4-0s and the S2 class renewals of the 1915 batch built by the same firm. They were classed as new building and had new and heavier frames. Both classes had 19″ x 26″ cylinders and 6′ 7″ diameter coupled wheels. These were probably the most successful of the numerous classes of inside cylinder 4-4-0s even in their original 1913–15 state they put up some very creditable performances. The boiler pressure was increased to 200 lb/sq. in. in the late 1920s.

In 1939 Mr Howden's place was taken by Mr H. R. McIntosh. No further locomotives were built at Dundalk, all new locomotives which included a batch of three cylinder simple locomotives in 1948 (Class VS) and two cylinder simples (Class U) in the same year, all from Beyer Peacock.

Description of the locomotives built at Dundalk will not give a complete picture of locomotive development on the GNR, but it will be realised that a considerable contribution was forthcoming from the various locomotive engineers of the company.

The five VS 4-4-0s must have a short reference. In outline they were similar to the V Class compounds. They had three cylinders 15¼″ x 26″ and 8½″ piston valves with three sets of Walschaerts valve gear. The boilers had Belpaire fireboxes with rocking grates. Self cleaning smokeboxes and spark arresters were fitted. A difference in detail was the brass number plates on the cab-sides and in the centre of the buffer beam.

The U class were a modified version of the 1915 U class with a raised boiler pressure of 200 lb/sq. in. The cabs had side windows. They were for mixed traffic, and in a short while the pressure was reduced to 175 lb/sq. in.

In common with other railways, experiments were carried out by fitting a number of locomotives with oil burning equipment.

The earlier locomotives bore a striking resemblance to those on the English Great Northern Railway, particularly the chimney and general styling. J. C. Park came from the GNR in England which makes explanation simple.

Although surviving as an independent company longer than the Belfast & County Down

Fig. 193
GNR(I)
20/1898
Clifford's 5′7″
2-4-2T No. 95
CROCUS at
Amiens Street

Fig. 194
GNR(I)
30/1905
Class P
4-4-0 No. 105
formerly FOYLE

Fig. 195
GNR(I) 36/1937
G. B. Howden's
Class UG
0-6-0 No. 79 at
Adelaide shed

Railway, which was acquired by the newly formed Ulster Transport Authority in 1948, and the Northern Counties Committee absorbed by the UTA in 1949, the GNR (I) carried on with a struggle until the position got so bad that the company announced that as from 1 January, 1951 it proposed to cease running all services. The two governments stepped in and in 1953 a joint board was set up consisting of the Coras Iompair Eireann and the Ulster Transport Authority. Finally in 1958 the CIE and UTA shared the locomotive stock.

Dundalk was not a large works and it had limited facilities for building, hence the large amount of locomotives supplied by builders in England. New boilers were in some cases bought out – the Belpaire boilers used in the rebuilding of the Class V 4-4-0 were supplied by Messrs Harland & Wolff. Dundalk produced a total of forty-seven locomotives from 1887 to 1939, repairs and rebuilding proceeding after this date. By 1959 the works were in the hands of a private engineering company.

GREAT NORTHERN OF IRELAND RAILWAY

Wks No	Date	Type	Cylinders inches	D.W. ft in	Class	GNR No	Name	W'drawn
1	1887	4-4-0T	14 x 18	4 6	BT	1	was No 100 (a)	1935
2	1887	4-4-0T	15 x 18	4 7	BT	2		1921
3	1888	4-4-0T	15 x 18	4 7	BT	3		1921
4	1888	4-4-0T	15 x 18	4 7	BT	4		1921
5	1889	4-4-0T	15 x 18	4 7	BT	5		1921
6	1889	4-4-0T	15 x 18	4 7	BT	6		1921
7	1890	0-6-0	17 x 24	4 7¼	A	60	Dundalk	1959
8	1891	0-6-0	17 x 24	4 7¼	A	33	Belfast	1959
9	1891	4-4-0T	15 x 18	4 7	BT	7		1920
10	1892	4-4-0T	15 x 18	4 7	BT	8		1921
11	1893	4-4-0T	15 x 18	4 7	BT	91		1921
12	1893	4-4-0T	15 x 18	4 7	BT	92		1921
13	1894	0-6-0	17 x 24	4 7¼	AL	32	Drogheda	1960
14	1895	0-6-0	17 x 24	4 7¼	AL	29	Enniskillen	1959
15	1895	0-6-0	17 x 24	4 7¼	AL	55	Portadown	1961
16	1895	2-4-2T	16 x 22	5 7	JT	93	Sutton (b)	1955
17	1896	2-4-2T	16 x 22	5 7	JT	94	Howth	1956
18	1896	0-6-0	17 x 24	4 7¼	AL	56	Omagh	1960
19	1897	2-4-2T	16½ x 22	5 7	JT	90	Aster	1957
20	1898	2-4-2T	16½ x 22	5 7	JT	95	Crocus	1955
21	1899	0-6-0	17½ x 24	4 7	PGs	151	Strabane was No 78	1961
22	1900	0-6-0	17½ x 24	4 7	PGs	100	Clones	1961
23	1901	2-4-2T	16½ x 22	5 7	JT	91	Tulip was No 13	1963
24	1902	2-4-2T	16½ x 22	5 7	JT	92	Violet was No 14	1956
25	1903	0-6-0	17½ x 24	4 7	PGs	10	Bessbrook	1964
26	1903	0-6-0	17½ x 24	4 7	PGs	11	Dromore	1960
27	1904	4-4-0	18 x 24	5 6	PS	88	Victoria (c)	1956
28	1904	4-4-0	18 x 24	5 6	PS	89	Albert (c)	1956
29	1905	4-4-0	18 x 24	5 6	PS	104	Ovoca	1956
30	1905	4-4-0	18 x 24	5 6	PS	105	Foyle	1959
31	1908	0-6-0	19 x 26	4 7	LQGs	78	Pettigo Reno No 119	1959
32	1908	0-6-0	19 x 26	4 7	LQGs	108	Pomeroy	1959
33	1911	4-4-0	18 x 24	6 7	PPs	25	Liffey	1957
34	1911	4-4-0	18 x 24	6 7	PPs	43	Lagan	1960
35	1937	0-6-0	18 x 24	5 1	UG	78		1965
36	1937	0-6-0	18 x 24	5 1	UG	79		1963
37	1937	0-6-0	18 x 24	5 1	UG	80		
38	1937	0-6-0	18 x 24	5 1	UG	81		1960
39	1937	0-6-0	18 x 24	5 1	UG	82		1965
40	1938	4-4-0	19 x 26	6 7	S	173	Galtee More	1964
41	1938	4-4-0	19 x 26	6 7	S.2	192	Slievenamon	1965
42	1938	4-4-0	19 x 26	6 7	S	171	Slieve Gullion (d)	1965
43	1938	4-4-0	19 x 26	6 7	S	172	Slieve Donard	1965
44	1939	4-4-0	19 x 26	6 7	S.2	191	Croagh Patrick	1960
45	1939	4-4-0	19 x 26	6 7	S	170	Errigal	1965
46	1939	4-4-0	19 x 26	6 7	S.2	190	Lugnaquilla	1965
47	1939	4-4-0	19 x 26	6 7	S	174	Carrantuohill	1965

(a) Derry Shunter rebuilt as 0-6-0T 1920 Reno 119. (b) Belfast Transport Museum. (c) Rebuilt from 4-2-2. (d) Railway Preservation Soc. of Ireland.

Official Works Nos used only up to 1911

GNR (I) Locomotive classifications came into use in the late 1890s, being a development of an earlier system of classifying crank axles by letter.

GREAT NORTH OF SCOTLAND RAILWAY

Kittybrewster

Locomotive Superintendents: 1853–5 Daniel Kinnear Clark; 1855–7 J. F. Ruthven; 1857–83 W. Cowan; 1883–90 James Manson; 1890–4 James Johnson; 1894–1902 William Pickersgill.

Up to 1887 all locomotives required were supplied by locomotive building contractors. Mr D. K. Clark, better known perhaps for his book *Railway Machinery*, was the first superintendent and designed two 2-4-0s; one for passenger work with 5′ 6″ diameter coupled wheels and the other for goods traffic with 5′ diameter coupled wheels. In addition, although delivered after his resignation, two 0-4-0WTs were designed for banking purposes. They were all equipped with Clark's patent smoke consuming apparatus consisting of tubular holes in the sides of the firebox passing air over the fire bed, and steam jets were fitted in these openings to induce the air into the firebox. The results were good with an economy made in fuel.

W. Cowan produced the first 4-4-0 design for the company in 1862. All GNSR locomotives of the period had outside cylinders and domeless boilers, the combined dome and safety valves being fitted over the firebox. The 4-4-0 then became the standard passenger type right up to the end of the company's existence.

James Manson succeeded Cowan, coming from the Kilmarnock works of the Glasgow and South Western Railway in 1883. Manson favoured inside cylinders and set about modernising the locomotive stock and replacing the older locomotives especially those inherited from the various smaller lines. Up to this time the principal classes were embellished with plenty of polished brass and copper work but Manson did not favour these extravagances. He also introduced the domed boiler with Ramsbottom safety valves over the firebox.

Kittybrewster works which were situated on the north side of Aberdeen could only take under cover four locomotives at a time and consequently much of the repairs and rebuilding were done in the adjacent sidings. Despite this two new locomotives were erected in the works in 1887. There is no doubt that the major parts were supplied by a locomotive building contractor probably Neilson & Co. or Kitson & Co.

Whether the saving in money approached £300 to £400, claimed to be possible by Manson, is doubtful and no others were built at Kittybrewster, after the two in 1887.

These were 4-4-0s, with inside cylinders 17½″ x 26″ and coupled wheels 5′ 7″ diameter. The boiler barrel was 4′ 6″ diameter by 10′ 6″ long with a total heating surface, including the firebox, of 1159 sq. ft and a grate area of 16·5 sq. ft. The boiler working pressure was 165 lb/sq. in. and the weight in working order was 42 tons 5 cwt, and with tender 71 tons 5 cwt. No. 5 named KINMUNDY was completed in the February, and No. 6 THOMAS ADAM in the May, of 1887, and they were known as Class M.

The names were subsequently removed and both were taken in L&NER stock. No. 5 was renumbered 6805 and was withdrawn in 1936, and No. 6 became 6806 and lasted until 1932.

In 1890 James Manson returned to the Glasgow & South Western Railway as the locomotive superintendent.

Due to the poor facilities and cramped conditions it was decided to build a new works with room for expansion and building commenced in 1898 at Inverurie. In 1901 the carriage works were ready and in 1902 locomotive repairs commenced.

GREAT NORTH OF SCOTLAND RAILWAY

Inverurie

Locomotive Superintendents: 1902–14 William Pickersgill; 1914–22 Thomas E. Heywood.

The works were built on a twenty-five acre site beside the village of Inverurie sixteen miles north west of Aberdeen. A new railway town came into being, increasing the population initially by 1200 people.

At the time of the transfer from Kittybrewster, William Pickersgill was in charge, and by the addition of new machinery to supplement that moved from Kittybrewster repairs and rebuilding were carried out in better circumstances.

No new locomotives were built at Inverurie until 1909, the majority of new stock having come from Neilson & Co. (later Neilson Reid & Co.)

The first to be built were inside cylinder 4-4-0s, the first being completed in April 1909. Eight were built, the last one appearing in March 1915. Known as class V they were built after five of the class were delivered by Neilson & Co. in 1899. Traffic was rapidly increasing and more modern classes were required to replace some of the older stock.

GREAT NORTH OF SCOTLAND
RAILWAY Inverurie

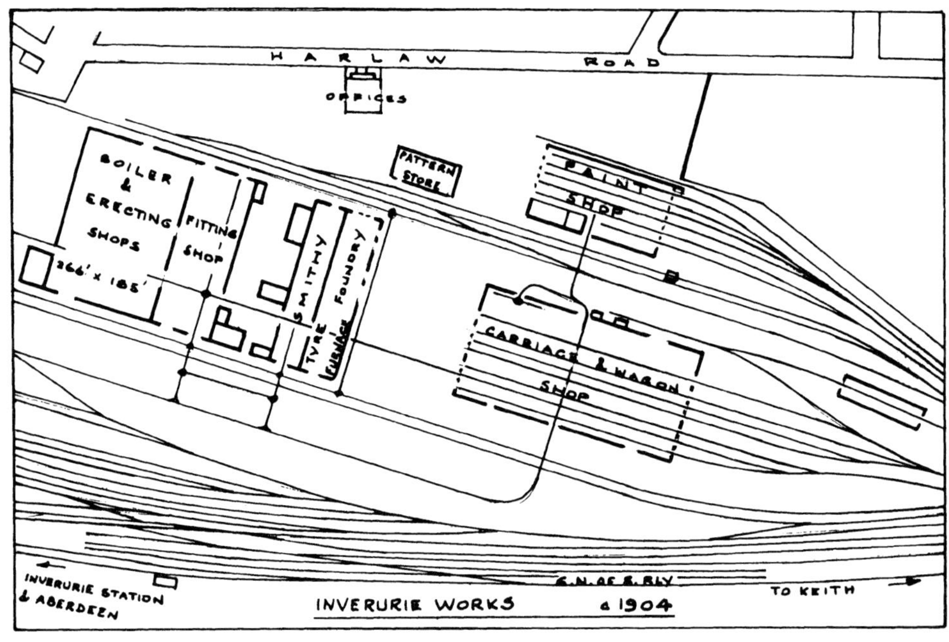

Fig. 196
GN of S Inverurie
1921
T. E. Heywood's superheated 6'1" 4-4-0 No. 46 BENACHIE as LNER 6846

Date	GNS No	LNER No		BR No	GNS Class	LNER Class	Name	W'drawn
		1st	2nd					
4/1909	27	6827	2265	62265	V	D40		12/1956
7/1909	29	6829	2267	62267	V	D40		8/1956
6/1910	31	6831	2268	62268	V	D40		7/1956
8/1910	36	6836	2272	62272	V	D40		3/1955
3/1913	28	6828	2266	—	V	D40		1/1947
9/1913	33	6833	2269	62269	V	D40		9/1955
9/1914	35	6835	2271	62271	V	D40		11/1956
3/1915	34	6834	2270	62270	V	D40		9/1953
6/1921	45	6845	2273	62273	F	D40	George Davidson	1/1955
9/1921	46	6846	2274	62274	F	D40	Benachie	9/1955

All above are 4-4-0, IC 18" x 26", 6' 1" D.W.

GREAT NORTH OF SCOTLAND RAILWAY Inverurie

The cylinders were 18″ x 26″ with unbalanced slide valves between the cylinders and Stephenson valve gear. A boiler of moderate proportions was fitted 4′ 6″ diameter and 10′ 6″ long. The total heating surface including the round topped firebox was 1207 sq. ft with a grate area of 18·24 sq. ft. The working pressure was 165 lb/sq. in. A new feature was the larger cab fitted with two windows each side and a raised ventilator on the roof. Ramsbottom safety valves were fitted over the firebox. The coupled wheels were 6′ 1″ diameter. The new locomotives were employed on all the major expresses and excursion trains.

No further new locomotives were built by Pickersgill at Inverurie, and in 1914 he left the GNSR to become locomotive superintendent of the Caledonian Railway.

He had been responsible for the new works lay-out at Inverurie and the locomotives he had added to the stock were well built and in good condition.

He was succeeded by Thomas Edward Heywood who came from the Taff Vale Railway and was principally involved in superheating various classes. In 1921 two further 4-4-0s were built at Inverurie, being superheated versions of the class V. They were fitted with balanced slide valves above the cylinders and operated by rocking shafts. The boiler gave a heating surface of 860 sq. ft with an additional 140 sq. ft from the superheater. The number of tubes was cut from 213 to 104. The weight in working order was 48 tons 13 cwt as compared with 46 tons 7 cwt for the V class.

They were known as the F class and were dual fitted with Westinghouse brake and vacuum ejector for working the Highland Railway stock between Aberdeen and Inverness.

The total number of locomotives built at Inverurie was ten, and all went into the L&NER stock, nine passing into British Railways' hands.

The works, although not building further new locomotives, continued to be used and are at present engaged on diesel locomotive repairs.

The original erecting shop had 20 pits including one through-road: 16 pits were in one bay which was served by a 60 ton travelling crane with two 30 ton hoists; 3 pits in the second bay were used for locomotives, the remainder for tenders and boilers.

In 1955–6 the erecting, boiler and tender shops were reorganised, and a new 100 ton capacity travelling crane installed to deal with the heavier types of locomotives.

GREAT SOUTHERN & WESTERN RAILWAY

Inchicore, Dublin

The railway was incorporated in 1844 and was of comparative small mileage until the various railways radiating from Waterford were absorbed in 1901.

Inchicore works were established in 1846 and the erecting shop had 18 pits on one side and 16 pits on the other with a traverser in between. This original erecting shop was replaced in 1934 by a large new shop, with through roads, which is still in operation.

The following locomotive superintendents were in charge: 1845–7 J. Dewrance; 1847–64 Geo. Miller; 1864–83 A. McDonnell; 1883–6 J. A. F. Aspinall; 1886–96 H. A. Ivatt; 1896–11 R. Coey; 1911–13 R. E. L. Maunsell; 1913–22 E. A. Watson; 1922–4 J. R. Bazin.

This railway had the advantage of starting with a fleet of locomotives comprising fifty-five and of these fifty-three of them were 2-2-2 of three classes so that their repair and maintenance was not fraught with the problems of so many other railways with an infinite variety of classes.

It was not until 1852 that the first new locomotive was steamed at Inchicore and the first passenger 2-2-2 came in July 1853 numbered 43, and later renumbered 59.

During George Miller's term of office the posts of civil and locomotive engineer were both held by him but his locomotive running superintendent Mr. W. Wakefield took over most of the work in the workshops and in this period some 2-2-2s were produced with 15″ x 22″ inside cylinders and 6′ 6″ diameter driving wheels. These worked the bulk of the main line trains. They were very much like McConnell's 'Bloomers' which were built at Wolverton from 1857 onwards, with inside frames and dome on the front ring.

The goods engines were 0-6-0 and 0-4-2 types, the latter being a common form of wheel arrangement on many Irish railways.

Mr Alexander McDonnell became locomotive superintendent in 1864. He had been in charge of locomotives on the Newport, Abergavenny and Hereford Railway, and from 1860 on the West Midland Railway when the NA&HR was absorbed. In 1862 he went abroad to superintend the Danube & Black Sea Railway's locomotive department.

His previous experience of locomotive running and maintenance showed its effect in the way he organised the works at Inchicore and the economies shown in producing the required parts for the locomotives being built and under repair.

When designing locomotives he wisely adapted, improved and copied existing types or left it to outside contractors to put forward a design to his requirements.

GREAT SOUTHERN & WESTERN RAILWAY

Fig. 197
GS&WR 118/1876
McDonnell's
Class 21
2-4-0 No. 66

Fig. 198
GS&WR
171/1882
Maid of all work
101 class
0-6-0 No. 101

The most important instance of this was the famed '101' class 0-6-0s originating from a Beyer Peacock design, the same firm building the first batch for McDonnell. A total of 119 were built, 12 by Beyer Peacock, 8 by Sharp Stewart and the remainder at Inchicore from 1866 to 1903. These locomotives became 'maids of all work' and were extremely efficient and one of the best classes to run on this railway.

From time to time modifications were made to cylinders, frames and boilers and it is pleasing to note that one has found a resting place in the Belfast museum (No. 186).

McDonnell's locomotive had quite distinctive feaures. His footplating was low and curved to clear the crank bearings; smokebox doors were double folding, on early types; chimneys were cast iron, mostly tapered; boilers were parallel with round type fireboxes with Ramsbottom safety valves mounted on them. His early express passenger locomotives were 2-4-0 which he built in considerable numbers, the last four in 1876 (Nos 66–69). These had 16″ x 20″ cylinders and 5′ 8″ diameter driving wheels and were used on semi-fast and light express passenger trains.

The next step was to increase the size of passenger locomotives and the first 4-4-0 appeared in May 1877 having the same cylinders and driving wheels as the previous 2-4-0s, but the bogie was of the swing link, or American pendulum, type and was the first example in the British Isles. It was subsequently used on the North Eastern, Lancashire & Yorkshire & Great Northern Railways.

0-4-4Ts and 0-6-4Ts were built during McDonnell's period. The GS&WR had few tank locomotives and the first 0-4-4Ts were of the Single Fairlie type with inside cylinders; later locomotives of the same wheel arrangement were equipped with orthodox frames but the trailing bogie had outside frames and back tanks. They were used between Dublin and Kildare, and Cork and Queenstown.

The larger 0-6-4T introduced in 1876 was

the first of this wheel arrangement in the British Isles for main line use. The trailing bogies were equipped with compensating beams.

An interesting replacement which took place in 1873 was that of the old 0-2-4T inspection carriage. The new tank was a 0-4-4WT named SPRITE and was attached to a separate four wheeled carriage. Its official use was the 'pay carriage'. It had 5 ft driving wheels and inside cylinders 8″ x 12″ with a diminutive boiler 2′ 9″ diameter and a grate area of 10 sq. ft. SPRITE was rebuilt as a 0-4-2T with additional side tanks and in 1894 a further pay carriage/locomotive was built bearing the name FAIRY.

Cast iron number plates were introduced by McDonnell in 1866 similar to those later introduced on the London & North Western Railway by F. W. Webb in 1873.

In 1883 McDonnell was appointed locomotive superintendent of the North Eastern Railway and his place was taken by the Inchicore works manager John A. F. Aspinall. In the following year he put in hand a number of 4-4-0s intended for the mail and express passenger trains and in 1885 a larger class of 4-4-0s was built with 18″ × 24″ inside cylinders, 6′ 7″ diameter driving wheels and a boiler giving a heating surface of 1051·5 sq. ft and working at a pressure of 150 lb/sq. in.

Whilst in office he substituted white metal for the number plates and totally enclosed the splashers, instead of the slotted type favoured by McDonnell.

Nevertheless, due to McDonnell's policy of standardisation, Aspinall did not depart radically from the designs he took over. He did, however, fit many locomotives with a vacuum brake, which at first was of the non-automatic type but after various experiments an automatic vacuum brake was designed and fitted.

In 1886 Aspinall was appointed by the Lancashire and Yorkshire Railway to take charge of their locomotive department and it was here that his best work was carried out.

To take his place H. A. Ivatt was duly ap-

Fig. 199
GS&WR
190/1884
Aspinall's 0-4-4WT
Class 47
No. 72

Fig. 200
GS&WR
198/1885
Aspinall 6′7″
4-4-0 No. 96

GREAT SOUTHERN & WESTERN RAILWAY

Fig. 201
GS&WR
249/1892
33 class
2-4-2T No. 34
for Kerry branches

Fig. 202
GS&WR
258/1895
60 class
4-4-0 6'7" No. 65
at Cork

Fig. 203
GS&WR
280/1901 Coey's
37 class No. 318
4-4-2T at Inchicore

Fig. 204
GS&WR
328/1907
2-cylinder 4-6-0
No. 366

pointed and here the locomotive policy was repeated in that no specific new designs were produced for the passenger and goods tender locomotives but in 1892 Ivatt introduced a 2-4-2T with 16″ x 20″ inside cylinders and 5′ 8½″ diameter driving wheels and side tanks and radial axleboxes on the leading and trailing wheels. The boilers had raised fireboxes and were of modest dimensions. They were used on various branch lines. In 1894 a modified design was brought out incorporating a leading bogie instead of the radial truck but dimensionally similar. They rode steadier and were not so hard on the flanges. Two were built, Nos 37 and 38, the former not being scrapped until 1954.

A 0-6-0T was also introduced, the first appearing in December 1887. Inside cylinders were 18″ x 24″, with wheels 4′ 6½″ diameter, and a weight in working order of 44 tons 10 cwt. On some duties they replaced the earlier 0-6-4Ts.

In 1891 one of the '101' class 0-6-0s was converted to a compound, one of the 17″ cylinders being replaced by one of 26″ diameter. The valves were placed on top of the cylinders, and the boiler heating surface modified. In 1894 one of McDonnell's 4-4-0s No. 93 was also converted to a compound. The goods locomotive was apparently more successful than the 4-4-0, showing an economy in coal and water, but at the expense of pulling power. Both were subsequently reconverted to simple expansion.

Inchicore seemed to be a training ground for locomotive engineers, McDonnell going to the North Eastern, Aspinall to make his mark on the Lancashire & Yorkshire and then in 1896 Ivatt left to take up the reigns at Doncaster on the Great Northern Railway.

Mr R. Coey succeeded Mr Ivatt and after adding ten '101' class 0-6-0s with modified details a new class of 4-4-0 was designed and four were built in 1900 with larger cylinders (18″ x 26″) and driving wheels 6′ 7″ diameter. The boiler heating surface was 1220 sq. ft and pressure 160 lb/sq. in. The tender was also enlarged and carried 3345 gallons of water and 7 tons of coal. The weight with tender was 77 tons 5 cwt in working order.

Apart from sundry tank locomotives it had not been the practice of the GS&WR to name their larger types but the four 4-4-0s were an exception and they bore names as follows: 301 VICTORIA; 302 LORD ROBERTS; 303 ST. PATRICK; 304 PRINCESS ENA.

No doubt onlookers anticipated that this was a change of policy and further new locomotives would be turned out bearing names, but it was not until 1913 that one solitary further named locomotive appeared. The 301 class were built to cope with the increasing loads on the major expresses and the mail trains.

Some additional 4-4-2Ts were put in hand; four were similar to those built by Aspinall in 1894 with 16″ x 20″ cylinders and four others were slightly larger with 17″ x 22″ cylinders. Both classes had side tanks with the sand boxes as an extension to the front ends of the tanks. The smokebox doors were of the standard pattern in two halves with a vertical joint. They were built in 1900 and 1901.

Four additional 0-6-0Ts were put in hand and were to Ivatt's '201' class design. They were numbered 217–220, and were followed by four more 4-4-0s similar to the 1900 batch, but not named. The weight in working order was 49 tons 16 cwt as compared with 47 tons for the previous batch.

They were numbered 305 to 308 and in 1904 No. 307 was fitted with Marshall's valve gear as an experiment.

The '101' class 0-6-0s continued to be built. Coey introduced various modifications including marine type big ends, cast steel wheels and single slide bars. He also rebuilt two earlier '101's with larger boilers 4′ 4″ diameter and the last twelve, built 1902–1903, had these boilers fitted and over a long period the majority of this class were rebuilt with the 4′ 4″ boiler. These were the last '101' class to be built, the period extending from 1866 to 1903.

Three further classes of 4-4-0s were built to supplement the existing classes for increased traffic. Two had 18″ x 26″ inside cylinders but the driving wheels were 5′ 8½″ and 6′ 8″ and the other class had 17″ x 26″ cylinders and 5′ 8½″ coupled wheels. One of the 17″ class was fitted with a Schmidt superheater from 1912–1916 (No. 326) and another of the same class (No. 332) was fitted with piston valves in 1926.

Taper boilers were fitted from 1903 on all passenger types but when rebuilding took place at a later stage they were replaced by parallel boilers.

The first 4-6-0s were introduced in 1905. They had inside cylinders 19¼″ x 26″ and 5′ 1¾″ diameter coupled wheels. The boiler was high-pitched and the heating surface of the tubes was 1466·75 sq. ft; the firebox was 133 sq. ft making a total of 1599·75 sq. ft. The grate area was 24·8 sq. ft. No superheater was fitted and the boiler pressure was 160 lb/sq. in. The weight in working order was 56 tons 19 cwt and the tender 35 tons, carrying 4 tons coal and 3340 gallons of water. The boiler diameter was 4′ 10½″ and the firebox was round topped.

Four were built at Inchicore (Nos. 362–365) in 1905 and two (Nos. 366 and 367) in 1907.

R. Coey left Inchicore in 1911 and his position was filled by R. E. L. Maunsell for a short period, and the only new locomotives built during his stay were four of Coey's larger 0-6-0s. Maunsell left at the end of 1913 to take up his appointment as locomotive superintendent of the South Eastern & Chatham Railway.

In 1914 Mr E. A. Watson came to Inchicore from the Great Western Railway at Swindon.

His first passenger locomotive, not surprisingly, was a four cylinder 4-6-0. Having had experience with the GWR 4 cylinder *Star* class, it seems logical that this type should have been introduced on to the GSWR the largest type running being two cylinder 4-4-0s.

The four cylinders were 14″ x 26″ and the coupled wheels 6′ 7″; that diameter being the accepted standard for express work. The large boiler and 1772 sq. ft of heating surface, a grate area of 28 sq. ft and a pressure of 180 lb/sq. in. The weight in working order was 72 tons 10 cwt. It was numbered 400 and appeared in August 1916. No more were built until 1921 when three more came out of Inchicore and six were built by Armstrong Whitworth & Co.

They were not entirely successful in their four cylinder form and the four Inchicore 4-6-0s were rebuilt with two outside cylinders 19½″ x 26″, and two were fitted with Caprotti valve gear. The results were much more satisfactory.

Previous to the 4-6-0s, a 4-8-0T was built in September 1915. It had inside cylinders 19¼″ x 26″ and 4′ 6½″ diameter coupled wheels. Its weight in working order was 80 tons 15 cwt. Another one followed in December 1924. They were the only eight coupled 5′ 3″ gauge locomotives in Ireland.

In accordance with the 1924 Railways Act the Great Southern & Western Railway merged with the Midland & Great Western Railway together with a number of smaller lines on 12 November, 1924 into a company designated the GREAT SOUTHERN RAILWAY COMPANY and on 1 January, 1925 the Dublin & South Eastern Railway was also amalgamated with the GSR and the title became the GREAT SOUTHERN RAILWAYS COMPANY and all railways whose lines ran entirely in the Free State were absorbed.

The chief mechanical engineer on amalgamation was J. R. Bazin who had replaced E. A. Watson in 1922 on the GS&WR.

Inchicore became the headquarters of the GSR locomotive department and the only other works which carried on building new locomotives was the old Midland Great Western Railway workshops at Broadstone who built some 2-6-0s from 1925 to 1927.

Mr Bazin's first design was a two cylinder 4-6-0. Three were built, one in 1924 (No. 500) and two in 1926 (Nos. 501 and 502). The outside cylinders were 19½″ x 28″; boilers had top feed, and Belpaire fireboxes,a pressure of 180 lb/sq. in. and Ross pop safety valves were fitted. In working order they weighed 73 tons 2 cwt.

During 1927 to 1929 some 2-6-0 tender locomotives were assembled from parts purchased from Woolwich Arsenal. The first eight had the standard 5′ 6″ diameter coupled wheels and the remaining six had 6 ft diameter wheels. Presumably the latter had new wheels supplied and all converted to 5′ 3″ gauge at Inchicore.

For Inchicore services Bazin designed a 2-6-2T and the prototype (No. 850) was built in September 1928. It had outside cylinders and motion. The boiler was parallel with a Belpaire firebox. The cylinders were 17″ x 28″, 5′ 6″ diameter coupled wheels and the weight in working order 71 tons 10 cwt.

In 1929 W. H. Morton took over the post of chief mechanical engineer. He had previously been in charge of locomotive affairs at the Broadstone works of the MGWR. In the same year the '700' class of 0-6-0 was put in hand at Inchicore. Five were built, and were a direct development of the '101' class and the wheels, motion, and cylinders were identical, but the boiler was slightly larger, 4′ 5¾″ diameter with a heating surface of 1013 sq. ft including the Belpaire firebox. The pressure was 160 lb/sq. in. No superheater was fitted. Their performance compared favourably with the '101' class, but when the latter was fitted with superheated boilers, they could not compete.

In 1932 A. W. Harty became chief mechanical engineer and in the following year built five 0-6-2Ts with 18″ x 24″ inside cylinders and 5′ 6″ diameter coupled wheels. They were indeed a tank version of the '710' class 0-6-0s which followed and were intended for the Dublin & South Eastern section.

This series of ten 0-6-0s were built in 1934–5. Known as the '710' class they had superheated boilers, 8″ piston valves, and detail features of Broadstone and Inchicore. They were not successful locomotives, being heavy on coal and not strong pullers. The chief reason for their poor performance seems to have been the inclination of the piston valves in relation to the cylinders.

Five further 4-4-0s were built in 1936; twenty-three years had elapsed since the last 4-4-0 was built at Inchicore.

They had inside cylinders 18″ x 26″ with 5′ 8½″ diameter coupled wheels. An interesting feature of these locomotives was the outside frames bogie with laminated springs above each wheel on the bogie.

The boiler was fitted with Belpaire firebox, a larger cab with side windows and a tender with

Fig. 205
GS&WR
333/1908
Class 333
4-4-0 No. 340.
Note outside frame bogie

Fig. 206
GS&WR
352/1915
E. A. Watson's 4-8-0T
Only two built

Fig. 207
359/1926
2-cylinder 4-6-0
No. 501

the top sides canted inwards. They were intended for mixed traffic duties.

In 1937 yet another chief mechanical engineer took office. He was Mr E. C. Bredin and in 1939–40 three 4-6-0s were built and were the last GSR locomotives to be built at Inchicore and undoubtedly the most magnificent.

In appearance they were very similar to the rebuilt *Royal Scot* class of the LM&SR of England. They had three cylinders 18″ x 28″ and 6′ 7″ diameter coupled wheels. The taper boiler with Belpaire firebox had a heating surface of 1670 sq. ft; superheater 468 sq. ft, firebox 200 sq. ft, giving a total of 2338 sq. ft. The grate area was 33·5 sq. ft. The three locomotives were named: 800 MAEDHBH; 801 MACHA; 802 TAILTE.

The weight in working order was 84 tons. The use of these locomotives on the Cork mail trains enabled the timings to be considerably accelerated. Even so they were never extended to their limits and, with war conditions following, their capabilities were never utilised to any degree.

With the completion of the 4-6-0s over 400 new locomotives had been built at Inchicore. Since 1915 only 50 had been built due mainly to the finances of the railways, increased competition and the incidence of two major wars making fuel and material in short supply.

In 1945 it was decided that all rail and road transport in Eire should be governed and controlled by the Transport Company of Ireland with the name of CORAS IOMPAIR EIREANN.

GREAT SOUTHERN & WESTERN RAILWAY

Because of the coal shortage many locomotives were converted to oil firing in 1945–6 including the 'Woolwich' 2-6-0s. The burners were made at Inchicore and large white circles were painted on the smokebox doors and tender sides for identification. In the winter of 1946–7 coal was practically unobtainable, after which the situation eased and in 1948 the oil burners were converted back to coal. The oil burning scheme had cost a lot of money in conversions and installations and for such a short period meant a serious loss.

Apart from painting green the passenger class locomotives, including the '372', '393', '400' and '500' classes, a period of austerity ensued. The growing number of heavy repairs required, and obsolescent types needing replacement, meant that by 1948 the financial state of the CIE was in such poor shape that a team of experts was asked to investigate and report on the whole transport structure.

Mr O. V. S. Bulleid was responsible for the report on locomotive matters and formed part of the whole Milne report. Many recommendations were put forward by Mr Bulleid, including proposals for repairing certain classes of locomotives, replacing others by new work, and the manufacture of quantities of standard boilers.

In 1949 Mr Bulleid joined the staff of the CIE as the chief mechanical engineer in an advisory capacity, but in 1952 he achieved full status.

During this perod it was proposed to build a quantity of locomotives specially adapted for burning turf and a 2-6-0 was experimentally adapted for this purpose as a preliminary. In 1957, as a result of these trials, Mr Bulleid built a 0-6-6-0 (No. CC1) akin to his ill-fated *Leader* class of the Southern Railway. It had four cylinders 12" x 14", 3' 7" diameter wheels, weighed 118 tons and had a boiler pressure of 250 lb/sq. in. After many tests it was withdrawn in 1958.

Various modifications took place with existing locomotives. The '800' class 4-6-0s were fitted with single chimneys. Two of the smaller two cylinders '400' class had larger boilers fitted. By now a major board decision had been made that steam locomotive building was to be abandoned in favour of diesel/electric and diesel/mechanical locomotives.

Inchicore works were to be adapted for this new work, and Limerick was to undertake the majority of steam repairs until such propulsion was phased out. In fact, Inchicore continued to carry out a small proportion of steam repairs until 1962.

One minor point on 'austerity' was that from 1946 the cast iron number plates were removed as locomotives went through the shops and large yellow figures painted on instead – a doubtful saving. In 1953 it was decided to standardise on grey for locomotive painting.

Fig. 208 367/1928 Sole example of Bazin's 2-6-2T No. 850

GREAT SOUTHERN & WESTERN RAILWAY

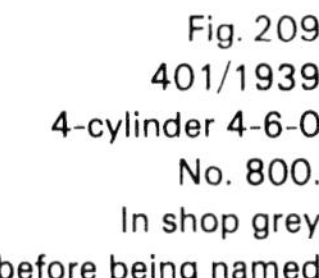

Fig. 209
401/1939
4-cylinder 4-6-0
No. 800.
In shop grey
before being named

P No	Built	Type	Cylinders inches	D.W. ft in	1st No.	1864 No			W'drawn
1	11/1852	0-4-2	15 x 24	5 1	57	117			
2	11/1852	0-4-2	15 x 24	5 1	58	118			
3	7/1853	2-2-2	15 x 24	6 0	59	43			
4	4/1854	0-6-0	17 x 24	5 0	60	139			
5	5/1854	0-6-0	17 x 24	5 0	61	140			
6	10/1854	0-4-2	16 x 24	5 2	62	122			
7	11/1854	0-4-2	16 x 24	5 2	63	123			
8	7/1855	2-4-0	14½ x 22	5 2	64	119			
9	7/1855	2-4-0	14½ x 22	5 2	65	120			
10	7/1855	2-4-0	14½ x 22	5 2	66	121			
11	2/1856	0-6-0	16 x 24	4 6	67	141			
12	3/1856	0-6-0	16 x 24	4 6	68	142			
13	4/1856	0-6-0	16 x 24	4 6	69	143			
14	10/1856	0-6-0	16 x 24	4 6	72	144			
15	10/1856	0-6-0	16 x 24	4 6	73	145			
16	12/1856	0-6-0	16 x 24	4 6	74	146			
17	1857	0-2-4T	10½ x 12	4 6			Sprite (a)		
18	3/1858	2-2-2	15 x 22	6 6	77	45			
19	4/1858	2-2-2	15 x 22	6 6	78	46			
20	4/1858	2-2-2	15 x 22	6 6	79	49			
21	12/1859	2-2-2	15 x 22	6 6	76	44			
22	12/1859	2-2-2	15 x 22	6 6	80	48			
23	12/1859	2-2-2	15 x 22	6 6	81	49			
24	8/1860	0-4-2	16 x 24	5 2	82	124			
25	8/1860	0-4-2	16 x 24	5 2	83	125			
26	10/1860	0-4-2	16 x 24	5 2	84	126			
27	2/1861	0-4-2	15 x 24	4 6	85	127			
28	3/1861	0-4-2	15 x 24	4 6	86	128			
29	3/1861	0-4-2	15 x 24	4 6	87	129			
30	8/1861	2-2-2	15 x 22	6 6	88	50			
31	8/1861	2-2-2	15 x 22	6 6	89	51			
32	10/1861	2-2-2	15 x 22	6 6	90	52			
33	3/1862	0-4-2	15 x 24	4 6	91	130			
34	4/1862	0-4-2	15 x 24	4 6	92	131			
35	4/1862	0-4-2	15 x 24	4 6	93	132			
36	10/1862	2-2-2	15 x 22	6 6	94	53			
37	12/1862	2-2-2	15 x 22	6 6	95	54			
38	1/1863	2-2-2	15 x 22	6 6	96	55			
39	5/1863	0-4-2	16 x 24	5 0	97	133			
40	7/1863	0-4-2	16 x 24	5 0	98	134			
41	8/1863	0-4-2	16 x 24	5 0	99	135			
42	4/1865	0-4-2	16 x 24	5 0	136				
43	5/1865	0-4-2	16 x 24	5 0	137				
44	5/1865	0-4-2	16 x 24	5 0	138				
45	4/1866	2-4-0			16		(b)		
46	6/1866	0-6-0	16 x 24	5 0	112			J15	1929
47	9/1866	2-4-0	16 x 20	5 7½	1				
48	10/1866	2-4-0	16 x 20	5 7½	9		Renewed 1874		
49	12/1866	0-6-0	17 x 24	5 0	113			J15	1930

GREAT SOUTHERN & WESTERN RAILWAY

P No	Built	Type	Cylinders inches	D.W. ft in	1st No			W'drawn
50	1/1867	2-4-0			27	Renewed 8/1870 & Reno 62		
51	1/1867	2-4-0			30	Renewed 5/1871 & Reno 63		
52	5/1867	0-6-0	17 x 24	5 0	118			1890
53	8/1867	0-6-0	15 x 24	5 0	103			1886
54	9/1867	0-6-0	15 x 24	5 0	111			1888
55	10/1867	2-4-0	15 x 20	5 7½	4			
56	10/1867	2-4-0			11			
57	1/1868	2-4-0			14			
58	2/1868	2-4-0			18			
59	5/1868	0-6-0	17 x 24	5 1½	105			1895
60	7/1868	0-6-0	17 x 24	5 1½	110			1890
61	11/1868	2-4-0	16 x 22	6 6	56			
62	12/1868	2-4-0	16 x 22	6 6	57			
63	2/1869	2-4-0	16 x 22	6 6	58			
64	3/1869	2-4-0	16 x 22	6 6	59			
65	5/1869	2-4-0	16 x 20	5 8	3			
66	6/1869	2-4-0	16 x 20	5 8	12			
67	8/1869	0-6-0	16 x 24	5 1½	114			1885
68	10/1869	0-6-0	16 x 24	5 1½	115		J15	1929
69	12/1869	0-4-4T	15 x 20	5 7½	33	Fairlie Patent		
70	2/1870	2-4-0	15 x 20	5 7½	20			
71	5/1870	0-4-4T	15 x 20	5 7½	34	Fairlie Patent		
72	6/1870	0-4-4T	15 x 20	5 7½	31			
73	6/1870	0-4-4T	15 x 20	5 7½	32			
74	10/1870	2-4-0	16 x 22	6 6	60			
75	11/1870	2-4-0	16 x 22	6 6	61			
76	1/1871	0-4-4T	15 x 20	5 7½	27			
77	2/1871	0-6-0	17 x 24	5 1½	155		J15	1929
78	4/1871	0-6-0	17 x 24	5 1½	156		J15	1961
79	9/1871	0-6-0	17 x 24	5 1½	159		J15	1949
80	9/1871	0-4-4T	15 x 20	5 7½	30			
81	10/1871	0-6-0	17 x 24	5 1½	160		J15	1955
82	10/1871	0-6-0	17 x 24	5 1½	161		J15	1963
83	12/1871	0-6-0	17 x 24	5 1½	162		J15	1963
84	3/1872	0-6-0	17 x 24	5 1½	157		J15	1963
85	4/1872	0-6-0	17 x 24	5 1½	158		J15	1957
86	6/1872	2-4-0	15 x 20	5 7½	17			
87	7/1872	2-4-0	15 x 20	5 7½	19			
88	1/1873	0-6-0	17 x 24	5 1½	102		J15	1962
89	1/1873	2-4-0	15 x 20	5 7½	16			
90	1/1873	0-4-4WT	8 x 12	5 0		Sprite (a)	L4	1927
91	1/1873	0-6-0	17 x 24	5 1½	104		J15	1965
92	3/1873	0-6-0	17 x 24	5 1½	167		J15	1960
93	4/1873	0-6-0	17 x 24	5 1½	168		J15	1962
94	6/1873	2-4-0	16 x 20	5 8	21		G4	1928
95	7/1873	2-4-0	16 x 20	5 8	22		G4	1928
96	7/1873	2-4-0	16 x 20	5 8	23			
97	9/1873	2-4-0	16 x 20	5 8	24			
98	10/1873	2-4-0	16 x 20	5 8	25			
99	11/1873	2-4-0	16 x 20	5 8	26		G4	1928
100	1/1874	0-6-0	17 x 24	5 1½	106		J15	1964
101	3/1874	0-6-0	17 x 24	5 1½	117		J15	1930
102	3/1874	0-6-0	17 x 24	5 1½	169		J15	1928
103	4/1874	0-6-0	18 x 24	5 1½	170		J15	1963
104	7/1874	0-6-0	18 x 24	5 1½	171		J15	1961
105	8/1874	0-6-0	18 x 24	5 1½	172		J15	1964
106	9/1874	0-6-0	18 x 24	5 1½	173		J15	1933
107	11/1874	0-6-0	18 x 24	5 1½	174		J15	1953
108	2/1875	0-6-0	18 x 24	5 1½	108		J15	1959
109	3/1875	0-6-0	18 x 24	5 1½	142		J15	1928
110	4/1875	0-6-0	18 x 24	5 1½	179		J15	1963
111	5/1875	0-6-0	18 x 24	5 1½	180		J15	1928
112	7/1875	2-4-0	17 x 22	6 6	64			

GREAT SOUTHERN & WESTERN RAILWAY

P No	Built	Type	Cylinders inches	D.W. ft in	No			W'drawn
113	8/1875	2-4-0	17 x 22	6 6	65			
114	10/1875	0-4-4T	15 x 20	5 7½	35			
115	10/1875	0-4-4T	15 x 20	5 7½	36			
116	10/1875	0-4-4T	15 x 20	5 7½	37			
117	10/1875	0-4-4T	15 x 20	5 7½	38			
118	3/1876	2-4-0	16 x 20	5 8	66		G4	1928
119	3/1876	2-4-0	16 x 20	5 8	67		G4	1928
120	5/1876	2-4-0	16 x 20	5 8	68		G4	1928
121	6/1876	2-4-0	16 x 20	5 8	69			
122	9/1876	0-6-4T	10 x 18	3 6		(c)	J30	1959
123	9/1876	0-6-4T	18 x 24	4 6½	201	Negro	J13	1957
124	9/1876	0-6-4T	18 x 24	4 6½	202	Jumbo, named 1897 (r)		
125	1/1877	0-6-0	18 x 24	5 1¾	109		J15	1964
126	1/1877	0-6-0	18 x 24	5 1¾	119		J15	1962
127	2/1877	0-6-0	18 x 24	5 1¾	120		J15	1955
128	2/1877	0-6-0	18 x 24	5 1¾	121		J15	1963
129	5/1877	4-4-0	16 x 20	5 8½	2		D19	1953
130	6/1877	4-4-0	16 x 20	5 8½	5		D19	1949
131	8/1877	4-4-0	16 x 20	5 8½	6		D19	1952
132	9/1877	4-4-0	16 x 20	5 8½	7		D19	1953
133	12/1877	0-6-0	18 x 24	5 1¾	143		J15	1960
134	2/1878	0-6-0	18 x 24	5 1¾	144		J15	1954
135	2/1878	0-6-0	18 x 24	5 1¾	145		J15	1926
136	2/1878	0-6-0	18 x 24	5 1¾	146		J15	1955
137	6/1878	4-4-0	16 x 20	5 8½	43		D19	1945
138	7/1878	4-4-0	16 x 20	5 8½	44		D19	1950
139	10/1878	4-4-0	16 x 20	5 8½	45		D19	1945
140	11/1878	4-4-0	16 x 20	5 8½	46		D19	1935
141	5/1879	0-4-4T	16 x 20	5 7½	29			
142	5/1879	0-4-4T	16 x 20	5 7½	35			
143	6/1879	0-4-4T	16 x 20	5 7½	39			
144	7/1879	0-4-4T	16 x 20	5 7½	40		E3	1936
145	8/1879	0-6-0	18 x 24	5 1¾	181		J15	1959
146	9/1879	0-6-0	18 x 24	5 1¾	182		J15	1962
147	12/1879	0-6-4T	18 x 24	4 6½	203		H1	1940
148	12/1879	0-6-4T	18 x 24	4 6½	204	REB 0-6-0T 1914	J12	1950
149	1/1880	0-6-0	18 x 24	5 1¾	183		J15	1965
150	2/1880	0-6-0	18 x 24	5 1¾	184	(d)	J15	1962
151	4/1880	0-6-4T	18 x 24	4 6½	205		H1	1928
152	5/1880	0-6-4T	18 x 24	4 6½	206		H1	1928
153	7/1880	4-4-0	16 x 20	5 8½	8		D19	1945
154	7/1880	4-4-0	16 x 20	5 8½	10		D19	1951
155	9/1880	4-4-0	16 x 20	5 8½	15		D19	1951
156	10/1880	4-4-0	16 x 20	5 8½	13		D19	1953
157	2/1881	0-6-0	18 x 24	5 1¾	107		J15	1957
158	2/1881	0-6-4T	18 x 24	4 6½	91	REB 0-6-0ST 1924 (e)	J29	1930
159	4/1881	0-6-4T	18 x 24	4 6½	92	(e)	H2	1945
160	4/1881	0-6-0	18 x 24	5 1¾	139		J15	1961
161	5/1881	0-6-0	18 x 24	5 1¾	140		J15	1961
162	5/1881	0-6-0	18 x 24	5 1¾	141		J15	1959
163	7/1881	0-6-0	18 x 24	5 1¾	123		J15	1963
164	9/1881	0-6-0	18 x 24	5 1¾	124		J15	1965
165	10/1881	0-6-0	18 x 24	5 1¾	125		J15	1965
166	11/1881	0-6-0	18 x 24	5 1¾	126		J15	1959
167	2/1882	0-6-0	18 x 24	5 1¾	127		J15	1963
168	3/1882	0-6-0	18 x 24	5 1¾	128		J15	1963
169	3/1882	0-6-0	18 x 24	5 1¾	187		J15	1957
170	7/1882	0-6-0	18 x 24	5 1¾	188	To Traffic 12/82	J15	1959
171	9/1882	0-6-0	18 x 24	5 1¾	101		J15	1962
172	9/1882	0-6-0	18 x 24	5 1¾	122		J15	1963
173	12/1882	0-6-0	18 x 24	5 1¾	130		J15	1965
174	12/1882	0-6-0	18 x 24	5 1¾	131		J15	1963
175	5/1883	0-4-4T	16 x 20	5 7½	47		E3	1945

GREAT SOUTHERN & WESTERN RAILWAY

P No	Built	Type	Cylinders inches	D.W. ft in	No		*	W'drawn
176	7/1883	0-4-4T	16 x 20	5 7½	48		E3	1930
177	8/1883	0-4-4T	16 x 20	5 7½	49		E3	1945
178	9/1883	0-4-4T	16 x 20	5 7½	50			
179	11/1883	4-4-0	17 x 22	6 7	52		D17	1949
180	12/1883	4-4-0	17 x 22	6 7	53			1925
181	12/1883	4-4-0	17 x 22	6 7	54		D17	1959
182	1/1884	0-4-4T	16 x 20	5 7½	81		E3	1934
183	1/1884	0-4-4T	16 x 20	5 7½	82		E3	
184	1/1884	0-4-4T	16 x 20	5 7½	83		E3	1928
185	2/1884	4-4-0	17 x 22	6 7	55		D17	1955
186	4/1884	0-4-4T	16 x 20	5 7½	51		E3	1934
187	6/1884	0-4-4T	16 x 20	5 7½	70		E3	1940
188	12/1884	0-4-4T	16 x 20	5 7½	84			
189	12/1884	0-4-4T	16 x 20	5 7½	71			
190	12/1884	0-4-4T	16 x 20	5 7½	72		E3	1940
191	6/1885	4-4-0	18 x 24	6 7	93		D14	1959
192	6/1885	0-6-0	18 x 24	5 1¾	133		J15	1963
193	6/1885	0-6-0	18 x 24	5 1¾	134		J15	1961
194	8/1885	4-4-0	18 x 24	6 7	94		D14	1959
195	9/1885	0-6-0	18 x 24	5 1¾	135		J15	1957
196	10/1885	0-6-0	18 x 24	5 1¾	191		J15	1962
197	10/1885	4-4-0	18 x 24	6 7	95		D14	1955
198	11/1885	4-4-0	18 x 24	6 7	96		D14	1959
199	9/1886	0-4-4T	16 x 20	5 7½	77		E3	1931
200	9/1886	0-4-4T	16 x 20	5 7½	78		E3	1945
201	10/1886	0-4-4T	16 x 20	5 7½	79			
202	10/1886	0-4-4T	16 x 20	5 7½	80		E3	1931
203	1886	4-4-0	18 x 24	6 7	85		D14	1959
204	1886	4-4-0	18 x 24	6 7	86		D14	1957
205	1886	4-4-0	18 x 24	6 7	87		D14	1957
206	1886	4-4-0	18 x 24	6 7	88		D14	1957
207	1886	4-4-0	18 x 24	6 7	89	(f)	D13	1960
208	12/1886	4-4-0	17 x 22	6 7	9		D17	1955
209	12/1886	4-4-0	17 x 22	6 7	16		D17	1959
210	2/1887	4-4-0	17 x 22	6 7	97		D17	1930
211	2/1887	4-4-0	17 x 22	6 7	98		D17	1954
212	6/1887	0-4-4T	16 x 20	5 7½	73		E3	1928
213	6/1887	0-4-4T	16 x 20	5 7½	74		E3	1930
214	9/1887	0-4-4T	16 x 20	5 7½	75		E3	1931
215	9/1887	0-4-4T	16 x 20	5 7½	76		E3	1931
216	12/1887	0-6-0T	18 x 24	4 6½	207		J11	1959
217	12/1887	0-6-0T	18 x 24	4 6½	208		J11	1959
218	12/1887	0-6-0T	18 x 24	4 6½	209		J11	1949
219	12/1887	0-6-0T	18 x 24	4 6½	210		J11	1959
220	4/1888	4-4-0	17 x 22	6 7	4		D17	1957
221	4/1888	4-4-0	17 x 22	6 7	11		D17	1949
222	5/1888	4-4-0	17 x 22	6 7	14		D17	1957
223	6/1888	4-4-0	17 x 22	6 7	18		D17	1959
224	9/1888	4-4-0	17 x 22	6 7	56		D17	1951
225	10/1888	4-4-0	17 x 22	6 7	57		D17	1957
226	1888	4-4-0	17 x 22	6 7	58		D17	1953
227	10/1888	4-4-0	17 x 22	6 7	59		D17	1955
228	12/1888	0-6-0	18 x 24	5 1¾	132		J15	1965
229	12/1888	0-6-0	18 x 24	5 1¾	136		J15	1962
230	12/1888	0-6-0	18 x 24	5 1¾	137		J15	1960
231	12/1888	0-6-0	18 x 24	5 1¾	138		J15	1962
232	8/1889	0-6-0	18 x 24	5 1¾	103		J15	1957
233	8/1889	0-6-0	18 x 24	5 1¾	114		J15	1961
234	9/1889	0-6-0	18 x 24	5 1¾	129		J15	1940
235	3/1890	4-4-0	17 x 22	6 7	1		D17	1955
236	3/1890	4-4-0	17 x 22	6 7	3		D17	1957
237	5/1890	4-4-0	17 x 22	6 7	12		D17	1949
238	6/1890	4-4-0	17 x 22	6 7	20		D17	1959

GREAT SOUTHERN & WESTERN RAILWAY

P No	Built	Type	Cylinders inches	D.W. ft in	No		*	W'drawn
239	12/1890	0-6-0	18 x 24	5 1¾	110		J15	1963
240	12/1890	0-6-0T	12 x 18	3 8½	99		J30	1931
241	1/1891	0-6-0T	12 x 18	3 8½	100		J30	1959
242	3/1891	0-6-0	18 x 24	5 1¾	111		J15	1963
243	5/1891	0-6-0	18 x 24	5 1¾	118		J15	1966
244	10/1891	4-4-0	18 x 24	6 7	60		D14	1957
245	11/1891	4-4-0	18 x 24	6 7	61		D14	1955
246	12/1891	4-4-0	18 x 24	6 7	62		D14	1959
247	12/1891	4-4-0	18 x 24	6 7	63		D14	1955
248	10/1892	2-4-2T	16 x 20	5 8½	33		F6	1957
249	12/1892	2-4-2T	16 x 20	5 8½	34		F6	1957
250	12/1892	2-4-2T	16 x 20	5 8½	41		F6	1958
251	1/1893	2-4-2T	16 x 20	5 8½	42		F6	1963
252	2/1894	2-4-2T	16 x 20	5 8½	35		F6	1959
253	4/1894	2-4-2T	16 x 20	5 8½	36		F6	1957
254	7/1894	4-4-2T	16 x 20	5 8½	37		C7	1954
255	10/1894	4-4-2T	16 x 20	5 8½	38		C7	1950
256	12/1894	0-4-4T	9 x 12	5	–	Fairy (g)	L4	1927
257	10/1895	4-4-0	18 x 24	6 7	64		D14	1959
258	10/1895	4-4-0	18 x 24	6 7	65		D14	1959
259	12/1895	0-6-0T	18 x 24	4 6½	201		J11	1963
260	12/1895	0-6-0T	18 x 24	4 6½	202		J11	1955
261	6/1896	0-6-0	18 x 24	5 1¾	105		J15	1963
262	8/1896	0-6-0	18 x 24	5 1¾	116		J15	1964
263	9/1898	0-6-0	18 x 24	5 1¾	192		J15	1956
264	9/1898	0-6-0	18 x 24	5 1¾	193		J15	1963
265	11/1898	0-6-0	18 x 24	5 1¾	194		J15	1959
266	12/1898	0-6-0	18 x 24	5 1¾	195		J15	1965
267	5/1899	0-6-0	18 x 24	5 1¾	196		J15	1961
268	6/1899	0-6-0	18 x 24	5 1¾	197		J15	1961
269	6/1899	0-6-0	18 x 24	5 1¾	198		J15	1965
270	11/1899	0-6-0	18 x 24	5 1¾	199		J15	1954
271	4/1900	4-4-0	18 x 26	6 7	301	Victoria	D11	1962
272	4/1900	4-4-0	18 x 26	6 7	302	Lord Roberts	D11	1957
273	6/1900	4-4-0	18 x 26	6 7	303	St. Patrick	D11	1959
274	6/1900	4-4-0	18 x 26	6 7	304	Princess Ena	D11	1959
275	11/1900	4-4-2T	17 x 22	5 8½	27		C4	1953
276	12/1900	4-4-2T	17 x 22	5 8½	30		C4	1950
277	12/1900	4-4-2T	17 x 22	5 8½	31		C4	1953
278	6/1901	4-4-2T	17 x 22	5 8½	32		C4	1951
279	6/1901	4-4-2T	16 x 20	5 8½	317		C7	1955
280	6/1901	4-4-2T	16 x 20	5 8½	318		C7	1953
281	6/1901	4-4-2T	16 x 20	5 8½	319		C7	1950
282	1901	0-6-0T	18 x 24	4 6½	217		J11	1961
283	1901	0-6-0T	18 x 24	4 6½	218		J11	1959
284	1901	0-6-0T	18 x 24	4 6½	219		J11	1955
285	1901	0-6-0T	18 x 24	4 6½	220		J11	1959
286	6/1902	4-4-0	18 x 26	6 7	305		D12	1957
287	6/1902	4-4-0	18 x 26	6 7	306		D12	1959
288	6/1902	4-4-0	18 x 26	6 7	307	(h)	D12	1959
289	6/1902	4-4-0	18 x 26	6 7	308		D12	1933
290	6/1902	4-4-2T	16 x 20	5 8½	320		C7	1954
291	10/1902	0-6-0	18 x 24	5 1¾	240		J15	1957
292	12/1902	0-6-0	18 x 24	5 1¾	241		J15	1957
293	12/1902	0-6-0	18 x 24	5 1¾	242		J15	1957
294	1/1903	0-6-0	18 x 24	5 1¾	243		J15	1955
295	2/1903	0-6-0	18 x 24	5 1¾	200		J15	1960
296	2/1903	0-6-0	18 x 24	5 1¾	223		J15	1960
297	3/1903	0-6-0	18 x 24	5 1¾	229		J15	1960
298	3/1903	0-6-0	18 x 24	5 1¾	232		J15	1963
299	1903	0-6-0	18 x 24	5 1¾	253		J15	1963
300	1903	0-6-0	18 x 24	5 1¾	254		J15	1961
301	1903	0-6-0	18 x 24	5 1¾	255		J15	1964

GREAT SOUTHERN & WESTERN RAILWAY

P No	Built	Type	Cylinders inches	D.W. ft in	No		*	W'drawn
302	1903	0-6-0	18 x 24	5 1¾	256		J15	1959
303	11/1903	0-6-0	18 x 26	5 1¾	351		J9	1964
304	12/1903	0-6-0	18 x 26	5 1¾	352		J9	1955
305	12/1903	0-6-0	18 x 26	5 1¾	353	(i)	J9	1931
306	12/1903	0-6-0	18 x 26	5 1¾	354		J9	1963
307	8/1904	0-2-2T			RM1			
308	12/1904	4-4-0	18 x 26	6 7	321		D2	1957
309	1/1905	4-4-0	18 x 26	6 7	322		D2	1960
310	1/1905	4-4-0	18 x 26	6 7	323		D2	1955
311	2/1905	4-4-0	17 x 26	5 8½	324		D3	1928
312	5/1905	4-4-0	17 x 26	5 8½	325		D3	1928
313	6/1905	4-4-0	17 x 26	5 8½	326		D3	1927
314	6/1905	4-4-0	17 x 26	5 8½	327		D2	1959
315	6/1905	4-4-0	17 x 26	5 8½	328		D2	1959
316	12/1905	4-6-0	19¼ x 26	5 1¾	362		B3	1928
317	12/1905	4-6-0	19¼ x 26	5 1¾	363		B3	1928
318	12/1905	4-6-0	19¼ x 26	5 1¾	364		B3	1928
319	12/1905	4-6-0	19¼ x 26	5 1¾	365		B3	1928
320	10/1906	4-4-0	17 x 26	5 8½	331		D2	1959
321	10/1906	4-4-0	17 x 26	5 8½	332		D2	1959
322	11/1906	4-4-0	17 x 26	5 8½	329		D2	1960
323	12/1906	4-4-0	17 x 26	5 8½	330		D2	1957
324	6/1907	4-4-0	18 x 26	5 8½	333		D4	1955
325	6/1907	4-4-0	18 x 26	5 8½	334		D4	1955
326	6/1907	4-4-0	18 x 26	5 8½	335		D4	1955
327	6/1907	4-4-0	18 x 26	5 8½	336		D4	1957
328	6/1907	4-6-0	19¼ x 26	5 1¾	366		B3	1931
329	6/1907	4-6-0	19¼ x 26	5 1¾	367		B3	1928
330	6/1908	4-4-0	18 x 26	5 8½	337		D4	1955
331	6/1908	4-4-0	18 x 26	5 8½	338		D3	1959
332	6/1908	4-4-0	18 x 26	5 8½	339		D4	1959
333	6/1908	4-4-0	18 x 26	5 8½	340		D4	1955
334	9/1909	2-6-0	19 x 26	5 1¾	368		K4	1928
335	9/1909	2-6-0	19 x 26	5 1¾	369		K4	1957
336	9/1909	2-6-0	19 x 26	5 1¾	370		K4	1957
337	9/1909	2-6-0	19 x 26	5 1¾	371		K4	1928
338	4/1912	0-6-0	18 x 26	5 1¾	249		J9	1965
339	5/1912	0-6-0	18 x 26	5 1¾	250		J9	1963
340	5/1912	0-6-0	18 x 26	5 1¾	251		J9	1965
341	6/1912	0-6-0	18 x 26	5 1¾	252		J9	1961
342	5/1913	4-4-0	19 x 26	6 7	341	Sir William Goulding (j)	D1	1928
343	10/1913	0-6-0	19 x 26	5 1¾	257		J4	1960
344	10/1913	0-6-0	19 x 26	5 1¾	258		J4	1963
345	11/1913	0-6-0	19 x 26	5 1¾	259		J4	1959
346	6/1914	0-4-2ST	16 x 20	4 6½		Sambo	L2	1963
347	7/1914	0-6-0	19 x 26	5 1¾	260		J4	1962
348	11/1914	0-6-0	19 x 26	5 1¾	261		J4	1966
349	11/1914	0-6-0	19 x 26	5 1¾	262		J4	1965
350	12/1914	0-6-0	19 x 26	5 1¾	263		J4	1962
351	12/1914	0-6-0	19 x 26	5 1¾	264		J4	1960
352	9/1915	4-8-0T	19¼ x 26	4 6½	900		A1	1928
353	8/1916	4-6-0	4/14 x 26	6 7	400		B2	1929
354	4/1921	4-6-0	4/14 x 26	6 7	401	(k)	B2	1961
355	8/1921	4-6-0	4/14 x 26	6 7	402	(l)	B2	1961
356	11/1921	4-6-0	4/14 x 16	6 7	406	(k)	B2	1957
357	11/1924	4-6-0	19½ x 28	6 7	500	(m)	B1	1955
358	12/1924	4-8-0T	19¼ x 26	4 6½	901		A1	1931
359	1926	4-6-0	19½ x 28	6 7	501		B1	1955
360	4/1926	4-6-0	19½ x 28	6 7	502		B1	1957
361	1927	2-6-0	19 x 28	5 6	384	(n)	K1	1960
362	1927	2-6-0	19 x 28	5 6	385		K1	1960
363	1927	2-6-0	19 x 28	5 6	386		K1	1959
364	1928	2-6-0	19 x 28	5 6	387		K1	1959

GREAT SOUTHERN & WESTERN RAILWAY

P No	Built	Type	Cylinders inches	D.W. ft in	No		*	W'drawn
365	1928	2-6-0	19 x 28	5 6	388		K1	1962
366	1928	2-6-0	19 x 28	5 6	389		K1	1955
367	10/1928	2-6-2T	17½ x 28	5 6	850		P1	1955
368	6/1929	2-6-0	19 x 28	5 6	390		K1	1955
369	6/1929	2-6-0	19 x 28	5 6	391		K1	1957
370	11/1929	0-6-0	18 x 24	5 1¾	700		J15A	1963
371	1929	0-6-0	18 x 24	5 1¾	701		J15A	1959
372	1929	0-6-0	18 x 24	5 1¾	702		J15A	1955
373	1929	0-6-0	18 x 24	5 1¾	703		J15A	1961
374	1929	0-6-0	18 x 24	5 1¾	704		J15A	1961
375	1930	2-6-0	19 x 28	6 0	393		K1A	1954
376	1930	2-6-0	19 x 28	6 0	394		K1A	1959
377	1930	2-6-0	19 x 28	6 0	395		K1A	1957
378	1930	2-6-0	19 x 28	6 0	396		K1A	1959
379	1930	2-6-0	19 x 28	6 0	397		K1A	1957
380	1930	2-6-0	19 x 28	6 0	398		K1A	1955
381	1933	0-6-2T	18 x 24	5 6	670		I3	1959
382	1933	0-6-2T	18 x 24	5 6	671		I3	1959
383	1933	0-6-2T	18 x 24	5 6	672		I3	1959
384	1933	0-6-2T	18 x 24	5 6	673		I3	1962
385	1933	0-6-2T	18 x 24	5 6	674		I3	1959
386	1934	0-6-0	18 x 24	5 1¾	710		J15B	1959
387	1934	0-6-0	18 x 24	5 1¾	711		J15B	1962
388	1934	0-6-0	18 x 24	5 1¾	712		J15B	1959
389	1934	0-6-0	18 x 24	5 1¾	713		J15B	1959
390	1934	0-6-0	18 x 24	5 1¾	714		J15B	1959
391	1935	0-6-0	18 x 24	5 1¾	715		J15B	1959
392	1935	0-6-0	18 x 24	5 1¾	716		J15B	1961
393	1935	0-6-0	18 x 24	5 1¾	717		J15B	1959
394	1935	0-6-0	18 x 24	5 1¾	718		J15B	1959
395	1935	0-6-0	18 x 24	5 1¾	719		J15B	1959
396	1936	4-4-0	18 x 26	5 8½	342	(o)	D4	1959
397	1936	4-4-0	18 x 26	5 8½	343		D4	1959
398	1936	4-4-0	18 x 26	5 8½	344		D4	1959
399	1936	4-4-0	18 x 26	5 8½	345		D4	1959
400	1936	4-4-0	18 x 26	5 8½	346		D4	1960
401	4/1939	4-6-0	3/18½ x 28	6 7	800	Maedhbh (p)	BIA	1962
402	1939	4-6-0	3/18½ x 28	6 7	801	Mácha	BIA	1962
403	6/1940	4-6-0	3/18½ x 28	6 7	802	Táilte	BIA	1957
C I E								1958
404	1957	0-6-6-0	4/12 x 14	3 7	CC1	(q)		

(a) Engine and carriage rebuilt 7/1871 on separate 4 wheel frame 'L5' until 1926. (b) Renewal of Sharp 'Single' Frames renewed 1/1873. (c) Built for Castle Island Rly. Absorbed 1879 and numbered 90. Rebuilt 0-6-0T. (d) Preserved by C I E. (e) Built as inspection carriage. (f) New frames and '700' boiler in 1925. 'Z' boiler in 1933 and reclassified D14. (g) Pay carriage Reb 0-4-2T. (h) Fitted with Marshall's valve gear 1904. (i) Scrapped after Monasterevan collision. (j) Inside Walschaert's valve gear. (k) Rebuilt two outside cylinders 19½" x 26" and Caprotti valve gear 1930 Class B2A. (l) Rebuilt two outside cylinders 19½" x 26". (m) Feed water heater. (n) Locos 384–391, 393–398 built from Woolwich Arsenal Parts. (o) Outside frame bogie side window cab No 800 preserved. (p) Originally double chimney. (q) Turf burning prototype. (r) Converted to 0-6-0T.

One additional Loco (0-4-0VB) built in 1884 for Cork Coal gantry probably from scrap parts.

*Column shows final classification.

GREAT SOUTHERN RAILWAY & GREAT SOUTHERN RAILWAY CO.

See MIDLAND GREAT WESTERN RAILWAY, and GREAT SOUTHERN & WESTERN RALWAY.

GREAT WESTERN RAILWAY
Swindon

BROAD GAUGE The original gauge of the GWR was 7′ 0¼″, recommended by the company's engineer Isambard Kingdom Brunel from London to Bristol. With its gentle gradients and gradual curves it was suggested that high speeds could easily be attained with the advantage of wider carriages and wagons giving increased accommodation and pay loads. It was unfortunate that the majority of the earlier locomotives built were very poor performers, mainly because of Brunel's insistence upon his own specifications being rigidly adhered to by the locomotive builders. Apart from the *Star* class built by Robert Stephenson & Co. the locomotive department were always in trouble and Daniel Gooch who had been appointed locomotive superintendent in 1837 with Thomas R. Crampton as his chief draughtsman was faced with the unenviable task of trying to keep such locomotives at work. Gooch's first designs were based on this successful *Star* class of Stephenson's.

The decision to build the main repair shops at Swindon was taken in 1840. Swindon was a strategic spot between London and Bristol, and the nearby Wilts and Berks Canal was used to bring in materials. A start was made shortly afterwards in building a new railway town between the village and the railway.

Construction of the workshops commenced in 1842 and was partly in use at the beginning of 1843 and completed in 1846. Archibald Sturrock was transferred from Westbourne Park to Swindon and made assistant locomotive superintendent and the first works manager.

Fig. 210 **GWR** Broad Gauge Swindon 1876 8′ single TARTAR

The first locomotive to be built was a 0-6-0 of the *Premier* class which was completed in February 1846 and a total of twelve were built in 1846–47. They were not entirely built in the workshops, the boilers with haycock fireboxes, being supplied by an outside contractor.

In April 1846 the prototype 2-2-2 named GREAT WESTERN was built with outside slotted sandwich frames, haycock firebox, inside cylinders 18″ x 24″ and 8 ft diameter driving wheels. It was found, after a leading axle was fractured, that there was excessive weight at the front end, and it was brought back into the shops for conversion to 4-2-2 and in this form was the basic design for future 8 ft singles, including the *Iron Dukes*.

The haycock firebox was now abandoned and further 0-6-0s of the *Pyracmon* and *Ariadne* classes were provided with the raised top variety with spring balance safety valves fitted above, surrounded by a squat brass cover. The boiler was domeless.

Six smaller 2-2-2s were built (1846–47); five with 7 ft diameter driving wheels, and one, WITCH, which had 7′ 6″ diameter drivers. They had sandwich frames, but the wheels were outside, similar to the 0-6-0 classes. They were known as the *Prince* class.

After initial troubles the *Great Western* had proved successful and in 1847 a start was made in building an improved version known as the *Iron Duke* class of 4-2-2s. The first six were built that year and altogether twenty-two were built up to 1851. The 8 ft diameter driving wheels were flangeless. The sandwich frames were outside the wheels and in addition plate frames were fitted between the rear end of the cylinders and the front of the firebox. There were three of these plate frames between the wheels, the middle one acting as a centre stay. The driving axle had no less than five bearings; the leading and trailing axles had bearings in the sandwich frames only. The sandwich frames had been fitted to Stephenson's *Star* class, and Daniel Gooch, probably due to their suitability for the longitudinal sleeper track on the GWR, adopted them as standard for all his locomotives. The plate frames were subject to far more vibration and rough riding. The sandwich frame comprised a thick wooden slab, usually ash, with iron plates bolted each side. The wood was later changed to oak, but this corroded the bolts which also gave trouble by becoming loose. Teak was then introduced at a later stage.

A type of balanced slide valve was used, but

Fig. 211 **GWR** Broad Gauge Swindon 1876 Convertible Type 0-6-0ST No. 1236

later unbalanced valves were fitted, giving less trouble. They were operated by Gooch's stationary link motion.

The domeless boiler had the regulator box on the front tubeplate and steam was collected by means of a perforated pipe and a mid feather was fitted inside the firebox. One of this class built in 1851 and named LORD OF THE ISLES was sent to the Hyde Park Exhibition that year.

Four saddle tanks were built. They were similar to the standard 0-6-0 goods locomotives being built except for the saddle tanks fitted over the boiler barrel. The front sandbox was fitted on the front of the smokebox above the door. Compensating levers were fitted linking all these axles and their springs.

After the first twenty-two *Iron Dukes*, Gooch brought out a 2-4-0 passenger type and eighteen were built between 1856–64. These also had compensating levers between all axles. Six smaller versions of this class were built in 1865 and tweny more by the Avonside Engine Co.

For work on the South Devon line a 4-4-0ST was designed; only two were built at Swindon, and thirteen by R. & W. Hawthorn & Co. They had inside sandwich frames, and the bogie was fitted with a ball and socket joint and a rigid pivot. The frames terminated behind the bogie, the boiler making the connection between the cylinders and main frames. A circular sandbox was fixed on top of the saddle tank, and another interesting feature was the 'sledge' brake between the coupled wheels. This type of brake was later dispensed with in favour of the conventional brake hangers and blocks. The sledge brake, acting on the rails vertically, tended to lift the wheels off the rails and when the joints were generally slack the brake fouled the points and check rails.

For working on the Metropolitan lines ten 2-4-0WTs were built (1863–4); twelve were previously delivered by outside contractors in 1862. The standard Gooch design of inside cylinders was not used for these tanks, nor were the sandwich frames. The outside cylinders were at the base of the smokebox and steeply inclined. They were the first in the country to be equipped with condensing gear which was fitted under the boiler, and two water tanks were sited between the frames, one under the boiler barrel and a smaller one under the footplate. Steam was exhausted into the main tank in which was fitted a perforated pipe – non-return valves were fitted to the exhaust pipes and flap valves operated from the footplate, diverted the exhaust up the chimney or into the tanks. They were not successful and later had the gear removed and some of the class converted to tender locomotives.

Gooch resigned in September 1864 and became chairman of the company. He had improved the locomotive stock considerably; the earlier locomotives were a strange lot and although he was not responsible for their ordering or design, he had to try and keep them running. His narrow gauge or standard gauge locomotives will be described later.

He was succeeded by Joseph Armstrong who came from the 'narrow' gauge stronghold at the Stafford Road works, Wolverhampton. He was now made responsible for the whole of the system, including Wolverhampton and Worcester.

Few broad gauge locomotives were built under Armstrong – the conversion of the South Wales lines to standard gauge was imminent which meant that the Welsh broad gauge stock would supplement the existing stock on the Western lines – and none were built after 1866.

In 1865–66 six 2-4-0s were built as 'renewals' and twenty more were built by Avonside Engine Co. in the same period. They were all named after engineers except the Swindon six. One of them was named after the principal

Fig. 212
GWR
Broad Gauge
Swindon 1888
0-4-4WT
No. 3549
originally 0-4-2ST

of the Avonside Engine Co. whose previous title had been Slaughter Gruning & Co. To be hauled by a locomotive named SLAUGHTER rather put off the travelling public and the name was tactfully changed to AVONSIDE.

From 1871 to 1888 the famous *Iron Duke* class were officially renewed and although a few parts of the former 4-2-2s were used on the first three the remainder were new locomotives entirely, and a total of twenty-four formed the class which was known as the *Rover* class and these gradually replaced the original twenty-two. They worked the main line express passenger trains between London and Bristol and some very fast runs were recorded – indeed they were without doubt performing the fastest running in the country in this period.

In 1866 six more tank locomotives were designed for the Metropolitan lines. They were the only side tanks to be built for the broad gauge and were known as the *Sir Watkin* class. After the conversion to standard gauge of the Metropolitan Railway in 1869, three were sold to the South Devon Railway, returning to the GWR when the South Devon Railway was amalgamated with the GWR.

The broad gauge of course was the odd man out and caused chaos at exchange points with the standard or 'narrow' gauge. From the broad gauge point of view it was the narrow gauge which caused all the confusion.

Up to 1892 when it was abolished, many fine locomotives ran on the 7 ft gauge – perhaps if that gauge had been standardised the trend and scope of steam locomotive design might have altered the whole policy of motive power. We might be still in the position of being able to see and hear real steam giants at work.

Swindon works built a total of 241 broad gauge locomotives. Later some broad gauge convertible locomotives were built, but these will be dealt with in the standard gauge section.

BROAD GAUGE–SWINDON GWR

Lot No	Type	First Built	No Built	Class	Cylinders inches	D.W. ft in	Heating Surface sq ft	Grate Area	WP lb/sq in	Weight WO ton cwt
1st Goods	0-6-0	1846	12	Premier (F)	16 x 24	5 0	982·18	13·67	100	26 15
2nd Goods	0-6-0	1847	6	Pyracmon (F)	16 x 24	5 0	1255·73	18·44	115	28 3
3rd Goods	0-6-0	1851	8	Caesar (F)	16 x 24	5 0	1255·73	18·44	120	32 9
4th Goods	0-6-0	1852	4	Ariadne (F)	17 x 24	5 0	1574·0	19·2	120	31 2
5th Goods	0-6-0ST	1852	2	Banking	17 x 24	5 0	1574·0	19·2	120	38 0
5th Goods	0-6-0	1852	22	Ariadne	17 x 24	5 0	1574·0	19·2	120	31 2
6th Goods	0-6-0	1854	40	Caliph	17 x 24	5 0	1574·0	19·2	120	31 2
6th Goods	0-6-0ST	1854	2	Banking	17 x 24	5 0	1574·0	19·2	120	38 0
7th Goods	0-6-0	1857	6	Caliph	17 x 24	5 0	1574·0	19·2	120	31 2
8th Goods	0-6-0	1859	6	Caliph	17 x 24	5 0	1574·0	19·2	120	31 2
9th Goods	0-6-0	1861	12	Caliph	17 x 24	5 0	1574·0	19·2	120	31 2
10th Goods	0-6-0	1862	12	Caliph	17 x 24	5 0	1574·0	19·2	120	31 2
11th Goods	0-6-0	1865	2	Swindon	17 x 24	5 0				
12th Goods	0-6-0	1865	12	Swindon	17 x 24	5 0				
	0-6-0T	1865	6	Sir Watkin	17 x 24	4 6				40 16
	2-2-2	1846	1	Great Western (F)	18 x 24	8 0		22·64	100	
1st Pass.	2-2-2	1846	6	Prince (F)	16 x 24	7 0	982·18	13·67		25 16
2nd Pass.	4-2-2	1847	6	Iron Duke (F)	18 x 24	8 0	1944·99	21·66	100	35 10
3rd Pass.	4-2-2	1848	12	Iron Duke (F)	18 x 24	8 0	1919·47	25·47	120	38 4
4th Pass.	4-2-2	1850	4	Iron Duke (F)	18 x 24	8 0	1919·47	25·47	120	38 4
5th Pass.	2-4-0	1856	8	Victoria (F)	16 x 24	6 6	1263·87	13·5		31 4
6th Pass.	2-4-0	1863	10	Victoria (F)	16 x 24	6 6	1263·87	13·5		31 4
	2-4-0	1865	6	Hawthorn	16 x 24	6 0	1201·0	19·0	130	29 10
	4-2-2	1871	24	Rover	18 x 24	8 0	1793·4	24·0	140	41 14
1st Bogie	4-4-0ST	1849	2	Corsair (F)	17 x 24	6 0	1255·73	19·0		35 15
1st Metro	2-4-0WT	1863	10	Metropolitan	OC 16 x 24	6 0		18·0		

During D Gooch's time, the tube heating surface was calculated for the 'fire side' of the tubes ie the inside diameter, whereas standard practice was to calculate the heating surface on the water side, or outside diameter. Heating surface on 'fire side' are marked (F).

STANDARD GAUGE The first standard gauge locomotives to be built at Swindon were twelve 0-6-0s to Gooch's designs and were a scaled down version of the broad gauge type. Slotted sandwich frames were used, but outside the wheels instead of inside as most of the broad gauge 0-6-0s. The boiler was domeless with raised firebox with typical Gooch's safety valves and casing over the firebox. Compensating beams were fitted to the springs between axles. The inside cylinders were 15½" x 22", with Gooch's stationary link motion, and the coupled wheels were 5' diameter. Boiler pressure was 120 lb/sq. in. The first two, numbered 57 and 58 were completed in May 1855 and were the forerunners of the standard goods 0-6-0 built up to 1876.

Gooch's series were as follows: '57' class 1855–56 15½" x 22" cylinders 5' 0" diameter wheels 12 built; '79' class 1857–62 16" x 24" cylinders 4' 6" diameter wheels 24 built; '131' class 1862 16" x 24" cylinders 5' 0" diameter wheels 16 built.

All three classes had boilers 4' 0" diameter and 11' 0" barrel varying slightly in heating surface. They were all initially sent to Wolverhampton for the northern division. They were despatched there by loading them on special broad gauge wagons until 1869 when they were off-loaded at Didcot and it was not until 1872 that standard gauge access was available from Swindon direct.

In 1860 the first two six coupled side tanks were built, Nos. 93 and 94. They had well tanks in addition and inside frames were used. The boiler was the standard type of Gooch's domeless raised fireboxes and Gooch's valve gear. The majority of six coupled tank locomotives were being built at Wolverhampton.

Fig. 213
GWR
Swindon 466/1873
2-4-0 No. 810
Last Armstrong
2-4-0 design

Fig. 214
GWR
Swindon
1391/1894
Dean 7'8"
single ACHILLES
The first 4-2-2

Two outside cylinder 2-4-0WTs were built in 1864 for the Metropolitan Railway's lines with condensing gear and their careers followed a similar pattern to the broad gauge locomotives built for the same purpose, being converted to tender locomotives and the condensing gear removed.

As mentioned in the broad gauge section Joseph Armstrong became the first locomotive, carriage and wagon superintendent of the company in 1864. In 1866 he put in hand the first passenger tender type namely the 2-2-2 *SIR DANIEL* class. Ten were built in 1866 followed by twenty more in 1869.

They had inside cylinders 17" x 24" and 7' 0" diameter driving wheels. The double frames were of plate; the boiler was domed, with a flush firebox casing with safety valves above, and with large brass covers. The driving wheel splashers were open. Four were named, No. 378 bearing the name SIR DANIEL as a compliment by Armstrong to his predecessor.

An interesting conversion took place in 1900–2 when the class became too small to cope with express work, and all but three were converted to 0-6-0 goods locomotives. A very rare metamorphosis!

In the same year, as the first batch of 2-2-2s were built, Armstrong introduced his 'standard' goods and up to 1876 a total of 280 were built at Swindon. They had outside double plate frames, 17" x 24" cylinders and 5' diameter wheels. The boiler was domed and the total heating surface was 1203 sq. ft with a diameter of 4' 2" and barrel length of 11' 0". The boiler pressure was 140 lb/sq. in. and the locomotive in working order weighed 29 tons 18 cwt. Dimensions varied of course particularly in heating surface and firebox details.

A 4' 6" wheel version was built in 1874 known as 'coal engines'. A total of twenty were built and the majority were used on coal trains between Pontypool Rd and Birkenhead.

In 1868 six 2-4-0s were built, Nos. 439–444. Known as 'bicycles', they had all the characteristics of the broad gauge 2-4-0s; inside frames, 6' 1" diameter coupled wheels with 'mudguards' only and no closed in splashers. The inside cylinders were 16" x 24" and they were indeed extremely ungainly looking machines but when renewed at Wolverhampton in 1885–86 they reappeared in a much improved form.

In the following year Armstrong designed a 2-4-0T known as *METRO* tanks, the first batch having condensing gear for Metropolitan work and they were as ugly as his 'bicycles'. The next batch (Lot 25), built in 1871, were slightly larger. In fact this class, which was built up to 1899, was of infinite variety, having different size tanks, wheelbases and boilers and when rebuilt a variety of cabs and bunkers were fitted. Up to 1906 as many as fifty of this class were working on the Metropolitan lines operating an astonishing number of routes but on the conversion of most of the lines to electric haulage the gear was removed except for a few whose main duties were the goods trains to and from Smithfield.

Whilst Swindon were turning out these 2-4-0Ts Wolverhampton were building 0-4-2Ts; the first sixty as saddle tanks and later converted to side tanks.

In 1870 Armstrong commenced building a tank version of his 'standard' goods, but they were far from standard, the first six having side tanks, the next fifty having saddle tanks extending over the boiler and firebox, and the remainder full length saddle tanks over smokebox as well. They all had 17" x 24" cylinders and 4' 6" diameter wheels.

The building of this class went on until 1881 with a total of 266. Ten were built in 1876 as broad gauge convertibles (Nos. 1228–37) and a further twenty (Nos. 1238–57) were converted to broad gauge in 1887–88.

For working the London to Swindon expresses Armstrong brought out the *Queen* class which were larger than the *Sir Daniel* class with 18" x 24" cylinders and 7' 0" diameter driving wheels. It was the first to be fitted with two-bar slide bars instead of the customary four-bar type, and valve spindle guides were used instead of suspension links. The boiler was domed but the further twenty built in 1875 had domeless boilers.

Joseph Armstrong died on 5 June, 1877 from nervous exhaustion – which was not to be wondered at. He had been involved with broad and standard gauge locomotives, the broad gauge conversion to standard of the South Wales and Weymouth lines, the taking over of South Devon and Bristol and Exeter stock and providing convertible stock. It was a period fraught with difficulties and he carried out the requirements with quiet efficiency.

His place was taken by William Dean who had served his apprenticeship at Wolverhampton becoming works manager there. He was transferred to Swindon in 1868 and was appointed chief assistant to Joseph Armstrong.

The building of Armstrong's 'standard' 0-6-0STs and 2-4-0Ts continued. Dean's first design was an experimental 4-4-0T with double frames with underslung springs for the coupled wheels and an outside frame bogie with Mansell wheels. The bogie was not pivotted, the connection was made via the spring shackles which were connected to an auxiliary framing fastened to the outside frames. The inside

Fig. 215
GWR
Swindon
1596/1898
No. 3296
CAMBRAI
Badminton class.
First class
with Belpaire
firebox

Fig. 216
GWR
Swindon
1666/1898
0-6-0PT 2725
as rebuilt from
saddle tank in
1922

Fig. 217
GWR
Swindon
1723/1899
4-6-0 2601
KRUGER class.
Note combustion
chamber and
sandboxes

Fig. 218
GWR
Swindon
1882/1902
Suburban 2-4-2T
3617

cylinders were 17″ x 26″ and coupled wheels 5′ 6″ diameter. The locomotive was numbered 1 and in its initial form was not successful, the bogie giving a lot of trouble. It was converted to a 2-4-0T in 1882, the long side tanks being shortened and as such was far more successful, lasting until 1924.

A second experimental locomotive No. 9, a 4-2-4T, was built in 1881. This too was unsuccessful in its original form. No. 9 had side tanks projecting past the smokebox with built-in sandboxes, the valve gear was outside, with eccentrics fixed to the driving axle and operated the inside valves which were placed on top of the cylinders by means of links and rocking shafts. The boiler was non-standard, being 4′ 2″ diameter and 11′ 6″ long barrel in two sections only. Mansell wheels were used on the two bogies. Derailments were frequent, due probably to the peculiarities of the bogies. In 1884 it was stripped down and rebuilt as a 2-2-2 tender locomotive.

The same cylinders were used and the 7′ 8″ diameter driving wheels. A shorter boiler was fitted with a total heating surface of 1250 sq. ft and a grate area of 19·23 sq. ft.

Another 2-2-2 (No. 10) was built in 1886 but with double frames and inside motion. Both were rebuilt with 7′ diameter driving wheels, and both had double frames after rebuilding as more conventional 'singles'.

2-4-0s, 2-4-0Ts and six coupled tanks continued to be built with little modifcation to Armstrong's designs.

In 1883 the remarkable Dean 'goods' first made their appearance. They differed considerably from Armstrong's 'standard' 0-6-0. They had inside single plate frames, and the first twenty (numbered 2301–2320) had domeless boilers, flush fireboxes and in addition the boiler cleading came flush to the smokebox diameter instead of the usual step up. The next twenty had a boiler with a dome on the front ring. From 1883 to 1899 two hundred and sixty were built. All were fitted with 17″ x 24″ cylinders and had 5′ 0″ diameter wheels. Boilers of varying heating surfaces were fitted from 1192·6 sq. ft to 1370·71 sq. ft. Later ones had larger domes on the second ring and larger smokeboxes.

A batch of twenty similar 0-6-0s were built known as the '2361' class in 1885–6. They were notable in having outside frames, underslung springs to all axles and the stroke increased to 26″. They were built as possible broad gauge convertibles but none of them were ever transformed.

At the same time a saddle tank version of these outside framed 0-6-0s were built. Actually they used the frames intended for a further batch of tender locomotives, but due to the success of previous classes of tank locomotives on the Aberdare and Pontypool Road coal trains they were built with saddle tanks to a total of forty in 1886–7. This class could also be converted to broad gauge if required but none were.

In addition five 2-4-0s and twenty 2-4-0Ts were built 1884–5 as a stage in Dean's standardisation plan. Cylinders, wheels and motion were common to both types, as well as for the '2361' class 0-6-0 and the '1661' 0-6-0ST class. The boilers were not exactly the same but similar. Ten of the 2-4-0Ts were built as 'convertibles' and in this state the coupled wheels were outside the frames and additional framing was provided for the leading wheels.

In 1886 Dean built his first tandem compound No. 7 followed by No. 8 a similar type but as a 'convertible'.

They were 2-4-0s with two low pressure cylinders in front of the two high pressure with a stroke of 21″. The LP valve chests were above the cylinders and the HP below and were actuated in the normal manner. To accommodate the drive the distance between the leading axle and driving axle was 10′ 2″. Double frames were used with curved footplating to clear the cranks driving the 7′ 0″ diameter wheels. The springs were underhung for the coupled wheels and above the axle of the leading wheels.

The HP cylinders for No. 7 were 23″ diameter and 22″ diameter for No. 8 and the LP cylinders 15″ diameter and 14″ diameter respectively. The boiler pressure was 180 lb/sq. in.

No. 7 ran for about two years but was always in trouble mainly due to the congested arrangement of cylinders and valve gear and No. 8 was a worse performer, never actually handed over to the running department. Both were withdrawn and then rebuilt as 4-4-0s for the standard gauge.

Much more successful were two larger 2-4-0s (Nos. 14 and 16) for the broad gauge with sandwich frames, 20″ x 24″ cylinders and 7′ 0″ diameter coupled wheels. These too were rebuilt as standard gauge 4-4-0s, all four forming one class, and it was in this form that they did their most successful work.

All the locomotives mentioned with low running numbers were experimental types and the practice of using low numbers persisted when prototypes were built.

At this period it seemed that Dean had changed his mind about frames after the designing of the 2301 class. The majority of types had the well tried sandwich frames and others were fitted with outside frames.

The most successful 2-4-0 design was built in 1889 known as the *Barnum* class and numbered 3206–3225. They were well proportioned and very handsome machines. All the springs were underhung, the footplating curved over the cranks; they had slotted curved frames, boiler with flush top firebox with polished dome cover over the first barrel ring and typical brass safety valve cover over the firebox with easing levers passing into the cab. The whistles were fixed above the cab roof (no shields required) and parallel built up chimney with polished copper band. Brass beaded splashers and individual brass numbers rivetted on the trailing splasher finished off their pleasing appearance.

The inside cylinders were 18″ x 24″ and coupled wheels 6′ 1½″ diameter, the heating surface was 1468·82 sq. ft and grate area 19·01 sq. ft. The boiler pressure was increased to 150 lb/sq. in. The class went through many changes including boiler changes and superheating and two received piston valves.

In 1891–2 appeared the final 'single' design. Thirty 2-2-2s were built with 20″ x 24″ cylinders, double plate frames, boiler 4′ 3″ diameter 11′ 6″ long and giving a total heating surface of 1459 sq. ft and a grate area of 20·8 sq. ft. The working pressure was 160 lb/sq. in.

A return was made to the raised firebox and a transverse water pocket was fitted. The driving wheels were 7′ 8½″ diameter.

Eight were built as 'convertibles' and were turned out as broad gauge locomotives in 1891 (Nos. 3021–3028), being converted to standard gauge in the following year which marked the end of the 7 ft gauge.

As with Gooch's *Great Western* of 1846, the same trouble of excessive weight at the front end was experienced. In addition the 20″ diameter cylinders were too big for the boiler output, the diameter of which could not be increased due to the large diameter of the driving wheels, hence the raised firebox and large dome. No. 3021 was derailed in Box Tunnel in September 1893 and it was decided that modifications were necessary. In 1894 all thirty had their frames lengthened and were converted to 4-2-2s. The cylinders were reduced to 19″ diameter.

At the same time new locomotives from No. 3031 were being built as 4-2-2s and a total of eighty were built up to 1899. In 1893 existing locomotives were named, continuing the previous policy but only passenger types were dealt with, except for one or two odd types. The majority of names chosen have never been excelled. In fact with the introduction later on of series names such as *Halls, Manors* etc., the glamour disappeared. A few examples should prove the point: FLYING DUTCHMAN, CORSAIR, WARLOCK, THUNDERBOLT or how about LORNA DOONE to haul you to the West Country?

In their rebuilt form, in appearance and performance, they could stand comparison with any other railways' 4-2-2s. They gleamed with brass and copper work; on the splashers were cast brass coat of arms, and the curved strip name plates with brass letters made them very handsome.

They were built for the west of England expresses and early in the present century took their share of the Birmingham and Wolverhampton expresses. That their life was comparatively short was due to the constant improvement in motive power and heavier train requirements. Some were fitted with Belpaire boilers and a few with the new standard domeless boilers from 1900 with the pressure increased to 180 lb/sq. in. Withdrawals started in 1908 and the last to disappear were 3050 ROYAL SOVEREIGN and 3074 PRINCESS HELENA in December 1915.

For the severe gradients west of Newton Abbot a locomotive with more adhesive weight and power was required and Dean put in hand in 1895 his well known *Duke of Cornwall* class of 4-4-0s. They had outside frames and inside cylinders 18″ x 26″, and 5′ 8″ diameter coupled wheels. The boiler had an extended smokebox, flush round top firebox, large dome on the back ring of the barrel and a total heating surface of 1398·18 sq. ft with a grate area of 19 sq. ft. Fifty were built up to 1899. The bogie had the wooden segmental Mansell wheels for the first forty built which were later changed to the normal spoked type. The nameplates were rectangular and most were fitted on the boiler barrel and some on the side of the firebox. The last four built had raised Belpaire fireboxes and No. 3312 BULLDOG was built with a much larger boiler with raised Belpaire firebox with a total heating surface of 1712·3 sq. ft and a grate area of 23·5 sq. ft. The pressure was raised to 180 lb/sq. in. The cab was also different with a longer roof supported by pillars.

From 1904 the nameplates were removed and replaced by the standard curved pattern over the splashers.

The first 4-6-0 to run on the GWR was built in August 1896. It had double frames, the inside pair terminating at the front of the firebox. The cylinders were 20″ x 24″ and coupled wheels 4′ 6″. The boiler was 4′ 6″ diameter and 14 ft long with a total heating surface of 1517·89 sq. ft and a grate area of 34 sq. ft. The firebox was raised and the boiler pressure was 165 lb/sq. in. The bogie was the usual Dean outside framed type with Mansell wheels, controlled by swing links. It was designed for goods traffic and often worked through to South Wales. Very little is known about its performance and it was withdrawn in 1905.

In 1899 a further 4-6-0 was built with a bigger boiler with 10′ 6″ length of barrel and a combustion chamber 3′ 6″ long giving a total heating surface of 1879·68 sq. ft and a grate area of 32·19 sq. ft. Cylinders were 19″ x 28″ and coupled wheels 4′ 7½″ diameter. An unusual feature was the single slide bars and sharply inclined cylinders complete with 8½″ diameter piston valves. Sandboxes like a miniature saddle tank were fitted over the front of the boiler barrel. Instead of the customary laminated springs, nests of volute springs were used. The bogie had inside frames with outside swing links. It was numbered 2601 and weighed 60 tons 16 cwt in working order. A second locomotive was built in 1901, but as a 2-6-0 with the same boiler, and eight more were built in 1903. Yet another 2-6-0 was built in 1900 (No. 33) with Belpaire boiler and parallel domeless boiler with safety valves on the barrel similar to the previous 2-6-0s but with a combustion chamber. This was the forerunner of the standard *Aberdare* class and eighty were built with various sized boilers and eventually all were fitted with the standard No. 4 boiler.

William Dean seems to have left the later boiler designs to his chief assistant G. J. Churchward and a number of standard sizes were brought out with Belpaire fireboxes, and parallel domeless barrels.

For suburban passenger traffic a prototype 2-4-2T was built in 1900 (No. 11). It had inside cylinders 17″ x 24″, coupled wheels 5′ 2″, and a domeless Belpaire boiler was fitted, resting on a saddle above the cylinders. This was successful and thirty more were built up to 1903. Many went to the Birmingham area, where they replaced the smaller 2-4-0T and 0-4-2Ts on many services. Their acceleration was good which helped considerably where the suburban stations had short distances between them.

Two 0-6-4 Crane tanks were built in 1901, Nos. 17 and 18, the former being sent to the Stafford Road works and No. 18 remaining at Swindon.

They were the second and third to be fitted with Pannier tanks which were to become such a characteristic of GWR tank locomotives. The first Pannier tank was No. 1490, a 4-4-0T built in 1898 as an experiment with tanks the length of the barrel and firebox only, whereas Nos. 17 and 18 had full length tanks.

William Dean's finale was the first GWR express passenger 4-6-0, No. 100, which was more Churchward than Dean, particularly the boiler which had the raised Belpaire firebox and domeless barrel. The outside cylinders were 18″ diameter and had a long stroke of 30″, the coupled wheels were 6′ 8½″ diameter. The total heating surface was 2410·31 sq. ft, grate area 27·62 sq. ft and boiler pressure of 200 lb/sq. in.

Dean retired at the end of April 1902, and during his term of office had built many successful classes. In most cases he had favoured double frames, with four bearings on the driving axle, and foundations had been laid for a programme of standardisation which was to be carried out with precision by his successor.

George Jackson Churchward had been a very active partner during the last years of William Dean's superintendency and so continuity of design and building was assured, a great advantage over some of the other major railways where changes in personnel meant radical changes in design and practice. There were changes in design and practice at Swindon but they were logical steps taken from a firm foundation.

Churchward commenced his new duties on 1 June, 1902 and in 1903 Swindon turned out three prototypes No. 97 a 2-8-0, No. 98 a 4-6-0 and No. 99 a 2-6-2T.

No. 98 was a development of Dean's No. 100 with outside cylinders 18″ x 30″ made in two halves, with half the smokebox saddle cast in each, forming the joint for the drum-head smokebox when bolted together, and bolted also to an extension of the main frames. Large 10″ diameter piston valves were fitted with a 5⅞″ travel and Stephenson's valve gear, between the frames. To combat priming, which had been experienced with the new parallel domeless boilers, No. 98 received a boiler with the rear section of the boiler barrel coned, coming up flush to the Belpaire firebox. The total heating surface was 2143·04 sq. ft with a grate area of 27·22 sq. ft and a pressure of 200 lb/sq. in. The 2-8-0 had the same size boiler fitted, known as Standard No. 1, the same cylinders, and coupled wheels were 4′ 7½″ diameter. It was the first 2-8-0 in the country.

No. 99 (the 2-6-2T) also had cylinders the same size, and boiler of similar design known as Standard No. 4 with a total heating surface of 1517·89 sq. ft and a grate area of 20·35 sq. ft. The barrel was 11′ 0″ long and 4′ 5⅛″ diameter rising to 5′ 0½″ to join the firebox.

An experimental 0-4-0T which was not repeated, was No. 101 with outside cylinders 13″ x 22″, outside Joy's Valve gear, 3′ 8″ diameter wheels and was equipped with Holden's oil burning equipment. It was built in 1902 and received a new boiler with corrugated firebox in 1905, but was withdrawn in 1911.

Double framed 4-4-0s continued to be built with domeless boilers of various sizes.

The taper boiler series of standard sizes gradually developed and the coned portion in-

Fig. 219
GWR
Swindon 2204/1906
2-cylinder
Saint class
LADY OF LYNN

Fig. 220
GWR
Swindon 2210/1906
3802
COUNTY CLARE
The first
County class

Fig. 221
GWR
Swindon 2279/1908
No. 111
THE GREAT BEAR
rebuilt in 1924
as a Castle class.
The first *Pacific*
for a British railway

Fig. 222
GWR
Swindon 2370/1909
Churchward's 4-cylinder
4-6-0 No. 4026
KING RICHARD

creased. The firebox was waisted and tapered in plan from front to back. The regulator was housed on the front tube plate.

Standard boilers had been introduced, which were used on new locomotives as shown in the table.

Churchward had adopted a number of American ideas including his two piece cylinders and front bracing bars and was very interested in American testing methods. In 1903 he built a stationary test bed in the new erecting shop. It consisted of a large cast iron bed with five pairs of bearings, arranged to slide longitudinally so that adjustment could be made for any centres, for the locomotive's wheels to be tested. The axles in these bearings carry wheels fitted with steel tyres and on these tyres the locomotive ran. Drums fitted to the axles on which band brakes were applied to absorb all or part of the power developed by the locomotive. The power thus absorbed was transmitted by means of connecting belts to an air compressor to avoid wasting the energy.

Hydraulic brakes actuated by water pumps were so arranged to absorb just the necessary energy from the locomotive to keep it at a required speed.

The carrying wheels were 4′ 1½″ diameter with bearings 9″ diameter x 14″ long.

The locomotive was run on to the test bed on an elevated frame. The carrying wheels were then adjusted to suit the wheel centres and the elevated frame would then be lowered and the locomotive wheels engaged onto the carrying wheels. All wheels were engaged in addition to the coupled wheels. The footplate came up to a firing stage where the coal bunker was situated complete with weighing machine. Water tanks were sited over the firing platform with the necessary calibrations. Beneath the platform was a dynamometer to enable the drawbar pull to be measured. The test bed was completely equipped with all the necessary instruments for a complete programme of testing. The chimney was shrouded by an adjustable hood for the exhaust and also enabled ash etc. ejected from the chimney to be trapped.

It was originally intended to use this test bed for running-in repaired and newly built locomotives besides experimentally testing them, but the former idea was abandoned.

The excellent work of the French De Glehn compounds attracted Churchward's attention and he persuaded the board to buy one. This was agreed and in 1903 LA FRANCE (No. 102) appeared, which was similar to the Nord's *Atlantic* type. The object was to compare the performance of this compound *Atlantic* with his own designs. A further two cylinder 4-6-0 (No. 171) was built with a standard No. 1. boiler but as the French locomotive had a working pressure of 227 lb/sq. in. the No. 1 boiler was up rated to 225 lb/sq. in, the boiler dimensions being very similar. As an added comparison Churchward converted No. 171 to a 4-4-2 so that the adhesive weight was brought nearer to that of No. 102, although the compound had four cylinders: two HP 13⅜″ x 25″ and two LP 22″ diameter and the same stroke. Separate sets of Walschaert's valve gear were fitted. Two larger compounds were ordered, and were delivered in 1905. They were No. 103 PRESIDENT and No. 104 ALLIANCE.

Later all three were fitted with No. 1 standard boilers and this entailed fitting new outside pipes – which became a feature of the *Castle* and *King* classes in the 1920s.

The conclusions arrived at after extensive trials were that as far as fuel consumption was concerned, there was little difference. Oil consumption was heavier on the compounds, but the compounds were far superior in riding qualities due to the divided drive and the balanced reciprocating masses.

Further two cylinder 4-4-2s and 4-6-0s were built but the 4-4-2s were subsequently converted to 4-6-0s.

Churchward was so impressed with the smooth running of the compounds that he designed a four cylinder simple 4-4-2. The inside cylinders drove the leading axle and the outside cylinders, which were placed well back to avoid too long connecting rods, drove the rear pair of coupled wheels. They were all 14¼″ x 26″. Two sets of valve gear between the frames were used, rocking levers connected the outside valve spindles. There was no room for eccentrics and the gear took the form of a modified Walschaerts gear similar to that employed on Deeley's Midland Railway '999' class.

With protests from Derby, when the 4-6-0 version of No. 40 appeared in 1907, the special 'scissors' valve gear was not fitted and the Walschaert's type of gear was used, with the same type of rocking levers as for others being built.

No. 40 ran as a 4-4-2 until it was converted to a 4-6-0 in 1909. Ten four cylinder 4-6-0s had already been built in 1907 and ten more in 1908 until seventy-two completed the class in 1923. They were very successful locomotives, incorporating the best from the French *Atlantics* and equipped with the efficient No. 1 boiler. Superheating was introduced 1909–10 for the class, although the motion was between the frames they were very light on maintenance and the criticism levelled against these and other GWR four cylinder types regarding inaccessibility was unwarranted.

In 1904 an outside cylinder 4-4-0 was put in hand and altogether thirty-one were built up to 1912. They were fitted with the standard No. 4 boiler, standard cylinders and wheels, 18" x 30" and 6' 8½" diameter respectively. Like all the outside cylinder locomotives built during this period, they were very angular with apparently no regard for appearance, but when the front and rear platform ends were curved in more graceful lines it made all the difference. Front bogie type brakes were fitted at first but were later discarded as were the equalising beams. The 4-4-0s were named after counties, and a tank version was brought out in 1905 and was allocated to the London area and used on fast and semi-fast trains between London and Reading. They were known as 'County Tanks' and had a fine turn of speed. The tender versions were very rough riders with plenty of fore and aft movement.

In February 1908 the first and only GWR *Pacific* was completed at Swindon with a lot of publicity.

This was the first *Pacific* built in the British Isles and created intense interest. The arrangement of the cylinders, motion, bogie and driving wheels were similar to the 4-6-0 *Star* class. It had four cylinders 15" x 26" and the standard 6' 8½" diameter driving wheels. The large boiler produced a total heating surface of 3376·59 sq. ft and a grate area of 41·79 sq. ft. The boiler barrel was 5' 6" coned up to 6' 0" diameter and 23 ft long. The trailing end was carried on a pair of radial wheels with inside boxes. The tender was a departure from standard practice and was carried on two four wheeled bogies with bar frames and inside bearings. An improved water pick-up gear was fitted. It carried 3500 gallons of water and 6 tons of coal and weighed 45 tons 15 cwt in working order and the engine weighed 97 tons. Two sets of Walschaert's valve gear were fitted, with combining levers actuated by links from the inside crossheads. The outside cylinder valves were connected to the inside valve spindles by horizontal rocking levers.

The firebox was a special type of Belpaire, with four arch tubes extending from front to back and supporting the brick arch.

A Swindon superheater was fitted comprising eighty-four 1⅜" diameter x 21' 4" long tubes, giving a heating surface of 545 sq. ft.

In service its performance was disappointing and its route availability very restricted, being limited to the Paddington/Bristol line, having an axle load of 20 tons 9 cwt. Its weakest point was the trailing radial truck having inside bearings which frequently overheated. Steaming troubles also occurred with such lengthy tubes. Various modifications were made to the superheater and adjustments to the weight distribution. In 1924 it was converted to a *Castle* 4-6-0 with very little of the original locomotive incorporated.

Two smaller 2-6-2Ts were introduced by Churchward. The first, No. 115, was built in 1904 with 16½" x 24" outside cylinders and 4' 1½" diameter coupled wheels. What was to be the standard No. 5 boiler was fitted with the second ring of the boiler barrel coned. Only the prototype was built at Swindon and a further ten on order, were built at Wolverhampton in 1905–06. When renumbered in 1912 they became Nos. 4400–4410. The second class had 17" diameter cylinders and 4' 7½" diameter coupled wheels and was fitted with the same boiler class. One hundred and seventy-five were built from 1906 to 1929, the last hundred having large tanks and were fitted with superheated boilers from the start. Both types were very successful, the 44XX with their smaller wheels were used mainly on the hilly Cornish branch lines and the 45XX were used on branch and suburban trains and were noted for their good acceleration.

In 1908 a further class of outside framed 4-4-0s were built known as the *Flower* class. These were similar to the *Atbara* class after they had been rebuilt with coned boilers, and had 18" x 26" cylinders and 6' 8½" diameter coupled wheels. The standard No. 2 boiler was fitted. For a time they were engaged in express

Fig. 223 GWR Swindon 1922 Mixed traffic 5'8" 2-8-0 No. 4705

passenger work until replaced by succeeding types.

For dock shunting work, five 0-6-0STs were built in 1910 following closely the design of the Cornwall Mineral Railway tanks which had been absorbed much earlier. The boiler was designed to fit the older type and had raised round top fireboxes. The saddle tanks did not cover the smokebox and the bunker had a sloping back joined to an all over cab. These were the last locomotives to be built with saddle tanks.

Previous classes with saddle tanks, which were receiving Belpaire boilers during overhauls, were fitted with pannier tanks supported by brackets on the smokebox and firebox sides and supported on top by tee-irons passing over the boiler. Saddle tanks were not practicable for Belpaire fireboxes.

A 2-8-0T was built in 1910 – a rare wheel arrangement, designed for the heavy short haul coal trains in the coalfield areas of South Wales, it was in this area that most of them worked throughout their existence. They had two outside cylinders 18½″ x 30″ and coupled wheels 4′ 7½″ diameter and were fitted with the standard No. 4 superheated version boiler. They were very powerful locomotives and 205 were built, the last ten in 1940 as replacements for ten earlier ones which had been converted to 2-8-2Ts.

In 1911 Churchward introduced his top feed apparatus, up to this time the clack boxes had been fitted in various positions varying from the back of the firebox to the front of the boiler barrel, low down near the smokebox. Whatever position they were in, corrosion was set up varying in intensity according to position resulting in pitting of the internal barrel surfaces. By introducing his top feed, all this trouble was eliminated, the feed water passing through the clack valves fitted each side of the safety valves in the case of domeless boilers or as a separate casting fitted on a stool between the dome and chimney. The feed water was discharged onto trays in the steam space, allowing the water to disperse and at the same time causing the precipitated solids to deposit over a wide area.

His last standard class was a tender version of his large 2-6-2T with the standard No. 4 boiler. This was a mixed traffic locomotive with 5′ 8″ diameter coupled wheels and 18½″ x 30″ outside cylinders and they were so successful that a total of 342 were built from 1911 to 1932. This was one of the few instances when an outside contractor built locomotives for the GWR, and Robert Stephenson & Co. supplied thirty-five of these 2-6-0s in 1921–22. They were extremely adaptable, handling passenger trains comfortably at high speed and hauling heavy freight with ease.

Standard classes had now been provided for all types of traffic and building went ahead with his 2-6-0s, 2-8-0Ts, 2-8-0s, 4-6-0s and 2-6-2Ts.

Nevertheless Churchward was not one to sit back, as well he might. Two further 'experiments' were put forward. In 1913 a scaled down version of the *County* tank was built, No. 4600 with 5′ 8″ coupled wheels, a standard No. 5 boiler and 17″ x 24″ outside cylinders. Intended for light suburban work, its use was found to be limited and no more were built.

The other project was a large 2-8-0 mixed traffic locomotive built in 1919 using a standard No. 1 boiler, having outside cylinders 19″ x 30″ and 5′ 8″ coupled wheels. It was not successful in this form.

Eight more were built in 1922–23 with a new standard boiler No. 7 which increased the total heating surface from 2171·43 sq. ft to 2556 sq. ft and the grate area from 27·07 to 30·28 sq. ft. With this larger boiler which tapered from 5′ 6″ to 6′ 0″ the weight in working order was 82 tons with a maximum axleload of 19 tons 12 cwt. They were very impressive machines and although only nine were built (Nos 4700–4708) they took over the fitted freight trains between London and Plymouth and Wolverhampton and later took over a number of passenger turns when required and handled them with ease. Their route availability was limited at the time and this was the reason for not multiplying the class.

In 1916 the title of Chief Mechanical Engineer was introduced. Churchward assumed the title and in 1921 he officially retired but remained in the same railway residence adjacent to the works and practically every day crossed the line and walked round to see how things were progressing until that terrible moment on 19 December 1933 when fog and his impaired hearing drove him into the path of a down express with fatal results. Swindon had lost its father figure.

Churchward was undoubtedly the foremost locomotive engineer of his time and much has been written about his technical achievements. The tapered boiler, top feed, stream lined steam passages to get rid of the exhaust quickly, piston valves used with long travel, valve liners and other design details are all of interest. His locomotives were on top of their jobs, but he was always looking for improvements. The efficent use of steam was the main preoccupation, he favoured high pressures and the economical use of the steam using early cut-offs. For the two cylinder classes Stephenson's valve gear was favoured and Walschaert's gear was used for the four cylinder classes. A great help was the jumper blast pipe top which prevented excessive pressure by varying the orifice according to the blast pressure.

Charles B. Collett took over the reins and traditions at Swindon and carried on the good work. The Railway Act of 1921 unlike other companies, did not affect the GWR except that a multitude of locomotives were accumulated, mainly from South Wales and the classifying and sorting out of the various types was the main preoccupation. Many were rebuilt with standard taper boilers where this was possible and many were scrapped. In August 1923 the first of the 4-6-0 *Castle* class came out, No. 4073 CAERPHILLY CASTLE. This was an enlarged *Star* class with 16″ diameter cylinders and with a new boiler, standard No. 8. It was the first standard class to have outside steam pipes, and a larger side-window cab. The boiler pressure was the standard 225 lb/sq. in. and when built was the most powerful express passenger locomotive in the country. Bogie brakes were still fitted but these were removed from all classes soon afterwards. 8″ piston valves were fitted and the front end was very similar to the *Stars.* This class turned out to be one of the most successful passenger types in the country. The publicity and partisanship of the 1925 trials with the L&NER *Pacific* class is history and the results astonished the locomotive departments of many railways who had frankly doubted C. B. Collett's claim that the *Castle* class coal consumption under test was 2·83 lb per dhp hour; this, when a figure anywhere from 4 lb to 7 lb was considered quite acceptable. Careful design of valves, valve events, large lap and port openings were the principal contributors to the success of the 2 and 4 cylinder passenger locomotives.

A 0-6-2T was designed to replace old stock needing replacement on the South Wales branches. This type was prolific and popular with the Welshmen and 200 were built from 1924 to 1928, fifty of which were ordered from Armstrong Whitworth & Co. 8″ piston valves were used, driven indirectly by inside Stephenson's valve gear. Standard cylinders, boiler and wheels were used and they were well liked in the 'Valleys'. Most of them spent their entire life in South Wales, and Caerphilly works, which had been modernised with a new erecting shop, was capable of dealing with this class together with the 2-8-0Ts and all the amalgamated railway's locomotives, which enabled West Yard works at Cardiff to be closed down.

More 45XX 2-6-2Ts were built together with additional mixed traffic 2-6-0s and 2-8-0Ts.

Resulting mainly from the rivalry of the other three groups, Collett embarked on the final development of the 4-6-0 and four years after the appearance of the first *Castle*, No. 6000

Fig. 224 **GWR** Swindon 1927 Collet *King* class **6005** KING GEORGE II with indicator shelter

Fig. 225 **GWR** Swindon 1930 2-8-0T 5287. Later converted to 2-8-2T (No. 7212)

KING GEORGE V appeared. To achieve this 'ultimate' many standards had to be put aside. The coupled wheels which had for so long been kept at 6′ 8½″ diameter, were reduced to 6′ 6″ diameter. The maximum cylinder diameter possible was 16¼″ and the stroke was increased from 26″ to 28″ and the bogie wheels reduced by 2 in to 3 ft diameter with outside bearings for the leading axle and inside for the rear bogie wheels to clear the inside cylinders. A new boiler was designed with a 16′ 0″ long barrel, with 250 lb/sq. in. pressure and 11′ 6″ firebox. It was a majestic and massive machine and represented the final development of Churchward's NORTH STAR.

Twenty were built in 1927–8 and a further ten in 1930. No. 6007 which was wrecked in the Shrivenham disaster of 1936 was replaced making an actual total built of thirty-one.

No. 6000 was sent to America in 1927 for the Baltimore and Ohio Railroad's centenary and was well admired. They were put on the West Country and Wolverhampton expresses.

One of the two cylinder *Saint* class 4-6-0s was fitted with 6 ft diameter wheels in 1924 and proved so successful in the mixed traffic category that a new class was created known as the *Hall* class. These were built with modifications from time to time right up to 1950, to a total of 330.

In 1926 a new 4000 gallon high sided tender was introduced which greatly improved the general appearance of the larger classes.

The period 1928 to 1941 was a period of renewal. Improved types were introduced to replace older locomotives but the same standards were adhered to in most cases, and interchangeability was not lost sight of.

A new 0-6-0PT was designed to replace earlier tanks and this class proved extremely versatile in handling goods, shunting, and passenger traffic, so much so that 863 were built from 1929 to 1950. During the period of trade depression 250 were built by outside locomotive builders which was the biggest departure from Swindon building policy. A feature was the high boiler pressure of 200 lb/sq. in. Ten were built with condensing apparatus for the London area and eighty (67XX series) had steam brakes only. An improved cab was fitted from 1933.

For replacing the older tanks fitted for operating auto-trains and for branch line working three classes were built: the 54XX with standard 21 boiler and 5′ 2″ wheels; 64XX with standard 21 boiler and 4′ 7½″ wheels; the 74XX with standard 21 boiler and 4′ 7½″ wheels.

The 74XX was not auto fitted but all had 16½″ x 24″ cylinders.

A taper boiler 0-6-0 was then designed particularly for Central Wales and the old Midland and South Western Junction lines. They were fitted with a No. 10 boiler which had been specially designed for rebuilding various classes of Welsh tanks. 120 were built from 1930 to 1948. The motion was interchangeable with the new 0-6-0PTs.

To perpetuate a 0-4-2T design seemed archaic but the old Wolverhampton '517' class were still pulling and pushing auto trains, but were needing replacement. Seventy-five auto fitted 0-4-2Ts were built from 1932 to 1936 and twenty not auto fitted in 1933. The basic design was not altered but various details were modernised. Criticism was not long in forthcoming but why build heavier and more powerful machines with limited route availability and surplus power when the 0-4-2T would do the job?

To replace the *County* tanks a standard 2-6-2T was built but with an increased boiler pressure of 225 lb/sq. in and to keep the same dimensions high tensile nickel steel was used for the boiler plates. The London suburban services were then accelerated and the 61XXs easily coped with the faster trains. Seventy were built 1931–35.

W. A. Stanier who had been principal assistant to C. B. Collett was appointed CME of the London Midland and Scottish Railway in 1932.

Two further 2 cylinder 4-6-0s were built. The *Grange* class was built in 1936, parts from the mixed traffic 2-6-0s being used, and they were a 5′ 8″ version of the *Hall* class. The other, the *Manor* class was a lighter class intended for the Cambrian section with another standard boiler (No. 14) and they too used parts from the 2-6-0s.

Various combinations were tried with the rebuilding of some of the larger 2-6-2Ts but the outbreaak of war in 1939 curtailed this programme. For additional heavy freight traffic a modernised 2-8-0 was put in hand in 1938 with outside steam pipes and side window cabs. Eighty-three were built up to 1942 when a series of LMS 2-8-0s were put through, ordered by the government – so the Stanier touch returned to Swindon.

C. B. Collett retired in 1941 having carried on the Churchward tradition but not allowing standardisation to become hide-bound.

As a result of experiments with cab signalling apparatus, carried out in 1906, the GWR developed a type of automatic train control and the lines between Paddington and Slough were fitted with the necessary ramps. Over 150 locomotives were fitted with the apparatus, but it was not until the 1930s that it was fully utilised and by 1939, two thousand six hundred and fifty miles of track were so equipped. The majority of classes, except various absorbed companies locomotives, had the apparatus installed and Collett was instrumental in this programme of safety measures. His *Castle* and *King* 4-6-0s were acknowledged as notable designs and the latter, after additional developments, especially the modifications to the superheating, made them fine performers, but it was left to a *Saint* class to make the inaugural run of the 'Cheltenham Flyer' on 3 July 1923. This was, for a long time, the fastest start-to-stop run in Great Britain – running the 77·3 miles from Swindon in 75 minutes.

C. B. Collett instigated many improvements in building and repair techniques and two important features was the Zeiss optical gear for lining up the frames and cylinders, and the horncheek grinding machine which cut out hours of labour and did a more accurate job in grinding the horn cheeks parallel and square. More up to date machines were installed in the boiler shop for drilling, reaming and tapping the firebox shells. The auxiliary works at Caerphilly and Wolverhampton were also modernised.

F. W. Hawksworth was Collett's successor and, unlike previous members holding this position, had not spent all his life on the GWR, but on the other hand he had long been associated with Collett and had been responsible for the detail design of the *King* class and subsequent classes.

In 1944 he introduced a number of modifications to the *Hall* class. The bogie had plate frames; the main frames were in one piece which necessitated the redesigning of the cylinders. The standard No. 1 boiler superheater was altered to a 3 row type giving a larger element surface area. The series commenced at No. 6959.

Hawksworth was responsible for the final development of the two cylinder 4-6-0. A new standard No. 14 boiler was fitted, with a boiler pressure of 280 lb/sq. in., non-standard coupled wheels at 6′ 3″ diameter, and 10″ diameter piston valves. The boiler had twenty-one flues with a total of eighty-four superheater elements (1¼″) giving a higher degree of superheat. The 18½″ x 30″ cylinders were bolted to each frame instead of sitting over the frame extensions as on previous two cylinder classes. The first, built in 1945, had an experimental double chimney of large dimensions which was later removed and the whole class of thirty built up to 1947 were all fitted with more slender double chimneys in BR days. The tractive effort of 32,580 lb at 85% boiler pressure made them more powerful than the *Castle* class, on paper, but for economy reasons the pressure was reduced to 250 lb/sq. in. later on. A new flat-sided 4000 gallon tender was introduced at the same time and was fitted to all thirty.

Other new construction proceeded with *Castles, Halls, Manors*: taper boiler 0-6-0s, 0-6-0PTs and 2-6-2Ts.

Only two further new designs were put in hand, one in 1947 for a taper boiler 0-6-0PT. This class was a logical development of the 57XX domed boiler 0-6-0PT. A standard No. 10 boiler was fitted but only ten were built at Swindon, a further 200 being built by outside contractors. Those built at Swindon were superheated but the remainder were not. An im-

Fig. 226
GWR
Swindon 1937
Castle class
No. 5057
EARL WALDEGRAVE
on test plant

Fig. 227
GWR
Swindon 1013
COUNTY OF DORSET

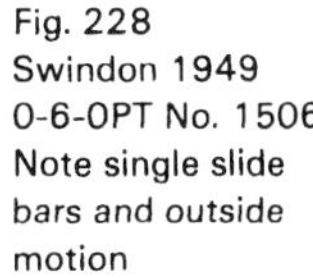

Fig. 228
Swindon 1949
0-6-0PT No. 1506
Note single slide bars and outside motion

Fig. 229 Swindon 1950 The last *Castle* No. 7037 SWINDON

portant alteration was the underhung springs on the rear axle, instead of the coil springs above the wheels in previous six coupled tanks which had been located in the cab and reduced the working space. They were numbered 3400–09, 8400–99, 9400–99.

The other, also a six coupled pannier tank had outside cylinders and outside Walschaert's valve gear to provide access to the motion and working parts. Welding was used to a great extent and no side or front platform were provided. A No. 10 taper boiler was provided. Although designed before the birth of British Railways they were not built until 1949.

On 31 December, 1947 the Great Western Railway was no more, having survived longer than any other British railway – a period of 110 years, with locomotive practice unbroken in tradition to the end.

The number of steam locomotives built at Swindon were as follows: Broad gauge 241; Standard gauge 5124; Narrow gauge 2; Rail car units 92; LMS 2-8-0 80; making a total of 5539 locomotives.

British Railways, Swindon

Standard classes continued to be built right up to 1955. Two new classes were introduced after nationalisation both were 0-6-0PTs, and the first of each type were built in 1949. The 15XX tank was a radical departure from usual GWR practice in that outside Walschaert's valve gear was fitted, no platforms were fitted and to provide the maximum accessibility the pannier tanks were shortened back beyond the end of the smokebox. A non-superheated standard No. 10 boiler was fitted and outside steam pipes were taken to the outside cylinders. The wheelbase was reduced to 12′ 10″ to enable a curve with a radius of three chains to be negotiated. They were heavier than the previous standard classes and weighed 58 tons 4 cwt in working order. They were mainly used for yard and carriage shunting. Only ten were built.

The 16XX tank was a modernised version of the old '2021' class and were intended to replace the older tanks where weight and clearance restrictions were in operation, such as the old Burry Port and Gwendraeth Valley line where they dealt with all traffic. One, No. 1646, travelled far afield and ended up on the Dornoch branch of the old Highland Railway. A total of seventy were built. Both types were designed by F. W. Hawksworth.

The first BR design to be put in hand was the Class 4, 4-6-0s, existing LMS flanging blocks being used for the boilers. However in the case of the Class 3 2-6-2Ts no suitable LMS design boiler could be found and so the flanging blocks of the Swindon No. 2 boiler were adopted.

The 2-10-0s were started in 1957, the boilers being supplied by Crewe and production proceeded very slowly because of a difficult labour situation and having to wait for parts.

The last steam locomotive to be built was No. 92220 which was steamed on 25 March, 1960. A competition had been organised for an appropriate name for this locomotive and the name selected was EVENING STAR. The first

GREAT WESTERN RAILWAY

Swindon Works

GW EVENING STAR had been built by Robert Stephenson & Co. and delivered in July 1839, and the second was a *Star* class of G. J. Churchward built in March 1907 at Swindon, so the name was full of history.

Locomotive repairs were carried out vis-à-vis with new diesel locomotive building and repairs.

No new building has taken place since 1965 but diesels are still repaired.

The estimated number of steam locomotives built at Swindon from 1846–1960 are: Broad Gauge (1846–1876) 242 including renewals; Narrow Gauge (1923) 2; Standard Gauge (1856–1960) 5720; making a total of 5964 locomotives.

Locomotive Superintendent: Daniel Gooch 1837–64.

Locomotive, Carriage and Wagon Superintendents: Joseph Armstrong 1864–77; William Dean 1877–1902; George Jackson Churchward 1902–16.

Chief Mechanical Engineers: George Jackson Churchward 1916–21; Charles B. Collett 1922–41; F. W Hawksworth 1941–48.

It was decided on 6 October, 1840, after a special visit of Brunel and Gooch to the environs of Swindon, a hamlet of less than 2000 people, that the principal locomotive station and repairing shops 'were to be built near the junction with the Cheltenham and Great Western Union Railway at Swindon', and work was

Fig. 230 Swindon 1954 BR *Class 3* 2-6-0 No. 77001

Fig. 231 Swindon 1954 BR *Class 3* 2-6-2T No. 82030

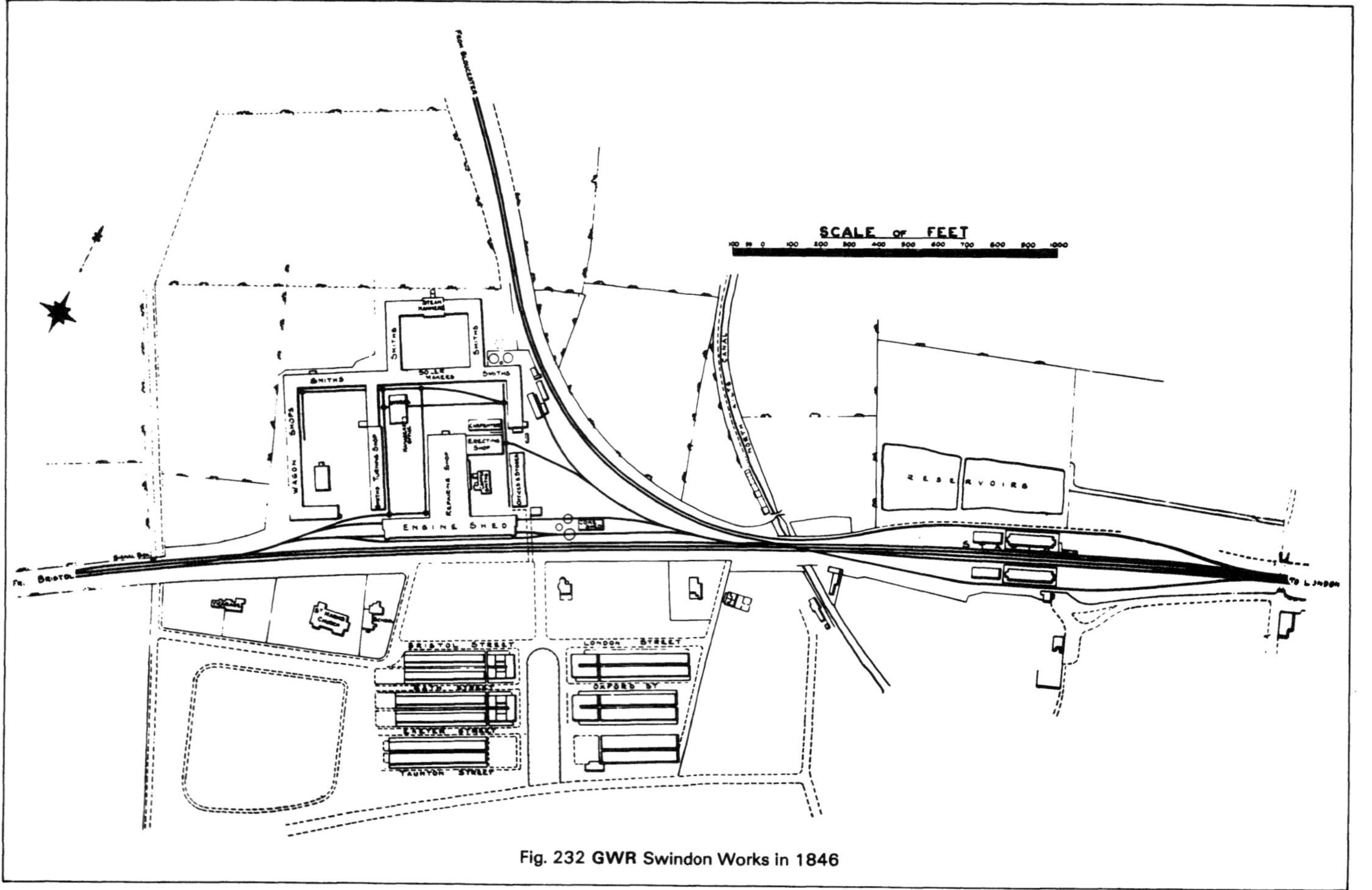

Fig. 232 GWR Swindon Works in 1846

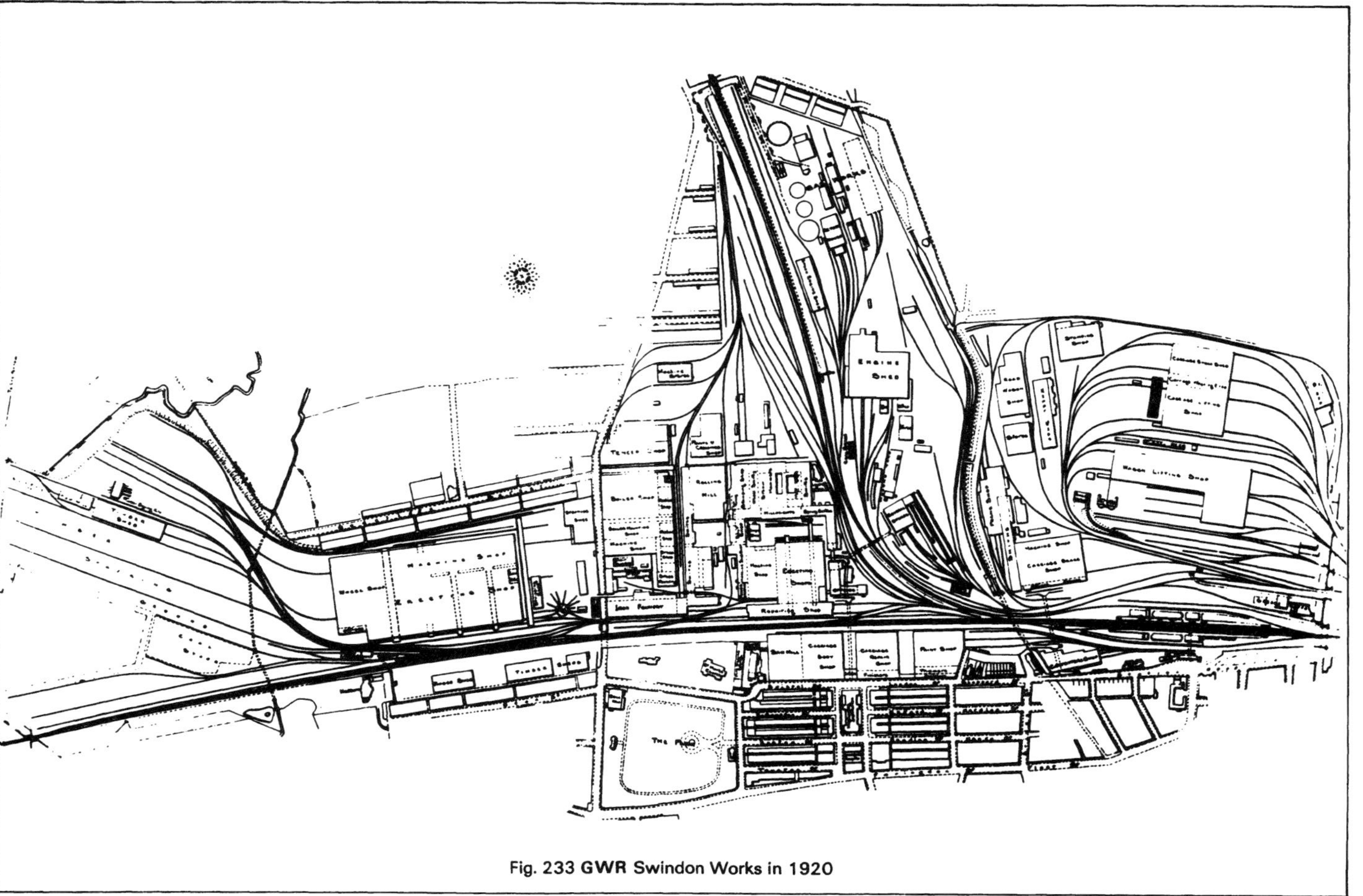

Fig. 233 GWR Swindon Works in 1920

Fig. 234 **Swindon** 1960 B.R. *Class 9* 2-10-0 No. 92220 EVENING STAR—last steam locomotive built by British Railways

Fig. 235
Original erecting shop,
'B' shed Swindon

Fig. 236
Swindon works plate
off 0-6-0ST 1286

started soon afterwards in building the works and the township to house the work people. The original erecting shop took eighteen locomotives. The works then occupied 14½ acres. Extensions and new shops were soon required and in 1854 a start was made, including new erecting shops. At first broad gauge locomotives were built and repaired, and the first standard gauge locomotives built in 1855 had to be sent on wagons to the northern division. Mixed gauge rails were then laid and one of the biggest problems was the reception of all broad gauge stock in May 1892 at Swindon. Temporary sidings were laid down west of the works to receive them and those that were convertible were dealt with as quickly as possible to get them into service again. The site of these sidings is now covered by the large 'A' shop constructed in 1902. By this time the locomotive, carriage and wagon works employed over 12,000 work people, compared with 4,500 in 1876.

The 'A' Shop consisted of fitting, boiler, wheel, and erecting shops, and had an area of half a million square feet, and the test plant was also located in this area. The 'B' shop at the other end of the works was still used for the repair of the smaller types. In the 'A' shop five new locomotives were laid down simultaneously. Hand fitting was eliminated where possible by machining parts to finer limits (such as the boring of axleboxes, grinding horns etc.) cutting down the tedious scraping operations previously necessary.

The disposition of locomotive pits favoured the traverser type of shop, the traverser feeding pits on both sides. Two such traversers were installed in 'A' shop and one smaller edition in 'B' shop.

GWR SWINDON

Date	Type	Class	Cylinders inches	D.W. ft in	Heating Surface sq ft			Grate Area sq ft	WP lb/sq in	Weight WO ton cwt
					Tubes	F' Box	Total			
J. ARMSTRONG										
1866	2-2-2	Sir Daniel	17 x 24	7 0	1105	98	1203	16 6	140	29-13
1866	0-6-0	Std Goods	17 x 24	5 0	1105	98	1203	16 6	140	29-18
1869	2-4-OT	Metro	16 x 24	5 0	988	92	1080	15 75	140	33-4
1869	2-4-0	481'	16 x 24	6 1	971	92	1063	15 97	140	27-4
1870	0-6-OST	1076'	17 x 24	4 6	1062	98	1160	16 85	140	37-14
1873	2-2-2	Queen	18 x 24	7 0	1155	128	1283	18 75	140	33-10
W. DEAN										
1878	2-4-OT	1401	16 x 24	5 0	1064·5	88·3	1152·8	14 52	140	36-14
1878	2-2-2	157 (a)	18 x 24	7 0	1099·27	115·0	1214·27	19 3	140	34-14
1883	0-6-0	2301	17 x 24	5 0	1079	113·6	1192·6	16 4	140	33-1
1884	2-4-0	3201 (b)	17 x 26	5 1	1106·57	103·29	1209·86	15 2	140	36-0
1890	0-6-OST	1854	17 x 24	4 6	1262·9	107·81	1370·71	17 2	150	43-8
1894	4-2-2	3031 (c)	19 x 24	7 8½	1434·27	127·06	1561·33	20 8	160	49-0
1895	4-4-0	3252 (d)	18 x 26	5 7½	1285·58	112·61	1398·18	19 0	160	46-0
1897	4-6-0	36	20 x 24	4 6	1402·06	115·83	1517·89	30 5	165	59-10
1897	4-4-0	3292 (e)	18 x 26	6 8	1175·32	121·58	1296·9	18 32	180	52-3
1899	4-4-0	3332 (f)	18 x 26	5 8	1538·06	124·96	1663·02	21 45	180	49-16
1899	4-6-0	2601	19 x 28	4 7½	1712·88	166·8	1879·68	32 19	180	60-14
1900	4-4-0	3373 (g)	18 x 26	6 8½	1540·18	124·10	1664·28	21 28	180	51-12
1902	2-4-2T	11 (h)	17 x 24	5 2	1436·76	124·89	1561·65	21 35	180	64-13
1902	4-6-0	100 (i)	18 x 30	6 8½	2252·37	157·94	2410·31	27 62	200	67-16
1903	2-6-0	2611	18 x 26	4 7½	1689·82	128·3	1818·12	20 56	200	56-15
1903	4-4-0	3710 (j)	18 x 26	6 8½	1689·82	128·3	1818·12	20 56	200	55-6

(a) Known as 'Cobham' class. (b) Known as 'Stella' class. (c) Known as 'Achilles' class. (d) Known as 'Duke of Cornwall' class. Old number. (e) Known as 'Badminton' class. Old number. (f) Known as 'Bulldog' class. Old number. (g) Known as 'Atbara' class. Old number. (h) Renumbered 3600. (i) Renumbered 2900. (j) Known as 'City' class.

Date	Type	1st No	Cylinders inches	D.W. ft in	Heating Surface—sq ft: Tubes	S'heat	F'box	Total	Grate Area sq ft	WP lb/ sq in	Weight WO ton cwt	*	
G. J. CHURCHWARD													
1903	2-8-0	97	18 x 30	4 7½	1988·65		154·39	2143·04	27·22	200	68 6	1	(a)
1903	4-6-0	98	18 x 30	6 8½	1988·65		154·39	2143·04	27·22	200	68 6	1	(b)
1903	2-6-2T	99	18 x 30	5 8	1396·58		121·31	1517·89	20·35	195	72 3	2	(c)
1904	4-4-0	3473	18 x 30	6 8½	1689·82		128·3	1818·12	20·56	200	55 6	4	(d)
1904	2-6-2T	115	16½ x 24	4 1½	1176·9		95·7	1272·6	16·83	165	55 15	5	
1905	4-4-2T	2221	18 x 30	6 8½	1396·58		121·31	1517·89	20·35	195	75 0	2	
1906	4-4-2	40	4/14¼ x 26	6 8½	1988·65		154·39	2143·04	27·07	225	74 10	1	(e)
1906	2-6-2T	3150	18½ x 30	5 8	1692·14		128·21	1820·35	20·56	200	78 16	4	
1906	2-6-2T	2161	17 x 24	4 7½	1176·9		95·7	1272·6	16·83	180	57 0	5	(f)
1907	4-6-0	4001	4/14¼ x 26	6 8½	1988·65		154·26	2142·91	27·07	225	75 12	1	
1908	4-6-2	111	4/15 x 26	6 8½	2697·67	545	158·14	3400·81	41·79	225	97 0	6	
1910	2-8-0T	4201	18½ x 30	4 7½	1228·02	215·8	122·92	1566·74	20·56	200	81 12	4	
1911	2-6-0	4301	18½ x 30	5 8	1228·02	215·8	122·92	1566·74	20·56	200	62 0	4	
1919	2-8-0	4700	19 x 30	5 8	1686·6	330·05	154·78	2171·43	27·07	225	77 14	7	(g)

(a) 2-8-0 Prototype Renumbered 2800. (b) 'Saint' prototype renumbered 2998. (c) '3111' prototype renumbered 3111. (d) 44xx prototype renumbered 4400. (e) Rebuilt 4-6-0 renumbered 4000. (f) 45xx prototype renumbered 4500. (g) 4701-8 built with new Std No. 7 boiler 4700 fitted with same boiler in 1921.

* Standard superheated boiler subsequently fitted.

C. B. COLLETT

Date	Type	Class	Cylinders inches	DW ft in	Heating Surface Sq Ft: Tubes	S'heat	F'Box	Total	Grate Area sq ft	WP lb/sq in	Weight WO ton cwt
1922	2-8-0	47xx	19 x 30	5 8	Standard No 7					225	82 0
1923	4-6-0	Castle	4/16 x 26	6 8½	Standard No 8					225	79 17
1924	0-6-2T	56xx	IC 18 x 26	4 7½	Standard No 2					200	68 12
1927	4-6-0	King	4/16¼ x 28	6 6	Standard No 12					250	89 0
1928	4-6-0	Hall	18½ x 30	6 0	Standard No 1					225	72 10
1929	2-6-2T	5101	18 x 30	5 8	Standard No 2					200	78 9
1930	0-6-0	2251	IC 17½ x 24	5 2	Standard No 10					200	43 8
1929	0-6-OPT	57xx	IC 17½ x 24	4 7½	Standard Domed PH					200	47 10
1930	0-6-OPT	54xx	IC 16½ x 24	5 2	Standard No 21 Domed					165	46 12
1931	2-6-2T	61xx	18 x 30	5 8	Standard No 2					225	78 9
1932	0-6-OPT	64xx	IC 16½ x 24	4 7½	Standard No 21 Domed					165	45 12
1932	0-4-2T	48xx	IC 16½ x 24	5 2	Standard Domed SS					165	41 6
1934	0-6-OPT	1366	16 x 20	3 8	715	—	73	788	10·7	165	35 15
1934	2-8-2T	72xx	19 x 30	4 7½	Standard No 4					200	92 12
1936	4-4-0	Earl	IC 18 x 26	5 8	1001	81·2	108	1190·2	17·0	180	49 0
1936	4-6-0	Grange	18½ x 30	5 8	Standard No 1					225	74 0
1938	2-6-2T	31xx	18½ x 30	5 3	Standard No 4					225	81 9
1938	2-6-2T	81xx	18 x 30	5 6	Standard No 2					225	76 11
1938	4-6-0	Manor	18 x 30	5 8	Standard No 14					225	68 18

F. W. HAWKSWORTH

Date	Type	Class	Cylinders inches	DW ft in	Tubes	S'heat	F'Box	Total	Grate Area sq ft	WP lb/sq in	Weight WO ton cwt
1944	4-6-0	6959	18½ x 30	6 0	1582·6	314·6	154·9	2052·1	27·07	225	75 16
1945	4-6-0	County	18½ x 30	6 3	Standard No 15					280	76 17
1946	4-6-0	Castle	4/16 x 26	6 8½	1799·5	313	163·5	2276	30·28	225	79 17
1947	0-6-OPT	94xx	IC 17½ x 24	4 7½	Standard No 10					200	55 7
† 1949	0-6-OPT	15xx	17½ x 24	4 7½	Standard No 10 (Sat)					200	58 4
† 1949	0-6-OPT	16xx	IC 16½ x 24	4 1½	Standard No 16 Domed					165	41 12

† Built under BR ownership.

GWR – STANDARD TAPER BOILERS

Std No	*	Boiler Barrel Min Dia ft in	Boiler Barrel Max Dia ft in	Boiler Barrel Length ft in	Heating Surface sq ft Tubes	Heating Surface sq ft S' Heat	Heating Surface sq ft F' Box	Heating Surface sq ft Total	Grate Area sq ft	Max WP lb/sq in	Classes Fitted To	Notes
1	AH	4 10¾	5 6	14 10	1686.6	262·62	154.78	2104	27·07	225	28xx 29xx 40xx 49xx 68xx	Std Nos 1 to 7 were introduced by Churchward, others by his successors
2	BA	4 5⅛	5 0½	11 0	1144·95	82.2	121.8	1348·95	20.4	200	3111 Class 41xx 5101 Class 56xx 61xx 81xx 2221 Class 33 xx	
3	CA	4 5⅛	5 0½	10 3	1069·42	75·68	121.8	1266.9	20.35	200	33xx 36xx	
4	DG	4 10¾	5 6	11 0	1349·64	191·79	128.72	1670.15	20.6	200	26xx 3150 Class 37xx 38xx 42xx 43xx 72xx	
5	EB	4 2	4 9½	10 6	992·51	77·64	94·25	1164.4	16.6	200	39xx 44xx 45xx	
6	FA	5 6	6 0	23 0	2596·97	398·52	158.51	3154	41.79	225	111 The Great Bear	
7	GA	5 6	6 0	14 10	2062·35	289.6	169.75	2521.7	30.28	225	47xx	
8	HA	5 2	5 9	14 10	1885·62	262.62	163.76	2312	30.28	225	4073 Class	* Each Std boiler had a key letter ie. Std No 3 was C. A second letter starting with A was added to indicate variations in No of tubes etc.
	HC	5 2	5 9	14 10	1670·1	393.2	163.32	2226.62	30.28	225	5098 Class 4 row S'heater	
10	KA	4 5⅛	5 0½	10 3	1069·42	75·68	102.4	1247.5	17.4	200	2251 Class 94xx (1st 10)	
	KB	4 5⅛	5 0½	10 3	1245·7	–	101.7	1347.4	17.4	200	84xx 94xx	
12	WA	5 6¼	6 0	16 0	2007·5	313	193.5	2514	34.3	250	60xx	
	WB	5 6¼	6 0	16 0	1818	489	194.5	2501.5	34.3	250	60xx 4 row S'heater	
14	ZA	4 7⅝	5 3	12 6	1285·5	190	140	1615.5	22.1	225	78xx	
15	OA	5 0	5 8⅜	12 7	1545	265	169	1979	28·84	280	10xx	

STANDARD DOMED BOILERS

Std No	*	Boiler Barrel Min Dia ft in	Boiler Barrel Max Dia ft in	Boiler Barrel Length ft in	Heating Surface sq ft Tubes	Heating Surface sq ft S' Heat	Heating Surface sq ft F' Box	Heating Surface sq ft Total	Grate Area sq ft	Max WP lb/sq in	Classes Fitted To	Notes
9	JA	4 5		10 3	960·85	75·3	103·85	1140	19·65	160	Absorbed Locos (1922)	
	JD	4 5		10 3	1074·7	–	103·85	1178·55	19·7	160	Absorbed Locos (1922)	
16	QA	3 10		10 1	877.2	–	79·5	956.7	14·9	165	16xx	
21	FA	4 3		10 6	1004.2	–	81·8	1086	16·76	165	54xx 64xx 74xx	
	SS	3 10		10 0	869.8	–	83·2	953	12·8	165	48xx (Later 14xx) 58xx	
	PH	4 5 & 4 3⅞		10 3	1075.7	–	102·3	1178	15·3	200	57xx	

BRITISH RAILWAYS–SWINDON

Date	2-10-0	4-6-0	2-6-0	0-6-0	2-6-2T	0-6-OPT(IC)	0-6-OPT(OC)
1948		6981-95 7008-17		3218-3219	4160-69	6760-65 7430-39 9662-72	
1949		6996-99 7900-06 7018-27			4170-79	1600-19 6766-69 9673-82	1500-09
1950		7028-37 7820-29 7907-29				1620-29 6770-79 7440-49	
1951		75000-15				1630-49	
1952		75016-19	46503-14		82000-19		
1953		75020-24 75030-49	46515-27				
1954		75025-29	77000-19		82020-31	1650-54	
1955		75065-77			82032-44	1655-69	
1956	92087-92	75050-52 75078-79					
1957	(92178-83 (92093-96	75053-64					
1958	92184-202						
1959	92203-17						
1960	92218-20						

69xx = Modified 'Hall' class also 79xx
70xx = 'Castle' class
78xx = 'Manor' class
16xx = 4' 1½'' class

67xx = 4' 7½'' steam brake only
74xx = 4' 7½'' passenger branch tanks
96xx = '8750' class

GREAT WESTERN RAILWAY
Wolverhampton

Locomotive superintendents: Joseph Armstrong 1854–64; George Armstrong 1864–96.

The Shrewsbury and Birmingham Railway had their locomotive repair shops and running sheds on the west side of Stafford Road where they were established in 1849. It was not until 1854 that the Great Western Railway arrived at Wolverhampton with broad gauge track from Birmingham. Mixed gauge track was provided for interchange traffic with the S&BR. In 1854 the S&BR was taken over by the GWR together with the Shrewsbury and Chester Railway, adding fifty-six locomotives to the company's stock.

Joseph Armstrong was in charge of the locomotives of the S&CR at Saltney, but upon the amalgamation he was kept on by the Great Western and assumed charge of all locomotive maintenance with headquarters at Stafford Road.

The policy at Swindon was the building of their own locomotives, and the same policy was extended to the Stafford Road works. This entailed considerable reorganisation, and re-equipping, and during this interval Swindon supplied new locomotives to the 'standard' gauge.

By 1859 the old running shed had been converted to an erecting shop and the old repair shop became a machine and fitting shop, with two pits for the erection of new locomotives remaining. The boiler shop and foundry were built on the site of the old goods yard. The first to be built were two of the 2-2-2 type to replace the old S&CR locomotives. The leading and trailing wheels had outside bearings and the 6′ 2″ driving wheels had inside bearings. The cylinders were 14½″ x 24″, and the boiler was domeless.

With the acquisition of twenty-one locomotives of the Birkenhead Railway in 1860 and the West Midland Railway stock in 1863 amounting to 131, the capacity of the works, even with the assistance of Worcester, was such that extensions to the works were essential and a further erecting shop was built adjacent to the broad gauge running sheds on the opposite side of the road together with fitting and machine shops and a smithy.

The erecting shop (for repairs only) was built on similar lines to the old erecting shop ('B' shed) at Swindon, with pits on each side of a central traverser.

Two more of the 2-2-2 type were built in 1860 and 1862, with running numbers 30 and 110, and then Joseph Armstrong designed his first 2-4-0 with double frames, domeless boilers, 16″ x 24″ cylinders and 6′ coupled wheels. Six were built in 1863–4. A further batch had domed boilers and chimneys with copper capping. They bore a strong resemblance to the 1862 2-4-0s built by Geo. England & Co. and designed by Daniel Gooch, the locomotive superintendent at Swindon, who resigned in 1864. Joseph Armstrong was appointed locomotive, carriage and wagon

Fig. 237 **GWR Wolverhampton** 4/1862 6′ 2-4-0 No. 110 as rebuilt in 1887

Fig. 238 **GWR Wolverhampton** 84/1868 0-4-2T No. 530 auto fitted. Built as saddle tank

Fig. 239 **GWR Wolverhampton** 700/1901 0-6-0ST No. 2096 domeless boiler

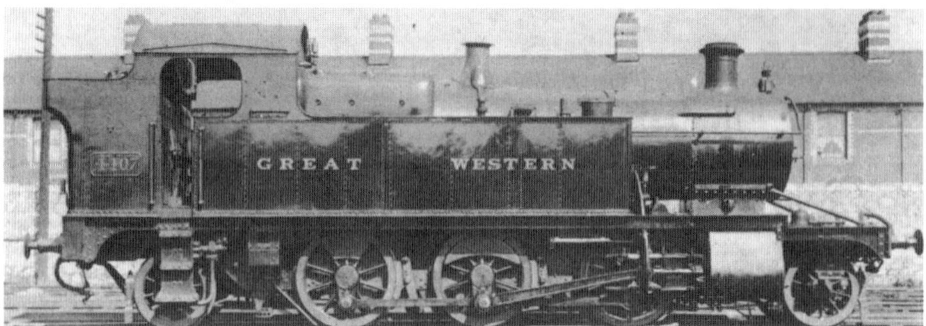

Fig. 240 **GWR Wolverhampton** 771/1906 2-6-2T No. 4407 at Newton Abbot

superintendent of the GWR. He was first to bear this title as J. Gibson the carriage and wagon superintendent had resigned at the same time as Gooch.

The vacancy at Wolverhampton was filled by Joseph Armstrong's brother George as the northern divisional locomotive superintendent. The independence of Wolverhampton works was carried on, in that, although controlled by Swindon, the design and construction of locomotives for the northern division was their own concern.

With such a variety of locomotive types and makers, a large proportion of the early work carried out was the extensive rebuilding of as many as possible.

In 1868 George Armstrong brought out the first 0-4-2T which was the forerunner of the well-known '517' class. The first fifty-four had saddle tanks and later ones side tanks, but all had side tanks eventually. There were many varieties and their construction continued until 1885 with a total of 156. They were not confined to the northern division but were dispersed all over the system. They were built in batches of twelve, as were most standard classes at Stafford Road. The other type built in quantity over the years was the 0-6-0ST; the first were double frame saddle tanks introduced by Joseph Armstrong. The tanks covered the boiler barrel only and the boilers were domeless with raised fireboxes. George Armstrong continued building similar locomotives but with domed boilers.

In 1871–2 twelve inside frame side tanks were built, some with condensing apparatus for the London area. A peculiarity of this class was that the clack boxes for the boiler feed water were both positioned on one side of the boiler.

In 1872 came the first of the 'large' saddle tanks with inside frames, 16″ x 24″ cylinders and a weight in working order of 34 tons 14 cwt, the tank holding 980 gallons.

It is interesting to note that five of these locomotives were sold direct (ex works) to the South Wales Mineral Railway.

Perhaps the best known class was the six coupled saddle tanks with inside frames, 15″ x 24″ cylinders, and weighing 30 tons 16 cwt. in working order. The 4′ 0″ wheels had cast iron centres. These 'small' saddle tank locomotives were built solely at Wolverhampton. They were known as the '850' class and a total of 144 were built between 1874 and 1895, followed by 140 more of the '2021' class which had 16½″ x 24″ cylinders, 4′ 1½″ diameter wheels and weighed 40 tons 13¾ cwt in working order with 1000 gallon tanks. These locomotives were extremely successful and, in the main, long lived; very few were withdrawn before the 1950s.

George Armstrong retired in 1897 at the age of 75 after a period of thirty-three years at Stafford Road. After this the independence of Wolverhampton vanished. The works continued of course, but under the control of the works manager. Even the '2021' class being built were now modified to take a Swindon domeless Belpaire boiler.

After the finish of the saddle tank orders, came an order to build some standard 2-6-2Ts as part of Churchward's standardisation plan. There were two classes with 4′ 1½″ diameter and 4′ 7½″ diameter coupled wheels, otherwise the two classes had the same boiler, side tanks, leading and trailing pony trucks. Ten 4′ 1½″ diameter class were built and twenty 4′ 7½″ diameter class being completed in 1908. These were the last locomotives built here although rebuilding and heavy repairs continued.

After the death of Sir Daniel Gooch in 1889 the subject was debated as to whether Swindon or Wolverhampton should become the principal works for the locomotive department. Wolverhampton had decided advantages, in practically sitting on coalfields, iron and steel manufacturers and with an abundance of the right type of labour. The big drawback was the congested area in which the works were placed with little room for expansion and so Swindon remained the headquarters. Nevertheless the large amount of locomotives requiring maintenance in the area resulted in a new set of shops being built in 1935 to replace the worn out and inconvenient layout as it was.

New locomotives were turned out in difficult conditions. This was particularly felt during the construction of the 2-6-2Ts, which were easy enough to build, but the problem was getting them out of the erecting shop when completed. To carry this out the two pony trucks and the buffers were removed, and temporary track was put down over the traverser into the wheel shop opposite. The locomotive had then to be barred through this shop. The trucks and buffers were then replaced.

During its term of independence the painting of the locomotives differed from Swindon. A darker shade of green was used with a bluish tinge, and the frames and wheels were painted a reddish brown. Chimney, Domes and Safety Valve covers were also of distinctive design and finish.

Stafford Road locomotives carried a standard GWR number plate, except the earlier ones, and the works number was cast in small numerals above the running number and the month and year built below.

In all 794 new locomotives and six rail motor units were turned out.

The works were completely closed on 1 June, 1964.

Lot No	Works Nos	Date	Type	Cylinders inches	DW ft in	Original numbers
	1-2	1859	2-2-2	14½ x 24	6 2 6 6	7 & 8
	3-4	1860 & 1862	2-2-2	15 x 22	6 (110)	30 110
	5-10	1863-4	2-4-0	16 x 24	6 0	111-114 115A 116A
	11-13	1864	2-4-0BT	14½ x 22	5 0	1A 2A 3A
	14-21	1864-5	0-6-0ST	16 x 24	4 6	302-309
	22-30	1865-6	2-4-0BT	14½ x 22	5 0	4A 11 177 344-349
	31-32	1866	0-6-0	15 x 24	4 6	34 35
	33-34	1866	2-4-0	15 x 24	5 0	108 109
A	35-40	1866	2-4-0	16 x 24	6 0	5A-8A 1010-1011 Renewals
A	41-46	1866-7	2-4-0	16 x 24	6 0	372-377
B	47-58	1867	0-6-0ST	16 x 24	4 6	238 1017-1027
C	59-70	1867-8	0-6-0ST	16 x 24	4 6	1028-1039
D	71-82	1868	0-4-2ST	15 x 24	5 0	1040-1051
E	83-94	1868-9	0-4-2ST	15 x 24	5 0	1052-1063
F	95-106	1869	0-4-2ST	15 x 24	5 0	1064-1075
G	107-118	1869	0-4-2ST	15 x 24	5 0	1076-1087
H	119-124	1869-70	0-4-2ST	15 x 24	5 0	1100-1105
H	125-130	1870	0-4-2T	15 x 24	5 0	1106-1111
J	131-142	1870	0-6-0ST	16 x 24	4 6	1040-1051
K	143-154	1870-1	0-6-0ST	16 x 24	4 6	1052-1063
L	155-166	1871	0-6-0ST	16 x 24	4 6	1064-1075
M	167-178	1871-2	0-6-0T	16 x 24	4 6½	633-644
N	–	–	–	–	–	Not used
O	179-190	1872	0-6-0ST	16 x 24	4 6	645-655 SWM1
P	191-202	1872-3	0-6-0ST	16 x 24	4 6	656 757 SWM2 CC2 SWM3 CC1 758-763
–	203-204	1872	2-2-2	17 x 24	6 6	70 74 Beyer Peacock Renewals
Q	205-216	1873	0-6-0ST	16 x 24	4 6	764-775
R	217-228	1873-4	0-4-2T	15 x 24	5 0	826-837
S	229-240	1874-5	0-4-2T	15 x 24	5 0	838-849
T	241-252	1874	0-6-0ST	15 x 24	4 0	850-861
U	–	–	–	–	–	Not used
V	253-264	1874-5	0-6-0ST	15 x 24	4 0	862-873
W	277-288	1875-6	0-4-2T	15 x 24	5 0	1154-1165
X	265-276	1875-6	0-6-0ST	15 x 24	4 0	987-998
Y	289-300	1876-7	0-6-0ST	15 x 24	4 0	1216-1227
Z	301-312	1876	0-4-2T	15 x 24	5 0	202-205 215-222
I	313-324	1877	0-4-2T	15 x 24	5 0	1421-1432
A2	325-336	1878	0-6-0ST	16 x 24	4 6	1501-1512
B2	337-348	1878-9	0-6-0ST	16 x 24	4 6	1513-1524
C2	349-360	1877-8	0-4-2T	15 x 24	5 0	1433-1444
D2	361-372	1879	0-6-0ST	16 x 24	4 6	1525-1536
E2	373-384	1879-80	0-6-0ST	16 x 24	4 6	1537-1548
F2	385-396	1880	0-6-0ST	16 x 24	4 6	1549-1560
G2	397-408	1881	0-6-0ST	16 x 24	4 6	1801-1812
H2	409	1880	0-4-0ST	15 x 24	4 1½	45
I2	Not used	–	–	–	–	–
J2	410-421	1881-2	0-6-0ST	16 x 24	4 1	1901-1912
K2	–	1877-87	0-6-0	16 x 24	5 0	132 135-6 139 141-6 310-13 315
L2	422-433	1882	0-6-0ST	16 x 24	4 1	1913-1924
M2	434-445	1883	0-4-2T	16 x 24	5 0	1465-1476
N2	Not used	–	–	–	–	–
O2	446-457	1883-4	0-6-0ST	16 x 24	4 1	1925-1936
P2	458-469	1884-5	0-4-2T	16 x 24	5 0	1477-1488 5' 1"DW 1483 onwards
Q2	470-481	1886-7	0-6-0ST	16 x 24	4 1	1937-1948
R2	482-493	1888	0-6-0ST	16 x 24	4 1	1949-1960
S2	494-499	1889	2-4-0	17 x 24	6 1	104 3227-3231
T2	500-511	1889-90	0-6-0ST	16 x 24	4 1	1961-1972
U2	Not used	–	–	-	–	–
V2	512-523	1890-1	0-6-0ST	16 x 24	4 1	1973-1984
W2	524-526	1890-1	0-6-0	17 x 24	5 1	316-318
X2	527-538	1891	0-6-0ST	16 x 24	4 1	1985-1996

Lot No	Works Nos	Date	Type	Cylinders inches	DW ft in	Original numbers
Y2	539-550	1891-3	0-6-0ST	16 x 24	4 1	1997-2008
Z2	551-562	1893-5	0-6-0ST	16 x 24	4 1	2009-2040
A3	563-574	1892	0-6-0ST	17 x 24	4 6	655 767 1741-1750
B3	575-594	1892-4	0-6-0ST	17 x 24	4 6	1771-1790
C3	595-604	1895	0-4-2T	16½ x 24	5 2	3571-3580
D3	625-634	1897	0-6-0ST	16½ x 24	4 1½	2021-2030
E3	605-624	1896-7	0-6-0ST	17 x 24	4 6	2701-2720
F3	635-644	1897-8	0-6-0ST	16½ x 24	4 1½	2031-2040
G3	645-664	1898-9	0-6-0ST	16½ x 24	4 1½	2041-2060
H3	665-684	1899-1900	0-6-0ST	16½ x 24	4 1½	2061-2080
I3	Not used	—	—	—	—	
J3	685-704	1900-1	0-6-0ST	16½ x 24	4 1½	2081-2100
K3	705-724	1902-3	0-6-OST	16½ x 24	4 1½	2101-2120
L3	725-744	1903-4	0-6-OST	16½ x 24	4 1½	2121-2140
M3	745-764	1904-5	0-6-OST	16½ x 24	4 1½	2141-2160
N3	775-794	1906-8	2-6-2T	17 x 24	4 7½	2161-2180 (a)
*	765-774	1905-6	2-6-2T	16½ x 24	4 1½	3101-3110 (b) (d)
1088	—	1905	0-4-0	OC 12 x 16	4 0	0853-0858 (c)

(a) Became 4500-4519 12/1921 45xx class. (b) Became 4401-4410 12/1912 44xx class. (c) Rail motor units car lot no. (d) Order transferred from Swindon (Lot 147).

CC: Carmarthen & Cardigan Rly. SWM: South Wales Mineral Rly.

GWR WOLVERHAMPTON

Date	Type	Class	Frames	Cylinders inches	DW ft in	Heating Surface sq ft			Grate Area sq ft	WP lb/sq in	Weight WO ton cwt
						Tubes	F'box	Total			
J. ARMSTRONG											
1859	2-2-2	7 & 8		14½ x 24	6 2	987	85·86	1072·86	13·58		25 18
1864	2-4-0BT	17		14½ x 22	5 0	641·81	71·21	713·02	12· 1	120	27 5
G. ARMSTRONG											
1867	0-6-0ST	1016	DF	17 x 24	4 7½	1045·25	91·75	1137	16·25	140	37 0
1868	0-4-2ST	517		15 x 24	5 0	734	74	808	12·33	140	27 2
1871	0-6-0T	633		16 x 24	4 6½	1208	92	1300	16·0	140	34 12
1874	0-6-0ST	1901		16 x 24	4 1	838	76	914	12·33	140	32 13
1878	0-6-0ST	1501		17 x 24	4 6½	1051	94	1145	16·0	140	39 8
1889	2-4-0	3226	DF	17 x 24	6 1	1181	99	1280	16·5	140	36 14
1895	0-4-2T	3571		16½ x 24	5 2	926·25	92·5	1018·75	15·0	140	40 5
1897	0-6-0ST	2021		16½ x 24	4 1½	926·25	92·5	1018·75	14·5	150	40 14
G. J. CHURCHWARD											
1905	2-6-2T	4401		16½ x 24	4 1½	1176·9	95·7	1272·6	16·83	165	55 15
1906	2-6-2T	4500		17 x 24	4 7½	1176·9	95·7	1272·6	16·83	180	57 0

All inside cylinders except 2-6-2Ts

GREAT WESTERN RAILWAY
Bristol

See BRISTOL & EXETER RAILWAY

GREEN, THOMAS & SON
Smithfield Ironworks, North Street, Leeds

Thomas Green purchased a site in 1848 in Leeds and founded the Smithfield Ironworks which initially was equipped for the manufacture of agricultural machinery.

Locomotive building commenced in the early 1880s when steam trams were in demand, and the potential was realised. The first tram locomotives were of a design combining the Kitson inside cylinder layout with Falcon type air condensers and a cab similar to the Wilkinson locomotive. Many were manufactured for William Wilkinson to his patent – thirty-nine in all. Later steam tram locomotives were built to Green's own design. The firm became a limited company about 1880 and were known as: THOMAS GREEN & SON LTD. and the manufacture of 157 tram locomotives was

THOS. GREEN

Order No	Date	Type	Cylinders inches	DW ft in	Gauge ft in	Customer	No	Name
1272	c1884	Tram	7¼ x 11	2 4	3 6	Mr Lee Australia (a)		
132	1889	0-4-0WT	OC 8½ x 14		3 6	Mr Lee Australia (b)		
161	1891	0-4-0ST	OC 10 x 18		Std	Mr Mathewson (c)		
170	1892				3 0	Mr Connell		
179	1892	2-4-2T	IC 13 x 20	3 6	5 3	Dublin & Blessington Tramway	7	
180	1892	0-4-4T	OC 14 x 20	3 6	3 0	Cork & Muskerry Rly	5	Donoughmore
192	1892	0-4-OST	OC 10 x 18		Std	Société Génerale de Paris (d)		
200	1893	0-4-4T	OC 14 x 20	3 6	3 0	Cork & Muskerry Rly	6	The Muskerry
201	c1894		8 x 12		2 6	Birch & Co		
218	1896	0-4-2T	IC 13 x 18	3 3	5 3	Dublin & Blessington Tramway	8	
219	c1896		7 x 12		2 6	Birch & Co		
220	c1896	0-4-0ST	OC 6 x 10		2 6	R Douse & Sons		
223	c1896	0-4-0ST	OC 10 x 18		Std	Lima Rlys Co		
224	c1896	0-4--0ST	OC 10 x 18		Std	Lima Rlys Co		
227	c1896	0-4-0ST	OC 10 x 16		Std	Griffiths Wharf Co London		IDA
228	c1897	0-4-0ST	OC 10 x 18		Std	Sociéte Génerale de Paris		
229	c1898	2-6-2T	OC 15 x 20	3 6	3 0	West Clare Rly	9	Fergus
234	1900	2-6-2T	OC 15 x 20	3 6	3 0	West Clare Rly	2	Ennis
236	1901	2-6-2T	OC 15 x 20	3 6	3 0	West Clare Rly	4	Liscannor
277	1901	0-4-0ST	OC 10 x 16		Std	Lima Rlys Co		
300	1902	0-4-0ST	OC 10 x 16		Std	Société Générale de Paris		
301	1902	0-4-2ST	OC 6 x 10		1 11½	Harrogate Waterworks Dept		Harrogate
312	1903	0-4-2ST	OC 7 x 12		1 11½	Harrogate Waterworks Dept		Claro
358	1903	0-4-0ST	OC 12 x 18		Std	Société Générale de Paris		
365	1904	0-4-0ST	OC 12 x 18		Std	Société Générale de Paris		
366	1904	0-4-2ST	9½ x 14		1 11½	Harrogate Waterworks Dept		Masham
367	1906	2-4-2T	IC 12 x 18	3 0¼	5 3	Dublin & Blessington Tramway	2	
439	1906	0-4-OST	OC 12 x 18		Std	Société Générale de Paris		
440	1907	0-4-OST	OC 5½ x 10		Std	Wm Cory & Son Ltd Kent		Hornchurch
441	1908	0-6-2ST	OC 10 x 16	2 6	2 0	Harrogate Gas Co		Barber
482	c1908	0-4-OST	OC 12 x 18		Std	Société Générale de Paris		
483	1908	0-4-OST	OC 12 x 18		Std	Société Générale de Paris		
484	c1908	0-4-OT	OC 5 x 7½		2 0	Robert Hudson & Co (g)		John
489	1908	0-6-OST	OC 8 x 10		1 11½	Robert Hudson & Co (e)		
490	c1913	0-4-OT	OC 5 x 7½		Std	Contracts Supply Co (f)		
625	1920	0-4-OST	OC 10 x 18		Std	Société Générale de Paris		

(a) Wilkinson Patent tram locomotive (b) Probably tram locomotive (c) Subsequently 'El Chiquito No 24' (d) Agents. Locos went to Peru (e) Had Robert Hudson Plates (f) At Kirton-in-Lindsey (g) for Rangoon

effected between 1885 and 1898. Most of them went to the tramways in Lancashire and Yorkshire, others to Birmingham, Coventry, Ireland, Dundee – as the table shows. Construction of this type of locomotive ceased in 1898, but a number of industrial and light railway tank locomotives were proceeded with up to 1920. These included three circular railway locomotives in 1895. Three interesting 2-6-2Ts were built (1898–1901), for the 3′ gauge West Clare Railway, with outside cylinders and inside frames and with a weight in working order of about 37 tons. They had long side tanks extending to the smokebox. They were the heaviest locomotives built by this firm.

Two 2-4-2Ts were built for the Dublin & Blessington Tramway which, although not true tram engines, had cabs and controls fore and aft; the fireman remaining in the rear cab, the front end cab only being used by the driver when travelling forward.

In all thirty-eight locomotives were built, from 1′ 11½″ to 5′ 3″ gauge, and all had tanks and no tenders. The last was built in 1920.

Over two hundred tram locomotives were built including the initial thirty-nine to Wilkinson's orders.

GREENWOOD & BATLEY LTD.
Albion Works, Leeds

After Loftus Perkins had his patent tramway locomotive (which was not a success) built by the Yorkshire Engine Company in 1874 for the Brussels tramways a larger one was built by this firm in 1878. It had a vertical boiler with a working pressure of 500 lb/sq. in. The triple expansion engine was geared to a jackshaft with a reduction of 4:1. The crankpins operated stirrups fitted to the outside coupling rods driving 2 ft diameter wheels.

Vertical condenser tubes were fitted on each side of the locomotive.

It was tried out on the Leeds tramways but apparently was no more successful than the previous one.

Fig. 241 Thomas Green 161/1891 for Mr. Matthewson EL CHIQUITO No. 24

GRENDON, THOMAS & CO.

Drogheda Iron Works

The foundry was probably established in 1835 and was an extremely versatile and prosperous concern, eventually extending its activities to shipbuilding, weighing machines, locomotive turntables, bridges and general engineering. At its busiest it employed between 600 and 700 work people.

After a number of years the foundry passed into the hands of a Scotsman named Mr McCoy who had risen from the ranks to become foreman under Thomas Grendon. From Mr McCoy the ownership was passed on to Messrs Frederick Smith & Co., and work was carried out until about 1885 when orders gradually declined and the works ultimately closed down.

Locomotive work commenced in 1844 when a standard gauge locomotive named FIREFLY, apparently built by the Butterley Co., was converted to 5′ 3″ gauge for the Dublin & Drogheda Railway. Grendons received their first order for three new locomotives from the same railway and the first was completed in January 1845, the second in May of the same year, followed by a third at the same time. They were D&DR No. 13 McNEILL; No. 14; and No. 15 HIBERNIA, all with 2-2-2 wheel arrangement, driving wheels 5′ 6″ and cylinders 14″ x 18″. Two more of the same type were ordered by the same company and the Midland & Great Western Railway ordered another two. They were D&DR No. 16 and No. 17 DROGHEDA and MGWR No. 7 DUNSANDLE and No. 8 VESTA, all delivered in 1847.

D&DR No. 16 was built as a tank locomotive, otherwise the remainder were generally similar to McNEILL. All had inside cylinders and inside frames, the fireboxes were copper and the tubes made of brass. Six wheeled tenders were built for six of them (with 1200 gallons of water). They had wrought iron wheels. Three more were ordered by the MGWR which were delivered in the same year.

They met with varying fortunes, most acquitting themselves well, and compared favourably with their English counterparts. Others were not so well put together and caused a considerable amount of trouble, a general weakness being that the frames were not substantial enough.

An order for four more of the same type came from the Dundalk & Enniskillen Railway but subsequently the third and fourth were built as 0-4-2s, together with two more of the same type for the GS&WR – all built in 1849. The latter were larger locomotives with 16″ x 24″ cylinders and 5′ 0″ diameter coupled wheels. Boilers were domed with round topped fireboxes and raised casings.

During this year (1849) the first mention of Mr McKay is made, which could also have been McCoy or MacKay – this is quoted because some sources state a change of name of the firm at some time to Grendon & McKay (or MacKay). This is not so, the name Thomas Grendon & Co being kept to the end.

In 1850 appeared the first of a class of 2-2-2Ts. This was for the Newry and Warrenpoint Railway, named and numbered VICTORIA No. 3. This was intended for light passenger traffic with cylinders 9″ x 14″ and 4 ft diameter driving wheels. It was awarded a Gold Medal (value £5)

Fig. 242 **Thomas Grendon** 2-2-2 VICTORIA for Newry, Warrenpoint & Rostrevor Rly, probably completed as a tank locomotive

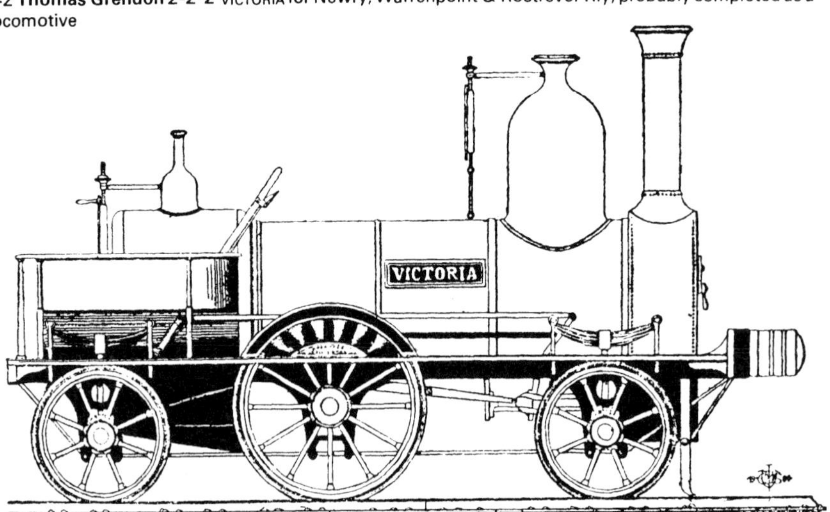

at the Dublin Exhibition of 1850. It is probable that VICTORIA was intended to be a tender locomotive but was altered to a tank before completion.

Grendons when they had no definite orders usually built locomotives for stock. One of these was sold to the MGWR in 1852, and was most likely built in 1851. It was a standard 2-2-2 and became No. 30 PALLAS. One other was disposed of in 1854 which at some later date was converted into an inspection vehicle as a 2-2-4T with accommodation at the rear. It was named FALCON.

In June 1855 a 2-4-0 was completed by the order of Robert Stephenson & Co. who had subcontracted this order to Grendons. It was Stephenson's works number 930 and believed to be a 'long boiler' type. It was sent to the Maua Railway of Brazil and could have been the only locomotive exported from Ireland although it is possible that one or two of Grendon's locomotives which were built but not traced could have been sent abroad.

Six 0-6-0s with 16" x 24" cylinders and 5 ft diameter wheels were built 1855–56; four went to the MGWR and two to the D&ER. They were the 'long boiler' type and were the largest built by Grendons. Copper fireboxes, brass tubes and wrought iron wheels were used and the framing was substantial being 22' 10" x $10\frac{1}{4}$" x $1\frac{1}{4}$" thick. Perhaps they had learned a lesson from the weak framing on earlier locomotives. One of these, supplied to the D&ER as No. 8 became GNR (I) No. 50 was fitted with a new boiler at the Dundalk workshops and was the last surviving Grendon locomotive, being withdrawn in November 1902 and broken up in December the following year.

From 1856 to 1860 a total of nine 2-4-0s were built and another was built in 1868. Eight were supplied to the MGWR, one to the D&DR and the last one to the BNCR, although the latter was officially a rebuild of a *Bury* 2-2-2 built in 1847 and converted to a 2-4-0 in 1855. The only existing parts re-used were the wheels, axles and parts of the motion.

Since 1860 orders had been falling off. One reason was that the railways were building in their own workshops, and also, Grendon's prices were usually far higher than their English competitors. Finally they were unable to meet the deliveries required, their layout preventing locomotives being laid down in any quantity. The usual time taken to build their locomotives was twelve months after receiving the orders.

The class of workmanship varied considerably; some were splendid machines and gave good service whereas others were the reverse.

Only one locomotive was completed in 1859, a 0-4-2 for the L&CR having 14" x 20" cylinders and 5 ft coupled wheels. The boiler pressure was 120 lb/sq. in. It became BNCR No. 33 and lasted until 1889.

Grendons, besides building new locomotives, did a considerable amount of repairs and rebuilding and supplied parts to the various railways. Their particular speciality was close grain iron casting and cylinder castings of

Fig. 243 **Thomas Grendon** VICTORIA in her final state

GRENDON, THOMAS & CO.

Date	Type	Cylinders inches	DW ft in	Railway	Rly No	Name		W'drawn
1/1845	2-2-2	14 x 18	5 6	D & DR	13	McNeill		1860
5/1845	2-2-2	14 x 18	5 6	D & DR	14			1859
5/1845	2-2-2	14 x 18	5 6	D & DR	15	Hibernia		1859
4/1847	2-2-2	14 x 18	5 7	MGWR	7	Dunsandle		1871
5/1847	2-2-2	14 x 18	5 7	MGWR	8	Vesta		1872
5/1847	2-2-2T	14 x 18	5 6	D & DR	16			1861
6/1847	2-2-2	14 x 18	5 6	D & DR	17	Drogheda		1891
7/1847	2-2-2	14 x 18	5 7	MGWR	9	Venus		1872 or 1869
12/1847	2-2-2	14 x 18	5 7	MGWR	10	Luna		1872 or 1869
12/1847	2-2-2	14 x 18	5 7	MGWR	11	Juno	S	1867
6/1848	2-2-2	14 x 18	4 0	D & ER	1		S	1888?
6/1848	2-2-2	14 x 18	4 0	D & ER	2		S	1874
3/1848	0-4-2	14 x 18	4 4	D & ER	3		S	1885
4/1848	0-4-2	14 x 18	4 4	D & ER	4		S	1850
1848	0-4-2	16 x 24	5 0	GSWR	51			1865
1848	0-4-2	16 x 24	5 0	GSWR	52			1865
7/1850	2-2-2T	9 x 12	4 0	NWRR	3	Victoria		*c* 1885
1851	2-2-2	14 x 18	5 6	MGWR	30	Pallas	(a)	1875
c 1851	2-2-2	14 x 18	5 6					
c 1851	2-2-2	14 x 18	5 6					
c 1851	2-2-2	10 x 12	4 0					
c 1853	0-6-0	15 x 24	4 7½					
1/1854	2-2-2T	14 x 18	5 6	MGWR	33	Falcon	(b)	1875
6/1855	2-4-0		4 10				(c)	
10/1855	0-6-0	16 x 24	5 0	MGWR	36			1881
10/1855	0-6-0	16 x 24	5 0	D & ER	7		GNR 49	1885
12/1855	0-6-0	16 x 24	5 0	MGWR	37			1881
2/1856	0-6-0	16 x 24	5 0	MGWR	38			*c* 1881
4/1856	0-6-0	16 x 24	5 0	MGWR	39			*c* 1881
4/1856	0-6-0	16 x 24	5 0	D & ER	8		GNR 50	1902
8/1856	2-4-0	15 x 20	5 7	MGWR	40			1880
9/1856	2-4-0	15 x 20	5 7	MGWR	41			*c* 1883
1856	2-4-0	15 x 20	5 6	D & DR	4			1888
1857	2-4-0	15 x 20	5 7	MGWR	2			1880
1857	2-4-0	15 x 20	5 7	MGWR	3			5/1880
1857	2-4-0	15 x 20	5 7	MGWR	4			*c* 1883
1857	2-4-0	15 x 20	5 7	MGWR	5			*c* 1883
1857	2-4-0	15 x 20	5 7	MGWR	6			*c* 1883
1858	T			NNWR	1	Rostrevor?		1892
1859	T			NNWR	2	Mourne?	S	*c* 1886
1/1859	0-4-2	14 x 20	5 0	L & CR			BNCR 33	1889
1860	2-4-0	15 x 20	5 7	MGWR	1			*c* 1883
1865	0-4-2?			NNWR	4	Drogheda	(d) S	1885
9/1868	2-4-0	15 x 20	5 6	BNCR	3		(e)	1902

BNCR Belfast & Northern Counties Rly D & DR Dublin & Drogheda Rly D & ER Dundalk & Enniskillen Rly L & CR Londonderry & Coleraine Rly MGWR Midland Great Western Rly NWRR Newry, Warrenpoint & Rostrevor Rly GSWR Great Southern & Western Rly

(a) Bought ex stock 1852 (b) Probably built as 2-2-2 or 2-2-2T and later converted to 2-2-4T (c) Sub-contracted from R. Stephenson & Co to Maua Rly of Brazil (R S No 930) (d) Dublin Exhibition 1865, bought 1866. (e) Rebuild of Bury 2-2-2 built in 1847 (S) Sold.

the highest quality were supplied to all railways located in the Dublin area in addition to those for the Drogheda & Dundalk lines. All cylinders required at the Inchicore works of the GSWR from 1862 to 1874 were supplied by Grendons and when Inchicore built their own iron foundry about 1874–5, the loss of orders was considerable. Cylinders were cast also for the MGWR until they lost the contract in 1886 to a rival firm, but the GNR continued to buy them until around 1893. Locomotive boilers were also built in numbers over the years.

In 1885 the firm lost the Guinness contract which was their mainstay, as at one time they had no less than 30 fitters and 50 assistants on this work alone. It is generally agreed that the cessation of this lucrative contract caused the firm's eventual closure, although the cancellation of contracts for the foundry as previously mentioned must have weakened the economy of the firm. About this date they closed down through lack of orders and after a number of years the works restarted on a much smaller scale by Mr Patrick Murphy, a local man, and about 1913 the works were acquired by Messrs Telford and Burn.

It is believed that Grendon's drawings were taken over by William Spence and no doubt those relating to the Guinness contract were included together with the general engineering business of Grendon & Co. apart from the foundry.

As will be gathered the precise number of locomotives built by Thomas Grendon & Co. cannot be obtained but the total must lie between forty-four and fifty. An added complication lies in the fact that they hired some locomotives from time to time to railway contractors, which might be returned, or sold and re-sold.

Thanks are due to Mr R. N. Clements of Celbridge, Co. Kildare who placed his records at my disposal and which made possible a more comprehensive list than my own previous attempts.

GROOM & TATTERSHALL LTD.
Station Works, Towcester, Northants

Although in some records this firm is credited with building one or two tank locomotives, there is nothing to substantiate this, and although they reconditioned some steam locomotives principally for the Heyford Iron Company no new engines were forthcoming.

GRYLLS & CO.
Llanelly

One locomotive was built for the Monmouthshire Railway and Canal Co. in 1847. It was probably built as a 0-8-0 with 3 ft diameter wheels and cylinders 16″ x 16″. It did not prove successful as it was up for sale in 1850. At some time it became a six coupled locomotive, the alteration probably made necessary by the original long wheelbase.

The Llanelly Railway and Dock Co. was opened in 1839 and the responsibility of providing locomotives was contracted for, first by Messrs Waddle & Wargrave and in 1847 Messrs Grylls & Co. took over.

Their terms of reference was for maintaining the existing locomotive stock and for building a new locomotive for the company.

This contract was rather suspect because W. Stubbs, the man in charge of locomotives, had a financial interest in Grylls & Co. and the railway Company dispensed with his services and shortly afterwards Grylls & Co. went bankrupt at the end of 1848.

Some reports state that the firm never completed a locomotive, although at the sale of Grylls & Co's effects a new boiler was purchased by the Llanelly Railway and fitted to the locomotive PRINCE OF WALES originally built by the railway company in 1843.

GUEST KEEN & NETTLEFOLDS LIMITED

Dowlais Works, Cardiff, Glamorgan

Construction of the first works on this site started in 1888 and of the second in 1934. The latter, subject to a good many additions and alterations since 1955 is the works as it now stands. It did not become known as East Moors Works until shortly after Nationalisation in 1967.

Two locomotives were built at these works GKN No. 8 a 0-4-0ST in 1916 and GKN No. 10 a similar locomotive in 1919. They had outside cylinders 15" x 22" and weighed 27 tons, and both were rebuilt, one in 1935.

GUEST KEEN & NETTLEFOLDS LIMITED

Cwmbran Ironworks, Monmouthsire

The following locomotives were built at Cwmbran. 1911 0-6-0ST ic GKN No. 1; 1914 0-6-0ST ic GKN No. 3; 1928 0-4-0ST oc GKN No. 2.

These works were originally the Patent Nut and Bolt Co. Ltd. which became Guest Keen & Nettlefolds Ltd. in 1900.

It is interesting to note that, at nearby Oakfield, a works was built in 1842 to produce railway locomotives and trucks, but due to the discovery of a rich seam of fireclay, firebricks were produced instead!

These works were owned by Mr Cyrus Hanson and bought by GKN in 1900.

Mr Robson of Dowlais Works designed a class A 0-6-0T for Cwmbran in 1909 noted as No. 1 CWMBRAN. It was not built at Dowlais and would probably be GKN No 1 of 1911. If this was the case it would have been a Side Tank similar to Dowlais No. 38 ARTHUR KEEN but the cylinders were $17\frac{1}{2}$" x 26", wheels 4' 0" diameter, heating surface 1285 sq. ft, grate area 20 sq. ft, wheelbase 12' 2" and boiler pressure 200 lb/sq. in.

Fig. 244 **Guest Keen & Nettlefolds** Cardiff 1916(?) at East Moors, Cardiff – No. 8

GUEST KEEN & NETTLEFOLDS LIMITED

Ifor Works, Dowlais, Glamorgan

The Ifor Works were built in 1839 by the Dowlais Iron Co. as an expansion to the original works and consisted of blast furnaces, forges, mills, foundries, and fitting shops and all maintenance work including some on locomotives was carried on here.

Until 1901 all the main overhauling and rebuilding of the many locomotives owned by the Dowlais Co. were sent outside to private locomotive builders, the majority going to Peckett & Sons of Bristol. From 1901 it was decided to carry out all major maintenance work and rebuilding at the Ifor Works and also to design and build new locomotives for their own use. It was about this time that Mr George Robson was appointed Locomotive Superintendent whose terms of reference were to carry out the new policy of the company. Robson came from the Great Western Railway's Swindon works having gone there from the Taff Vale Railway.

His first new locomotive design was for a 0-6-0 side tank with inside cylinders, Class A No. 38 ARTHUR KEEN which was built in 1906 for marshalling all incoming purchased material for weighing at Upper Branch. It was scrapped in 1937. In the following year an outside cylinder 0-4-0T (Class D) No. 7 was built for the Morlais Limestone Quarry traffic and Ifor Works. When the steelworks closed in 1930 it was transferred to Cardiff works and renumbered 5. In the same year a 0-4-0T with inside cylinders Class D was built for taking hot metal from the Blast Furnaces and cupolas to the metal mixers and another of the same class was built in 1909 for the same duties. They were No. 40 and No. 42 respectively.

No. 41 SANDYFORD a 0-6-0T with inside cylinders class A.1 built in 1908 was a larger and more powerful edition of the previous Class A and was used for all main line traffic from Upper to Lower works.

The Class A 0-6-0T had a round top firebox but the SANDYFORD and the 0-4-0Ts had Belpaire fireboxes which rested on the top of the frames. All Robson's locomotives were of distinctive design, the chimney incorporated an exaggerated rim almost twice the diameter of the chimney itself. The Class A.1 had coupling rods with a hinged joint near the centre crank pin enabling the locomotive to traverse the sharp curves of the system. A remarkable feature was the high boiler pressure ranging from 200 to 175 lb/sq. in. It is believed that SANDYFORD had a boiler pressure of 200 lb/sq. in later reduced to 180 lb/sq. in. and must have been one of the most powerful six coupled tanks in this country or elsewhere. Comparing its dimensions with the standard 57XX 0-6-0PT of the GWR reveals some interesting figures.

	Class A.1	57XX
Cylinders (inside)	19" x 26"	17½" x 24"
Wheels	4' 0"	4' 7½"
Boiler pressure (lb/sq.in)	200	200
Total heating surface (sq.ft)	1562	1178
Grate area (sq.ft)	21	15.3
Water (Gallons)	940	1200
Coal	15 Cwt	3 tons 6 cwt
Tractive effort 85% B.P.	33250lb	22510lb
Overall Length	29' 6"	31' 2"
Height	12' 5"	12' 3"
Weight W.O. tons	55	47½

Further Class D 0-4-0Ts were built in 1912 for charging the blast furnace bunkers with iron ore, coke and limestone and one in 1915 for attending to the finished steel traffic, and in this year George Robson left Dowlais leaving behind these well-built and impressive machines. His successor was David Hicks who had been Robson's assistant and further locomotives were built to Robson's well-tried designs.

In 1917 a 0-4-0T was built for the 3 ft gauge lines and used for attendance at the Bessemer Steel plant for taking bogies of cast ingot moulds to the ingot stripper.

The last new locomotive was built in 1920 No. 46 a Class D 0-4-0T making a total of nine new locomotives built at these works.

From this date until the steelworks closed in 1930 many locomotives were repaired and rebuilt including one of the Dowlais built 0-4-0Ts No. 44 which in 1927 was rebuilt as a 0-6-0T.

Considering that all material was manufactured at Ifor Works for the construction of these locomotives, and that the works main business was iron, steel and coal and not locomotive building the results were a magnificent testimonial to the engineers and craftsmen who were responsible for their building and maintenance.

Thanks are due to Mr John Owen of British Steel Corporation, Dowlais and Mr G. W. Kelland of GKN Group Services Ltd., London for much of the information this section contains.

GUEST KEEN & NETTLEFOLDS LIMITED

Works No	Built	Type	Cylinders inches	DW ft in	Total HS sq ft	Grate Area sq ft	Press lb/sq in	Coal ton cwt	Water galls	Weight tons	W' Base ft in	Class	No	Name	W'drawn
1	1906	0-6-0T	IC 18 x 26	4 3			180			55	12-2	A	38	Arthur Keen (a)	1937
2	1907	0-4-0T	OC 17½ x 22	3 6	1137	17·5	175	1-12	500	42	7-6	C1	7	(b)	
3	1907	0-4-0T	IC 14 x 18	3 6	1535	15·0	190	0-17	476	35		D	40	King George V (c)	1937
4	1908	0-6-0T	IC 19 x 26	4 0	1562	21·0	180	0-15	940	57½	12-2	A1	41	Sandyford	1938
5	1909	0-4-0T	IC 14 x 18	3 6	1535	15·0	200	0-17	476	35	7-6	D	42	Queen Mary (c)	
6	1912	0-4-0T	IC 14 x 18	3 6			200	0-17	476	35	7-6	D	43		
7	1915	0-4-0T	IC 14 x 18	3 6			200	0-17	476	35	7-6	D	44	Pant (d)	
8	1917	0-4-0T	OC 10½ x 16				200			13½	4-9	H	45	(e)	
9	1920	0-4-0T	IC 14 x 18	3 6			200	0-17	476	35	7-6	D	46		

(a) Reboilered 1914 to conform to class A1 (b) Transferred to Cardiff Steelworks *c* 1930 and renumbered 5 (c) Named 6/1912
(d) Rebuilt 0-6-0T in 1927 (e) 3' gauge.
Locomotive superintendents: c 1900–15 George Robson 1915–24 David Hicks 1924–30 Peter Jones. Steelworks closed 1930.

Fig. 245 **Dowlais** 1908 0-6-0T SANDYFORD

Fig. 246 **Dowlais** 1909 No. 42 QUEEN MARY with No. 40 KING GEORGE V (Built 1907)

HACKWORTH, TIMOTHY

Wylam Colliery

Timothy Hackworth served his apprenticeship in the engine shops at Wylam Colliery, terminating in 1807. His appointment to foreman blacksmith followed and he came under the control of William Hedley who was viewer at the colliery.

He took part in Hedley's experiments in 1811 concerning the adhesion of wheels to rails compared with tha cog and rack rail favoured by Blenkinsop.

The first locomotive at Wylam was built in 1813 by Thomas Waters who had a workshop at Gateshead but Christopher Blackett the owner of the colliery agreed to have the work carried out in the workshops at Wylam. Waters was assisted by Hackworth, and so the latter obtained first hand experience in locomotive construction. The locomotive itself had a single through fire tube in the boiler, one cylinder with the usual gearing and layout of Trevithick's locomotive of 1803. Water's locomotive was tried out and various modifications carried out but it was not successful and was removed in the following year. Undaunted Blackett gave orders for a further locomotive to be built. The responsibility for it rested on Hedley's shoulders but there is no doubt that Hackworth carried out most of the practical work.

This locomotive had a boiler with a single return flue, the two cylinders were vertical above the rear wheels and power was transmitted through longitudinal levers to which were attached the connecting rods, coupled to a geared shaft which through further gears turned the rail wheels. It was known as the 'Grasshopper' type and began work in 1813. It was found far superior to the Water's eegine, but due to its weight the cast iron plate rails were constantly broken. It was obvious that the engine weight had to be better distributed and an eight wheeled locomotive was put in hand and completed in 1813. Apart from the two pairs of bogies it was similar to its predecessor. It has been stated that one bogie was fixed and the other pivoted to assist in traversing curves. This locomotive ran until the plateways were converted to edge rails in the 1830s.

Although the above locomotives are ascribed to Hedley, Hackworth played a major part in their construction.

In 1815 he left Wylam (one reason given is that he was a devout preacher and refused to work on Sundays) and obtained a position at Walbottle Colliery as foreman smith. No locomotives were built here and numerous offers were received by Hackworth, including one from Robert Stephenson to look after his newly established Forth Street Locomotive works in Newcastle, whilst Stephenson was busy on other projects. Hackworth finally consented, moved to his new job in 1824 but only stayed a few months and returned to Walbottle. This period was one of unrest and indecision – he first agreed to contract to build some boilers for the Tyne Iron Company at Newcastle which he did, and then he set up his own works in Newcastle to make boilers, and stationary engines. In 1825 he was asked to accept the position on the newly formed Stockton and Darlington Railway as resident engineer including the maintenance of their locomotives and stationary engines. He did accept and at first was primarily concerned in the number of stationary engines necessary for incline haulage.

The S&DR works were built at New Shildon and were completed in 1826.

For details of the work carried out here please see the section on STOCKTON & DARLINGTON RAILWAY (see also WILLIAM HEDLEY).

HACKWORTH & DOWNING

Shildon

Timothy Hackworth entered into a contract with the Stockton and Darlington Railway in 1833 to work the railway and provide the necessary locomotives and to maintain them.

He had already built workshops on land acquired in Shildon and his brother Thomas was put in charge and remained with the firm until 1840, when Thomas entered into partnership with George Fossick to found the firm of Fossick and Hackworth.

The works at Shildon were then simply called TIMOTHY HACKWORTH.

The works were kept busy repairing the locomotives and other equipment for the S&DR. The first new locomotive built there was S&DR No. 24 MAGNET in 1835. It was a six wheeled locomotive with tender. Timothy Hackworth favoured six wheels, primarily to spread the weight on the track.

Other eminent builders still built four wheeled vehicles despite the advantages of the larger type. In the following year he built the first locomotive to run in Russia, for the Russian government. This locomotive was a big advance in design. Until this period he had stuck to the return tube boiler, but for the Russian order the boiler was of the conventional loco-type with raised firebox, dome on boiler, $1\frac{1}{2}''$ diameter tubes, outside frames and a wheel arrangement of 2-2-2. The cylinders were placed inside the frames beneath the smokebox.

In 1837 a similar locomotive was built for the S&DR, their No. 26 ARROW.

For the S&DR Hackworth reverted to outside vertical cylinders and No. 11 BEEHIVE and No. 12 BRITON were built in 1837 and 1842 respectively.

In 1838 he designed a new class, the first of which No. 15 TORY had outside inclined cylinders at the rear end. Others of this class were No. 17 WHIG, No. 10 AUCKLAND.

In May 1840 he terminated his contract with the Stockton & Darlington Railway, but continued building locomotives, stationary engines and other work. He built for the Clarence Railway and received an order for twelve single driver express locomotives from the London & Brighton Railway. They were designed by John Gray and the last one was completed in 1848. In 1849 he designed and built a 2-2-2 with inside cylinders, and outside frames to the leading and trailing wheels. The cylinders were 15″ x 22″. It represented what was to be Hackworth's last design, as he died suddenly in 1850.

There is no doubt that Hackworth was in the forefront of locomotive design at this time and the number of features which came from his inventive mind were many. The more important are worth listing: (1) Blast pipe, although the invention of this has been claimed by others; (2) Six coupled locomotives; (3) Spring balance safety valve; (4) Cylinders placed under the smokebox; (5) Eccentrics used for working the feed pumps; (6) used waste steam for preheating feed water; (7) Steam drying chamber in boiler; (8) Separate crankshaft hung in bearings fixed to frames; (9) Equalising beams introduced.

There are gaps in this list and it is probable that between forty and fifty locomotives were built from 1835 to 1849. The order for five increased to fourteen 'singles', placed by the London and Brighton Railway, and was the largest obtained, but building was very slow – so slow in fact that the company had the first one despatched to Brighton in an incomplete form and this one was finished in the company's own works. More of the order may have been dealt with in similar fashion. Their performance was far from satisfactory and, even when rebuilt at Brighton in various wheel arrangements, they gave a lot of trouble. Hackworth, as previously mentioned, was extremely inventive but most of the work carried out in his works was not of a high standard.

John Hackworth who had been works manager left to set up a works in Darlington. The Soho works were subsequently used as a gas works by the North Eastern Railway.

Much information has been obtained or verified from Robert Young's classic *Timothy Hackworth and the Locomotive*.

Fig. 247 **T. Hackworth, Shildon** 1838 SAMSON – note tender and position of fireman. Albion Mining Co. Nova Scotia, active until 1882

Fig. 248 **T. Hackworth, Shildon** 1842 Stockton & Darlington Rly. No. 8 LEADER – compare cylinders with Fig. 247

THE TWO HACKWORTH SANSPAREILS

On 25 April, 1829 came the publication of the proposed locomotive competition which was sponsored by the Liverpool and Manchester Railway with a prize of £500 for the best locomotive engine to be built in accordance with rules and regulations laid down by the railway company. The competition was open to anyone and the actual contest was to be carried out on 1 October in the same year, giving only five months to design and build such an engine.

Hackworth was at this time the resident engineer on the Stockton and Darlington Railway and although busy, building and repairing his company's locomotives, he was extremely keen to build a locomotive for this competition. He asked the committee for permission to build such a machine in the company's workshops and this was granted, but he had to finance the project himself. However, his resources were very meagre.

He spent all his spare time on the design and ordered the major parts from other locomotive builders – the cylinders from Stephenson's, the boiler from Longridge's, and the prolonged delivery of some of the parts resulted in a hurried assembly of the locomotive with no time left for a proper trial trip. It was taken by road to Liverpool. In these circumstances it was small wonder that in a very short time troubles were encountered during the trial of SANSPAREIL, as it was named, in the form of boiler leaks and then one of the cylinders fractured. These unfortunate occurrences were no fault of Hackworth's but of his suppliers, and, to crown all, his locomotive was over the stipulated weight of four and a half tons, which it exceeded by five hundredweights and was therefore disqualified. Nevertheless the judges agreed to let it run to see its capabilities, giving it a load of twenty-five tons, five tons more than for the others, including the ROCKET.

Before misfortune fell upon it SANSPAREIL hauled the load for two hours without any trouble with a maximum speed of about 15 miles per hour. In fact it was the most powerful locomotive at the trials. It was sold to the Liverpool and Manchester Railway and in 1831 again, to John Hargreaves who furnished the motive power on the Bolton and Leigh Railway.

The boiler of the SANSPAREIL was 6 ft long by 4′ 2″ in diameter and was made of wrought iron. The fire tube was in the shape of a 'U' with the firegrate at the beginning about 5 ft long with the tube taken to the other end of the boiler and returning to the same end as the grate and terminating in the chimney.

The vertical cylinders 7″ x 18″ were fitted at the rear end and connected to the trailing wheels. The two axles were coupled and the wheels were 4′ 6″ diameter, and were built with wooden spokes.

Hackworth's second SANSPAREIL was built twenty years later in 1849 and was the last one built by him, but circumstances had altered and this one was built in his own workshops and he endeavoured to produce his ideal locomotive as a culmination of all his past experience and endeavour. It was a 2-2-2 with inside cylinders 15″ x 22″, driving wheels 6′ 6″ diameter and a total weight in working order of 23¾ tons. It was full of his patents, and the boiler had all the longitudinal seams welded, as also were the firebox and smokebox flange plates. The total heating surface was 1188 sq. ft.

It was a handsome locomotive and its performance was such that Hackworth issued a challenge to Robert Stephenson to match his latest 2-2-2, built for the York, Newcastle and Berwick Railway, against SANSPAREIL but nothing came of it.

The 1829 locomotive ran until 1844 and is now in the Science Museum, Kensington.

HACKWORTH & DOWNING–SHILDON

Date	Type	Cylinders inches	DW ft in	Customer	No	Name	
1835	0-6-0	RVC 15 x 16	4 0	S & DR	24	Magnet	
1836				Deanery Colly			
1836	2-2-2	IC 17 x 9	5 0	Russian Govt			
5/1837	2-2-2	IC 17 x 9	5 0	S & DR	26	Arrow	Reno 51
6/1837	0-6-0	RVC 14 x 16	4 0	S & DR	11	Beehive	
8/1838	0-6-0	RVC15⅝ x 18	4 0	Nova Scotia		Samson	(a)
1838	0-6-0	RVC15⅝ x 18	4 0	Nova Scotia		Hercules	
1838	0-6-0	RVC15⅝ x 18	4 0	Nova Scotia		John Buddle	
11/1838	0-6-0	ROC/I 12½ x 18	4 0	S & DR	15	Tory	
12/1838	0-6-0	ROC/I 12½ x 18	4 0	S & DR	17	Whig	
1839	0-6-0	RVC14¼ x 16	4 0	Llanelly Rly & Dock Co		Victoria	
1839	0-6-0	RVC14¼ x 16	4 0	Llanelly Rly & Dock Co		Albert	
6/1839	0-6-0	ROC/I 14 x 18	4 0	S & DR	6	Despatch	Reno 10
6/1839	0-6-0	ROC/I 12 x 22	4 0	S & DR	10	Auckland	Reno 6
TIMOTHY HACKWORTH							
1840	0-6-0	14 x 22	4 0	Clarence Rly		Coxhoe	
1840	0-6-0	14 x 22	4 0	Clarence Rly		Evenwood	
c 1840	0-6-0	FOC/I 15 x 18	4 0	S Hetton Coal Co		Buddle	(b)
1840	0-6-0	FOC/I 15 x 18	4 0	S Hetton Coal Co		Kellor	
1840	0-6-0	FOC/I 15 x 18	4 0	S Hetton Coal Co		Wellington	
1840	0-6-0	FOC/I 15 x 18	4 0	S Hetton Coal Co		Prince Albert	
1840	0-6-0						
4/1840	0-4-0	IC 13 x 16	4 0	S & DR	41	Dart	
4/1842	0-6-0	ROC/I 14 x 18	4 0	S & DR	8	Leader	
9/1842	0-6-0	RVC 14 x 22	4 0	S & DR	12	Briton	
1845	0-6-0	ROC/1		Seghill Colly		John	
1845	0-6-0	ROC/1		Seghill Colly		Samson	
9/1846	2-2-2	IC 15 x 24	6 0	London & Brighton Rly	52		
11/1846	2-2-2	IC 15 x 24	6 0	London & Brighton Rly	53		
3/1847	2-2-2	IC 15 x 24	6 0	London & Brighton Rly	54		
4/1847	2-2-2	IC 15 x 24	6 0	London & Brighton Rly	55		
6/1847	2-2-2	IC 15 x 24	6 0	London & Brighton Rly	56		
8/1847	2-2-2	IC 15 x 24	6 0	London & Brighton Rly	57		
9/1847	2-2-2	IC15 x 24	6 0	London & Brighton Rly	58		
12/1847	2-2-2	IC15 x 24	6 0	London & Brighton Rly	59		
1/1848	2-2-2	IC15 x 24	6 0	London & Brighton Rly	60		
3/1848	2-2-2	IC14 x 24	6 0	London & Brighton Rly	50		
5/1848	2-2-2	IC15 x 24	6 0	London & Brighton Rly	51		
7/1848	2-2-2	IC15 x 24	6 0	London & Brighton Rly	49		
9/1848	2-2-2	IC15 x 24	6 0	London & Brighton Rly	52		
11/1848	2-2-2	IC15 x 24	6 0	London & Brighton Rly	55		
1849	2-2-2	IC15 x 22	6 6	North Eastern Rly	2	Sanspareil	
	0-6-0		4 0	Clarence Rly	1		
	0-4-0		4 0	Clarence Rly	2		
	0-4-0		4 0	Clarence Rly	3		
	0-6-0		4 0	Clarence Rly	4		
	0-6-0		4 0	Clarence Rly	7		
	0-6-0		4 0	Clarence Rly	9	Seymour	
	0-6-0		4 0	Clarence Rly	10	Pilot	
	0-6-0		4 0	Clarence Rly	13		

Also supplied locos to: Earl of Durham Colliery, Seaton Delaval Colliery, Cramlington Colliery, Stanhope & Tyne Rly (a) Albion Coal Mining Co (b) Renamed BRADYLL

Cylinders: IC = inside, RVC = outside at rear, vertical, ROC/I = outside at rear, inclined, FOC/I = outside at front, inclined.

HAGUE, JOHN

Cable Street, London

One engine is credited to John Hague, this was for the Stockton & Darlington Railway, their No. 42 named LONDON which was turned out in July 1839. It was a 0-4-0.

Cable Street is in the Shadwell district of London in an industrial area, but it is more likely that John Hague, a pioneer of refrigeration, acted as an agent for a locomotive builder.

This locomotive had inside cylinders 11″ x 16″ and 4′ 6″ diameter wheels. It had a 'D' shaped firebox 2′ $7\frac{1}{2}$″ long, 3′ 3″ wide, 3′ 7″ high. The boiler was 2′ 7″ diameter and 7′ 0″ long and had 71 copper and brass tubes $1\frac{3}{4}$″ diameter 7′ 4″ long. It had Gab motion. The tender had four cast iron wheels 2′ $10\frac{1}{2}$″ diameter with two wooden brake blocks, the brake being applied by a long hand lever. The water tank contained 491 gallons.

HAIGH FOUNDRY

Wigan

This foundry was established in 1810 for the purpose of manufacturing equipment for the mining industry including winding engines and pumping equipment.

In 1835 Mr E. Evans and Mr T. C. Ryley formed the HAIGH FOUNDRY CO. and took the works on a twenty-one year lease. The first manager was Mr Eckersley followed by Mr W. Melling.

At the time of the formation of the new company it was decided to add locomotive construction to its business activities. Over 100 were built from 1835 to 1856.

The first locomotives built were subcontracted by Edward Bury and a number of 0-4-0 and 2-2-0 types were built to his standard designs.

In October 1837 a 0-4-2 was delivered to the Leicester and Swannington Railway named AJAX. The order was received in April of the same year and it was delivered two months ahead of the delivery quoted. The cost of this locomotive was £1320.

A larger 0-6-0 for the same railway was delivered in September 1839 named HECTOR. Its weight in working order was 17 tons and the boiler had a total heating surface of 688 sq. ft and a grate area of 11·64 sq. ft.

Two interesting 7 ft gauge locomotives were built for the Great Western Railway in 1838.

Fig. 249 **Haigh Foundry** 1837 Leicester & Swannington Rly 0-4-2 AJAX

They were 2-2-2s with 6′ 4″ driving wheels geared up in the ratio of 3:2, and the cylinders were $14\frac{3}{4}$″ x 18″. They were not a success and the gearing was soon removed, and in 1839 or 1840 they were rebuilt with larger cylinders, new outside framing, valve gear and regulator. After this they performed quite satisfactorily. They are shown as equipped originally with Melling's patent which was probably the form of gearing. Mr John Melling, who at that time had patented a friction roller drive whilst on the Liverpool & Manchester Railway, had no connection with the W. Melling of Haigh Foundry. Four other broad gauge locomotives were built from 1851 to 1853 for the South Devon Railway. These were of Gooch's design with inside sandwich frames, four coupled 5′ 9″ diameter wheels and a leading bogie. They had saddle tanks and were intended for passenger service and were similar to the GWR *Corsair* class. They were very ugly with saddle tanks over the boiler barrel only, and cut away to clear the driving wheels. A circular sand box was mounted on the tank, with safety valves over the firebox. No protection was given to the driver and fireman, not even a weatherboard.

Some long-boilered 4-2-0s were built for Jones and Potts, for the Chester and Holyhead Railway, and for France, and they were all generally to the same design.

Three others of the same wheel arrangement were built for T. R. Crampton to his patents with 7 ft diameter driving wheels, outside valve gear and cylinders. After trials they were also sent to France.

Many of the earlier locomotives built, except for the Bury designs were equipped with Haigh's four eccentric valve gear.

The locomotive built for Mr J. Smith was for testing some 'improvements', but what these were is not known. It was eventually bought by the Grand Junction Railway and named TANTALUS.

The 0-4-0s for the Haigh Collieries were to the Haigh Foundry's own design. The Earl of Balcarres owned the collieries and had an interest in the foundry.

In 1855 two 0-8-0s were built to the order of the War Dept, and were designed to haul guns up an incline of 1 in 10 and were sent to the Crimea.

The only details handed down were that they had horizontal circular furnace (not a square firebox), the boiler was fed by two force pumps, two pairs of wheels were flangeless and the third pair of wheels were driven by outside cylinders.

In 1838 Haigh Foundry built a 4-2-0 for Bourne Bartley & Company. It had an unequal bogie with wheels 3′ 8″ and 2′ 10″ diameter and driving wheels 5 ft diameter. It was originally intended for America but was probaby rejected and was eventually sold to the North Union Railway.

Works numbers 48 to 53 require investigation. 48 is recorded both for the Manchester & Leeds and for the Preston & Wyre Railways but is more likely to be M&L No. 8. Nos. 73–77 have no customer recorded but East Lancashire Railway was supplied with three 0-6-0s and one 2-2-2 in 1847–48. Locomotive repairs were also carried out and a number were fitted with new boilers.

The lease expired in 1856 and as orders at this time were few and far between it was decided to give up the works, and locomotive building ceased.

Reference is made to two rack locomotives built by arrangement with John Blenkinsop by Robert Daglish and which may have been built by the Haigh Foundry. (See ROBERT DAGLISH)

HAIGH FOUNDRY

Works No	Built	Type	Cylinders inches	DW ft in	No/Name	Customer	
1	1835	0-4-0				For E. Bury (Bury Type)	
2	1835	0-4-0				For E. Bury (Bury Type)	
3	2/1836	2-2-2	12½ x 16	5 0	43 Vesuvius	Liverpool & Manchester Rly	
4	6/1836	2-2-2	12½ x 16	5 0	45 Lightning	Liverpool & Manchester Rly	
5	1836	2-2-2	12½ x 16	5 0	46 Cyclops	Liverpool & Manchester Rly	
6	1836	2-2-0				For E. Bury (Bury Type)	
7	1836	2-2-0				For E. Bury (Bury Type)	
8	1836	2-2-0				For E. Bury (Bury Type)	
9	1837	2-2-2	12½ x 18	5 0	2 Hecla	Grand Junction Rly	
10	1837	2-2-2	12½ x 18	5 0	4 Hecate	Grand Junction Rly	
11	1837	2-2-2	12½ x 18	5 0	5 Falcon	Grand Junction Rly	
12	1837	0-4-0				Irish Contractor—Dublin	
13	1837	0-4-0				Irish Contractor—Dublin	
14	1837	2-2-0	12 x 18	5 6	22	London & Birmingham Rly	(a)
15	1837	2-2-0	12 x 18	5 6	23	London & Birmingham Rly	(a)
16	1837	2-2-0	12 x 18	5 6	24	London & Birmingham Rly	(a)
17	10/1837	0-4-2	14 x 18	4 6	9 Ajax	Leicester & Swannington Rly	
18	11/1837	2-2-2				J. Wilkie Contractor—Preston	
19	12/1837	2-2-0	12 x 18		8	North Union Rly	(a)
20	1838	2-2-0	12 x 18		9	North Union Rly	(a)
21	1838	2-2-0	12 x 18		10	North Union Rly	(a)
22	1838	2-2-0	12 x 18		11	North Union Rly	(a)
23	6/1838	2-2-2				Paris & St Germain Rly	
24	7/1838	2-2-2				Paris & St Germain Rly	
25	9/1838	2-2-2	14¾ x 18	6 4	Snake	Great Western Rly	(b)(m)
26	8/1838	2-2-2	14¾ x 18	6 4	Viper	Great Western Rly	(b)(m)
27	1838	2-2-2				Paris & St Germain Rly	(b)
28	1838	2-2-2				Paris & St Germain Rly	(b)
29	1838	2-2-2				Birmingham & Gloucester Rly Contractor	
30	1838	2-2-2		6 0		Trial locomotive	(c)
31	1838	4-2-0		5 0		Bourne Bartley & Co	(d)
32	1839	2-2-0	12 x 18	5 6		Midland Counties Rly	(a)
33	1839	2-2-0	12 x 18	5 6		Midland Counties Rly	(a)
34	1839	2-2-0	12 x 18	5 6		Midland Counties Rly	(a)
35	9/1839	0-6-0	16 x 20	4 6	Hector	Leicester & Swannington Rly	
36	1839	2-2-2	11 x 18	5 0		Mr Day London	(e)
37	1840	2-2-2				Rennie & Co	(f)
38	1840	2-2-2				Rennie & Co	(g)
39	1840	2-2-2				For E. Bury	(a)
40	1840	2-2-2				For E. Bury	(a)
41	1840	0-4-0				For E. Bury	(a)
42	1840	0-4-0				For E. Bury	(a)
43	1840	2-2-2				Mr J Smith Bradford	(h)
44	3/1841	0-4-2	14 x 18	5 0	33	Manchester & Leeds Rly	
45	4/1841	0-4-2	14 x 18	5 0	34	Manchester & Leeds Rly	
46	4/1841	0-4-2	14 x 18	5 0	35	Manchester & Leeds Rly	
47	5/1841	0-4-2	14 x 18	5 0	36	Manchester & Leeds Rly	
48	4/1841	2-2-0		5 0	11	North Union Rly	
49	1841	2-2-2					
50	1841	2-2-2					
51	1841	2-4-0					
52	1841	2-4-0					
53	1842	0-6-0					
54	1842	0-4-0				Irish Contractor—Dublin	
55	1842	0-4-0				Irish Contractor—Dublin	
56	1842	0-4-0				Irish Contractor—Dublin	
57	1842	0-4-0				Irish Contractor—Dublin	
58	1842	0-4-0				Irish Contractor—Dublin	
59	1842	0-4-0				Irish Contractor—Dublin	
60	1842	0-4-0	12 x 18	4 0		Haigh Collieries	
61	1843	0-4-0	13¾ x 18	4 0		Haigh Collieries	

Works No	Date	Type	Cylinders inches	DW ft in	No/Name	Customer	
62	1843	0-4-0	14 x 18	4 0		Haigh Collieries	
63	1843					Built to Order	
64						Built to Order	
65						Built to Order	
66						Built to Order	
67						Built to Order	
68						Built to Order	
69						Built to Order	
70						Built to Order	
71						Built to Order	
72	1844	0-4-0				Haigh Collieries	
73	1845					No record	
74						No record	
75						No record	
76						No record	
77	1846					No record	
78	9/1847	4-2-0	12 x 18	5 0		Preston & Wyre Rly (L & Y5)	(a)
79	1847	4-2-0	OC 15 x 22	6 6		For Jones & Potts	(b)
80	1847	4-2-0	OC 15 x 22	6 6		For Jones & Potts	
81	1847	4-2-0	OC 15 x 22	6 6		For Jones & Potts	
82	1847	4-2-0	OC 15 x 22	6 6		For Jones & Potts	
83	1847	4-2-0				Paris	
84	1847	4-2-0				Paris	
85	1847	4-2-0				Paris	
86	1848	4-2-0		7 0		For T. R. Crampton	(j)
87	1848	4-2-0		7 0		For T. R. Crampton	
88	1848	4-2-0		7 0		For T. R. Crampton	
89	1849	2-2-2			202	L & NWR (5 DIV)	(k)
90	1849	2-2-2			203	L & NWR (5 DIV)	
91	1849	2-2-2			204	L & NWR (5 DIV)	(k)
92	1850	2-2-2				Liverpool Crosby & Southport Rly	
93	1850	2-2-2				Liverpool Crosby & Southport Rly	
94	1850	2-2-2				Liverpool Crosby & Southport Rly	
95	1850	2-2-2				Liverpool Crosby & Southport Rly	
96	1850	2-4-0		5 0		Kendal & Windermere Rly	
97	1851	2-4-0		5 0		Kendal & Windermere Rly	
98	1851	2-2-2				For Stock	
99	1851	0-6-0				For Stock	
100	1851	2-4-0				For Stock	
101	1852	2-2-2				Birkenhead, Lancs and Ches J Rly	
102	1852	0-6-0				Birkenhead, Lancs and Ches J Rly	
103	11/1851	4-4-0ST	IC 17 x 24	5 9	Priam	South Devon Rly	
104	2/1852	4-4-0ST	IC 17 x 24	5 9	Damon	South Devon Rly	
105	9/1852	4-4-0ST	IC 17 x 24	5 9	Falcon	South Devon Rly	
106	2/1853	4-4-0ST	IC 17 x 24	5 9	Orion	South Devon Rly	
107	1854	0-4-0				For G & J Rennie	
108	1855	0-4-0				HM War Dept	
109	1855	0-8-0	OC 15 x 20			HM War Dept	(l)
110	1855	0-8-0	OC 15 x 20			HM War Dept	(l)
111	1856	0-4-2					
112	1856	0-4-2					
113	1856	6W				For Mr Crampton	
114	1856	4-4-0		6 6		For Mr Crampton	

(a) To E. Bury's designs (b) Melling's Patent (c) On loan to Grand Junction North Midland & other railways, then sold to Birmingham & Gloucester Rly (d) Intended for America but sold to North Union Rly. (e) Sold to London & Greenwich Rly (f) To London & Brighton Rly BRIGHTON (g) To London & Brighton Rly SHOREHAM (h) For testing improvements sold to GJR TANTALUS (i) 79 to 85 'Long Boiler' Passenger type. (j) 86 to 88 sent to France. (k) 89 to 91 'Long Boiler' Passenger type. (l) For hauling guns, incline 1 in 10 in the Crimea. (m) 7' gauge.

HANCOCK, WALTER
High Road, Stratford, Essex

Hancock made many road steam carriages and patent boilers during the years 1824–36 and in 1840, he entered the field of railway locomotives, designing and building for the Eastern Counties Railway an engine after the style of his road carriages. It included his patent multi-chambered boiler, two vertical cylinders driving an independent shaft with a final drive by endless link chains running over graduated pulleys on both crankshaft and driving axles. The graduated pulleys acted as variable speed gear, having a neutral position which enabled the engine to work its feed pumps.

HARRINGTON IRON & COAL COMPANY
Cumberland

The original title of this firm was H. C. Plevins, changing to Bain & Patterson and again to Sir James Bain & Company until 1876 when it became the Harrington Iron & Coal Co Ltd., Harrington Ironworks about 1905.

A 0-4-0ST was built somewhere between 1876 and 1882 and was No. 8.

The works were closed in 1921 after being taken over by the Workington Iron & Steel Co. Ltd. in 1909.

No other information is available.

HARRIS, JOHN
Albert Hill Foundry, Darlington

John Harris leased a section of William Lister's works and built about twelve locomotives between 1863 to 1869.

Very little information has been found regarding Harris's products. They were rather primitive in design at the front end, inclined cylinders being fitted high up with frames shaped for bolting on. Two 0-4-0STs are illustrated in 'The Chronicles of Boulton's Sidings' in diagram form.

Joicey's locomotives although built much later were very similar in some details.

Both four and six coupled were built. The earlier ones had 'Hopetown Foundry' on the maker's plates and some of the later ones had 'Albert Hill Foundry' on the plates. The establishment was quite distinct from Messrs W. & A. Kitching.

When John Harris ceased manufacturing, this part of the works was taken over by Messrs Summerson, makers of railway switches, points and crossings.

The following have been identified:
1863 0-4-0ST VICTORY*;
1865 0-4-0ST DERWENT*;
1868 0-4-0ST BYRON*;
Also supplied Derwent Iron Co, with one or more.

* Contractors for Team Valley branch railway, subsequently all three sold to Pelaw Main Colliery.

Fig. 250 J. Harris *c*1865. Typical 0-4-0ST

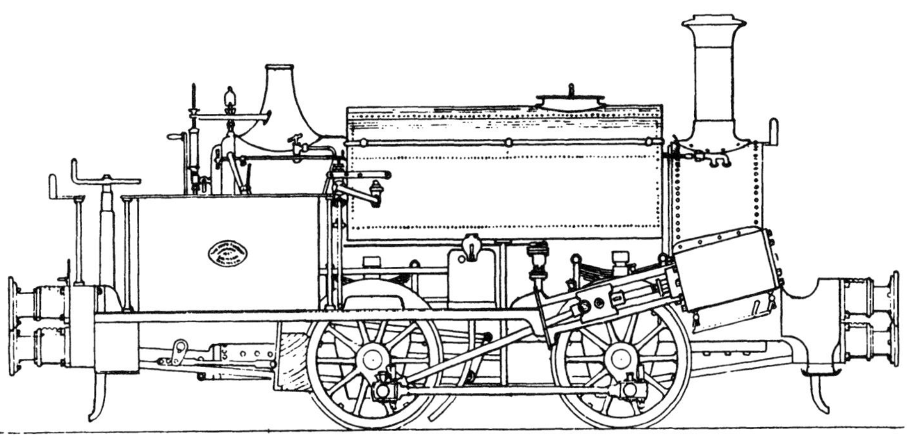

HARRISON & SON (HANLEY) LTD

Victoria Mill, Stanley, Endon, Stoke-on-Trent

This old established family business, making ceramic colours, glazes and other pottery material, built their own steam locomotive around 1885/6 and was probably constructed by Mr Harrison, grandfather of the present chairman of the company. The locomotive had a vertical boiler mounted on four wheels but details of the actual method of transmitting the drive to the track wheels are not known. It was subsequently rebuilt as an electrically driven locomotive for overhead wire traction probably about 1910 and was eventually scrapped in 1925 and replaced by a battery locomotive. The conversion could not have included many of the existing steam locomotive parts.

The railway was built from the factory, running across fields and across a swing bridge over the canal to a siding at Endon station where traffic was shunted from the main line from Stoke to Leek. This was closed down about 1958 and the track and bridge now removed.

The present company is Harrison Mayer Ltd.

Thanks are due to Mr B. Harrison who furnished the details.

HARTLEPOOL IRON WORKS

See THOMAS RICHARDSON & SONS LTD.

HARTLEY ARNOUX & FANNING

Stoke-on-Trent

This firm's main line of business appears to have been in the manufacture of machinery for the pottery and brickmaking industries and there is a record of some pottery machinery dated 6 December, 1877. They also undertook a certain amount of general engineering and millwrighting.

Locomotive building and other railway work was carried out on behalf of Kerr Stuart & Company of London and other firms, and the first record of work carried out for Kerr Stuart & Company appears as a drawing of a tipping box wagon dated 12 October, 1888.

The first locomotive built was named JOHN BULL on their order No. 4723. It was a 0-4-0 side tank with 8″ x 16″ cylinders and built to a gauge of 3′ 9″. The works number was E51 the prefix only being used for locomotives. It was despatched on 8 March, 1891.

It is thought that all locomotives built up to E71 were by this firm and although numbers E73 – 76 are recorded as built by H. A. & F. they were probably completed after the firm was taken over by Kerr Stuart & Company. Nos. 77 and 78 were ordered on 10 April, 1893 and recorded as 'K.S. & Co.' and No. 72 which was ordered on 18 April, 1893 would definitely have been by K. S. & Co.

In October 1891 six metre-gauge 0-6-0Ts were ordered by Kerr Stuart and had H.A. & F. works numbers 60–62 and 66–68. It is possible that all locomotives ordered by Kerr Stuart had their own work plates fitted.

The total number of locomotives built would appear to be works numbers E51 to E71, that is 21.

Fig. 251 **Hartley Arnoux & Fanning** 51/1891 JOHN BULL to the order of Dick Kerr – destination probably St. Lucia, British West Indies

Order No	Class	E No	Ordered	Despatched	Type	Cylinders inches	DW ft in	Gauge ft in	Customer/Destination	No/Name
		51		8/3/1891	0-4-0T	8 x 16	2 6	3 $9\frac{11}{16}$	For Dick Kerr & Co	John Bull
		52	18/2/1891	16/7/1891	0-4-0ST	5 x 10	1 7	2 5½	Pernambuco Brazil (a)	No 3
	Midge	53	18/2/1891	–	0-4-2T	8½ x 15	2 6	2 5½	Pernambuco Brazil (a)	561 Arabella
		54	–	1891	0-4-0ST	5 x 10	1 6	1 8	For Dick Kerr & Co (c)	Wasp
1181	Midge	55	–	–	0-4-2T	7 x 12	2 3	Metre	Not known	
1181	Midge	56	–	–	0-4-2T	7 x 12	2 3	Metre	Not known	
1182	Midge	57	–	1891	0-4-2T	6 x 10	1 9	2 5½	Pernambuco Brazil	
1182	Midge	58	–	1891	0-4-2T	6 x 10	1 9	2 5½	Pernambuco Brazil	
		59	–	1892	0-4-0ST	9½ x 16	3 0	Metre	For Dick Kerr & Co	Zara
1187		60	19/10/1891	–	0-6-0T	12 x 20	3 6	Metre	Spain	Turis
1187		61	19/10/1891	–	0-6-0T	12 x 20	3 6	Metre	Spain	Torrente
1187		62	19/10/1891	–	0-6-0T	12 x 20	3 6	Metre	Spain	Picassent
	Midge	63	1892	–	0-4-2T	7 x 12	2 3	Metre	Turis Spain	
1192		64	2/2/1892	–	0-4-0ST	8 x 12	2 3	Metre	Santander Spain	Solia No 1
1192		65	2/2/1892	–	0-4-0ST	8 x 12	2 3	Metre	Santander Spain	Obregon No 1
1187		66	19/10/1891	–	0-6-0T	12 x 20	3 6	Metre	Spain (d)	Monserratt
1187		67	19/10/1891	–	0-6-0T	12 x 20	3 6	Metre	Spain	Carlet
1187		68	19/10/1891	–	0-6-0T	12 x 20	3 6	Metre	Spain	Catadau
1197		69	4/3/1892	–	0-4-0TM	8 x 14	2 6	Metre	Avilles Steam Tramway Co Gijon Spain (b)	Avilles No 1
1197		70	4/3/1892	–	0-4-0TM	8 x 14	2 6	Metre	Avilles Steam Tramway Co Gijon Spain (b)	Salinas No 2
		71	11/11/1892	–	0-4-0TM	8 x 12	2 0	2 6	Great Yarmouth Port & Harbour Commissioners	Sir Harry Bullard
	Midge	72	18/4/1893	–	0-4-2T				Casa Lupton (b)	
						7 x 12	2 3	1 $11\frac{5}{8}$	São Paulo Brazil (e)	Chaperin
1218		73	9/2/1893	–	0-4-0ST	4¾ x 8	1 6	1 $11\frac{5}{8}$	Antofagasta	Autlagas
1218		74	9/2/1893	–	0-4-0ST	4¾ x 8	1 6	1 $11\frac{5}{8}$	Antofagasta	Colquechaca
1218		75	9/2/1893						Antogafasta	Consuelo
	Midge	76	24/3/1893	30/10/1893	0-4-2T	8½ x 15	2 3	2 5½	Fry Miers & Co Pernambuco (f)	

(a) Originally for Danube Coal & Minerals Co Ltd Austria (b) Tram Locos (c) KS Register shows 2′ 5½″ gauge but loco built for l′ 8″ gauge (d) Photographed as 'Roseberry' with KS Maker's plate numbered 66. (e) Probably ordered after KS had taken over (f) Register states built by Hartley as engine no 53. All Locos had inside frames except E Nos 53, 54 and 76. Thanks are due to Mr G Horsman for much of this information.

HAWKS & THOMPSON
Gateshead

This firm were general engineers and built beam engines, stationary engines and in 1837–38 built three locomotives for the Newcastle and Carlisle Railway. The first was an outside framed 0-4-0 with cylinders set low under the smokebox, the pistons passing under the leading axle.

The throw of the outside cranks was much greater than half the stroke – an unusual feature. The wheelbase was 5′ 2″ and the tender bore the name LIGHTNING (Fig 252). It was later converted to a 0-4-2. The maker's plate has Newcastle-on-Tyne inscribed although it is fairly certain the works were at Gateshead. It is difficult to understand the reason for this.

In the same year a 0-6-0 was built with the same size cylinders and coupled wheels. The boiler barrel was identical but with a small tube variation.

The third locomotive was a 2-4-0 built in 1838 with 6 ft coupled wheels which one would think unsuitable for the hilly route it had to traverse. It was claimed to be the swiftest on the line, but its performance up the banks has not been stated.

Some N&CR orders were then placed with Thompson Bros of Wylam in the following year, but no definite connection can be traced between the two firms.

Date	Type	Cylinders inches	D.W. ft in	Htg Surface sq ft	Pressure lb/sq in	Newcastle & Carlisle Rly		NER No	W'drawn
						No	name		
1/1837	0-4-0	14 x 15	4 6	440	65	10	Lightning	459	1863
6/1837	0-6-0	14 x 15	4 6	440	65	12	Carlisle	(a)	1853
6/1838	2-4-0	15 x 18	6 0	530	80	14	Victoria	463	1866

(a) Sold to R. Stephenson & Co.

Fig. 252 **Hawks & Thompson** 1837 Newcastle & Carlisle Rly LIGHTNING

HAWTHORN, R. & W. LTD

Forth Banks Works, Newcastle

The firm was founded in 1817 by Robert Hawthorn trading as marine and steam engine builders and engineers. In 1820 the firm became R. & W. Hawthorn Ltd.

The two brothers were at the Rainhill trials and no doubt it was this event that gave them the incentive to build railway locomotives themselves, and this they did and in 1831 appeared the first recorded locomotive built by them. It was a 2-2-2 for Vienna with 10½" x 18" cylinders and quoted as '50 HP Condensing'. It was named MODLING.

This was followed by six locomotives for the Stockton and Darlington Railway, three *Majestic* class, and three *Wilberforce* class 0-6-0s. These all had vertical cylinders, fastened to an extension of the framing in front of the boiler, driving a jack shaft which in turn was coupled to the leading coupled wheels. Hawthorn's earlier locomotives were fitted with their own design of valve gear which had no eccentrics, but consisted of a swivelling block attached mid-way along the connecting rod working in a slotted frame. From an arm on this frame, motion was derived through a rocking arm to the valve. This was superseded in the 1840s by a link motion.

Another feature of some earlier locomotives was a steam dryer and superheater apparatus consisting of a chamber in the upper part of the smokebox and extending partly round it. Hawthorn's were early users of the perforated steam collector pipe extending in the steam space primarily to combat the foaming and priming common to early locomotive boilers.

In 1835 the COMET was built for the Newcastle & Carlisle Railway and it was equipped with four fixed eccentrics instead of the two loose ones which were commonly used. It was not the first to be so fitted; George Forrester & Company started fitting them in 1834.

An unusual locomotive built for the Stockton & Darlington Railway, No. 27 SWIFT was a 0-4-0 with vertical cylinders driving a jack shaft between the two axles and on which were mounted the eccentrics and the motion link fulcrums were attached to the boiler by means of pads rivetted to the barrel.

Perhaps the most startling locomotives were two built in 1838 for the broad gauge Great Western Railway, under the patents of T. E. Harrison who was to become chief engineer on the North Eastern Railway. The arrangement was that the boiler was placed on a separate carriage, another carrying the cylinders and driving gear, the steam and exhaust between the two carriages being connected by flexible couplings. The THUNDERER had 16" x 20" cylinders driving a train of gears giving a ratio of 3:1. The driving wheels were 6 ft diameter, giving an equivalent diameter of 18 ft. The boiler was carried on three axles and the engine on two axles, one driving and the other trailing but not coupled together.

The other locomotive was the HURRICANE which had a similar boiler unit, but the engine itself had no spur gearing, and the driving wheels had their diameter increased to 10 ft. The object of this arrangement was to spread the axle weight and by doing so, to be able to place on a carriage as large a boiler as possible. The obvious disadvantage was that the driving axle had a very small proportion of the total weight available for adhesion. It was claimed that on a test trip the THUNDERER rode very smoothly and touched sixty miles per hour, but nevertheless both locomotives were failures, the cost of constant repairs being out of all proportion to the mileage run. The motion and the flexible couplings gave a great deal of trouble and they were withdrawn. Nevertheless it was a courageous experiment and could be termed a forerunner of the Garratts, Fairlies, etc. Besides supplying the home railways R. & W. Hawthorn did a considerable export business supplying locomotives to: Altona & Kiel, Leipzig & Dresden, Cologne Minden, Sweden, Denmark, India, Australia, S. America, and S. Africa.

In 1854 four 'Crampton' 4-2-0s were built for the Zealand Railway of Denmark and four 0-4-0s for the Glasgow & South Western Railway with dummy crankshafts. The wheelbase for these 0-4-0s was no less than 12' 1", the dummy crankshaft being fitted between the two axles. They had a brief life. The Cramptons on the other hand were more successful; in fact all Cramptons that went abroad and were built abroad had successful careers compared with the few built for home railways which were not viewed with favour and had short lives. In 1870 St. Peter's works were built on the north bank of the Tyne for marine engineering extensions, adjoining the works of Robert Stephenson & Co. Ltd.

A standard range of four and six coupled industrial tank locomotives were designed and built, variations being Crane and Fireless locomotives to the same wheel arrangement. Some double and single ended Fairlie's were built in the 1870s for abroad, and in 1879 a tram locomotive, presumably for demonstration purposes, was built in keeping with other firms about this time. One was built for the Palma-Majorca tramways in 1880. The only other 'tram' type were six coupled units for Belgium in 1917.

In 1884 the firm amalgamated with A. Leslie & Company of Hebburn who were shipbuilders and the firm became HAWTHORN LESLIE & CO. LTD.

A unique crane locomotive was built in 1888

Fig. 253 2113/1888 2-2-2CT 'Cross' Patent 10 ton crane. Palmers Shipbulding & Iron Co. Jarrow

for Palmer's Shipbuilding & Iron Co. Ltd. of Jarrow. It was a 2-2-2T with inside cylinders 14″ x 18″ outside frames with a Cross's patent 10 ton crane mounted above the boiler.

In 1893 a 4-2-2-0 'simple' was built to the order of Westwood & Winby for the Chicago World Fair. It had four cylinders, the two 17″ x 22″ inside driving the front pair of 7′ 6″ diameter driving wheels and the two 16½″ x 24″ outside, driving the rear pair. The driving wheels were not coupled. The large boiler, working at 175 lb/sq. in., had a total heating surface of 2000 sq. ft and a grate area of 28 sq. ft. F. C. Winby, who had served his apprenticeship at Crewe, designed this locomotive but unfortunately it was not road tested before being shipped and was found to be badly deficient in steam, which put paid to a wager that it could match any American locomotive.

1904 brought an order for a 0-8-0T for the Kent & East Sussex Railway – a most unlikely locomotive for a 'light' railway.

In 1932 it was sold to the Southern Railway, who in turn furnished the K&ESR with an old Beyer Peacock 0-6-0ST, which was far more suitable for the duties required.

One railmotor was built for the Port Talbot Railway in 1906. This was the largest one built in this country. It was six coupled with 12″ x 16″ outside cylinders and 3 ft diameter wheels. Its length was 77 ft and the carriage portion was built by Hurst Nelson.

In 1907 Hawthorn Leslie acquired ground adjoining their Forth Banks works for much needed extensions – this land formed the chief part of the loco works of R. S. & Co. Ltd before they removed to Darlington.

In 1915 six 4-6-0s were ordered by the Highland Railway; these were large machines with two outside cylinders 21″ x 28″ and 6 ft driving wheels, to the designs of Mr F. G. Smith.

Unfortunately the engineer-in-chief of the time banned them as being too heavy – whether he had been consulted in the first place is problematical, but it was a first rate blunder.

They were taken off the Highland Railway's hands by the obliging Caledonian Railway.

During these years a great variety of locomotives were sold to overseas railways of a great variety of types but mostly to the designs of the Crown Agents or the railways supplied.

Along with other firms a number of fireless locomotives were built. These were supplied to works where fire and explosive risks made it impossible for the normal coal or oil fired locomotive to be employed. This type could therefore be used in powder factories, magazines, chemical works, paint, cotton, paper and wood factories.

In lieu of the ordinary boiler and firebox, a cylindrical heat storage chamber was provided, the water in which was heated to a sufficient temperature to form the steam required for working the locomotive. This was effected by charging it with water and high pressure steam in the proportion of 3 or 4 to 1 from a stationary boiler plant outside the danger zone. The formation of fresh steam from the heated water was effected as the result of the reduction of pressure in the reservoir as the steam was used from the upper portion of the reservoir, allowing fresh steam to be liberated at a correspondingly reduced pressure. A reducing valve fitted between the reservoir and the cylinders ensured

Fig. 254 2671/1906 0-6-0ST Backworth Colliery No. 4

Fig. 255 3659/1926 0-6-0T Frodingham Iron & Steel Co. Ltd. No. 23

Fig. 256 3900/1935 Phoenix Colliery Ltd. No. 1 Johannesburg 3′ 6″ gauge cylinders 20″ x 24″ 3′ 9″ D.W.

a constant low pressure for as long a period as possible. The cylinders were of relatively large diameter to permit a useful range of steam pressure. Whilst sufficient pressure remained the locomotive must be able to return to the stationary boiler plant for recharging.

The advantages of this type of locomotive were: low initial cost, cheap maintenance due to absence of boiler deterioration, no danger of fire, no stoking required, no fuel bunkers or water tanks required. The main disadvantage was the limited range of operation.

A further refinement was the fitting of a steam drier and superheater.

Marine and other types of engines were included in the work's number sequence. In all 2611 steam locomotives were built, the last one being No. 3953 of 1938, a 0-4-0ST for the Consett Iron Company.

From 1 January, 1937 the locomotive department of Hawthorn Leslie & Co. Ltd. (excluding the Boiler dept.) was purchased by Robert Stephenson & Co. Ltd. and the title of the firm became R. STEPHENSON & HAWTHORN LTD. For further history see R. Stephenson & Co. Ltd.

HAWTHORNS & COMPANY

Leith Engine Works, Great Junction Street, Leith.

These works were bought or built by R. & W. Hawthorn of Newcastle in 1846 for the erection of locomotives built by them at Newcastle for those ordered for use in Scotland to avoid transporting the completed units by sea. Later complete locomotives were built at Leith, the first in 1847 for the Edinburgh, Leith & Granton Railway.

After the Royal Border bridge was opened on 29 August, 1850 – it linked Tweedmouth and Berwick – the works were sold to a firm with the title Hawthorns & Company who proceeded to build new locomotives into the 1880s. Rebuilding and repairing were carried out until a much later date.

A number of outside cylinder 2-2-2s, and 2-4-0s of various classes and to the design of William Barclay were built for the Inverness & Nairn, Inverness & Aberdeen Junction, and Inverness & Rosshire railways. The cylinder layout and framing were typical of the 'Allan' type.

Naturally many orders came from Scottish railways and industry. Various four coupled tanks were built and an unusual feature of many of them was the outside Stephenson's valve gear, and the well tanks used the frames as sides of the tank (S. D. Davidson's patent) very similar to the Borrows-built locomotives.

Locomotives were built for the Deeside, Dundee & Arbroath, and Scottish North Eastern railways and other were sent abroad to the Chilean Government Railway, South Africa, Ireland, Germany and India.

The works list is practically non-existent but it is estimated that approximately 425 locomotives were built up to 1872, and many went overseas.

Fig. 257
Hawthorns, Leith 0-4-0WT *c.* 1860 at Bourtriehill Brickworks – outside Stephenson valve gear

HAWTHORN LESLIE
EXAMPLES OF SIX-COUPLED INDUSTRIAL TANK LOCOMOTIVES

Date	Class	Type	Cylinders inches	DW ft in	Heating Surface sq ft			Grate Area sq ft	WP lb/sq in	Tank gallons	Weight WO ton cwt
					Tubes	F'box	Total				
1906	Anzac	0-6-0ST	OC 16 x 24	3 8	746	64·3	810·3	12·6	180	1200	43 6
1906	Kersley	0-6-0ST	OC 17 x 24	3 10	848	87	935	15·5	170	1100	44 5
1906	Backworth	0-6-0ST	OC 17 x 26	4 1	909	86	995	15·5	160	1100	44 10
1926	Scunthorpe	0-6-0T	OC 18 x 24	4 0	1002	90	1092	18·0	180	1200	52 12
1926	Oriental *	0-6-0T	OC 18 x 24	3 8	1025	90	1115	18·0	180	1255	56 0

* 5′ 6″ gauge.

HAZELDINE & RASTRICK
Bridgnorth

The only locomotive built by this firm was to the designs of Richard Trevithick. Known as CATCH-ME-WHO-CAN, it ran on a circular track erected in the place now occupied by Euston Square. It was built in 1808.

John Hazeldine of Bridgnorth was joined by William Foster *c.* 1807 who stayed with the firm until 1817 when he joined J. U. Rastrick at Stourbridge to form another locomotive building establishment: Foster Rastrick & Co. The main occupation of the firm was stationary engines and bridges, the best example of the latter being the cast iron bridge at Chepstow over the River Wye. CATCH-ME-WHO-CAN was a single cylinder locomotive, the cylinder being mounted vertically on the centre line of the boiler and sunk into it. The rear wheels were driven by connecting rods and a pair of wheels fixed forward to carry the front end.

Trevithick had wagered that this engine would travel farther in 24 hours than any horse, but there did not appear to be any challengers. For a short period it pulled passengers around the track in a converted road carriage but despite publicity and an admission charge of two shillings (10p) the exhibition was short lived.

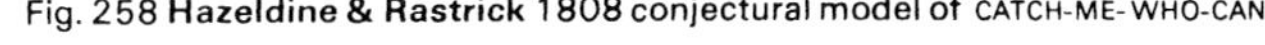

Fig. 258 **Hazeldine & Rastrick** 1808 conjectural model of CATCH-ME-WHO-CAN

HEAD, T. H.

90 Cannon Street, London.

Thomas H. Head is credited with having built in 1871 at least one vertical boilered 0-4-0T for the Dorking Greystone Lime Co. Ltd. The locomotive was known as COFFEEPOT. It was cab-less when built, with a 2:1 geared drive, subsequently altered to 3:1 ratio.

The frame was of cast iron and was made as one casting. The link motion was actuated by small cranks at the end of the crankshaft. The wheels were 2′ 4″ in diameter and the cylinders 6″ x 12″.

This locomotive was probably built by Head Wrightson & Co. Ltd of Thornaby, although the characteristics of the T. H. Head COFFEEPOT differ from those of the Thornaby firm.

HEADLY, J. & E.

Eagle Foundry, Newmarket Road, Cambridge

This engineering firm was engaged in manufacturing steam pumps, beam and stationary engines and work allied to the agricultural industry.

Also quoted as HEADLY BROS. the firm was run by two brothers, James and Edward.

Their only incursion into the locomotive building field was completed in 1849 for the Norfolk Railway. It was intended for tours of inspection of the line and as there were no couplings provided at either end, scanty accommodation was available on the footplate. EAGLE as it was named was a 2-2-0WT with inside cylinders 7″ x 14″, driving wheels 4′ 6″ diameter and 2′ $6\frac{1}{2}$″ diameter leading wheels. The boiler was domeless with a raised firebox, and a safety valve above. The heating surface was about 22 sq. ft with a grate area of 2·6 sq. ft. Wooden lagging was used on the boiler barrel and firebox with brass bands. The well tank was fixed beneath the footplate. A tall chimney gave it a height of 8 ft and the length overall was 12 ft. The driving wheels were cast, which due to the slender spoke section must have been quite an achievement. It weighed approximately two tons empty.

Notoriety was soon earned in 1850, by EAGLE being involved in a fatal accident near Norwich when Mr Newall, the district superintendent was reported to have jumped off the footplate for some unknown reason and was run over and killed by the locomotive.

Fig. 259 J. & E. Headly 1849

The Norfolk Railway had been operated by the Eastern Counties Railway since 1848 and the rolling stock was purchased in 1850.

EAGLE was subsequently rebuilt, probably due to the fatality, as a 2-2-2WT, the frames extended rearwards to house the trailing wheels and a small inspection saloon, thus separating the officials from the footplatemen, but a speaking tube was installed for communication. The boiler was rebuilt with a new raised firebox, and compensating beams linked the leading and driving springs.

It was then used by James Samuel, the Eastern Counties Railway Engineer, for his own inspection saloon and was kept at Stratford. In association with W. Bridges Adams, Samuel had taken a keen interest in steam carriages and a number had been built at Bow including EXPRESS and ENFIELD for the ECR.

EAGLE was withdrawn in 1868.

The title of the firm changed later to HEADLY & MANNING and then to HEADLY & EDWARDS and ceased trading in the early 1920s.

An account of this locomotive may be found in R. H. Clark's *Steam Engine Builders of Suffolk, Essex and Cambridgeshire.*

HEAD WRIGHTSON CO.

Teesdale Ironworks, Stockton-on-Tees

Formerly known as Head, Ashly & Co. this firm of well-known engineers, ironfounders, bridge and roof builders and makers of hydraulic machinery built a number of locomotives in the 1860s and 1870s.

The majority were vertical boilered four coupled locomotives for collieries, iron works and contractors, and it is thought that approximately forty were built.

In advertisements in 1866 and 1869 appeared drawings of four coupled saddle tanks but whether any of this type actually were built is debatable, although in the 1866 advertisement it stated that three engines of that type were now ready for delivery and others to be finished shortly.

Two types were built with vertical boilers. The geared type had two vertical cylinders bolted to the boiler, the connecting rods which had forked ends drove a crankshaft whose bearings were bolted on to the firebox. A gear wheel was fitted to the crank shaft and meshed with a gear wheel on the axle. The ratio was usually 2:1 or 3:1 and the cylinders varied from 6″ x 12″ to 9″ x 14″. A link motion was operated by small cranks on either side of the main cranks. The wheels were coupled and the frames, axlebox horns, buffers and cokebox were cast in one piece. The other type had conventional outside cylinders which were sharply inclined and well forward of the leading wheels.

The boiler for both types had a conical crown with vertical tubes. The pressure was 120 lb/sq. in.

Fig. 260 **Head Wrightson** 21/1870 Geared type preserved at maker's works

HEAD WRIGHTSON & CO.

One went to Hanover and for a time was used as works shunter in the Hanomag locomotive works.

Fortunately one example has been preserved by the makers where it stands in the works yard, with wheels lifted off the rails so that the engine can be turned with the aid of compressed air.

Fig. 261 **Head Wrightson** Advertisement for improved tank locomotive 1869

Works No	Date	Type		Cylinders inches	DW ft in	Gear Ratio	Customer
21	1870	0-4-0T	VB G	6¼ x 14	2 6½	2:1	Londonderry Estates Co (a)
	1871	0-4-0T	VB G	6 x 12	2 4	2:1	Dorking Greystone Lime Co (b)
33	1873	0-4-0T	VB OC	9 x 14	2 5½	—	Londonderry Estates Co (a)
25?	1876	0-2-2	VB				Chatterley Iron Co

(a) to Seaham Harbour (b) named COFFEEPOT gear ratio altered to 3:1

Fig. 262 **Head Wrightson** 33/1873 outside cylinder type at Seaham Harbour

HEATH, ROBERT & SONS LTD.

Biddulph, near Stoke on Trent

Built approximately eleven locomotives for their own use between 1888 and 1926. They were 0-4-0ST and 0-6-0ST with outside cylinders.

Subsequently (*c* 1920) Robert Heath and Low Moor.

1888	0-4-0ST	R. Heath & Sons Ltd	No. 8
1915	0-6-0ST	R. Heath & Sons Ltd	No. 15
1916	0-4-0ST	R. Heath & Sons Ltd	No. 7
1924	0-6-0ST	R. Heath & Low Moor	No. 16
1925	0-4-0ST	R. Heath & Low Moor	No. 1
1926	0-4-0ST	R. Heath & Low Moor	No. 9
	0-4-0ST	R. Heath & Low Moor	No. 2
	0-4-0ST	?	No. 4
	0-4-0ST	?	No. 5
	0-4-0ST	?	No. 6
	0-4-0ST	?	No. 12
	0-4-0ST	?	No. 14

See also COWLISHAW WALKER & CO. LTD.

HEDLEY, WILLIAM

Wylam Colliery

William Hedley was appointed viewer at Wylam Colliery in 1805, the colliery being owned by Christopher Blackett. Coal was conveyed from the pit to Lemington-on-Tyne on a track made of wood, by means of horses hauling wagons.

The track was relaid about 1808 with cast iron plate rails and according to accounts Hedley conducted experiments in 1812 with a geared trolley operated by hand to test simple adhesion of a smooth wheel tread on a smooth rail. Blenkinsop of Middleton had already taken out a patent in 1811 for a locomotive engine with toothed wheels engaging in a rack rail. In addition Mr Chapman in 1812 patented a different method of propulsion using a chain in between the rails on which gearing on the locomotive engaged.

No confidence had been placed on simple adhesion and gravity by Blenkinsop or Chapman but Hedley after carrying out his tests was quite satisfied and took out his patent in March 1813.

Thomas Waters of Gateshead built the first locomotive for Wylam which is described in that section and known at Wylam as BLACK BILLY.

The second was put in hand at Wylam and Hedley discarded the single cylinder and the single flue cast iron boiler. Instead the boiler was made of malleable iron with a return flue which positioned the chimney at the same end as the fire hole. It had two vertical cylinders driving a beam arrangement with central rods driving a geared jackshaft engaging a train of gears on the two axles. This locomotive known as PUFFING BILLY was much more successful

Fig. 263 **Robert Heath & Sons** 1888 Norton Colliery No. 8

than Waters' locomotive. It began work in 1812 and in 1865 was sent to the South Kensington Museum and has the distinction of being the oldest locomotive in existence.

A third locomotive was built in 1813, and due to the excessive weight on the cast iron rails which were having to be replaced constantly, the new locomotive was mounted on eight wheels to spread the load. The wheels were on two separate bogies all wheels being driven through gears from a central geared jackshaft.

It is not known whether either or both bogies were pivoted; if they were, problems with the lateral movement of the gearing would have been experienced. On the other hand if the whole wheelbase was rigid the track would have suffered.

From about 1828 to 1830 the track was replaced by wrought iron edge rails and the four wheeled PUFFING BILLY type were used and the eight wheeled locomotive was probably converted to the four wheeled type.

Besides PUFFING BILLY at least two other similar locomotives were built known as WYLAM DILLY and LADY MARY.

These may have been eight wheeled drives and converted back to four wheelers when the track was reconstructed. Most carried on for 50 years. WYLAM DILLY is also preserved, in the Royal Scottish Museum, Edinburgh.

See also WATERS, THOMAS and HACKWORTH, TIMOTHY.

HETTON COAL COMPANY

Hetton Lyons, Co. Durham

Between 1820 and 1822 five locomotives were built in the colliery workshops under the guidance of George Stephenson.

The Hetton Colliery railway was the first entirely new railway to be constructed by George Stephenson and was opened officially on 18 November 1822 some three years before the Stockton & Darlington Railway.

The locomotives were all 0-4-0 type with 3 ft driving wheels. The cylinders were vertical, one at each end of the boiler and recessed into it. Each drove one pair of wheels by means of a connecting rod and crankpin. The two axles were connected with spur gearing and chains. Patent steam springs were fitted. At first they were not entirely successful and were removed from a section of the line being replaced by fixed engines. At least three were dispensed with, but the one built in 1822 had a long career, being rebuilt in 1857 and 1882 and eventually preserved at York in 1912.

Hetton Coal Company No. 1 was built in 1820, Nos 2, 3, and 4 in 1821, and No. 5 in 1822. Four of them carried names HETTON, DART, TALLYHO and STAR.

Hetton Coal Company became HETTON COAL COMPANY LTD. in 1884, and about the year 1900, two more locomotives were built.

These were 0-4-0Ts with vertical boilers named LYONS and EPPLETON, they had a geared drive and the two axles were connected by means of endless chains around gear wheels.

Fig. 264 **William Hedley** 1813 (?) geared drive known as PUFFING BILLY

The entire length was roofed over together with side panels and end cab with large spectacle windows, quite well protected for the period.

An illustration and description of LYONS may be found in the *Locomotive Magazine* for April 1901.

The colliery line was closed down in Sept. 1959 after becoming the property of the National Coal Board in 1947.

HEYWOOD, SIR ARTHUR P.

Duffield Bank, Derbyshire

Although not strictly within the limits of this book, the building of locomotives at Duffield Bank to a gauge of 15″ was an interesting achievement. Sir Arthur Percival Heywood was a pioneer of 'miniature' locomotives and built six from 1874 to 1916.

All the boilers were built by Abbott & Co. of Newark and had cylindrical fireboxes of the launch pattern. The disadvantage of this type of firebox was that although the grate area was proportional to the boiler heating surface the firebox volume was too small and trouble was experienced in maintaining steam.

The first locomotive EFFIE was a 0-4-0T with outside cylinders 4″ x 6″ and 1′ 3½″ diameter wheels. The boiler, working at a pressure of 125 lb/sq. in. had a heating surface of 23 sq. ft and a grate area of 1·25 sq. ft. Stephenson's valve gear was fitted between the frames.

The second, built in 1881 named ELLA was a six coupled tank with 4⅞″ x 7″ cylinders and 1′ 1½″ diameter wheels. A much larger boiler was fitted with 70 sq. ft of heating surface and grate area 2·12 sq. ft. Boiler pressure was increased to 160 lb/sq. in., which was used on all subsequent locomotives. To enable it to traverse

Fig. 265 **Hetton Coal Co.** 1822 Geo. Stephenson – in service until 1908, preserved at York.

the sharp curves on the Duffield Bank railway an ingenious method was used. The axles of the leading and trailing wheels were housed in conventional axleboxes in the outside frames. In the centre of the two axles ball joints were used; the actual wheels were mounted on sleeves surrounding the axles which could swivel on the axis of the ball joints. The valve gear was of necessity outside and was specially designed and evolved, from that used by Mr Brown of Brown-Boveri, Winterthur; vibrating links being used instead of eccentrics. The cranks were counterbalanced instead of the wheels and this method was used by many locomotive builders of small gauge types.

His third locomotive for Duffield was the largest built, it was named MURIEL and was a 0-8-0T. The cylinders were $6\frac{1}{4}''$ x 8″ and 1′ 6″ diameter wheels. The boiler had 91 sq. ft of heating surface and 3 sq. ft grate area. The weight in working order was 5 tons. The layout was practically the same as ELLA and it was completed in 1894.

All three locomotives were built with the object of proving the efficiency of this sort of transport on large estates and he not only built the locomotives, but carriages and wagons too. Another objective was to prove the suitability of this gauge and motive power for military railways behind the lines for ammunition etc.

The Duke of Westminster who resided at Eaton Hall was the only owner of a large estate who showed any practical interest in the Duffield Bank railway and in 1896 KATIE a 0-4-0T was built for the Eaton Hall Railway, which had been laid under Sir Percival's supervision. This locomotive was larger than EFFIE and had $4\frac{5}{8}''$ x 7″ cylinders, 1′ 3″ diameter wheels. The boiler had 53 sq. ft of heating surface and a grate area of 2·12 sq. ft. Stephenson's valve gear was again used. For heavier loads a six coupled tank was ordered, and completed in 1904. Named SHELAGH, it was a little larger than the earlier 0-6-0T. The outside cylinders were $5\frac{1}{2}''$ x 8″, wheel diameter 1′ 4″. The boiler had 80 sq. ft of heating surface and 3·0 sq. ft grate area. This was probably the most successful of those built and in 1916 a second one with identical dimensions was completed and named URSULA. The same year Sir Percival died and the works were closed.

So ended a fascinating chapter in the development of light railway rolling stock and no small contribution came from Duffield Bank.

Fig. 266 **Sir A. P. Heywood** 1894 0-8-0T MURIEL

HICK, BENJAMIN

Soho Ironworks, Bolton

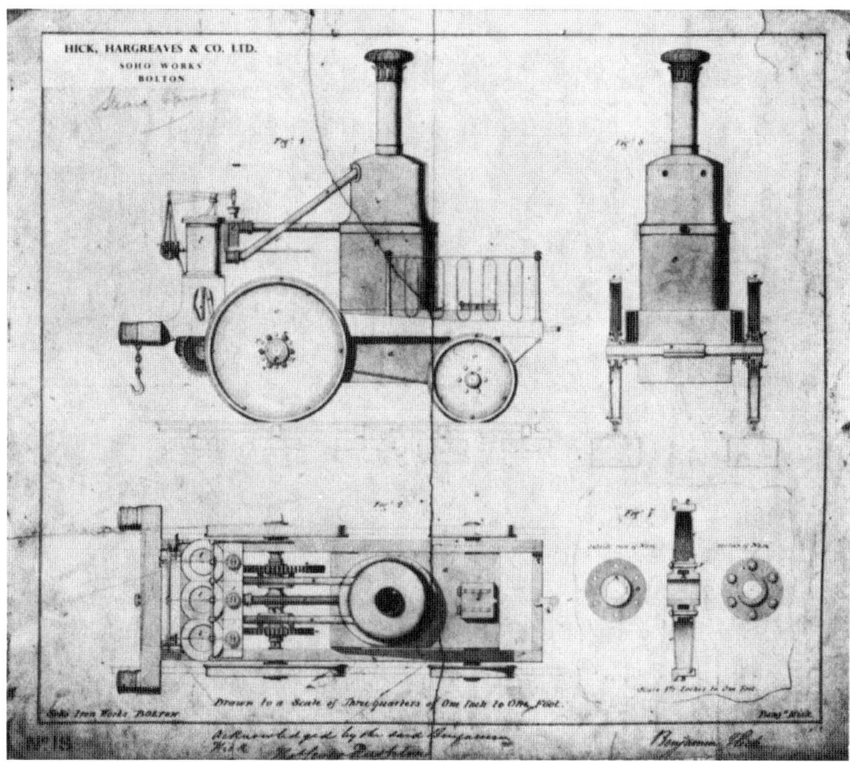

Fig. 267 **Benjamin Hick** 1833 steam carriage with geared drive

Benjamin Hick founded the firm in 1833 having left Rothwell, Hick and Rothwell in the previous year.

His two sons, John and Benjamin were also brought into the firm, but Benjamin only stayed for twelve months. John carried on with his father until the latter died in 1842.

Some years after his father's death, John Hick took William Hargreaves into partnership and the name of the company became HICK, HARGREAVES & COMPANY.

Locomotive building commenced in 1833 but never became the major part of the business, the principal products being stationary engines, boilers, mill gearing, hydraulic machinery, water wheels and marine engines. Nevertheless nearly 100 locomotives were built and although some were short lived, the workmanship in their manufacture was excellent.

In 1833 a four wheeled steam rail carriage was built for a Mr Thomas Lever Rushton. The drawing shows a vertical boiler centrally placed and a three cylinder unit at one end and driving a crankshaft on which were two gear wheels engaged with two gears on the driving axle. Two different ratios were used, the appropriate gear on the driving axle being engaged by dog clutches. The built-up wheels are interesting, and also the simple valve gear. This was followed by a 2-2-0 for the Pontchartrain Railroad in 1834 with inclined outside cylinders fed by outside steam pipes from a dome just behind the chimney. Two eccentrics had adjustable rods and operated the valves through a rocking shaft. Plunger feed pumps were driven off the crossheads. The buffers appear to be spring loaded – a very early date for this type.

In the same year a 0-4-2 was built for Mr John Hargreaves, the lessee of the Bolton, Kenyon and Liverpool Railway (Bolton and Leigh) and this was a more conventional design with inside frames, two domes with spring balance safety valves and a steam collector between the chimney and first dome. A man-hole cover was fitted to the barrel behind the first dome. The cylinders were low-set under the smokebox with the piston rods beneath the leading axle. Two further 0-4-2s were built for this railway; UTILIS had 14″ cylinders and 4′ 8″ coupled wheels and CASTLE had 14″ cylinders and 4′ 6″ coupled wheels.

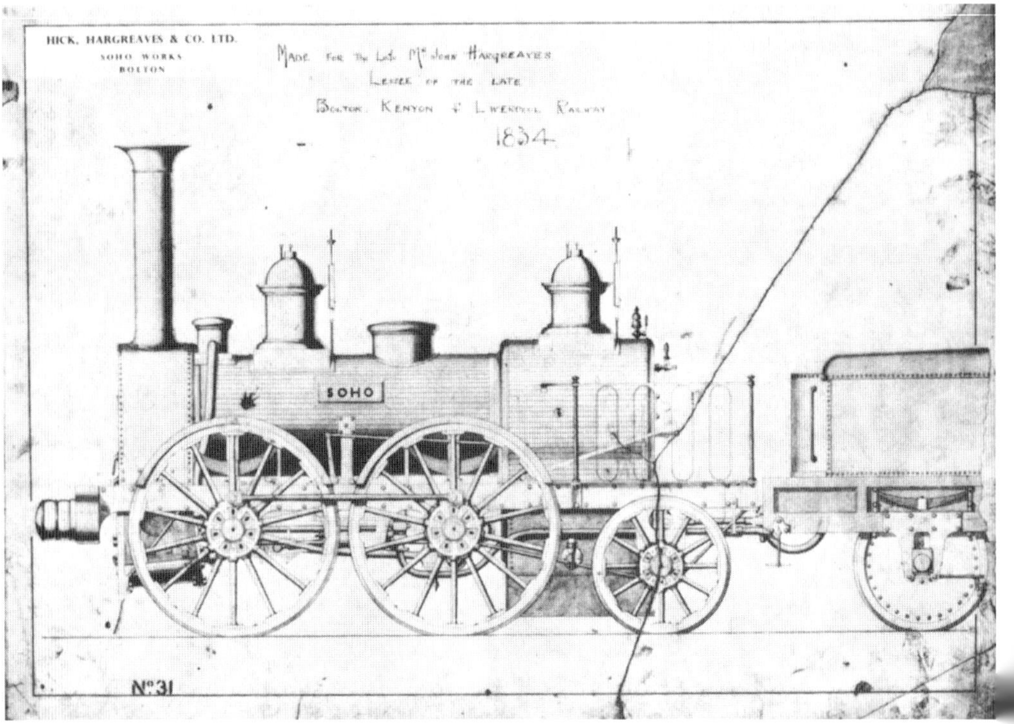

Fig. 268 **Benjamin Hick** 1834, for John Hargreaves-Bolton, Kenyon & Liverpool Rly SOHO

Fig. 269 **Benjamin Hick** 1840 Birmingham & Gloucester Rly–similar to American 'Norris' type

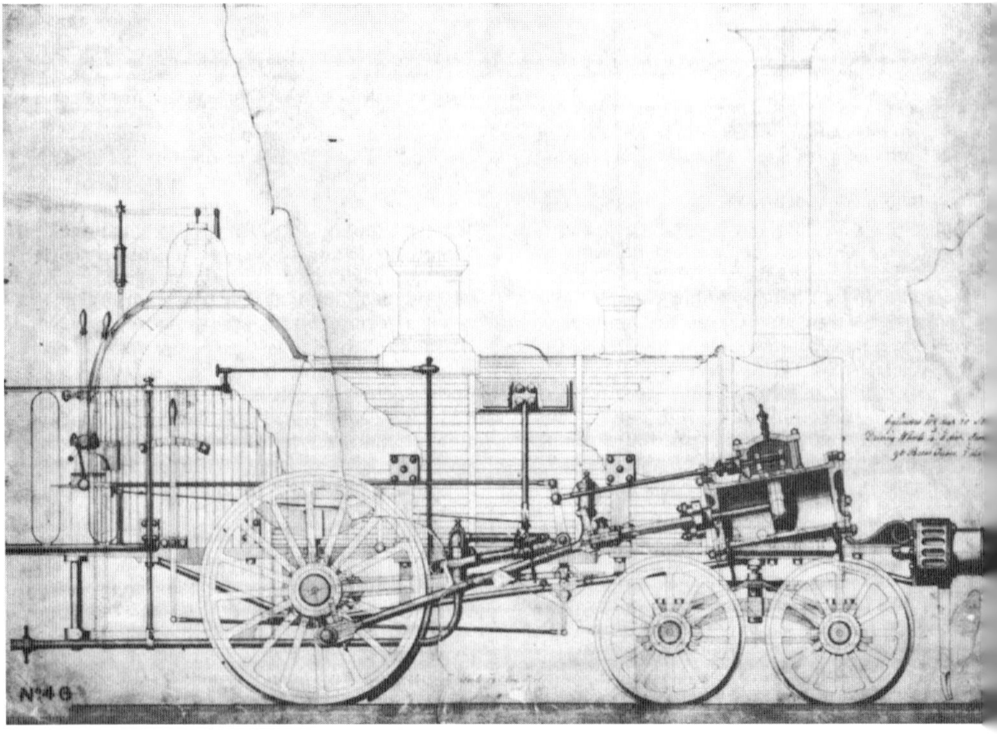

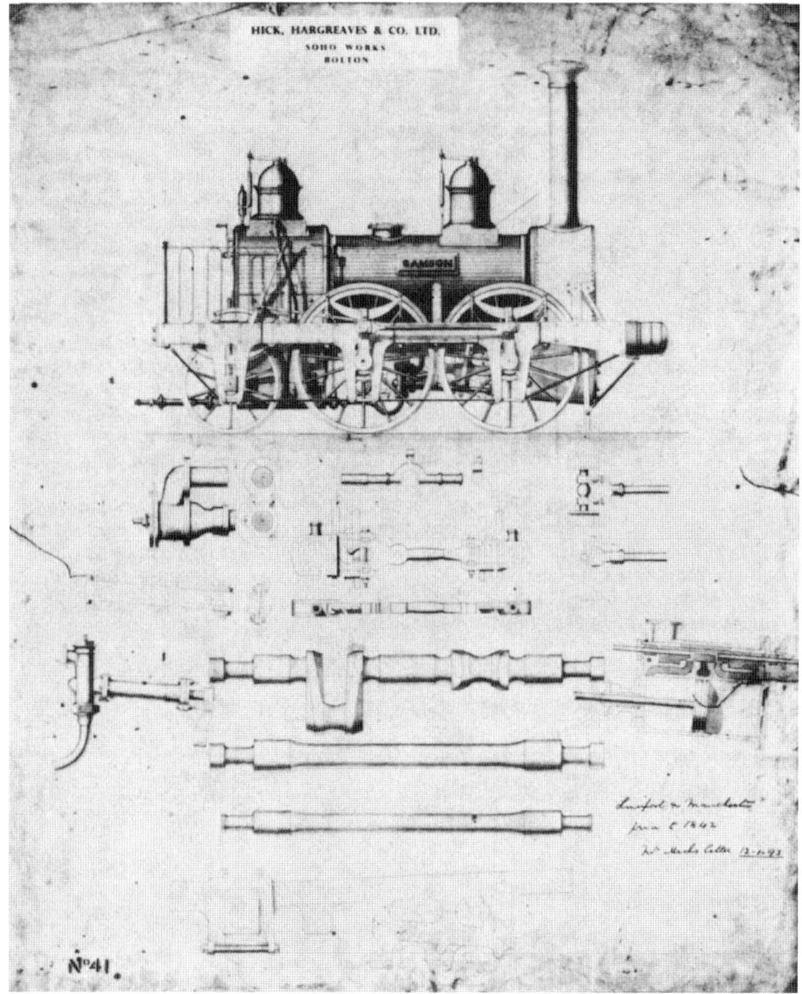

Fig. 270 **Benjamin Hick** 1839 Liverpool & Manchester Rly No. 66 SAMSON

At some unknown date the UTILIS was rebuilt either as a 2-4-2 or 0-4-4 tender locomotive with 2′ 6″ and 1′ 8″ carrying wheels.

Further orders were fulfilled for American railways but no details are known.

Benjamin Hick became one of the sub-contractors for Edward Bury and a number of his bar-framed 2-2-0s were built from 1837 to 1841 for the London & Birmingham, North Union, Midland Counties, and Manchester & Leeds railways.

Two 0-4-2s for the Liverpool & Manchester Railway were built in 1839 (Fig. 270) and the sketches shown beneath the locomotive give details of various parts. The inside frame which was bolted on to the firebox at one end and the smokebox at the other would indicate that the leading axle was passed through the framing before the wheels were pressed on!

In 1840 two 'singles' were built for the Paris and Versailles Railway. The boiler had a raised firebox on which was mounted the combined dome and safety valves. The regulator was positioned in the smokebox and the blast pipe extended to the base of the chimney.

In 1840 two 4-2-0s were built for the Birmingham & Gloucester Railway (Fig. 269) and closely followed the prototypes built by Norris in America and shipped over for this railway. They had bar-frames outside inclined cylinders, the front bogie was pivoted and the firebox was of the 'haystack' type. The slide bars were square in section fixed on the horizontal plane with the diagonal of the square in the vertical plane. They were reported to have been built with workmanship superior to the Norris locomotives. The importation of the latter raised quite an outcry, but they were indisputably superior in performance to any other type trying to ascend the Lickey incline.

Some 'long-boiler' 0-6-0s were built in 1844 and 1846 with haystack fireboxes and domeless boilers; the steam collector in the dome incorporated the regulator.

The inclined cylinders were low set, with the pistons beneath the leading axle. The stool and cover fitted to the barrel was utilised for the addition of another safety valve in the case of the three Taff Vale Railway locomotives and may have been fitted at a later date at Cardiff.

Two outside cylinder 2-2-2s built for the Edinburgh & Glasgow railway had large domes near the chimney incoporating the regulator valve. The raised firebox had spring balance safety valves, exhausting through a miniature flared chimney. The driving wheels were fitted with balance weights.

A similar type of boiler was fitted to two 0-6-0s built in 1848 for the Birkenhead railway (Fig 271). On these, four eccentrics and forked connecting rods were fitted. They were quite modern for the period and were some of the earliest six coupled type with inside frames. Forked gab motion was used. The most unusual feature was the inverted springing used on the leading wheels to clear the smoke-box, the springs being placed to the rear of the axle centre line. One transverse spring was used on the trailing wheels beneath the footplate.

Four outside framed 2-4-0s for the North Staffordshire railway were built in 1848. They had inside cylinders 14" x 21" and 5 ft diameter coupled wheels. Inside frames were attached to the rear of the cylinders and the front of the firebox and the driving axle had four bearings. The boiler was domeless with a very high raised firebox with dome and safety valves fitted on top.

Fig. 271 **Benjamin Hick 1849 Birkenhead Rly 0-6-0—one of four ordered**

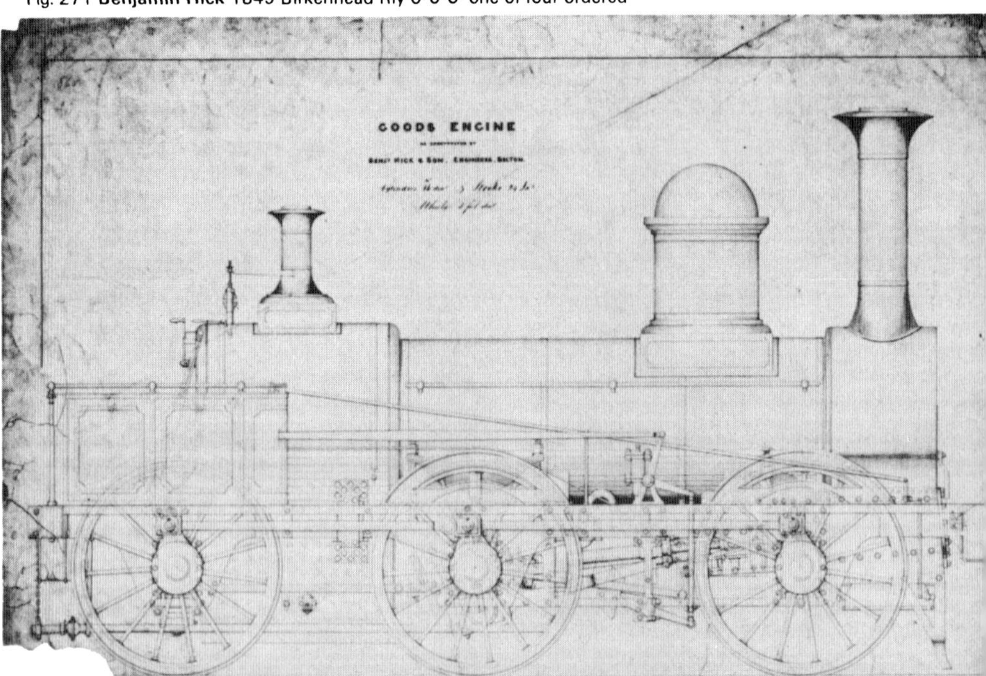

Six further 2-4-0s were built for the Birkenhead railway which was formed by the amalgamation of the Birkenhead, Lancashire and Cheshire Junction railway with the Chester & Birkenhead railway in July 1847. They were all delivered in 1851, but had been built in 1849/50 for stock. Outside cylinders, inside frames, and a boiler similar to the North Staffordshire 2-4-0s were fitted.

These were the last locomotives built by the firm, the policy being to concentrate on stationary engines and other mill work.

The number of locomotives built was approximately between 90 and 100.

John Hick retired from the business in 1868 and after William Hargreaves death in 1889 it was formed into a limited company, and under the title of Hick Hargreaves & Co. Ltd. is one of the foremost engineering establishments in the country.

Thanks are due to Messrs. Hick Hargreaves & Co. Ltd for information and drawings willingly made available.

Date	Type	Cylinders inches	DW ft in	No/Name	Customer
1833	0-2-2	3VC			Thomas L. Rushton (a)
1834	2-2-0	OC/inc		Fulton	Pontchartrain RR
1834	0-4-2	13	5 8	Soho	John Hargreaves Junr.
1836				Potomac	Richmond Fredericksburg & Potomac RR
1836	0-4-2	14 x 18	4 8	Utilis	Bolton & Leigh Rly
1837				Virginia	Raleigh & Gaston RR
1837				Louisa	Richmond Fredericksburg & Potomac RR
1837				New Orleans	Carrolton RR
1837	0-4-2	14 x 18	4 8	Victoria	Bolton & Leigh Rly (b)
7/1837	2-2-0	12 x 18	5 6	10	London & Birmingham Rly (c)
1838	2-2-0	12 x 18	5 6	5	London & Birmingham Rly
1838	2-2-0	12 x 18	5 6	11	London & Birmingham Rly
1838	2-2-0	12 x 18	5 6	12	London & Birmingham Rly
1838	2-2-0	12 x 18	5 6	13	London & Birmingham Rly
1838	2-2-0	12 x 18	5 6	14	London & Birmingham Rly
1838	2-2-0	12 x 18	5 6	15	London & Birmingham Rly
1838	0-4-2	14 x 18	4 6	Castle	Bolton & Leigh Rly
1838				Phoenix	Richmond Fredericksburg & Potomac RR
1838	2-2-0	12 x 18	5 6	4	North Union Rly (c)
1838	2-2-0	12 x 18	5 6	5	North Union Rly
1838	2-2-0	12 x 18	5 6	6	North Union Rly
1839	2-2-0	12 x 18	5 6	7	North Union Rly
1839	0-4-2	11 x 20	5 0	66 Samson	Liverpool & Manchester Rly
1839	0-4-2	11 x 20	5 0	68 Goliath	Liverpool & Manchester Rly
1840	0-4-2	14 x 18	5 0	62	North Midland Rly
1840	0-4-2	14 x 18	5 0	65	North Midland Rly
1840	0-4-2	14 x 18	5 0	66	North Midland Rly
6/1840	2-2-0	12 x 18	4 0	Dragon	Midland Counties Rly (c) (l)
7/1840	2-2-0	12 x 18	4 0	Scorpion	Midland Counties Rly
1840	2-2-0	12 x 18	4 0	Hornet	Midland Counties Rly
9/1840	2-2-0	12 x 18	4 0	Wyvern	Midland Counties Rly
10/1840	2-2-2	12 x 18	4 0	Vampire	Midland Counties Rly
9/1840	2-2-0	12 x 18	4 0	Lynx	Midland Counties Rly
8/1840	2-2-0	13 x 18	5 6	Centaur	Midland Counties Rly
8/1840	2-2-0	13 x 18	5 6	Hydra	Midland Counties Rly
8/1840	2-2-0	13 x 18	5 6	Harpy	Midland Counties Rly
1840	2-2-2	13 x 18	5 6		Paris & Versailles Rly
1840	2-2-2	13 x 18	5 6		Paris & Versailles Rly
1840	4-2-0	OC 11½ x 20	4 0	18 Breedon	Birmingham & Gloucester Rly
1840	4-2-0	OC 11½ x 20	4 0	24 Spetchley	Birmingham & Gloucester Rly
1841	2-2-2	$14\frac{1}{8}$ x 18	5 8	1	West Hartlepool Harbour & Rly Co
1841	2-4-0	$14\frac{1}{8}$ x 18	5 4	2	West Hartlepool Harbour & Rly Co
4/1841	2-2-0	12 x 18	5 0		North Union Rly (c)

Date	Type	Cylinders inches	DW ft in	No/Name	Customer
11/1841	2-2-0	12 x 18	5 0	2	Manchester & Leeds Rly (c)
1841	2-2-0	13 x 18	5 6	50	London & Birmingham Rly (c)
1841	2-2-0	13 x 18	5 6	51	London & Birmingham Rly (c)
1841	2-2-0	13 x 18	5 6	52	London & Birmingham Rly (c)
	2-2-0	13 x 18	5 0		
1842	4-2-0	OC 11½ x 20	4 0	25 Eckington	Birmingham & Gloucester Rly
1843	2-4-0	14⅛ x 18	5 7	3	West Hartlepool Harbour & Rly Co
1844	0-6-0	15 x 24	4 9	171	Midland Rly (d)
1844	0-6-0	15 x 24	4 9	172	Midland Rly (d)
10/1844	2-2-2	13 x 18	5 6	101	Joint Committee (e)
10/1844	2-2-2	13 x 18	5 6	102	Joint Committee
12/1845	2-2-2	OC 15 x 22	5 6	1	South Eastern Rly (f)
2/1846	2-2-2	OC 15 x 22	5 6	2	South Eastern Rly
3/1846	2-2-2	OC 15 x 22	5 6	3	South Eastern Rly
3/1846	2-2-2	OC 15 x 22	5 6	4	South Eastern Rly
5/1846	0-6-0	15 x 24	4 8	Cambrian	Taff Vale Rly (f)
1846	0-6-0	15 x 24	4 8	Llandaff	Taff Vale Rly
1846	0-6-0	15 x 24	4 8	Newbridge	Taff Vale Rly
2/1847	2-4-0	15 x 22	4 6	6	South Eastern Rly (f)
4/1847	2-4-0	15 x 22	4 6	7	South Eastern Rly
5/1847	2-4-0	15 x 22	4 6	8	South Eastern Rly
1847	2-2-2	15 x 22	5 6	42	Edinburgh & Glasgow Rly
1847	2-2-2	15 x 22	5 6	43	Edinburgh & Glasgow Rly
10/1847	2-4-0	OC 15 x 22	5 6	182	Eastern Counties Rly (g)
10/1847	2-4-0	OC 15 x 22	5 6	183	Eastern Counties Rly
1847	2-2-2	OC 15 x 22	5 6	42	Edinburgh & Glasgow Rly (h)
1847	2-2-2	OC 15 x 22	5 6	43	Edinburgh & Glasgow Rly (h)
5/1848	2-4-0	OC 15 x 22	5 6	184	Eastern Counties Rly
6/1848	2-4-0	OC 15 x 22	5 6	185	Eastern Counties Rly
7/1848	2-4-0	OC 15 x 22	5 6	186	Eastern Counties Rly
8/1848	2-4-0	OC 15 x 22	5 6	187	Eastern Counties Rly
1848	0-6-0	16 x 24	5 0	11 Birkenhead	Birkenhead Railway (j)
1848	0-6-0	16 x 24	5 0	12 Chester	Birkenhead Railway (j)
1848	2-4-0	OC 14 x 21	5 0	5	North Staffordshire Rly (f)
1848	2-4-0	OC 14 x 21	5 0	6	North Staffordshire Rly
1848	2-4-0	OC 14 x 21	5 0	7	North Staffordshire Rly
1848	2-4-0	OC 14 x 21	5 0	8	North Staffordshire Rly
1849	0-6-0	16 x 24	5 0	13 Mersey	Birkenhead Rly
1849	0-6-0	16 x 24	5 0	14 Dee	Birkenhead Rly
1849	2-2-2	OC 15 x 22	6 0	15	Birkenhead Rly (k)
1859/50	2-2-2	OC 15 x 22	6 0	16	Birkenhead Rly
1849/50	2-2-2	OC 15 x 22	6 0	17	Birkenhead Rly
1849/50	2-2-2	OC 15 x 22	6 0	18	Birkenhead Rly
1849/50	2-2-2	OC 15 x 22	6 0	19	Birkenhead Rly
1850	2-2-2	OC 15 x 22	6 0	20	Birkenhead Rly

(a) Steam Carriage. (b) Or 'Victory'. (c) 'Bury' type. (d) Long Boiler type, ordered by Midland Counties Rly. (e) Joint Locomotive Committee of the London and Croydon, South Eastern and London & Brighton Rlys. Bought January 1845. (f) 'Long Boiler' type. (g) Sub-contracted from R. Stephenson & Co E C R Nos 182-187. (h) Sold to Stockton & Darlington Rly 1855—93 Uranus and 94 Neptune. (j) Delivered 1849. (k) Delivered 1851 Birkenhead Rly Nos 15 to 20. (l) Subsequently numbered.

HIGHLAND RAILWAY

Lochgorm, Inverness

Locomotive Superintendents: William Stroudley 1865–69; David Jones 1869–1896; Peter Drummond 1896–1911; Frederick G. Smith 1912–1915; Christopher Cumming 1915–March 1922; David C. Urie April 1922 to December 1922.

The Highland railway was formed in 1865 by the amalgamation of the Inverness and Perth Junction railway and the Inverness and Aberdeen Junction railway, and at that time William Barclay was the locomotive superintendent of both these constituent companies. He was a nephew of Alexander Allan and the typical 2-2-2 and 2-4-0 types of the 'Allan Crewe' pattern were well in evidence with the inclined outside cylinders well supported by double frames, leading wheels with underslung springs, domeless boilers with safety valves on the boiler barrel and over the firebox. They had all been built by outside contractors principally by Hawthorns of Leith, Neilsons and Sharp Stewart.

When William Stroudley took over in 1865 he spent most of his time rebuilding the older locomotives and it was not until his last year in office that Lochgorm works turned out the first official new locomotive. Nevertheless the boiler was second hand from old No. 3 ST MARTINS, a 2-2-2 built by Hawthorns in 1856, and was shortened to enable it to be accommodated in its new frames. The 'new' locomotive was a 0-6-0T designed for shunting duties and three were sanctioned. The first No. 56 BALNAIN appeared in February 1869 and its chief claim to fame was that it was the forerunner of the famous *Terrier* tanks. No 56 had the characteristic cab, bunker and rounded top side tanks, and although the other two were built after Stroudley had left, the three locos were similar. Stroudley left in the same year to take up the reins at Brighton Works on the London Brighton & South Coast railway.

Mr David Jones took his place, having come from Longsight, Manchester where he had served his apprenticeship under John Ramsbottom, locomotive superintendent of the North Eastern division of the LNWR.

Except for the other two Stroudley tanks which appeared in 1872 and 1874, the existing locomotive stock was deemed adequate and no new design was forthcoming until 1874 when a 4-4-0 was put in hand. Mr Jones was also steeped in the Allan tradition and this class appeared still with the 'Crewe' front end. The cylinders were 18" x 24" and coupled wheels 6' $3\frac{1}{2}$" diameter, and Allan's link motion used as on previous types.

David Jones introduced a novel type of chimney comprising an outer casing with louvres which produced a current of air around the chimney proper, the object being to lift the exhaust above the cab especially when the locomotive was coasting down the numerous gradients of the line. It was quite successful and the majority of locomotives built in this period were so fitted. He also introduced the counter pressure brake on to the Highland Railway for use on these formidable down gradients.

The 4-4-0s were known as the '60' class and the first batch built by Dubs & Co. Seven were built in the company's workshops at Lochgorm, the first being No. 4 ARDROSS in July 1876, the others appearing from 1883 to 1888. They were more familiarly known as *Skye Bogies.*

Fig. 272 **Highland Rly** 1877 D. Jones No. 29 RAIGMORE– note Stroudley cab roof and louvred chimney

Date	Type	Class (1901)	Cylinders inches	DW ft in	1st HR No	Name	Later Nos	LMS No	Notes	W'drawn
2/1869	0-6-0T	R	IC 14 x 20	3 7	56	Balnain	56A/$_{20}$ 56B/$_{21}$	16118	(a)	1928
11/1872	0-6-0T	R	IC 14 x 20	3 7	57	Lochgorm	57A/$_{20}$ 57B/$_{21}$	16119		1932
10/1874	0-6-0T	R	IC 14 x 20	3 7	16	St Martins	49/$_{01}$ 49A/$_{12}$	16383	(b)	1927
7/1876	4-4-0	F	OC 18 x 24	6 3	4	Ardross	31/$_{99}$ 31A/$_{11}$	–	(c)	1913
7/1877	2-4-0	H	OC 16 x 24	6 3	3	Ballindalloch	30/$_{98}$	–		1910
11/1877	2-4-0	H	OC 16 x 24	6 3	1	Raigmore	29/$_{98}$ 29A/$_{10}$	–		1912
12/1878	2-4-0T	O	OC 16 x 24	4 9	58	Burghead	58A/ 58B/	15011	(d)	1928
6/1879	2-4-0T	O	OC 16 x 24	4 9	59	Highlander	59A/ 59B/	15010	(e)	1932
12/1879	2-4-0T	O	OC 16 x 24	4 9	17	Breadalbane	50/$_{01}$ 50A/$_{12}$ 50B/$_{20}$	15012	(f)	1929
5/1882	4-4-0	L	OC 18 x 24	5 3	70		70A/$_{16}$ 67/$_{13}$	14277		1930
12/1883	4-4-0	F	OC 18 x 24	6 3	71	Clachnacuddin	71A/$_{12}$	–		1915
6/1884	4-4-0	F	OC 18 x 24	6 3	72	Bruce	72A/$_{16}$	–	(g)	1923
2/1885	4-4-0	F	OC 18 x 24	6 3	73	Thurlow	73A/$_{16}$	–	(h)	1925
9/1885	4-4-0	F	OC 18 x 24	6 3	74	Beaufort	74A/$_{16}$	–		1915
10/1886	4-4-0	F	OC 18 x 24	6 3	75	Breadalbane	75A/$_{17}$	–		1925
12/1888	4-4-0	F	OC 18 x 24	6 3	84	Dochfour	84A/$_{17}$	–		1925
5/1890	0-4-4ST	S	IC 14 x 20	4 3	13	Strathpeffer	53/$_{99}$ 53A/$_{19}$	15050	(i)	1929
8/1892	4-4-0	L	OC 18 x 24	5 3	85		85A/$_{19}$	–		1923
2/1893	4-4-0	L	OC 18 x 24	5 3	86			14279		1927
12/1893	4-4-0	L	OC 18 x 24	5 3	87			(14280)		1926
4/1895	4-4-0	L	OC 18 x 24	5 3	88			(14281)		1926
8/1897	4-4-0	L	OC 18 x 24	5 3	5		32/$_{99}$	14282		1929
11/1897	4-4-0	L	OC 18 x 24	5 3	6		33/$_{99}$	14283		1929
7/1898	4-4-0	L	OC 18 x 24	5 3	7		34/$_{99}$	14284		1930
7/1899	4-4-0	C	IC 18¼ x 26	6 0	9	Ben Rinnes		14405		1944
8/1899	4-4-0	C	IC 18¼ x 26	6 0	10	Ben Slioch		14406		1947
11/1899	4-4-0	C	IC 18¼ x 26	6 0	11	Ben Macdhui		14407		1931
4/1900	4-4-0	C	IC 18¼ x 26	6 0	12	Ben Hope		14408		1947
6/1900	4-4-0	C	IC 18¼ x 26	6 0	13	Ben Alisky		14409		1950
8/1900	4-4-0	C	IC 18¼ x 26	6 0	14	Ben Dearg		14410		1949
2/1901	4-4-0	C	IC 18¼ x 26	6 0	15	Ben Loyal		14411		1936
2/1901	4-4-0	C	IC 18¼ x 26	6 0	16	Ben Avon		14412		1947
2/1901	4-4-0	C	IC 18¼ x 26	6 0	17	Ben Allighan		14413		1933
12/1901	4-4-0	L	OC 18 x 24	5 3	48			14285		1928
10/1903	0-6-0T	V	OC 18 x 24	5 2½	22			16380		1930
12/1903	0-6-0T	V	OC 18 x 24	5 2½	23			16381		1932
5/1904	0-6-0T	V	OC 18 x 24	5 2½	24			16382		1930

3/1905	0-4-4T	W	IC 14 x 20	4 6	25			15051	(j)	1956
9/1905	0-4-4T	W	IC 14 x 20	4 6	40			15052		1930
12/1905	0-4-4T	W	IC 14 x 20	4 6	45			15053	(j)	1957
2/1906	0-4-4T	W	IC 14 x 20	4 6	46			15054		1945

(a) Renamed DORNOCH 1902 (b) Renamed FORT GEORGE 1899 (c) Renamed AUCHTERTYRE 1900
(d) Rebuilt 4-4-0T 1882 (e) Rebuilt 4-4-0T 1881 (f) Rebuilt 4-4-0T 1882 renamed ABERFELDY 1886
(g) Renamed GRANGE 1886 (h) Renamed ROSEHAUGH 1898 (i) Renamed LYBSTER 1903. Rebuilt side tanks 1901 (j) Renumbered British Railways 55051 and 55053 respectively.

HR renumberings shown thus $56A/_{20}$: renumbered 56A in 1920

LMS numbers allocated but loco withdrawn before they were used: (14280)

Total locomotives built 1869 to 1906: 41

W. STROUDLEY

Date	No Built	Type	Cylinders inches	DW ft in	Heating Surface sq ft			Grate Area sq ft	WP lb/sq in	Weight WO ton cwt	1901 Class
					Tubes	F'box	Total				
1869	3	0-6-0T	IC14 x 20	3 7	608	63·25	671·25	12·25	120	23-10	R

D. JONES

Date	No Built	Type	Cylinders inches	DW ft in	Tubes	F'box	Total	Grate Area sq ft	WP lb/sq in	Weight WO ton cwt	1901 Class
1876	7	4-4-0	OC 18 x 24	6 3	1132	96	1228	16·25	140	41 0	F
1877	2	2-4-0	OC 16 x 22	6 3	1004	93	1097	16·2	160	35 0	H
1878	3	2-4-0T	OC 16 x 24	4 9	820	93	913	16·2	140	36 11	O (a)
1882	9	4-4-0	OC 18 x 24	5 3	1123	93	1216	16·2	150	43 0	L (b)
1890	1	0-4-4ST	IC 14 x 20	4 8	578	62	640	12·5	100	32 0	S (c)

P. DRUMMOND

Date	No Built	Type	Cylinders inches	DW ft in	Tubes	F'box	Total	Grate Area sq ft	WP lb/sq in	Weight WO ton cwt	1901 Class
1899	9	4-4-0	IC 18 x 26	6 0	1060	115	1175	20·3	175	46 0	C (d)
1903	3	0-6-0T	OC 18 x 24	5 3½	1093	93	1186	16·2	160	47 18	V
1905	4	0-4-4T	IC 14 x 20	4 6	652	67·5	719·5	13·0	150	35 15	W

(a) Rebuilt as 4-4-0T 1881-1882 (b) Skye Bogies (c) Named STRATHPEFFER (d) Small Bens

Number built given for Lochgorm works only.

A goods version of this design was also put in hand with 5′ 3″ coupled wheels, the first one from Lochgorm being No. 70 in May 1882. It was ten years before further building of this class took place when eight more were built between 1892 and 1901. Their chief duties were on the Dingwall to Strome Ferry line – the 'Skye Line' and for pilot duties assisting the heavy passenger traffic over the banks.

The *Strath* class introduced in 1892 were similar to the *Skye Bogies* but had larger boilers and were built by Neilson & Co. These were the last locomotives to have the familiar 'Crewe' front end.

It is interesting to note that at the commencement of David Jones' term of office the locomotive stock consisted of 2-2-2, 2-4-0, 0-4-0T and 0-6-0T types. No six coupled goods tender locomotives had been thought of, or thought necessary – the four coupled type being in favour on most Scottish railways. Mr Jones however thought the time was ripe for introducing a six coupled class and in 1894 designed the first 4-6-0 to run in the British Isles. They were built by Sharp Stewart & Co., being too large to handle at Lochgorm. Indeed they were his masterpiece, and unfortunately he met with a serious accident on one of them which led to his retirement in 1896.

Peter Drummond was appointed in Mr Jones place. He had served his apprenticeship with Forrest & Moor of Glasgow and was the younger brother of Dugald Drummond. He had also served under William Stroudley on the LBSC then at Cowlairs on the North British Railway under Dugald, following him to St. Rollox.

His first design was the *Ben* class 4-4-0 of which nine were built at Lochgorm between 1899 and 1901. They had two inside cylinders 18¼″ x 26″ and 6′ diameter coupled wheels. They were the first tender locomotives to be turned out at Lochgorm with inside cylinders and had the Drummond family characteristics being very allied to the *290* class of the L&SW Railway designed by brother Dugald. The *Bens* had comparatively small boilers and the second batch built in 1908–9 had larger boilers initially with 1516·2 sq. ft as compared with 1060 sq. ft on the first design. These were known as *Big Bens* and the *Bens* became *Small Bens*.

As far as Lochgorm was concerned, after building the nine *Small Bens* and the last of the outside cylinder *L class*, three 0-6-0T were built in 1903–4 and four 0-4-4Ts in 1905–6. After this no more new locomotives were built in the Company's workshops but all rebuilding and repair work was continued, the works eventually closing under British Railways' ownership on 18 July, 1959.

All other new designs by Peter Drummond and his successors were built by outside contractors which are amply described in a number of works on the Highland Railway.

In all forty-one locomotives were built at Lochgorm, although the first three could not be called brand new productions as parts from replaced locomotives were used. It was a small works with an erecting shop with two parallel roads, 526′ x 40′ 6″ and pits 330′ long. The class of workmanship was acknowledged as very high as was the maintainance of the locomotives.

Fig. 273 **Highland Rly** 1890 0-4-4ST No. 13 STRATHPEFFER *Class S*

HOLT, HENRY P.
Leeds

Patented a design for a tramway engine in 1872 with compound inside cylinders and horizontal boiler, the most remarkable feature of which was an auxiliary chimney or flue constructed in the crown of the firebox and fitted with a self acting damper controlled by a cylinder connected with the boiler, by means of which the heat of the firebox was regulated. Steam was partly exhausted into the firebox flue and partly condensed in a chamber under the roof through which passed longitudinal air tubes.

However, it does not appear that the tramway engine was ever actually built.

HOPKINS GILKES & COMPANY
Middlesborough

See GILKES WILSON & COMPANY.

HOPPER, JOHN
Britannia Foundry, Fencehouses, Co. Durham

One locomotive was built for the Merrybent and Darlington Railway in 1870.

It was a six coupled saddle tank with outside cylinders 13″ x 20″ named MERRYBENT.

Major repairs and rebuilding were carried out at the North Eastern Railway's Darlington workshops, but when the Merrybent and Darlington Railway was taken over by the NER in 1890 it was sold, and eventually scrapped in 1905.

HORLOCK, A. & COMPANY
Northfleet Ironworks, Kent

Two four coupled locomotives with tenders were built in 1848 for the 4 ft gauge Dinorwic Slate Quarry lines: No. 1 FIRE QUEEN and No. 2 JENNY LIND.

These were two unusual locomotives with some interesting features (see Figs 275 & 276): a long wheel base with trailing wheels behind the firebox on T. Crampton's principle, screw reversing gear which is usually attributed to William Stroudley (1871) and valve gear driven off the front axle patented by Fletcher of Fletcher Jennings in 1864.

Fig. 275 **A. Horlock & Co**
View of underside of FIRE QUEEN

Fig. 274 **John Hopper** 1870 0-6-0ST Merrybent & Darlington Rly.

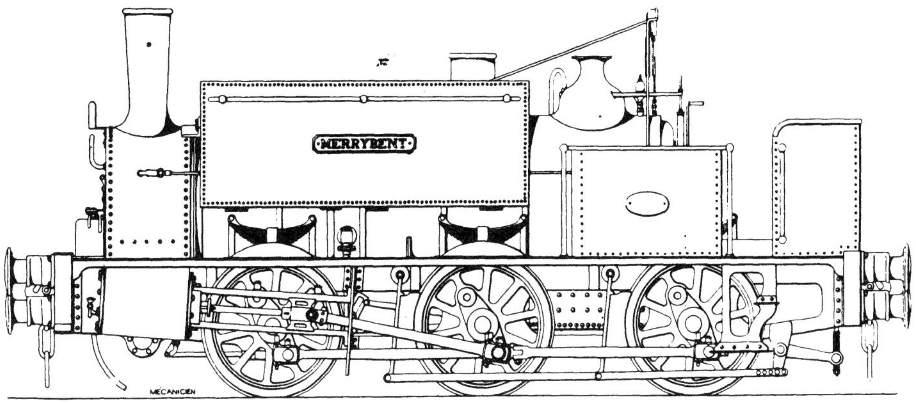

HORLOCK, A. & COMPANY

The firebox had no foundation ring and gussets were fixed between the barrel and throat plate. The cylinders and boiler had wood lagging and the regulator was of the pull-out type with separate rods to the right and left hand valves. Brakes were on the tender and sanding was behind the rear wheels.

At some time the cabs were removed. Both were withdrawn about 1886 but FIRE QUEEN is still intact.

Thanks are due to Mr C. R. Weaver for information and photographs.

HORNER & WILKINSON

It is extremely doubtful whether this firm built any locomotives, but they did build a number of passenger coaches for early railways in the north of England.

HORNSBY, RICHARD & SONS LIMITED

Spittlegate Ironworks, Grantham

Richard Hornsby established a foundry in Grantham in 1815. In 1851 the firm's title was altered to RICHARD HORNSBY & SONS and in 1880 it became a limited company.

Their main product was agricultural machinery including road traction engines and an adaptation of this design was carried out, retaining the two-speed gear and offering sizes from 8 nominal horse power to 12 n.h.p. The wheel formation was 2-2-0 similar to the many Aveling & Porter locomotives, with plate frames, solid wheels and geared drive.

No particulars are available as to whether any were actually built.

In 1918 the firm amalgamated with Messrs Ruston Proctor & Co. Ltd. to form the firm of Ruston & Hornsby Ltd.

An illustration of the tram engine can be found in *Steam Engine Builders of Lincolnshire* by R. H. Clark.

Fig. 276 **A. Horlock & Co.** 1848 *Crampton* type JENNY LIND 4′ gauge

HORSLEY COAL & IRON COMPANY

Tipton, near Birmingham

In 1833 a locomotive, built to the design of Isaac Dodds, was produced for trial on the Liverpool and Manchester Railway, but after being damaged in an accident in 1835 the L&M decided not to purchase the locomotive, and in September of the same year it was sold to the Dublin & Kingstown Railway.

It was named STAR and, as built, was a 2-2-0 with outside frames and 11″ x 18″ outside cylinders. From reports the workmanship of this engine was of a low standard and within three years it had been rebuilt with a pair of trailing wheels added. Notwithstanding this expense, so soon after purchase, the locomotive was never a success and in 1840 it was dismantled and a few usable parts incorporated in a 'new' engine built at the D&K shops.

Three locomotives for coal firing were also built for the St. Helens Railway. They were four coupled with 13″ x 20″ cylinders and 4′ 6″ wheels. The cost was £450 each and building took place between 1833–5.

HUDSON, ROBERT

Gildersome Foundry, Morley, Leeds

This firm, established in 1865, were specialists in light railway equipment but never manufactured any steam locomotives. When steam orders were placed with Hudson's they were always sub-let to either the Hunslet Engine Company and their associates or Hudswell Clarke & Company.

HUDSWELL & CLARKE

Railway Foundry, Leeds

The firm was founded in 1860 as Engineers, by W. S. Hudswell, the son of a local pastor who had served his apprenticeship with Kitson & Company, and Mr John Clarke, Kitson's works manager for many years.

Heavy engineering machinery, steam hammers and the like were built from the outset but their principal manufacturing talents were from 1861 concentrated on locomotive building. They realised that although they were neighbours to Kitson & Company, and Manning Wardle, both locomotive builders, that the home and overseas markets still held opportunities for additional output and orders.

That their faith was justified was amply borne out by the steady flow of orders received, particularly for four coupled and six coupled industrial saddle tank locomotives.

Besides these standard tanks a number of 0-6-0 tender locomotives were built for northern collieries, Cockermouth & Workington, North Staffordshire and North Eastern railways from 1862 to 1870.

From 1865–7 the firm undertook the rebuilding of Jones and Potts 2-4-0s and R. Stephenson's 0-6-0s for the North Staffordshire Railway, some of the latter being converted to saddle tanks.

Mr Rodgers joined the firm in1866, as did a Mr Clayton (a 'sleeping partner') but it was not until 1870 that the firm's name was altered to: HUDSWELI, CLARKE & RODGERS.

Not many locomotives were sent overseas; a few small tanks went to New Zealand, Greece, and elsewhere but the home market was the mainstay.

Some handsome 4-4-0Ts were supplied to the Lynn & Fakenham, and Yarmouth and North Norfolk railways from 1878 to 1881. A total of seven were built, the first with outside cylinders 14″ x 20″ and 4′ 6″ diameter coupled wheels. The boiler gave a total heating surface of 565·8 sq. ft and a grate area of 9 sq. ft. A large cab was fitted with side windows affording more than usual protection for the period.

Subsequent locomotives had the cylinders increased to 15″ and had a slightly larger boiler.

In 1880 Mr Rodgers left and in 1881 the firm became: HUDSWELL CLARKE & COMPANY.

In December 1899, after the death of the original partners, it was formed into a private limited liability company.

Most standard tank locomotives of this period had unbalanced wheels, springs above the axleboxes on the leading and trailing axleboxes and a transverse laminated spring on the trailing axle on the six coupled types. Chimneys were built up in three sections with a pronounced radius and reverse radius on the

Fig. 277
Hudswell Clarke & Rogers
60/1866 North Staffordshire Rly 2-4-0 No. 25

Fig. 278
Hudswell Clarke
216/1881
Atterbury & Shaw
10″ x 16″ 0-4-0ST

top, some however were fitted with stove-pipe chimneys. The firebox which was raised, housed the regulator and safety valves were fitted over the top.

Three standard 0-6-0STs with outside cylinders 12″ x 18″ and 3 ft diameter wheels were supplied to the Guimarães Railway, Portugal in 1882. Tasmania received a 3′ 6″ gauge 0-6-0T in 1884. In the same period the Royal Arsenal was supplied with a number of 0-4-0STs for the 1′ 6″ gauge lines. They had outside cylinders 7″ x 12″ and 2′ 1″ diameter wheels and nine were built in 1884–9.

T. A. Walker, the civil engineering contractor, received seven 0-4-0STs, nine 0-6-0STs and two 0-6-0Ts in 1887–9.

In 1889–91 the firm built five 0-6-0Ts for the Barry Railway. They were small locomotives of the builder's own design with inside cylinders 14″ x 20″ and 3′ 3½″ diameter wheels. They weighed 27½ tons in working order.

Three 0-4-2STs were supplied to the Labuan Colliery, Borneo in 1890–1 to 2′ 5″ gauge, one 0-4-0ST to Spain in 1890 and a 4-4-0T with outside cylinders 11″ x 16″ to the Selangor State Railway in the same year.

A small compound 4-4-0 was supplied to the Santander and Bilbao Railway in 1892, and in 1897 three neat 2-4-2Ts were supplied to the Barry Railway. They were comparatively large locomotives weighing 60 tons 18 cwt in working order and had inside cylinders 18″ x 26″ and 5′ 7½″ diameter coupled wheels. They were finished off with brass caps to the chimneys, domes and safety valve covers also of polished brass; and they were most successful.

Five 0-6-2STs were supplied to the Rhymney Railway in 1899 and two interesting 4-6-2Ts for the Londonderry and Lough Swilly Railway in the same year followed by two similar ones in 1901. All wheels had outside bearings and inside Stephenson's link motion was fitted. The outside cylinders were 15″ x 22″ and coupled wheels 3′ 9″ diameter. The 1901 pair were slightly larger and unique in having Allan's straight link motion – the only locomotives in Ireland so fitted.

Hudswell, Clarke & Company also built two 4-8-0 tender locomotives in 1905 for the same railway. These two were unique in being the only tender locomotives on the narrow gauge railways of Ireland and they were the first in the country to have eight coupled wheels. They had outside cylinders 15½″ x 22″ and 3′ 9″ diameter coupled wheels and all wheels had outside bearings and outside Walschaert's valve gear.

By 1900 approximately 575 locomotives had been built, averaging between fourteen and fifteen locomotives per annum, the majority being standard four coupled and six coupled industrial tanks.

A good deal of business was done with various Welsh railways at the turn of the century; six 16″ 0-6-0STs supplied to the Port Talbot Railway in 1900 and six 0-6-2Ts with 17½″ x 26″ cylinders for the Taff Vale Railway in 1902.

The Manchester Ship Canal Company were supplied with twenty-four 0-6-0Ts, with inside cylinders 15½″ x 20″ and 3′ 4½″ diameter wheels – a considerable order from a private firm, and a useful standard class.

Two steam rail motor units were built in 1907 for the Rhymney Railway, Cravens supplying the bodies. They were powerful four coupled units with loco type boilers. They were afterwards converted as separate tank locomotives with six coupled wheels.

In 1904 a 2-6-0T had been built for the Castlederg and Victoria Bridge Tramway (No. 4) with coupled wheels 3′ 1″ diameter and a rigid wheelbase of 7′. Apparently this was too much for the curves and in 1912 a 0-4-4T was built with an outside framed bogie and inside frame for the coupled wheels. Stephenson's valve gear was used. It was numbered 5 and had inside cylinders 12½″ x 18″ and 3′ 1″ diameter coupled wheels on a 5′ 6″ wheelbase. A large sandbox was placed on top of the boiler between the chimney and the dome and as was customary on Tramways the motion was shrouded with hinged plates.

In 1913 the Londonderry and Lough Swilly Railway received two 4-8-4Ts – the only two ever built to this wheel arrangement to run in the British Isles, and the most powerful for the Irish narrow gauge railways. They weighed 51 tons, had outside bearings to all wheels. The boilers were fitted with Belpaire fireboxes and gave a total heating surface of 1063 sq. ft and a grate area of 17·6 sq. ft. They were numbered 5 and 6, and were a tank version of the two 4-8-0s supplied in 1905.

A further batch of 15½″, 0-6-0Ts were supplied to the Manchester Ship Canal Company 1912–14 and the Port of London Authority had ten 16″ 0-6-0Ts and five 14″ 0-4-0STs both types with outside cylinders, in 1915–17.

Fig. 279 **Hudswell Clarke** 1332/1918 Shelton Iron & Steel Co. MYRESIDE

During the war years the company was busily occupied on building their standard tank designs, and a number of 0-4-0WTs for the Air Ministry. Steel works, collieries and other manufacturers engaged on products for the war effort were supplied. The War Dept also received a number of 0-6-0WT mostly for overseas work.

After the war they were busily engaged in fulfilling orders which had been put on one side and besides industrial tank locomotives some 0-6-2Ts were built for the Rhymney Railway, seven 0-6-0Ts for the Rhymney and Cardiff railways; and eight for the Takoradi Harbour Works, 3′ 6″ gauge with outside cylinders 11½″ x 16″ and 2′ 9½″ diameter wheels.

In 1923 a 0-4-0CT was built for Murray and Patterson Ltd of Coatbridge, the crane being supplied by Thomas Smith and Sons, Rodley. The crane had a capacity of 2 tons at 20′ 6″ radius.

By 1927 approximately 1600 locomotives had been built both for narrow and standard gauges for all kinds of duties and conditions and were dispersed to most parts of the world.

From time to time the standard range of tank locomotives were modified and in 1929 an improved version of their 14″ four coupled saddle tanks appeared. The tanks were lengthened to cover smokebox, boiler barrel and firebox, the stroke increased from 20″ to 22″ and a standard diameter of wheels 3′ 3½″ used. Boiler pressure was 160 lb/sq. in. and this version weighed 28 tons 10 cwt in working order. The result was a locomotive with neater lines and a useful addition to the company's range.

During the trade depression orders were fewer and the output in the early 1930s was scanty and this recession continued up to the 1939 period. During the war years industrial demands were heavy and in 1943 an order was commenced for fifty Hunslet designed 'austerity' 0-6-0STs. These were completed in 1946, and formed the bulk of locomotive construction during these years. After the war the main portion of production was still four and six coupled industrial tanks. An order for twenty British Railways Western region 0-6-0PTs with taper boilers was subcontracted to Robert Stephenson and Hawthorn Ltd.

In 1950 four 2-6-2 outside cylinder tender locomotives were built to the 2′ 6″ gauge for the Dehri Rotas Light Railway of India; and in 1951–2, three 2-8-4Ts to metre gauge for the Iraq Petroleum Co.

Diesels had been built by Hudswell Clarke & Company from the 1920s and orders for this type of locomotive gradually outstripped those for steam.

The last steam locomotive was built in 1961 being the last of an order placed by the National Coal Board for their standard 14″ 0-4-0STs.

The works number was 1893, and taking into consideration locomotives which were rebuilt only, and the fact that in early days some works numbers were given to stationary engines and marine work, plus tenders for main line locomotives for a few years, the total number of steam locomotives built totalled 1807 for a period of 101 years.

Diesel locomotives only are now built by the firm.

Fig. 280 **Hudswell Clarke** 1523/1923 Holditch Colliery, Chesterton, Staffs. CORNIST

HUGHES, HENRY & COMPANY

Falcon Works, Loughborough

In 1865 Henry Hughes, an engineer and timber merchant founded the Falcon Works, having bought seven acres of land alongside the Midland Railway. He commenced by building rail coaches, wagons and horse-drawn tram-cars, principally of wooden construction following on the timber merchant's business. There was a ready sale for the company's products and within a few years, some two hundred people were employed there.

Locomotive building started probably in 1865. In *Engineering* for 8 March, 1867 is a description of a 0-4-0ST shown at the Paris Exhibition. It had outside cylinders 12" x 20" and 3′ 0" diameter on a 5′ 9" wheelbase. The saddle tank held 500 gallons and a Giffard injector was fitted. In the same article it was stated that nearly one hundred locomotives had already been constructed to gauges varying from 2′ 3" to 7′ 0". That such a statement was correct is extremely doubtful and well nigh impossible. Many were supplied to contractors in this period, but it is strange that only a few can be traced and indentified.

As with the other firms in the same business, the Tramways Act of 1870 attracted Hughes attention and the construction and development of steam locomotives was proceeded with. In 1875 a tram engine was built to the order and design of John Downes of Birmingham, running public trials between Handsworth and West Bromwich in January 1876, but did not go into regular service. A further tram engine patented by Hughes commenced running on the Leicester Tramways on 6 March 1876, where it performed for a few months coupled to a trailer car. This tram engine weighed approximately four tons and was equipped with a device for consuming its own smoke and condensing the steam (which was a stipulation of the Act). It did not win favour in Leicester and was transferred to Govan near Glasgow where in the company of eight others, built between 1877 and 1879 worked regularly until 1881 when they were all withdrawn and replaced by new tram engines of another maker.

Other similar tram engines were supplied to the Swansea and Mumbles Railway Co, Bristol, Guernsey, Paris, Wantage and Lille.

These engines were mostly conventional saddle tanks with a body completely enclosing the engine, they were known as dummies, as distinct from some of the earlier tram engines which were constructed with the engine and boiler integral with the passenger car. The tram engine supplied to the Swansea and Mumbles Railway Co. was used on the inaugural run of steam hauled trains on 16 August, 1877; it was named THE PIONEER. In addition a number of 0-4-0 saddle tanks were built and principally used as contractors' engines.

In the late 1870s and early 1880s the firm got into financial difficulties due to a trade recession and in 1883 the firm was

Fig. 281 **Henry Hughes**–Typical 0-4-0ST for contractor John Aird No. 75

Fig. 282
137/1887
(Falcon)
Cork & Muskerry
Lt. Rly. No. 1
CITY OF CORK
at St. Annes

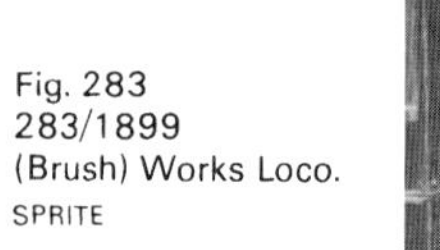

Fig. 283
283/1899
(Brush) Works Loco.
SPRITE

Fig. 284
314/1906–
last owner
Berry Wiggins & Co.
presented to
Leicester Museum

reconstructed under the title of FALCON RAILWAY PLANT WORKS. Control was taken over by Mr Norman Scott Russell, Henry Hughes having left the company. It is said he emigrated to New Zealand where he died in 1896.

Trade improved and the factory became busy on the manufacture of locomotives and rolling stock for home and abroad. Unfortunately there are no records of Henry Hughes works list and numbers, and it is assumed that two series were used, one for Tram engines and one for conventional locomotives. Russell developed an air condensing tram locomotive which had also a speed regulating governor driven by a fifth wheel running on the rails and kept in position by a spring. The air condenser consisted of 250 thin copper tubes arched transversely across the roof of the cab and communicating with each other through a box chamber running along each side.

About forty-two Hughes and sixty-one Russell tram loomotives were built between 1874 and 1888. A number of four coupled saddle and well tanks were built during this period, 2-4-0Ts for the Cork and Muskerry Railway, 4-4-0s for the Bolivar Railway Co. and four coupled tanks for Spain and the Azores.

The assets of the Falcon Railway Plant Works were acquired by the Anglo-American Brush Electric Light Corporation in 1889 and a new Company formed entitled BRUSH ELECTRICAL ENGINEERING CO., and although the activities of the firm were of a more varied nature the tradition of railway rolling stock continued, including steam locomotives until 1914, the last big contract being received in 1910 for seven six-coupled side tanks from the Royal Siamese State Railways.

After the 1914–18 war no more steam locomotives were built due to foreign competition, and the concentration of building electric tramcars, buses, wagons and battery operated vehicles. The firm is now one of the principal suppliers of diesel-electric locomotives, many of which have gone to overseas customers.

No records of Hughes earlier locomotives have been discovered, but before the Falcon list was commenced about thirty to fifty 0-4-0STs were produced so that altogether approximately 250 steam locomotives were built in addition to the tram engines.

HUGHES, OWEN

Valley Foundry, Anglesey

This general engineering firm is credited with building two vertical boilered locomotives but it is doubtful whether all the material was new.

One was supplied to Brundrit & Co of Penmaenmawr for their 3 ft gauge lines, named MONA and built in the 1870s. It was taken by De Winton & Co in part exchange for one of their new locomotives PENMAEN delivered in 1878. No doubt MONA was reconditioned by De Winton and re-sold.

The second was used by Parry & Co contractors who built the Penrhyn Railway, completed in 1876. It was named COETMOR and would have been built in the early 1870s and was sold by the contractors to Lord Penryhn's Slate Quarries in 1876 on completion of the railway. It was re-named BRONLLWYD.

No further details are available, but it has been suggested that these two locomotives could have been of Chaplin manufacture reconditioned by Owen Hughes.

HULL & SELBY RAILWAY

Hull

Two six coupled tender engines with 16″ x 24″ cylinders and 5 ft wheels were built in 1844. John Gray was the Locomotive Superintendent and the locomotives built had a number of characteristics. He had designed and fitted an expansive valve motion, and the boilers were pitched higher than was usual at this time. Gray left the Hull & Selby to take up a similar position on the London Brighton & South Coast Railway.

HUNSLET ENGINE COMPANY
Leeds

The Hunslet Engine Company was founded in 1864 and workshops were built on the plot of land on which had been built the locomotive works of E. B. Wilson & Company, and named the Railway Foundry. These buildings had been pulled down and notwithstanding the fact that there were already three locomotive builders established around this site – Manning Wardle, Hudswell Clarke and Kitson, the founder of the Hunslet Engine Company thought there was enough scope for a fourth works in the neighbourhood.

The founder John Towlerton Leather was a railway contractor and civil engineer, and had built the firm for his son Arthur Leather hoping that orders could be obtained for contractors' and industrial locomotives. The manager appointed was James Campbell, son of Alexander Campbell the manager of Manning Wardle. Although building the same type of product there was a close relationship between the two firms and Hunslet were assisted in the building of their first locomotives by Manning Wardle supplying material. Close examination of the designs of each firm shows great similarity in main features.

The first locomotive built by Hunslet was a 0-6-0ST with inside cylinders 14″ x 18″, tank over the boiler barrel only, and combined dome and safety valves over the firebox. The cab was a weatherboard bent over a small amount towards the rear. Compensating beams were fitted between the leading and driving wheels, and the driving and trailing wheels had hand brakes with wooden blocks. The 3′ 4″ diameter wheels had the double crank pin bosses as balance weights and additional weights were added between the spokes each side of the dummy bosses. A very plain parallel chimney with a sharp flare at the top was used and the smoke box door was semi-circular with a horizontal hinge. Laminated springs were used on the leading and driving wheels and one transverse spring on the trailing wheels. It was completed on 18 July, 1865 and went to Brassey & Ballard working at Ampthill. Its name was LINDEN. A large number of these 0-6-0STs were built for contractors and industry and also 0-4-0STs for similar customers. The four coupled saddle tanks had outside cylinders 10″ x 15″ and 2′ 9″ diameter wheels. Sizes varied according to gauge and requirements.

Saddle tank locomotives were sold to the Bengal-Nagpur Railway, and Dominion Coal Company, Nova Scotia.

A unique locomotive was constructed in 1871 (Wks No. 52) for the Oudh and Rohilkund Railway. It was a vertical boilered 2-2-0WT inspection car and was turned out with a steel framing surrounding the boiler and the space in front of it where the seating and windows were fitted on arrival in India. The outside cylinders were $5\frac{1}{2}$″ x 12″ and the single drivers 3′ 9″ diameter. It was the only 'single' built by Hunslet.

In 1872 six 4-4-0STs were built for the Prince Edward Island Railway's 3″ 6″ gauge and in 1874 some 4-6-0Ts and 4-4-0Ts for Tasmania.

Although Hunslet never seriously developed the 0-4-0T steam tram a fireless loco was built about 1879 for L. Francq of Paris. The cylinders were 9″ x 12″, four coupled wheels 2′ 6″ diameter and a boiler pressure of 225 lb/sq. in. A patent condensing arrangement was used with 950 tubes 1″ outside diameter with a cooling surface of 954 sq. ft. The volume of water was 400 gallons and the diminutive

Fig. 285
Hunslet Engine Co.
1/1865 for
Brassey & Ballard
contractors–
cylinders 11″ x 18″
3′ 4″ D.W.

Fig. 286
Hunslet Engine Co.
254/1880
Carthagena & Herrerias Steam Tramway Co. Ltd.
Designed to work back to back with No. 253
ESCOMBRERA

Fig. 287
Hunslet Engine Co.
299/1882
Hodbarrow Mining Co.
Millom, Cumberland.
Worked until 1968

locomotive weighed 7 tons empty and 8 tons 15 cwt in working order. It was not very successful compared with the more conventional tram locomotives made by neighbours Manning Wardle, Kitson, Thomas Green and others farther afield.

Narrow gauges were catered for from 1870 and many saddle tanks were supplied to ironstone and slate quarries particularly in Wales. Larger types were supplied to the North Wales Narrow Gauge railway (1′ 11½″) a 0-6-4ST BEDDGELERT in 1878 (Wks No. 206), a 2-6-2T RUSSELL in 1906 (No. 901) and a 0-6-4T GOWRIE (No. 979) in 1908.

The first eight coupled locomotives were eleven saddle tanks for the F. C. Los Blancos, Cartagena and Herrerias Steam Tramway.

They were built from 1880 to 1907 and had outside cylinders 13″ x 20″, weighed 30 tons in working order and were to 3′ 6″ gauge.

In 1887 appeared three of the most novel locomotives built by Hunslet. They were for the Listowel and Ballybunnion Railway for Lartigue's patent monorail. The locomotives had twin boilers of locomotive type and were suspended on a rail fastened on to trestles. There were three coupled wheels 2 ft diameter, and two cylinders 7″ x 12″. The tender was also powered by two cylinders 5″ x 7″ and drove two wheels 2 ft diameter through a clutch and shaft. It was a very early type of 'booster', but the scheme proved superfluous as the locomotive itself was capable of all duties required.

Control of the Hunslet Engine Company had earlier passed into the hands of James Campbell. Young Arthur Leather was not interested in locomotive matters and so the Leathers' interests were bought and all financial and general policies were governed by Campbell. The works were extended from time to time and the production of four coupled and six coupled tank locomotives remained the mainstay of the firm, each type being improved and enlarged at intervals.

Fig. 288 **Hunslet Engine Co.** 1155/1914 Gold Coast Rly No. 83– superheated 3′ 6″ gauge

In Ireland the Tralee & Dingle Light Railway was supplied by Hunslet with its first locomotives. The first three were 2-6-0Ts built in 1889 and were fitted with cabs at either end and were first used by the contractor building the railway which was opened in 1891. They had outside cylinders 13″ x 18″ and 3′ 0½″ diameter coupled wheels. The fourth, built in 1890 was a 0-4-2T with the two cabs and wheels and motion enclosed. Subsequently the front cab and platework around the wheels and motion were removed. In 1892 a larger 2-6-2T was built on the same lines as the 2-6-0Ts with 13½″ diameter cylinders. It was initially equipped for oil burning to Holden's patent – the first to be so equipped in Ireland. Two further 2-6-0Ts were built by Hunslet for this 3 ft gauge line, one in 1898 and one in 1910. Other Irish railways and contractors were supplied, including three 4-6-0Ts for the West Clare Railway (1912 and 1922).

In 1898 three 2-6-2Ts were built for the Sierra Leone Government Railway to the order and inspection of the Crown Agents. They had outside frames, cylinders, cranks, and motion. The tanks went the full length of the boiler and smokebox, the base being cut away at the front for access purposes. The boiler was domed and the firebox protruded well into the cab and the Ramsbottom safety valves were fitted on the firebox in the cab with exhaust pipes through the cab-roof. The cylinders were 10″ x 15″ and the coupled wheels 2′ 4″ diameter. The boiler with round top firebox gave a heating surface of 312 sq. ft and had a grate area of 6·25 sq. ft. These locomotives were typical of many Hunslet types in that they obtained many repeat orders extending over many years with few major modifications. A total of thirty-two were built, the last two as late as 1954.

Three curious articulated locomotives were built in 1901 for the Lagos Steam Tramway to 2′ 6″ gauge. The units consisted of a conventional outside cylinder four coupled locomotive with the frames cut down at the rear. Between the frames was fitted a casting with a ball joint on which the rear portion of the unit rested, the frame passing either side of the firebox and resting on a bogie at the rear. This formed a platform on which the coal bunkers and water tanks were fitted and a small covered compartment behind the footplate. This gave a flexible unit. The cylinders were 6¼″ x 8″ and coupled wheels 1′ 6½″ diameter. Two more were built in 1910.

In 1902 the firm was reconstituted as a private limited company and became THE HUNSLET ENGINE CO. LTD.

The export side of the business steadily increased for a variety of gauges and types, and locomotives were sent to South America, Africa, India, Federated Malay States, Italy, Spain, Portugal, Australia, Tasmania and New Zealand including some 4-6-0 tender locomotives for the Federated Malay States with outside cylinders 14½″ x 20″ and 4′ 3½″ diameter coupled wheels, and 5′ 3 gauge 4-6-0s for the Ceylon Government Railway in 1908.

The 1000th locomotive was built in 1909 – a 2-4-2T with outside cylinders 9½″ x 14″ for the Baraset-Basirhat Light railway, via T. A. Martin & Co., Calcutta.

In 1912 Edgar Alcock joined the firm as Works Manager, having received his training on the Lancashire and Yorkshire Railway and subsequently becoming assistant works manager at Beyer Peacock and Company. He became a director in 1917 and in this preliminary period the works were extended and re-organised, new methods introduced and a policy of building much larger locomotives commenced after the 1914–18 war.

During the Great War, Hunslet built a large quantity of 4-6-0Ts for the War Department for use in the forward areas of the battle fronts in Europe. A total of 155 were built from 1916 to 1919 to a gauge of 60 cm. The outside cylinders were $9\frac{1}{2}$" x 12" and coupled wheels 2' 0" diameter. They were built to a strict specification with a maximum permitted axle load of $3\frac{1}{2}$ tons and the bogie to have sufficient side play to traverse a radius of 20 metres. For such a wheel arrangement they were very small locomotives. The boiler gave a total heating surface of 205 sq. ft and a grate area of 3·95 sq. ft. The dome was combined with safety valves of the 'pop' variety and this combination became a characteristic on some of Hunslet's locomotives. Behind the dome was a large cylindrical sand box.

The factory was also engaged on munition of all descriptions.

After the war the first large order was for twenty-six superheated 4-6-0s for the metre gauge Indian Railways. This was the beginning of the manufacture of larger orders of larger locomotives and steps were taken to cope with this sort of order by extending the erecting shop and other departments and installing new equipment. Some of the Indian 4-6-0s had 16" x 22" outside cylinders and 4 ft diameter coupled wheels and others had $16\frac{1}{2}$" x 22" cylinders and 4' 9" diameter wheels. The order was completed in 1921.

Similar locomotives were built for Ceylon in 1922. In the same period three large metre gauge 4-6-4Ts were built for the Burma Railways with outside cylinders 16" x 22" and 4 ft diameter coupled wheels and in 1923 two very similar tank locomotives were built for the British Guiana Railways, and in the same year appeared a new Hunslet standard 0-6-0ST with inside cylinders 16" x 22" and 3' 9" diameter wheels.

During 1924–28 Hunslet built ninety 0-6-0T for the London Midland & Scottish Railway, they were a development of the Midland Railway's Johnson *Class 3* tanks built by the Vulcan Foundry in 1899–1902. The current ones had two inside cylinders 18" x 26" and 4' 7" diameter wheels, Belpaire firebox, and side tanks holding 1200 gallons. Working pressure was 160 lb/sq. in. and the weight in working order 49 tons 10 cwt, a total of 422 were built by various firms up to 1931.

In 1927 Manning Wardle went into voluntary liquidation, their equipment and shops had not been modernised in any way and as a consequence they could not compete on equal terms with firms such as Hunslet which had endeavoured to keep abreast of development and to invest in new plant from time to time. The goodwill was purchased by Kitson & Co, and the Hunslet Engine Co. bought a portion of Manning Wardle's works which amounted to about five acres. This enabled them to extend their own workshops at a time when space was badly needed.

In the same year Edgar Alcock's son John joined the firm and as a result of his father's insistance that he should make a special study of internal combustion engines, Hunslet have always kept in the forefront of diesel locomotive development and construction. 1930 witnessed the closure of Kerr Stuart's works at Stoke-on-Trent. The goodwill of this firm was obtained by the Hunslet Engine Co.

Kerr Stuart had considerable experience in building diesel locomotives and it was Kerr Stuart No. 4428, completed at Hunslet, that became the first of many diesels to emerge from this works.

Kerr Stuart had a working agreement with the Leeds firm of Robert Hudson & Co. who were manufacturers of light railway equipment and this relationship was maintained by the Hunslet Engine Co.

In 1930 eight 0-8-0Ts were built for the Gold Coast Railway's 3' 6" gauge. These powerful tanks weighed nearly 50 tons and had two outside cylinders 18" x 23" and wheels 3' $6\frac{3}{4}$" diameter. Further orders gave a total of fifty-seven locomotives up to 1955, some going to the Nigerian Railways.

Two other 0-8-0Ts built 1933–4 were built for the Chinese Government for marshalling trains on to the Nanking ferry. They had outside cylinders $22\frac{1}{2}$" x 26", 4' 3" diameter wheels and weighed nearly 85 tons. The boiler pressure was 200 lb/sq. in. giving a tractive effort of 38,710 lb which must have made them the most powerful eight coupled standard gauge tank locomotives in the world.

A similar locomotive was built for the Tata Steel Co. India in 1935 but with a tender instead of side tanks.

Following on the closure of Kerr Stuart in 1930, a further casualty of the trade depression was the failure of the Avonside Engine Co., Bristol in 1934 which went into voluntary liquidation. In the following year Hunslet bought the goodwill of the firm together with the drawings, patterns etc, but did not acquire the premises.

Standard four and six coupled tanks continued to be built for industry at home and abroad – interspersed with diesel locomotives.

During the Second World War orders were completed where possible and a great amount of work was done for the Ministry of Supply including anti-tank guns, shell cradles, aircraft parts etc.

Fig. 289
Hunslet Engine Co.
3902/1971
0-4-2ST–similar to Kerr Stuarts *Brazil* class for Indonesia

In the later years of the war two types of locomotives were required by the Ministry of Supply for the anticipated moving of huge quantities of supplies before and after D-Day. One was for a main line freight and the other a powerful shunting tank. The LMS standard 3F 0-6-0T was put forward for the latter but eventually it was decided on a design based on Hunslet's 18″ 0-6-0ST which had first appeared in 1937.

The final design was a saddle tank locomotive with the tank extended over the smokebox, 18″ x 26″ cylinders, 4′ 3″ diameter wheels, 170 lb/sq. in. boiler pressure and a weight in working order of 48 tons 5 cwt. A total of 484 were built from 1943 to 1964, 217 by Hunslet. The life of those built during the war was to be two years, changes in specifications and manufacture were introduced due to the shortage of many materials. Considering that their life expectation was two years, and some are still working over twenty-five years after a parallel can be drawn with our pre-fab houses many of which are with us still.

In 1944 John Alcock was made joint managing director and on the death of his father in 1951 became chairman and managing director.

Fifty more *Austerity* class were ordered by the National Coal Board and these were built in the 1950s. Others have been rebuilt, others tried with mechanical stokers, oil burners and Giesl ejectors. Many of the original WD batches were sold to the NCB Port of London Authority, the London & North Eastern Railway and to other industries.

Other steam locomotives built during this period were 0-6-2Ts for the 5′ 6″ gauge Calcutta Port Commissioners with 16″ x 24″ outside cylinders and 3′ 10″ diameter coupled wheels. The main feature of this class was the high working pressure of 210 lb/sq. in. Repeat orders were received for 0-8-0Ts for the Gold Coast & Nigerian Railways.

2-8-0s and 2-8-2s were built, the latter for the Barsi Light Railway 2′ 6″ gauge and a 2 ft gauge of the same wheel arrangement for South Africa.

2-6-4Ts and 2-8-4Ts appeared in the 1950s for India and thirty 0-6-0PTs were ordered for the Western region of British Railways. The latter locomotives were sub-contracted to the Yorkshire Engine Company and were built in 1954.

No. 3409, the last of the order, steamed in October 1956 and was the last pannier tank to be built of the long line of GWR tanks – a design which was so characteristic of that railway. It seems almost unbelievable that three of this batch 3404, 3407 and 3408 were scrapped in 1962 their age being six years. If that is not inept planning to put it mildly – what is?

In 1960 Hunslet acquired the goodwill and drawings for Kitson and Manning Wardle, industrial locomotives from Robert Stephenson & Hawthorns Ltd, and this of course brought in additional work, especially for spares.

Many repairs and rebuilding took place of locomotives by Hunslet and other makers. At least four of the Snowdon Mountain Railway's locomotives originally built by the Swiss Locomotive Works at Winterthur on the Abt rack system, have been rebuilt and modified by Hunslet.

The restored and repainted ex LMS *Jubilee* class 4-6-0 No. 5596 BAHAMAS for the Stockport (Bahamas) Locomotive Society, was handed over on 11 March, 1968. The company still has the plant and capacity to build steam locomotives, and are still manufacturing spares

HUNSLET ENGINE COMPANY

for many locomotives at home and abroad.

The Hunslet Engine Company was entrusted with an order by the Indonesia Forestry Commission to build a steam locomotive for their 2′ 6″ gauge railway. This (Works No. 3902) was completed in November 1971. (Fig. 289).

This 0-4-2 Saddle Tank engine will be employed on the Commission's Railway hauling logs from the forest to the saw mills and the sawn timber to the river for further transporting by boat to the cities. An interesting aspect of this application is that this steam locomotive is replacing a diesel powered locomotive. The reason is basically one of economics.

Fuel oil for use by the diesels on this particular railway has to be transported a considerable distance up river by barge and this results in the high cost of over £1 per gallon. This steam locomotive is designed to burn waste timber as fuel.

The design was developed in 1919 and with the exception of fitting a steam driven generator for the electric lights, the locomotive was built to these fifty year old drawings. It must be a unique accomplishment for a locomotive manufacturer to have under construction modern locomotives ranging in size from the tiny 6HP battery electric units, through the 300 and 400 HP class of standard shunting locomotives and the latest 1124 HP *Bo-Bo* diesel-electric heavy industrial locomotives, to this vintage steam locomotive and all in building at the same time. The total number of steam locomotives built to date is 2236. In 1972 the Hunslet group of companies purchased the equity of Andrew Barclay Sons & Co. Ltd. the two factories being fully occupied in a variety of work. Thanks are due to The Hunslet Group of Companies for much of the information in this section.

I'ANSON, C & COMPANY
Hope Town Foundry, Darlington

Charles I'Anson was Alfred Kitching's cousin. William and Alfred Kitching had been partners in a locomotive building establishment (see W. & A. KITCHING). The Lister Foundry was bought by the brothers and passed on to Charles I'Anson.

The shops were rebuilt and among its activities was the production of about five tank locomotives between 1875 and 1881. The works were later known as the 'Whessoe Foundry' and I'Anson ceased activities about 1885.

According to correspondence in *The Engineer* in 1920, six locomotives were built between 1875 and 1885 possibly five by C. I'Anson and Son and one by A. E. & H. Kitching. The gentleman responsible for these engines was Thomas Hudson who was originally with John Harris at the Albert Hill Foundry.

INCE FORGE COMPANY
Wigan

This firm built one locomotive for the South Devon Railway and supplied material for three more. The completed locomotive was a 2-4-0ST with inside cylinders 12″ x 17″, a domeless boiler and 4 ft diameter coupled wheels. It was built in June 1871 to the 7 ft gauge and named PRINCE, and was converted to the standard gauge in 1893 and afterwards became a portable engine. It was not until 1935 that it was cut up at Swindon.

Fig. 290
Ince Forge
Diagram of proposed 0-6-0ST

INCE FORGE COMPANY

In 1878 three sets of parts for 2-4-0STs were despatched to Newton Abbot for erection but as the South Devon Railway had been taken over by the GWR in 1876, the parts were despatched to Swindon and completed as side tanks and numbered 1298 to 1300. No. 1299 was fitted with a crane in 1881.

The Ince Forge Company became WM. PARK & COMPANY.

INGLIS, GEORGE & COMPANY
Albert Works, Airdrie

This firm built locomotives from the early 1900s. One was delivered to the Lanarkshire Steel Company in 1900. All were probably 0-4-0STs and in 1904 the type was redesigned with 14″ diameter cylinders. Five or six were built for local collieries and another was supplied to G. & T. Earle at Hope. It was named PINDALE and bore the works number E.800 of 1928. The cylinders were outside and horizontal, wheel spokes were I section and a stove pipe chimney fitted. Few details are available but about seven locomotives were built. PINDALE was the last.

Fig. 291 **Geo. Inglis** *c*1900
Redding Colliery, Falkirk

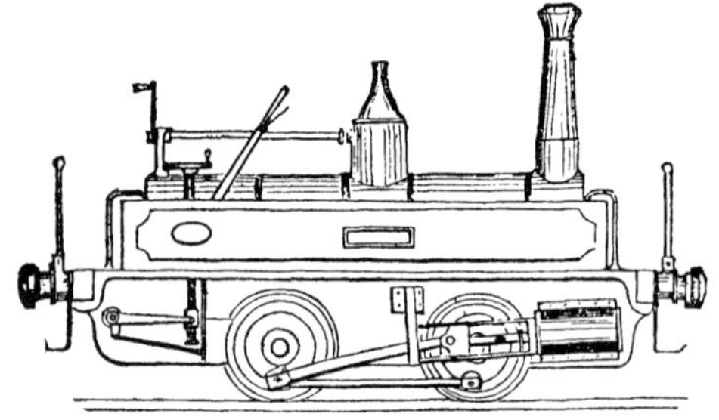

Fig. 292 **Bryan Johnson**
Advertisement
showing 0-4-0ST
in 1862

IRISH NORTH WESTERN RAILWAY

Barrack Street, Dundalk

The railway was opened in 1862 and Charles Clifford was the Locomotive Engineer. Three locomotives were built at Barrack Street; the first in 1873 was more truly 'erected'. It was No. 31 a 2-4-0 with 15″ x 21″ cylinders and 5′ 6″ diameter coupled wheels. The boiler was supplied by Courtney & Stephens of Dublin, the frames were built by Sharp Stewart & Co of Manchester, the wheels and axles by Monkbridge, and the crossheads and connecting rods by Pemberton. A strange procedure and one wonders what advantage if any was gained by diversifying the orders for various parts. The heating surface of the tubes was 901 sq. ft, the firebox 86 sq. ft making a total heating surface of 987 sq. ft. The weight in working order was 29 tons.

Built	Type	Cylinders inches	D.W. ft in	Weight WO tons cwt	I N W No	G N (I) No	W'drawn
1873	2-4-0	IC15 x 21	5 6	31 0	30	73	1892
1874	2-4-0	IC16 x 21	6 0	30 0	31	72	1921
11/1876	0-6-0	IC17 x 24	5 0	35 0	(2)	44	1948

Nos 31 and (2) had strong Dübs characteristics

The second built in 1874 was another 2-4-0 No. 30 with 16″ x 21″ cylinders and 6 ft diameter coupled wheels. How much of this locomotive was actually made at the Barrack Street works is problematical. The boiler was slightly larger with tubes giving 930 sq. ft of heating surface, the firebox 94 sq. ft giving a total of 1024 sq. ft. The weight in working order was 31 tons.

A 0-6-0 was completed in 1876. The number intended was No. 2 but by the time it was finished the INWR had became part of the Northern Railway of Ireland taking effect from 1 January of that year. Cylinders were 17″ x 24″ and wheels 5 ft. diameter. Tube heating surface was 1007 sq. ft, firebox 102 sq. ft giving a total heating surface of 1109 sq. ft and a grate area of 16·5 sq. ft. The weight in working order was 35 tons. The Northern Railway running number allocated was 44. The works were closed in 1880.

JOHNSON, BRYAN

Engineer, Chester

Advertisements around 1862 indicated that this firm was engaged in manufacturing machinery and castings for mines and collieries, lead works etc. Also special machinery for Shot, Red-lead and Pipe making and machines for compressing peat and patent fuel. Mill work in general and a good collection of wheel patterns was advertised. The illustration is a copy from an advertisement showing a drawing of a 0-4-0T with outside cylinders and combined safety valve and dome.

The Festiniog Railway were quoted on 29 October 1862 for two engines of Johnson's own design and again on 1 November, 1862 for three locomotives. If the tenders had been accepted it is certain that they would have been sub-contracted to a locomotive building firm.

JOHNSON C. G. & COMPANY

Exchange Place, Middlesbrough

This firm of engineers advertised as makers of 8″, 10″, 12″ and 14″ locomotives but no other details are available. No locomotives can be traced.

H. W. JOHNSON

See E. BORROWS & SONS.

JOHNSTON & McNAB
Port Dundas

This firm built a 0-4-0 in February 1832 named GLASGOW for the Garnkirk and Glasgow Railway and was their No. 2. It had inside cylinders 11″ x 18″ and 4 ft coupled wheels. Robert Stephenson & Co built the first locomotive (No. 1) and drawings were supplied to Johnston & McNab and Murdoch & Aitken and GLASGOW had the same dimensions as No. 1. The boiler was 3 ft diameter with 104 tubes 1$\frac{5}{8}$″ diameter giving a heating surface of 319·17 sq. ft. The boiler pressure was 50 lb/sq. in. The locomotive weighed 9 tons and the tender was built at the railway shops at St. Rollox. The gauge of the railway was 4′ 6″.

The firm was principally engaged in coach building and built a number for the early Scottish railways.

JOICEY, J. & G. & COMPANY
Forth Banks West Factory, Newcastle

Jacob Gowland Joicey was a partner in this firm and during this period it is reported that between 1867 and 1894 about twenty-four locomotives were built for colliery and factory use.

The firm at some date became JAMES JOICEY & CO. LTD.

Only a small proportion of locomotives built can be traced, and they were all located in the north east. Four coupled and six coupled saddle tanks were built, although the reproduction from an advertisement shows small side tanks fitted, and comparing the photograph of the Gateshead Gas Co's locomotive, the shape of the safety valve cover is very similar although this alone cannot confirm that it was a Joicey locomotive. The wheels, rods and slide bars were quite different.

The works numbers were not kept separately from other engineering work carried out relating to the collieries. The company amalgamated with Lambton & Hetton collieries in 1924 becoming LAMBTON, HETTON, & JOICEY COLLIERIES LTD.

Date	Type	Cylinders inches	DW ft in	Gauge ft in	Railway	No	Name
2/1832	0-4-0	11 x 16	4 0	4 6	Garnkirk & Glasgow	2	Glasgow

Fig. 293 Jas. Joicey & Co. drawing of 0-4-0T

JONES, JOHN
William Street, Liverpool

John Jones who had been a partner in the firm of Jones Turner & Evans (later becoming Jones & Potts) started up a locomotive business at the above address after the dissolution of the Viaduct Foundry in 1852. He commenced building about 1853 and continued rather erratically until 1863. The old series of Viaduct Foundry works numbers were continued and commenced about No. 292 and terminated at No. 342.

The title of the firm at an unknown date became JONES, JOHN & SON. Many of the locomotives built, went abroad to Central Argentine, and Langres railways.

JONES, TURNER & EVANS
Viaduct Foundry, Newton-Le-Willows

This Lancashire firm commenced operations 1837–8. There appeared to be strong business connections with other locomotive building firms especially Edward Bury and Robert Stephenson & Co and they obtained a number of orders from these firms as sub-contracts. Orders from Bury were for locomotives for the North Union, and Midland Counties railways; four 2-2-0s and three 2-2-2s respectively. These were followed by orders from the Great Northern of England, London & Brighton and Grand Junction railways. Six broad gauge 2-2-2s were supplied to the Great Western Railway in 1840, part of a class of sixty-two locomotives

Fig. 294 John Jones 338/1863 2-4-0T LA PLATA Buenos Ayres Northern Rly.

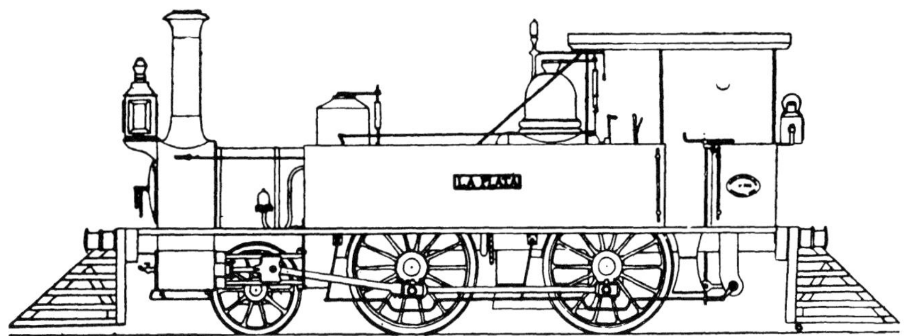

JOICEY

Works No	Date	Type	Cylinders inches	Gauge ft in	No/Name	Customer
210	1869	0-4-0ST	12 x 20		4	Weardale Iron & Coal Ltd (a)
	1870	0-4-0ST	12 x 20		No 8 Middridge	Weardale Iron & Coal Ltd
	1872	0-4-0ST	12 x 20		No 7 Spawood	Weardale Iron & Coal Ltd
	1874 (b)	0-4-0ST	12 x 20		No 9 Sedgefield	Weardale Iron & Coal Ltd
	1878	?				
305	1883	0-6-0ST	IC		7	South Hetton Colly
350	1882	0-4-0T	OC 4½ x 9	1 8		
377	1885	0-4-0ST	12 x 20		Tanfield	Jas Joicey Tanfield Lea
		0-4-0ST	12 x 20		Alma	Jas Joicey Tanfield Lea
		0-4-0ST	12 x 20		Leigh	Jas Joicey Tanfield Lea
388	1890	0-4-0ST	OC 14 x		Sadlers No 1	Sadler & Co Middlesboro
429	1894	0-4-0ST			Harparley	Jas Joicey Tanfield Lea
		0-4-0ST			16	Acklam Iron Co
					4	Mickley Coal Co Ltd (c)
					5	Mickley Coal Co Ltd
					6	Mickley Coal Co Ltd
					7	Mickley Coal Co Ltd
					8	Mickley Coal Co Ltd
					9	Mickley Coal Co Ltd

(a) Rebuilt to 0-6-0ST may have been works no 240 (b) Possibly two more to Weardale about this date (c) Two of these were 3′ 6″ gauge for the Coke Ovens lines.

Fig. 295
Jones, Turner & Evans
1840 for
Great Western Rly
broad gauge
FIRE FLY

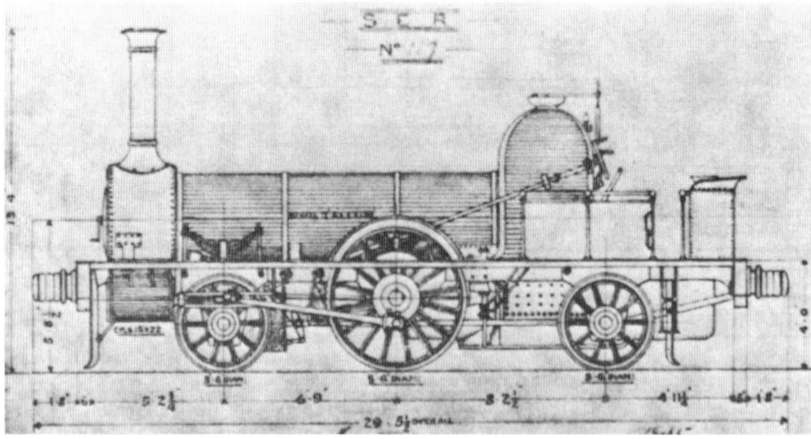

Fig. 296
1845–for South Eastern Rly
No. 117 2-2-2WT

all built by contractors. They had 7 ft diameter driving wheels, outside sandwich frames with 3″ oak or ash planks between ½″ thick iron plates with bolts through holding the 4″ thick frames together. The inside frames consisted of iron beams bolted to the inside cylinders at the front and the firebox casing at the rear so that the crank axle had no less than six bearings. The first of the class delivered named FIREFLY attained a speed of 56 mph shortly after delivery.

The first locomotives for abroad were ten for the Northern Railway of Austria. They were of two types, 2-2-2 and 0-4-2 of Stephenson's designs and built in 1841 together with four 2-2-0s, with bar frames, for the Eastern Counties Railway ordered by Braithwaite and Milner.

In 1842 few orders were obtained and for more than two years no orders materialised. About 1844 a change of management took place and a Mr Arthur Potts joined the firm which became known as JONES & POTTS.

Mr Jones was the practical man looking after the actual building, whereas Mr Potts was the man who got the orders, which he certainly did to good effect from 1845 to 1850. It was in 1845 that the firm became associated with Alexander Allan and many locomotives were built following the Allan pattern with outside cylinders and slotted frames which became to a large extent Jones & Potts standard types.

Many locomotives were built for the Eastern Counties Railway, the first order, being for ten

2-2-2s, three 2-4-0s and a 4-2-0 in February 1845 and due to some pretty fast work in the shops the first 2-2-2 was completed in July. One of this class is reported to have reached a speed of 70·5 mph near Waltham. A total of thirty-five locomotives were built for the ECR.

At this time the 4-2-0 became popular and Jones & Potts turned out quite a number. Seven were supplied to the London & Birmingham Railway, nine for the Chester & Holyhead, and six for the Eastern Counties Railway. They were long boiler type with the driving wheels behind the firebox. This design resulted in very unsteady running and frequent derailments.

A large number of locomotives were built for Scottish railways including the Scottish Central, Aberdeen, Scottish Midland, Glasgow, Dumbarton & Helensburgh, Glasgow, Paisley & Greenock railways. Many were of Allan's *Crewe* type singles with outside cylinders. Others were 0-4-2s and five 0-6-0s which were converted subsequently to 0-4-2s falling in line with Scottish practice of no more than four wheels coupled.

Those locomotives built for Robert Stephenson had the firm's characteristic wheels of the period with large cast iron centres with spokes shaped as two tee-iron sections back to back, the opposite to many built by other firms and H section spokes the flanges being parallel to the frames.

Seven small 2-2-2WTs were built in 1848 for the London & Blackwall Railway to Allan's outlines and had 13½" x 18" outside cylinders, except the seventh which had 15" x 18" cylinders. The driving wheels were 5′ 6" diameter. They did good work in the London area for many years.

In the early 1850s orders tailed off and Jones & Potts ceased trading in 1852. The London & North Western Railway took a lease on the works from 1 March, 1853, and purchased them outright on 11 May, 1860 to form the well known Earlestown carriage and wagon works.

One partner John Jones decided to carry on building locomotives, and transferred to Liverpool. At the time of the dissolution of Jones & Potts the works numbers had reached 290 approximately. John Jones continued the series and terminated at 342.

KENWORTHY, TAYLOR & COMPANY

Barnsley Foundry, Barnsley

Built tank locomotives. No further information.

KERR, JAMES & CO.

Glasgow

See KERR STUART & CO. LTD.

KERR, MITCHELL & NEILSON & KERR NEILSON & COMPANY

Glasgow

See NEILSON & MITCHELL.

KERR STUART & CO. LTD.

California Works, Stoke-on-Trent

The firm was founded in Glasgow in 1881 by James Kerr and was known as James Kerr & Company. In Glasgow the firm were dealers in railway plant of all descriptions and when orders for new locomotives were received these were sub-contracted to the Stoke-on-Trent firm of Hartley, Arnoux & Fanning, John Fowler and Falcon/Brush.

In 1893 the firm of Hartley, Arnoux and Fanning were taken over by James Kerr & Co and became known as Kerr Stuart & Co. Ltd. of California Works. They commenced building locomotives in the same year and presumably previous locomotives built here bore James Kerr & Company or Kerr Stuart works plates.

They specialised in industrial locomotives both for standard and narrow gauge, more particularly the latter for home and abroad. Later main line types were built and some 1500 were built in all, up to the year 1930 when the works closed and the goodwill was bought by the Hunslet Engine Co. Ltd.

Some standard gauge saddle and side tanks were built but the bulk of the early built locomotives were for narrow gauge lines at home and abroad. In 1899 some 0-6-4 tender locomotives were supplied, to the Gwalior Light Railway's 2 ft gauge, with outside cylinders 8¼" x 15" and 2′ 3" diameter coupled wheels and in the following year twelve 0-4-0Ts were supplied to the Sao Paulo Railway 5′ 3" gauge with special brakes for service on the New Serra inclines. They had outside cylinders 13½" x 16"

Fig. 297
Kerr Stuart & Co.
754/1902 *Skylark*
class 0-4-2T for
Hadfields
2′ gauge

Fig. 298
Kerr Stuart & Co.
786/1904
Argentine Gov. Rly.
No. 185 metre
gauge 0-6-2T

Fig. 299
Kerr Stuart & Co.
2368/1915
Metropolitan Water
Board SUNBURY
2′ gauge

and 3 ft diameter wheels. Other railways supplied were Londonderry & Lough Swilly (4-6-0T) British North Borneo Railway (0-4-2T), Buenos Ayres Midland Railway (0-6-0T) Tralee & Dingle (2-6-0T), Anglo-Chilean Nitrate (2-6-2) Interoceanic & Mexican Eastern and many others abroad. At the same time standard and narrow gauge tank locomotives were supplied to collieries and quarries in this country.

A Meyer articulated 0-6-6-0 was supplied to the Anglo-Chilean Nitrate Railway (3′ 6″ gauge) in 1903 having cylinders 14″ x 18″ and 2′ 10¾″ diameter wheels.

Kerr Stuart built a number of engine units for rail motors, including four 2′ gauge for the Gwalior Light Railway, three 2-2-0s for the Buenos Ayres Great Southern Railway with outside cylinders 9″ x 15″ and 3′ 5″ diameter driving wheels. Other rail motors were six 0-2-2 for the Taff Vale, two 2-2-0 and twelve 0-4-0 for the Great Western, two for the Italian State Railways and two for the Mauritius Railways.

A comparatively large order for thirty-two 0-4-4Ts was carried out in 1909–10 for the South Indian Railway (metre gauge) with outside cylinders 11″ x 18″ and 3′ 6½″ diameter coupled wheels. The tank was in the bunker.

Probably the strangest order for this firm was for four 4-2-2 locomotives for the Shanghai-Nanking Railway with outside cylinders 18″ x 26″ and 7 ft diameter driving wheels.

During the 1914 war Kerr Stuart built seventy 0-6-0Ts for the French Government Artillery Railways to 600 mm gauge (Works Nos. 2402–2416, 2428–2457 in 1915 and 2995–3019 in 1916) and all delivered to Nantes.

They had outside cylinders, 8½″ x 11″ and outside valve gear, the side tanks were carried to the front of the smokebox and brought down in two levels to the low level footplate. The chimney was fitted with a spark arrester.

During the same period a number of 0-4-0WTs were built similar to those built by E. Borrows for Brunner Mond's works. Ten smart looking 0-6-0Ts were built for the Government for the Inland Waterways and Docks and subsequently becoming ROD Nos. 601–610. The outside cylinders were 17″ x 24″ and wheels 4′ 0″ diameter and weight in working order about 49 tons.

Another War Dept order was for eighteen of their standard *Wren* class for the Air Ministry for use on various aerodromes.

After the war eight 4-4-4Ts were built for the Metropolitan Railway (MR Nos. 103–110) they were to the design of Charles Jones, the Chief Mechanical Engineer of the railway, and were intended for fast passenger services. They had outside cylinders 19″ x 26″, 5′ 9″ diameter driving wheels, a moderate boiler pressure of 160 lb/sq. in., a total heating surface of 1178 sq. ft and grate area of 21·2 sq. ft. The boilers were interchangeable with the 'G' class 0-6-4Ts, and were fitted with Robinson superheaters. They were good looking machines and apart from their tendency to slip were excellent locomotives.

Thirty 4-6-0s were then put in hand for various Indian State railways of Metre gauge.

Kerr Stuart were used to small quantity orders and a great variety of types were built, interspersed with four coupled industrial tank locomotives. Examples in 1921 to 1924 were

2-8-4T	2′ 6″ gauge	Dholpur State Railway
2-4-0CT	5′ 6″ gauge	Tatanagar India Peninsula Loco Company
2-8-2	2′ gauge	Gwalior Light Railways
4-6-2	2′ 6″ gauge	Mysore Railways
4-6-0	2′ 6″ gauge	Mysore Railways
2-4-2T	Std gauge	Takapuna Railway. N.Z.
2-6-4T	5′ 6″ gauge	Eastern Bengal Railway

Very little opportunity occurred for quantity production until in 1925 they received an order for fifty standard *Class 4* 0-6-0s from the London Midland & Scottish Railway. The first twenty-five were built in 1925 (LMS Nos. 4082–4106) and the remainder (LMS Nos. 4332–4356) were built in 1926 and 1927.

In 1930 a six coupled narrow gauge shunter was built with a water tube boiler fitted with superheater and operating at a working pressure of 300 lb/sq. in.

A high speed vertical geared engine was fitted, the cylinders were 6″ x 8″ with piston valves and Marshall's valve gear and the driving wheels were 2 ft diameter.

The final large order came from the Great Western Railway in 1930 for twenty-five of the *57xx* 0-6-0PTs (GWR Nos. 7000–7724).

Large numbers of narrow gauge 0-4-0ST, 0-4-0T and 0-4-2ST types were built over the whole period and they went to plantations, quarries, light railways and industry in many parts of the world. Each type received a class name and many parts were standardised where possible.

The most prolific type was the *Wren* class of 0-4-0ST, and were typical of all Kerr Stuart's small locomotives.

The outside cylinders were 6″ x 9″ and wheels 1′ 8″ diameter with a 3 ft wheelbase. The gauge varied mainly between 2′ and 3′. They had inside plate frames with no splashers or running plates. The saddle tank (87 gallons) rested on the boiler only, the dome had the safety valves on top and auxiliary valved steam supplies to the injectors and blower. Early tank locomotives had solid cast steel wheels with

DIMENSIONS OF STANDARD CLASSES

Class	Type	Cylinders inches	DW ft in	Weight Empty ton cwt	Weight WO ton cwt
Egypt	0-4-0T	6 x 10	2 0		
Sirdar	0-4-0T	6 x 10	2 0	5 3	
Dundee	0-4-0WT	6½ x 9	1 8	3 4	
Priestley	0-4-0WT	14½ x 20	3 4	22 14	
Buya	0-4-0ST	6 x 9	1 8	3 6	
Moss Bay	0-4-0ST	15 x 20	3 6	22 15	29 0
Southend	0-4-0ST	6 x 9	1 8	3 8	
Witch	0-4-0ST	12 x 16	3 1	15 19	
Wren	0-4-0ST	6 x 9	1 8	3 7	4 3
Huxley	0-4-2T	11 x 16 12 x 16	2 9	15 17	20 10
Midge	0-4-2T	6 x 10	1 9	6 0	
		7 x 12	1 9		
Midge	0-4-2T	7½ x 12	2 3	to	
		8½ x 15	1 9		
		9½ x 15	2 3	10 13	
Skylark	0-4-2T	7 x 12 7½ x 12	2 3	8 8 8 16	
Waterloo	0-4-2T	9½ x 15	2 9	11 11 to 13 7	
Brazil	0-4-2ST	9 x 15	2 6	10 10	12 15
Darwin	0-4-2ST	8 x 12	2 3	8 13	10 15
Tattoo	0-4-2ST	7 x 12	2 0	6 10	8 10
Tattoo Special	0-4-2ST	7 x 12	2 0	6 13	
Argentina	0-6-0T	15 x 20	3 9	27 0	34 3
Haig	0-6-0T	8½ x 11	1 11⅝	8 10	10 10
Triana	0-6-0T	14 x 20	3 6	23 13	
Matary	0-6-2T	10 x 15	2 3	12 16	16 15
Orinoco	2-6-2T	15 x 20	3 3	29 2	

Fig. 300
Kerr Stuart & Co.
4092/1920
Metropolitan Rly.
C. Jones 4-4-4T
No. 107

cored holes, but later ones had cast iron centres with steel tyres. The valve gear was Stephenson's link motion but was changed to Hackworth's motion about 1915.

The crossheads were of the single bar type and connecting and coupling rods were circular. Many detailed variations were built but the main dimensions were kept as standard as possible.

Very few light tank locomotives were built with outside frames, the *Sirdar* class was one example. Accessibility was important for maintenance especially in the 'back of beyond!'

Referring to the section on Hartley, Arnoux and Fanning, the firm was taken over by Kerr Stuart in 1893 and works numbers commenced at No. 72. The series ran from 72 to 120 and from No. 121 five hundred was added to the numbers, i.e. becoming 621. This series of numbers ran from 621 to 1360.

Then from 22 January one thousand was added to the numbers making the next series 2361 to 2494, after which five hundred was added to the next series which was 2995 to 3130. Nos. 3027–3046 were blanks, probably a cancelled order.

The next series commenced at 4000 and continued to the final locomotive built which was No. 4483 a 0-4-2PT of the *Tamar* class with 9″ x 15″ outside cylinders and built to 3′ 0″ gauge for the Anglo-Persian Oil Co. Ltd. and despatched on 23 July, 1930.

In the 500 series of works numbers were a number of locomotives built by other firms in the period when Kerr Stuart were in business as factors. Some were built by Henry Hughes and others by John Fowler.

Taking into account some gaps in the sequences Kerr Stuart built approximately 1500 steam locomotives. The goodwill was taken over by the Hunslet Engine Company in 1930 and at least three completed locomotives and a boiler were transferred to Leeds.

KILMARNOCK ENGINEERING CO.

Britannia Works

This firm was successor to Dick, Kerr and Co. Ltd. in 1919 and building took place in 1920, with works numbers starting at 500. About twenty-seven locomotives were built probably all in the same year. Production was carried on at Britannia Works on a number of 0-4-0STs to 3′ and 3′ 6″ gauges. They were probably designed by Dick Kerr & Co. Ltd. Four 3′ 6″ gauge locomotives went to the Swanscombe works of Associated Portland Cement Manufacturers Ltd. and two 3′ gauge to the Islip Iron Co. Ltd. at their Islip Quarries. Dick Kerr & Co. Ltd. had supplied a similar locomotive in the previous year. All had outside cylinders. As will be seen from the table there are many gaps in the works numbers, some of which may have been filled by other types of machinery. The APCM locomotives were either scrapped or sold after the narrow gauge track working ceased in 1928. The two Islip Iron locomotives were scrapped in 1953 when the 3 ft gauge track was lifted.

Fig. 301 **Kilmarnock Eng. Co. 500/1920** A.P.C.M. Ltd. Swanscombe, Kent HUSTLER

KINGSBURY IRONWORKS

Ball's Pond, London

See MATTHEWS, JAMES.

K'LMARNOCK ENGINEERING CO. LTD.

Wks No	Date	Type	Cylinders inches	D.W. ft in	Gauge ft in	No/Name	Customer
500	1920	0-4-0ST	OC10 x 16	2 8	3 6	Hustler	APCM Ltd. Swanscombe
501	1920	0-4-0ST	OC10 x 16	2 8	3 6	Swanscombe	APCM Ltd. Swanscombe
507	1920	0-4-0ST	OC10 x 16	2 8	3 6	Northfleet	APCM Ltd. Swanscombe
508	1920	0-4-0ST	OC10 x 16	2 8	3 6	Greenhithe	APCM Ltd. Swanscombe
509	1920	0-4-0ST	OC		3 0	7 Slipton	Islip Iron Co. Ltd.
510	1920	0-4-0ST	OC		3 0	8 Slipton	Islip Iron Co. Ltd.
526	1920	0-4-0ST	OC		3 6	Alkerden	APCM Ltd. Swanscombe

KINMOND HUTTON & STEEL

Wallace Foundry, Blackness, Dundee

This firm commenced locomotive building in 1838 and apart from three locomotives supplied to Canada in 1847 all their output was for Scottish railways.

The first locomotive built, one of three, was the WALLACE for the 5′ 6″ gauge Dundee and Arbroath Railway, with 5 ft diameter driving wheels, inside frames and outside cylinders, and a boiler working at 50 lb/sq. in. They weighed twelve tons and cost £1012 each. Three more were built 1840–1 but with 5′ 6″ diameter driving wheels, weighing 13 tons 13 cwt and costing £1270 each. They were of similar design to those supplied to the Arbroath and Forfar Railway by the firm's neighbour Stirling & Co.

Six 2-2-2s were built in 1840 for the Glasgow, Paisley, Kilmarnock and Ayr Railway. They had 5′ 6″ diameter driving wheels and a boiler with raised firebox. The dome, placed in a forward position on the boiler, had a safety valve fitted on top, a further valve being fitted on top of the firebox. The locomotives were sent by sea to Ayr in a dismantled state and re-erected on delivery. They evidently proved satisfactory and six similar but more powerful 2-2-2s were built, taking up to 1846 to complete the order.

Little is known about the three 2-2-2s that were sent to Canada apart from the fact that they had outside cylinders.

In 1847 three 0-4-2s were built for the Dundee Perth and Aberdeen Junction Railway, with inside cylinders 15″ x 20″, with outside frames and 5 ft diameter coupled wheels for goods traffic. These were followed by three 2-2-2s for the same company with double frames and inside cylinders the same size as for the 0-4-2s.

In 1848 nine 2-2-2s were delivered to the Glasgow Dumfries and Carlisle Railway. They had inside cylinders and 6 ft diameter driving wheels. One named QUEEN was used by D. Kinnear Clarke in various experiments relating to boiler efficiency in February 1850.

In his report the boiler figures given for this locomotive were: Heating surface of tubes 570 sq. ft; Heating surface of firebox 64 sq. ft; Total 634 sq. ft. Grate area was 13·4 sq. ft. It is interesting to note that coal not coke was burned.

In the locomotive list a gap appears between 1848 and 1855 but locomotives may have been built during this period. Four 2-2-2s were built for the Dundee and Arbroath Railway in 1855. They were built to 4′ 8½″ gauge, the railway having been converted from 5′ 6″ gauge in 1847. In the early days this railway instituted running on the right hand track.

In 1853 Kinmond & Co established a branch locomotive building works in Montreal which built eleven locomotives for the Grand Trunk Railway and one other, but the works were closed down in 1857 and no doubt manufactured other machinery besides locomotives.

Mr James Steel who had been a partner of the firm left the Wallace Foundry about 1850 and it would be then that the firm became KINMOND & CO. After cessation of locomotive building in the late 1850s the firm lost its identity and in 1861 the Wallace Foundry belonged to Robertson Orchar.

No definite output can be assessed for Kinmond Hutton & Steel but would probably be about fifty.

Fig. 302 **Kinmond Hutton & Steel** 1847 2-2-2 for Dundee and Arbroath Rly WALLACE

Date	Type	Cylinders inches	D.W. ft in	Railway	No	Name
1838	2-2-2	OC 13 x 18	5 6	Dundee & Arbroath		Wallace (a)
1838	2-2-2	OC 13 x 18	5 0	Dundee & Arbroath		Fury
1838	2-2-2	OC 13 x 18	5 0	Dundee & Arbroath		Griffin
1839	2-2-2	OC 13 x 18	5 0	Dundee & Arbroath		Rapid
1840	2-2-2	OC 13 x 18	5 6	Dundee & Arbroath		Dart
1840	2-2-2			GPK & A	6	Bruce (b)
1840	2-2-2			GPK & A	9	Kelburne
1840	2-2-2			GPK & A	11	Eglinton
1840	2-2-2			GPK & A	12	Portland
1840	2-2-2			GPK & A	13	Ailsa
1840	2-2-2			GPK & A	14	Loudoun
1841	2-2-2	OC 13 x 18		Dundee & Arbroath		Queen
1843	2-2-2			GPK & A	21	Burns
1843	2-2-2			GPK & A	26	Mars
1845	0-4-0	13 x 18	4 0	Aberdeen	25	
1846	2-2-2			GPK & A	38	North Star
1846	2-2-2			GPK & A	39	Meteor
1846	2-2-2			GPK & A	40	Planet
1846	2-2-2			GPK & A	41	Comet
1847	2-2-2	OC		Montreal & Lachine		James G. Ferrier
1847	2-2-2	OC		Montreal & Lachine		Montreal
1847	2-2-2	OC		Champlain & St. Lawrence		John Molson
1847	0-4-2	IC 15 x 20	5 0	DP & AJ	11	(c)
1847	0-4-2	IC 15 x 20	5 0	DP & AJ	12	
1847	0-4-2	IC 15 x 20	5 0	DP & AJ	13	
1847	2-2-2	IC 15 x 20	5 0	DP & AJ	14	Vulcan
1847	2-2-2	IC 15 x 20	5 0	DP & AJ	15	Lucifer
1848	2-2-2	IC 15 x 20	5 0	DP & AJ	16	Dundee
1848	2-2-2	IC 15 x 20	6 0	GD & C		Dumfries (d)
1848	2-2-2	IC 15 x 20	6 0	GD & C		Glasgow
1848	2-2-2	IC 15 x 20	6 0	GD & C		Carlisle
1848	2-2-2	IC 15 x 20	6 0	GD & C		Solway
1848	2-2-2	IC 15 x 20	6 0	GD & C		Afton
1848	2-2-2	IC 15 x 20	6 0	GD & C		Queen
1848	2-2-2	IC 15 x 20	6 0	GD & C		Albert
1848	2-2-2	IC 15 x 20	6 0	GD & C		Princess
1848	2-2-2	IC 15 x 20	6 0	GD & C		Nith

KINMOND & CO.

Date	Type	Cylinders inches	D.W. ft in	Railway	No	Name
1855	2-2-2	IC 15 x 18	5 6	Dundee & Arbroath	1	
1855	2-2-2	IC 15 x 18	5 6	Dundee & Arbroath	2	
1855	2-2-2	IC 15 x 18	5 6	Dundee & Arbroath	7	

(a) 5′ 6″ Gauge. (b) Glasgow, Paisley, Kilmarnock & Ayr Rly. (c) Dundee, Perth & Aberdeen Junction Rly. Nos 11-16 Reno. 1-6. (d) Glasgow, Dumfries & Carlisle Rly. Became G & SWR Nos 77-85.

KIRKLESS HALL COAL & IRON CO.
Wigan

Some 0-4-0STs were built by this company. Little information remains of any details except that they had dummy crankshafts and were probably to 4′ 1½″ gauge. In 1865 an amalgamation took place of various local collieries and the title became: WIGAN COAL & IRON CO. LTD.

At the Kirkless workshops a number of 0-4-0STs and 0-6-0STs were built between 1865 and 1912. The six coupled tanks bore a number of Crewe characteristics, due to the colliery engineer Mr Percy having spent a few months in Crewe Works with the object of obtaining technical information to help in building further locomotives at Kirkless for internal use.

The first, built in 1865 named BESSEMER had typical LNWR pattern cast iron wheels with H section spokes, lift up horizontal smoke box door, elongated lower half cab sides, and other small details such as square socket lamp irons.

Later built locomotives varied somewhat in detail and at least two, SOL and CRAWFORD, had saddle tanks over the boiler and firebox only. The cylinder diameter was also enlarged from 14″ to 16″.

Nasmyth Wilson built four locomotives for the company between 1892 and 1902 to this design. They must have been well built as many survived in the 1950s and LINDSAY still exists and it is hoped, may be preserved.

Built	Type	Cylinders inches	D.W. ft in	Name	W'drawn
1865	0-6-0ST	IC 14 x 24	4 3	Bessemer	*c*1919
1866	0-6-0ST	IC 14 x 24	4 3	Vulcan	
1867	0-6-0ST	IC 14 x 24	4 3	Ajax	1954
1867	0-6-0ST	IC 14 x 24	4 3	Hector	1959
1869	0-6-0ST	IC 16 x 24	4 3	Vesta	*c*1938
1870	0-6-0ST	IC 16 x 24	4 3	Ludovic	*c*1939
1874	0-6-0ST	IC 16 x 24	4 3	Shah	1957
1876	0-4-0ST			Venus	
1878	0-6-0ST	IC 16 x 24	4 3	Sol	*c*1962
1883	0-6-0ST	IC 16 x 24	4 3	Crawford	1964
1887	0-6-0ST	IC 16 x 24	4 3	Lindsay	
1892	0-6-0ST	IC 16 x 24	4 3	Balcarres	1957
1893	0-6-0ST	IC 16 x 24	4 3	Emperor	1948
*c*1897	0-4-0ST			Berkune	Sold 1898
1902	0-4-0ST			Minnie	1934
1905	0-6-0ST	IC 16 x 24	4 3	Nelson	1954
1908	0-6-0ST	IC 16 x 24	4 3	Manton	1959
1912	0-6-0ST	IC 16 x 24	4 3	Siemens	1959

Fig. 303 1883 0-6-0ST CRAWFORD with modified cab

KIRTLEY & CO.
Dallam Foundry, Warrington

This firm under William Kirtley built main line locomotives for various railways between 1837 to 1841. No complete history is available as to the full extent of their output. The 2-2-0 locomotives for the Leeds and Selby Railway were probably of the *Bury* type with bar frames. The COMET lasted until 1875. SWALLOW and SWIFT cost £1200 without tenders, the latter being supplied by E. Bury.

KITCHING, W. & A.
Hope Town Foundry, Darlington

William and Alfred Kitching took over a family business established as an iron foundry in 1790. They extended the works in 1832 with the object of building main line locomotives, which they did, commencing in 1835 and finishing in 1860.

Kitching's principal customer was the Stockton and Darlington Railway – William Kitching was a member of the SDR committee for many years.

Most of the locomotives for the SDR were to Hackworth's designs and consisted of thirteen 0-6-0 type, one 0-4-0 type, two 0-4-2 type, one 2-2-2 type, seven 2-4-0 type.

The first 0-6-0 was the ENTERPRISE which was copied from the MAGNET built by Hackworth at Shildon. This had rear vertical outside cylinders transmitting power to a crankshaft with cranks set at 90° and were connected to the coupled wheels by a short con-

KIRTLEY & CO.

Built	Type	Cylinders inches	D. Wheels ft in	Customer	No/Name
1837	2-2-2			Leipzig & Dresden	Renner
	2-2-2			Leipzig & Dresden	Sturm
	2-2-2			Leipzig & Dresden	Elephant
	2-2-2			Leipzig & Dresden	William Kirtley
1838	2-2-2			Leipzig & Dresden	Greif
1839	2-2-0			Leeds & Selby	Prince
	2-2-0			Leeds & Selby	Queen
	2-2-0			Leeds & Selby	Swift
	2-2-0			Leeds & Selby	Swallow
	2-2-0			Leeds & Selby	
	2-2-0			Leeds & Selby	
	2-2-0			Leeds & Selby	
	2-2-0			Leeds & Selby	
1840	2-2-0			Leeds & Selby	
2/1841	2-2-2	IC 14 x 18	5 6	Stockton & Darlington	52. Comet

Fig. 304 **W. & A. Kitching** 1845 Stockton & Darlington Rly. No. 25 DERWENT Preserved at Bank Top Station, Darlington

necting rod on each side. The valves were operated by two eccentrics to each slide valve and the eccentrics operated the feed pumps as well. The boiler had a large fire tube around which were a series of smaller tubes. The locomotive was equipped with two 'tenders', one carrying coal at the fire tube end and the other carrying water in a large wooden cask. This was also the driving end, so that the driver and fireman were separated from each other by the length of the boiler. The next 0-6-0 TEES had inclined outside cylinders, still at the rear but driving the front pair of coupled wheels. In 1838 a four coupled locomotive was built to Kitching design. It had inside frames and cylinders with conventional 'loco' type boiler. This was followed by a 2-2-2 with outside frames and inside cylinders also to Kitching design.

The 2-4-0 HACKWORTH was quite a step forward in design. It had outside frames, a domed boiler with the raised round top firebox between the coupled wheels. The inside cylinders were 17″ x 24″, and the wheels had springs above the running plates. It ran for fifty years.

The reason for the extension to the works in 1832 was that the early locomotives of the Stockton and Darlington Railway were giving a lot of trouble, so much so that the directors of the railway had seriously considered their abandonment. William Kitching from the outset had great faith in the future of the steam locomotive and, ably supported by Timothy Hackworth, the two Kitching brothers were determined to produce locomotives that could be relied upon. As will be seen from the list, apart from one locomotive supplied to the Clarence Railway and two to the Whitehaven & Furness Junction Railway, all their products went to the SDR. They not only built steam locomotives but produced numbers of wagons, cranes, turntables, gas retorts, and corn grinding machinery. They designed the first hopper wagon with bottom discharge doors.

The *Derwent* class of 0-6-0s did excellent work and could haul eighty chaldron wagons.

The QUEEN (SDR 40) was the first locomotive made for that railway with a crankshaft. The crank axle was $5\frac{3}{4}$″ diameter the inside cylinders 12″ x 18″ and the boiler 3′ 8″ diameter x 8′ 6″ long with eighty-six copper tubes 2″ diameter. A copper firebox was fitted

Date	Type	Cylinders inches	D.W. ft in	Railway	Rly No	Name	W'drawn
7/1835	0-6-0	15 x 16	4 0	Stockton & Darlington	25	Enterprise (a) (c) (f)	
1837	0-6-0	15 x 16	4 0	Stockton & Darlington	14	Tees (a) (c)	1871
11/1838	0-4-0	IC 12 x 18	4 6	Stockton & Darlington	29	Queen	1858
12/1838	0-6-0	12½ x 18	4 0	Stockton & Darlington	17	Whig (b) (e)	1870
10/1839	2-2-2	IC 12 x 18	5 0	Stockton & Darlington	30	Raby Castle	1858
1840	0-6-0		4 0	Clarence	8		
1840	0-6-0	14 x 20	4 0	Stockton & Darlington	26	Pilot (c)	1868
7/1841	0-6-0	15 x 20	4 0	Stockton & Darlington	5	Hope Town (c)	1869
10/1844	0-4-2	14 x 18	5 6	Stockton & Darlington	48	Active	
1845	0-6-0	15 x 24	4 0	Stockton & Darlington	16	Stanhope (c)	1870
11/1845	0-6-0	15 x 24	4 0	Stockton & Darlington	25	Derwent (c)	
5/1846	0-6-0	15 x 24	4 0	Stockton & Darlington	24	Skelton Castle (c)	1870
5/1847	0-4-2	13 x 18	4 6	Stockton & Darlington	55	Walsingham	1879
1847	0-4-2			Whitehaven & Furness Jc			
1847	0-4-2			Whitehaven & Furness Jc			
1/1848	0-6-0	15 x 24	4 0	Stockton & Darlington	22	Alert (c)	1869
8/1848	2-4-0	15 x 20	5 2	Stockton & Darlington	58	Woodlands (d)	1877
10/1848	2-4-0	15 x 20	5 2	Stockton & Darlington	59	Hallgarth (d)	1881
3/1849	0-6-0	15 x 20	4 0	Stockton & Darlington	3	Times	1865
7/1849	0-6-0	15 x 20	4 0	Stockton & Darlington	2	Graham	1865
3/1851	2-4-0	17 x 24	5 9	Stockton & Darlington	71	Hackworth	
5/1855	2-4-0	15 x 20	5 2	Stockton & Darlington	98	Pierremont (d)	1876
9/1855	2-4-0	15 x 20	5 2	Stockton & Darlington	101	Marske (d)	1882
7/1857	2-4-0	16 x 19	5 2	Stockton & Darlington	118	Elmfield	1880
8/1858	0-6-0	17 x 20	4 3	Stockton & Darlington	132	Appleby	1903
9/1858	0-6-0	17 x 20	4 3	Stockton & Darlington	133	Kirby Stephen	1880
11/1860	2-4-0	16 x 19	5 2	Stockton & Darlington	166	Oswald Gilkes	1877

(a) Same as MAGNET built at Shildon. (b) Same as TORY built at Shildon. (c) Rear vertical cylinders driving leading wheels. (d) Long boiler type. (e) Cylinders on rear of boiler, inclined, and driving leading wheels. (f) Built 1836 according to Hackworth List.

2′ 7″ long, 3′ 6″ wide and 4′ 5″ high. The tender tank was carried on wooden frames and malleable iron horns. The wheels too were of malleable iron four in number, with springs and screw brake. The tank carried 654 gallons. The locomotive was subsequently converted to a 0-4-2.

The only 'single' built was RABY CASTLE designed for passenger traffic with inside cylinders 12″ x 18″ and 5 ft diameter driving wheels. Subsequently the 2-4-0 was standardised as the most suitable wheel arrangement for the SDR passenger work and seven were built from 1848 to 1860 all had 5′ 2″ diameter driving wheels except No. 71 HACKWORTH with 5′ 9″ diameter. The latter also differed from the other 2-4-0s, having the firebox between the coupled wheels.

In 1858 the 0-6-0s built had inside cylinders, the designers breaking away from the outside inclined cylinders which were so characteristic of the SDR early mineral locomotives. The works were purchased by the Stockton & Darlington Railway in 1862 and used for carriage building until 1886.

KITSON & LAIRD
KITSON, THOMPSON & HEWITSON
KITSON & CO.

See TODD, KITSON & LAIRD.

LANCASHIRE & YORKSHIRE RAILWAY
Bury

The Bury Shops had been built by the East Lancashire Railway for the maintenance of their locomotives and it was was not until after the ELR was taken over by the Lancashire & Yorkshire Railway in 1859 that any new locomotives were built in the workshops.

In 1862 two 2-4-0s were built. They had curved outside frames and inside cylinders 15″ x 20″ and 5′ 6″ diameter coupled wheels. The boiler had a raised firebox with safety valve on top; the barrel had a large dome placed close to the chimney. A weatherboard only, was fitted. In 1867 three 2-4-0Ts were built with the main dimensions the same as the two tender locomotives, but were rebuilt in the following year to conform with the two 2-4-0s.

In 1871 a larger 2-4-0 was built with inside frames and inside cylinders 16″ x 24″ and 6 ft diameter coupled wheels. It was designed by William Hurst. A further four were built in 1873 and differed from those built at Miles Platting in that they had domed boilers, the domes being on the centre of the barrels.

From 1875 to 1877 six additional 2-4-0s were built. These reverted to outside curved frames but with a substantial cab, boiler with flush top firebox and dome on centre ring of the barrel. The boiler pressure for all locomotives built at Bury was 140 lb/sq. in.

New building ceased after the two 2-4-0s built in 1877, and work transferred to Miles Platting. Repairs were continued at the Bury workshops until 1888 after which they were used as stores.

Fig. 305 **L&YR Bury** 1876 2-4-0 PHANTOM as L&YR No. 638

LANCASHIRE & YORKSHIRE RAILWAY

Miles Platting, Manchester

The Manchester & Leeds Railway in 1847 assumed the title of the Lancashire & Yorkshire Railway. In 1846 the locomotive workshops at Miles Platting were completed under the supervision of Sir John Hawkshaw, the Locomotive Superintendent to the M&LR. The first locomotives to be constructed at Miles Platting were to Hawkshaw's designs and, strictly speaking, the first three bearing M&LR Nos. 53, 56 and 59, were their build before the title was changed.

Thirty two 2-2-2s were put in hand and were built from 1846 to 1849. They had outside cylinders 15″ x 20″ and driving wheels varying between 5′ 9″ and 5′ 10″ diameter. The boilers had raised round top fireboxes with the dome placed above. Four wheeled tenders were fitted and tender brakes only were used.

The locomotive superintendents at Miles Platting works were: 1846 Sir John Hawkshaw; 1847–68 William Jenkins; 1868–76 William Hurst; 1876–81 William Barton Wright.

The 2-2-2s that had been built by Mr Jenkins proved too light for the loads and heavy gradients and to provide more adhesion they were all rebuilt as 2-4-0s between 1866 and 1873 but were all scrapped before 1880.

Twenty three 0-4-2s were then built with inside cylinders and frames with a similar type of boiler to the 2-2-2s. At least three were converted to 0-6-0s by Mr Hurst.

From 1854 a large number of 0-6-0s were built, most of them having 15″ x 24″ inside cylinders and 4′ 10″ diameter wheels. When rebuilt they received 16″ cylinders. Many were also rebuilt as saddle tanks.

Jenkins then introduced a new class of 2-4-0s in 1861 with inside frames and inside cylinders 15″ x 24″. The coupled wheels were 5′ 9″ diameter. These were intended for express passenger duties and twenty-two were built. The first eleven bore names of directors but later ones were nameless. Building took place from 1861 to 1867. The boilers on this class were domed on the boiler barrel whereas former classes were domeless on the barrel, the dome being fitted on the firebox. The shape of the dome which was of polished brass and very large was likened to a 'cottage loaf' by Mr E. L. Ahrons, and was quite unique. The chimneys had copper tops. Rebuilding commenced in 1875 and all received 16″ cylinders. They were withdrawn between 1887 and 1900.

Apart from a few 2-4-0STs with 5 ft

LANCASHIRE & YORKSHIRE RLY – BURY

Built	Type	Cylinders inches	D.W. ft in	LYR No	Name		W'drawn
3/1862	2-4-0	15 x 20	5 6	673	Blacklock		1882
9/1862	2-4-0	15 x 20	5 6	680	Craven		1877
1/1867	2-4-0T	15 x 20	5 6	601	Odin		1882
1/1867	2-4-0T	15 x 20	5 6	603	Clio		1882
9/1867	2-4-0T	15 x 20	5 6	618	Titan		1881
10/1871	2-4-0	16 x 24	6 0	713	Juno		1886
10/1873	2-4-0	16 x 24	6 0	650	Banshee		1897
10/1873	2-4-0	16 x 24	6 0	654	Bacchus		1893
11/1873	2-4-0	16 x 24	6 0	607	Reindeer		1893
11/1873	2-4-0	16 x 24	6 0	609	Vesta		1898
12/1875	2-4-0	16 x 22	5 6	746	Thor		1901
4/1876	2-4-0	16 x 22	5 6	638	Phantom		1892
5/1876	2-4-0	16 x 22	5 6	608	Jupiter		1893
6/1876	2-4-0	16 x 22	5 6	651	Centaur		1892
2/1877	2-4-0	16 x 22	5 6	680	Craven		1892
5/1877	2-4-0	16 x 22	5 6	662	–		1898

First Built	Type	Cylinders inches	D.W. ft in	Heating Surface sq ft Tubes	F'box	Total	Grate Area sq ft	WP lb/sq. in	Weight WO ton cwt
1862	2-4-0	IC 15 x 20	5 6	905.5	69.5	975	14.7	120	24 16
1867	2-4-0T	IC 15 x 20	5 6	905.5	69.5	975	14.7	120	25 11
1871	2-4-0	IC 16 x 24	6 0	832.2	72.5	904.7	14.4	140	31 0
1875	2-4-0	IC 16 x 22	5 6	905.5	69.5	975	14.7	140	26 2

LANCASHIRE & YORKSHIRE RAILWAY

HAWKSHAW

Date	Type	Cylinders inches	D.W. ft in	Heating Surface sq ft			Grate Area sq ft	WP lb/sq. in	Weight WO ton cwt
				Tubes	F'box	Total			
1846	2-2-2	OC 15 x 20	5 9	730	59.5	789.5	11.0	120	24 3
1849	0-4-2	15 x 24	4 9	730	59.5	789.5	11.7	120	24 6

JENKINS

Date	Type	Cylinders inches	D.W. ft in	Tubes	F'box	Total	Grate Area sq ft	WP lb/sq. in	Weight WO ton cwt
1854	0-6-0	18 x 24	5 0	696.6	86.0	782.6	14.4	130	24 3
1856	0-6-0ST	15 x 24	5 0	696.6	86.0	782.6	14.4	130	30 18
1861	2-4-0	15 x 22	5 9	809	86.0	895	14.4	130	27 3

HURST

Date	Type	Cylinders inches	D.W. ft in	Tubes	F'box	Total	Grate Area sq ft	WP lb/sq. in	Weight WO ton cwt
1868	2-4-0ST	15 x 20	5 0	697.2	50.5	747.7	12.7	140	29 4
1869	0-6-0	17 x 24	5 0	841	109	950	16.25	140	29 0
1870	2-4-0	16 x 24	6 0	832.2	72.5	904.7	14.4	140	31 0
1873	0-6-0ST	17 x 24	5 0	841	109	950	16.25	140	34 11

BARTON WRIGHT

Date	Type	Cylinders inches	D.W. ft in	Tubes	F'box	Total	Grate Area sq ft	WP lb/sq. in	Weight WO ton cwt
1878	0-6-0	17½ x 26	4 6	971	90	1061	19.0	140	37 3

diameter coupled wheels built in 1868 and continuing into 1869, Jenkins did not produce any further designs.

Mr Hurst took over in 1868 and the building of 0-6-0, 0-6-0ST, and 2-4-0 types continued with the addition of a few 0-4-2s and 2-4-0STs.

Mr Hurst's 2-4-0s had inside cylinders 16″ x 24″ and inside frames. The coupled wheels were 6 ft diameter. They had domeless boilers and flush fireboxes and were known as 'straightbacks' – as distinct from the raised fireboxes previously built. Building took place from 1870 to 1876 when thirty four were turned out, the last ten having 17″ cylinders. They were used on main line services for about ten years. Another difference in the ten built in 1875–6 was that domed boilers were fitted.

To supplement the deficiency of main line locomotives of suitable power ten LNWR *Newton* class were supplied by Crewe in 1873. In fact the L&YR ordered a total of 101 locomotives to be built at Crewe, the other types being 0-4-0ST (five) and 0-6-0 *DX* goods (eighty-six) and as described elsewhere the locomotive contracting business obtained an injunction prohibiting any one railway building for another, but as the selling of second hand locomotives was legitimate new locomotives could be run for a short period and then sold.

Regarding tank locomotives, Hurst preferred the 2-4-0ST to the six coupled saddle tanks built by Jenkins and a total of twenty-six were built. They had the usual Hurst domeless boiler and the saddle tank covered the boiler and firebox but not the smokebox. They were used on branch and local trains but were not very effective as far as load pulling was concerned. A cab was an improvement instead of the previous type of spectacle plates.

A large stud of 0-6-0s had been built and, with the *DX* goods, formed the larger part of the total locomotives. They were used for passenger traffic as well as freight.

In 1876 Mr William Barton Wright took over the locomotive department at Miles Platting. Just over 450 locomotives had been built up to this time. He was faced with the demand for larger locomotives but the workshops were reaching saturation point, and due to lack of space for expansion, being completely hemmed in by factories and houses the Board decided that a new site was becoming imperative. Repairs were very much in arrears and the new building programme was such that outside contractors had to provide most of the new locomotives so badly needed.

From the advent of Barton Wright all new locomotives built were six coupled apart from two 2-4-0s of Hurst, completed in August 1876. The locomotives built were 0-6-0STs and 0-6-0s and all of the latter were to be rebuilt as saddle tanks in the 1890s.

A total of 522 new locomotives were built at Miles Platting, the last one being a 0-6-0 No. 305 which left the works in June 1881.

These works were notable for building the first 0-6-2T, although it was really a rebuilt 0-6-0 tender type. It appeared in 1880 in this form and a total of nine were dealt with in this way from 1880 to 1883.

Another event happened in 1882 when Baldwin's 4-2-2 LOVATT EAMES, having been shipped over from America, was re-assembled at Miles Platting to demonstrate the Eames vacuum brake.

Between 1881 and 1888 all new locomotives required were ordered from outside contractors. Rebuilding and repair work was carried on until 1888 when the works were closed, four years after a Directors meeting of March 1884 had decided that the works could no longer cope with the work required.

Number plates were introduced in 1876; numbers before this had been brass figures attached to the buffer beam and numbers on tanks and tenders were painted on. In 1887 Horwich was ready to receive repair work.

LANCASHIRE & YORKSHIRE RAILWAY

Horwich

Horwich Works were completed by the end of 1887 but the erecting shop, which was 1520 ft long and 118 ft wide made up of five bays, commenced repair work in November 1887 and new building commenced in the following January.

The name Locomotive Superintendent was now dropped and Chief Mechanical Engineer became the new title. The L&Y were probably the first railway to adopt this title. Personnel involved were: 1886–1900 John Audley Frederick Aspinall; 1900–4 H. A. Hoy, 1904–21 George Hughes.

Mr Aspinall had taken over the superintendency from Barton Wright at Miles Platting. He had been a pupil under John Ramsbottom at Crewe and from 1883 to 1886 was Locomotive Superintendent at Inchicore on the Great Southern and Western Railway.

His first design was the well known 'radial' 2-4-2Ts and an order for ten was put in hand. The first one No. 1008, now happily preserved, did not appear until 20 February, 1889 due to the new works not being ready to produce all the material required immediately, and to some

Fig. 306
L&YR Miles Platting
308/1870 2-4-0T
ELK No. 623

Fig. 307
L&YR Horwich
95/1891
Aspinall 7′ 3″
4-4-0 No. 1097

LANCASHIRE & YORKSHIRE RAILWAY

Fig. 308
L&YR Horwich
792/1902 *Aspinall*
7′ 3″ *Atlantic*
1406 as built
with inside
trailing axleboxes

Fig. 309
L&YR Horwich
909/1905
2-4-2T No. 869
as LMS No. 10939
superheated and
with large
bunker

labour trouble. Once these had been overcome, production was improved and the tenth appeared in July 1889.

Twenty 0-6-0s followed, with inside cylinders 18″ x 26″, inside frames and boilers with round top fireboxes. Many further locomotives of this type were built at Horwich following the same general design with various modifications and some with 17½″ cylinders. Joy's valve gear was fitted.

In 1891 Aspinall brought out some 4-4-0s with 7′ 3″ diameter driving wheels and 19′ x 26″ inside cylinders. These were not the first 4-4-0s on the L&Y as Barton Wright had designed a 6 ft 4-4-0 but they were all built by outside contractors, and thirty Aspinall 4-4-0s with 6 ft coupled wheels and 18″ x 26″ cylinders were built by Beyer Peacock in 1888 and 1889. They followed the Barton Wright design with larger boilers and Joy's valve gear instead of Stephenson's motion.

Aspinall built twenty of the 7′ 3″ class and the second ten had 18″ cylinders and later the others had their cylinders lined to 18″ diameter. They handled the majority of expresses including the Bradford to St. Pancras trains which they worked as far as Sheffield. Twenty more were built in 1894.

The Horwich works had a network of 18″ gauge railways 6½ miles long interconnecting the various shops for the conveyance of material, similar in function to the Crewe Works layout. For motive power three very small 0-4-0STs were ordered from Beyer Peacock in 1887 with outside cylinders 5″ x 6″ and 1′ 4¼″ diameter wheels. They were named DOT, ROBIN and WREN. From time to time these were added to by similar locomotives built at Horwich, two in 1891 named FLY and WASP, two in 1899 MIDGET and MOUSE and one in 1901 BEE.

The boiler was 2′ 3″ diameter by 4′ 3″ long and had a circular firebox with a grate area of 1·78 sq. ft. Besides a saddle tank holding 50 gallons there was a small tank between the frames with 26½ gallons of water. Coal was carried in a small four wheeled tender. They were extremely useful and did a tremendous amount of work.

In 1886 the Vulcan Foundry built three standard gauge 0-4-0STs and these were supplemented by a number of similar saddle tanks built at Horwich from 1891 to 1910; known as 'Pugs' they were used in shunting the Dock areas and the goods yards around Manchester. They were a neat design with 13″

x 18″ outside cylinders and 3 ft diameter wheels. The domed boiler had a heating surface of 475·75 sq. ft and a grate area of 7·2 sq. ft. The boiler pressure was 160 lb/sq. in.

Strangely enough the first seventeen were not allocated works numbers, neither were the 18″ gauge locomotives. Some were fitted with spark arresters, and instead of Joy's valve gear Stephenson's motion was used, between the frames.

Two other interesting designs by Aspinall were built at Horwich. A 0-6-0 side tank was introduced in 1897 and was the first locomotive on the L&Y to have a Belpaire firebox; one of the first railways to introduce this type of firebox, the first being the Manchester, Sheffield & Lincoln Railway in 1891 on one of their 0-6-2Ts. It is difficult to understand why this type was built whilst the 0-6-0STs were giving such good service. They were slightly larger but the total wheelbase was reduced from 15 ft to 12 ft, the outside cylinders were smaller, being 17″ x 24″ compared with the 17½″ x 26″ cylinders of the saddle tanks. Boiler pressure was increased from 140 to 160 lb/sq. in.

They were not popular with the shed staff and were far from lively. The other type was the large inside cylinder 4-4-2s which created quite a stir when it first appeared in 1899. Inside cylinders were 19″ x 26″, coupled wheels 7′ 3″ diameter, a large domed Belpaire boiler had a total heating surface of 1721 sq. ft and a grate area of 26 sq. ft. The pressure was 180 lb/sq. in. The weight in working order with tender was 89 tons 8 cwt. The boiler centre line was pitched 8′ 11″ above the rails which heightened the impression of massiveness. At first the trailing wheels had inside bearings and spiral springs, but due to rough riding they were modified to outside bearings which improved their riding qualities. Unusual features were: access doors forming part of the cab front through which the footplate men could pass to the outside running plates, regulator operated from either side by a horizontal bar linked up to the regulator shaft and steam sanding gear. Twenty were built in 1899 and twenty more in 1902. They were a great success being reliable and fast running – a speed of over 100 mph was claimed on a trial run from Liverpool to Southport in July 1899 and there should be no reason to doubt it. A tendency to slip on starting was perhaps their main fault but in the hands of experienced drivers this could be overcome.

In June 1899 Sir John Aspinall, as he had now become, was appointed General Manager and Mr H. A. Hoy became CME. He carried on with the output of 2-4-2Ts and 0-6-0s as a matter of policy, but with the goods department needing something stronger than the 0-6-0s, the first eight-coupled freight locomotive appeared in April 1900. Its design was by Aspinall and at the time was probably the most powerful locomotive in Great Britain. They had inside cylinders 20″ x 26″ with 4′ 6″ diameter wheels. The total heating surface of the boiler was 1693·2 sq. ft and the grate area 23 sq. ft. Boiler pressure was 180 lb/sq. in. giving a tractive effort of 29,466 lb. They were a neat design with domed boiler and Belpaire firebox, the wheel splashers were a continuous channel from end to end. They were fitted with Richardson's balanced slide valves. Unfortunately the boiler of No. 676 built in July 1900 exploded in March 1901, killing the footplate crew, and ending up in a field 50 yards away. The findings of the enquiry established that the firebox stays were of unsuitable material and many were defective.

As a result Mr Hoy designed a special boiler with circular firebox eliminating the usual stays supporting the firebox. It was in some respects similar to the well-known industrial Cornish boiler with single flue (as distinct from the Lancashire industrial boiler which had two flues). The flue in Mr Hoy's design was corrugated and made of steel. He fitted this type of boiler to one of the early 0-8-0s for test purposes and twenty new 0-8-0s were put in hand in 1903 fitted with this type of boiler. Instead of being the answer to the usual troubles experienced with stayed fireboxes, these boilers gave a lot of trouble, including leaking tubes, priming, and maintenance difficulties, added to which they took an inordinate time to raise steam from cold conditions. Various modifications were tried out and they were all subsequently fitted with orthodox boilers but with fireboxes modified to give more water space between the inner and outer firebox shells. They were known as 'Sea Pigs'.

Hoy designed a 2-6-2T for working the Manchester, Oldham, Rochdale & Bury services. They had inside cylinders 19″ x 26″ and 5′ 8″ coupled wheels. The large side tanks held 2000 gallons and the weight in working order was 77 tons 9 cwt. The boiler was high pitched at 9 ft and had a Belpaire firebox, a large heating surface totalling 2038·6 sq. ft and a grate area of 26 sq. ft. They were not popular with the permanent way department, the long rigid wheelbase playing havoc by spreading the rails and, as the centre coupled wheels were flangeless, derailments were frequent. As they grew older, trouble was experienced with cracked frames and leaking tanks. Twenty were built in 1903–4.

In March 1904 Mr Hoy left the L&Y to become General Manager of Beyer Peacock & Co and was succeeded by George Hughes.

LANCASHIRE & YORKSHIRE RAILWAY

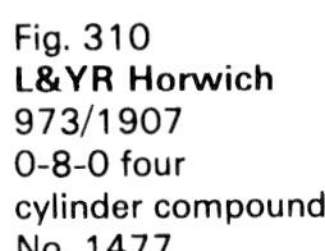

Fig. 310
L&YR Horwich
973/1907
0-8-0 four
cylinder compound
No. 1477

Fig. 311
L&YR Horwich
1169/1912
0-6-0 No. 237
Superheated with
Belpaire firebox

Fig. 312
L&YR Horwich
1175/1912
Class 31 0-8-0
as LMS 12841–
large cab, plug
superheater

LANCASHIRE & YORKSHIRE RAILWAY

Hughes had been premium apprentice under F. W. Webb at Crewe and then went straight to Horwich.

0-8-0s, 2-4-2Ts and 0-4-0STs continued to be built and in May 1906 a four coupled rail motor appeared. Kerr Stuart had already supplied two 0-2-2 rail motors in 1905 but the Horwich four coupled were larger. Eighteen were built between 1906 and 1911. They had outside cylinders 12" x 16" with outside Walschaerts valve gear and Richardson's balanced slide valves. The total heating surface of the boiler was 494·5 with a grate area of 9·4 sq. ft. The flush firebox was round topped.

They had short side tanks holding 550 gallons and unlike many other rail motors the engine part was quite divorced from the carriage except that the front portion of the carriage was pivoted on the rear of the engine. They were used on many branch lines including that from Blackrod to Horwich.

One of these units (No. 15) built in March 1907 was the 1000th locomotive to be built at Horwich although it was allocated works No. 983. The missing works numbers were: Lot 10 six 0-4-0ST 1891; Lot 20 six 0-4-0ST 1893 to 1894; Lot 26 five 0-4-0ST 1895. A total of seventeen locos added to 983 = 1000.

This total still does not include the five 18" gauge locos built 1891 to 1901 but they were never officially on the stock list, but for the purist the totals should have another five added on!

In 1908 a tank version of the 0-8-0s being built was designed by Hughes. It was 0-8-2T and five were built for banking and shunting duties. Their boilers were larger than their tender counterparts giving a total heating surface of 2204·8 sq. ft and a grate of 25·6 sq. ft. The cylinders were 21½" x 26". They were extremely powerful with a tractive effort of 34,052 lb at 85% of the boiler pressure of 180 lb/sq. in.

In 1906 one of the 0-8-0s was altered to a four cylinder compound with HP cylinders 15½" x 26" and LP 22" x 26", all driven off the second coupled axle. The valve gear was Joy's, and an interesting feature was that the LP valves were the normal 'D' type slide valves but the HP valves were piston type and were operated from the inside motion by rocking shafts.

The locomotive could be changed automatically from simple to compound working by a special valve. The locomotive (No. 1452) performed well and ten new 0-8-0 compounds were built, the only modification being to the drive to the outside high pressure cylinders driving the third coupled axle. They were built in 1907.

Regarding superheating on the L&Y, Aspinall was the first to introduce superheating to a locomotive in Great Britain and one of his 4-4-2s (No. 737) was fitted with a low degree superheater and the last five *Atlantics* were fitted with the same apparatus, which was in effect a steam drier, when they were built but it was Hughes who pursued a more positive line on superheating.

The principle of superheating was no new discovery, its advantages had been appreciated almost from the inception of the steam engine but applying it to a locomotive boiler presented a number of problems. The main advantage of superheating was to reduce cylinder losses and George Hughes ran his comparative trials combined with experiments on compounding at the same time as their advantages were synonymous. Low superheat had already been tried – 'low' being defined as up to 100°F above saturation point and 'high' as superheat over 200°F, which maintains steam in a dry condition throughout the stroke.

Condensation losses through the cylinder walls had been appreciated by Aspinall and he applied steam jacketted cylinders to his *Atlantics* passing steam for the injectors through the jacket, the other losses during steam expansion being combatted by superheating and compounding. Special attention was paid to lubrication of superheated locomotives especially those fitted with slide valves, oil being fed to the port faces by means of a mechanical lubricator and the valves themselves having several oil holes on the faces. It was found that the slide valve locomotive was not so economical as those fitted with piston valves, the saving by using superheaters was 15·3 per cent and 18·5 per cent per train mile respectively. These results were obtained on 4-4-0 locomotives the superheater surface being 209·9 sq. ft. in addition to the tubes and firebox total of 886·1 sq. ft.

The results of compounding on his 0-8-0s were interesting:

Average train load	*Simple*	*Compound*
400 tons & under	·209	·210
401–425 tons	·212	·202
426–450 tons	·188	·191
451–475 tons	·186	·186
476–500 tons	·190	·181
501–525 tons	·183	·169
526–550 tons	·184	·171

The figures given are the average amount of coal consumed per ton/mile. From the above it can be deduced that as the load increases consumption per ton/mile decreases and at higher loads the compound showed an advantage.

Fig. 313 **L&YR Horwich** 1344/1923 Hughes 4-6-0 No. 1674 four-cylinder, superheated, Walschaert's valve gear

Fig. 314 **Horwich** 1360/1924 *Baltic* Tank LMS No. 11116

Compounding had been neglected because comparatively good results were obtained from simple engines, and economy obtained by compounding must show a saving in maintenance costs as well as lower fuel consumption. Greater advantages were derived from compounding in low speed locomotives – when piston speeds were below 600 ft/min, cylinder condensation is lessened by compounding, but above that speed there was hardly any difference. This was amply shown in tests with the Hughes 0-8-0 freight locomotives.

In 1908 appeared the first Hughes 4-6-0 designed for the ever increasing loads on express passenger services. It was the forerunner of an order for twenty and had four cylinders 16″ x 26″ and coupled wheels 6′ 3″ diameter. The boiler was saturated, with a Belpaire firebox. Joy's valve gear was fitted. The total heating surface was 2507 sq. ft and a grate area of 27 sq. ft. Boiler pressure was the now standard 180 lb/sq. in. As built they had draughting problems and were very heavy on coal, so blast pipes and chimneys were modified more than once. Shed maintenance presented difficulties due to the inaccessible internal motion, glands and packings. The twenty were completed by March 1909, and an early decision was made to rebuild them all, with superheated boiler and replacing the Joy's valve gear by Walschaerts, new cylinders were to be fitted with piston valves. Unfortunately the war broke into these arrangements and all thought of rebuilding had to be postponed.

LANCASHIRE & YORKSHIRE RAILWAY

During the 1914–18 war locomotive work was reduced to a minimum, the only new locomotives built were forty 0-8-0s and ten 0-6-0s; the works were engaged manufacturing military equipment of all types.

After the war building 0-8-0s continued and in 1921 a new series of 4-6-0s commenced with the four cylinders enlarged to 16½" diameter and a superheated boiler with 304·9 sq. ft of superheated surface and a total heating surface of 2305·4 sq. ft. They were also equipped with a new type of 3000 gallon tender with 6 tons of coal and the weight in working order with tender was 119 tons 1 cwt. At the same time the origial batch of 4-6-0s were rebuilt with 16½" cylinders and superheaters; Walschaerts valve gear and long travel piston valves were also fitted and in this form their performance was transformed.

In December 1921 the amalgamation with the London & North Western Railway was ratified and from 1 January, 1922 the Lancashire & Yorkshire Railway lost its identity.

This lasted exactly a year when the L&NWR became part of the London Midland & Scottish Railway on 1 January, 1923.

During this one year, five 4-6-0s Nos. 1657 to 1661 were built and the last three, although carrying the standard L&Y number plates had L&NWR inscribed on them instead of their previous owners.

A more detailed mention should be made of the large amount of 2-4-2Ts that Horwich built from 1889 to 1911 with a grand total of 330, many more than any other British railway.

The 1889 tanks had small bunkers 18" x 26" cylinders and saturated boilers and flush fireboxes. All the 2-4-2Ts had 5' 8" diameter coupled wheels. The first modification was the reduction of the cylinder diameter to 17½" and to provide larger bunkers.

The first to be fitted with Belpaire fireboxes were those built in 1910, and the next step was the provision of superheated boilers with Schmidt type superheaters and cylinders 20½" diameter.

Vacuum operated water scoops were fitted to all 2-4-2Ts and these locomotives could be found all over the system working all classes of passenger trains from expresses to local suburban traffic.

Six, in Mr Hoy's time, were fitted with thermal storage heaters, patented by Druitt & Halpin, as an effort to save fuel which varied from 4 per cent to 12 per cent according to the duties performed. They were all removed after about six years use.

Up to the formation of the LMS Horwich works had built 1348 new locomotives to standard gauge and the five 18" gauge 0-4-0STs.

LONDON MIDLAND & SCOTTISH RAILWAY

Building continued with the four cylinder 4-6-0s and in 1924 a tank version was built as 4-6-4T *Baltic.* Ten were built although frames were cut for a total of twenty but it was thought that ten would be sufficient and the frames for the second ten were modified and became tender locomotives.

George Hughes was appointed CME of the whole LMS system which to the Midland division particularly was very unexpected, and contrary to expectations he refused to be moved from Horwich and during his term of office they automatically became the locomotive headquarters of the new system. He continued in office until 1925 and then Sir Henry Fowler took over.

The *Baltic* tanks may be said to be the last Lancashire & Yorkshire locomotive design although the *13,000* 2-6-0 mixed traffic locomotives were designed at Horwich under George Hughes' supervision. Sir Henry Fowler made certain alterations to the original design but the first batch of thirty was built at Horwich in 1926–7. They had two inclined outside cylinders 21" x 26" and coupled wheels 5' 6" diameter. The boiler had a parallel barrel and Belpaire firebox with a total heating surface of 1812 sq. ft including a superheater with 307 sq. ft. The grate area was 27·5 sq. ft. They had the large double window cab of the later 4-6-0s and a high running plate. Outside Walschaerts valve gear was fitted. They were extremely successful locomotives and a total of 245 were built, including forty more at Horwich from 1928 to 1930.

Fowler's policy was to concentrate building at four workshops: Crewe, Derby, Horwich and St Rollox.

Besides the seventy 2-6-0s built from 1926 to 1930 the following standard LMS locomotives were built:

20	4-4-0	compounds	1115–1134	1925–6
10	0-6-0	Class 4F	4457–4466	1928
15	0-6-0T	Class 3F	16750–16764	1931
75	2-8-0	Class 8F	8331–8399	1943–5
			8490–8495	1945
70	4-6-0	Class 5	4932–4999	1945–7
			4783–4799	1947
3	2-6-0	Class 4	3000–3002	1948

At the end of 1947 when the LMS became part of British Railways Horwich, the following new locomotives had been built:

Lancashire & Yorkshire 1343; London & North Western 5; London Midland & Scottish 315; a total of 1663.

It will be noticed that no new work was carried out from 1932 to 1942. The depression of 1931 was the beginning, and locomotive repairs only were carried out during this period and during the 1939–45 war a considerable

Fig. 315
Horwich
1452/1929 2-6-0
13118
with R.C. poppet
valve gear

Fig. 316
Horwich
1792/1953 0-4-0ST
47008–similar
to those supplied by
Kitson 1932

amount of armament work was carried out including the manufacture of 481 *Centaur, Matilda* and *A13 Cruiser* tanks, shells and other ordnance. In addition Horwich dealt with a number of American built 2-8-0s requiring modifications.

Building of new locomotives was resumed in 1943 and carried on as part of British Railways from 1 January, 1948.

Work proceeded in 1948 with the new Ivatt standard *Class 4* 2-6-0s, followed by some *Class 5P* two cylinders 4-6-0s. In 1953 five 0-4-0STs were put in hand for dock work.

The only British Railways design built at Horwich was the *Class 4* 2-6-0s which had many parts in common with the Ivatt 2-6-0s. Forty of this class were built and 76099 emerged from the works on 28 November, 1957 being the last steam locomotive built at Horwich.

The locomotives built during this period were:

Year	No.	Railway	Class	Type	Numbers
1948	20	LMS	Class 5	4-6-0	44698–44717
1948	20	LMS	Class 4	2-6-0	43003–43022
1949	2	LMS	Class 5	4-6-0	44668, 44669
1949	27	LMS	Class 4	2-6-0	43023–43049
1950	26	LMS	Class 5	4-6-0	44670–85/88–97
1951	2	LMS	Class 5	4-6-0	44686, 44687
1951	24	LMS	Class 4	2-6-0	43112–43135
1952	1	LMS	Class 4	2-6-0	43136
1952	6	BR	Class 4	2-6-0	76000–76005
1953	14	BR	Class 4	2-6-0	76006–76019
1953	4	LMS	Class OF	0-4-0ST	47005–47008
1954	1	LMS	Class OF	0-4-0ST	47009
1956	4	BR	Class 4	2-6-0	76075–76078
1957	21	BR	Class 4	2-6-0	76079–76099
	172	Total			

The total number of steam locomotives built at Horwich was 1835 and five 18″ narrow gauge works locomotives, from 1889 to 1957.

Diesel shunters were built from 1958 to 1962 when all new and repair work finished, the factory ceasing to be a locomotive works.

See LONDON MIDLAND & SCOTTISH RAILWAY & BRITISH RAILWAYS.

LYR – HORWICH

ASPINALL

Date	Type	Cylinders inches	D.W. ft in	Heating Surface sq ft				Grate Area sq ft	WP lb/sq in	Weight WO ton cwt	
				Tubes	S'heat	F'box	Total				
1889	2-4-2T	17½ x 26	5 8	1108.73	–	107.68	1216.41	18.75	160	55 19	
1889	0-6-0	18 x 26	5 0	953.17	–	107	1060.17	18.75	180	42 3	
1891	4-4-0	19 x 26	7 3	1108.73	–	107.7	1216.43	18.75	160	44 16	
1891	0-4-0ST	13 x 18	3 0½	432.7	–	43	475.7	7.2	160	21 5	
1897	0-6-0T	17 x 24	4 6	1081.5	–	85.9	1167.4	17.0	160	50 0	
1899	4-4-2	19 x 26	7 3	1767	–	161	1928	23.0	175	58 15	
1900	0-8-0	20 x 26	4 6	1767	–	147	1914	23.0	175	53 16	

HOY

Date	Type	Cylinders inches	D.W. ft in	Tubes	S'heat	F'box	Total	Grate Area sq ft	WP lb/sq in	Weight WO ton cwt	
1902	0-8-0	20 x 26	4 6	1775	–	125	1900	26.0	180	57 19	(a)
1903	2-6-2T	19 x 26	5 8	1877	–	161.6	2038.6	26.0	175	77 9	

HUGHES

Date	Type	Cylinders inches	D.W. ft in	Tubes	S'heat	F'box	Total	Grate Area sq ft	WP lb/sq in	Weight WO ton cwt	
1906	2-4-2T	18 x 26	5 8	1072	–	111.5	1183.5	18.75	180	60 4	
1906	0-4-0RM	12 x 16	3 7½	437.5	–	57	494.5	9.4	180	25 17	
1907	0-8-0	15½/22 x 26	4 6	1539	–	154.2	1693.2	23.0	180	60 16	(b)
1908	0-8-2T	21½ x 26	4 6	2008	–	196.8	2204.8	25.6	180	84 0	
1908	4-6-0	4/16 x 26	6 9	2317	–	190	2507	27.0	180	77 1	
1911	2-4-2T	20½ x 26	5 8	812.86	191.1	108.1	1112.06	18.75	180	66 9	
1912	0-6-0	20½ x 26	5 0	762.5	191.1	108.1	1061.7	18.75	180	46 10	
1912	0-8-0	21½ x 26	4 6	1768	396	195	2359	25.6	180	66 11	
1921	4-6-0	4/16½ x 26	6 3	1511	552	175	2238	27.0	180	72 12	
1925	4-6-4T	4/16½ x 26	6 3	1511	552	175	2238	27.0	180	99 19	(c)

(a) Corrugated Firebox. (b) 4 Cyl. Compound. (c) LMS Built.

LANCASHIRE & YORKSHIRE RAILWAY

LMSR – HORWICH

Date	2-8-0	4-6-4T	4-6-0	2-6-0	0-6-0	0-6-0T	4-4-0
1923			10433-54				
1924		11110-19	10455-70				
1925			10471-74				1115-30
1926				2700-06 (a)			1131-34
1927				2707-29 (a)			
1928				2810-16 (b)	4457-66		
1929				2817-29 (b)			
1930				2830-49 (b)			
1931						7667-81 (c)	
1943	8331-37						
1944	8338-81						
1945	8382-99 8490-95		4932-43				
1946			4944-66 4982-90				
1947			4991-99 4783-99	3000-02			

(a) Originally 13000-13029. (b) Originally 13110-13149. (c) Originally 16750-16764.

HORWICH WORKS

The works were designed to carry out all necessary locomotive repairs and the renewal of locomotive stock. In addition all mechanical, electrical and hydraulic engineering was located here. The land enclosed was 116 acres and a workshop area of twenty-two acres. This included a steel foundry fitted with Siemens-Martin regenerative melting furnaces. The erecting shop 1520 ft x 118 ft wide was equipped with twenty 30 ton overhead travelling cranes. Access to this shop for locomotives to the centre portion of the shop was effected by two traversers. Horwich in 1886 had a population of 4000 and in 1903 it had increased to 15,000. The narrow gauge railway had some 7½ miles of track in the works for transporting material from the stores to various shops and between shops. In the large erecting shop there was room for ninety locomotives and thirty tenders.

Orders for locomotives were mostly in units of ten or twenty with some exceptions and each order received a Lot Number, very similar to Swindon practice. Steam Locomotive lot numbers were from 1 (1889) to 107 (1957). Due to the conversion of many 0-6-0 tender locomotives to 0-6-0STs, there were many tenders rendered redundant and many new locomotives received second hand tenders.

Like many works Horwich had its own Gas Works and of course the indispensible Mechanics Institute.

Fig. 317
Horwich
1815/1957 BR
Class 4 2-6-0
No. 76096

LEEDS NORTHERN RAILWAY
Holbeck, Leeds

Mr Johnson and afterwards Mr Taylor were the Locomotive Superintendents at Holbeck.

At least one locomotive was built here – a 2-2-2 in 1854 with 15½" x 20" cylinders and driving wheels 5' 6" diameter. The same year the Leeds Northern Railway became part of the North Eastern Railway.

The locomotive was sold to the West Hartlepool Harbour and Railway Co. about 1865 and became their No. 37. In 1865 the WHH&R Co. became part of the North Eastern Railway and the locomotive became NER No. 610.

LENNOX LANGE
Glasgow

Despite information to the contrary, this firm did not build any locomotives but dealt in locomotives made by other firms.

LEWIN, STEPHEN
Poole Foundry, South Road, Poole

Stephen Lewin built a number of narrow gauge locomotives between 1868 and 1879 for industrial use and a tramway engine for Bilbao. Agricultural machinery was also built together with steam launches.

It is probable that he took over the works from a William Pearce who was an iron and brass founder, machinist and maker of agricultural implements, in the early 1860s.

The majority of locomotives built were for the narrow gauge. Details are very scanty but what were built found their way to the Isle of Man, Dublin, the West Country and the north and in 1878 a steam tram loco was sent to Bilbao and probably one to Guernsey the following year.

One of the first was a side tank with outside cylinders. The connecting rods were circular in section with single slide bars. It had no cab, but subsequently one was fitted which overshadowed the rest of the locomotive by its size. It was sold to B. Fayle & Co. Ltd. for their 3' 9" gauge line at the Clay Mines near Corfe Castle. It had an extraordinary long life of eighty years.

The Laxey Mines had two well tanks, each with one inside cylinder, plate frames and launch-type firebox which Lewin favoured. The cylinder was 4' x 6", the wheels 14" diameter and the gauge was 1' 6".

Fig. 318 **Lennox Lange** 1882 0-4-2ST with Lennox Lange plates

Fig. 319
Stephen Lewin
*c*1875 0-4-0WT
BEE Laxey Lead Mines

Fig. 320
Stephen Lewin
design for 0-4-0WT with inside frames

Fig. 321
Stephen Lewin
design for 0-4-2T with outside frames

Another type was a 'steam roller' adaption with cylinder on top of the boiler driving the coupled wheels through a series of gears. The gear was keyed on to the outside of the track wheel and the illustration shows both wheels similarly treated and it is supposed that one acted as a spare, should the teeth be stripped or worn. Another similar locomotive LINDHOLME had the gearing between the frames.

The earlier locomotives had eccentrics fixed to the leading axle; the cylinders were outside and inclined, with outside steam pipes. Stephenson's link motion was used.

According to local reports Stephen Lewin was a rich man and manufactured steam launches and locomotives more as a hobby than a serious business. He originally came from Lincolnshire and became a prominent person in Poole.

Despite his influence he is reported to have gone bankrupt, the works being taken over by another firm, about 1879. In 1878 Mr Lewin's works manager Mr Tarrant had left to go over to an adjacent general engineering works known as the Dorset Iron Foundry.

STEPHEN LEWIN

Works No	Date	Type	Cylinders inches	D.W. ft in	Gauge ft in	Customer	No/Name
	? 1868	0-4-0WT	OC 9 x 18	1 9	Std	Londonderry Rly	18
	1868	0-4-0T	OC 6½ x 9	2 6	3 9	B. Fayle & Co Ltd	Corfe
	*c.*1870	0-4-2T	OC 9 x 18				Lewin
	1871	0-4-0T	OC		3 0	Torrington & Marland Rly	Peter
		4WTG		2 6			Lindholme
	1875	0-4-0WT	OC 9½ x 18	3 0	3 5½	APCM Swanscombe (a)	Erith
	1875	0-4-0ST	OC 10 x 18	1 2	Std	Salt Union Ltd (b)	Cheshire
684	*c.*1875	0-4-0WT	IC 4 x 6	1 2	1 6	Laxey Leadmines I.O.M	Ant
685	*c.*1875	0-4-0WT	IC 4 x 6		1 6	Laxey Leadmines I.O.M.	Bee
	*c.*1875	0-6-0T					Edith
	*c.*1875	0-6-0T					Helen
	1876	0-4-0TG				Northampton Gas Light Co.	Crocodile
	1876	0-4-0TG	1/6¼ x 8		1 10	A. Guinness Son & Co. Ltd. Dublin	2 Hops
	1876	0-4-0TG	1/6¼ x 8		1 10	A. Guinness Son & Co. Ltd. Dublin	3 Malt
	1878	0-4-0TM					
	1879	0-4-0TM				Guernsey Steam Tramway Co.	(c)
		0-4-0T	OC		2 0	Bradley Langburgh Whinstone Quarry	Clara
		0-4-0T	OC		2 0	Winn Cliffe Ridge Quarry	

All Machinery made by Stephen Lewin, including Steam Launches Agricultural Machinery and Locomotives had Works Numbers in the same series.

(a) Outside eccentrics on front axle. (b) Originally WHARTON. (c) May have been the 1878 tram.

LILLESHALL CO.

St George's, Oakengates

This well known firm was established in 1764. As general engineers, the firm was particularly noted for its winding, pumping and blast engines. In the early 1860s they decided to build railway locomotives and the first ones were built at their Donnington Wood works.

On 1 January, 1889 the title became LILLESHALL CO. LTD.

In 1862 an exhibition locomotive was built, what type it was is not known, but in 1867 another was built for the Paris Exhibition. It was a 2-2-2 express passenger locomotive and it was evident that the intention of the firm was to build for the main line railways. A silver medal was awarded for this exhibit but no customers were forthcoming and eventually it was converted to a saddle tank, with six coupled wheels, and sold to the Cannock & Rugeley Colliery (No. 4 RAWNSLEY).

From 1862 the locomotives built were mainly four and six coupled saddle tanks with inside and outside cylinders. A number were built for their own internal use and others for local collieries, and a few went further afield.

Many of the earlier outside cylinder locomotives had a distinctive bolted framework forming the crosshead slide bars. Later ones had single slide bars. The long wheel base and small coupled wheels of some of the 0-6-0STs made them look a little out of proportion.

Fig. 322
Lilleshall Co.
1868 Cannock & Rugeley Colliery 0-6-0ST
ANGLESEY

Fig. 323
Lilleshall Co.
1886 Lilleshall Co. No. 2

LILLESHALL CO.

Date	Type	Cylinders inches	D.W. ft in	Gauge ft in	Customer	No/Name
1862	?				Exhibition Locomotive (London 1862)	
1862	0-4-0ST	OC 13 x 20	3 0	Std	Thos. Savin	Lilleshall
1865	0-4-0ST	OC 13 x 22	3 6	Std	Lilleshall Co.	4. Constance
1865	0-4-2T	OC		2 7½	C. Brain, Trafalgar Coll. Cinderford	Free Miner
1866	0-4-0ST			Std	Lilleshall Co.	5.
1867	0-6-0ST	IC 17 x 22	3 7	Std	Cannock & Rugeley Coll.	1. Marquis
1867	0-6-0ST			Std	W. Harrison Ltd. Grove Coll. Brownhills	Warrior
1867	2-2-2	*	*	Std	Exhibition Locomotive (Paris)	
1868	0-6-0ST	IC 17 x 22	3 8	Std	Cannock & Rugeley Coll.	2. Anglesey
1868	0-4-0ST	OC		Std	Cannock & Rugeley Coll.	3. Uxbridge
1869	0-4-0ST	OC		2 7½	C. Brain, Trafalgar Coll., Cinderford	Trafalgar
1869	0-6-0ST	13 x 22	3 6	Std	Lilleshall Co.	6.
1870	0-6-0ST	13 x 22	3 6	Std	Lilleshall Co.	7.
1870	0-4-0T	OC		2 7½	C. Brain, Trafalgar Coll., Cinderford	The Brothers
1870	0-6-0ST			Std	Cannock & Rugeley Coll.	Cannock Wood
1886	0-6-0ST	14 x 22	3 2	Std	Lilleshall Co.	2.
1888	0-4-0ST	OC		Std	Shelton Iron, Steel, & Coal Co.	Alice
1888	0-4-0ST	OC		Std	Shelton Iron, Steel, & Coal Co.	Althorp
	0-6-0ST				Cheslyn Hay	Emlyn then Sons
	0-4-0ST				Devon Basalt & Granite Co	Alderman
	0-4-0ST				Lilleshall Co.	1
	0-4-0ST				Lilleshall Co.	2
	0-4-0ST				Lilleshall Co.	3
	0-4-0ST	OC				Ravenhead
	0-4-0ST			2 3½	Earl of Dudley's Rly (Pensnett Rly) †	

* After rebuilding in 1872 sold to Cannock & Rugeley Coll. RAWNSLEY as 0-6-0ST IC 17" x 21", 3' 7" dia. wheels. † Quantity supplied – unknown.

LINGFORD GARDINER CO. LTD.
Railway Street, Bishop Auckland

The firm was established in the late 1850s as engineers, manufacturing mainly for local collieries. This included building and repairing locomotives, the building commencing about 1900 and ending in 1931. The closure of North Durham collieries in the 1920s severely affected their trading and the firm closed down in 1931.

Watts Hardy & Company who carried out the dismantling of the works sold a 0-4-0ST, which was probably the last built, to the Kettering Iron & Coal Co. Ltd. in 1933. No complete list available.

LINTON
Selby

Two locomotives are recorded as built by Linton for the York and North Midland Railway in 1842. They were: Y&NMR No. 21 ETNA 2-4-0 14" x 20" 5' 6" NER 265 and Y&NMR Nos. 22 SELBY 2-2-2 14" x 20" 5' 6" NER 266. Both were completed in February 1842 and both had outside frames. No further details are available.

LINGFORD GARDINER CO. LTD.

Built	Type	Cylinders inches	D.W. ft in	Customer	Name	W'drawn
*c.*1900	0-4-0ST			Pease & Partners Ltd	15 Miriam	
1900	0-4-0ST			Pease & Partners Ltd		
1900	0-4-0ST			Pease & Partners Ltd		
1904	0-4-0ST					1935
1906	0-4-0ST	OC 12 x 18	3 3	N. Bitchburn Coal Co.	Mostyn	1949
1914	0-4-0ST	OC 12 x 18	3 3	S. Derwent Coal Co. Ltd	Shield Row	1952
1931	0-4-0ST			Watts Hardy & Co. to Kettering Iron & Coal Co. Ltd	Kettering Furnaces No. 14	

Fig. 324 **Lingford Gardner** 1931 Kettering Coal & Iron Co. Ltd. KETTERING FURNACES No. 14

LISTER, WILLIAM
Hope Town, Darlington

These works were close to W. & A. Kitching's Hope Town Foundry. Four locomotives were built for the Stockton & Darlington Railway. The 0-6-0s were copies of the TORY designed by Timothy Hackworth. TORY was actually taken into Lister's works, measured up, drawings made and then returned into traffic. The outside cylinders were inclined and fitted to the rear end and driving the leading coupled wheels. These 0-6-0s were very powerful in their day and lasted until the late 1860s. The exact date of building the two for the Clarence Railway is not known but they were built before 1841. They both belonged to a Mr Norton, the lessee of the passenger traffic on the Clarence Railway.

LISTER, WILLIAM

Date	Type	Cylinders inches	D.W. ft in	Railway	No	Name
1838	0-4-0	$10\frac{3}{8}$ x	4 6	Stockton & Darlington Rly	3	Black Diamond
8/1839	0-6-0	12 x 22	4 0	Stockton & Darlington Rly	9	Middlesboro
6/1840	0-6-0	14 x 20	4 0	Stockton & Darlington Rly	13	Ocean
1840	0-6-0	14 x 22	4 0	Stockton & Darlington Rly	18	Etherley
	0-6-0		4 0	Clarence Rly	6	
	0-4-0	13 x 18	4 6	Clarence Rly		Victoria (a)

(a) Working pressure 55lb/sq. in. One large fire tube and 75 small tubes.

Fig. 325 **William Lister** 1839 Stockton & Darlington Rly No. 9 MIDDLESBORO built from Hackworth's drawing of *Tory* class

LIVERPOOL & MANCHESTER RAILWAY

Edge Hill, Liverpool

This famous railway was incorporated on 5 May, 1826 and opened on 15 September, 1830. In 1828 the question of what type of motive power should be used was investigated, with visits to the Stockton and Darlington Railway to compare the merits of fixed engines and locomotives. On 20 April, 1829 the Directors resolved to offer a sum of £500 for the best 'improved' type of locomotive – the competition to be held on 6 October, 1829. The successful design was Stephenson's ROCKET which did all that was required, and locomotives were decided upon for the railway apart from fixed engines at Edge Hill to haul trains up to Crown Street and from the Wapping terminus through the tunnel to Edge Hill.

Twenty-seven out of the initial total of thirty-six locomotives were built by Robert Stephen-

EDGE HILL – L & MR

Date	Type	Cylinders inches	D.W. ft in	No		Name
				L & M	GJR	
9/1841	2-2-2	12 x 18	5 0	69	128	Swallow
9/1841	2-2-2	12 x 18	5 0	71	131	Kingfisher
11/1841	2-2-2	12 x 18	5 0	72	130	Heron
12/1841	2-2-2	12 x 18	5 0	73	132	Pelican
1/1842	2-2-2	12 x 18	5 0	70	129	Martin
3/1842	2-2-2	12 x 18	5 0	74	133	Ostrich
3/1842	2-4-0	13 x 20	5 0	75	134	Owl
5/1842	2-2-2	12 x 18	5 0	77	136	Stork
6/1842	2-4-0	13 x 20	5 0	76	135	Bat
10/1842	2-2-2	12 x 18	5 0	78	137	Crane
1842	2-2-2	12 x 18	5 0	79	138	Swan
11/1842	0-4-2	12 x 18		81	140	Atlas (a)
12/1842	2-2-2	12 x 18	5 0	80	139	Cygnet
1/1843	2-2-2	12 x 18	5 0	82	141	Pheasant
4/1843	2-4-0	13 x 20	5 0	84	142	Bittern
6/1843	2-2-2	12 x 18	5 0	83	126	Partridge
10/1843	2-4-0	13 x 20	5 0	85	143	Lapwing
12/1843	2-4-0	13 x 20	5 0	86	144	Raven
1/1844	2-4-0	13 x 20	5 0	87	145	Crow
3/1844	2-2-2	12 x 18	5 0	88	146	Redwing
1844	2-2-2	12 x 18	5 0	89	147	Woodlark
7/1844	2-4-0	13 x 20	5 0	91	149	Petrel
1/1845	2-4-0	13 x 20	5 0	90	148	Penguin
1845	2-2-2	12 x 18	5 0	92	150	Linnet
1845	2-2-2	12 x 18	5 0	93	151	Goldfinch
1845	2-2-2	12 x 18	5 0	94	152	Bullfinch
6/1845	2-2-2	12 x 18	5 0	95	153	Chaffinch
7/1845	2-2-2	12 x 18	5 0	96	154	Starling

GJR

Date	Type	Cylinders inches	D.W. ft in	L & M	GJR	Name
5/1847	2-2-2	OC 13 x 20	6 0		164	Sun (b) Crewe Type
5/1847	2-2-2	OC 13 x 20	6 0		165	Star
5/1847	2-2-2	OC 13 x 20	6 0		166	Comet
8/1847	2-2-2	OC 13 x 20	6 0		167	Rhinoceros
8/1847	2-2-2	OC 13 x 20	6 0		168	Dromedary
9/1847	2-2-2	OC 13 x 20	6 0		118	Roderic
9/1847	2-2-2	OC 13 x 20	6 0		169	Huskisson
11/1847	2-2-2	OC 13 x 20	6 0		182	Roebuck
11/1847	2-2-2	OC 13 x 20	6 0		183	Theorem
2/1848	2-2-2	OC 13 x 20	6 0		184	Problem
2/1848	2-2-2	OC 13 x 20	6 0		185	Alderman
5/1848	2-2-2	OC 13 x 20	6 0		211	Onyx
5/1848	2-2-2	OC 13 x 20	6 0		212	Megatherium
5/1848	2-2-2	OC 13 x 20	6 0		213	Talbot

(a) Probably rebuild of old Stephenson Loco. (b) These 14 locos erected at Edge Hill Shops. Some parts probably supplied by Crewe. Some if not all were SFB (small firebox) with crooked frames.

son & Co, and it was not until 1841 that the main repair shops at Edge Hill, established in the early 1830s, began building new locomotives.

The first Locomotive Superintendent was John Dixon, quickly succeeded by John Melling, and in 1840 the post was taken over by John Dewrance.

2-2-2 passenger and 2-4-0 luggage locomotives were designed and the first 2-2-2 No. 69 SWALLOW was turned out in September, 1841.

Very little information is available about these two classes built at Edge Hill. The 2-2-2s weighed 15½ tons and the 2-4-0s, 16½ tons. Outside frames and inside cylinders were used. The boilers had raised fireboxes with domes fitted above. Two safety valves were fixed on the boiler barrels.

It has been stated that eleven additional locomotives were built during Melling's term of office taking the names and numbers of those they replaced, but no positive records are available. C. F. Dendy Marshall in his *Centenary History of the Liverpool & Manchester Railway* (1930) states that locomotives VESUVIUS, STAR and LIGHTNING were almost certainly replaced at Edge Hill.

However the total number built from 1841 to 1845 was twenty-eight plus the rebuilt ATLAS which may have been new.

By an act dated 8 August, 1845 the Liverpool & Manchester Railway together with the Bolton & Leigh Railway were incorporated in the GRAND JUNCTION RAILWAY.

As the GJR were about to transfer their own workshops at Edge Hill to Crewe, locomotive building ceased at the Liverpool & Manchester Railway shops and major repairs ceased by 1847.

LLANELLY RAILWAY & DOCK CO.
Tyissa, Llanelly

The railway company hired premises at Tyissa, Llanelly to build their locomotives which they commenced in 1840. The first one took a long time to build, not appearing until May 1843.

It was named PRINCE OF WALES and was a six coupled locomotive with outside cylinders 12" x 24" and weighed 17 tons. From these figures it must have been of modest proportions.

It was not successful and went into the workshops again at the end of 1844 for alterations and repairs. It remained in the shops for about two years when it was stated that it was ready for work again. Between 1847 and 1853 it probably did some work but in September 1853 it was reported as out of use and no use, and it was decided to dismantle it and use as much of the materials as possible in building a further locomotive.

It is of interest to note that Grylls & Co., local engineers, offered their services to contract for maintaining the company's locomotives and building a new one, but before this could be carried out the firm went bankrupt in 1848. At the sale of the firm's effects a boiler was purchased for the new locomotive which was put in hand in 1854 and was probably finished in 1856. It was named WALES and the wheels were given as 4 ft and were probably the ones originally fitted to the PRINCE OF WALES as the tyre wear could not have been very great. It must have been more successful than its predecessor as it was not withdrawn until 1871 and was actually sold to a Mr Richards the following year.

This railway served a number of collieries mining anthracite and this was used for firing WALES. It may have been one of the causes of trouble with the first locomotive as certain firing techniques and draughting had to be mastered and James McConnell of Wolverton had been the first to introduce coal instead of coke on his locomotives and his advice was sought and given regarding the burning of anthracite.

LLANELLY RAILWAY & DOCK CO.

Built	Type	Cylinders inches	Dia of Wheels		Name		GWR No	W'drawn
			Drive	Trail				
5/1843	0-6-0	OC 12 x 24		–	Prince of Wales		–	1853
3/1856	0-6-0		4 0		Wales	(a)	–	1872
1858	0-4-2	13 x 18	5 1½	3 6	Alice	(b)	896	1877
1863	0-4-2	13 x 18	5 1½	3 6	Alfred	(c)	897	1877
1865	0-6-0	16 x 24	4 9	–	Alexandra		907	1881
1873	0-6-0	17 x 24	4 9	–	Wales		908	1880

(a) Used parts of the 1843 locomotive with new boiler. (b) Rebuilt from LNWR 0-4-0 PRINCESS ALICE built by Bury 1839 and purchased 1847. (c) Rebuilt from LNWR 0-4-0 PRINCE ALFRED built by Bury 1839 and purchased 1853.

LLANELLY RAILWAY & DOCK CO.

Considerable rebuilding and 'renewals' took place in the workshops, two *Bury* four coupled locomotives were rebuilt as 0-4-2s and it was not until 1865 that an entirely new one was built. This was the ALEXANDRA another six coupled type with 16″ x 24″ cylinders, 4′ 8″ diameter wheels and a boiler with a heating surface of 1306·03 sq. ft including the firebox. This was followed by a similar design in 1873 named WALES once more, with cylinders increased to 17″ diameter bores.

It is likely that the appearance of these two locomotives was similar to those that had been supplied by Fossick & Blair with inside cylinders and frames with dome over the firebox surmounted by the safety valves. The earlier ones of 1843 and 1856 may have been similar to the Fossick & Hackworth locomotives supplied with inclined outside cylinders attached to the smokebox.

Locomotive Superintendents: W. Stubbs 1843–8; Joseph Hepburn 1850–71; Robert Hepburn 1871–3.

Joseph Hepburn was in charge of locomotives when I'Anson, Fossick & Hackworth had the contract of working the line and when their contract terminated in 1853 he remained in charge for the railway company. The Great Western Railway took over in 1873.

LONDON & BIRMINGHAM RAILWAY

Wolverton

The works were established in 1838 under the management of Edward Bury, the Locomotive Superintendent of the railway, who also had his own locomotive works at Liverpool and a large number (over one hundred) of 2-2-0 and 0-4-0 types were supplied. Wolverton Works were used initially for repairing and rebuilding, new construction not commencing until 1845 continuing until 1862, but repairs being carried on until 1877. After this the works became the centre for carriage and wagon building.

The first two new locomotives were two *Bury* type 2-2-0s with bar frames and circular fireboxes. The inside cylinders were 14″ x 18″ and driving wheels 5′ 9″.

On 16 July, 1846 the London and Birmingham Railway became part of the LONDON & NORTH WESTERN RAILWAY, and was known as the Southern Division.

Edward Bury moved on, his place being taken by James Edward McConnell who in 1836 had worked at Edward Bury's works, becoming his shop foreman. In 1840 he became engine superintendent at Bromsgrove on the Birmingham & Gloucester Railway.

It was not until 1848 that McConnell decided to continue new building at Wolverton. Designs were quite independent of Crewe and a variety of 2-2-2 passenger types were built incorporating many ideas instigated by McConnell. Prior to these being built, in 1849 a large 2-2-2 with outside cylinders and frames was turned out, with a large boiler for the period with a total heating surface of 1539 sq. ft. It was slightly out of gauge and various platforms had to be moved back for the cylinders to clear. Its nickname was 'Mac's Mangle'.

Fig. 326 **L&BR** 1861 2-2-2 No. 373 'Extra Large Bloomer' later named CAITHNESS

McConnell was interested in using coal instead of coke and his patent firebox was divided down the centre longitudinally for firing each side alternatively, with two fire doors. A combustion chamber formed an extension to the boiler barrel.

He designed a standard 0-6-0 goods and between 1854 and 1863 one hundred and seven were built by various locomotive contractors and Wolverton built no less than ninety-five of this total. They had inside cylinders 16″ x 24″, inside frames and 5′ 6″ diameter wheels.

His standard passenger classes were of three types, all with inside cylinders and inside frames: 'Small Bloomers' Cylinders 15″ x 21″ Drivers 6′ 6″ diameter; 'Large Bloomers' Cylinders 16″ x 22″ Drivers 7′ 0″ diameter; 'Extra Large Bloomers' Cylinders 18″ x 24″ Drivers 7′ 6″ diameter.

The 'Small Bloomers' had 15″ x 21″ cylinders which later were increased to 16″ x 22″. The original boiler pressure was 150 soon reduced to 120 lb/sq. in. Trouble was encountered with slipping and they were taken off main line services. 'Large Bloomers' with 16″ x 22″ cylinders were most successful and could cope with moderate loads with a good turn of speed. The most successful were the 'Extra Large Bloomers' which were heavier and more powerful, they had mid-feathers in the firebox, combustion chamber and a 'superheating' vessel in the smokebox through which steam passed from the boiler to the cylinders.

McConnell was considerably in advance of his time particularly regarding boiler design and combustion, and the majority of his locomotives included his combustion chamber and patent dual firebox. Fifteen 0-4-2Ts built at Wolverton were also fitted with this type of boiler.

McConnell had his own colour scheme – brick red – as opposed to the Crewe 'green'. He resigned in 1861 and the Southern Division was amalgamated with the Northern in 1862. Both locomotive departments then came under John Ramsbottom. Locomotives already in hand at Wolverton were completed and the building of the Standard 0-6-0s continued into 1863. Most of the machinery was then transferred to Crewe and the department closed in 1873. As far as can be ascertained the total output of new locomotives reached a total of 178.

The reader may be interested to note that the appellation 'Bloomer' was given to McConnell's 2-2-2s as their building coincided with the efforts of Mrs Amelia Bloomer in her dress reform campaign and, after all, with inside frames and cylinders they looked a bit bare under the splashers!

Fig. 327 **L&BR** Longitudinal section of McConnell's patent firebox

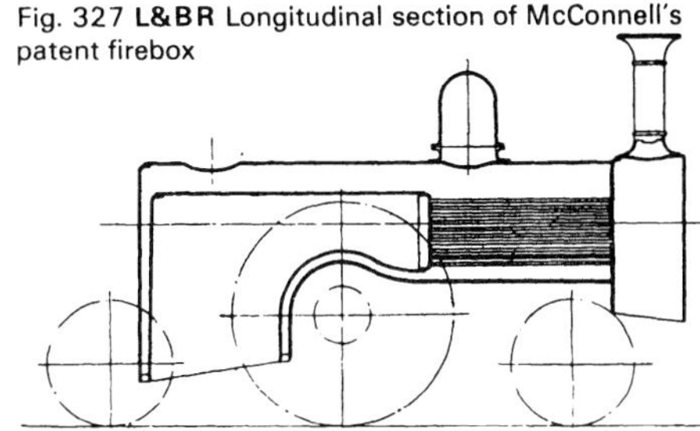

Date	Type	Cylinders inches	D.W. ft in	No	L & NW Name	Remarks
1845	2-2-0	14 x 18	5 9	92		Bury Type (a)
1845	2-2-0	14 x 18	5 9	9	No 95?	Bury Type
1848	0-6-0	18 x 24	5 0	228		Long Boiler Type (b)
1848	0-6-0	18 x 24	5 0	229		Long Boiler Type
1848	0-6-0	18 x 24	5 0	230		Long Boiler Type
1849	2-2-2	18 x 21	6 6	227		Outs, Frames & Cylinders
1852	0-4-2	16 x 24	5 0	28		
1852/3	0-4-2	16 x 24	5 0	124		
1852/3	0-4-2	16 x 24	5 0	148		
1852/3	0-4-2	16 x 24	5 0	164		
1852/3	0-4-2	16 x 24	5 0	236		
1853	0-4-2	16 x 24	5 0	192		
1853	0-4-2	16 x 24	5 0	193		
1853/4	0-4-2	16 x 24	5 0	194		
1853/4	0-4-2	16 x 24	5 0	195		
1853/4	0-4-2	16 x 24	5 0	231		
1856	0-6-0	16 x 24	5 6	11		Standard Goods
1856	0-6-0	16 x 24	5 6	19		Standard Goods
1856	0-6-0	16 x 24	5 6	20		Standard Goods
1856	0-6-0	16 x 24	5 6	49		Standard Goods

LONDON & BIRMINGHAM RAILWAY

WOLVERTON

Date	Type	Cylinders inches	D.W. ft in	No	L & NW Name	Remarks
1856	0-6-0	16 x 24	5 6	90		Standard Goods
1856	0-6-0	16 x 24	5 6	147		Standard Goods
1856	0-6-0	16 x 24	5 6	149		Standard Goods
1856	0-6-0	16 x 24	5 6	150		Standard Goods
1856	0-6-0	16 x 24	5 6	167		Standard Goods
1857	0-6-0	16 x 24	5 6	151		Standard Goods
1857	0-6-0	16 x 24	5 6	166		Standard Goods
1857	0-6-0	16 x 24	5 6	179		Standard Goods
1857	0-6-0	16 x 24	5 6	182		Standard Goods
1857	0-6-0	16 x 24	5 6	196		Standard Goods
1857	0-6-0	16 x 24	5 6	197		Standard Goods
1857	0-6-0	16 x 24	5 6	198		Standard Goods
1857	0-6-0	16 x 24	5 6	199		Standard Goods
1857	0-6-0	16 x 24	5 6	241		Standard Goods
1857	2-2-2	15 x 21	6 6	7	Inglewood	Small 'Bloomer'
1857	2-2-2	15 x 21	6 6	21	Bela	Small 'Bloomer'
1857	2-2-2	15 x 21	6 6	103	Osprey	Small 'Bloomer'
1857	2-2-2	15 x 21	6 6	140	St. David	Small 'Bloomer'
1857	2-2-2	15 x 21	6 6	238	Petrel	Small 'Bloomer'
1857	2-2-2	15 x 21	6 6	240	Lonsdale	Small 'Bloomer'
1858	2-2-2	15 x 21	6 6	2	Caliban	Small 'Bloomer'
1858	2-2-2	15 x 21	6 6	165	Herald	Small 'Bloomer'
1858	2-2-2	15 x 21	6 6	168	Glyn	Small 'Bloomer'
1858	2-2-2	15 x 21	6 6	180	Bucephalus	Small 'Bloomer'
1858	0-6-0	16 x 24	5 6	82		Standard Goods
1858	0-6-0	16 x 24	5 6	172		Standard Goods
1858	0-6-0	16 x 24	5 6	242		Standard Goods
1858	0-6-0	16 x 24	5 6	318		Standard Goods
1859	0-6-0	16 x 24	5 6	50		Standard Goods
1859	0-6-0	16 x 24	5 6	83		Standard Goods
1859	0-6-0	16 x 24	5 6	91		Standard Goods
1859	0-6-0	16 x 24	5 6	120		Standard Goods
1859	0-6-0	16 x 24	5 6	175		Standard Goods
1859	0-6-0	16 x 24	5 6	183		Standard Goods
1859	0-6-0	16 x 24	5 6	245		Standard Goods
1859	0-6-0	16 x 24	5 6	319		Standard Goods
1859	2-2-2	15 x 21	6 6	3	Langdale	Small 'Bloomer'
1859	2-2-2	15 x 21	6 6	66	Pheasant	Small 'Bloomer'
1859	2-2-2	15 x 21	6 6	117	Swift	Small 'Bloomer'
1859	2-2-2	15 x 21	6 6	189	Cadmus	Small 'Bloomer'
1860	2-2-2	15 x 21	6 6	317	Napier	Small 'Bloomer'
1860	0-6-0	16 x 24	5 6	5		Standard Goods
1860	0-6-0	16 x 24	5 6	51		Standard Goods
1860	0-6-0	16 x 24	5 6	100		Standard Goods
1860	0-6-0	16 x 24	5 6	101		Standard Goods
1860	0-6-0	16 x 24	5 6	104		Standard Goods
1860	0-6-0	16 x 24	5 6	105		Standard Goods
1860	0-6-0	16 x 24	5 6	107		Standard Goods
1860	0-6-0	16 x 24	5 6	169		Standard Goods
1860	0-6-0	16 x 24	5 6	170		Standard Goods
1860	0-6-0	16 x 24	5 6	171		Standard Goods
1860	0-4-2T	15 x 22	5 6	102		Double Firebox & Combustion Chamber
1860	0-4-2T	15 x 22	5 6	110		Double Firebox & Combustion Chamber
1860	0-4-2T	15 x 22	5 6	125		Double Firebox & Combustion Chamber
1860	0-4-2T	15 x 22	5 6	174		Double Firebox & Combustion Chamber
1860	0-4-2T	15 x 22	5 6	354		Double Firebox & Combustion Chamber
1861	0-6-0	16 x 24	5 6	52		Standard Goods
1861	0-6-0	16 x 24	5 6	72		Standard Goods

LONDON & BIRMINGHAM RAILWAY

Date	Type	Cylinders inches	D.W. ft in	No	L & NW Name	Remarks
1861	0-6-0	16 x 24	5 6	74		Standard Goods
1861	0-6-0	16 x 24	5 6	86		Standard Goods
1861	0-6-0	16 x 24	5 6	89		Standard Goods
1861	0-6-0	16 x 24	5 6	92		Standard Goods
1861	0-6-0	16 x 24	5 6	93		Standard Goods
1861	0-6-0	16 x 24	5 6	94		Standard Goods
1861	0-6-0	16 x 24	5 6	97		Standard Goods
1861	0-6-0	16 x 24	5 6	98		Standard Goods
1861	0-6-0	16 x 24	5 6	99		Standard Goods
1861	0-6-0	16 x 24	5 6	108		Standard Goods
1861	0-6-0	16 x 24	5 6	109		Standard Goods
1861	0-6-0	16 x 24	5 6	113		Standard Goods
1861	0-6-0	16 x 24	5 6	114		Standard Goods
1861	0-6-0	16 x 24	5 6	115		Standard Goods
1861	0-6-0	16 x 24	5 6	118		Standard Goods
1861	0-6-0	16 x 24	5 6	173		Standard Goods
1861	0-6-0	16 x 24	5 6	355		Standard Goods
1861	0-6-0	16 x 24	5 6	356		Standard Goods
1861	0-6-0	16 x 24	5 6	357		Standard Goods
1861	0-6-0	16 x 24	5 6	358		Standard Goods
1861	0-6-0	16 x 24	5 6	359		Standard Goods
1861	0-6-0	16 x 24	5 6	360		Standard Goods
1861	0-6-0	16 x 24	5 6	361		Standard Goods
1861	0-6-0	16 x 24	5 6	362		Standard Goods
1861	0-6-0	16 x 24	5 6	363		Standard Goods
1861	0-6-0	16 x 24	5 6	364		Standard Goods
1861	0-6-0	16 x 24	5 6	365		Standard Goods
1861	0-6-0	16 x 24	5 6	366		Standard Goods
1861	0-6-0	16 x 24	5 6	367		Standard Goods
1861	0-6-0	16 x 24	5 6	368		Standard Goods
1861	0-6-0	16 x 24	5 6	369		Standard Goods
1861	0-6-0	16 x 24	5 6	370		Standard Goods
1861	0-6-0	16 x 24	5 6	371		Standard Goods
1861	2-2-2	18 x 24	7 6	372	Delamere	Extra Large 'Bloomer'
1861	2-2-2	18 x 24	7 6	373	Caithness	Extra Large 'Bloomer' (c)
1861	2-2-2	18 x 24	7 6	375	Maberley	Extra Large 'Bloomer'
1861	2-2-2	15 x 21	6 6	377	Sultan	Small 'Bloomer'
1861	2-2-2	15 x 21	6 6	378	Mammoth	Small 'Bloomer'
1861	2-2-2	15 x 21	6 6	379	Wasp	Small 'Bloomer'
1861	2-2-2	15 x 21	6 6	380	Vandal	Small 'Bloomer'
1861	2-2-2	15 x 21	6 6	381	Councillor	Small 'Bloomer'
1862	0-4-2T	15 x 22	5 6	134		Double Firebox & Combustion Chamber
1862	0-4-2T	15 x 22	5 6	374		Double Firebox & Combustion Chamber
1862	0-4-2T	15 x 22	5 6	376		Double Firebox & Combustion Chamber
1862	0-4-2T	15 x 22	5 6	382		Double Firebox & Combustion Chamber
1862	0-4-2T	15 x 22	5 6	383		Double Firebox & Combustion Chamber
1862	0-4-2T	15 x 22	5 6	384		Double Firebox & Combustion Chamber
1862	0-4-2T	15 x 22	5 6	385		Double Firebox & Combustion Chamber
1862	0-4-2T	15 x 22	5 6	386		Double Firebox & Combustion Chamber
1862	0-4-2T	15 x 22	5 6	387		Double Firebox & Combustion Chamber
1862	0-4-2T	15 x 22	5 6	388		Double Firebox & Combustion Chamber
1862	2-2-2	16 x 22	7 0	389	Archimedes	Large 'Bloomer'

LONDON & BIRMINGHAM RAILWAY

Date	Type	Cylinders inches	D.W. ft in	No	Name	
1862	2-2-2	16 x 22	7 0	390	Alaric	Large 'Bloomer'
1862	2-2-2	16 x 22	7 0	391	Japan	Large 'Bloomer'
1862	2-2-2	16 x 22	7 0	392	Stork	Large 'Bloomer'
1862	2-2-2	16 x 22	7 0	393	Burmah	Large 'Bloomer'
1862	2-2-2	16 x 22	7 0	394	Ariel	Large 'Bloomer'
1862	2-2-2	16 x 22	7 0	395	Briareus	Large 'Bloomer'
1862	2-2-2	16 x 22	7 0	396	Raglan	Large 'Bloomer'
1862	2-2-2	16 x 22	7 0	397	Baronet	Large 'Bloomer'
1862	2-2-2	16 x 22	7 0	398	Una	Large 'Bloomer'
1862	0-6-0	16 x 24	5 6	409		Standard Goods
1862	0-6-0	16 x 24	5 6	410		Standard Goods
1862	0-6-0	16 x 24	5 6	411		Standard Goods
1862	0-6-0	16 x 24	5 6	412		Standard Goods
1862	0-6-0	16 x 24	5 6	413		Standard Goods
1862	0-6-0	16 x 24	5 6	414		Standard Goods
1862	0-6-0	16 x 24	5 6	415		Standard Goods
1862	0-6-0	16 x 24	5 6	416		Standard Goods
1862	0-6-0	16 x 24	5 6	417		Standard Goods
1862	0-6-0	16 x 24	5 6	418		Standard Goods
1862	0-4-2			601		
1862	0-4-2			606		
1862	0-4-2			616		
1862	0-4-2			688		
1862	0-4-2			695		
1862	0-4-2			696		
1862	0-4-2			741		
1862	0-4-2			778		
1862	0-4-2			784		
1862	0-4-2			787		
1862	0-4-2			790		
1862	0-4-2			811		
1862	0-4-2			1019		
1862	0-4-2			1020		
1862	0-4-2			1021		
1863	0-6-0	16 x 24	5 6	1056		Standard Goods
1863	0-6-0	16 x 24	5 6	1057		Standard Goods
1863	0-6-0	16 x 24	5 6	1058		Standard Goods
1863	0-6-0	16 x 24	5 6	1059		Standard Goods
1863	0-6-0	16 x 24	5 6	1060		Standard Goods
1863	0-6-0	16 x 24	5 6	1071		Standard Goods
1863	0-6-0	16 x 24	5 6	1072		Standard Goods
1863	0-6-0	16 x 24	5 6	1073		Standard Goods
1863	0-6-0	16 x 24	5 6	1074		Standard Goods
1863	0-6-0	16 x 24	5 6	1075		Standard Goods

(a) Built as London & Birmingham Rly Stock. (b) Built as L & NW Rly (Southern Div.) Stock. (c) Shown at the 1862 International Exhibition, London. (d) Built as L & NW Rly Stock with 600 added to S. Div. numbers.

First Built	Type	No	Cylinders inches	D.W. ft in	Boiler Heating Surface			Grate Area	WP lb/sq in	Weight WO ton cwt	
					Tubes	F'box	Total				
1845	2-2-0	2	IC 14 x 18	5 9							(a)
1848	0-6-0	3	IC 18 x 24	5 0			1102.0				(b)
1849	2-2-2	1	OC 18 x 21	6 6			1383.0				
1852	0-4-2	10	IC 16 x 24	5 0			1007.0				
1856	0-6-0	95	IC 16 x 24	5 6							(c)
1857	2-2-2	20	IC 15 x 21	6 6			1047.0		150		(d)
1860	0-4-2T	30	IC 15 x 22	5 6	1152	142	1294			28 15	
1861	2-2-2	3	IC 18 x 24	7 6	980.3	242.5	1222.8	25.0	150	34 14	(e)
1862	2-2-2	10	IC 16 x 22	7 0	753.5	241.5	995		150	27 0	(f)

(a) 'Bury' Type. (b) Long Boiler Type. (c) Standard Goods. (d) Small 'Bloomers'. (e) Extra Large 'Bloomers'. (f) Large 'Bloomers'. Numbers built are for Wolverton only.

LONDON BRIGHTON & SOUTH COAST RAILWAY

Brighton

Locomotive Superintendents: T. Statham 1840–1845; J. Gray 1845–7; S. Kirtley 1847; J. C. Craven 1847–69; W. Stroudley 1870–89; R. J. Billinton 1890–1904; D. E. Marsh 1905–11; L. B. Billinton 1912–22.

John Chester Craven was in charge of the locomotive department when Brighton Works became operational in 1852. The locomotive stock at that time consisted of a large number of different classes and makers ranging from Bury 2-2-0s to Hackworth and E. B. Wilson's 2-2-2s. The first two locomotives to be built at Brighton were two 2-2-2WTs with 13″ x 20″ cylinders and 5′ 7″ diameter driving wheels. They were followed by six 2-2-2s which were completed in 1854. To enumerate and describe all the types which Craven built at the company's works would be almost impossible, and take too long.

To summarise the types turned out from 1854 to 1869 the following list will give a good idea of a non-standardisation policy.

0-6-0	1854–65	16″ x 24″	4′ 9″ dia wheels	14 locomotives
0-6-0	1855–66	16″ x 24″	5′ 0″ dia wheels	10 locomotives
0-6-0	1864	16″ x 24″	5′ 1″ dia wheels	2 locomotives
0-6-0	1866	17″ x 24″	5′ 1″ dia wheels	4 locomotives
0-6-0	1867	17″ x 24″	5′ 2″ dia wheels	2 locomotives
2-2-2	1856–68	43 locomotives six sizes of cylinders and six sizes of driving wheels. 12 combinations.		
2-4-0	1854–68	34 locomotives, four sizes of cylinders and five sizes of coupled wheels. 9 combinations.		
0-4-2	1862–66	7 locomotives, two sizes of cylinders and two sizes of coupled wheels. 3 combinations.		

In 1870 William Stroudley was appointed in J. C. Craven's place. Stroudley had started his railway career on the GWR as a fitter in Swindon Works under Daniel Gooch. He left for a similar job at the Peterborough repair shops on the GNR and for a period attended to the locomotives on Lord Willougby de Eresby's Edenham and Little Bytham Railway. He became running shed foreman at Peterborough under W. Brown and when the latter obtained the post of Locomotive, Carriage and Wagon Superintendent on the Edinburgh and Glasgow Railway he was accompanied by Stroudley who was appointed manager at Cowlairs. In 1865 he took charge of the Lochgorm Works on the Highland Railway and left at the end of 1869 to take the LB&SCR position at Brighton in January 1870.

In addition to these types he built the following tank locomotives (quantity in brackets): 0-4-0WT (1); 2-2-2ST (1); 2-2-2T (4); 2-4-0T (5); 2-4-0WT (6); 0-4-2T (3); 0-4-2ST (6); 0-4-4T (1); 4-4-0ST (2); 0-6-0ST (2); 0-6-0PT (2).

The 0-6-0 goods locomotives had double frames and many lasted until the end of the century. The 2-2-2s and 2-4-0s, particularly the 'singles' took over the duties of the older Sharp singles, and worked the principal express services from Victoria & London Bridge to Brighton, Hastings & Portsmouth. Two interesting 2-2-2s No. 162 and 163 built in 1863 had driving wheels 7 ft. diameter with inside cylinders 17″ x 22″, double fireboxes similar to Mr Cudworth's designs built on the South Eastern Railway. They worked the fastest timed expresses to Brighton and back and were very successful. The 7 ft diameter wheels were the largest to be ever used on the LB&SCR.

Despite the great variety of shapes and sizes, and it should be borne in mind that the aforementioned locomotives were the products of Brighton Works, many more were built by outside contractors to increase the variety. This railway could well have said at the time 'you name it – we've got it'!

Altogether 150 locomotives had been built at Brighton when Stroudley took over. None were built in 1870 and two of Craven's 0-4-2Ts were completed in 1871 when the first two Stroudley 'C' class 0-6-0s were also built. They had inside $17\frac{1}{2}$″ x 26″ cylinders and 5′ diameter wheels. The boiler was domed with flush round-top firebox and inside frames were used.

In the following year he built his first batch of 0-6-0Ts, later known as *Terriers*. They had 13″ x 20″ inside cylinders and 4′ diameter wheels, the tanks had rounded top edges which Stroudley featured on all his tank locomotives. The boiler had a combined dome and safety valve on the last ring adjacent to the firebox. They were intended for working the suburban traffic of south London. They were known as *Class A* and they proved so useful that fifty were built. A larger six coupled tank was introduced in 1874 which became Stroudley's standard goods tank *Class E*. The *Class D* 0-4-2T wnich appeared in the following year had interchangeable boilers and cylinders with the *Class E* tank. The cylinders were 17″ x 24″ and the six coupled tanks had 4′ 6″ diameter wheels and the 'D' class 5′ 6″ diameter coupled wheels.

LONDON BRIGHTON & SOUTH COAST RAILWAY

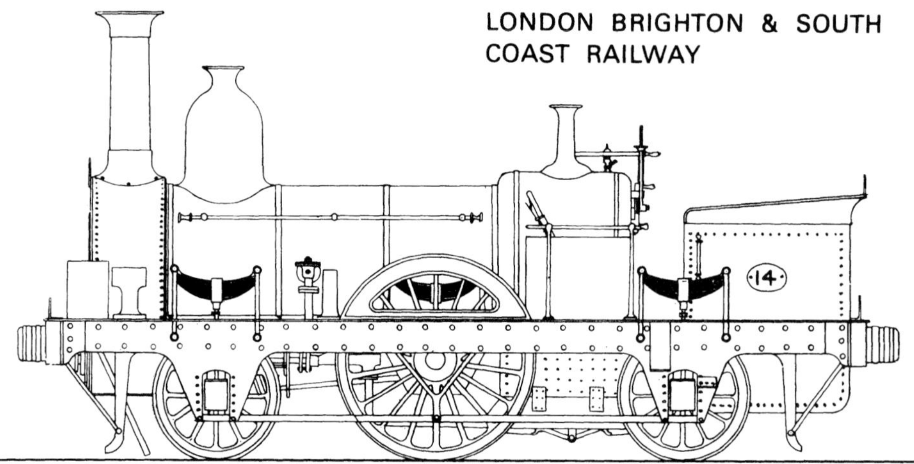

Fig. 328 **LB&SCR** 1852 2-2-2WT No. 14—first locomotive built at Brighton

Fig. 329 **LB&SCR** 1873 0-6-0 *Class C* W. Stroudley design No. 404

Fig. 330 **LB&SCR** 1874 *Terrier* class 0-6-0T No. 65 TOOTING

For passenger traffic, after building a few 2-4-0s and a 2-2-2 he designed the *Class D2* 0-4-2 the first one appearing in September, 1876, No. 300 LYONS, with inside cylinders 17" x 24" and 5′ 6" diameter coupled wheels. They were for mixed traffic and fourteen were built between 1876 and 1883. In conjunction with the *D2* class, an express version was built as *Class D3* with 17½" x 26" cylinders and 6′ 6" diameter coupled wheels. Stroudley favoured this wheel arrangement preferring the larger wheels to take the heavy weight at the front end. Despite this he continued developing the 2-2-2 and a most successful class was built as *Class G* commencing with No. 325 in January 1877 bearing the name ABERGAVENNY and from 1880 to 1882 twenty-three more were built at Brighton.

Stroudley was in favour of naming his locomotives, and all his locomotives had names bestowed upon them, with the exception of the 0-6-0 tender types. Unfortunately they were practically all place names and some must have looked very odd, such as TOOTING, FRESHWATER, TROCADERO, CRAWLEY to name a few. Even some of Craven's locomotives which Stroudley rebuilt were subsequently named.

The 0-4-2 was further enlarged, resulting in the famous *Class B Gladstone.* The cylinders were increased to 18¼" x 26", the coupled wheels remaining at 6′ 6" diameter. The boiler pressure was at first 140 lb/sq. in. which was the standard pressure, but was increased to 150 lb/sq. in. The success of this class was mainly due to the large boiler fitted, having a total heating surface of 1485·4 sq. ft. The weight in working order was 38 tons 14 cwt.

Stroudley used steam water feed pumps instead of injectors, claiming a saving in fuel, two water gauges instead of a single gauge with shut off cocks. Westinghouse automatic brakes were employed on the LB&SCR and the pump was a standard fitting. Some of the *Gladstones* were also fitted with vacuum brake pumps and fittings.

Stroudley's tenders were also unusual in that they had all inside frames and the wheels were larger than the usual tender wheel diameters.

For passenger locomotives each driver had his own, and was responsible for its running maintenance, cleanliness and fuel consumption and he was encouraged to get the best out of his charge. His name was inscribed in the cab, and although the locomotive could not be utilised to its full availability, considerable advantages accrued from this scheme.

Stroudley was indeed a perfectionist and great attention was paid to details in his designs and although his locomotives were expensive to build compared with some of the other railways, the additional expense was justified.

One of the *Gladstone* class was sent to the Paris Exhibition of 1889 this was No. 189

Fig. 331 **LB&SCR** 2-2-2 *Class G* No. 338 BEMBRIDGE 6′ 6" D.W.

LONDON BRIGHTON & SOUTH COAST RAILWAY

Fig. 332
LB&SCR
1896 4-4-0 No. 317
GERALD LODER

EDWARD BLOUNT. There were thirty-six in the class numbered 172 to 214 and curiously were built in the reverse order of numbering. They were used on the principal expresses between London, Brighton and Eastbourne, which they pulled with ease, keeping time into the bargain. Most trains however on this railway were not noted for their punctuality, neither was its neighbour the South Eastern Railway.

In June 1882 Stroudley produced an enlarged version of his *Class C* 0-6-0s known as *Large C* class. Cylinders were increased from 17½" to 18¼" diameter with the same stroke, and in 1884 he modified the design of the *Class E* 0-6-0T in the same way.

Unfortunately, Stroudley who went with his EDWARD BLOUNT to the Paris Exhibition, caught a chill whilst testing the locomotive on the Paris, Lyons and Mediterranean Railway after the exhibition, and died in Paris in December 1889. Whilst in office he had endeavoured to bring some order out of chaos and had designed and built a minimum of classes with a great degree of standardisation. At the time of his departure 478 locomotives had been built at Brighton.

His successor was R. J. Billinton and in 1890 and 1891, Stroudley's *Class B* 0-4-2 and *Class E1* 0-6-0Ts continued to be built and in the latter year a 0-6-2T version of the *E1* was built, having previously been planned by Stroudley. This was the prototype for a long line of 0-6-2Ts. It was No. 158 with the same cylinders and coupled wheels as the 0-6-0T, and was named WEST BRIGHTON.

In 1892 appeared Billinton's first design – a 0-4-4T intended for suburban passenger traffic and semi-fast trains; thirty-six were built from 1892 to 1896 with 18" x 26" inside cylinders, 5' 6" diameter coupled wheels and a boiler giving a total heating surface of 1203·5 sq. ft. Many worked on the London–Tunbridge Wells trains.

The 0-6-2Ts were built with 4' 6" diameter coupled wheels (*Class E3*) 5' 0" diameter (*Class E4*) and 5' 6" diameter (*Class E6*). The *E3s* were mainly for shunting, *E4s* shunting and pick-up goods and the *E6s* for passenger work.

Billinton was content to keep to the lines of his predecessor although of course to meet traffic needs his locomotives were larger and Brighton's first 4-4-0 (*Class B2*) appeared in 1895. They had inside cylinders 18" x 26" and 6' 9" diameter coupled wheels. The boiler had a round-top firebox with dome and safety valve combined and on a man-hole over the firebox the whistle was mounted. Curved continuous splashers were fitted over the bogie wheels. The boiler pressure was raised to 160 lb/sq. in. the total heating surface was 1227·3 sq. ft and grate area 18·73 sq. ft. Twenty-five were built from 1895 to 1898. The last one No. 213 BESSEMER had a larger boiler with 1464·8 sq. ft heating surface and a grate area of 20·6 sq. ft. It was classed *B3*.

In 1900 another class (*B4*) of 4-4-0 was built with 19" x 26" cylinders 6' 9" diameter coupled wheels, a heating surface of 1635 sq. ft and the working pressure increased to 180 lb/sq. in. The weight with tender in working order was 77 tons 7 cwt.

Lock-up safety valves were now fitted above the firebox. The first came out in January 1900, No. 53 SIRDAR and eight were built at Brighton 1900–5 and twenty-five by Sharp Stewart & Co. Ltd. in 1901.

Billinton also brought out a standard *Class C2* 0-6-0 with 18″ x 26″ cylinders and 5 ft wheels but all were built by outside contractors to a total of fifty-five. He died in 1904.

He had followed well in Stroudley's footsteps and during his fifteen years in office had built at Brighton: 125 0-6-2T Classes E3, E4, E5, E6; thirty-six 0-4-4T Class D; thirty-three 4-4-0 Classes B2, B3, B4; six 0-6-0T Class E1.

Douglas Earle Marsh succeeded Billinton in 1905. He was a Doncaster man and, as was to be expected, his first design was for an *Atlantic* tender locomotive which, when they appeared, showed a great similarity to the Great Northern Railway counterparts of Ivatt's *251 Class.*

Comparing the two classes dimensionally and in external lines it is very apparent that Marsh used the GNR drawings.

	LBSC No. 37	*GNR No. 251*
Cylinders	18½ x 26	18¾ x 24
Driving Wheels dia	6′ 7½″	6′ 8″
Tubes	246 x 2¼″	248 x 2¼″
Heating Surface Tubes	2337·1 sq ft	2359 sq ft
H Surface Firebox	136·4 sq ft	141 sq ft
Total	2473·5	2500 sq ft
Firebox Width,*	6′ 7½″	6′ 8″
Grate Area	30·9 sq ft	31 sq ft
Wheelbase	26′ 4″	26′ 4″
Weight W.O.	67 t	68 t – 6 c
Working pressure	200 lb. psi.	175
Built	1905	1902
Builder	Kitson & Co	Doncaster

*Wootten

Billinton had introduced the outside frame type of tender and Marsh carried on with this type.

In 1906 ten 0-6-0s, were built and then another GNR counterpart in the form of two classes of 4-4-2T. The *Class I1* had 17½″ x 26″ cylinders and 5′ 6″ diameter coupled wheels and the *I3* had 19″ x 26″ cylinders and 6′ 7½″ diameter wheels.

The *I1s* were the least successful of Marsh's designs but the *I3s* were amongst the first of this country's locomotives to be superheated. Two only, Nos. 22 and 23 were so equipped when built but the remainder were soon converted. No. 23 was used on the trials carried out between Brighton and Rugby in 1909 with the L&NWR's No. 7 TITAN of the *Precursor* type which was not superheated. From the results of this trial, in which the tank locomotive was superior, the L&NWR commenced fitting their boilers with superheaters soon afterwards.

As a result of the advantages shown by superheated boilers six 4-4-2s were put in hand at Brighton. They were generally similar to the five built by Kitson & Co. but the boilers all had Schmidt superheaters giving a heating surface of 461 sq. ft and a total heating surface of 2511 sq. ft. The cylinders were altered to 21″ diameter and boiler pressure to 170 lb/sq. in. Mechanical lubricators were fitted, and the smokebox was supported by a cast iron saddle. They were classed *H2.*

Marsh's final design was a 4-6-2T, one was built in December 1910 and the other in February 1912. Both had 21″ x 26″ outside cylinders and 6′ 7½″ diameter coupled wheels and round topped flush fireboxes. The first, No. 325 ABERGAVENNY had Stephenson's valve gear and a total heating surface of 1865 sq. ft including Schmidt superheater.

Fig. 333 LB&SCR 1907 4-4-2T *Class I*[1] No. 2 on trial before painting

LONDON BRIGHTON & SOUTH COAST RAILWAY

The second, No. 326 BESSBOROUGH, had outside Walschaerts valve gear – the first of this type of valve gear on the LB&SC. The heating surface was increased to 1944 sq. ft of which 357 sq. ft was from the Schmidt superheater. It weighed 87 tons in working order with 2000 gallons of water and 3 tons of coal. No. 325 was a ton lighter.

Boiler feed water was heated in the tanks by means of exhaust steam and fed into the boiler by a horizontal steam pump and hot-water injectors were also fitted.

The connecting rods were fitted with marine type big ends which Stroudley had introduced on his locomotives. They both worked between Brighton and London.

Unhappily Mr Marsh died in 1911 before the *Atlantics* and the second 4-6-2T were completed.

He had certainly brought Brighton locomotive practice to an advanced state. His successor was Lawson B. Billinton, son of R. J. Billinton.

Not content to increase the number of the 4-6-2Ts, he introduced in 1914 his *Baltic* 4-6-4Ts. The main features were a Belpaire firebox, Robinson superheater, Weir feed pump and feed water heater, piston valves and Walschaerts valve gear. The first, No. 327 CHARLES C. MACRAE appeared in March 1914. The cylinders were 22″ x 28″, 6′ 9″ coupled wheels, boiler heating surface total 2070 sq. ft including 383 sq. ft of superheater. Weight in working order was 98 tons. The trailing bogie enabled a large bunker to be built carrying 3½ tons of coal. Seven were built from 1914 to 1922 and were classed J.

A need at this time was for a more powerful heavy goods locomotive and Billinton decided on a 2-6-0 *Mogul* with outside cylinders 21″ x 26″ and 5′ 3″ diameter coupled wheels. They too had Belpaire fireboxes and superheaters. Seventeen were built, 1913–21. No locomotives were built at Brighton between 1917 and 1919.

Fig. 334
LB&SCR 1911
4-4-2 *Class H*²
D. E. Marsh
No. 421

Fig. 335
LB&SCR 1914
4-6-4T *Class L*
No. 327
CHARLES C. MACRAE

LONDON BRIGHTON & SOUTH COAST RAILWAY

For shunting the *Class E2* was built, to replace the earlier *E1s* of Stroudley origin. They had 17½" x 26" cylinders and 4' 6" diameter wheels. A Weir feed pump was fitted in front of the tanks. Ten were built and at the end of 1922 locomotives built at Brighton totalled 807.

Much rebuilding was done there, many older types received new and improved boilers and in some cases superheaters. Early locomotives were small but on the whole efficient and it was not until the advent of the *Atlantics* that large locomotives were seen on the Brighton line.

A duplicate list was in operation from early days the first series being in the 200s then as the locomotive stock increased numbers were given in the 300, 400, 500 and 600 series.

Naming of locomotives introduced by Stroudley was carried on by R. J. Billinton but discontinued by D. E. Marsh. From a publicity point of view names were popular and although never reaching the proportions of early days the *Atlantics* and the 4-6-2T locomotives received names and three only of the 4-6-4Ts. The last 4-6-4T No. 333 was the last built for the LB&SCR at Brighton and bore the appropriate name REMEMBRANCE, dedicated to the 532 men of the railway who lost their lives in the 1914–18 war.

On 1 January, 1923 the London Brighton and South Coast Railway lost its identity becoming the *Central Division* of the newly formed SOUTHERN RAILWAY.

At the amalgamation L. B. Billinton was rebuilding twelve of his father's *B4* 4-4-0s. New boilers, interchangeable with the 2-6-0 *K class* were fitted. These had superheaters, Belpaire fireboxes and new cylinders 20" x 26" with piston valves. Very little of the originals were incorporated and they were virtually new locomotives. The boilers had the top feed device introduced by L. B. Billinton on his 2-6-0s.

At the amalgamation R. E. L. Maunsell became the first Chief Mechanical Engineer and Eastleigh Works became the headquarters. Apart from the major rebuilding of the *B4s* now classified *B4X* no new locomotives were built until 1926 when ten 2-6-4Ts were 'erected' at Brighton, the components being built at other works and the boilers by the North British Locomotive Co. They were numbered A800–A809 and were *Class K*. Each bore the name of a river.

Due to a number of mishaps with these locomotives culminating in the Sevenoaks disaster in 1927 when thirteen lives were lost, the whole class totalling twenty were withdrawn from traffic. They had unstable characteristics at speed. It was decided to rebuild them as 2-6-0 tender locomotives and six were thus dealt with at Brighton in 1928.

Ten new *U class* 2-6-0s using Woolwich Arsenal parts were also erected in 1928 and these were originally intended to be another batch of 2-6-4Ts but the above mentioned events forced a change of policy.

In 1929 came an order for eight 0-8-0Ts (*Class Z*). These were intended for marshalling yard duties. They had three cylinders, the outside pair driving the third axle and the inside cylinder with an inclination of 1 in 8 drove the second axle; they were 16" x 28", the wheels 4' 8" diameter. The boiler had a round-top firebox and no superheater. The tanks were short and were inclined at the front to provide a better view for the driver and fireman. They weighed 71 tons 12 cwt and had a tractive effort of 29,376 lb at 85 per cent boiler pressure. Steam reversing gear was fitted.

Fig. 336
1929
0-8-0T *Class Z*
No. A951

LONDON BRIGHTON & SOUTH COAST RAILWAY

Routine repairs, rebuilding and maintenance was carried on at Brighton up to 1941 and no new building took place in this period. Brighton was to become redundant was the general impression and this would probably have happened if the war years had not intervened. In 1942 came an order for twenty *Class Q1* 0-6-0s.

R. E. L. Maunsell had retired at the end of October 1937 and his place was taken by Mr O. V. S. Bulleid coming from the LNER.

The *Class Q1* was his first design for the Southern Railway, and it was essential that this type should be able to work over the whole system without restriction yet be sufficiently powerful for any duty required. Its weight was restricted to 52 tons and what emerged from Brighton was indeed like nothing that had been built there before. All surplus non-essential fittings and refinements were discarded. The inside cylinders were 19″ x 26″ with piston valves above, wheels 5′ 1″ diameter, boiler heating surface 1472 sq. ft including 218 sq. ft superheater, a grate area of 27 sq. ft and a working pressure of 230 lb/sq. in. These dimensions made the class the most powerful 0-6-0s in the country, or probably anywhere else. Multiple-jet blast pipes were fitted and the wheels were of the Boxpok type. The weight was 51 tons 5 cwt without tender. The design was of course criticised as unnecessarily ugly, but to be fair Bulleid did the only sensible thing in cutting down the weight in this manner to obtain a tractive effort of 30,000 lb, a wonderful achievement for a 0-6-0 of limited weight.

They were numbered C1 to C16 and C37 to C40, Bulleid having introduced a new numbering scheme for his new locomotives. The next order was for Stanier's 2-8-0 which had been adopted by the War Dept as a standard locomotive. Ninety-three were built, sixty-eight for the London Midland and Scottish Railway and twenty-five for the London and North Eastern Railway, all in 1943 and 1944.

Considering that the maximum output at Brighton Works up to 1942 had been thirty locos, it speaks volumes for the reorganisation and planning that must have taken place to build these 2-8-0s which were being turned out at the rate of one a week towards the end of the order, and Brighton was in the front line of fire at this time.

In 1945 Bulleid introduced his 4-6-2 *West Country* class and from this date to the end of 1947 fifty of this class were built, and twenty very similar *Battle of Britain* class which differed from the *West Country* class in having modified cab fronts.

Fig. 337
1942 0-6-0
Bulleid's
Austerity Class Q
No. C1

Fig. 338
1948 Bulleid's
six-cylinder
Leader class

They had three cylinders 16½" x 26" and 6' 2" diameter coupled wheels; a large taper boiler gave a total heating surface of 2122 sq. ft including a large superheater surface of 545 sq. ft. Boiler pressure was 280 lb/sq. in. and the weight in working order was 86 tons. They were numbered 21C 101 to 21C 170. The designation meaning two front axles, one rear axle and C stood for three coupled axles. They were completely streamlined, had electric lighting (the first in this country) Boxpok wheels, welded steel firebox with two thermic syphons and the valve gear was completely new, chain driven with oil bath casings. They were used principally on the hilly routes of the West Country.

On 1 January, 1948 the Southern Railway became part of the nationalised *British Railways*. Up to 31 December, 1947 Brighton Works had built 1018 steam locomotives in 96 years and after the lull between 1929 and 1941 the works had seemed to take on a new lease of life.

Mr Bulleid had designed another class of locomotive, three of which were in the process of being built at the end of 1947. Five were originally ordered but only three were completed in 1948 and two of these were not run.

They were 0-6-6-0T. The two six wheeled powered bogies were each driven by three cylinders. The boiler was between the units and at a slight angle. No trial work was seriously pursued and all were scrapped in 1951.

Up to and during 1951 building of *West Country* and *Battle of Britain* Pacifics continued. The 'BB' or *Battle of Britain* class were similar to the West Country class but the cab front was cut away at the sides and windows fixed at an angle to improve the drivers visibility.

The LMS *Class 4* 2-6-4Ts had been found eminently suitable for suburban and semi-fast traffic, particularly in and out of London, and Brighton built twenty-four in 1950.

The only BR design that was built was the *Class 4* 2-6-4Ts which were the same as the LMS *Class 4* but incorporating standard BR fittings.

The following steam locomotives were built for BR:

19	SR	Class BB	4-6-2	34071–34089	1948
1	SR	Class BB	4-6-2	34090	1949
7	SR	Class WC	4-6-2	34091–34094/96/98 34100	1949
5	SR	Class WC	4-6-2	34103, 34105–34108	1950
1	SR	Class BB	4-6-2	34109	1950
24	LMS	Class 4	2-6-4T	42066–42078, 42096–42106	1950
1	SR	Class BB	4-6-2	34110	1951
17	LMS	Class 4	2-6-4T	42079–42095	1951
17	BR	Class 4	2-6-4T	80010–80026	1951
27	BR	Class 4	2-6-4T	80027–80053	1952
18	BR	Class 4	2-6-4T	80059–80076	1953
22	BR	Class 4	2-6-4T	80077–80098	1954
22	BR	Class 4	2-6-4T	80099–80105, 80116–80130	1955
20	BR	Class 4	2-6-4T	80131–80150	1956
4	BR	Class 4	2-6-4T	80151–80154	1957
205 Total					

Fig. 339
Bulleid's *West Country* class No. 34107
BLANDFORD

LONDON BRIGHTON & SOUTH COAST RAILWAY

After the last steam locomotive No. 80154 had been steamed on 20 March, 1957 locomotive repairs continued sporadically until the end of 1958 when no further locomotive work was carried out. Early in 1957 part of the works were taken over for the production of BMW Isetta cars. 1211 steam locomotives were built at Brighton from 1852 to 1957.

See also SOUTHERN RAILWAY and BRITISH RAILWAYS.

W. STROUDLEY

First Built	Class	Type	Cylinders inches	D.W. ft in	Heating Surface sq ft Tubes	F'box	Total	Grate Area sq ft	WP lb/sq in	Weight WO ton	cwt	
1871	C	0-6-0	17½ x 26	5 0	1312	102	1414	19.5	140	38	12	Domeless
1872	A	0-6-0T	13 x 20	4 0	463	55	518	10.0	140	24	7	'Terriers'
1873	D	0-4-2T	17 x 24	5 6	952	91	1043	15.0	150	38	10	
1874	G	2-2-2	17 x 24	6 9	1100	110	1210	19.3	150	33	0	(a)
1875	B	2-4-0	17 x 24	6 6	1022	110	1132		150	39	3	OF (b)
1874	E	0-6-0T	17 x 24	4 6	772	80	852	15.0	150	39	10	
1876	D^2	0-4-2	17 x 24	5 6	971	103	1074	17.0	150	34	7	
1878	D^3	0-4-2	17½ x 26	6 6	1074	109	1183	17.0		36	0	
1880	G	2-2-2	17 x 24	6 6	1084.5	100	1184.5	17.0	140	33	8	
1882	New C	0-6-0	18¼ x 26	5 0	1312	101	1413	20.95	150	40	7	'Jumbos'
1882	B	0-4-2	18¼ x 26	6 6	1374	112.5	1486.5	20.65	150	38	14	'Gladstones'

R. J. BILLINTON

First Built	Class	Type	Cylinders inches	D.W. ft in	Tubes	F'box	Total	Grate Area sq ft	WP lb/sq in	Weight WO ton	cwt	
1891	E Spl	0-6-2T	18¼ x 26	4 6	1082.2	91.7	1173.9	17.0				
1891	D Bogie	0-4-4T	18 x 26	5 6	1106.4	99	1205.4	17.0	160	51	14	Later E^3
1894	E^4	0-6-2T	18 x 26	4 6	1106.4	93	1199.4	17.4		48	9	Later D^3
1894	E^2	0-6-2T	18 x 26	5 0	1106.4	93	1199.4	17.4		51	0	
1895	B^2	4-4-0	18 x 26	6 9	1227.3	114.7	1342	18.73				
1899	B^4	4-4-0	19 x 26	6 9	1509	126	1635	24	180	42	16	
1902	E^5	0-6-2T	18 x 26	5 6	1106.4	105.2	1211.6	19.32	160	49	0	'Siemens' (c)
1904	E^6	0-6-2T	18 x 26	4 6	1106.4	105.2	1211.6	19.32	160	58	0	

(a) GROSVENOR was Class B. (b) CARISBROOKE & FRESHWATER. (c) Eight built at Brighton and twenty-five by Sharp Stewart & Co.

D. E. MARSH

First Built	Class	Type	Cylinders inches	D.W. ft in	Heating Surface sq ft Tubes	F'box	Total	Grate Area sq ft	WP lb/sq in	Weight WO ton	cwt	
1905	H^1	4-4-2	18½ x 26	6 7½	2337.1	136.4	2473.5	30.9	200	58	10	(e)
1906	C^3	0-6-0	17½ x 26	5 0	1183.4	101.4	1284.8	18.64	170	47	10	
1906	I^1	4-4-2T	17½ x 26	5 6	939.05	96.17	1035.22	17.43	170	68	6	
1907	I^2	4-4-2T	19 x 26	6 9	1499	126	1625	24.0	170	73	0	(f)
1907	I^2	4-4-2T	17½ x 26	5 6	1003.14	96.89	1100.33	17.35	170	68	10	
1908	I^4	4-4-2T	20 x 26	5 6	789	79	1118 (a)	17.35	160	70	10	PV
1908	I^3	4-4-2T	21 x 26	6 7½	850	126	1281 (g)	23.8	160	76	0	8" PV
1909	I^3	4-4-2T	19 x 26	6 7½	1507	120	1627 (h)	23.8	160	73	0	
1910	J	4-6-2T	21 x 26	6 7½	1461.69	124.4	1943.09(b)	25.16	170	89	0	
1911	H^2	4-4-2	21 x 26	6 7½	1895	136	2031	30.9	170	68	10	10" PV

L. B. BILLINTON

First Built	Class	Type	Cylinders inches	D.W. ft in	Tubes	F'box	Total	Grate Area sq ft	WP lb/sq in	Weight WO ton	cwt	
1913	E^2	0-6-0T	17½ x 26	4 6	1003	97	1100	17.35	170	52	15	
1913	K	2-6-0	21 x 26	5 6	1155.4	139	1573.4 (c)	24.8	170	63	15	
1914	L	4-6-4T	22 x 28	6 9	1534.92	152.08	2070 (d)	26.68	170	98	0	10" PV
1921	L	4-6-4T	22 x 28	6 9	1664.47	152.08	2199.55(d)	26.68	170	98	5	10" PV

(a) Includes 250 sq ft Superheating. (b) Includes 357 sq ft Superheating. (c) Includes 279 sq ft Superheating. (d) Includes 383 sq ft Superheating. (e) Built Kitson & Co. was Class B.5. (f) Altered to I^3. (g) Includes 305 sq ft Superheating. (h) Saturated 1^3s.

LONDON CHATHAM & DOVER RAILWAY

Longhedge, London

Originally all repair work was handled at Dover but when Longhedge Works were built in 1862 all repair work was transferred. The site of seventy-five acres had been acquired just south of the London & South Western Railway, and on it were built the locomotive workshops and running shed, and a carriage and wagon works. The locomotive erecting shop was 200 ft long and 100 ft wide, divided down the centre by a traverser which served ten pits on each side. An overhead crane served each set of pits. The length of the shop was considerably increased in 1880–1 and twenty-two additional pits were provided, eleven each side. The factory at the time of building was very well equipped with foundry, forge, machine shop, boiler shop and brass finishing shop. Considering its facilities, the erection of new locomotives occurred spasmodically and most of the labour employed did more repair and rebuilding work.

The locomotive superintendents of the LC&DR were: 1860–74 William Martley and 1874–98 William Kirtley.

The first new locomotive was completed in March 1869, a 2-4-0 with 6 ft driving wheels and 16″ x 24″ cylinders. It was followed by two more in 1870 with 7′ diameter coupled wheels. They were named ENIGMA, MERMAID and LOTHAIR respectively. The first had this name bestowed, it is alleged, because it was a puzzle how ever it was completed at all, principally due to lack of funds – a state of affairs that was with the railway for many years and was the main reason why so few locomotives were built at Longhedge, although the Board always seemed to be able to pay for new ones built by outside contractors.

Two similar 2-4-0s appeared in 1876 but with 17″ x 24″ cylinders. Previous to these, four 2-4-0Ts had been built in 1872–3 using the boilers off some 1860 4-4-0s built by R. & W. Hawthorn. These can not be termed new locomotives although in some lists they are designated thus.

William Kirtley designed a 0-6-0T with 17″ x 24″ inside cylinders and 4′ 6″ wheels and ten were built from 1879 to 1893 (Nos. 141–150). They were neat and robust and the last one disappeared in 1951. The side tanks had rounded tops and the cab and bunkers were narrower than the tanks. The Westinghouse pump was fitted on the right hand side, on the cab side. The boiler was domed with round topped firebox surmounted by Ramsbottom safety valves. They were *Class T*.

The next class were 4-4-0s *Class M1* and *M2*, the *M2*s having tenders with a water capacity of 2470 gallons compared with 2550 gallons for the *M1 Class*. They had 17½″ x 26″ inside cylinders and 6′ 6″ diameter coupled wheels. A domed boiler and round firebox gave a total heating surface of 1071 sq. ft and a grate area of 16·5 sq. ft. They were used on the Continental boat trains, until relegated to secondary

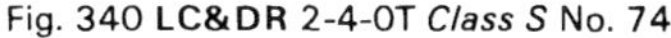
Fig. 340 **LC&DR** 2-4-0T *Class S* No. 74

Date	Type	Cylinders inches	D.W. ft in	Class	No		Name	W'drawn
					LCDR	SE & CR		
8/1865	2-4-0T	15 x 18	5 6	(a)	62	521	Lake	8/1909
9/1865	2-4-0T	15 x 18	5 6	(a)	60	519	Sittingbourne	7/1909
9/1865	2-4-0T	15 x 18	5 6	(a)	64	523	Chatham	9/1909
10/1865	2-4-0T	15 x 18	5 6	(a)	59	518	Sondes	7/1909
11/1865	2-4-0T	15 x 18	5 6	(a)	61	520	Crampton	8/1909
11/1865	2-4-0T	15 x 18	5 6	(a)	63	522	Faversham	9/1909
3/1869	2-4-0	16 x 24	6 0	L	50	509	Enigma	8/1906
6/1870	2-4-0	16 x 24	6 6	L	51	510	Mermaid	7/1906
9/1870	2-4-0	16 x 24	6 6	L	52	511	Lothair	12/1905
9/1872	2-4-0T	16 x 22	5 6	(b)	72	531	Bacchus	11/1909
1/1873	2-4-0T	16 x 22	5 6	(b)	74	533	Comus	7/1905
2/1873	2-4-0T	16 x 22	5 6	(b)	73	532	Vulcan	4/1906
5/1873	2-4-0T	16 x 22	5 6	(b)	71	530	Aeolus	7/1905
8/1876	2-4-0	17 x 24	6 6	C	57	516		5/1908
11/1876	2-4-0	17 x 24	6 6	C	58	517		2/1909
12/1879	0-6-0T	17 x 24	4 6	T	141	600		12/1936
12/1879	0-6-0T	17 x 24	4 6	T	142	601		6/1933
12/1880	4-4-0	17½ x 26	6 6	M1	175	634		5/1914
12/1880	4-4-0	17½ x 26	6 6	M1	176	635		3/1923
6/1881	4-4-0	17½ x 26	6 6	M1	177	636		5/1912
6/1881	4-4-0	17½ x 26	6 6	M1	178	637		11/1914
11/1885	4-4-0	17½ x 26	6 6	M2	179	638		2/1914
12/1885	4-4-0	17½ x 26	6 6	M2	180	639		1/1913
12/1889	0-6-0T	17 x 24	4 6	T	149	608		12/1936
4/1890	0-6-0T	17 x 24	4 6	T	146	605		7/1932
12/1890	0-6-0T	17 x 24	4 6	T	148	607 (c)		11/1949
3/1891	0-6-0T	17 x 24	4 6	T	145	604 (d)		11/1950
6/1891	0-6-0T	17 x 24	4 6	T	147	606		12/1936
9/1891	0-6-0T	17 x 24	4 6	T	150	609		12/1936
5/1892	4-4-0	18 x 26	6 6	M3	14	473		8/1926
7/1892	4-4-0	18 x 26	6 6	M3	25	484		8/1926
6/1893	0-6-0T	17 x 24	4 6	T	143	602 (e)		7/1951
10/1893	0-6-0T	17 x 24	4 6	T	144	603		1/1936
11/1893	4-4-0	18 x 26	6 6	M3	16	475		1/1927
4/1894	4-4-0	18 x 26	6 6	M3	17	476		1/1927
11/1894	4-4-0	18 x 26	6 6	M3	20	479		11/1927
5/1895	4-4-0	18 x 26	6 6	M3	15	474		4/1926
12/1895	4-4-0	18 x 26	6 6	M3	12	471		10/1926
5/1896	4-4-0	18 x 26	6 6	M3	13	472		11/1927
11/1896	4-4-0	18 x 26	6 6	M3	23	482		1/1927
3/1897	4-4-0	18 x 26	6 6	M3	5	464		4/1926
6/1897	4-4-0	18 x 26	6 6	M3	19	478		7/1926
10/1897	4-4-0	18 x 26	6 6	M3	3	462		3/1926
6/1898	4-4-0	18 x 26	6 6	M3	7	466		1/1927
8/1898	4-4-0	18 x 26	6 6	M3	24	483		1/1926
12/1898	4-4-0	18 x 26	6 6	M3	6	465		12/1925
5/1899	4-4-0	18 x 26	6 6	M3	4	463		6/1928
11/1899	4-4-0	18 x 26	6 6	M3	8	467		8/1927
2/1900	4-4-0	18 x 26	6 6	M3	(26)	485		7/1927
2/1901	4-4-0	18 x 26	6 6	M3	(9)	468		8/1925
5/1901	4-4-0	18 x 26	6 6	M3	(10)	469		10/1927
2/1902	0-6-0	18½ x 26	5 2	C		592	31592	(f)
5/1902	0-6-0	18½ x 26	5 2	C		593	31593	2/1958
10/1902	0-6-0	18½ x 26	5 2	C		460	31460	2/1949
12/1902	0-6-0	18½ x 26	5 2	C		461	31461	8/1958
12/1902	0-6-0	18½ x 26	5 2	C		486	31486	4/1953
6/1903	0-6-0	18½ x 26	5 2	C		580	31580	8/1953
6/1903	0-6-0	18½ x 26	5 2	C		583	31583	8/1961
1/1904	0-6-0	18½ x 26	5 2	C		480	31480	7/1961
4/1904	0-6-0	18½ x 26	5 2	C		481	31481	11/1961

(a) Known as Second Sondes Class. Boilers and other parts off 'Crampton' tanks. (b) Known as New Aeolus Class. Replacements. (c) Renumbered BR No. 31607. (d) Renumbered BR No. 31604. (e) Renumbered BR No. 31602. (f) Transferred to Service Dept 1963 as DS.239. W'drawn finally in 1966 and preserved at Ashford.
(a) and (b) are not usually included as new locomotives.

duties. Four *M1s* and two *M2s* were built from 1880 to 1885. They had followed six *Class M* 4-4-0s built by Neilson & Co. and were very similar in appearance, not unlike the Cambrian Railway 4-4-0s apart from the sandboxes which were incorporated in the leading driving splashers.

As the working of the Kent expresses became heavier and harder a further 4-4-0 design *Class M3* was brought out by Kirtley in 1892 with larger 18″ x 26″ cylinders and the boiler pressure increased by 10 lb to 150 lb/sq. in. The boiler was 3″ longer giving a total heating surface of 1110·25 sq. ft and a grate area of 17 sq. ft, yet notwithstanding these increases the official weight in working order was 42 tons 9 cwt compared with 42 tons 16½ cwt of the *M class.* Their performance however showed a great improvement and from 1892 to 1898 fifteen were built.

On 1 January, 1899 the LC&DR and the South Eastern Railway formed the SOUTH EASTERN & CHATHAM RAILWAY.

William Kirtley retired at the end of 1898 and Harry S. Wainwright was appointed Locomotive and Carriage Superintendent of the newly formed SE&CR. He had previously been the South Eastern Railway's Carriage & Wagon Superintendent.

Five more *M3s* were built at Longhedge from 1899 to 1901.

In 1900 Ashford works, being the chief locomotive works of the new company, designed a 0-6-0 *Class C* and this design was originally evolved by Kirtley at Longhedge to provide replacements for earlier 0-6-0s.

One hundred and nine of this class were built of which nine came from Longhedge in 1902–4, and were the last new locomotives to come from this factory. They had inside cylinders 18½″ x 26″ and 5′ 2″ diameter wheels. The boiler had a round top firebox, Ramsbottom safety valves and a total heating surface of 1200 sq. ft and a grate area of 17 sq. ft. An unusual feature was the sandboxes integral with the centre driving wheels splashers. They were operated by steam, as was the reversing gear and Westinghouse pumps were fitted. They were painted green and all brass and copper work was polished. Though primarily for goods traffic, they were frequenty used satisfactorily on various passenger duties.

Although no new work was carried out after 1904, Longhedge continued to be busy on rebuilding and repair work and it was not until 1911 that the works were officially closed, and most of the machinery transferred to the Ashford works in the following year after the latter works had been sufficiently extended.

A total of fifty locomotives had been built; thirty-six by the LC&DR and fourteen by the SE&CR.

Fig. 341 **LC&DR** 1885 4-4-0 *Class M*² 6′6″ No. 179

LONDONDERRY RAILWAY

Seaham Engine Works, Seaham, Co. Durham

This railway was owned for nearly fifty years by the Marquis of Londonderry until it was taken over by the North Eastern Railway on 6 October, 1900. Built principally for mineral traffic it became operative on 3 August, 1854 and started carrying passengers on 2 July, 1855.

The railway took over the working of the Seaham harbour and docks and continued to do so until 1900, when the Seaham Harbour Dock Co. became a separate entity.

There were extensive workshops at Seaham and besides doing colliery work, the steamships employed by the Marquis were also repaired there. Locomotive building commenced in 1889 although the repairing and rebuilding of locomotives was started much earlier.

Three locomotives are known to have been built at Seaham, with the possibility of a fourth.

The 2-4-0T was for passenger services and equipped with steam, air and hand brakes, and had inside cylinders and frames. The 0-6-0 had inside cylinders and outside frames, the boiler having a round topped firebox. The 0-4-4T was quite modern in appearance, and had inside frames and cylinders.

In general the main characteristcs of locomotives built at Seaham were: side window cabs, in the majority of cases; square front spectacle plates; smokebox fronts with wings; double buffers; and built-up chimneys.

All had to work without turning so that the tank locomotives had all over cabs and rear weatherboards and the tender locomotive was equipped with cabs on both with a flexible cover connecting them.

Passenger and goods locomotives were painted green, lined out in orange, with the railway initials LR on tank and tender sides with a coronet or coat of arms between the letters. Number plates were brass.

The 1901 0-6-0T was rebuilt and laid aside until 1904 when it was sold to the Seaham Harbour Dock Co. Ltd.

Built	Type	Cylinders inches	D.W. ft in	LR No	NER No
1889	2-4-0T	14½ x 24	4 11	2	1113
1891	0-6-0	17½ x 26	4 9	20	1335
1895	0-4-4T	17 x 24	5 4½	21	1712

LOCOMOTIVES REBUILT AT SEAHAM

Rebuilt	Type	Cylinders inches	D.W. ft in	LR No	NER No	Rebuilt from			
						Mkr	Works No	Built	Type
?	2-4-0T	14½ x 22	4 11	8	2269	RS	1075	1856	2-4-0
1877	0-6-0ST	16¾ x 24	4 7	12	2273	RS	1327	1860	0-6-0
1880	2-4-0T	14½ x 22	4 9	9	2270	RS	1096	1856	2-4-0
1883	0-6-0ST			6					0-4-0ST
1887	0-6-0	16¾ x 24	4 9	15	2276			*c* 1868	0-6-0
1901	0-6-0T	16¾ x 26	4 6	7		RS	1073	1856	0-6-0

RS: Robert Stephenson & Co.

Fig. 342
Londonderry Rly
1891 0-6-0 No. 20
as Hartley Main Colliery
No. 7

LONDON MIDLAND & SCOTTISH RAILWAY

As a result of the Railway Act of 1921, this railway became the largest of the four main companies, with the main locomotive works situated at Bow, Crewe, Derby, Horwich, Kilmarnock, Stoke and Inverness. Smaller repairs shops were at Barrow and Maryport.

George Hughes became the first Chief Mechanical Engineer. He remained at Horwich, making these works his headquarters, for two years. The following were his successors: 1925–31 Sir Henry Fowler; 1931–2 Ernest H. J. Lemon; 1932–44 William Stanier, (Later Sir William); 1944–5 Charles E. Fairbairn; 1945–7 H. G. Ivatt.

Building continued at Horwich with Hughes' four cylinder 4-6-0s derived from the rebuilt version of the 1908 4-6-0s which had received boilers with Horwich superheaters, new cylinders with piston valves and Walschaerts instead of Joy's valve gear. In their rebuilt form they were far more successful and the newly built ones equally so.

A tank version was also put in hand, huge *Baltic* machines – twenty were planned but only ten built. The frames for the other ten were shortened and built as tender locomotives.

At Derby some 4-4-2Ts were built to replace old locomotives of the same type on the London Tilbury and Southend section, and a start was made to multiply the standard MR *Class 4* 0-6-0s and 4-4-0 compounds.

The only LNW design that Crewe built after amalgamation was a modification to the 0-8-2T designed by Bowen Cooke. The new tanks were 0-8-4Ts with 4′ 3″ coupled wheels, inside cylinders 20½″ x 24″, and a superheated boiler giving a total heating surface of 2046 sq. ft, and grate area 23·6 sq. ft. They weighed eighty-eight tons in working order and gave a lot of trouble to the permanent way department.

After these had been completed a batch of *Class 4* 0-6-0s were put in hand, these having been designated a 'standard' class by Derby and a total of 580 were built from 1924 to 1941 and, with those already built by and for the Midland Railway, made a grand total of 772 which could be found practically all over the system in England, Scotland and Wales.

Although Hughes had a number of large passenger and goods locomotive designs on the board when he retired in 1925, the 'small locomotive' policy of the Midland was too firmly embedded at Derby for any radical changes to take place and so numerous 4-4-0 compounds, and *Class 2* 4-4-0s were built. The compounds were built at Derby, Horwich and contractors, and the *Class 2* 4-4-0s at Derby and Crewe.

In 1926 the two-cylinder 2-6-0 designed by Hughes was built at Horwich and Crewe. Their appearance was striking, with outside cylinders set high at an angle, outside Walschaerts valve gear and parallel Belpaire boiler. The running plate was set high over the cylinders. They were built for mixed traffic duties and were undoubtedly the most successful of locomotives built during the 1920s.

Although not built in the company's workshops, an important type, Beyer Peacock Garratts, was introduced in 1927. Three were built and they were, as far as the engine part was concerned, two 2-6-0 units. They were required for the heavy coal traffic principally between Toton and Brent and were designed by Sir Henry Fowler in collaboration with Beyer Peacock. The tractive effort at 85 per cent boiler pressure was 45,600 lb and their weight in working order was 148 tons 15 cwt. They were spoilt by the exclusion of long travel valves, and by using the inadequate Derby journal diameters and lengths, to which they obstinately clung. Nevertheless they did the work of two *Class 4* 0-6-0s and in 1930 thirty more were built which were subsequently fitted with revolving bunkers.

In September 1926 the GWR sent on loan one of their 4-6-0 *Castle* class No. 5000 LAUNCESTON CASTLE for a series of trials between Euston, Crewe and Carlisle. The desperate situation on the LMS regarding adequate locomotive power prompted these trials, especially as in 1927 the proposed new *Royal Scot* train was to be introduced. The *Castle* performed all the required duties with ease. The projected *Pacific* which had been put in hand was abandoned, and a three-cylinder simple 4-6-0 was designed with the co-operation of the North British Locomotive Co. who built fifty, the first appearing in August 1927 and all being finished by the end of the year – a remarkable feat in design and construction. There was little similarity between these locomotives and the *Castles,* in fact they bore a strong likeness to the Southern Railway's *Lord Nelson* class. The larger diameter of the boiler 5′ 7⅜″ increasing to 5′ 9″ necessitated a very short chimney, squat dome and horizontal whistle. When built they were equipped with bogie brakes and a vacuum pump driven off the LH crosshead. These were subsequently removed, and smoke deflectors fitted from 1947.

It was not until 1930 that the company built any themselves, when the order for a further twenty was put in hand in the Derby works – they were numbered 6150–6169.

As originally built they had three cylinders 18″ x 26″, coupled wheels 6′ 9″ diameter. The boiler with Belpaire firebox gave a heating surface of 2081 sq. ft with an additional 300 sq. ft of superheater. The grate area was 31·2 sq. ft and the working pressure 250 lb/sq. in. The cranks were set at 120° to each other and the inside cylinder drove the leading coupled wheels, and three independent sets of Walschaerts valve gear were fitted. The last to be withdrawn was 46115 in December 1965.

In 1928 Fowler introduced the first of a long line of 2-6-4Ts. They were fitted with a standard G8AS Belpaire superheated boiler and two outside cylinders 19″ x 26″. Walschaerts valve gear was standard. They were used on suburban and semi-fast trains. A lighter version 2-6-2T was put in hand, in 1930, originally numbered fom 15500, but later renumbered 1 onwards. Twenty were fitted with condensing apparatus and vacuum brake trip cock gear for working over the Metropolitan Railway.

Fowler had planned a standardisation programme which had to cover the replacement eventually of some 9000 locomotives of many classes and varied conditions. As building of the new standard classes progressed an extensive scrapping programme was enforced, and as and when various pre-grouping classes required extensive repairs or rebuilding they were withdrawn.

A standard *Dockyard* 0-6-0T was built in 1928, ten in number, with a short wheelbase, outside cylinders and valve gear. They were built at Derby and originally numbered 11270–11279.

Two *Claughton* class 4-6-0s were rebuilt in 1930 with a new parallel Belpaire boiler, and three cylinders. In appearance they resembled a scaled down *Royal Scot*, in fact they were popularly known as *Baby Scots*. Very little of the original was left, but the coupled wheels were used. Some of this class had already been reboilered and ten were fitted with Beardmore-Caprotti valve gear in 1928.

Forty further *Claughtons* were rebuilt in 1932–3 and ten new locomotives to the same design were built at Crewe in 1934 (Nos. 5542–5551).

Another standard design was the 0-6-0T based on Johnson's 1900–59 class introduced in 1900. They were mainly built by outside contractors; out of a total of 422 (including seven for the Somerset and Dorset Joint Railway) only ten were built in the company's workshops – those at Horwich in 1931.

An 'improved' type of 0-8-0 freight locomotive was built at Crewe, 1930–2 reaching a total of 175 most of which were allocated to the Lancashire and Yorkshire section. They did not compare favourably with the well-tried and successful LNW *Class G2* 0-8-0s.

In October 1930 Sir Henry Fowler was appointed assistant to the Vice-President for Works and Mr E. J. H. Lemon succeeded him for short period.

It was evident that pre-grouping rivalry was having a serious effect on the running of the Locomotive department. Crewe, and St. Rollox particularly, were strongly critical of all that emanated from Derby and it was mainly for this reason that someone was required without any particular leanings to any of the pre-grouping railways.

So it came about that Mr William Stanier, who was principal assistant to Mr C. B. Collett, the Chief Mechanical Engineer of the Great Western Railway, was approached by the LM&SR and asked to take over the reins as CME of that company. On 1 January, 1932 Mr Stanier took office – a difficult task of great magnitude requiring both tact and strong leadership. Stanier had both.

To thrust Swindon practice on the LMS design staff could not be done overnight and Stanier's first project was the introduction of a taper boiler on the *2700* class 2-6-0s. Commencing with 2945, forty were built with this boiler.

The first *Pacific* No. 6200 was built in 1932 and as expected showed many Swindon characteristics and the basic dimensions were the same as the GWR *King* class 4-6-0, namely four cylinders 16¼″ x 28″, 6′ 6″ diameter coupled wheels and a boiler pressure of 250 lb/sq. in., but the boiler was much larger.

Unfortunately the Swindon practice from Churchward's days was for a relatively moderate degree of superheat, which was quite satisfactory on Swindon boilers as long as they were in good condition and pressure maintained.

It was soon appreciated that the boiler for 6200 would have to be modified and this was carried out and a comparison of the heating surface figures is interesting:

	As built	*As modified*
Small tubes	2078	1272
S'heater tubes	445	825
S'heater elements	370	653
Firebox	190	217
Total	3083 sq ft	2967 sq ft

This modification, carried out in 1935, made all the difference in performance and No. 6201, which had been built in 1933 had its boiler modified in 1936 but in this case the heating surface was: tubes 2429, superheater 594, and firebox 190 giving a total of 3213 sq. ft.

Nos. 6203–6212 were put in hand with modified boilers as for 6201 and were all completed in 1935. All were named after royal personages.

No. 6202 was built in July 1935. It was a 4-6-2 and had a boiler similar to the second batch (6203–6212) of *Pacifics*. A double chimney was fitted and the arrangement of the blast pipe modified. Turbine propulsion was used and the unit was designed in collaboration with Ljüngstrom and Metropolitan–Vickers.

In 1928 trials were carried out with a Beyer-Ljüngstrom turbine locomotive on the Midland section between St. Pancras and Manchester. On the new locomotive there were two turbines, one for forward and one for reserve. They were non-condensing type and the forward running turbine drove through double-helical triple reduction gearing on to the leading coupled axle and was housed on the left hand side, with the reversing turbine on the right hand side. It was far more successful than any previous turbine locomotive tried out in this country; very smooth running, no slipping or hammer blow. It ran principally on London to Liverpool expresses until the outbreak of war. It was stored at Crewe for a number of years, as maintenance could not be carried out as previously, due to the specialised nature of the transmission which had not given an abnormal amount of trouble up to this time. Had it not been for the war there is no doubt that the development of this type of motive power would have made it more reliable and competitive with the orthodox *Pacifics,* but due to the war time conditions the attention it required was not available. It ran again at the end of the war until 1952 when it was converted to a conventional reciprocating locomotive and was named PRINCESS ANNE. Her life was indeed short being completely wrecked in the Harrow accident in October of the same year.

The primary need was for a mixed traffic locomotive, and a two-cylinder 4-6-0 was brought out, the first coming from the Vulcan Foundry in 1934, and the first from the company's workshops at Crewe in 1935. No prototype was built, as in the case of the *Pacifics.* They were straight off the drawing board, and proved to be one of the most successful classes built. The moderate degree of superheat was later modified as in the previous case and in this form large orders were placed with outside contractors, as this type was urgently required throughout the system. By 1947 a total of 742 of this class had been built. The boiler was modified from time to time, with varying numbers of superheat elements. They were known as *Black Fives* and the numbers were from 4758 to 5499.

In 1937, Coronation year, both the LM&SR and L&NER were preparing for a high-speed express between London and Scotland – Glasgow on the LMS and Edinburgh on the LNE. Stanier's design was for a *Pacific* with enlarged cylinders, wheels and boilers and fully streamlined. The LNE *Pacifics* which were semi-streamlined, with a wedge-type casing, had appeared in 1935.

Particular attention was paid to the design of the cylinders and steam chests. The piston valve diameter was increased to 9″ and the steam passages enlarged and streamlined. Instead of four sets of valve gear the new *Duchess* class had two for the four cylinders, the motion for the two inside valve spindles being transmitted by rocking levers. Reference to the table of dimensions will show the increase of boiler capacity.

The first locomotive of this class was built at Crewe in June 1937 No. 6220 CORONATION followed by four more in the same year, ten in 1938, five in 1939, and five in 1940, all streamlined except five built in 1938 (6230–6234).

This class was a great improvement on the earlier *Pacifics,* extremely free-running and of tremendous haulage power. One (No. 6229 disguised as No. 6220) was sent to the New York World's Fair in 1939 where it had to remain until the end of hostilities.

During the war, production was concentrated on 2-8-0s at Horwich, whilst Derby built *Class 5* 4-6-0s and 2-6-4Ts. Crewe also built 2-8-0s and *Class 5* 4-6-0s and managed to build eight more *Pacifics.*

In 1942 Stanier had been seconded to the Ministry of Production and during this period C. E. Fairbairn acted as Deputy. In 1943 Stanier appeared in the New Year's Honours list receiving a Knighthood for distinguished services in the locomotive and mechanical engineering field – an honour richly deserved.

The first *Royal Scot* to be rebuilt with a taper boiler was 6170 BRITISH LEGION which had started life as the ill-fated high pressure locomotive No. 6399 FURY. This was rebuilt in 1935.

For purely express passenger duties a three-cylinder 4-6-0 class was built, the first, No. 5552 coming from Crewe in July 1934 (later renumbered 5642). They were not an immediate success but modifications to the boiler and blast pipe improved their performance considerably.

Stanier next turned his attention to the 2-6-4Ts of the *2300* class which had been one of Fowler's standard classes with the Derby standard parallel boiler with straight sided Belpaire firebox.

The first to be built by Stanier had three cylinders and taper boiler and were exclusively used on the London Tilbury and Southend lines.

Thirty-seven were built in 1934. These were followed by a two cylinder class which were dispersed all over the system and a total of 206 were built from 1935 to 1943, seventy-three of which were ordered from the North British Locomotive Co, the remainder being built at Derby.

For lighter axle load requirements the 2-6-2Ts designed by Fowler were redesigned with taper boilers and Nos. 71 to 209 were built at Derby and Crewe. In 1940–2 four of this type were rebuilt with larger taper boilers.

There was an urgent need for more powerful main line freight locomotives and following the experience of the L&NER and GWR, a 2-8-0 was designed. The first, No. 8000 was steamed at Crewe in July 1935, and like the *Black Fives* the first batch had domeless boilers, but subsequently the same boiler was fitted to these 2-8-0s as on the 4-6-0s and the cylinders and motion were common to both classes. Many were built at Crewe and outside contractors and in 1939 at the outbreak of the war the type was chosen by the Ministry of Supply with certain modifications and 240 were built to this specification.

During the war they were built by other railway workshops at Eastleigh, Brighton, Ashford, Darlington, Doncaster and Swindon. Crewe and Horwich also built a quantity.

It was not until 1942 that the first *Royal Scot* class No. 6103 ROYAL SCOTS FUSILIER received a standard No. 2A taper boiler and as each of the class came in for general repairs it received a taper boiler, the whole class being dealt with, the last, No. 6137, being rebuilt in 1955. Two *Jubilee* class, Nos. 5735 & 5736 also received these boilers.

The rebuilt *Scots* were transformed, and their steaming characteristics improved considerably. As rebuilt, the boiler and firebox gave a heating surface of 1862 sq. ft and the superheater 357 sq. ft giving a total heating surface of 2219 sq. ft and a grate area of 31·2 sq. ft. The boiler was type 2A, top feed apparatus was fitted between the dome and chimney – which was elongated for the double blast pipe. The dimensions for cylinders, wheels and wheelbase remained the same for both rebuilt *Scots* and *Jubilees*.

In 1944 Sir William retired from the LMS and C. E. Fairbairn became CME. There is no doubt that from the outset Sir William Stanier had created a completely different concept of LMS locomotive design and construction – he introduced many well proven practices from Swindon – they did not all work under such variable conditions with which the LMS had to contend, but he was versatile enough to modify as required. He had established a basis of standardisation though well nigh impossible and five types of main line locomotives were all that were necessary to carry out the work: 4-6-2 *Duchess* class, 4-6-0 Rebuilt *Scot*, 4-6-0 *Black Fives*, 2-8-0 *Class 8F* and 2-6-4T *Class 4*. Other secondary classes were of course necessary for shunting, banking etc.

The 2-6-4T in 1945 came under review and Fairbairn carried out certain design modifications to the batch of forty-five sanctioned, beginning with No. 2673. The coupled wheelbase was reduced from 16′ 6″ to 15′ 4″ so that a five chain radius could be safely negotiated. All unnecessary plating was eliminated and more welded fabrication utilised. The same standard boiler was used.

Unfortunately Mr Fairbairn died suddenly in October 1945 and H. G. Ivatt became Acting CME and in January he was appointed CME.

A reappraisal of standardisation was carried out by H. G. Ivatt and the number increased, by establishing eleven types to cover all requirements. They were:

1	Class 7	*Pacific* 4-6-2
2	Class 6	Rebuilt *Royal Scots* 4-6-0
3	Class 5	Mixed traffic (*Black Fives*) 4-6-0
4	Class 8F	Freight 2-8-0
5	Class 4	Freight 2-6-0
6	Class 2	Mixed traffic 2-6-0
7	Class 4	Passenger 2-6-4T
8	Class 2	Mixed traffic 2-6-2T
9	Class 3	Freight 0-6-0T
10	Class 2	Freight 0-6-0T
11		Diesel Shunter 0-6-0

Classes 1, 2, 3, 4, 7, 9, 10 and 11 were all well tried designs and entirely new designs were made for the Class 4 and 2, 2-6-0s and the Class 2, 2-6-2T. Both Class 2 designs were put in hand in 1946. They were identical in design apart from the tanks and tenders and both intended for mixed traffic services but the tank version more for branch lines and push/pull working. The boiler was of conventional design having a taper of 5″ and a Belpaire firebox. Top feed valves were fitted between dome and chimney and instead of the 'Churchward' type of perforated trays to disperse the water, sloping plates were used. A rocking grate was provided and the cylindrical smoke box was self cleaning. The hornblocks for the driving axleboxes were 'horseshoe' type, the others being separate.

Two outside cylinders 16″ x 24″ with 8″ piston valves were fitted and outside Walschaerts valve gear. The tender was quite a new design with special emphasis on good look-out for tender-first running, and had a cab and rear windows.

The rebuilt *Claughton* class known as *Patriots*, commencing with 5530 SIR FRANK

REE were next taken in hand and were subjected to a further rebuilding by the introduction of the class 2A taper boiler which had already been fitted to two of the *Silver Jubilee* class Nos. 5735 and 5736. New cabs, cylinders, smokebox and saddle were fitted together with the standard 4000 gallon tender. When rebuilt these locomotives were almost identical to the two rebuilt *Jubilees*. The new cylinders were 17″ x 26″ (3) and the coupled wheels remained at 6′ 9″ diameter.

In 1947 various non-standard features were fitted to a batch of two cylinder 4-6-0s being built at Crewe viz:

4758–4764	Single chimney, Walschaert's V. G. and Timken bearings.
4765 & 4766	Double chimney, Walschaert's V.G. and Timken bearings.
4767	Double chimney, Stephenson's V. G. and Timken bearings.

4767 was the only main line locomotive in England to have outside Stephenson's valve gear, with the exception of William Dean's experimental 2-2-2 No. 9 rebuilt from a 4-2-4T in 1884. In the case of 4767 the motion was by fly cranks whereas outside eccentrics were used on No. 9.

Ivatt's first *Class 4* standard 2-6-0 was completed at Horwich at the end of 1947 and numbered 3000. Everything was designed for accessibility; a high running plate well above motion and cylinders, and outside Walschaert's valve gear. The cylinders were cast steel, and made at Crewe, and large 10″ diameter piston valves were intended to provide free running. A double chimney was fitted with the exhaust from the twin blast pipes diverging into a common stream. The regulator handles were duplicated and fixed to a common shaft. Many features of the *Class 2* locomotives already built were also incorporated, and the tender had the same features but was larger.

To compare with the 3200 HP diesel locomotives two further *Pacifics* were put in hand by H. G. Ivatt in 1947 and, although to the general design of the pevious *Pacifics*, Ivatt introduced a number of additional features to facilitate maintenance and availability. All axles had roller bearings and the boiler had the usual self cleaning smokebox, double chimney, new top feed design, and the trailing pony truck was a one piece steel casting and side check springs were incorporated. The first was completed at the end of 1947 numbered 6256 and named SIR WILLIAM STANIER F.R.S. a fitting tribute to the man who had designed the first LMS *Pacific*.

On the 1 January, 1948 the LM&SR became part of the nationalised British Railways.

See BRITISH RAILWAYS; CALEDONIAN RAILWAY – ST. ROLLOX, GRAND JUNCTION RAILWAY – CREWE, LANCASHIRE & YORKSHIRE RAILWAY – HORWICH, MIDLAND RAILWAY – DERBY.

See BELFAST & NORTHERN COUNTIES RAILWAY.

STANDARD TAPER BOILERS

Class	Loco Type	Loco Class	Distance Between Tube Plates ft in	Heating Surface sq ft				Grate Area sq ft
				Tubes	S'heat	F'box	Total	
1	4-6-2	Princess Royal	19 3	2516	536	217	3269	45.0
1X	4-6-2	Coronation	19 3	2807	822	230	3859	50.0
2	4-6-0	6170	14 3	1968	367	195	2530	31.25
2A	4-6-0	Royal Scot	13 0	1851	367	195	2413	31.25
3A	4-6-0	Jubilee	13 2⅞	1641	313	181	2135	31.0
3B	4-6-0	Mixed Traffic	13 2⅞	1650	365	171	2186	28.65
3C	2-8-0	8F	12 2⅞	1649	241	171	2061	28.65
3D	2-6-0	5MT	12 2⅞	1633	244	135	2012	27.6
4C	2-6-4T	3 Cyl.	12 3	1172	209	143	1524	26.7
4C	2-6-4T	2 Cyl.	12 3	1369	240	143	1752	26.7
4D	2-6-0	4 MT	10 10½	1212	247	131	1590	23.0
6A	2-6-2T	3	10 10½	1045	74	107	1226	19.2
6B	2-6-2T	3	10 10½	1105	147	111	1363	19.2
7	2-6-0 2-6-2T	2	10 10½	1025	124	101	1250	17.5

LONDON MIDLAND & SCOTTISH RAILWAY

First Built	Type	Power Class	Cylinders inches	D.W. ft in	Heating Surface sq ft Tubes	S'heater	F'box	Total	Grate Area	WP lb/sq in	Weight WO ton cwt
G. HUGHES, H. FOWLER											
1923	4-4-2T	3P	OC 18 x 26	6 6	929	–	117	1046	19.77	170	68 2
1923	0-8-4T	7F(a)	20½ x 24	4 5½	1545	360	147	2052	23.6	185	88 0
1923	4-6-0	5P	4/16½ x 26	6 3	1837	394.9	180	2411.9	27.0	180	79 1
1923	4-6-4T	5P	4/16½ x 26	6 3	1837	394.9	180	2411.9	27.0	180	99 19
1924	0-6-0	4F	20 x 26	5 3	909.9	252.7	123.8	1286.4	21.1	175	48 15
1924	4-4-0	4P	19/21 x 26	6 9	1169.5	290.75	147.25	1607.5	28.4	200	61 14
1924	0-6-0T	3F	18 x 26	4 7	967.5	–	97	1064.5	16.0	160	49 10
1925	4-6-0	4P	OC 20½ x 26	6 1	1529.5	258.2	146.5	1934.2	25.5	175	74 15
1926	2-6-0	4F 5P	OC 21 x 26	5 6	1361	307	160	1828	27.5	180	66 0
1927	2-6-6-2		4/18½ x 26	5 3	1954	466	183	2603	44.5	190	155 10
1927	2-6-4T	4P	OC 19 x 26	5 9	1083	246	138	1467	25.0	200	86 5
1927	4-6-0	6P	3/18 x 26	6 9	1892	399	189	2480	31.2	250	84 18
1928	0-6-0T	2F	OC 17 x 22	3 11	923	–	85	1008	14.5	160	43 12
1928	4-4-0	2P	19 x 26	6 9	909.9	252.7	123.8	1286.4	21.1	180	54 11
1929	0-8-0	7F	19 x 26	4 8½	1434	352.5	149.8	1936.3	23.6	200	60 15
1930	2-6-2T	3P	17½ x 26	5 3	692.7	172.7	103.5	968.9	17.5	200	71 16
1932	4-6-0	5XP	3/18 x 26	6 9	1450	365	183	1998	30.5	200	80 15
W. STANIER											
1932	0-4-4T	2P	18 x 26	5 7	902.7	–	103	1005.7	17.5	160	58 1
1932	4-6-2	7P	4/16¼ x 28	6 6	2523	370	190	3083	45.0	250	104 10
1933	2-6-0	5F	OC 18 x 28	5 6	1633	244	135	2012	27.6	225	65 0
1934	2-6-4T	4P	3/16 x 26	5 9	1126	198	139	1463	26.7	200	92 5
1934	4-6-0	5	18½ x 28	6 0	1460	307	171.3	1938.3	28.65	225	70 12
1934	4-6-0	5XP	3/18 x 26	6 9	1460	307	181.1	1948.1	31.0	225	79 11
1935	4-6-2	7P	Turbine	6 6	1951	577	217	2745	45.0	250	110 11
1935	2-8-0	8F	OC 18½ x 28	4 8½	1649	241	171	2061	28.65	225	70 10
1935	2-6-4T	4P	OC 19⅝ x 26	5 9	1223	230	143	1596	26.7	200	87 17
1935	2-6-2T	3P	OC 17½ x 26	5 3	859.1	72.8	106.9	1038.8	19.2	200	71 5
1937	4-6-2	7P	4/16½ x 28	6 9	2577	830	230	3637	50.0	250	105 5
1943	4-6-0	6P	3/18 x 26	6 9	1667	348	195	2210	31.2	250	83 0
H. G. IVATT											
1946	2-6-0	2F	OC 16 x 24	5 0	924.5	134	101	1159.5	17.5	200	47 2
1946	2-6-2T	2P	OC 16 x 24	5 0	924.5	134	101	1159.5	17.5	200	63 5
1946	4-6-0	5XP	3/17 x 26	6 9	1667	348	195	2210	31.2	250	82 0
1947	2-6-0	4	OC 17½ x 26	5 3	1090	231	131	1221	23.0	225	59 2

(a) designed by H.P.M. Beames.

LONDON & NORTH EASTERN RAILWAY

This railway came into being on 1 January, 1923, the main railways joining forces being: North Eastern, Great Northern, Great Eastern, North British, Great Central, Great North of Scotland, (Hull & Barnsley amalgamated with NER in 1922).

Locomotive building up to the date of amalgamation had been carried out at Darlington (NER) Doncaster (GNR) Stratford (GER) Cowlairs (NBR) and Gorton (GCR). Inverurie (GN of SR) and Hull-Springhead (H&BR) carried out repairs and rebuilding only, but whereas Springhead had never built any new locomotives, Inverurie had built ten 4-4-0s from 1909 to 1921.

Herbert Nigel Gresley became the Chief Mechanical Engineer of the new company and to assist him each area had a District Mechanical Engineer. No change was made on policy at first, and the building which was being carried out at each works was allowed to continue. Darlington had several types in the course of erection. Thirty-two NE *Class S3* three cylinder 4-6-0s, ten NE *Class P3* superheated 0-6-0s and five small NE *Class H* inside cylinder 0-4-0Ts with 3′ 5″ diameter wheels were completed. The latter was an old Worsdell design of 1888 with domeless boiler and open back cab.

Three Raven *Pacifics* were completed in 1924 making a total of five in service, the last three differing from the previous two by having outside frames and bearings to the trailing truck. This type was of course subjected to comparative tests with Gresley's own design and although one was rebuilt with a Gresley boiler, the cost involved in improving the front end design resulted in their withdrawal in 1936 and 1937. A more successful NE design was the three-cylinder *Class T3* 0-8-0 and ten were built in 1924 using the same boiler as the *S3* 4-6-0s. They were extremely successful, probably the best of any 0-8-0 design. Finally the last NE design to be built was the *Class X* 4-8-0T; ten were put in hand in 1925. They had three cylinders 18″ x 26″ and 4′ 7″ diameter coupled wheels, the cylinders and valve chests being a single casting.

At Doncaster, work continued with the building of Gresley's *A1 Pacifics, N2* 0-6-2Ts, *J23* (LNE *Class J50*) 0-6-0Ts, which were to be a standard class, and *Class O2* three-cylinder 2-8-0s.

In 1925 ten of Worsdell's *E1* 0-6-0Ts were built and thirty-two of Mr Hill's very successful G.E. *Class L77* 4′ 10″ 0-6-2Ts, but whereas those built elsewhere had Belpaire fireboxes those built at Doncaster had round topped fireboxes.

The most interesting locomotives built in this period were two 5′ 2″, 2-8-2 tender locomotives with boosters. They were the first *Mikados* to run in Great Britain and were designed for hauling the heavy coal trains betwen New England and Ferme Park. A booster had already been tried on one of Ivatt's large *Atlantics* by rebuilding the trailing truck and equipping it with two 10″ x 12″ cylinders. The type was a development of Gresley's *Pacific*, but for freight traffic. The trouble was that although they could haul 100 wagon trains of 1600 tons, special traffic arrangements had to be made because of the abnormal length of the trains with few refuge loops that could take them. This was an experiment in freight movement and proved rather an embarrassment. The boosters were subsequently removed. No attempt was made to rebuild them and due to their low availability they were withdrawn.

At Stratford the last ten 4-4-0s were laid down in 1923, GER Nos. 1780-1789 and were known as *Super Clauds*.

These were followed by ten 0-6-0Ts (LNE *Class J68*) with side window cabs. Twenty had been built 1912–14 as GE *Class C72* and finally the very successful 0-6-2Ts to *Class L77* (LNE *N7*).

At Gorton ten 4-6-0s were completed in 1923–4. They were additions to the *Class 9Q* (LNE *B7*) four cylinder mixed traffic with 5′ 7″ diameter coupled wheels. The works were then engaged on building the GE 0-6-2Ts and a total of forty were built from 1925–8.

Cowlairs contribution was twenty of Reid's 4′ 6″ 0-6-2Ts and these were the last new locomotives built there.

Gresley had introduced his large three-cylinder 5′ 8″ 2-6-0s with 6 ft diameter boilers in 1920 and both Darlington and Doncaster continued to build this type – improved by the addition of side window cabs – and a total of 113 were produced from 1924 to 1937 and others were built by outside contractors. The ninety-three built at Darlington differed by having NE style cabs and boiler mountings. They were mixed traffic locomotives and extensively used on passenger duties.

The maximum axle load was 20 tons which kept them off most of the Great Eastern section.

Many 0-6-0s required replacing and two classes were designed, the *J38* with 4′ 8″ diameter wheels and the *J39* with 5′ 2″ diameter. Apart from the wheels they were of similar dimensions. The boiler was superheated, with a working pressure of 180 lb/sq. in. and there was the usual Gresley round topped firebox. The *J38s* were intended for Scotland and the *J39s* were mostly allocated to English sheds and were far more successful than the *J38s*. They were established as a standard class, being equally at home on freight and passenger turns. They became one of the most numerous classes, a total of 289 being built from 1926 to 1941 and all except twenty-eight (built by Beyer Peacock & Co.) were built at Darlington and it was seldom that 0-6-0s were not in the process of building between 1923 and 1941. The whole *J38* class were also built at Darlington.

As a result of experiments carried out on two of the *A1 Pacifics*, with increased boiler pressure and long lap valves a new class of *Pacifics* was built from 1928–30 and in 1934–5. They were known as *Class A3*. The exchange trials between the Great Western Railway and the LNER have been written about ad nauseam, but from a practical point of view the more economical working of the GWR *Castle* class and the Swindon front end design showed that long lap valves and a higher boiler working pressure of 225 lb/sq. in. were the key to their success. The *A3s* had a working pressure of 220 lb/sq. in. and the cylinders increased in diameter to 19″. Their performance warranted the conversion of the earlier *A1s* to the same specification. The later *A3s* had a modified dome which was shaped like a banjo, housing a perforated steam collector.

For the London–Edinburgh non-stop trains it was necessary to change the engine crew, enroute and to avoid stopping, a corridor tender was introduced so that the second crew rode on the train until it was time to change over which they did by traversing the tender – and the other crew left by the same means.

Whilst all this experimenting and innovation was being carried out at Doncaster, Darlington in 1929 were engaged in building a 4-6-4 locomotive of a striking design resulting from Gresley's wish to try out a high pressure water tube boiler whose efficiency was far superior on land and marine applications. Yarrow & Co. Ltd. of Glasgow had a vast experience in this field and a locomotive boiler was designed and built by them with a boiler pressure of 450 lb/sq. in. The engine part was built at Darlington in the form of a four-cylinder compound with HP cylinders 12″ x 26″, and LP cylinders 20″ x 26″. The HP cylinders, after preliminary trials, were reduced to 10″ diameter. The coupled wheels were 6′ 8″ diameter. The boiler itself consisted of an upper drum and two lower drums connected by a large number of tubes, with the furnace gases impinging on the outside of the tubes. It was numbered 10,000 and completed in December 1929. Two sets of valve gear were used, the cut-off to the HP and LP cylinders being independent. Due to the shape of the boiler cladding, it was streamlined in form and the front portion was curved with the outside plating brought forward. Initially the locomotive was fairly consistent in performance but the early hopes of success were diminished by increased maintenance troubles including leaking tubes. In 1937 No. 10,000 was rebuilt as a three-cylinder Simple, with a conventional boiler and streamlined casing. Strangely enough it never bore a name, although all major classes were named.

For the Great Eastern section Gresley designed a three-cylinder 4-6-0 *Class B17*. The cylinders were 17½″ x 26″, the inside cylinder driving the leading coupled wheels which was a departure from Gresley's standard practice of driving one axle only. The first ten were built by the North British Locomotive Co. and Darlington proceeded with the next fifty-two, many of which had short GE pattern tenders to enable them to be turned on the limited length of GE turntables. Robert Stephenson & Co. completed the class by building eleven in 1937. Two were streamlined in 1937 for working the *East Anglian* express.

Their haulage power was greater than the inside cylinder Holden 4-6-0s and became popular with the GE crews. When more were built some were sent to the GC section and some outstanding runs were made.

A three-cylinder 4-4-0 was also built at Darlington, from 1927, intended for the North Eastern region carrying 'Shire' names and were *Class D49*. Later ones were sent to Scotland. They were the most powerful 4-4-0s in the country when they were built. The cylinders were 17″ x 26″ and the coupled wheels 6′ 8″ diameter. They were fitted with Walschaerts valve gear and Gresley's motion for the inside cylinder operated from the rear and not the front of the valve chest as in the case of other three cylinder types. Six were fitted with Walschaerts valve gear and oscillating-cam Lentz poppet valves instead of piston valves. A number of different GE locomotives had already been fitted with this arrangement and reduced coal consumption resulted from the experiments. The poppet valves were operated by rotary cams. Later *D49s* were then fitted with the rotary cam variety and this sub-division of the class was named after 'Hunts'. A total of 75 *D49s* were built, the last being completed in 1935 and all were built at Darlington.

A series of 2-6-2Ts were built at Doncaster from 1930, *Class V1*, which were in all essentials a tank version of Gresley's *K2* mixed traffic tender locomotives, except that three cylinders were used, 16″ x 26″, instead of the *K2s* two 20″ x 26″ cylinders. The round topped boiler, with a working pressure of 180 lb/sq. in. had a total heating surface of 1609 sq. ft. including 284 sq. ft. of superheater elements. The grate area was 22·0 sq. ft.

They were fine looking locomotives and did excellent work wherever they went. The first batch went to Scotland to replace Reid's 4-4-2Ts on some duties. In 1939 ten further 2-6-2Ts were built the same as the *V1s* but having an increased boiler pressure of 200 lb/sq. in. Eighty two *V1s* were built, all at Doncaster.

In 1934 Doncaster built the second 2-8-2 design, for passenger trains and expected to pull 550 tons on the main line north of Edinburgh. This route is one of the toughest, not only for severe gradients but sharp curves as well. This enormous machine had a rigid wheelbase of 19′ 6″ with 6′ 2″ diameter coupled wheels. The engine weighed 110 tons 5 cwt and with tender 165 tons 11 cwt in working order. The boiler had a total heating surface of 3490·5 sq. ft including 776·5 sq. ft of superheater elements. The grate area was 50 sq. ft. The three cylinders were 21″ x 26″ and steam was admitted by 8″ diameter poppet valves and exhausted through 9″ diameter valves operated by rotary cams. A double chimney was fitted of the Kylchap variety with double blastpipe. All this resulted in a free running locomotive. In December 1934 it was sent to Vitry locomotive testing station near Paris. Numbered 2001 it was *Class P2* and the second, No. 2002 also built in 1934 differed from the prototype in having ordinary piston valves and no feed water heater which had been incorporated on the first one.

As a result of the Vitry tests the next four (2003–2006) built in 1936 had modified front ends and 2006 was first built with a combustion chamber and larger firebox. As in the case of the *P1* 2-8-2s the runs were short for such large locomotives and, due mainly to inexperienced firing, the coal consumption was heavy; overheating was prevalent. They were transferred south of the border and during Thompson's regime were transformed into *Pacifics*. The complete class was: 2001 COCK O' THE NORTH; 2002 EARL MARISCHAL; 2003 LORD PRESIDENT; 2004 MONS MEG; 2005 THANE OF FIFE; and 2006 WOLF OF BADENOCH. They were the first eight coupled locomotives in the country designed for passenger work.

A much more successful type were the *Class V2* 2-6-2 tender locomotives which was more a development of the *K3* 2-6-0s than a modified *Pacific*. They were designed primarily for fast freight traffic with fitted vehicles, but they soon found their way on to fast and semi-fast passenger trains where they were equally at home. They were a very popular and successful type and a total of 184 were built at Doncaster and Darlington, 1936–44.

They had the usual three cylinders and for this class were 18½″ x 26″ and 6′ 2″ diameter coupled wheels. Boilers were not standardised to the same extent as the Great Western and LMS railways, Gresley preferring to design a particular locomotive for a particular duty, without being tied down to specific dimensions and being forced to build round a specified boiler.

In 1935 appeared the first of the steamlined *Pacifics*. Four were built specially for the *Silver Jubilee* express and on a trial run with No. 2509 SILVER LINK (later No. 14) a speed of 112·5 mph was reached on two occasions with a sustained speed of 100 mph for an uninterrupted 43 miles. The streamlining embraced the cylinders and part of the motion, the front end was wedge shaped and the cab fronts set at an angle. The appearance of these locomotives had a mixed reception but for high speed work a saving of at least 10 per cent in power output could be achieved running at 90 mph.

These *Pacifics* were classed *A4* and the boiler pressure was increased to 250 lb/sq. in. and the cylinders reduced to 18½″. Great attention was paid to the design and streamlining of the steam passages and the piston valve diameter increased from 8″ to 9″. The motion was manufactured of nickel chrome steel giving strength with comparative lightness as for the previous *Pacifics*.

Thirty-five were built from 1935 to 1938, the first four being painted silver for working the *Silver Jubilee*. A new train was introduced in 1937 named *The Coronation* and it was on this train that No. 4468 MALLARD (later No. 22) attained a maximum speed of 126 mph, the world's record for steam locomotives. The improved performance of those that had been fitted with Kylchap double blast pipe and chimney led to the others being fitted with them. The *A4s* were undoubtedly Gresley's masterpiece and it is pleasing to note that at least six have been preserved, one going to Canada and one to the USA and MALLARD of course in the York Museum.

For the West Highland section Gresley designed a 2-6-0 with 5′ 2″ coupled wheels instead of the standard 5′ 8″ diameter for the *K1, K2* and *K3* classes. Known as *K4* class, a higher tractive effort was obtained by reducing the wheel diameter and increasing the boiler pressure to 200 lb/sq. in. They were an immediate success and six were built at Darlington in 1937 and were all named.

A light weight 2-6-2 was then designed with the intention of making it a general utility locomotive with high route availability and a maximum axle load of 17 tons. The result was the building of two *Class V2* 2-6-2s at Doncaster. They had three cylinders 15″ x 26″ and

5′ 8″ coupled wheels and a superheated boiler giving a total heating surface of 1800 sq. ft and a grate area of 28·7 sq. ft. The second was built with a Thermic syphon and steel firebox for comparison purposes. They were both sent to Scotland.

Unfortunately before further developments could take place Sir Nigel Gresley died on 6 April, 1941 thus cutting short a career of achievement which would be hard to match. He had been created a Knight Bachelor in 1936 and many other distinctions and awards had been conferred upon him.

Gresley was a staunch three-cylinder proponent claiming that he could get more mileage out of them between overhauls, less coal consumption, earlier cut-off in full gear and reduced hammer blow and by early trials with two cylinder types, he had no cause to change his policy. Many books and papers have been written about his locomotives and their performances.

Besides his new locomotives Gresley put in motion the re-boilering of many classes, trials in feed-water heaters, poppet valves, and boosters.

Although Cowlairs and Stratford built some 0-6-2Ts in 1923 and 1924 the ensuing years were spent on repairs and rebuilding only. Gorton however was used to a greater if limited extent and building proceeded until 1939 with small batches. The largest locomotives built were two additional 0-8-4Ts similar to the four designed by Robinson and built by Beyer Peacock in 1908, they were used for the increased traffic dealt with at the hump yard at Wath.

Edward Thompson was appointed Chief Mechanical Engineer in succession to Sir Nigel Gresley in 1941. He was a North Eastern man having had experience with carriages and wagons besides locomotives. In 1927 he was in charge at Stratford, then Darlington and finally in 1938, Doncaster.

At Doncaster twenty-five standard *O2* 2-8-0s were built 1942–3. New tenders were not built for this order, existing stock being used. These were followed by a Ministry order for fifty Stanier LMS type *8F* 2-8-0s, thirty bore LMS numbers (8510–8539) and twenty LNER numbers (3148–3167). They were completed in 1945. No. 8510 was the first locomotive to be built at Doncaster with a Belpaire firebox.

Thompson's first design was for a two-cylinder general service 4-6-0 and was completed at Darlington in December 1942. With the exception of fitting two outside cylinders instead of the three-cylinder practice of Gresley, these 4-6-0s followed, in general, LNER practice.

It is well known that Gresley and Thompson had many opposing views on locomotive design which unhappily developed on a personal level. Thompson was, and had been, preoccupied on the maintenance side of locomotives, and he set out to eliminate some of the troubles he had encountered, especially Gresley's conjugated valve gear and the fixation for three-cylinder layouts. His personal dislike for all things 'Gresley' should have had some restraint, but by the efforts he made in this direction it was obvious that little or no opposition was encountered. The two-cylinder 4-6-0 built in 1942, No. 8301 SPRINGBOK (later renumbered 1000), was the forerunner of a class totalling 410 locomotives by 1950. Cylinders were 20″ x 26″, coupled wheels 6′ 2″ diameter, the boiler, with a working pressure of 225 lb/sq. in. had a round topped firebox, and a total heating surface of 2005 sq. ft including 344 sq. ft of superheater elements. The grate area was 27·9 sq. ft 10″ piston valves were fitted with a maximum travel of 6½″. They were comparable with Stanier's mixed traffic *Black Fives* and were free running and fast. Many were built by outside contractors and could be found all over the system.

Thompson kept the works busy on modifying previous designs. The larger 2-8-2s were converted to *Pacifics* from 1943 and were classed *A2/2*. In 1944–5 four new *Pacifics* appeared which were a development of the rebuilt 2-8-2s, and were originally intended to be the final four 2-6-2 *V2s* and the *V2* boilers were used on the modified design.

The Robinson *04* 2-8-0s were rebuilt with the same cylinders, motion and boilers as his *Class B1* and this type was to be the standard heavy freight locomotive. His standard 2-6-0 was a rebuilt *K4* to be known as the new *K1* class, and in 1945 a *K3* Gresley 2-6-0 was rebuilt with two cylinders and higher boiler pressure to be known as the *K5* class, which made the classes rather complicated.

A new tank design was built in 1945 – 2-6-4T type *Class L1* No. 9000 (later renumbered 7701). The same cylinders and motion were used as for *B1* 4-6-0s and the coupled wheels were 5′ 2″. This was another successful design and one hundred of them were built, all except the prototype, after nationalisation.

Vincent Raven's three-cylinder 4-6-0s also came in for Thompson's attention, and although Gresley had rebuilt some of this type with separate cylinder castings instead of the monobloc casting originally used, the Thompson rebuilding included three separate sets of Walschaerts valve gear instead of the conjugate 2:1 motion.

One of the *D49* 'Hunt' class was rebuilt with

two 20″ x 26″ inside cylinders, and the Lentz poppet valves were replaced by Stephenson link motion.

Perhaps the deed that caused the greatest uproar – whether it was deliberate or not is debatable – but the prototype Gresley *Pacific* GREAT NORTHERN was rebuilt as a prototype for Thompson's new standard *Pacific.* The drive was divided with three sets of Walschaerts valve gear, the wheelbase was lengthened from 35′ 9″ to 38′ 5″ made necessary by the repositioning of the outside cylinders well back from the rear wheels of the front bogie. The smokebox was lengthened to accommodate the inside cylinder. The cylinders were 19″ x 26″, coupled wheels 6′ 8″ diameter and boiler pressure 250 lb/sq. in. A particularly ugly double chimney was fitted which had smoke deflector plates. It was indeed particularly unhandsome at the front end and, what made matters worse, it showed no striking improvement on running performance or coal consumption. Nevertheless Thompson's main motives must have been standardisation and simplification of maintenance and the *B1s* and *L1s* were excellent machines. Further *Pacifics* were built to *Class A2/3* at Doncaster following closely the 'Great Northern' rebuild design.

Thompson did an excellent job when he brought out his renumbering scheme. Before this the numbering of locomotives was in the old tradition of haphazard numbering, giving new locomotives the same number of a recently scrapped one, but with the new scheme each type had a set of numbers and identification was made much easier. Summarised, the scheme was as follows:

1–999	4-6-2 and 2-6-2
1000–1999	4-6-0 and 2-6-0
2000–2999	4-4-2, 4-4-0 and 2-4-0
3000–3999	0-8-0 and 2-8-0
4000–5999	0-6-0
6000–6999	Electric locomotives
7000–7999	2-4-2T, 0-4-4T, 4-4-2T, 2-6-2T, 4-4-4T
8000–8999	0-6-0T, 0-4-0T, 0-4-2T
9000–9998	2-6-4T (ex GC & Met) 0-6-4T, 0-6-2T, 4-6-2T, 0-8-4T, 4-8-0T, 0-8-0T
9999	2-8-8-2 Garratt
10000	4-6-6

There were modifications from time to time but the scheme was established and all current locomotives renumbered.

Class	Type	Cylinders inches	D.W. ft in	Heating Surfaces sq ft				Grate Area sq ft	WP lb/sq in	Weight WO ton cwt	First Built
				Tubes	S'heat	F'box	Total				
GRESLEY											
K3	2-6-0	3/18½ x 26	5 8	1719	407	182	2308	28.0	180	71 14	1925
P1	2-8-2	3/20 x 26	5 2	2715	525	215	3455	41.25	180	99 19	1925 (a)
J38	0-6-0	20 x 26	4 8	1283	289	171.5	1743.5	26.0	180	58 19	1926
J39	0-6-0	20 x 26	5 2	1226	271	171.5	1668.5	26.0	180	57 17	1926
D49	4-4-0	3/17 x 26	6 8	1226	271	171.5	1668.5	26.0	180	65 11	1927
A3	4-6-2	3/19 x 26	6 8	2521.6	635.5	215	3372.1	41.25	180	92 9	1928
B17	4-6-0	3/17½ x 26	6 8	1508	344	168	2020	27.5	200	77 5	1928
W1	4-6-4	2/12 x 26 2/20 x 26	6 8		140		2126	34.95	450	103 12	1930
V1	2-6-2T	3/16 x 26	5 8	1198	284	127	1609	22.08	180	84 0	1930
P2	2-8-2	3/21 x 26	6 2	2477	776.5	237	3490.5	50.0	220	107 3	1934 (a)
P2	2-8-2	3/21 x 26	6 2	2477	635.5	237	3349.5	50.0	220	110 5	1934
A4	4-6-2	3/18½ x 26	6 8	2345.1	748.9	231.2	3325.2	41.25	250	102 19	1935
V2	2-6-2	3/18½ x 26	6 2	2216.07	679.67	215	3110.74	41.25	220	93 2	1936
K4	2-6-0	3/18½ x 26	5 2	1253	310	168	1731	27.5	180	68 8	1936
V3	2-6-2T	3/16 x 26	5 8	1198	284	127	1609	22.08	200	84 0	1939
V4	2-6-2	3/15 x 26	5 8	1292.5	355.8	151.6	1819.4	28.7	250	70 10	1941
THOMPSON											
B1	4-6-0	20 x 26	6 2	1508	344	168	2020	27.9	225	71 3	1942
A2/1	4-6-2	3/20 x 26	6 2	2216.07	679.67	245.3	3141.04	50.0	225	98 0	1944
L1	2-6-4T	20 x 26	5 2	1198	284	138.5	1620.5	24.74	225	89 9	1945
A2/3	4-6-2	3/19 x 26	6 2	2216.07	679.67	245.3	3141.04	50.0	250	101 10	1946
K1	2-6-0	20 x 26	5 2	1240	300	168	1708	27.9	225	66 17	1949
PEPPERCORN											
A2	4-6-2	3/19 x 26	6 2	2216.07	679.67	245.3	3141.04	50.0	250	101 0	1947
A1	4-6-2	3/19 x 26	6 8	2216.07	679.67	245.3	3141.04	50.0	250	104 2	1948

(a) Booster fitted – 2 Cylinders 10″ x 12″

LNER – DONCASTER

Date	2-8-2	2-8-0	4-6-2	2-6-2	2-6-0	2-6-2T	0-6-2T	0-6-0T
1923		02: 3932-39	A1: 103-112					
1924		02: 3940-46	A1: 44-55					J50: 8930-39
1925	P1: 2393-94†		A1: 56-63				N2: 9562-67	J72: 8745-54
1926								J50: 8940-66
1927							N7: 9702-12	J50: 8967-71
1928			A3: 89-94				N7: 9713-33	
1929			A3: 95-98		K3: 1870-89			
1930			A3: 84-88, 99-101			V1: 7600-08		J50: 8972-77
1931						V1: 7609-27		
1932		02: 3947-54						
1933		02: 3955-58						
1934	P2: 2001-02	02: 3959-62	A3: 35-42			V1: 7628		
1935			A3: 43 A4: 14-17			V1: 7629-49		
1936	P2: 2003-06		A4: 23, 24	V2: 800-804		V1: 7650-61		
1937			A4: 3,4,7-13, A4: 18-20,25-31					
1938			A4: 1,2,5,6,21,22 A4: 32-34, 4469†			V1: 7662-76		
1939				V2: 872-875		V1: 7677-81 V3: 7682-87		
1940				V2: 876-881		V3: 7688-91		
1941				V2: 928-933 V4: 1700-1701				
1942		02: 3963-85		V2: 934-937				
1943		02: 3986-87 LMS: 8510						
1944		LMS: 8511-27 08: 3148-55						
1945		08: 3156-65 LMS: 8528-39				L1. 7701*		
1946		08: 3166-67	A2/3 500,511-518					
1947			A2/3 519-525					

All loco running numbers are as the 1946 renumbering scheme except the P1 and P2 classes.
* Type 2-6-4T †Withdrawn before renumbering

LONDON & NORTH EASTERN RAILWAY

Arthur H. Peppercorn succeeded Thompson in July 1946 and he was a Great Northern trained man. Building of Thompson's *B1s* continued at Darlington, and the existing orders for the *Class A2/3 Pacifics* were allowed to be completed in 1947 and a total of twenty Thompson *Pacifics* had been built altogether.

Peppercorn redesigned the Thompson *Pacifics* and incorporated Thompson and Gresley practice. They had three 19" x 26" cylinders. The cylinders were brought practically in line, with the outside ones back in the normal place on the bogie centre. They had three sets of Walschaerts valve gear. The boiler had a total heating surface of 3141·04 sq. ft of which 679·67 sq. ft was superheater. The grate area was 50 sq. ft and working pressure 250 lb/sq. in. The weight in working order without tender was 104 tons 2 cwt with 66 tons for adhesion. Seven were built in 1947 and others built after nationalisation.

A. H. Peppercorn remained Chief Mechanical Engineer on British Railways (E&NE Regions) until the end of 1948 when the post was abolished.

See BRITISH RAILWAYS; EASTERN COUNTIES RAILWAY – STRATFORD, GREAT NORTHERN RAILWAY – DONCASTER, MANCHESTER SHEFFIELD & LINCOLNSHIRE RAILWAY – GORTON, NORTH EASTERN RAILWAY – DARLINGTON.

LONDON & NORTH WESTERN RAILWAY

See GRAND JUNCTION RAILWAY and LONDON & BIRMINGHAM RAILWAY.

LONDON & SOUTH WESTERN RAILWAY

Eastleigh

The works were completed in 1910 and there was ample room for the various shops which, at the time, were the most up-to-date and advanced in the country.

The first new locomotive to appear was one of two 0-4-0T *S14* class for motor train working. The original order was for five, but three were cancelled. Following these were the *P14* 4-6-0 class of five which were identical to the *G14s* built at Nine Elms.

Drummonds's last 4-6-0 design was for the Bournemouth line and had 6′ 7″ diameter driving wheels and four 15″ x 26″ cylinders. They had the same boiler as the *G14* and *P14* classes with a total heating surface of 1920 sq. ft but the pressure was increased to 200 lb/sq. in. and the weight in working order 70 tons 19 cwt, 119 tons 10 cwt with the 4500 gallon eight wheeled tender. The outside cylinders were moved forward and were located between the bogie wheels and a casing sweeping down from the smokebox encased these cylinders. The outside Walschaerts gear operated the inside cylinders through rocking levers.

The smokebox housed a steam drier – Drummond was not interested in superheaters – and consisted of a bank of interconnected tubes placed in line with the tubes so that the hot gases passed around them. Their performance was an improvement on the *G14* and *P14* classes but they did not come up to expectations, due principally to bad steaming. They were classified *T14*.

It was considered at one time to lay down water troughs on the main lines but this did not materialise, hence the large tenders fitted to the express passenger classes.

The last class designed and built by Drummond was a reversion to the 4-4-0. These were the *D15s*. Ten were built in 1912 and were larger than any of his previous 4-4-0s. They had 19½″ x 26″ cylinders, 6′ 7″ diameter driving wheels, 10′ 0″ coupled wheelbase and a boiler giving a total heating surface of 1724 sq. ft and a grate area of 27 sq. ft. The working pressure was 200 lb/sq. in. and they weighed 59 tons 15 cwt, 108 tons 15 cwt with the 4500 gallon tenders which were fitted with feed water heating pipes. The feed water was fed to the boiler by pumps. They were equipped with steam reversing gear and the water tube firebox as standard. These were superb locomotives and a far cry from Drummond's first 4-4-0 design of 1876 for the North British Railway, the *476* class for the Waverley route which acquitted themselves with distinction.

Unfortunately Drummond had to have his leg amputated as a result of a bad scald at Eastleigh works and he died on 8 November, 1912. So ended the career of a distinguished locomotive

LONDON & SOUTH WESTERN RAILWAY

Fig. 343
L&SWR
Eastleigh 1912
Drummond
Class T14
6'7" 4-6-0 No. 460

Fig. 344 **L&SWR**
Eastleigh 1919
R. Urie *Class N15*
6'7" 4-6-0 No. 740

Fig. 345 **L&SWR**
Eastleigh 1921
R. Urie *Class G16*
4-8-0T No. 493
for Feltham Yards

Fig. 346 **L&SWR**
Eastleigh 1922
R. Urie
Class H16 4-6-2T
No. 519

engineer, with locomotives to his design working in the far south and far north of Great Britain.

His works manager at Eastleigh was Robert Wallace Urie – a fellow Scot and one who had worked in close co-operation with Drummond. He had been with the company almost as long as his chief and he was the obvious choice to succeed as Chief Mechanical Engineer and in December 1912 he took office. It was not until February 1914 that his first 4-6-0 design materialised, known as the *H15* class. They had two outside cylinders 21″ x 28″ with 11″ diameter piston valves and outside Walschaerts valve gear. Drummonds cross water tubes were not fitted to the firebox, neither were the feedwater heating arrangements and pumps. The running boards were much higher providing better accessibility to the motion. Safety valves and cover, cab, chimney, and dome cover were Drummond type. The tenders were eight wheeled but had outside bearings. Urie adopted a standard boiler pressure of 180 lb/sq. in. The coupled wheels were 6 ft diameter and the distance between driving and trailing coupled wheels was increased to make room for a 9′ 0″ long firebox with sloping grate. All these features indicated that Urie was determined not to perpetuate features that had given trouble in the past.

Superheaters were fitted to eight of these locomotives, four having 360 sq. ft of superheat surface and the other four 333 sq. ft. The former were Schmidt superheaters and the latter Robinson type. Comparisons were made and the Robinson type preferred and formed the basis for Urie's own design of superheater.

They were an immediate success and were popular with the running staff. One of Drummond's not so successful, 4-6-0 No. 335, was converted to the same class with small detail differences.

Most of the 1914–18 war period was taken up with repairs, rebuilding and Government work and it was August 1918 before Urie's next 4-6-0 design appeared. Twenty were built as *Class N15.* With two outside cylinders 22″ x 28″ and coupled wheels 6′ 7″ diameter they were similar in appearance to the *H15* class but the boiler was tapered, on the first ring only, from 5′ $1\frac{1}{2}$″ to 5″ $5\frac{3}{4}$″. The stove pipe chimney was fitted with a capuchon and the dome cover was more in proportion. They gave good service, were cheap to maintain and reliable. The 22″ cylinders were the largest used in this country at that time. The first ten were completed by November 1919 and the second batch in 1922–3, the last three being built after amalgamation with the Southern Railway.

Urie's third 4-6-0 design was for goods traffic with 21″ x 28″ cylinders and 5′ 7″ wheels and were known as the S15 class. Twenty were built in 1920–1. A similar boiler to the *N15s* was fitted with a taller stove pipe chimney. They too acquitted themselves well and were used on relief passenger trains when required proving a useful mixed traffic type.

Two tank locomotive designs were also put in hand, the first was a 4-8-0T *Class G16* of which four were built and were intended for hump duties at Feltham Yard. They were basically a tank version of the *S15* 4-6-0s with a smaller boiler as fitted to Drummond's last 4-4-0s the *D15* class but with a modified firebox and grate. The two outside cylinders were 22″ x 28″ and the coupled wheels were 5′ 1″ diameter.

These massive locomotives weighed 95 tons 2 cwt of which 72 tons 18 cwt were available for adhesion. Steam brakes and steam reversing were fitted which were a great help in the marshalling yards.

The other tank design was a 4-6-2T with boiler, motion, bogie, cylinders (bored to 21″), as for the *G16s.* They were used for transfer goods traffic and also for shunting at Feltham. Five were built in 1921–2 and weighed 98 tons 8 cwt in working order.

At the end of 1922 the L&SWR lost its identity and on 1 January, 1923 became the *Western Section* of the SOUTHERN RAILWAY.

Ninety-four new locomotives were built, the last official locomotive being No. 752, a *N15* 4-6-0 under L&SWR ownership. Robert Urie retired on 31 December, 1922. His locomotives were all of massive proportions and his invaluable previous experience as Works Manager of the locomotive workshops was indicated in the alterations he carried out in his designs with an eye to easy maintenance and accessibility. He lifted the running boards well above the motion, injectors were used instead of steam pumps, he increased the cross section of axleboxes giving more bearing surfaces and keeping hot boxes to a minimum – everything was done for a purpose. Careful trials with various superheaters were carried out and the 'Eastleigh' superheater resulted.

Insufficient credit is given to Urie's work – he was a true locomotive engineer blending theory with a great deal of practice.

The table overleaf lists steam locomotives built and rebuilt at Eastleigh during this period.

LONDON & SOUTH WESTERN RAILWAY

1923	3	4-6-0	Class N15	753–755
1924	10	4-6-0	Class H15	473–478, 521–524
1925	10	4-6-0	Class N15	448–457
1926	12	4-6-0	Class N15	793–804
1926	1	4-6-0	Class LN	850
1927	2	4-6-0	Class N15	805–806
1927	14	4-6-0	Class S15	823–836
1928	1	4-6-0	Class S15	837
1928	8	4-6-0	Class LN	851–858
1928	7	4-6-0	Class LN	859–865
1930	10	4-4-0	Class V	900–909
1931	20	2-6-0	Class U.1.	A891–900, 1901–1910
1932	5	2-6-4T	Class W	1911–1915
1932	5	4-4-0	Class V	910–914
1933	10	4-4-0	Class V	915–924
1934	6	4-4-0	Class V	925–931
1935	8	4-4-0	Class V	932–939
1936	10	4-6-0	Class S15	838–847
1938	11	0-6-0	Class Q	530–540
1939	9	0-6-0	Class Q	541–549
1941	6	4-6-2	Class MN	21C1–21C6
1942	4	4-6-2	Class MN	21C7–21C10
1943	21	2-8-0	Class LMS	8600–8609, 8650–8660
1944	2	2-8-0	Class LMS	8661 and 8662
1944	2	4-6-2	Class MN	21C11 and 21C12
1945	8	4-6-2	Class MN	21C13–21C20

Rebuilds

1924	4	4-6-0	Class H15	330–333 rebuilt from class F13 4-6-0
1925	1	4-6-0	Class H15	334 rebuilt from class F13 4-6-0
1928	7	2-6-0	Class U	A790–A796 converted from Class K 2-6-4Ts.

Fig. 347
Eastleigh 1925
Class N15
4-6-0 No. 455
SIR LANCELOT

Fig. 348
Eastleigh 1928
Class LN No. 852
SIR WALTER RALEIGH
– K.C. blast pipe

LONDON & SOUTH WESTERN RAILWAY

Eastleigh became the chief Locomotive Works in the group and was kept busy on building Robert Urie's *H15, N15* and *S15*, 4-6-0s modified to Mr Maunsell's requirements.

Among the *H15* class were Nos. 330–334 which had already been fitted with superheated boilers by Urie. They were rebuilt in 1924–5, but to all intents and purposes were new locomotives with new frames, cylinders, coupled wheels and motion. The old boilers were modified. They had not been successful in their original form for any traffic, but after becoming almost identical with the new *H15s* were far more useful on passenger and goods duties.

The ten Drummond *Class T14* 4-6-0s had also been fitted with superheated boilers by Urie, but they were still heavy on coal and suffered from hot boxes. Maunsell decided to rebuild the whole class, and this was done in 1930 and 1931. The large 'paddlebox' splashers were removed and separate ones provided and the superheaters were removed and Maunsell's pattern fitted. The rebuilding did not improve their performance to any extent, and the boilers were still shy of steam and fuel consumption more than average. Their numbers were 443 to 447 and 458 to 462.

Eastleigh works at the end of 1947 had built 304 steam locomotives from 1910, a modest figure for such an up to date works. All requirements were met until 1925 when thirty *King Arthur* class were built by the North British Locomotive Co. to meet the urgent demand for passenger locomotives.

From 1 January, 1948 the Southern Railway became the *Southern Region* of BRITISH RAILWAYS.

Eastleigh built a further sixteen locomotives all of Southern Railway design:

8	SR	4-6-2	Class MN	35021–35028 in 1948;
3	SR	4-6-2	Class WC	34095, 34097, 34099 in 1949;
2	SR	4-6-2	Class MN	35029, 35030 in 1949, and
3	SR	4-6-2	Class WC	34101, 34102, 34104 in 1950.

Three hundred and twenty locomotives were built at Eastleigh from 1910 to 1950. Normal repair work was carried out and in 1956 the first *Merchant Navy Pacific* was rebuilt. A design had been prepared in 1950 at Brighton for a two-cylinder version of the *West Country* class, but being more pre-occupied in the BR designs for the *Class 4* 4-6-0s, 2-6-4Ts and *Class 9* 2-10-0s, it was not until 1955 that the designs were finalised for the rebuilding of the two *Pacific* classes.

Fig. 349 Eastleigh 1930 *Schools Class V* No. 904 LANCING

LONDON & SOUTH WESTERN RAILWAY

Type	Class	First Built	No	Cylinders inches	D.W. ft in	Boiler: Tubes & Flues	Boiler: Water Tubes	Boiler: Super	Boiler: F'box	Boiler: Total	Grate Area	Press lb/sq in	Weight WO ton cwt	
4-6-0	P14	1910	5	4/15 x 26	6 0	1580	200	—	140	1920	31.5	175	70 19	(a)
0-4-0T	S14	1910	2	OC 12 x 18	3 8	521	—	—	82	603	10.5	175	28 1	
4-6-0	T14	1911	10	4/15 x 26	6 7	1580	200	—	140	1920	31.5	200	74 10	(a)
0-4-4T	M7	1911	10	IC 18½ x 26	5 7	1067.7		—	123.9	1191.6	20.36	150	54 13	
4-4-0	D15	1912	10	IC 19½ x 26	6 7	1406	170	—	148	1172.4	27.0	200	59 15	
4-6-0	H15	1914	11	OC 21 x 28	6 0	1759	—	360	167	2286	30.0	180	79 2	
4-6-0	N15	1918	20	OC 22 x 28	6 7	1716	—	308	162	2186	30.0	180	77 17	
4-6-0	S15	1920	20	OC 21 x 28	5 7	1716	—	308	162	2186	30.0	180	77 8	
4-8-0T	G16	1921	4	OC 22 x 28	5 1	1267	—	231	139	1637	27.0	180	95 2	
4-6-2T	H16	1921	5	OC 21 x 28	5 7	1267	—	231	139	1637	27.0	180	98 8	

(a) Some previously built at Nine Elms.

The *Merchant Navy* class was rebuilt in its entirety and the main alterations were the removal of the streamlined casing, the replacement of the chain driven valve gear by three sets of Walschaerts valve gear, the outside cylinder piston valves having outside admission and the new cast steel inside cylinder with the valve chest slightly offset, inside admission. This inside cylinder design enabled a deeper smokebox to be fitted. The original boiler was retained and the oval smokebox door remained, but the boiler pressure was dropped to 250 lb/sq. in. The thirty locomotives comprising this class were completed in 1959.

In 1957 a start was made on rebuilding the *West Country* and *Battle of Britain* class and a total of sixty were dealt with, the last in 1961. The remaining fifty were not rebuilt.

Both classes were most successful and the abandoning of the chain driven valve gear was welcomed by the maintenance staff.

Withdrawal of the unrebuilt *Pacifics* commenced as early as 1963 and in the same year no less than thirteen of Bulleid's *Q1* 0-6-0s were also withdrawn. Withdrawals were accelerated by the electrification in 1959 of the Kent coast lines and the delivery of diesel locomotives.

Eastleigh then gradually changed over to electric and diesel repairs, and electric locomotive building.

The erecting shop comprised three two-road bays wih a centre service road to each bay. Each line of pits were 765′ long and were served by two 30 ton overhead cranes.

Situated 5$\frac{1}{2}$ miles from Southampton, Eastleigh was formerly known as Bishopstoke. See also BRITISH RAILWAYS and SOUTHERN RAILWAY.

LONDON & SOUTH WESTERN RAILWAY

Nine Elms, London

The first workshops were built on ground near the old passenger terminus of the London and Southampton Railway at Nine Elms Lane and were completed at the end of 1839 after the company had changed its name to the London & South Western Railway.

New locomotive building commenced in 1844, although four singles had been erected in 1843–4 as replacements for three earlier locomotives built in 1839 by Nasmyth Gaskell, and one built by Chas. Tayleur in 1838. They could hardly be classified as entirely built at Nine Elms as the major parts were supplied by William Fairbairn and other second-hand material was used.

The first locomotive superintendent was Joseph Woods (April 1838 to December 1840) followed by Joseph Viret Gooch (January 1841 to June 1850).

Building commenced with ten double framed 0-6-0s with inside cylinders and 4′ 9″ diameter wheels. As in many new Works these took a long time to complete. The first four had 15″ x 22″ cylinders, and the remainder 16″ x 22″. No. 49 BISON appeared in December 1845 and the last one No. 106 PANTHER in October 1848. A numbering scheme was adopted in 1845, all previous locomotives bore names only.

Meanwhile ten 2-2-2s had been ordered and all appeared in 1847 between the two batches of four and six 0-6-0s. The frames and outside cylinders were similar to those of the 2-2-2s built by Christie, Adams & Hill in 1848–9, but the boilers had a combined dome and safety valves were on the first ring of the boiler with a further safety valve on the raised round topped firebox. This class was very satisfactory and gave little or no trouble. Ten further 2-2-2s were ordered and again building was extremely slow

Fig. 350
Eastleigh 1941
Bulleid's *Class MN*
No. 35029
ELLERMAN LINES

and only one, No. 40 ETNA was completed before Gooch had left to take up the position of Locomotive Superintendent on the Eastern Counties Railway.

On 1 July, 1850 Joseph Hamilton Beattie replaced Gooch at Nine Elms and a further eight of the 2-2-2s were completed by 1853, but the tenth was completely redesigned by Beattie.

The initial nine had double frames with outside bearings to the leading and trailing wheels only. The outside cylinders were 15" x 20" and driving wheels 7 ft diameter. Beattie lost no time commencing his experiments on feedwater heating and fireboxes adapted for coal burning. One of the 2-2-2s No. 122 BRITANNIA was fitted with a firebox designed to burn coal and coke. It was divided by a mid-feather. There were two fire-holes and coke was burned in the main firebox and coal in the inner firebox, the hot gases passing into the outer firebox through a series of short tubes, thus improving the combustion of the volatile gases and eliminating smoke.

The redesigned tenth 2-2-2 had inside frames to the leading and driving wheels and outside for the trailing wheels. The outside cylinders were 15" x 21" and the driving wheels 6' 6" diameter. Numbered 123, it was named THE DUKE and had a similar firebox to No. 122 and was completed in December 1853.

The first Beattie designed 2-2-2 appeared in 1856 known as the *Canute* class. Twelve were built with the customary outside cylinders, small boiler with square based safety valve mounting over the firebox (introduced by Gooch) and a dome and safety valve on the first ring of the boiler. One innovation was the underslung springs and floating outside axleboxes on the leading wheels. The valves were placed over the cylinders and were actuated by rocker arms from the inside motion. Beattie's jet condenser was also fitted in front of the chimney, exhaust steam being used to heat the feed water. This apparatus was subsequently modified as trouble was experienced with the exhaust steam condensing and mixing with the feed water which became contaminated with oil and grease. The new condenser was a heat exchanger whereby the exhaust steam was not in direct contact with the feedwater. The *Canute* class were not entirely made at Nine Elms, Beattie having ordered some boilers, frames and other parts from private locomotive builders. He continued this practice for a long period.

A batch of twelve 2-4-0 goods locomotives were built between 1851 and 1854 with inside cylinders 15" x 22" and 5' 6" diameter coupled wheels and known as the *Hercules* class. A peculiarity of this class was the frames which were girder section, deepened between the wheels, but decreasing in depth at the horns, a bad feature contributing to cracked frames.

2-4-0s were also built for passenger services from 1854, Beattie favouring this wheel arrangement instead of the 'singles' giving more adhesive weight to the locomotives. No. 45 TITAN was the first built in 1856 with 6' 1" diameter coupled wheels and outside cylinders $14\frac{1}{4}$" x 22", then six larger locomotives were built of the *Tweed* class in 1858 and 1859.

Two further classes of 2-4-0s were built 1855–7 for goods traffic. The first, *Saxon* class with 15" x 24" inside cylinders of which twelve were built, the last six having $15\frac{1}{2}$" diameter cylinders. The other class known as the *Gem* class had 16" cylinders and six were built.

The original works were getting far too cramped, various shops being added from time to time, but it was realised that a new site would have to be found and in 1861 it was agreed to build a new works on the opposite side of the main line. These were completed in 1865. The new erecting shops were 750 ft long and comprised two bays, each 57 ft wide with three sets of rails. One hundred locomotives could be accommodated.

Meanwhile Beattie continued to build his inside and outside cylinder 2-4-0s and from 1851 to 1875 fifteen different classes were built with a total of 141 locomotives, six of which were built by Beyer Peacock in 1866 and it was probably from this class that Beattie took note of the beautiful lines of these locomotives with brass safety valve and dome covers, and chimney caps. Beattie's dome covers were ugly cast iron pieces which were actually painted yellow in an endeavour to make them look like brass. Early locomotives had rectangular brass name plates attached to the boiler barrel; later ones had varieties of names and number plates, the neatest being a combined name and number plate with date of construction included. They were bolted to the cab sides and were informative but unobtrusive.

One characteristic that stood out was the flywheel of the small donkey engine feedwater pump. The pump was usually mounted on the left-hand side of the firebox on the footplating.

For suburban traffic Beattie built some 2-2-2 well tanks at Nine Elms, eight *Sussex* class and three *Chaplin* class. Both classes had outside cylinders, 14" and $14\frac{1}{2}$" x 20", 5' 6" driving wheels, domeless boilers having a total heating surface of 750 sq. ft and a small grate with an area of 8·9 sq. ft. A large dome and spring balance safety valves were mounted on the firebox. The driving splashers had radial slots. The *Chaplin* class differed in trailing wheel diameter, increased bunker capacity and smaller tanks.

The natural development of the 2-2-2WT followed in the form of 2-4-0WTs introduced by Beattie in 1856 when three *Minerva* class were built with 14" x 21" outside cylinders and 5′ 6" diameter coupled wheels. The boiler was practically the same as for the 2-2-2WTs. Three *Nelson* and three *Nile* class followed in 1858 and 1859 being developments of the *Minerva* class, and in 1863 the final design was produced when Beyer Peacock built a total of eighty-two from 1862 to 1875 and three were built at Nine Elms in 1872. Such a large class had dimensional variations, the outside cylinders 15" diameter to 16½" diameter and the stroke 20" and 22". The coupled wheels were 5′ 6" diameter and the domeless boiler had a total heating surface of 795·2 sq. ft. Tank capacities were from 500 to 600 gallons and the weight in working order from 29 tons 17 cwts to 34 tons 9 cwt. Beattie's divided firebox was fitted with dome and spring balance safety valves fitted on top, and a small lock-up safety valve fixed on the boiler barrel. The valve motion was Allan's straight link and they were indeed very handsome in appearance which subsequent rebuildings did not spoil.

For goods traffic Beattie built some inside frame, inside cylinder 0-6-0s from 1863 to 1873 at Nine Elms with 16½" x 22" cylinders and 5′ 0" diameter wheels.

Beattie was very friendly with Charles Beyer of Beyer Peacock and the majority of locomotives ordered from outside contractors were made by this firm.

Suffering from ill health for quite a long time Beattie died in October 1871 leaving behind a stud of useful locomotives which in the main were very successful and designed for medium weight traffic which at this time was on the increase.

His place as Mechanical Engineer was filled by his son William George Beattie. He was content to carry on building locomotives to his father's designs with few alterations, and what alterations there were did not effect any improvements. Beyer Peacock were relied upon

Fig. 351 **L&SWR** Nine Elms 1894 Adams' *Class G6* 0-6-0T No. 257 with Drummond chimney

Fig. 352 **L&SWR** Nine Elms 1899 Drummond *Class T9* 6′7" 4-4-0 No. 116 as rebuilt in 1925

for their designs and advice. Traffic was increasing and the 2-4-0s were hard put to cope with the heavier trains. It was obvious that something larger was required and W. G. Beattie ordered twenty outside cylinder 4-4-0s which, to put it mildly, did not come up to expectations and in December 1877 he resigned.

In January 1878, William Adams was appointed the new Mechanical Engineer. Coming from the Great Eastern Railway he was a very experienced engineer, and he set about reorganising the locomotive department in no uncertain fashion.

Up to this time 228 locomotives had been built at Nine Elms. Two series of works numbers had been used: Nos. 1–111 from 1843–60 and Nos. 1–126 from 1862–75. For some peculiar reason Nos. 1–9 in the second series of works numbers were also Nos. 103–111 in the first series.

No locomotives were built at Nine Elms after the above series, until 1887. William Adams introduced Works Order numbers, similar to those in use on the GER at Stratford. Thus the first order for ten new locomotives was A12 and this primary order became the class number for that particular class of locomotives. This practice continued for all subsequent building at Nine Elms and was continued at Eastleigh.

Nine Elms were heavily committed on repairs and rebuilding and Adams' initial designs were all built by outside contractors.

From 1879 to 1886, 210 locomotives were built as follows:

12	5' 7"	4-4-0T	46 class	1879	Beyer Peacock
12	5' 7"	4-4-0	380 class	1879	Beyer Peacock
12	6' 7"	4-4-0	135 class	1880-1	Beyer Peacock
12	7' 1"	4-4-0	445 class	1883	R. Stephenson
10	6' 7"	4-4-0	460 class	1884	Neilson & Co
11	6' 7"	4-4-0	460 class	1884/7	R. Stephenson
28	5' 7"	4-4-2T	415 class	1882-5	R. Stephenson
12	5' 7"	4-4-2T	415 class	1882	Beyer Peacock
11	5' 7"	4-4-2T	415 class	1883/5	Neilson
20	5' 7"	4-4-2T	415 class	1884-5	Dubs & Co
70	5' 1"	0-6-0	395 class	1881-6	Neilson & Co

The introduction of the above types succeeded in meeting the increased traffic demands, and besides these new additions, Adams prepared and carried out extensive rebuilding of older classes and in the process removed Joseph Beattie's special fireboxes, feedwater heaters and feed pumps declaring that the cost of upkeep of these refinements cancelled out any saving in fuel costs.

The first Adams design to be built at Nine Elms was a 0-4-2 mixed traffic locomotive which appeared in May 1887, the first for twelve years. It bore all the characteristics of Adams' designs, stove pipe chimney, dome on boiler barrel, flush round top firebox surmounted by Ramsbottom safety valves neatly cased in. Inside cylinders were 18" x 26" and coupled wheels 6' 1" diameter. Outside frames were provided for the trailing wheels only. A large boiler gave a total heating surface of 1248 sq. ft and a grate area of 17 sq. ft. Tenders were

Fig. 353 **L&SWR** Nine Elms 1900 Drummond *Class M7* 0-4-4T No. 112

not built, older ones being coupled to this class. They were very successful and a total of fifty were built at Nine Elms from 1887 to 1895 and Neilson & Co. built five in 1892. They were known as *A12* class.

For London surburban work some 0-4-4Ts were put in hand and fifty were built 1888–96. They had similar cylinders to the *A12s* and 5′ 7″ diameter coupled wheels. These were the *T1s* and a smaller version known as *02s* were built to replace the smaller 2-4-0s and 2-4-0WTs. Sixty were built and the coupled wheels were 4′ 10″. They were used on branch lines mainly; and their smaller wheels were well suited to the gradients on many branches. Some of this class found their way to the Isle of Wight and did fine service there for many years.

Following GER practice Adams introduced a similar design of cast number plate although earlier Adams locomotives carried separate brass numerals on tank or cab side.

For dock work twenty 0-4-0Ts were built, ten at a time. The second ten built in 1893 were to replace older tank locomotives previously owned by the Southampton Dock Co, which the London & South Western Railway took over in 1891. Known as the *B4* class they had outside cylinders 16″ x 22″ and 3′ 9¾″ diameter wheels. The valve gear was inside the frames and the wheelbase was only 7′ 0″ which gave them access to the quays and dockyards.

For general shunting ten *G6* 0-6-0Ts were put in hand in 1894 and the class was subsequently increased to a total of thirty-four. They had the same boiler as the *02* class 0-4-4Ts and had 17½″ x 24″ cylinders and 4′ 10″ diameter wheels.

The need for more powerful passenger locomotives began to be felt in the late 1880s and Adams designed four more classes of 4-4-0s. They were the *X2, T3, T6,* and *X6* classes and were built from 1890 to 1896.

The main dimensions were as follows:

Class	Cylinders	D.W.	Heating Surface sq ft	Grate Area sq ft	Weight W.O.	Press. lb/sq in	No Built
X2	19″ x 26″	7′ 1″	1358·75	18·14	48 tons 12½ cwt	175lb	20
T3	19″ x 26″	6′ 7″	1315·8	19·75	48 tons 11 cwt	175lb	20
T6	19″ x 26″	7′ 1″	1263·8	19·65	50 tons 2½ cwt	175lb	10
X6	19″ x 26″	6′ 7″	1263·8	19·65	49 tons 13 cwt	175lb	10

The 6′ 7″ diameter driving wheel locomotives of classes *T3* and *X6* were intended for the West of England main lines with their attendant banks whereas the 7′ 1″ class were for the easier gradients from Waterloo. Timekeeping improved and when the forty had been completed, all the major expresses were being hauled by them. They were well built and robust and gave little trouble. Tail rods were fitted to these classes, and boilers and cylinders were interchangeable between the *T6* and *X6* classes.

Adams became interested in compounding at the time that Webb was turning out his various classes of compounds and by arrangement No. 300 COMPOUND was loaned to the L&SWR and although it did not distinguish itself at all, Adams converted one of his earlier *445* class 4-4-0s No. 446 using the Worsdell-Von Borries system. One outside 18″ x 24″ cylinder was used at high pressure and a new 26″ x 24″ cylinder fitted outside on the opposite side and strangely enough the necessary platform clearance was obtained. The conversion was carried out in 1888 and trials lasted until 1891 when it was rebuilt to a 'simple', very little saving in fuel having been accomplished.

Due to ill health William Adams had to retire in July 1895 leaving behind an efficient stock of locomotives and workshops well run and in good spirit.

This friendly atmosphere at the works was unfortunately dissipated by Adams' successor Dugald Drummond who, after a very varied and chequered career, had left his newly formed Glasgow Railway Engineering Company, to take up this post on the L&SWR.

Despite his intolerant and dour attitude, he was a very able engineer and he realised that the present state of affairs in the locomotive department was good, and he was wise enough to mark time. It was not until January 1897 that the first Drummond designed express locomotive appeared. This novel 4-2-2-0 had inside cylinders driving the leading wheels, and two outside cylinders driving the rear pair, but there were no coupling rods. Stephenson's link motion was employed for the inside cylinders and Joy's valve gear for the outside cylinders. All cylinders were 15″ x 26″ and the driving wheels 6′ 7″ diameter. Firebox water tubes were fitted giving a heating surface of 215 sq. ft out of a total heating surface of 1664 sq. ft.

There were also heating tubes in the tender which was carried on eight wheels and inside frames. A handsome Drummond chimney, combined dome and safety valves, organ pipe whistle and winged front to the smokebox made it quite different from anything that had appeared before at Nine Elms. It was classed *T7* and numbered 720 but it gave endless trouble

LONDON & SOUTH WESTERN RAILWAY

Fig. 354 **L&SWR** Nine Elms 1901 Drummond *Class E10* 4-cylinder 4-2-2-0 No. 370–water tube firebox

Fig. 355 **L&SWR** Nine Elms 1902 *Class K10* 5′ 7″ mixed traffic 4-4-0 No. 136

with shortage of steam, wheel slip and unbalanced valve motion and spent lengthy periods in the shops. It was reboilered in 1905 but no attempt was made to rebuild it as a coupled engine. Five similar locomotives were built in 1901.

His 4-4-0s however were far more successful and his *T9* class were, at the time, the most efficient four coupled passenger locomotives on the L&SWR. They had inside cylinders 18½″ x 26″ and 6′ 7″ diameter coupled wheels. The boiler gave a total heating surface of nearly 1500 sq. ft and a grate area of 24 sq. ft. The firebox was fitted with Drummond's water tubes and the feed water was passed through the smokebox, the clack valves being sited just behind the smokebox on the underside of the boiler barrel. No. 773 of this class, built by Dubs & Co, was shown at the Glasgow Exhibition in 1901, thirty-five were built at Nine Elms from 1899–1901 and thirty by Dubs & Co.

As could be expected they had marked Scottish lines with strong St. Rollox characteristics. The same applied to his *M7* 0-4-4Ts which had 18½″ x 26″ cylinders and 5′ 7″ diameter coupled wheels. The first series had conical smokebox doors and altogether ninety-five were built between 1897 and 1911 with various detail alterations from time to time.

The reason for the conical doors was to provide extra volume to compensate for the fitting of Drummond's spark arresters which took up a great deal of room but, when subsequently modified, the conical doors became unnecessary.

Later locomotives of this class had feed water heaters in the tanks through which part of the exhaust steam was passed. Feed pumps were fitted to cope with the hotter water.

In 1899 Drummond designed and built a novel inspection saloon. It was a combined locomotive and carriage intended for Drummond's frequent inspection tours on the system. The wheel arrangement was a 4-2-4T the trailing bogie being under the carriage portion. The two outside cylinders were 11½″ x 18″ and the driving wheels 5′ 7″ diameter.

In common with other railways he designed various rail motors all with leading single driving wheels the outside cylinders being fitted behind the drivers adjacent to the trailing wheels. No spare engine units were built and when any repairs were necessary the whole unit had to be taken out of service. To get over this problem ten independent units were built in 1906–7. They were 2-2-0 tanks with outside cylinders 10″ x 14″ and 3 ft diameter driving wheels. They were designed for push and pull motor services but their power was very limited and the extra coach required for peak traffic usually proved too much for such light-weights. Four were rebuilt as 0-4-0Ts and were used for other duties.

In 1901–2 Drummond built forty useful mixed traffic 4-4-0s with 5′ 7″ diameter wheels. The boilers were similar to the *M7* 0-4-4Ts and cylinders and motion were interchangeable. The boiler pressure was the standard 175 lb/sq. in. They were known as the *K10* class. A larger version followed and forty were built from 1903–7 (*Class L11*). The main difference was an enlarged firebox and the coupled wheelbase increased from 9 ft to 10 ft. They were more successful than the *K10s* and were more widely dispersed.

In addition, in 1903, ten 6′ 1″ diameter versions of the *L11s* were built and in 1904–5 twenty 6′ 7″ versions appeared. They were known as *S11* and *L12* respectively.

Apart from Drummond's early 4-4-0s these various classes proved excellent machines but inevitably loads were increasing faster than the pulling power of the locomotives and as the later 4-4-0s had almost reached the limits in size, Drummond designed a four cylinder 4-6-0, five of which were built in 1905 and carried numbers 330–334, and were classified *F13*. The cylinders were 16″ x 24″ and it had 6 ft diameter driving wheels. The large boiler was 5′ 6″ diameter and was fitted with no less than 340 1¾″ tubes giving a heating surface of 2210 sq. ft. The firebox was Drummond's patent water tube type with 2¾″ cross tubes (112 in number) giving 357·0 sq. ft and the firebox itself 160 sq. ft of heating surface, giving a total of 2727 sq. ft.

The inside valves were driven by Stephenson's link motion whilst the outside valves were operated by Walschaerts valve gear. The inside big ends were balanced pattern, obviating balance weights on the driving wheels. The crank-pins for the outside cylinder connecting rods and coupling rods were turned down to different diameters in the conventional manner but the two diameters had 2″ eccentricity.

The throw of the coupling rods was 10″ and 12″ for the connecting rods. Steam pumps were fitted instead of injectors and exhaust steam was used for preheating the water. The weight in working order was 73 tons for the engine and 44 tons 17 cwt for the eight wheeled tender which carried 5 tons of coal and 4000 gallons of water.

They were indeed massive locomotives and all the motion built on the grand scale. Despite their impressive looks they were not a success being very sluggish, heavy on coal and water and were frequent visitors to the shops.

A further 4-6-0 was built in 1907 with larger cylinders 16½″ x 26″ and the outside pair had piston valves fitted above the cylinders, but no appreciable improvement was obtained in performance and the coal consumption was astronomic. Like the other five they were eventually relegated to freight duties.

At the same time building commenced of five further 4-6-0s with four 15″ x 26″ cylinders and 6 ft diameter coupled wheels. They were numbered 453 to 457 and classed *G14*. In general outline they were similar to their predecessors and were used on the heavy West of England expresses where they performed quite creditably, but the same problem of heavy firing occurred and it is well known that on all these 4-6-0s when engaged on heavy trains two firemen were essential.

In 1905 the title of Mechanical Engineer was changed to Chief Mechanical Engineer.

Five additional *B4* Dock tanks were needed and Drummond put them in hand in 1908. They were William Adams design, originally built in 1891–3 and apart from smaller boilers and Drummond fittings they differed little from the original batch. In June 1908, No. 84 of this class was completed at Nine Elms and was the last new locomotive to be built at Nine Elms. In all 815 were built and all major repairs and rebuilding were carried out at the factory.

Again, the factory was proving too small for the increased locomotive stock and enlargement of the works was impossible. It was decided to build a new locomotive works at Eastleigh – the carriage and wagon department were already there, and the new buildings were ready for building new locomotives in 1910.

Nine Elms carried on with repairs until the end of 1909 when all useful plant had been transferred to Eastleigh and the former was finally closed on 31 December, 1909.

The standards of workmanship at Nine Elms had been excellent, the locomotives being robust and substantial and in the main well maintained.

See L&SWR – EASTLEIGH.

LONDON & SOUTH WESTERN RAILWAY

SOME EXAMPLES OF EARLY CLASSES

Date	Class	No Built	Cylinders inches	D.W. ft in	Heating Surface sq ft	Grate Area sq ft	WP lb/sq in	Weight WO ton cwt	
	2-4-0								
1851-4	Hercules	15	15 x 22	5 6	781.0		110	28 10	
1855	Saxon	6	15 x 24	5 0	1167.8	15.6	130	29 4	
1857	Saxon	6	15½ x 24	5 0	1167.8	15.6	130	29 4	
1855	Titan	1	14¼ x 22	6 1	769.0	16.0	130	27 8	
1858	Tweed	3	15 x 22	6 0	1167.8	15.6	130	26 10	
1859	Tweed	3	15½ x 22	6 0	1167.8	15.6	130	26 10	
1859-60	Undine	12	16½ x 22	6 6	904.4	18.0	130	33 3	
1859-68	Clyde	13	15 x 22	7 0	1149.0	16.1	130	28 10	
1862	Eagle	3	17 x 22	6 0	904.4	18.0	130	33 2	
1862-3	Gem	6	16 x 22	5 0	1167.8	15.6	130	29 4	
1863-6	Falcon	14	17 x 22	6 6	904.4	18.0	130		
1865	Falcon	3	16 x 22	6 6	904.4	18.0	130		
1866	231	6	16½ x 22	6 0	906.25	18.0	130	32 3	Beyer Peacock
1866-73	Volcano	18	{17 x 21 17 x 22 16 x 22}	6 1	904.4	18.0	130	33 1	
1869-75	Vesuvius	32	17 x 22	6 6	904.4	18.0	130	34 11	
		All outside cylinders except 'Hercules' class							
	2-2-2								
1843/4	Eagle	4	14½ x 21	6 6	767.0	9.4		17 0	
1847	Mazeppa	10	15½ x 22	6 6	898.5	12.4	100	26 13	
1850-3	Vesuvius	8	15 x 20	7 0	1057				
1855-9	Canute	6	15 x 21	6 6	769.0	16.0	130	25 9	
		All outside cylinders							
	0-6-0								
1865-73	Lion	38	16½ x 22	5 0	1039	16.6	130	33 2	Ins. Cyls

WILLIAM ADAMS

Type	Class	First Built	No	Cylinders inches	D.W. ft in	Boiler: Tubes	Boiler: Water Tubes	Boiler: F'box	Boiler: Total	Grate Area	WP lb/sq in	Weight WO ton cwt	
0-4-2	A12	1887	90	IC 18 x 26	6 1	1131.4	–	116.7	1248.1	17.0	160	43 8	(a)
0-4-4T	T1	1888	50	IC 18 x 26	5 7	1121	–	110	1231	17.0	160	53 0	
0-6-0	O2	1889	60	IC 17 x 24	4 10	898	–	89	987	13.83	160	44 15	
4-4-0	X2	1890	20	OC 19 x 26	7 1	1246.25	–	112.5	1358.75	18.14	175	48 12	
0-4-0T	B4	1891	25	OC 16 x 22	3 9¾	766	–	57	823	10.75	140	33 9	
4-4-0	T3	1892	20	OC 19 x 26	6 7	1193.7	–	122.1	1315.8	19.75	175	48 11	
0-6-0T	G6	1894	34	IC 17½ x 24	4 10	898	–	89	987	13.83	160	44 3	
4-4-0	T6	1895	10	OC 19 x 26	7 1	1141.7	–	122.1	1263.8	19.65	175	50 2	
4-4-0	X6	1895	10	OC 19 x 26	6 7	1141.7	–	122.1	1263.8	19.65	175	49 13	

DUGALD DRUMMOND

Type	Class	First Built	No	Cylinders inches	D.W. ft in	Tubes	Water Tubes	F'box	Total	Grate Area	WP lb/sq in	Weight WO ton cwt	
0-4-4T	M7	1897	105	IC 18½ x 26	5 7	1067.7	–	123.9	1191.6	20.36	175	54 13	(c)
4-2-2-0	T7	1897	1	4/15 x 26	6 7	1307	215	142	1664	27.4	175	54 11	
4-4-0	C8	1898	10	IC 18½ x 26	6 7	1067.7	–	123.9	1191.6	20.36	175	46 16	
4-2-0T	F9	1899	1	OC 11½ x 18	5 7	500	–	50	550.0	11.3	175	37 8	
4-4-0	T9	1899	66	IC 18½ x 26	6 7	1186.4	–	148.3	1334.7	24.0	175	46 4	(b)
4-2-2-0	E10	1901	5	4/14 x 26	6 6	1344	190	156	1690	27.4	175	57 15	
4-4-0	K10	1901	40	IC 18½ x 26	5 7	1067.7	100	123.9	1291.6	20.36	175	46 14	
4-4-0	L11	1903	40	IC 18½ x 26	5 7	1187	165	148	1500	24.0	175	51 1	
4-4-0	S11	1903	10	IC 19 x 26	6 1	1222	165	163	1550	24.0	175	52 0	
4-4-0	L12	1903	20	IC 19 x 26	6 7	1222	165	163	1550	24.0	175	53 19	
2-2-0T	C14	1906	10	OC 10 x 14	3 0	379	119	73	571	9.5	150	24 0	
0-4-0T	S14	1910	2	OC 12 x 18	3 8	521	–	82	603	10.5	175	28 1	

(a) Includes 40 built by Neilson & Co. (b) Includes 31 built by Dubs & Co. (c) Includes ten built at Eastleigh

LONGBOTHAM

Railway Foundry, Barnsley

The only locomotive reported to have been built here was a 0-6-0T about 1847. It had inside frames and cylinders, and the haystack firebox was surmounted by a combined dome and safety valves. Its early history is unknown, but in 1861 it was repaired at Swindon after an accident. Whether it acquired a saddle tank here is not known, but when it appeared on the construction of the Hundred of Manhood and Selsey tramways (later West Sussex Railway) it had a straight sided saddle tank. The coupling rods were removed from the trailing end and it worked as a 0-4-2ST and subsequently the Avonside Engine Company fitted smaller trailing wheels. It was scrapped in 1913. The cylinders were 11″ x 18″, coupled wheels 3′ 6″ diameter and when at Selsey it was named CHICHESTER. The weight in working order was 20 tons.

LONGRIDGE, R. B. & COMPANY

Bedlington, Northumberland

These works, established in 1785, were situated in the valley of the River Blyth about two miles up river from Blyth. They were known as the Bedlington Ironworks and at their peak employed 1500 workers and also took on a number of seven-year apprentices. They were well known as producers of tyres for locomotives in competition with Low Moor Company and others.

Locomotive building commenced in 1837 when a 0-6-0 was built for the Stanhope and Tyne Railway and named MICHAEL LONGRIDGE.

Michael Longridge had been one of the partners of Robert Stephenson & Company's works when they were founded in 1823. Whilst Robert Stephenson was abroad, he managed the works in his absence. He was also proprietor of the Bedlington Iron Works where the first malleable iron rails were rolled. Boilers and plates were supplied to the Newcastle works from Bedlington.

Although the Stephensons knew that Longridge was preparing to build locomotives at Bedlington in 1836 he did not dissolve his partnership until 1842. It is doubtful whether any orders were subcontracted by Robert

Fig. 356 **Longbotham** *c*1847 CHICHESTER during construction of line–Hundred of Manhood & Selsey Tramways

Stephenson & Company to Longridge although a number of 'long boilered' locomotives were built.

From 1838 to 1840 at least twenty locomotives were built for European countries. 2-2-2s were supplied to Belgium, Holland, Germany, France and Austria and two 0-4-2s to Germany.

The first locomotive to run on the Holland Railway was supplied in 1839. It was a 6 ft 'single' with 12½" x 18" inside cylinders. Some Great Western Railway 7 ft gauge 'singles' were built in 1841 to Daniel Gooch's designs with outside slotted sandwich frames and additional bearings for the driving axle. The driving wheels were flangeless. The boilers were domeless with huge 'gothic' type fireboxes on which spring balance safety valves were fitted.

In 1841–2 some further 2-2-2s were supplied to the 5 ft gauge Northern & Eastern Railway.

There was a gap of two years (1843 and 1844) when no recorded locomotive building took place.

In 1845 locomotives were built for Newcastle & Darlington, Hull & Selby, West Hartlepool Harbour & Railway Co. and Cologne & Minden railways.

Business now increased considerably and in the years 1846 and 1847 at least sixty locomotives were built. Useful orders came from the London and Birmingham Railway for twelve 0-6-0s, from the Midland Railway for twenty outside cylinder long boilered 2-4-0s and eight 0-6-0s for the London Brighton & South Coast Railways.

The overseas market disappeared apart from a number of various types sent over to Ireland from time to time including eight 2-4-0s for the Midland Great Western Railway.

In 1846 two 0-4-2s were built for the Eastern Counties Railway and were the first outside cylinder locomotives to be built by the firm.

The Bristol & Exeter Railway had ordered ten 4-2-2 passenger locomotives for the 7 ft gauge. These were large but not so large as Gooch's *Iron Duke* class. Nevertheless they had 7′ 6″ diameter driving wheels and 16½" x 24" inside cylinders and a large domeless boiler with raised firebox giving a total heating surface of over 1600 sq. ft and a grate area of 24 sq. ft. They were completed in 1849.

Next were two 2-2-2WTs for the Bristol & Exeter Railway (Nos. 30 & 31) designed by James Pearson. They had well and back tanks holding 480 gallons and were built in 1851 for branch line service.

In the same year Longridge built ten 4-2-0 *Cramptons* for the Great Northern Railway. The driving wheels 6′ 6″ diameter were driven from a jack-shaft in front of the firebox. The two leading axles had compensating beams between the springs. The boiler was domeless. They were subsequently converted to 2-2-2s by Archibald Sturrock.

Another order from another 7 ft gauge line was completed in 1851 for the South Devon Railway. Five 4-4-0STs were built and had inside sandwich frames and a saddle tank covering the boiler barrel and firebox. A circular sandbox was fitted on the saddle tank, the boiler was domeless and had a total heating surface of 1323 sq. ft with a grate area of 22 sq. ft.

Fig. 357 R. B. Longridge & Co. 1847 London Brighton & South Coast Rly No. 101

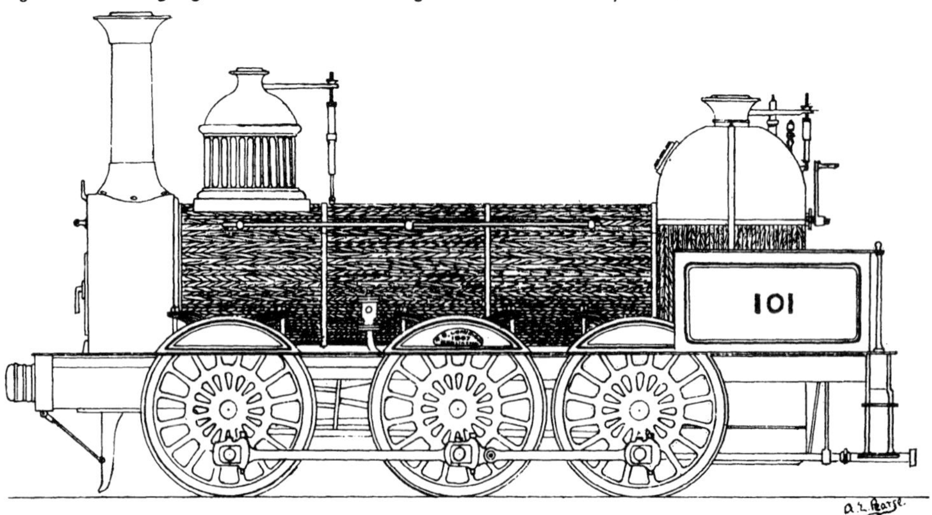

Wks No	Date	Type	Cylinders inches	D.W. ft in	Customer	No	Name	Gauge ft in
101	1837	0-6-0	IC 14 x 18	4 0	Stanhope & Tyne Rly		Michael Longridge	
102	3/1838	0-6-0	IC 14 x 18	4 6	Brandling Junc. Rly		Gateshead	
103	5/1838	2-2-2	IC 12½ x 16	5 6	Brandling Junc. Rly		Sunderland	
104	8/1838	2-2-2	IC 12 x 18	5 0	Brandling Junc. Rly		Newcastle	
105	11/1839	0-6-0	14 x 18	4 0	Brandling Junc. Rly		Brandling	
106	6/1839	0-6-0	14 x 18	4 6	Hartlepool Dock & Rly		NE 132	
107	6/1839	0-6-0	14 x 18	4 6	Hartlepool Dock & Rly		NE 133	
	1839	2-2-2			Naples—Portici—Italy		Bayard	
	1839	2-2-2			Naples—Portici—Italy		Vesuvio	
					Naples—Portici—Italy		Impavido	
					Naples—Portici—Italy		Impetuoso	
113	11/1838	2-2-2	12½ x 18	5 6	Belgian Rlys	54	Ortelins	
114	1/1839	2-2-2	12½ x 18	5 6	Belgian Rlys	56	De Lannay	
115	1/1839	2-2-2	12½ x 18	5 6	Belgian Rlys	57	Le Lion Belge	
116	9/1839	2-2-2	12½ x 18	6 0	Holland Rly	1	Snelheid	
117	5/1839	2-2-2	12½ x 18	5 0	Rhenish Rly		Pluto	
118	5/1839	2-2-2	12½ x 18	5 0	Rhenish Rly		Phoenix	
119	9/1839	2-2-2	14 x 18	6 0	Holland Rly	2	De Arend	
120	5/1839	0-4-2	14 x 18	5 0	Rhenish Rly		Atlas	
121	1839	0-4-2	14 x 18	5 0	Berlin—Potsdam Rly			
122	10/1839	2-2-2	12⅛ x 18	6 0	Holland Rly	3	Hoop	
123	2/1840	2-2-2	12⅛ x 18	5 6	Belgian Rlys	92	Hennepin	
124	2/1840	2-2-2	12⅛ x 18	5 6	Belgian Rlys	93	Aquilon	
125	12/1839	2-2-2	14 x 18	6 0	Holland Rly	4	Leeuw	
126	5/1840	2-2-2	12½ x 18	5 6	Belgian Rlys	107	Gramaye	
127	1840	2-2-2	12½ x 18	5 6	Belgian Rlys	108	Philippe Le Bon	
128	1839				Austria			
129	1839				Alais & Beaucaire Rly			
130								
131	5/1840	2-2-2	12½ x 18	5 6	Belgian Rlys	109	Rubruquis	
132								
133	8/1840	2-2-2	12½ x 18	5 6	Belgian Rlys	112	Verheyen	
134	9/1840	2-2-2	12½ x 18	5 6	Belgian Rlys	116	Baudouin De Constantinople	
	*c.*1840				Midland Counties Rly			
	9/1840	2-2-2	14 x 18	6 0	Eastern Counties Rly	23		5 0
	9/1840	2-2-2	14 x 18	6 0	Eastern Counties Rly	24		5 0
	11/1841	2-2-2	14 x 18	6 0	Eastern Counties Rly	25		5 0
	12/1841	2-2-2	14 x 18	6 0	Eastern Counties Rly	26		5 0
	1/1842	2-2-2	14 x 18	6 0	Eastern Counties Rly	28		5 0

Works No	Date	Type	Cylinders	D.W.	Customer	No	Name	Gauge
	2/1842	2-2-2	14 x 18	6 0	Eastern Counties Rly	29		5 0
	4/1842	2-2-2	14 x 18	6 0	Eastern Counties Rly	30		5 0
	4/1841	2-2-2	15 x 18	7 0	Great Western Railway		Jupiter	7 0
	6/1841	2-2-2	15 x 18	7 0	Great Western Railway		Saturn	7 0
	7/1841	2-2-2	15 x 18	7 0	Great Western Railway		Mars	7 0
	8/1841	2-2-2	15 x 18	7 0	Great Western Railway		Lucifer	7 0
	10/1841	2-2-2	15 x 18	7 0	Great Western Railway		Venus	7 0
	10/1841	2-2-2	15 x 18	7 0	Great Western Railway		Mercury	7 0
	1841				North Midland Rly			
156	11/1841	2-2-2	14 x 18	6 0	Northern & Eastern Rly	29		5 0
157	12/1841	2-2-2	14 x 18	6 0	Northern & Eastern Rly	30		5 0
142	11/1840	2-2-2	14 x 18	6 0	Northern & Eastern Rly	31		5 0
158	1/1842	2-2-2	14 x 18	6 0	Northern & Eastern Rly	35		5 0
159	2/1842	2-2-2	14 x 18	6 0	Northern & Eastern Rly	36		5 0
160	4/1842	2-2-2	14 x 18	6 0	Northern & Eastern Rly	23		5 0
		2-2-2	14 x 18	6 0	Northern & Eastern Rly	25		5 0
	1845	2-4-0	15 x 24	4 9	Chester & Birkenhead Rly	21		
	1845	2-4-0	15 x 24	4 9	Chester & Birkenhead Rly	22		
	1845				Cologne & Minden Rly		Weser	
	1845				Cologne & Minden Rly			
	3/1845	0-6-0	15 x 24	4 8	Newcastle & Darlington Junc. Rly		NER 46	
	4/1845	0-6-0	15 x 24	4 8	Newcastle & Darlington Junc. Rly		NER 47	
	5/1845	0-6-0	15 x 24	4 8	Newcastle & Darlington Junc. Rly			
	5/1845	0-6-0	15 x 24	4 8	Newcastle & Darlington Junc. Rly			
	5/1845	0-6-0	15 x 24	4 8	Newcastle & Darlington Junc. Rly.		NER 48	
	11/1845	0-6-0	15 x 24	4 8	Newcastle & Darlington Junc. Rly			
	1845	2-4-0 (?)	14 x 18	4 6	W. Hartlepool Harbour & Rly Co	8	NER 589	
	9/1845		15½ x 24	4 6	Hull & Selby Rly		Y & NM No 52 NER 284	
	9/1845		15 x 24	4 6	Hull & Selby Rly		Y & NM No 53 NER 285	
	1846	0-4-2	15 x 24	4 9	Eastern Counties Rly	104		
	1846	0-4-2	15 x 24	4 9	Eastern Counties Rly	105		
	2/1846	2-4-0	15 x 22	6 0	York & North Midland Rly	60	NER 292	
	1846	2-4-0	16½ x 24	4 7	W. Hartlepool Harbour & Rly Co	32	NER 607	
	1846	0-6-0			London & Birmingham Rly	125		
	1846	0-6-0			London & Birmingham Rly	126		
	1846	0-6-0			London & Birmingham Rly	127		
	1846	0-6-0			London & Birmingham Rly	128		

	1846	0-6-0			London & Birmingham Rly	129		
	1846	0-6-0			London & Birmingham Rly	130		
	1846	2-4-0			Londonderry & Enniskillen Rly	1		
		2-2-2			Londonderry & Enniskillen Rly	2		
		2-4-0			Londonderry & Enniskillien Rly	3		
	10/1846	2-4-0	OC 15 x 22	5 6	Midland Rly	143		
	10/1846	2-4-0	OC 15 x 22	5 6	Midland Rly	144		
	10/1846	2-4-0	OC 15 x 22	5 6	Midland Rly	145		
	11/1846	0-6-0	15 x 24	4 9	Shrewsbury & Chester Rly	1	*	
	11/1846	2-4-0	15 x 24	4 9	Shrewsbury & Chester Rly	2	*	
	11/1846	2-2-2	15 x 24	5 9	Shrewsbury & Chester Rly	3	*	
	11/1846	2-4-0	15 x 24	4 9	Shrewsbury & Chester Rly	4	*	
	11/1846	2-4-0	14½ x 18	4 6	Newcastle & Darlington Junc. Rly		NER 78	
	11/1846	2-4-0	14½ x 18	4 6	Newcastle & Darlington Junc. Rly			
	11/1846	2-4-0	14½ x 18	4 6	Newcastle & Darlington Junc. Rly			
	11/1846	2-4-0	14½ x 18	4 6	Newcastle & Darlington Junc. Rly			
	11/1846	2-4-0	14½ x 18	4 6	Newcastle & Darlington Junc. Rly			
	11/1846	2-4-0	14½ x 18	4 6	Newcastle & Darlington Junc. Rly			
	11/1846	2-4-0	14½ x 18	4 6	Newcastle & Darlington Junc. Rly			
	11/1846	2-4-0	14½ x 18	4 6	Newcastle & Darlington Junc. Rly			
	12/1846	2-4-0	14½ x 24	5 0	Shrewsbury & Chester	5		
	1/1847	2-4-0	14½ x 24	5 0	Shrewsbury & Chester	6		
	12/1846	2-4-0	OC 15 x 22	5 6	Midland Rly	128		
	12/1846	2-4-0	OC 15 x 22	5 6	Midland Rly	146		
	1/1847	2-4-0	OC 15 x 22	5 6	Midland Rly	139		
	1/1847	2-4-0	OC 15 x 22	5 6	Midland Rly	140		
	1/1847	2-4-0	OC 15 x 22	5 6	Midland Rly	141		
	1847	2-4-0	OC 15 x 22	5 6	Midland Rly	142		
	1/1847	2-4-0	14½ x 18	4 6	Newcastle & Darlington Junc. Rly		NER 101	
	1847	2-4-0	14½ x 18	4 6	Newcastle & Darlington Junc. Rly			
	1847	2-4-0	14½ x 18	4 6	Newcastle & Darlington Junc. Rly			
	1847	2-4-0	14½ x 18	4 6	Newcastle & Darlington Junc. Rly			
	1847	2-4-0	14½ x 18	4 6	Newcastle & Darlington Junc. Rly			
	1847	2-4-0	14½ x 18	4 6	Newcastle & Darlington Junc. Rly			
	1847	2-4-0	14½ x 18	4 6	Newcastle & Darlington Junc. Rly			
	2/1847	2-4-0	OC 15 x 22	5 6	Midland Rly	147		
	3/1847	2-4-0	OC 15 x 22	5 6	Midland Rly	129		
	3/1847	2-4-0	OC 15 x 22	5 6	Midland Rly	127		
	4/1847	2-4-0	OC 15 x 22	5 6	Midland Rly			
	5/1847	2-4-0	OC 15 x 22	5 6	Midland Rly	148		
	1847	2-4-0	OC 15 x 22	5 6	Midland Rly	149		
	6/1847	2-4-0	OC 15 x 22	5 6	Midland Rly	150		
	6/1847	0-6-0	15 x 24	4 9	London, Brighton & Sth. Coast Rly	70		
	7/1847	0-6-0	15 x 24	4 9	London, Brighton & Sth. Coast Rly	71		

* Long Boiler Type

Works No	Date	Type	Cylinders inches	D.W. ft in	Customer	No	Name etc	Gauge ft in
	8/1847	0-6-0	15 x 24	4 9	London, Brighton & Sth. Coast Rly	72		
	10/1847	0-6-0	15 x 24	4 9	London, Brighton & Sth. Coast Rly	75		
	7/1847	0-6-0	15 x 24	4 8	York, Newcastle & Berwick Rly		NE 146	
	9/1847	0-6-0	15 x 24	4 8	York, Newcastle & Berwick Rly		NE 154	
	9/1847	0-6-0	15 x 24	4 8	York, Newcastle & Berwick Rly		NE 155	
	9/1847	0-6-0	15 x 24	4 9	York, Newcastle & Berwick Rly		NE 158	
	10/1847	0-6-0	15 x 24	4 9	York, Newcastle & Berwick Rly			
	11/1847	0-6-0	15 x 24	4 8	York, Newcastle & Berwick Rly			
	11/1847	2-4-0	15 x 24	4 8	York, Newcastle & Berwick Rly			
	1847	2-4-0	14½ x 18	4 6	York, Newcastle & Berwick Rly			
	11/1847	2-4-0	15 x 22	5 6	Midland Rly	86		
	12/1847	2-4-0	15 x 22	5 6	Midland Rly	87		
	12/1847	2-4-0	15 x 22	5 6	Midland Rly	79		
	1847	2-4-0	16 x 20	5 6	Dundalk & Enniskillen Rly			
	1847	2-2-2	15 x 22	5 6	Dundalk & Enniskillen Rly			
	1847		15 x 24	5 6	Belfast & Ballymena Rly	27		
	1847	0-6-0	15 x 24	4 9	London & Birmingham Rly	131		
	1847	0-6-0	15 x 24	4 9	London & Birmingham Rly	132		
	1847	0-6-0	15 x 24	4 9	London & Birmingham Rly	133		
	1847	0-6-0	15 x 24	4 9	London & Birmingham Rly	134		
	1847	0-6-0	15 x 24	4 9	London & Birmingham Rly	135		
	1/1848	0-6-0	15 x 24	4 9	London, Brighton & Sth. Coast Rly	100		
	5/1848	0-6-0	15 x 24	4 9	London, Brighton & Sth. Coast Rly	105		
	7/1848	0-6-0	15 x 24	4 9	London, Brighton & Sth. Coast Rly	96		
	7/1848	0-6-0	15 x 24	4 9	London, Brighton & Sth. Coast Rly	97		
	1848	2-4-0			Midland Rly	81		
	1848	2-4-0			Midland Rly	82		
	1848	2-4-0			Midland Rly	83		
246	1848	2-4-0	OC 15 x 22	5 6	Midland Rly	84		
247	1848	2-4-0	OC 15 x 22	5 6	Midland Rly	85		
	2/1848	0-6-0	15 x 24	4 8	York, Newcastle & Berwick Rly		NE 169	
	4/1848	0-6-0	15 x 24	4 8	York, Newcastle & Berwick Rly		NE 172	
	6/1848	0-6-0	15 x 22	4 8	York, Newcastle & Berwick Rly		NE 182	
	1848	2-4-0			Albion Mines. Nova Scotia		Vulcan	
	1848	2-4-0			Albion Mines. Nova Scotia		Hibernia	
	2/1849	0-4-2	OC	4 4	Shrewsbury & Birmingham Rly	6	*	
	1849	0-4-2	IC 15 x 24	5 0	Shrewsbury & Birmingham Rly	7		
	1849	0-4-2	IC 15 x 24	5 0	Shrewsbury & Birmingham Rly	8		
	1849	0-4-2	IC 15 x 24	5 0	Shrewsbury & Birmingham Rly	9		
	1849	0-4-2	IC 15 x 24	5 0	Shrewsbury & Birmingham Rly	10		
	3/1849	0-6-0	IC 15 x 24	4 9	Shrewsbury & Birmingham Rly	11	*	
	1849	0-6-0	IC 15 x 24	4 9	Shrewsbury & Birmingham Rly	12	*	
	1849	0-6-0	IC 15 x 24	4 9	Shrewsbury & Birmingham Rly	13	*	

	1849	0-6-0	IC 15 x 24	4 9	Shrewsbury & Birmingham Rly	14		
	1849	0-6-0	IC 15 x 24	4 9	Shrewsbury & Birmingham Rly	15	*	
	1849	4-2-2	IC 16½ x 24	7 6	Bristol & Exeter Rly	11		7 0
	1849	4-2-2	IC 16½ x 24	7 6	Bristol & Exeter Rly	12		7 0
	1849	4-2-2	IC 16½ x 24	7 6	Bristol & Exeter Rly	13		7 0
	1849	4-2-2	IC 16½ x 24	7 6	Bristol & Exeter Rly	14		7 0
	1849	4-2-2	IC 16½ x 24	7 6	Bristol & Exeter Rly	15		7 0
	1849	4-2-2	IC 16½ x 24	7 6	Bristol & Exeter Rly	16		7 0
	1849	4-2-2	IC 16½ x 24	7 6	Bristol & Exeter Rly	17		7 0
	1849	4-2-2	IC 16½ x 24	7 6	Bristol & Exeter Rly	18		7 0
	1849	4-2-2	IC 16½ x 24	7 6	Bristol & Exeter Rly	19		7 0
	1849	4-2-2	IC 16½ x 24	7 6	Bristol & Exeter Rly	20		7 0
	1851	2-2-2WT	IC 12½ x 18	5 6	Bristol & Exeter Rly	30		7 0
	3/1851	2-2-2WT	IC 12½ x 18	5 6	Bristol & Exeter Rly	31		7 0
	10/1851	4-4-0ST	IC 17 x 24	5 9	South Devon Rly		Comet	7 0
	10/1851	4-4-0ST	IC 17 x 24	5 9	South Devon Rly		Lance	7 0
	10/1851	4-4-0ST	IC 17 x 24	5 9	South Devon Rly		Rocket	7 0
	11/1851	4-4-0ST	IC 17 x 24	5 9	South Devon Rly		Meteor	7 0
	1/1852	4-4-0ST	IC 17 x 24	5 9	South Devon Rly		Aurora	7 0
	1851	4-2-0	16 x 21	6 6	Great Northern Rly	91	Crampton Patent	
	1851	4-2-0	16 x 21	6 6	Great Northern Rly	92	Crampton Patent	
	1851	4-2-0	16 x 21	6 6	Great Northern Rly	93	Crampton Patent	
	1851	4-2-0	16 x 21	6 6	Great Northern Rly	94	Crampton Patent	
	1851	4-2-0	16 x 21	6 6	Great Northern Rly	95	Crampton Patent	
	1851	4-2-0	16 x 21	6 6	Great Northern Rly	96	Crampton Patent	
	1851	4-2-0	16 x 21	6 6	Great Northern Rly	97	Crampton Patent	
	1851	4-2-0	16 x 21	6 6	Great Northern Rly	98	Crampton Patent	
	1851	4-2-0	16 x 21	6 6	Great Northern Rly	99	Crampton Patent	
	1851	4-2-0	16 x 21	6 6	Great Northern Rly	200	Crampton Patent	
	11/1851	2-4-0	15 x 20	5 7	Midland Great Western Rly	18	Vulcan	5 3
	11/1851	2-4-0	15 x 20	5 7	Midland Great Western Rly	19	Childers	5 3
	11/1851	2-4-0	15 x 20	5 7	Midland Great Western Rly	20	Arabian	5 3
	12/1851	2-4-0	15 x 20	5 7	Midland Great Western Rly	21	Eclipse	5 3
	2/1852	2-4-0	15 x 20	5 7	Midland Great Western Rly	23	Cygnet	5 3
	5/1852	0-6-0	16 x 24	5 0	Midland Great Western Rly	25	Cyclops	5 3
	1852	0-6-0	16 x 24	5 0	Midland Great Western Rly	26		
	1851				York & North Midland Rly			
	4/1852	2-2-2WT	12 x 22	6 6	Eastern Counties Rly	4		
	10/1852	2-2-2WT	12 x 22	6 6	Eastern Counties Rly	5		
	10/1852	2-2-2WT	12 x 22	6 6	Eastern Counties Rly	6		
309	1852	0-4-0WT	12 x 18	3 2	Holyhead Breakwater		Prince Albert	7 0
	1852	0-4-0WT	12 x 18	3 2	Holyhead Breakwater		Queen	7 0
	1852	0-4-0WT	12 x 18	3 2	Holyhead Breakwater		London	7 0
	1852	0-4-0WT	12 x 18	3 2	Holyhead Breakwater		Holyhead	7 0
	1852	0-4-0WT	12 x 18	3 2	Holyhead Breakwater		Cambria	7 0
	1852	0-4-0WT	12 x 18	3 2	Holyhead Breakwater		?	7 0

* 'Long Boiler' Type

LONGRIDGE, R. D.

All South Devon Railway locomotives bore names only – as indeed the Great Western Railway did at this time.

The last locomotives built by the firm were six 0-4-0WTs for the 7 ft gauge railway for the Holyhead Breakwater. They had inside cylinders with a 3 ft diameter boiler working at 110 lb/sq. in. and the weight in working order was 14 tons. One found its way to South America about 1873.

These interesting tanks were completed in 1852. Orders had fallen off by then and the works ceased production, and were sold to James Spence in 1853 who closed down two years later.

Works numbers commencing at 101 were used for locomotives and the last known number was 309. Not all of the locomotives built by this firm have been traced but assuming all numbers were filled, 209 locomotives were built from 1837 to 1852.

LOWCA ENGINEERING CO. LTD.

Whitehaven

See TULK & LEY.

LYSAGHT, JOHN, LTD.

Normanby Park Steel Works, Scunthorpe, Lincs.

Due to shortage of labour and the urgent need for locomotives, the following were extensively rebuilt by John Lysaght Ltd using parts supplied by Messrs Peckett & Sons Ltd.

1942	3' gauge	No. 2	0-4-0ST		
1945	Std gauge	No. 22	0-4-0ST	16" x 24"	3' 10"
1945	Std gauge	No. 23	0-4-0ST	16" x 24"	3' 10"
1946	Std gauge	No. 24	0-4-0ST	16" x 24"	3' 10"
1946	3' gauge	No. 1	0-4-0ST		

Each locomotive was in effect a 'Replacement' having new frames, horns, cylinders, wheels, axles, axleboxes, cab, tank and boiler. Therefore the replacement locomotives were not built but erected only. Some extensive rebuilding was also carried out by this firm.

Fig. 358 J. Lysaght Ltd 1945 for own use No. 22–Peckett parts supplied

MANCHESTER & BIRMINGHAM RAILWAY

Longsight, Manchester

The works were established in 1842 and it is probable that the combined repair and new engine building shop took the form of a roundhouse with adjacent ancillary shops. After the amalgamations of July 1846 of which the Manchester & Birmingham Railway was a constituent, the locomotive department became the North Eastern division of the L&NWR with John Ramsbottom as Locomotive Superintendent until 1857 when Francis Trevithick retired from the superintendency of the Northern Division with headquarters at Crewe.

Fifteen passenger 'singles' were built and were based on the current Sharp Stewart & Co's design of that type. One 0-6-0 was built in 1858 and shortly afterwards the works were closed and the work transferred to Crewe.

MANCHESTER & LEEDS RAILWAY

Miles Platting, Manchester

See LANCASHIRE & YORKSHIRE RAILWAY – MILES PLATTING.

MANCHESTER, SHEFFIELD & LINCOLNSHIRE RAILWAY

Gorton, Manchester

The actual site was decided upon and the land purchased by the Sheffield and Manchester Railway to build the necessary workshops for the railway. On amalgamation the size of land was increased to twenty acres, and the buildings were completed in 1848, and consisted of a circular locomotive shed 150 ft in diameter with accommodation for seventeen locomotives and tenders, and an erecting shop 150 ft long and 60 ft wide with nine pits with travelling cranes. Each pit took two locomotives. The carriage and wagon departments were also catered for at Gorton.

The circular shed was noted for its construction,having a central pillar only, thus leaving the working area clear. The 40 ft diameter turntable had two sets of rails passing either side of the pillar. The programme of locomotive repairs and rebuilding was extremely heavy and it was not until 1858 that new building commenced.

Richard Peacock was Locomotive Superintendent, and had graduated from being the first locomotive driver of the Sheffield, Ashton-under-lyne and Manchester Railway to superintendent of the SA&MR in 1841. He was responsible for the layout of the Gorton workshops and carried out an extensive amount of rebuilding of some of the older locomotives. Before any new locomotives appeared Peacock left in 1854 to become partner with Charles Beyer to found the famous locomotive works of Beyer Peacock & Co. Ltd.

LNWR LONGSIGHT

Date	Type	Cylinders inches	D.W. ft in	Running Numbers			Rebuilt Saddle Tank	Renumbered	W'drawn
				NED	ND	1857 No			
1854	2-2-2	15 x 20	5 6	2		402	1868	1856/74	1879
1854	2-2-2	15 x 20	5 6	3		403	–	1858/73	1874
1854	2-2-2	15 x 20	5 6	5		405	1857	1839/74	1875
1854	2-2-2	15 x 20	5 6	6		406	1868	1830/74	1878
1854	2-2-2	15 x 20	5 6	12		412	?		1874
1854	2-2-2	15 x 20	5 6	14		414	1870	1274/68, 1932/72	1876
1854	2-2-2	15 x 20	5 6	24		424	–	1895/73	1874
1855	2-2-2	15 x 20	5 6	68		468	1871	1929/74	1877
1857	2-2-2	15 x 20	5 6	13		413	1870	1866/74	1875
1857	2-2-2	15 x 20	5 6	33		433	–		1873
1857	2-2-2	15 x 20	5 6	34		434	1870	1922/74	1879
1857	2-2-2	15 x 20	5 6	85		485	1871	1931/74	1878
1857	2-2-2	15 x 20	5 6	86		486	1873	1933/74	1879
1857	2-2-2	15 x 20	5 6	87		487	1870	1934/74	1878
1858	2-2-2	15 x 20	5 6	(69)		469	1868	1930/74	1879
1858	0-6-0	16 x 24	5 0		214		–		

NED: North Eastern Division. ND: Northern Division

In 1857 the NE Division was absorbed by the Northern Division and NED Locos had 400 added to their numbers. This date coincided with the retirement of F. Trevithick, and J. Ramsbottom took charge of the enlarged N. Division.

MANCHESTER, SHEFFIELD & LINCOLNSHIRE RAILWAY

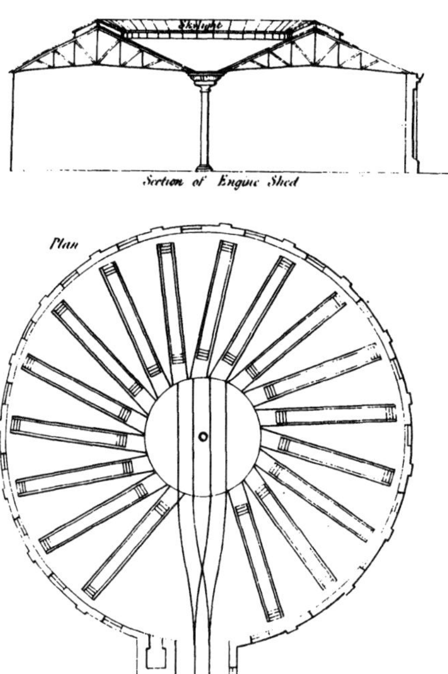

Fig. 359 **MS&LR** Gorton showing double entry turntables, angled pits and central pillar

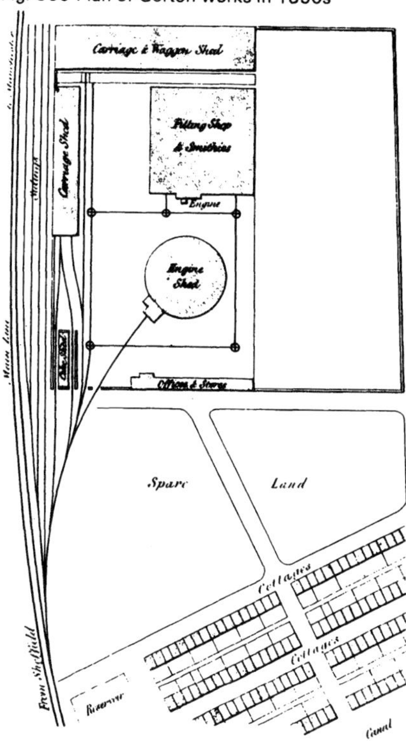

Fig. 360 Plan of Gorton works in 1850s

His successor was William Grindley Craig and his first locomotives were two double framed 0-6-0s replacing two earlier Sharp 'singles'. No. 6 ARCHIMEDES appeared in March and No. 7, PHLEGON in April 1858. They had 17" x 24" cylinders and 5 ft diameter wheels and were followed by two more, Nos. 36 & 53 in the following year. Domeless boilers were fitted with combined dome and safety valves above the firebox. Allan's straight link motion was fitted and the smokebox had a sloping front.

Craig was interested in the experiments of Beattie on the London & South Western Railway and Stephenson's on the Midland Railway. He followed Beattie's idea of a double compartment firebox for coal and coke, and the final judgement was that coal was better than coke especially from the boiler maintenance point of view and economy of fuel.

W. G. Craig was replaced by Charles Reboul Sacré in 1859 coming from the Great Northern Railway's Peterborough workshops where he had been in charge.

Two 2-4-0WTs were built in 1860 to his designs with inside cylinders 15½" x 20" and 5' 6" diameter coupled wheels. They were numbered 23 & 24 and were *Class 1*, Craig having introduced a system of class numbers for each type earlier on. Two more appeared in 1863. They had double frames, and the footplating above the frames were curved to clear the cottered crank pin brasses, in the Beyer Peacock fashion. The boiler had the dome in the centre and the safety valves placed over the raised firebox. A total of ten were built the last one in 1866. The formation of the double framing varied on some of this class.

Two classes of 0-6-0s, *Class 13* and *Class 23* were built in this period. The *Class 23s* had 16" x 24" cylinders and the other class had a shorter stroke of 22". Both had double frames and the *Class 23* was derived from a series supplied by Kitson & Co. (1859–61). Various classes of 2-4-0s were also built, classes *8, 1B, 24,* and *10* from 1862 to 1871. Three of these classes had double frames but the *Class 10* had inside frames and were built in 1869. Weatherboards only were fitted for the footplate crew's protection.

The locomotive requirements at the time Sacré took over, comprised many classes of 2-

2-2, 0-4-2, 2-4-0 and 0-6-0 types of varying efficiency, and although in this first period he put in hand various 2-4-0s and 0-6-0s of much sturdier build, his main preoccupation was in rebuilding the existing stock. Money was short at this time, a circumstance which bedevilled development quite considerably.

The mainstay for goods traffic was the *Class 23*, which were not only built at Gorton but were supplied by Beyer Peacock, Neilson, Kitson and Sharp Stewart companies, eventually reaching the large total of ninety-eight locomotives by 1867.

Sacré kept in touch with his old master Archibald Sturrock and when the latter had introduced his 0-6-0 with steam tender to act as a booster, Sacré followed suit; six being built by Neilson & Co. in 1865 were thus equipped. The locomotive cylinders were 16½″ x 24″ with 5′ 3″ diameter wheels and the tenders had 11½″ x 17″ cylinders and 4′ 6″ diameter wheels. As with the Great Northern Railway steam tenders they could haul additional loads, but at the cost of heavy maintenance and extra fuel.

2-4-0s continued to be used on express passenger duties and were built up to 1884. The last two designs were the *Class 12A* (1875) with 16″ x 24″ cylinders and 6 ft diameter coupled wheels and the *Class 6D* (1887) comprising three locomotives only, with 18″ x 26″ cylinders and 6′ 9″ diameter coupled wheels. One of the most successful class of this wheel arrangement was the ten built in 1873 to *Class 12* with 17″ x 24″ cylinders and 6′ 3″ diameter wheels (Nos. 311 to 320). The boiler pressure was 130 lb/sq. in. and they turned out extremely useful machines.

Some 0-6-0STs for shunting duties and light goods traffic were built; ten in 1871, six in 1873 and various batches up to 1880. They had inside frames and cylinders 17″ x 24″, 4′ 9″ diameter wheels and saddle tanks covering the full length of boiler and smokebox holding 907 gallons. The weight in working order was 41 tons 13 cwt. A total of forty-one were built.

Passenger traffic loads and timings were making it difficult for some classes of 2-4-0s to cope, and it was obvious that a more powerful class of locomotive was needed. To meet this requirement Sacré designed a double framed, inside cylinder, 4-4-0. The bogie had inside bearings but a peculiarity was that the coupled wheels frame extended to the front buffer beam. The frames were shallow but had a stiffener under the driving axleboxes which was continued also to the front end. The boiler was domed with a flush firebox casing as distinct from most of the previous boilers which had raised firebox casings. The pressure was increased to 140 lb/sq. in., the cylinders were 17″ x 26″ and the coupled wheels 6′ 3″ diameter. They were built from 1877 to 1880 and totalled twenty-seven and were *Class 6B*. The cab had small oval side windows fitted. They bore the brunt of main line traffic in the 80s and 90s. No. 434 of this class was involved in the Penistone accident and among the twenty-four killed was Mr Massey Bromley who had retired only three years earlier from the position of Locomotive Superintendent of the Great Eastern Railway.

The MS&LR were working in conjunction with the Great Northern Railway in providing a fast service from King's Cross to Manchester, stopping at Grantham and Sheffield only. The

Fig. 361 **MS&LR Gorton** 1868 Sacré *Class 1B* 2-4-0 No. 31

Midland Railway was competing for this traffic and to speed up the overall time Sacré introduced his well known *14* class 2-2-2s with outside cylinders $17\frac{1}{2}$" x 26" and 7' 6" diameter driving wheels. These savoured very much of the Allan 'Crewe' type and the Scottish equivalents. They were very fast runners but extremely unsteady due to the outside cylinders and the meagre wheelbase of 15' 9". Nevertheless with their introduction the time was cut from $4\frac{1}{2}$ to $4\frac{1}{4}$ hours. In 1885 six inside frame 0-6-0Ts were built with 15" x 20" inside cylinders, 3' 9" diameter wheels and weighing 38 tons 2 cwt in working order (*Class 7*). The MS&LR had very few tank locomotives in proportion to tender types. Sacré was quite content to use the smaller 2-4-0s for secondary and branch line duties. Unhappily in 1886 Sacré tendered his resignation to Sir Edward Watkin, the chairman of the company, and it was accepted. Although entirely found without blame at the enquiry into the Penistone acci-

Fig. 363 **MS&LR Gorton** 1882 Sacré *Class 14* 2-2-2 No. 104 7'6" D.W.

Fig. 364 **MS&LR Gorton** 1885 Sacré *Class 12A* 2-4-0 No. 539

dent, Sacré must have felt deeply on the matter and held himself responsible for the broken locomotive axle which caused the derailment. Six months after the first Penistone accident a second one occurred, caused by a broken wagon axle, not far from the scene of the previous derailment, and in 1887 another accident happened at Hexthorpe when an express from Manchester overran signals and crashed into the rear of another train. Charles Reboul Sacré took his own life in August 1889.

He had left behind a considerable stock of strongly built and soundly designed locomotives, most of which had double frames, low set boilers, distinctive chimneys and fittings.

Thomas Parker, his successor in 1886 had served his apprenticeship on the Caledonian Railway at the Greenock works under Robert Sinclair and since 1858 had been Carriage and Wagon Superintendent at Gorton.

Up to this time nearly 400 locomotives had been built and for a year no new designs were forthcoming, but in 1887 three 2-4-0 express passenger locomotives were built with double frames and the leading wheels had Webb pattern radial axleboxes with inside frames (*Class 6D*). In the following year appeared three 4-4-0s similar to the *Class 6D* 2-4-0s, except that the radial truck was replaced by a bogie, the inside cylinders were 18″ x 26″ and coupled wheels 6′ 9″ for both classes.

He used Stephenson's link motion for all express types, but Joy's valve gear for others. One minor alteration he made was to remove all the handsome brass number plates and use transfers instead which did not improve appearances.

In 1889 he introduced a 0-6-2T with his standard 18″ x 26″ cylinders and 5′ 1″ coupled wheels. Inside frames had now been established as standard and a stove-pipe tapered chimney was used on his classes, relieved a little by a radiused band around the top.

The 0-6-2Ts were *Class 9A* and weighed 61 tons 10 cwt and side tanks held 1300 galls. Some were later altered to *Class 9A alt* with 6″ longer tanks holding 1360 gallons. One of these locomotives had the distinction of being the first to be fitted with a Belpaire firebox – that is, for use in this country. It was No. 7 built in 1891 with 18½″ cylinders and classed *9c*. Belpaire fireboxes had been fitted to various locomotives since 1872 when Beyer Peacock built two 2-4-0s for the Malines-Terneuzen Railway.

A 2-4-2T was also designed, the first being built in 1889. Joy's valve gear was fitted; the boiler had a round top firebox with a total heating surface of 1278 sq. ft and a grate area of 18·85 sq. ft. The tanks were the same as for the 0-6-2Ts and some were similarly enlarged.

In 1892–3 six of the 2-4-0WTs which had been built for the Stockport services were converted to side tanks. The original design was credited to Samuel Waite Johnson as chief assistant to Sacré at Gorton from 1859 to 1864 and who went to the Great Eastern Railway in 1866 as Locomotive, Carriage and Wagon Superintendent.

Four classes of 0-6-0 were built, *6A1, 9, 9D*

and *9E*. The *6A1* were double frame locomotives developed from Sacré's previous *Class 6A* and were known as 'Bashers' with larger cylinders 18″ x 26″ and wheels 5′ 3″ diameter. Twelve were built in 1887 to 1888. The *Class 9* had inside frames and 4′ 9″ diameter wheels, and the *Class 9D* and *9E* had 5′ 1″ diameter wheels with the same size of cylinders but Stephenson's valve gear was fitted. Six *Class 9* (1889), six *Class 9D* (1893) and six *Class 9E* (1894–5) were built.

Parker also designed another 4-4-0 known as *Class 2* and *2A* with the usual 18″ x 26″ cylinders and 6′ 9″ diameter wheels. These had inside frames throughout and were based on No. 591 which had been supplied by Kitson & Co. in 1887. The boiler pressure was 160 lb/sq. in. The *Class 2* locomotives, of which twelve were built at Gorton, had round top fireboxes (1890–2) and six built in 1894 as *Class 2A* had Belpaire fireboxes, longer journals on the driving axles and helical springs for the driving wheels.

Parker retired at the end of 1893 having introduced larger locomotives which were badly needed.It was the transition period for frames and the single inside frame had been established by the time Parker had retired.

Mr Harry Pollitt was appointed from January 1894. He was a MS&LR man and also the son of Mr William Pollitt the General Manager of the company.

Parker's designs were adhered to at first and a new 4-4-0 appeared in 1895 purely Parker in design with 18½″ x 26″ cylinders and 7′ 0″ diameter wheels. The boiler pressure was increased to 170 lb/sq. in., Belpaire fireboxes were fitted and the total heating surface was 1318 sq. ft. Six were built (*Class 11*) in 1895 with slide valves, and Pollitt ordered no less than thirty-three *Class 11As* which were identical except that piston valves were fitted beneath the inside cylinders. Of these, thirteen were built at Gorton 1897–8. These locomotives were extensively used on the London extension when it was finally opened in March 1899.

Anticipating this, it was agreed by the Directors that from 1 August, 1897 the title of the railway would be GREAT CENTRAL RAILWAY.

Up to this time nearly 500 locomotives had been built at Gorton but this quantity was heavily supplemented by locomotives ordered from locomotive building firms, particularly Beyer Peacock & Co. who from the outset had maintained a very friendly and profitable association with their next door neighbours.

Pollitt continued to build Parker's 0-6-2Ts, all with Belpaire boilers, and in 1897 ten 0-6-0STs were built (*Class 5*) with outside cylinders 13″ x 20″ and 3′ 6″ diameter wheels.

They had single slide bars and the saddle tanks did not project over the smoke box. The boiler was 3′ 7″ diameter and the total heating surface 590 sq. ft with a grate area of 11 sq. ft. One was converted to a 0-6-2 Crane tank in 1905 but reverted to the standard type in 1918.

In January 1900 appeared Pollitt's famous batch of 4-2-2s, comprising six locomotives with inside cylinders 19½″ x 26″ and 7′ 9″

Fig. 365 **Gorton** 1906 Robinson *Class 8E* 4-4-2 as LNER No. 5365 SIR WILLIAM POLLITT

diameter driving wheels (*Class 13*). At first the boiler working pressure was 200 lb/sq. in. but this was soon reduced to 160 lb/sq. in. They had single frames throughout and the boiler centre line being 8′ 2″ above the rails made them appear large machines. They were numbered 967 to 972, and 971 was rebuilt with a much larger boiler 4′ 9″ diameter giving it a massive appearance. The standard 4′ 3″ diameter boiler was refitted in 1916. Piston valves were fitted to all six.

They were specially designed for express work on the new main line between Leicester and Marylebone and were Pollitt's 'Swan-song', and he resigned in June 1900.

John George Robinson was his successor and started in office in July, 1900. He was trained at Swindon on the Great Western Railway and had then gone to Ireland and at the time, just prior to his GCR appointment, was Locomotive, Carriage and Wagon Superintendent on the Waterford, Limerick and Western Railway.

In 1901 he commenced a series of 0-6-0s similar to Parker's *Class 9D* but with 18½″ diameter cylinders (*Class 9H*) followed by *Class 9J* popularly known as 'Pom-Poms' building continued until 1910 and a total of 174 were built, many by outside contractors. The boiler pressure was 180 lb/sq. in. with a total heating surface of 1426 sq. ft, the Belpaire firebox contributing 130 sq. ft. The grate area was 19·5 sq. ft. They became the mainstay of the heavy goods traffic throughout the system. The chimney was far more graceful than Parker's 'stove-pipes'; they were tapered with the smaller diameter at the top with a nicely moulded rim. The cab was much larger than hitherto.

For the heavy Manchester surburban duties Robinson brought out his *Class 9K*, 5′ 7″ 4-4-2Ts with inside cylinders 18″ x 26″. These were handsome locomotives, with Belpaire firebox and a domed boiler giving a total heating surface of 1143 sq. ft and a grate area of 19·25 sq. ft. The pressure was 160 lb/sq. in. The tank capacity was 1450 gallons. Twenty-eight were built at Gorton (1903–5) and another twelve by the Vulcan Foundry in 1903. A heavier version with larger tanks (*Class 9L*) carrying 1820 gallons was built in 1907 by Beyer Peacock & Co. Twelve were ordered, and were used at the southern end of the line.

Robinson's aim was to create standard classes to meet all the requirements of the company. He introduced the *Class 8A* 0-8-0 in 1902 and gave the first order for three locomotives to Neilson Reid & Co., followed by fifty-one to Kitson & Co. (1903–7), the remaining thirty-five being built at Gorton (1909–11). These were large powerful locomotives with outside cylinders 19″ x 26″ and wheels 4′ 7″ diameter and the 4′ 9½″ diameter boiler gave a total heating surface of 1764·96 sq. ft. The steam pressure was 180 lb/sq. in. and the total weight in working order with tender was 108 tons 5 cwt with 63 tons 15 cwt for adhesion. Fluted coupling and connecting rods were fitted as standard. The front splasher incorporated the sand box and a con-

Fig. 366 Gorton 1909 Robinson *Class 9J* 0-6-0 No. 16 superheated

Fig. 367 Gorton 1911 Robinson *Class 9N* 4-6-2T No. 170

tinuous splasher covered the two centre pair of wheels and the trailing wheels had a separate splasher. They were intended principally to work the heavily loaded coal trains in the Manchester & Sheffield areas.

For mixed traffic work the *Class 8* 4-6-0 was built, all by outside firms, with 6 ft driving wheels and 19″ x 26″ outside cylinders.

Regarding his proposals for express locomotives Robinson was undecided between 4-4-2s and 4-6-0s. He ordered four locomotives from Beyer Peacock two were 4-4-2s, capable of being converted to 4-6-0s if required – a parallel with the *Atlantics* and two-cylinder 4-6-0s on the GWR but in the case of GCR the *Atlantics* were not rebuilt as 4-6-0s.

All four locomotives had 6′ 9″ diameter coupled wheels, the same boiler with a total heating surface of 1911 sq. ft and grate area of 26 sq. ft and a working pressure of 180 lb/sq. in.

One gathers from the tests carried out that Robinson was more impressed with the *Atlantics*, as in 1904–6, seventeen were built by outside contractors and Gorton built eight. For the seventeen the pressure was increased to 200 lb/sq. in. The cylinders were 19½″ x 26″ and all were classed *8B*. He then turned his attention to compounds and built four *Atlantics*; No. 258 in 1905 and No. 259, 361 and 362 in 1906. They had one high pressure cylinder 19″ x 26″ and two low pressure 21″ x 26″. The HP cylinder had a 10″ piston and the LP cylinders slide valves. All bore names on a curved brass plate on the driving splasher. Three sets of Stephenson's valve gear were employed.

Experiments had been going on since 1909 on superheating. A Schmidt superheater had been fitted to one of the *Class 9J* 0-6-0s (No. 16). It was also fitted with piston valves and force-feed lubricator at the same time. Other locomotives were fitted with improved versions of superheater until Robinson brought out his own eighteen element type which was a modified Schmidt type with expanded elements into the header.

For the heavier London suburban traffic the first 4-6-2T to run on British metals was designed and the first two appeared in 1911. They were also the first GCR locomotives to be built new with superheaters, which had 145 sq. ft of heating surface out of a total of 1524 sq. ft. The grate area was 21 sq. ft and steam pressure 180 lb/sq. in. 10″ piston valves were fitted and were operated by Stephenson's valve gear. Inside cylinders were 20″ x 26″ and the coupled wheels 5′ 7″ diameter. They proved excellent machines, so much so, that after the initial twenty-one were built from 1911 to 1917, ten more appeared in 1923, and Hawthorn Leslie & Co built thirteen more in 1925–6 for the North Eastern area of the LNER.

Also in 1911 a new freight locomotive was designed in the form of a 2-8-0 (*Class 8K*) with two outside cylinders 21″ x 26″ and 4′ 7″ diameter coupled wheels, an obvious development of the *Class 8A* 0-8-0s. The boiler was in-

Fig. 368 Gorton 1912 Robinson *Class 1* 4-6-0 as LNER No. 5423 SIR SAM FAY

terchangeable with the *Atlantics*, and the coupled wheels and motion the same as for the 0-8-0s. The boiler was superheated with a total heating surface of 1691 sq. ft and a grate area of 26 sq. ft.

The splasher over the coupled wheels was continuous. It was a typical Robinson design 'simple', rugged with clean, well shaped lines. The success of this type was soon proven and it was adopted as a standard type for the War Department and hundreds were built during this period by many locomotive firms and, at the end of the 1914–18 war, were bought by a number of other British railway companies.

For goods, mixed traffic and passenger services, no less than nine classes of 4-6-0s were designed from 1903 to 1921. As previously mentioned, one class was built for comparative trials with the *Atlantics* (*Class 8C*).

In 1904 *Class 8* was built, with outside cylinders 19″ x 26″ and 6 ft diameter wheels. Fourteen were built for fish trains from Grimsby and other allied traffic. The boiler was interchangeable with the *Class 8A* 0-8-0. They were built by Beyer Peacock & Co., and Neilson Reid.

In 1906 *Class 8F* and *8G* was introduced, and ten of each class were built. Beyer Peacock & Co. built both classes, which had outside cylinders 19½″ x 26″. *Class 8F* had 6′ 7″ diameter, and *8G*, 5′ 3″ diameter coupled wheels. Standard parts were used where possible. In 1912–13 *Class 1* and *1A* with inside cylinders 21½″ x 26″, *Class 1* with 6′ 9″ diameter and *Class 1A* with 5′ 7″ diameter coupled wheels. The cylinders and motion were interchangeable. The splashers over the coupled wheels were continuous for both classes.

In 1917 and 1921 were classes *9P* and *9Q* respectively. These were four cylinder 16″ x 26″, *Class 9P* had 6′ 9″ diameter, and *Class 9Q* 5′ 8″ diameter coupled wheels; their cylinders and motion were interchangeable as also was the boiler. The cabs in most cases had two windows each side and an interesting feature was the extension of the piston rods towards the rear sliding in a bushed casting with the slide bars fixed well back between the rear bogie wheels and the leading coupled wheels. The object presumably was to shorten the connecting rods.

The last design to be dealt with was *Class 8N* of 1918 which had two outside cylinders 21″ x 26″ and 5′ 8″ diameter coupled wheels. The cylinders, motion and boiler were interchangeable with the *Class 8M* 2-8-0 introduced in 1911 which was a large boilered version of the prolific *Class 8K* 2-8-0.

A reference to the 4-6-0 table will illustrate the differences between these nine classes, but with Robinson's standardisation plans it is difficult to find reasons for such a medley of types and one wonders what would have been the outcome if the amalgamation had not happened when it did. Many of the 4-6-0s were named, some in complete classes, some without any relationship.

Robinson also built a number of 4-4-0s, and the *Director, Class 11E*, ten of which were built in 1913, were the best known and the most successful. Eleven more were built; one in 1919, four in 1920 and six in 1922. The last eleven had side window cabs, the boiler pitched 1½″ higher and were slightly heavier. They were probably the most successful inside cylinder 4-4-0s ever built and they competed easily with the *Atlantics*, were fast and economical with fuel. They were all superheated with a large boiler having a total heating surface of 1963 sq. ft and a grate area of 26 sq. ft. Their weight in working order was 109 tons 6 cwt with tender.

Once again their efficiency resulted in twenty-four being built after amalgamation by Kitson & Co. and Armstrong Whitworth & Co. for service in Scotland with cut-down boiler mountings. All had 6′ 9″ diameter coupled wheels and 20″ x 26″ cylinders.

In 1914 an unusual tank locomotive design was built at Gorton with a 2-6-4 wheel arrangement. It was intended for coal traffic from the Nottinghamshire coal fields to the docks at Grimsby and Immingham. They were *Class 1B* and had the same boiler as the *Director* class. Inside cylinders were 21″ x 26″ and coupled wheels 5′ 1″ diameter. A large railed coal bunker held 4½ tons of coal and the tanks 3000 gallons of water. A new feature was the top feed apparatus between the chimney and dome fitted as standard; a few 2-8-0s and 4-6-2Ts had been fitted previously. The principle was similar to Churchward's top feed apparatus on the GWR.

During the 1921 coal strike, in common with other railway companies, some locomotives were converted to oil burning as a temporary measure.

During Robinson's term of office many locomotives were rebuilt, boilers equipped with superheaters and at the time of the amalgamation the Great Central Railway had as fine a stock of locomotives as any in the country and credit was due to Robinson who ranks as one of the most capable of the locomotive engineers.

At the end of 1922 the Great Central Railway lost its identity and became part of the LONDON & NORTH EASTERN RAILWAY. Gorton had built 921 locomotives from 1858 to 1922 including three rail motor units and three ROD 2-8-0s which eventually arrived in Australia.

The output of new locomotives was moderate, many were bought from outside contractors, principally from their neighbour Beyer Peacock & Co. The works were enlarged only after the carriage and wagon department was moved to Dukinfield, the new workshops being completed in 1909–10.

The locomotive works had always been known as the 'Tank', the predominant feature of the works being a high level water storage tank. The erecting shop had been increased in size to a length of 480 ft with a width of 250 ft having five 50 ft bays which could take about 100 locomotives. New machine shops and stores were built in the late 1880s. Expansion was also limited by the fact that in this enclosed area was a twenty road locomotive running shed.

John Robinson was asked to consider the post of Chief Mechanical Engineer to the newly

Fig. 369
Gorton
1922 Robinson
Class 11F 4-4-0
as LNER No. 5503
SOMME

formed LNER – a title that was introduced by the Great Central in 1902. He refused on the grounds of age, and recommended H. N. Gresley for the post.

Work continued on building GCR designs and ten of the *Class 9N* 4-6-2Ts were built in 1923 followed by ten *Class 9Q*, 4-6-0s. This class was a mixed traffic version of the *Lord Faringdon Class 9P* with 5′ 7″ diameter coupled wheels and four cylinders, 16″ x 26″. The superheated boiler had a total heating surface of 2,382 sq. ft and a working pressure of 180 lb/sq. in. To increase their availability, they were fitted with cut down chimneys, domes and cabs.

Despite the fact that Gresley was building his own *N2* 0-6-2T design at Doncaster, the success of the Great Eastern 0-6-2Ts by A. J. Hill was such that building continued after the amalgamation, and Gorton built forty of this class (LNE *Class N7*) and many more came from outside contractors.

The Gorton order had Belpaire fireboxes, 18″ x 24″ cylinders and 4′ 10″ diameter coupled wheels, and were built from 1925 to 1928.

In 1932 two more 0-8-4Ts were built for Wath concentration yard. Four had been built in 1907–8 by Beyer Peacock & Co., and in 1932 one of these had been rebuilt with superheated boiler and a reversible booster fitted to the rear bogie. The wheels were coupled and this auxiliary engine increased the haulage power by 35 per cent. The two new 0-8-4Ts were also equipped with the booster bogie. The tanks were shortened to provide a better outlook for the footplate crew. They had three cylinders 18″ x 26″, 4′ 8″ diameter coupled wheels and 3′ 2″ diameter coupled wheels on the booster which gave a tractive effort of 46,896 lb making them the most powerful tank locomotives in the country and weighing in working order 104 tons 5 cwt.

In 1938–9 fourteen Gresley 0-6-0Ts to GN design (LNE *Class J50*) were built – they were known as 'Ardsley' tanks.

Almost 1000 steam locomotives had now been built at Gorton but orders were being concentrated more on Darlington and Doncaster.

On 1 January, 1948 nationalisation of the Railways, made the LNER the Eastern Region of BRITISH RAILWAYS.

Only ten more steam locomotives were built at Gorton 'Tank', and these of course were the final orders to be fulfilled. They were ten *Class B1* 4-6-0s of LNER design. Nos. 61340 and 61341 were built in 1948 and Nos. 61342–61349 in 1950.

Fig. 370 Gorton works in 1930s

MANCHESTER, SHEFFIELD & LINCOLNSHIRE RAILWAY

C. R. SACRÉ'S CLASSIFICATION 1859-1886 MS & LR

Class	Type	First Built	No Built	LNE Class	Cylinders inches	D.W. ft in	Remarks
1	2-4-0WT	1860	10		15½ x 20	5 6	DF
1B	2-4-0	1865	9		15 x 20	5 6	DF
1C	2-4-0	1864	6		15½ x 20	5 6	DF also as 'Class 9'
2	2-4-0	1861	3			5 6	Rebuilt at Sheffield from 2-2-2s
3	0-6-0	1861	10				Rebuilds of 2nd & 3rd Classes
4	2-2-2	1859	4		14 x 18		Built by W. Fairbairn
4	Misc	1876	13		–	–	Misc. 0-4-0ST & 0-6-0ST Ex Contractors
5	0-6-0	1858	4		17 x 24	5 0	DF
6A	0-6-0	1874	40		17 x 26	5 3	DF also as 'Class 6'
6B	4-4-0	1877	27	D12	17 x 26	6 3	DF
6C	0-6-0	1880	62	J12	17½ x 26	4 9	DF
7	0-6-0	1849	4		15 x 24	4 6	Ex South Yorkshire Rly
7	0-6-0T	1885	6	J64	15 x 20	3 9	
8	2-4-0	1862	7		16 x 20	5 6	DF Renewals of Old Sharps
10	2-4-0	1869	8		17 x 22	5 6	
11	0-4-2	1874	7				Built at Sheffield
12	2-4-0	1873	10		17 x 24	6 3	DF
12A	2-4-0	1875	28	E3	16 x 24	6 0	DF
12AT	2-4-0T	1880	8	E8	17 x 24	5 6	
13	0-6-0	1862	5		x 22	5 0	DF Renewals
14	0-6-0	1865	6		16½ x 24	5 3	DF with Steam Tenders
14	2-2-2	1882	12		17½ x 26	7 6	DF
15	2-4-0	1865	4		16 x 20	5 6	Beyer Peacock & Co.
16	2-4-0	1861	4				DF Rebuilt at Sheffield from 2-2-2s
18	0-6-0	1869	63		17 x 24	5 0	
18A	0-6-0	1871	7		17 x 24	5 0	Double Dome Boilers
18T	0-6-0ST	1871	41	J59	17 x 24	4 9	
19	2-4-0	1862	5		?	?	DF Rebuilt at Sheffield also as 'Class 20'
20	0-6-0	1849	7		16 x 24	5 0	Ex South Yorkshire Rly
22	0-6-0	1860	12			5 0	Rebuilds of RS & SS 1st Class
23	0-6-0	1859	99		16½ x 24	5 3	DF
24	2-4-0	1865	33		16 x 22	6 0	
25	2-4-0	1859	3		16 x 22	6 0	Rebuilt Fairbairn 2-2-2s

GENERAL DIMENSIONS FOR PASSENGER & MT LOCOMOTIVES GCR

Class	Type	First Built	No	Cylinders inches	D.W. ft in	Boiler - Heating Surface				Grate Area	WHG Press lb/sq in	Weight WO ton cwt	
						Tubes	F'box	S'htr	Total				
1	4-6-0	1912	6	IC 21½ x 26	6 9	2220	167	430	2817	26.3	180	76 0	(a)
1A	4-6-0	1913	11	IC 21½ x 26	5 7	2220	167	430	2817	26.3	180	75 0	(a)
8	4-6-0	1902	14	OC 19 x 26	6 0	1622	126	–	1748	23.6	180	65 0	
8C	4-6-0	1903	2	OC 19½ x 26	6 9	1778	133	–	1911	26.3	180	70 0	
8F	4-6-0	1906	10	OC 19½ x 26	6 7	1778	133	–	1911	26.3	180	71 0	
8G	4-6-0	1906	10	OC 19½ x 26	5 3	1778	133	–	1911	26.3	180	67 0	
8N	4-6-0	1918	3	OC 21 x 26	5 8	1640	174	308	2122	26.3	180	73 0	(c)
9P	4-6-0	1917	6	4/16 x 26	6 9	1881	163	344	2388	26.3	180	79 0	(b)
9Q	4-6-0	1921	28	4/16 x 26	5 8	1881	163	344	2388	26.3	180	80 0	(b)
8B	4-4-2	1903	27	19½ x 26	6 9	1778	154	–	1932	26.3	180	72 0	(d)
8D	4-4-2	1905	4	19/21 x 26	6 9	1778	154	–	1932	26.3	200	71 0	(d)
8J	4-4-2	1908	1	3/15½ x 26	6 9	1778	154	–	1932	26.3	180		(e)
11A	4-4-0	1898	33	IC 18½ x 26	7 0			–	1192	20.0	160		
11B	4-4-0	1901	40	IC 18½ x 26	6 9	1250	130	–	1380	21.0	180	53 0	
11E	4-4-0	1913	10	IC 20 x 26	6 9	1502	157	303	1962	26.5	180	61 0	
11F	4-4-0	1919	11	IC 20 x 26	6 9	1388	155	209	1752	26.5	180	61 0	

(a) Cylinders, Motion and Boiler Interchangeable. (b) Cylinders, Motion and Boiler Interchangeable
(c) Cylinders, Motion and Boiler interchangeable with 2-8-0 Class 8M (d) Boiler same as 2-8-0 Class 8K
(e) Rebuilt from Class 8B

MANCHESTER, SHEFFIELD & LINCOLNSHIRE RAILWAY

LATER LOCOMOTIVE CLASSIFICATION

Class	Type	First Built	No Built	LNE Class	Cylinders inches	D.W. ft in	
1	4-6-0	1912	6	B2	21½ x 26	6 9	Later LNE Class B19
1A	4-6-0	1913	11	B8	21½ x 26	5 7	
1B	2-6-4T	1914	20	L1	21 x 26	5 1	Later LNE Class L3
2	4-4-0	1890	25	—	18 x 26	6 9	Rebuilt to Class 2A
2A	4-4-0	1894	6	D7	18 x 26	6 9	Belpaire Fireboxes
3	2-4-2T	1889	39	F1/1&2	18 x 24	5 7	
3 ALT	2-4-2T			F1/3&4	18 x 24	5 7	Larger Tanks
5	0-6-0ST	1897	12	J62	13 x 20	3 6	
5A	0-6-0T	1906	7	J63	13 x 20	3 6	
6A1	0-6-0	1887	12	J8	18 x 26	5 3	DF
6D	2-4-0	1887	3	E2	18 x 26	6 9	DF
6DB	4-4-0	1888	3	D8	18 x 26	6 9	DF
8	4-6-0	1904	14	B5	19 x 26	6 0	
8A	0-8-0	1903	89	Q4	19 x 26	4 7	
8B	4-4-2	1903	26	C4	19½ x 26	6 9	
8C	4-6-0	1903	2	B1	19½ x 26	6 9	Later LNE Class B18
8D	4-4-2	1905	2	C5	19/21 x 26	6 9	From 8B before completion
8E	4-4-2	1906	2	C5	19/21 x 26	6 9	Compound
8F	4-6-0	1906	10	B4	19½ x 26	6 6	
8G	4-6-0	1906	10	B9	19½ x 26	5 4	
8H	0-8-4T	1907	4	S1	18 x 26	4 8	
8J	4-4-2	1908	1	—	3/15½ x 26	6 9	Rebuilt to Class 8B
8K	2-8-0	1911	129	04	21 x 26	4 8	5′ Boiler
8L	2-8-0	1918	1		21 x 26	4 8	6′ Boiler
8M	2-8-0	1918	18	05	21 x 26	4 8	5′ 6″ Boiler
8N	4-6-0	1918	3	B6	21 x 26	5 8	
9	0-6-0	1889	6	J13	18 x 26	4 9	
9A	0-6-2T	1889	41	N4/1&2	18 x 26	5 1	
9A ALT	0-6-2T	1892	14	N4/3&4	18 x 26	5 1	Larger Tanks
9B	0-6-0	1891	25	J9/1&2	18 x 26	5 1	
9C	0-6-2T	1891	3	N5/1&2	18½ x 26	5 1	
9D	0-6-0	1893	18	J10/1&2	18 x 26	5 1	
9E	0-6-0	1894	6	J9/3&4	18 x 26	5 1	
9F	0-6-2T	1893	126	N5/1&2	18 x 26	5 1	
9G	2-4-2T	1898	10	F2	18 x 26	5 7	
9H	0-6-0	1896	105	J10/3-6	18 x 26	5 1	
9J	0-6-0	1901	174	J11	18½ x 26	5 1	
9K	4-4-2T	1903	40	C13	18 x 26	5 7	
9L	4-4-2T	1907	12	C14	18 x 26	5 7	
9M	0-6-0	1908	1	—	18 x 26	5 1	9H with 9J Boiler. Rebuilt to 9H
9N	4-6-2T	1911	21	A5	20 x 26	5 7	
9O	0-6-2T	1915	1	N5/3	18 x 26	5 1	Longer Tanks
9P	4-6-0	1917	6	B3	4/16 x 26	6 9	
9Q	4-6-0	1921	28	B7	4/16 x 26	5 8	
10	2-6-0	1900	20	—	18 x 24	5 0	Or Class 15?
11	4-4-0	1895	6	D5	18½ x 26	7 0	
11A	4-4-0	1897	33	D6	18½ x 26	7 0	Piston Valves
11B	4-4-0	1901		—	18½ x 26	6 9	REB of 11 D 4′ 9″ Boiler 8′ Firebox
11C	4-4-0	1907	40	D9	18½ x 26	6 9	REB of 11D 5′ Boiler 8′ 6″ Firebox
11D	4-4-0	**			18½ x 26	6 9	REB of 11D 5′ Boiler 7′ Firebox
11E	4-4-0	1913	10	D10	20 x 26	6 9	
11F	4-4-0	1919	11	D11	20 x 26	6 9	
12AM	2-4-0T		6	E8	17 x 24	5 6	Class 12AT Motor Fitted
13	4-2-2	1900	6	X4	19½ x 26	7 9	
18 CONV	0-6-0ST	1902	14	J58	17 x 24	5 3	Rebuilt from Class 18
18 TALT	0-6-0ST		26	J59	17 x 24	4 9	Class 18T with Shorter Tanks
A	0-6-2T	1895	18	N6	18 x 26	4 9	Ex LD & ECR* Retained Same Class
B	0-6-0T	1897	4	J60	17 x 26	4 3	Ex LD & ECR* Retained Same Class
C	0-4-4T	1897	6	G3	17 x 24	5 6	Ex LD & ECR* Retained Same Class
D	0-6-4T	1904	9	M1	19 x 26	4 9	Ex LD & ECR* Retained Same Class

Unclassified : Rail Motors
: Ex Wrexham, Mold & Connah's Quay Rly

* Lancashire, Derbyshire & East Coast Rly
** Rebuild of Class 11B

Electric locomotives were built after this time and repairs to steam locomotives continued for a while.

The total of steam locomotives built from 1858 to 1948 was 1006.

Early Locomotive Classification Introduced by Mr W. G. Craig in 1858

1st Class	0-6-0	1852-4	18" x 24" cylinders 4' 4" and some 5' dia wheels RS & SB (20 locos)
2nd Class	0-6-0	1849	18" x 24" cylinders 5' 0" dia wheels built by SB (6 locos)
3rd Class	0-6-0	1846-8	18" x 24" cylinders 4' 6" dia wheels built by SB (5 locos)
4th Class	0-4-2 2-4-0	1841-50	Various dimensions various makes (41 locos)

RS: Robert Stephenson & Co.
SB: Sharp Bros. & Co.

Whether these classes signified age, condition or hauling power or other characteristics is not clear, but no serious steps were taken to provide a classification to identify different types. Some locomotives of the period such as the Crampton 2-2-2s, various 2-2-2, 2-4-0, and 2-2-2WT types were not classified.

MANCHESTER, SHEFFIELD & LINCOLNSHIRE RAILWAY

Sheffield, (Neepsend)

The locomotive running sheds at Sheffield were equipped to carry out major repairs but from 1874 to 1880 twelve locomotives were actually built in these sheds. Obviously all the necessary equipment for building new locomotives would not have been installed here and therefore boilers, frames, wheels and cylinders were either off earlier built types or they were supplied new by Gorton or locomotive building contractors.

Three classes were built as follows:

Seven 0-4-2 *Class 11* which almost certainly contained usable parts from earlier 0-4-2s built 1848–9. No dimensions of the 'new' *Class 11* are available but the older 0-4-2s had 5′ 0″ diameter driving wheels, 3′ 6″ diameter trailing wheels and cylinders varying in diameter from 15½″ to 16″ with a stroke of 22″. The wheels would be turned down or re-tyred and the cylinders bushed or bored out. The *Class 11* had inside cylinders and frames, domed boilers with round top fireboxes and were fitted with six wheeled tenders which would be off other locomotives. Most of them ran to the end of the century.

The second type was *Class 24* which was a double frame 2-4-0 with inside cylinders 16″ x 22″ and coupled wheels 6′ 0″ diameter. They were neat little locomotives but only two were built at Sheffield in 1877–8 but Gorton had started building them in 1866 and completed twenty-four by 1877 and the two from Sheffield completed the order. They had similar type boilers to the *Class 11s* and the cab had small oval side windows.

Fig. 371 Gorton 1923 4-cylinder *Class 9Q* 4-6-0 No. 5480–note slide bars

The final type was a double frame 0-6-0 *Class 6A* with inside cylinders 17″ x 26″ and 5′ 3″ diameter wheels. Three were built in 1879–80. The frames were slotted, boilers had round top fireboxes and were domed.

'Building' ceased after these 0-6-0s were built and Gorton took over the building of all new work, except when requirements necessitated obtaining them from outside contractors.

Throughout the period of building at Sheffield, Charles Reboul Sacré was in charge and was known at that time as Engineer and Superintendent of Locomotive and Stores Dept, with headquarters at Gorton.

In the case of classes *11* and *6A* the safety valves were enclosed in a tall brass cover which may point to the old boilers being used, as at the time of building, Gorton's new locomotives were fitted with Ramsbottom safety valves with a short base cover. The two *Class 24s* may have had new boilers as they were fitted with safety valves according to the current Gorton practice.

MANLOVE ALLIOTT & FRYER

Bloomsgrove Works, Ilkeston Road, Nottingham

Edward Perrett designed a combined steam car or railcar in 1876 comprising a tram type body with two vertical boilers, one at each end. They were 2′ 1½″ diameter and 6 ft high. A horizontal engine was fixed under the floor and supplied with steam from both boilers. The car which was double decked was mounted on eight wheels with the two centre axles coupled and a Bissel truck at each end. The driving wheels were 2′ 6″ diameter and the leading and trailing wheels 1′ 6″. It was designed in 1876 and was tried out on the Nottingham tramways.

This may have been the same car which appeared on the Dublin and Lucan Tramway in 1881, with cylinders 7″ x 9″ and a total wheelbase of 17′ 6″, and a weight of 9 tons.

More than one are reported to have been built, including one for Burnley and District Tramways Co. Ltd. in 1882 but this has not been confirmed. Only the Dublin car is mentioned in H. A. Whitcombe's *History of the Steam Tram*, and its life was very short, proving uneconomical due to prolonged standing time.

MSLR SHEFFIELD

Date	Type	MSL No	Class	Cylinders inches	D.W. ft in		W'drawn
4/1874	0-4-2	400	11				*c.* 1888
7/1875	0-4-2	401	11				
11/1875	0-4-2	403	11				1891
12/1875	0-4-2	402	11				
8/1876	0-4-2	404	11				1889
1/1877	0-4-2	405	11				
6/1877	0-4-2	406	11				1901
8/1877	2-4-0	409	24	16 x 22	6 0	Reno 214/1900	1904
6/1878	2-4-0	410	24	16 x 22	6 0	Reno 215/1900 215B	1905
5/1879	0-6-0	26	6A	17 x 26	5 3	Reno 26B	1913
9/1879	0-6-0	102	6A	17 x 26	5 3		1914
9/1880	0-6-0	69	6A	17 x 26	5 3	Reno 69B	1915

Fig. 372 **MS&LR** Sheffield 1880 *Class 6A* 0-6-0

MANNING WARDLE & CO.
Boyne Engine Works, Leeds

This firm was established in 1858 by Alexander Campbell and C. W. Wardle. Alexander Campbell had been Works Manager of the locomotive building department of Scott Sinclair & Co. of Greenock and had come south in 1856 to manage the affairs of E. B. Wilson & Co. at the Railway Foundry on the death of Mr Wilson, and when the firm closed down he set about establishing his own works with the financial assistance of Mr Wardle – a local vicar. More capital being required, Mr John Manning became a partner. Drawings and patterns from E. B. Wilson & Co. were obtained and the earlier locomotives of Manning Wardle & Co. bore the characteristics of Wilson's designs.

Manning Wardle & Co. concentrated on building contractors and industrial tank locomotives and built up an excellent reputation in this field. In proportion very few tender locomotives were built after the 1870s.

Their first locomotive was, significantly, a small 3 ft gauge 0-4-0ST for Dunston & Barlow Ltd., Sheepbridge followed by two 5′ 6″ gauge 2-4-0WTs for the Royal Portuguese Railway, all in 1859 and in the same year two 0-6-0Ts were built and two outside cylinder 0-4-0s for the Earl of Dudley. In 1860 they built their only examples of 2-2-2 passenger, for the New South Wales Government Railway. Four were built with cylinders ranging from 15″ x 20″ to 15″ x 22″ and driving wheels, 5′ 6″ to 5′ 9″ diameter and two 4-4-0s for the same railway were built in 1862. Many contractors working at home and abroad were furnished with 0-4-0STs and 0-6-0STs and these types predominated throughout the firm's existence. Their standard four-coupled saddle tanks had outside cylinders and the six-coupled classes had inside cylinders. At an early stage they appreciated the advantages of standardisation and standard classes were established for four-coupled and six-coupled saddle tanks.

These contractors locomotives had straight sided saddle tanks and were equipped with a water feed pump and one injector. A high crown round top firebox was generally used in which the steam collector was fitted. The safety valve cover was usually elongated to clear the cab of

Fig. 373 **Manning Wardle** 6/1859 Pensnett Railway BRANDON at Kingswinford Junction

Fig. 374 **Manning Wardle** typical early 0-6-0ST CONEYGRE

steam; the cabs were of course in early days very primitive being either a straight weatherboard, or bent over slightly towards the rear. Later a bent all-over cab in two pieces was used. Wooden brake blocks were fitted and the wheels had crank pin bosses cast on opposite sides, the unused one acting as a crude balance weight. In the case of the six-coupled tanks compensating beams linked the leading an driving springs with the intention of making the locomotives ride better over the uneven temporary tracks on which they were employed. This was an advantage at low speed but when rapid progress was attempted the locomotive became very unstable with yawing and pitching movements. The compensating beams were later left off.

By the end of 1880, 311 0-4-0STs and 298 0-6-0STs had been built out of a total of 760 locomotives built.

Other interesting types built during this period were nine 4-4-0STs with outside cylinders 12″ x 18″ and 4′ 6″ diameter coupled wheels for the Buenos Ayres Western Railway in 1869–70.

Three small 0-4-0Ts were built on the 'Fell' system for the Leopoldina Railway of Brazil. John Barraclough Fell had designed a locomotive suitable for steep gradients incorporating horizontal friction wheels gripping a central double-headed rail by spring pressure and driven by an additional set of cylinders on the locomotive, the normal outside cylinders being used on the easier sections of the railway.

For the 2′ 6″ gauge Pentewan Railway in Cornwall a six-coupled tender locomotive was built in 1873. Both locomotive and tender had outside frames, and outside cylinders 7″ x 12″ were fitted. The boiler was domeless, with a raised firebox and gave a total heating surface

Fig. 375 **Manning Wardle** 377/1872 Fell's patent central rail locomotives

Fig. 376 **Manning Wardle** 2047/1926 last to be built. Rugby Portland Cement Co. Ltd.

MANNING WARDLE DIMENSIONS OF STANDARD CLASSES

Class	Type	Drawing No.	Drawing Date	Cylinders	D.W.	W'base	Overhang LDG	Overhang TLG	(a) Buffer Beam	Boiler Dia.	Boiler L	Heating Surface sq. ft. Tubes	Heating Surface sq. ft. F'box	Heating Surface sq. ft. Total	Grate Area sq. ft.	(b) c. line from rail	Tank gallons	Weight Empty T–C	Weight W. O. T–C	
B	0-4-0ST	2	22.6.63	6 X 12	2' 3	4' 6	4' 0	4' 0	4½	1' 11	6' 9	104	21	125	3.0	3' 6	200			
C	0-4-0ST	2	22.6.63	7 X 12	2' 3	4' 6	4' 0	4' 0	4½	1' 11	6' 9	104	21	125	3.0	3' 6	200			
B	0-4-0ST	2	19.11.77	6 X 12	2' 6	4' 7	4' 0	4' 5	4½	1' 11	6' 9	82	18	100	2.9	4' 0	200			
C	0-4-0ST	2	19.11.77	7 X 12	2' 6	4' 7	4' 0	4' 5	4½	1' 11	6' 9	104	21	125	3.0	4' 0	200	7–5	8–10	
D	0-4-0ST	3	8.3.66	8 X 14	2' 8	4' 6	4' 4	4' 9	4½											
D	0-4-0ST	3		8 X 14	2' 8	4' 6	4' 4	4' 9	4½	2' 1	7' 0	153.5	23.5	177	3.5	3' 11½	200	8–13	9–18	
E	0-4-0ST	4	30.6.79	9 X 14	2' 9	4' 9	4' 6	5' 3	5⅛	2' 3	7' 3	217.5	30.5	248	5.0	3' 9½	250	10–3	12–5	(c)
F	0-4-0ST	5	4.6.69	10 X 16	3' 0	4' 9	5' 3	5' 2	5⅛	2' 7	7' 11	267.5	32.5	300	5.5	4' 7½	350	12–8½	15–2½	
F	0-4-0ST	5	27.11.78	10 X 16	2' 9	4' 9	5' 3	5' 2	5⅛	2' 7	7' 11	267.5	32.5	300	5.5	4' 6	350			
G	0-4-0ST	6	5.75	11 X 17	3' 0	5' 0	5' 9	5' 3	5⅝	2' 8	8' 3	315	35	350	5.5	4' 9	400			(d)
H	0-4-0ST	7	6.6.67	12 X 18	3' 0	5' 3	5' 6	5' 6	5⅝	2' 9	8' 3	335	39	374	6.5	4' 11	450	15–15½	19–10	
H	0-4-0ST	7	27.7.77	12 X 18	3' 0	5' 4	5' 6	5' 5	5 9/16	2' 9	8' 3	335	39	374	6.5	4' 11	450			
H	0-4-0ST	7	3.93	12 X 18	3' 0	5' 4	5' 6	5' 5	5 9/16	2' 9	8' 3	329	39	368	6.36	4' 11	450			
I	0-4-0ST	8	19.9.77	13 X 18	3' 0	5' 6	6' 3	5' 9	6⅛	2' 10	9' 4	428.25	43.5	471.75	6.5	5' 0	500			
I	0-6-0ST	8A	17.11.70	11 X 17	3' 1⅜	5' 5 + 4' 10	3' 7	3' 9	5½	2' 9	7' 3	305	40	345		4' 4½	420			
K	0-6-0ST	9	9.71	12 X 17	3' 1⅜	5' 5 + 5' 4	3' 7	3' 9	5 9/16	2' 9	7' 9	330	40	370	6.75	4' 4¾	450	15–0	18–10	
K	0-6-0ST	9	17.11.76	12 X 17	3' 1⅜	5' 5 + 5' 4	3' 7	3' 9	5 9/16	2' 9	7' 9	326	40	366	7.0	4' 4¾	450			(e)
L	0-6-0ST	10	23.6.79	13 X 18	3' 0	5' 9 + 5' 6	4' 0	4' 0	6⅛	3' 2	8' 2	418	47	465	7.25	5' 3	475	15–18¼	19–8½	
L	0-6-0ST	113	20.3.88	12 X 18	3' 0	5' 5 + 5' 4	4' 0	4' 0		3' 1	7' 9	402	46	448	7.0	5' 0½	450			
M	0-6-0ST	11	16.10.77	12 X 18	3' 0	5' 10 + 5' 8	4' 0	4' 3	6⅛	3' 4	8' 4	500	50	550	8.0	5' 4	550	17–12	22–10	
M	0-6-0ST	11	3.3.88?	12 X 18	3' 0	5' 10 + 5' 8	4' 0	4' 3	6⅛	3' 4	8' 4	491	51	542	8.0	5' 4	550			(f)
O	0-6-0ST	12	15.1.78	15 X 22	3' 9	5' 9 + 6' 3	4' 6	4' 7	6¾	3' 10	8' 6	692	70	752	10.0	6' 0½	700	23–18	29–13	
P	0-4-0ST	13	25.7.18?	14 X 18	3' 0	5' 6	6' 3	5' 9	6¼	3' 1	9' 4	503	47	550	7.0	5' 0	550			
Q	0-6-0ST	14	30.5.78	14 X 20		5' 11 + 6' 1	4' 4	4' 6	6⅛	3' 6	8' 8	600	60	660	8.8	5' 10½	600			
Q	0-6-0ST	14	24.7.91	14 X 20		5' 11 + 6' 1	4' 4	4' 6	6⅛	3' 6	8' 8	600	60	660	8.8	5' 10½	600			

0-4-0ST all outside cylinders   0-6-0ST all inside cylinders   Overhang excludes thickness of buffer beams

(a) Thickness of Buffer Beams (Wood with Flitch plates)
(b) Boiler centre line to rail level
(c) Includes works no. 60 (1862 exhibition)
(d) Although there is a G. A. DWG for Class G it was not a standard class and probably was to have been the basis for some 4' 4" gauge 0-4-0STs for the Pernambuco Steam Tramway with 11 x 17 outside cylinders. Any further 11 x 17 0-4-0STs were referred to as 'specials'
(e) 3' 0" diameter wheels after works no. 1000.
(f) Class M also built with 3' 6" diameter wheels and also as side tank version

Above applicable to locos built in 1860s up to c. 1910.

of 126·5 sq. ft. The wheels were 20″ diameter. The chimney and safety valve cover were very high and the whole locomotive had a very bizarre appearance.

A four-coupled crane tank was built in 1875 for the Kirkstall Forge Co., the crane parts being supplied by Jos. Booth & Bros, Rodley.

Regarding tram locomotives, Manning Wardle & Co. were the pioneers of the 'dummy' tram loco, where the unit was independent of the passenger cars.

In 1866 two 0-4-0STs, with suitable bodywork surrounding them, were built for the 4 ft gauge Pernambuco Tramway Co. in Brazil and a total of eight were built by 1870. The exhaust steam was passed back into the saddle tank.

In 1870 three tramway units were built for the Buenos Ayres tramways. At each end of the locomotive an articulated coach was attached.

When tram locomotives became popular in the 1880s the company supplied very few compared with other builders. Ten were supplied to the North Staffordshire tramways from 1880 to 1882. The first two had 8½″ x 14″ cylinders and the remainder 9″ x 14″, and water cooled condensers were fitted to the latter. Three were supplied to the standard gauge lines of the Manchester, Bury, Rochdale and Oldham Steam Tramway Co. in 1883 with 8½″ x 14″ cylinders. Standard practice was departed from in the last eleven tram locomotives as they were all fitted with inside cylinders.

In 1882 two 0-4-0STs were built for the Secretary of State for War and were used by Major General Gordon in his Sudan expedition. These were claimed to be the first armoured locomotives put into service.

In 1886 a second 0-6-0 named TREWITHEN, supplied to the Pentewan Railway, was very similar to the 1873 locomotive but the cylinders were 7½″ x 12″ and the boiler slightly larger.

In 1891–2 one 2-6-2T and one 2-6-4T were built for the Metre gauge of the Malta Railway. They both had outside cylinders 15″ x 20″, but types other than the standard saddle tanks were rarely built. Three 1′ 11½″ gauge 2-6-2Ts were built in 1897 for the newly built Lynton and Barnstaple Railway which was opened the following year. They were named YEO, EXE and TAW. They had outside frames and outside cylinders 10½″ x 16″ and coupled wheels 2′ 9¾″ and were not unlike the two Vale of Rheidol locomotives built by Davies & Metcalfe in 1903. They were very successful and a fourth was built in 1925.

By 1900 over 1500 locomotives had been built; the demand for these small tank locomotives did not seem to decrease.

Their dimensions inevitably increased with the consequent heavier loading required and some improvements to their design occurred from time to time.

In 1906 they entered the Rail Motor field and six 0-2-2 units were built for the Taff Vale Railway. The 10½″ x 14″ cylinders were outside, between the 3′ 6″ diameter driving wheels and the trailing wheels. Four similar units were supplied to the Great Northern Railway of Ireland in 1907 and finally two for the Dublin, Wicklow & Wexford Railway.

Seven large 0-6-2Ts were supplied to the Taff Vale Railway in 1907 with inside cylinders 17½″ x 26″ and 4′ 6½″ coupled wheels.

Additional Crane tanks were built in 1913 and 1915, also 2-4-0Ts for South America and Calcutta, a 2-6-0T for the Knott End & Garstang Railway, 4-4-2T for New Zealand, two 0-6-2Ts for Iquique and four 0-4-2Ts for the Great North of Scotland Railway in 1914–15.

During the 1914–18 war, locomotive production was reduced to supplying steelworks, collieries, etc. with locomotives, although quite a number were sent abroad to Spain, New Zealand, Australia, India and S. America. Ten M of M 0-6-0STs with outside cylinders 17″ x 24″ were supplied to the Inland Waterways and Docks department.

After the war Manning Wardle & Co., like everyone else suffered from the trade depression. In addition, although their locomotive design kept abreast of the times, their manufacturing methods did not and as a result the competition for contracts, with keen prices and deliveries required, brought the firm to a halt. In 1927 the firm went into voluntary liquidation. The goodwill of the company was bought by Kitson & Co., a near neighbour; and the Hunslet Engine Co. – its friendly rival – acquired part of the Boyne Engine Works.

It was an unfortunate end to a firm which had earned a reputation for excellent locomotives which could be found in all quarters of the globe.

The last works number was 2047 of 1926. Some numbers were blanks, others taken by petrol engines, and air engines so that the number of steam locomotives built from 1859 to 1926 was 2004.

MARKHAM & CO. LTD.

Broad Oaks Works, Chesterfield

See OLIVER & CO. LTD.

MARSDEN J. W.

Union Foundry

Quoted Festiniog Railway on 29 October, 1862 for three locomotives.

MARSHALL FLEMING & JACK
Dellburn Works, Motherwell

This works was established about 1890 principally for the construction of cranes of all types. In addition to the orthodox travelling railway cranes usually propelled by gearing on to one axle, this firm built a number of 0-4-0CTs with outside cylinders and vertical boilers. These were built from 1896 but the number built and customers are not known. In 1907 Mr Jack resigned and the firm became Marshall Fleming, and subsequently Marshall Fleming & Co. Ltd.

MARSHALL, SONS & CO. LTD.
Britannia Ironworks, Gainsborough

Commenced business as blacksmiths in 1848. In the early 1850s the firm started building portable engines and styled themselves as 'Millwrights and Engineers'. They were builders of traction engines and, like other builders of such vehicles, designed locomotives incorporating the layout of the road vehicles with flanged wheels for rail operation.

Only three were built:

1878	No. 6402	0-4-0TG for Pepper & Son Ltd 2 cylinders 7" x 10"
1898	?	? for ?
1902	No. 36741	4 WTG for own use.

They purchased the goodwill of Clayton Wagons Ltd. and in 1930 acquired the business of Clayton & Shuttleworth Ltd., Stamp End Works, Lincoln.

MARSHALL, WILLIAM
Gravesend

This firm, according to some sources, supplied locomotives to the London and Greenwich Railway. According to Mr D. L. Bradley's book *The Locomotives of the South Eastern Railway* (RCTS) access to the L&GR record books produced the following information:

In May 1834 a 15 HP engine was ordered from William Marshall, and one more in July, whilst in the same year two were ordered from Bury who sub-contracted to Forrester & Co. Two more were ordered from R. Stephenson. The Forrester pair were delivered but not the two from R. Stephenson. Marshall ultimately supplied four locomotives all of the *Planet* type but not identical: ROYAL WILLIAM, ROYAL ADELAIDE, DOTTIN, and TWELLS.

In 1838 they were numbered 1 to 4 and 1 to 3 became South Eastern Railway No. 127 to 129. TWELLS was sold to the Admiralty and used as an auxiliary engine in *HMS EREBUS*.

From Mr N. Wakeman's *The SECR Locomotive List 1842–1952* (Oakwood Press) SER Nos. 127 to 129 are given as ex L&GR locomotives built by Marshall in 1845.

In the *Railway Magazine* for August 1907 (Vol. 21) a series of articles on 'Early Locomotives of the SER' stated 'possessed nine engines the first six having names and Nos 1 to 3 were replacement engines built by Marshall in 1845. The original locomotives Nos 1 to 4 were given as built by C. Tayleur in 1836 but the works lists show only one supplied to the L&GR. (Works No. 25 of 1836).'

Fig. 377 Marshall Sons & Co. 6402/1878 Pepper & Son Ltd, Amberley, Sussex

A further source Mr A. R. Bennett's *First Railway in London* (Loco, Pub. Co.) shows No. 1 from Tayleur, Nos 2 to 4 from Braithwaite & Milner of London; No. 2 was named ROYAL ADELAIDE and No. 3 PRINCESS VICTORIA. Mr Bennett also stated that Marshall took over Braithwaites in 1840.

It seems on the weight of evidence that Marshall built at least two or probably three either in the 1830s or as replacements in the 1840s.

Thanks are due to Mr D. L. Bradley and Mr G. Woodcock for some of the information in this section.

MARTYN BROS

Chapelside Works, Airdrie

Built tank locomotives for collieries. The firm acquired Dick & Stevenson's patterns and drawings in 1890 and in the next ten years built at least two and possibly six from these designs. One is known to have gone to Devon Colliery, Alloa.

MARYPORT & CARLISLE RAILWAY

Maryport

Locomotive Superintendents: ? –1854 Mr Scott; 1854–70 George Tosh; 1870–8 Hugh Smellie; 1878–93 Robert Campbell; 1893–8 William Robertson; 1898–1904 William Coulthard; 1904–22 J. B. Adamson.

This railway, opened in 1840, was one of the oldest but smallest of English railways. The main line some 28 miles in length was from Maryport to Carlisle through Aspatria and Wigton with a branch to Brigham linking up with the L&NWR and the Cockermouth, Keswick and Penrith Railway.

Up to 1857 the railway workshops at Maryport were used for repairs and rebuilding. The stock of locomotives up to that year was of infinite variety and included a 4-2-0 Crampton built by Tulk and Ley at Whitehaven who also provided half the total stock of twelve.

M&C No. 5 was the first one built in the workshops and was a 2-2-2 with 14″ x 22″ cylinders and 6 ft driving wheels. It was the only 'single' built by the company although two had been previously supplied by Tulk & Ley, in addition to the Crampton

George Tosh was a progressive man and introduced coal firing at an early date. During his term the company's locomotives were fitted with steel tyres and, equally as important, the first steel boiler was fitted to a M&C locomotive in 1862, a year before the L&NWR fitted their first one.

Steel rails were also introduced as an experiment in 1867. The iron and steel works in the neighbourhood no doubt had the co-operation of the engineer in these experiments.

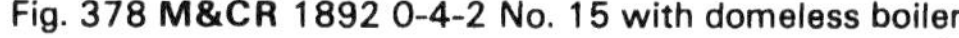

Fig. 378 **M&CR** 1892 0-4-2 No. 15 with domeless boiler

MARYPORT & CARLISLE RAILWAY

Most of the locomotives built by Tosh had domeless boilers with the safety valve mounted over the firebox. The wheel splashers had radial slots cut in them which was fashionable in the north country and Scotland.

Hugh Smellie took over in 1870, coming from Kilmarnock where he had been Works Manager under the Stirling brothers. One would have thought that being steeped in the domeless boiler tradition Smellie would have carried on in the same way, but this did not happen. The only feature he used was the Stirling rounded cab, but the boilers were domed.

Yet when he returned to Kilmarnock in 1878 the domeless boiler, Stirling's steam reversing gear and other traditional features were all incorporated in his Glasgow and South Western Railway designs. To make the situation more interesting still, Smellie's successor Robert Campbell promptly introduced the domeless boilers and new locomotives and rebuilds were thus equipped.

Hugh Smellie built various classes of 0-4-2, 2-4-0 and 0-6-0 wheel arrangements, all inside frames and cylinders.

The three 2-4-0s Nos 13, 8, and 10 were larger than the earlier ones, and had 17" x 24" cylinders with a boiler pressure of 135 lb/sq. in. and were used for the main line passenger trains. The 0-4-2s were built for mixed traffic duties but were mainly on passenger trains.

Two 0-6-0 goods tender locomotives built in 1875 (No. 9) and 1877 (No. 14) had inside cylinders 18" bore with an unusual stroke of 28". Few locomotives had 14" cranks in those days.

Smellie returned to Kilmarnock in 1878 and Robert Campbell filled the vacancy and was in charge until 1893. In 1897 William Robertson introduced a new wheel arrangement in the form of a 0-4-4T, one only being built for working the branch line, or the Derwent branch as it was called.

In 1900 the last locomotive was built at Maryport. It was a 0-6-0 similar to those built in

P No	Date	M & C No	Type	Cylinders inches	D.W. ft in			LMS No	W'drawn
1	1857	5	2-2-2	14 x 22	6 0				1872
2	1859	9	0-4-0	15 x 22					1875
3	1860	6	0-4-2	15 x 22	5 6				1872
4	1862	8	0-4-2	15 x 22	6 0				
5	1862	10	0-4-2	15 x 22	5 6	Reno R2/1878			1897
6	1864	4	0-4-2ST	15 x 22	4 9		Sold		1878
7	1865	7	0-4-2ST	14½ x 22	4 6	Reno R3/1882			1892
8	1865	16	0-6-0	16 x 22	4 6				1895
9	1865	17	0-4-2T	15 x 22	4 9	Rebuilt 0-6-0T		11563	1927
10	1866	1	0-6-0	15¾ x 24	5 0	Reno R4/1900		12077	1924
11	1867	18	0-4-2T	15 x 22	5 2	Reno R5/1908			1923
12	1867	19	2-4-0	15 x 22	6 0	Reno R1/1884			1921
13	1868	11	0-6-0	15¾ x 24	5 0	Reno R3/1881			1882
14	1869	2	0-6-0	15¾ x 24	5 0				1886
15	1870	12	0-6-0	17 x 24	5 1			12078	1924
16	1871	6	0-6-0	17 x 24	5 1½			12079	1928
17	1871	20	0-6-0	17 x 24	5 1½			12080	1929
18	1873	3	0-4-2	16 x 22	5 7½				1923
19	1873	13	2-4-0	17 x 24	6 1½			10005	1924
20	1873	5	0-4-2	16 x 22	5 7½				1921
21	1875	9	0-6-0	18 x 28	5 1½			12484	1930
22	1876	8	2-4-0	17 x 24	6 1½			10006	1925
23	1877	14	0-6-0	18 x 28	5 1½			12485	1925
24	1878	10	2-4-0	17 x 24	6 1½			10007	1925
25	1879	4	0-4-2	17 x 24	5 7½			10010	1928
26	1881	11	0-6-0	18 x 26	5 1½			12487	1928
27	1882	7	0-6-0	18 x 26	5 1½			12488	1930
28	1884	19	0-6-0	18 x 26	5 1½			12489	1930
29	1889	2	0-4-2	17 x 26	5 7½			10011	1928
30	1892	15	0-4-2	17 x 24	5 7½			10012	1928
31	1895	16	0-4-2	17 x 24	5 7½			10013	1928
32	1897	26	0-4-4T	16 x 22	5 1½			10618	1925
33	1900	1	0-6-0	18 x 26	5 1½			12492	1930

the early 1880s. Two more powerful six-coupled goods were to appear in 1921 with 19″ x 26″ cylinders but were supplied by the Yorkshire Engine Co.

The M&C locomotives were solidly built and had, in the main, quite long lives – out of a total of thirty-three built by the company, nineteen were taken over by the LM&SR.

A duplicate list was introduced in 1877 and six locomotives were transferred between 1877 and 1908 with numbers R1 to R5, R3 being used twice. After the last new engine was built in 1900 the works were confined to repairs and rebuilding and were eventually shut down.

One boiler at least was in evidence at Derby Works being used as a stationary boiler in the Carriage & Wagon works. It was built in 1920 and put on 0-4-2 No. 4 (LMS No. 10010) and when the locomotive was withdrawn in 1928 it served its secondary purpose until 1953.

Established in November 1826 this firm commenced as builders of marine and stationary steam engines and entered the field of locomotive building by completing a small four-coupled tank engine in February 1827, followed by two locomotive cranes and a steam traverser or 'traveller' for their own use.

Edward Bury, having many orders on his books at this time, sub-contracted an order for three engines received from the Petersburg Railroad. Two were four-coupled similar to Bury's LIVERPOOL, already sold to this customer, and the other was a four-wheeled single very similar to the LIVER on the Liverpool and Manchester Railway. These engines were despatched at the end of 1833. In 1834 came orders for four small tank engines for Liverpool docks, two steam cranes for Birkenhead docks, and a tank engine for Haydock Colliery. Four engines for the War Department followed in 1835, together with three similar ones for the Russian government, and three for collieries – Worsley Colliery (one) and Duke of Bridgewater (two) – and one for the North Union Railway (Parkside and Wigan Line).

Hargreaves of Bolton bought a 2-4-0 locomotive which was built in 1836. In this year, designs for four classes of engines were prepared. They were all six wheeled engines: 2-2-2, 0-4-2, 0-6-0, and 2-4-0 types. One of

Fig. 379 **Mather Dixon 1838 Broad gauge Great Western Rly.** AJAX - 10′ driving wheels built up from solid plates

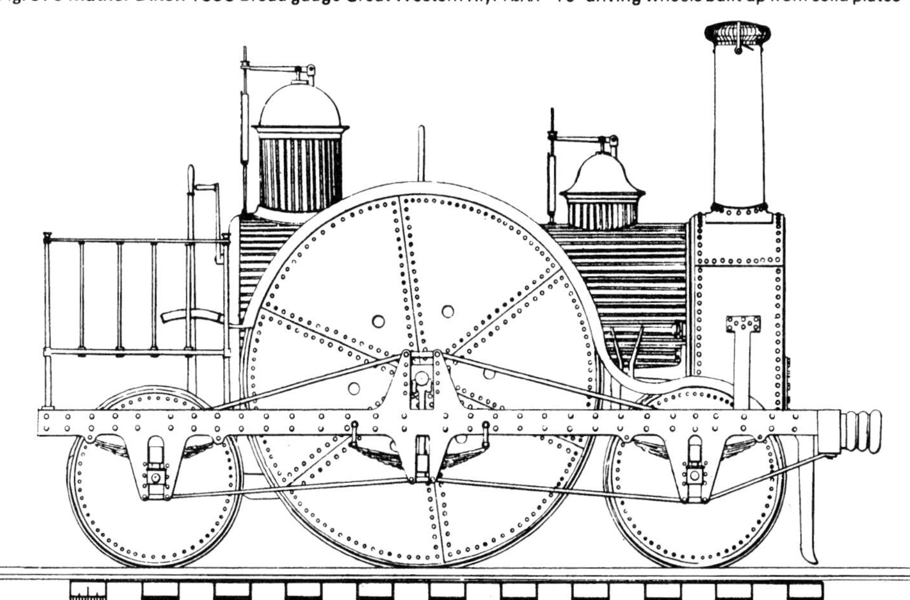

each type was built and was kept to show prospective customers but they were all finally sold to an agent in Manchester.

Further Bury type singles were supplied to the London and Birmingham Railway between 1836 and 1838 and two 0-4-2 and four 2-2-2 short stroke (12″) locomotives for the Liverpool and Manchester Railway between 1836 and 1837.

Next were the remarkable 7′ and 8′ broad gauge singles for the Great Western Railway built between 1837 and 1840, followed by one 6′ single for the Tsarskoe Selo Railway, near St Petersburg.

Three 5′ 6″ 2-2-2 locomotives for the Grand Junction Railway were again of the 'Bury' type and were delivered in 1838.

Early in 1839 the firm moved to new premises called the North Foundry, at William Street, Bootle and fifteen 2-2-2 locomotives were built as a batch and sold as follows: Grand Junction Railway two; London & Manchester Railway one; Birmingham & Derby Railway three; North Midland Railway three; Paris & Orleans two; Chester & Birkenhead Railway two; Chester & Birkenhead two not delivered.

In the early 1840s competition in this field was extremely acute and orders were hard to come by and no more engines were built until 1842, but ten boilers for the London and Manchester Railway were built in the interval and sent to the Edge Hill works, whilst a further ten were built for the Grand Junction Railway and were sent to Crewe.

Five 2-2-2s were built for stock in 1842 and also one 2-4-0, with the hope of better times to come.

At this time Mr John Grantham, the chief draughtsman, was taken into partnership, his name being added to the firm as Mather Dixon and Grantham. Unfortunately, while two six-coupled engines were in hand, it was decided to discontinue the business in 1843 due to the severe competition from other local firms of which there were seven.

In all seventy-five locomotives were built between 1827 to 1843.

John P. Mather, Dixon & Company were in the 1827 directory as Ironfounders and Engineers. In 1837 they were also Boilermakers. After the firm was dissolved in 1843 a new firm appeared at a different address called John P. Mather, Thomlinson (or Tomlinson) & Company, Iron and Gunpowder Manufacturers.

Ref No	Date	Type	Cylinders inches	D.W. ft in	Customer	No/Name
1	2/1827	0-4-0T		3 0	For Own Use	
2		4W C			Loco Crane	
3		4W C			Loco Crane	
4		Steam Traverser			For Own Use	
5	12/1833	0-4-0	IC 10 x 16	4 6	Petersburg RR (a)	New York
6	1833	0-4-0	10 x 16	4 6	Petersburg RR (a)	Philadelphia
7	1834	2-2-0	IC 12 x 20	5 6	Petersburg RR (a)	Petersburg
8	1834		OC		Liverpool Docks	
9	1834				Liverpool Docks	
10	1834				Liverpool Docks	
11	1834				Liverpool Docks	
12	1834	4WC			Birkenhead Docks	
13	1834	4WC			Birkenhead Docks	
14	1834	0-4-0			Richard Evans & Co Ltd	
15	1835	0-4-0			War Dept	
16	1835	0-4-0			War Dept	
17	1835	0-4-0			War Dept	
18	1835	0-4-0			War Dept	
19	1835	0-4-0			Russian Government	

MATHER DIXON

Ref No	Date	Type	Cylinders inches	D.W. ft in	Customer	No/Name
20	1835	0-4-0			Russian Government	
21	1835	0-4-0			Russian Government	
22	1835	0-4-0			North Union Rly	Wigan
23	1835	0-4-0			Duke of Bridgewater	
24	1836	0-4-0			Duke of Bridgewater	
25	1836	0-4-0			Worsley Colly	
26	1836	2-4-0			Hargreaves, Bolton	
27	1836	2-2-2			To Manchester Agent	
28	1836	0-4-2			To Manchester Agent	
29	1836	0-6-0			To Manchester Agent	
30	1836	2-4-0			To Manchester Agent	
31	1836	2-4-0			For Stock	
32	1836	0-4-2	15 x 16	5 0	Liverpool & Manchester Rly	39. Hercules
33	1836	0-4-2	15 x 16	5 0	Liverpool & Manchester Rly	44. Thunderer
34	1837	2-2-2	14 x 12	5 0	Liverpool & Manchester Rly	49. Dart
35	1837	2-2-2	14 x 12	5 0	Liverpool & Manchester Rly	52. Arrow
36	1837	2-2-2	14 x 12	5 0	Liverpool & Manchester Rly	54. Meteor
37	1837	2-2-2	14 x 12	5 0	Liverpool & Manchester Rly	55. Comet
38	1837	2-2-0	IC 12 x 18	5 6	London & Birmingham Rly	31. (a)
39	1837	2-2-0	IC 12 x 18	5 6	London & Birmingham Rly	32. (a)
40	11/1837	2-2-2	IC 14½ x 14½	7 0	Great Western Rly	Premier (b)
41	3/1838	2-2-2	IC 14 x 14	7 0	Great Western Rly	Ariel (b)
42	1838	2-2-2		6 0	Tsarskoe Selo Rly	England
43	1838	2-2-2	IC 13 x 18	5 6	Grand Junction Rly	37. Hawk
44	1838	2-2-2	IC 13 x 18	5 6	Grand Junction Rly	47. Vulture
45	1838	2-2-2	IC 13 x 18	5 6	Grand Junction Rly	50. Hornet
46	1838	2-2-0	IC 12 x 18	5 6	London & Birmingham Rly	33. (a)
47	1838	2-2-0	IC 12 x 18	5 6	London & Birmingham Rly	34. (a)
48	1838	2-2-0	IC 12 x 18	5 6	London & Birmingham Rly	35. (a)
49	1838	2-2-0	IC 12 x 18	5 6	London & Birmingham Rly	36. (a)
50	12/1838	2-2-2	IC 14 x 20	7 0	Great Western Rly	Ajax (b)
51	8/1839	2-2-2	IC 16 x 20	8 0	Great Western Rly	Planet (b)
52	9/1839	2-2-2	IC 16 x 20	8 0	Great Western Rly	Mercury (b)
53	4/1840	2-2-2	IC 16 x 20	7 0	Great Western Rly	Mars (b)
54	1839	2-2-2	13 x 18	5 6	Birmingham & Derby J Rly	Barton
55	1839	2-2-2	13 x 18	5 6	Birmingham & Derby J Rly	Tamworth
56	1839	2-2-2	13 x 18	5 6	Birmingham & Derby J Rly	Hampton
57	1839	2-2-2			North Midland Rly	
58	1839	2-2-2			North Midland Rly	
59	1839	2-2-2			North Midland Rly	
60		2-2-2		6 0	Paris–Orleans Rly	
61		2-2-2		6 0	Paris–Orleans Rly	
62	1840	2-2-2		5 6	Grand Junction Rly	
63	1840	2-2-2	13 x 18	5 6	Grand Junction Rly	
64	9/1840	2-2-2	13 x 18		Chester & Birkenhead Rly	2. Zillah
65		2-2-2			Chester & Birkenhead Rly	3. Touchstone
66					Chester & Birkenhead Rly	(c)
67					Chester & Birkenhead Rly	(c)
68 to 77 *Boilers only supplied*					London & Manchester Rly	
78	1842	2-2-2			For Stock	
79	1842	2-2-2			For Stock	
80	1842	2-2-2			For Stock	(e)
81	1842	2-2-2			For Stock	
82	1842	2-2-2			For Stock	
83	1842	2-4-0			For Stock	
84 to 93 *Boilers only supplied*					Grand Junction Rly	Del'd Crewe 1843
94	1843	0-6-0			For Stock	
95	1843	0-6-0			For Stock	
?					Liverpool & Manchester Rly	Victoria (d)

(a) Sub-contracted by E. Bury & Co. All were 'Bury' type with bar frames. (b) 7′ 0″ gauge. (c) Not delivered. (d) Some sources quote as Ref No 53. (e) Two out of Nos 78–82 sold to Holland Rly March 1843 *Salamander* and *Phoenix*
Ref No indicates *probable* works number.

MATTHEWS, JAMES

Broad Street, Bristol

James Matthews patented a tram engine design in 1879. It has been established that Matthews did not carry out the actual building, but as there are two schools of thought on who was the builder, it is convenient to put the matter under this heading.

Dr H. A. Whitcombe in his paper 'History of the Steam Tram' states that it was built 1879–80 probably by Fox Walker & Company of Bristol. It was tried out on the tramways of Bristol and Liverpool. It had outside cylinders 6" x 10¼", four wheels 2' 4" diameter, and a horizontal boiler pressed at 120 lb/sq. in. The exhaust steam was condensed in tanks placed along the inner sides of the cab. It was sold to the Wantage Tramway in 1885 becoming their No. 6 and was withdrawn on the cessation of passenger traffic on 31 July, 1925.

On the other hand, Mr S. H. Pearce Higgins in his book *The Wantage Tramway* (1958) gives the builder as Kingsbury Ironworks, Ball's Pond, London and the date of manufacture between 1879 and 1881. The three 'improvements' under Matthews' patent are described in detail, which included the arrangement of framing and cylinders and the condensing equipment.

This locomotive had the wheels between the frames and the cylinders and valve gear outside, but the wheels were not coupled so that it had the 2-2-0 wheel arrangement. The cylinders are given as 6¼" x 10".

MAUDSLAY, SONS & FIELD

Lambeth Marsh, London

This famous firm was established at the end of the 18th century by Henry Maudslay who produced stationary steam engines for marine and industrial use and various machine tools. He was a pioneer in the production of interchangeable screw threads.

Whitworth & Nasmyth both worked here for some time. Maudslay produced all his own equipment and created his own standards which were adhered to right up to the time the firm closed in 1900.

Henry Maudslay died in 1831 but his four sons carried on the business with Joshua Field who had joined the firm about 1805. To enter the locomotive market seems strange, although the manufacture of ships' engines probably influenced the decision.

MAUDSLAY, SONS & FIELD

Twelve 'Bury' type 0-4-0 tender locomotives were built for the London & Birmingham Railway, sub-contracted by Edward Bury. They were built as follows: 1838 L&B Nos 79–82 and 1839 L&B Nos 83–90.

They were the standard type of Bury bar-framed locomotives with 13" x 18" inside cylinders and 5' diameter wheels. Whether any further orders were fulfilled is not known, except that some type of steam carriage was built in 1836.

McCULLOCH, SONS & KENNEDY

Kilmarnock

This engineering firm was mainly concerned with colliery machinery, winding and stationary engines. They built a few 0-4-0STs with typical Kilmarnock characteristics, with stove pipe chimneys outside horizontal cylinders, combined dome and safety valves, and wheels with 'I' section spokes. A few were supplied to local collieries and New Zealand. One built in 1890, Works No. 324 was for Wemyss Coal Co. 11.

The firm closed down in 1890 and the patterns and drawings were acquired by Gibb & Hogg of Airdrie.

McFARLANE

Greenock

Although stated to have built locomotives no information is available as to whether this was so or not.

McHENDRICK & BALL

Glasgow

One 0-4-0T fitted with a vertical boiler was supplied for William Lee, Son & Co's 4' 3" gauge system at Halling, Kent in 1878 and was their No. 1.

No further details are known but this locomotive may have been built by Alexander Chaplin & Co. of Glasgow, McHendrick & Ball acting as agents.

McLAREN, J. & H.

Midland Engine Works, Leeds

Two locomotives of the traction engine type were built, one in 1896 and one in 1915. The 1896 locomotive was a 2-2-0 (Works No. 614) built for the Glanlivet Distillery, Scotland. The one built in 1915 was a four coupled 0-4-0 (Works No. 1547) for Hall & Company, Croydon, their No. 9. This was a compound with cylinders 6½"/10¾" x 12". Wheels were 3' 1" diameter.

The firm's principal products were oil and gas engines.

MELLING, JOHN

Rainhill

Mr John Melling set up as an engineer builder at Rainhill after working for the Liverpool and Manchester Railway as superintendent of locomotive repairs at the Manchester end. His services were dispensed with at the end of 1839.

He obtained an order for three locomotives in 1840 for the Grand Junction Railway at a price of £1400 each. The first to be delivered was GJR No. 70 SPHINX in 1841.

The other two were delivered in 1842 and a certain amount of mystery surrounds their GJR numbers and names. The railway board had in 1842 decided on the names PRINCE and PRINCESS, but according to Mr C. Williams, the authority of LNWR locomotive matters, they replaced No. 1 SARACEN and No. 7 SCORPION and took both numbers and names.

No. 73 was named PRINCE and No. 19 PRINCESS, both being built at Crewe in 1845, so that it is unlikely that the Melling locomotives bore these names, or if they did were re-named before 1845. In a request for a payment on account, Melling mentions the name PRINCE now delivered.

By 1842 the firm was known as Melling & Son, John probably being joined by his son Thomas after being superintendent of the Liverpool & Manchester Railway's locomotive shops at Manchester.

The three locomotives built were probably four coupled with 5 ft wheels but confirmation is not available. No further locomotives are known to have been built.

MERRYWEATHER & SONS LTD.

Tram Locomotive Works, Greenwich Road, London, S.E.

This well known firm was established in 1690 by Nathaniel Hadley in Cross Street. The firm then moved to Bow Street, Longacre and became Hadley & Simpkin and in 1791 Hadley, Simpkin & Lott.

About 1836 Moses Merryweather took over the business and in 1876 transferred to premises in Greenwich. The Lambeth branch which had opened in 1862 was closed down in 1879.

In the 1860s the firm was known as Merryweather & Field, Field specialising in the boiler section of the business which had concentrated on fire fighting equipment from its inception.

As an additional source of business they became interested in steam trams, having been given an order by John Grantham to manufacture the power unit for a combined locomotive and carriage which consisted of a four wheeled double deck car in the centre of which was placed two small vertical boilers 18" diameter and 4' 4" high, of the Field type. The engine was fixed under the floor and had two cylinders 4" x 10" driving on one pair of the 2' 6" wheels. Controls were such that the tram could be driven from either end.

It was tried out in London in this original form but was not satisfactory and the two Field boilers were replaced by one larger unit made by Shand Mason & Co. who were to provide a substantial number of boilers for the Merryweather trams.

After further trials it was sent to the Wantage Tramway but did not traverse the full length of the tramway until 30 June, 1876, although it had been at Wantage since 1875, but permission to use steam on the tramway had to be awaited. On 1 August it entered regular service.

Meanwhile orders from other tramways had been placed. In 1875 a small tram engine was delivered in Paris ordered by a Mr G. P. Harding who had the operating concession for the Southern Tramways of Paris. A second, slightly larger, appeared in 1876. Both had vertical boilers, but Merryweather decided that future tram engines should have horizontal boilers.

These tram engines were propulsion units only and were known as 'dummy locomotives' – a strange appellation.

Altogether forty-six tram engines were supplied to the Paris tramways from 1875 to 1877.

Merryweathers' designed six sizes of engines.

Type No	Cylinders	Weight Empty (tons) approx.
1	6 x 9	4
2	6½ x 10	5
3	7 x 11	6½
4	7½ x 12	7½
5	8½ x 14	8½
6	10 x 14	9½

In Merryweather & Sons' catalogue of 1882 they put their case forward for the replacement of horse haulage by steam traction as follows.

1 Considerable reduction in working expenses.
2 Applicability to Tramways with steep gradients, or of considerable length. There are many advantageous routes on which it would be quite impracticable to employ horse traction.
3 Utility for lines with a fluctuating traffic. Engines when not in use practically cost nothing to keep.
4 Increased power. This enables the use of extra cars, when required under special circumstances.
5 Greater speed when desirable.
6 Avoidance of cruelty to animals. This is inevitable in the case of Tramway horses on account of the overstraining to which they are constantly subject.
7 Special suitability for winter traffic. Tramway locomotives can run at times when it would be impossible to employ horse traction, through the slippery or snowed up state of the roadways.
8 Ability to start and stop more quickly.

D. Kinnear Clark in his book *Tramways – Their Construction and Working* gave the Merryweather Tram Engine a good reputation for economy, simplicity of construction and lack of noise, exhaust steam and smoke.

Continuing their introduction in their catalogue Merryweathers stated 'The engines we construct comply with all the conditions referred to as essential to a tramway locomotive, viz:

1 They are free from noise, either of blast or machinery.
2 They are smokeless, and free from any escape of steam.
3 There is no visible fire, and the working parts are excluded from observation.

Fig. 380 **Merryweather & Sons** *c*1880 tram engine under construction

4 They are perfectly safe, and explosion is practically impossible.
5 The machinery is boxed in, and protected as far as possible from dirt and dust.
6 They work both ways and are constructed to enable the driver at all times to have full command of his machine, as he has an uninterrupted view of the road-way both in front and on either side.
7 They start and stop very readily, both on the level and on steep gradients, and take sharp curves without difficulty.
8 The working parts are easily accessible for cleaning, and can be repaired with facility.
9 Whilst running they require no attention in firing and feeding, so that the Engineer can devote his whole attention to driving. One man is quite sufficient to work each Engine.
10 Parts of Engines of similar size are made interchangeable, and every piece being manufactured accurately to gauge, duplicate parts can be readily supplied.

All the mechanical difficulties which presented themselves when 'Steam on Tramways' was first discussed have therefore been entirely overcome by us, and, as regards the danger of frightening horses, experience has demonstrated that whenever tramway cars have been run by steam upon thoroughfares frequented by horses, the horses soon become familiar and indifferent to them.

The regulations for running steam trams on public highways was very stringent. The Acts of 1861 and 1865 provided for a maximum speed of 4 mph and a man to walk in front with a red flag. The act of 1870 cancelled these regulations but stipulated that an efficient braking system must be fitted, and a governor to shut off the steam supply when the speed reached 10 mph.

More important, from the manufacturers point of view, was that the issue of steam, smoke or hot air from the engine was forbidden and that no noise must be caused by the exhaust, or from the machinery.

When introduced, the steam trams had a very mixed reception. The protagonists hailed the event as a big step forward in passenger transport in townships, and despite the strong opposition the steam trams were used on many tramways until electric traction took over. Many steam trams operated for twenty-five years or more. The chief opposition came from those

Fig. 381 **Merryweather & Sons** 142/1885 North London Tramways No. 1

who had provided the horses, the stabling and the fodder. Large capital outlays were involved and with the introduction of steam trams, many had their livelihood threatened. The London General Omnibus Co. employed 8000 horses which were bought when they were about five years old. The average life of a horse drawing an omnibus was four and a half years, and when employed on pulling trams, four years. This of course depended on where the horses worked, as on some routes with severe gradients the horses were tried to the limit.

Bearing this in mind, and although the cost of steam trams was from £600 to £800 each, with this large turnover of horse flesh the steam tram was an economic proposition and resulted in cheaper fares in many instances.

After the few experimental locomotives, Merryweathers continued building the standard classes as shown above. They could be made to any gauge but the general specification remained the same. A typical tram engine as produced was as follows:

The cylinders were placed inside the frame above the leading axle and were inclined, to meet the centre line of the trailing axle. The cylinders were cast separately and bolted together forming the steam chest between them and the saddle on which the boiler was fixed. The cast steel crossheads worked between two steel slide bars and the crossheads were fitted with cast iron slippers. A motion plate between the frames provided support for the slide bars and the valve spindle guides were fixed on to it. The motion was of the shifting link type and the cast iron eccentrics were fixed on to the rear axle and the straps were of the same metal.

The boiler was horizontal and of the locomotive type with round top firebox made of Low Moor iron and double rivetted along the longitudinal seam. The firebox door was fitted on the side of the firebox. The position of the driver/fireman was on the side of the boiler so that only one set of controls was necessary as compared with other makers which had duplicate controls at either end. The reversing lever was of the conventional type situated about half way down on the left hand or driver's side.

The boiler was fed from a force feed pump operated by a separate eccentric on the rear axle. A Giffard's injector was also fitted. The wheels and axles were steel and brakes were fitted to engage all four wheels and operated by a foot pedal. The boiler feed tanks were placed on each side of the smokebox with an equalising pipe between the two. The coupling rods had solid phosphor bronze bushes, running on case hardened crank-pins.

Engines without roof condensers had the exhaust steam injected into a copper tank in the smoke box and then discharged through a rose jet into the chimney. It was claimed that this method was found to expand the steam sufficiently to prevent the blast from being audible and at the same time superheat the steam so rendering it invisible.

Those engines fitted with condensers had them placed on the roof and consisted of four horizontal layers of 1″ × 26 SWG copper tubes, slightly arched, laid transversely across the roof each approximately 6 ft long. There were sixty tubes in each layer, 240 tubes in all, coated with brown varnish to increase their radiating property. They were secured at the ends into 3″ diameter longitudinal pipes, four on each side 11 ft long. The exhaust steam was discharged by means of two copper pipes, one to each side, into the uppermost longitudinal pipe, whence it circulated through the transverse tubes. The condensate and remaining vapour were conducted into a separator at the front, the water running down to the feed water tank and the vapour passing away into the smokebox where it was exhausted together with the flue gases.

The engine could be worked all day with one charge of the feed water tank which held 100 gallons and the quantity consumed as uncondensed steam did not exceed 50 gallons per day.

Steam trams supplied to places where a hot climate predominated such as Barcelona and Rangoon were not fitted with condensers but with the previous mentioned smoke-box fitting. Tests carried out on the temperature of the feed water tank when condensers were used indicated a maximum temperature under 200°F.

Construction of tram engines was carried out from 1875 to 1892. The standard types were of robust build, particularly the motion which had to stand up to rough track, dirt and constant stopping and starting. The boiler and the semicircular part of the firebox were lagged with polished wooden strips. Below the footplate, the wheels and motion were covered with plating and so were the front and rear which were also strengthened. The average overall length was 12 ft, width 6′ 4″ and weight in working order 4 to 8½ tons.

In addition to the steam trams, a number of steam driven inspection cars were built, some going to South America. One is illustrated here and an illustration of a more elaborate one may be found in the *Locomotive Magazine* for 15 July, 1907. It was built for the Buenos Ayres and Pacific Railway.

The following table is not complete and many of the works numbers and sequences cannot be verified.

Fig. 382 Merryweather & Sons
Steam Tramway
Rangoon

Fig. 383 Merryweather & Sons
0-2-2 VB inspection car.
Some were supplied to South America

Works No	Date	Type	Customer	No		Gauge
1	1872		Wantage Tramway	3	Grantham Car	Std
2	1874		Vienna via Mr G. P. Harding		(a)	
3	1875		Paris Southern Tramway	1	(a)	
4	1876		Paris Southern Tramway	2	(a)	
5	1876		Paris Southern Tramway	3		
6	1876		Paris Southern Tramway	4		
7	1876		Paris Southern Tramway	5		
8	1876		Paris Southern Tramway	6		
9	1876		Wantage Tramway		(b)	Std
10	1876		Paris Southern Tramway	7		
11	1876		Paris Southern Tramway	8		
12	1876		Paris Southern Tramway	9		
13	1876		Paris Southern Tramway	10		
14	1876		Paris Southern Tramway	11		
15	1876		Paris Southern Tramway	12		
16	1876		Paris Southern Tramway	13		
17	1876		Paris Southern Tramway	14		
18	1876		Paris Southern Tramway	15		
19	1876		Paris Southern Tramway	16		

MERRYWEATHER & SONS LTD.

Works No	Date	Type	Customer	No		Gauge
20	1876		Paris Southern Tramway	17		
21	1876		Paris Southern Tramway	18		
22	1876		Paris Southern Tramway	19		
23	1876		Paris Southern Tramway	20		
24	1876		Paris Southern Tramway	21		
25	1876		Paris Southern Tramway	22		
26	1876		Paris Southern Tramway	23		
27	1876		Paris Southern Tramway	24		
28	1876		Paris Southern Tramway	25		
29	1876		Paris Southern Tramway	26		
30	1876		Paris Southern Tramway	27		
31	1876		Paris Southern Tramway	28		
32	1877	1	Wharncliffe Rifle Range	1	(c)	
33	1877	4	Kassel Tramways	1		
34	1877	4	Kassel Tramways	2		
35	1877	4	Kassel Tramways	3		
36	1877	4	Kassel Tramways	4		
37	1877	4	Kassel Tramways	5		
38	1877		Paris Southern Tramway	29	(d)	Metre
39	1877		Paris Southern Tramway	30		Metre
40	1877		Paris Southern Tramway	31		Metre
41	1877		Paris Southern Tramway	32		Metre
42	1877		Paris Southern Tramway	33		Metre
43	1877		Paris Southern Tramway	34		
44	1877		Paris Southern Tramway	35		
45	1877		Paris Southern Tramway	36		
46	1877	3	Barcelona à San Andrés	1		
47	1877	3	Barcelona à San Andrés	2		
48	1877	3	Barcelona à San Andrés	3		
49	1877	3	Barcelona à San Andrés	4		
50	1877	3	Barcelona à San Andrés	5		
51	1877		Paris Southern Tramways	37		
52	1877		Paris Southern Tramways	38		
53	1877		Paris Southern Tramways	39		
54	1877		Paris Southern Tramways	40		
55	1877		Paris Southern Tramways	41		
56	1877		Paris Southern Tramways	42		
57	1877		Paris Southern Tramways	43		
58	1877		Paris Southern Tramways	44		
59	1877		Paris Southern Tramways	45		
60	1877		Paris Southern Tramways	46		
61	1878	3	Wellington Tramways NZ	1		Metre
62	1878	3	Wellington Tramways NZ	2		Metre
63	1878	3	Wellington Tramways NZ	3		Metre
64	1878	3	Wellington Tramways NZ	4		Metre
65	1878	3	Wellington Tramways NZ	5		Metre
66	1878	3	Wellington Tramways NZ	6		
67	1878	3	Wellington Tramways NZ	7		
68	1878	3	Wellington Tramways NZ	8		
69	1878	3	Barcelona à San Andrés	6		
70	1878	3	Barcelona à San Andrés	7		
71	1878	3	Barcelona à San Andrés	8		
72	1878	3	Barcelona à San Andrés	9		
73	1878	3	Barcelona à San Andrés	10		
74						
75						
76						
77						
78						
79						
80	1878	2	Adelaide Tramways	1	Eureka (e)	5 6
81	1878	4	Kassel Tramways	6		
82	1878	3	Dutch Rhenish Rly	1	NRS 1 (f)	Std

Works No	Built	Type	Customer	No	Notes	Gauge ft in
83	1878		Dunedin Tramways NZ	1		
84	1879	3	Guernsey Steam Tramway Co	1		Std
85	1879	3	Guernsey Steam Tramway Co	2		Std
86	1879	3	Wellington Tramways NZ	9		
87	1879	3	Wellington Tramways NZ	10	(g)	
88	1879	3	Dutch Rhenish Rly	2	NRS 2	Std
89	1879	3	Dutch Rhenish Rly	3	NRS 3	Std
90	1879	3	Dutch Rhenish Rly	4	NRS 4	Std
91	1879	3	Dutch Rhenish Rly	6	NRS 6	Std
92	1879	3	Dutch Rhenish Rly	7	NRS 7	Std
93	1879	3	Dutch Rhenish Rly	5	NRS 5	Std
94	1880	2	Dewsbury, Batley & Birstal Tramways Co	1		Std
95	1880		Oporto Tramways			
96	1881		Dutch Rhenish Rly	8	NRS 8	Std
97	1881		Dutch Rhenish Rly	9	NRS 9	Std
98	1881		Dutch Rhenish Rly	10	NRS 10	Std
99	1881		Dutch Rhenish Rly	11	NRS 11	Std
100	1881		Dutch Rhenish Rly	12	NRS 12	Std
	1881	2	Dewsbury, Batley & Birstal Tramways Co Ltd	2	(h)	
	1881	2	Dewsbury, Batley & Birstal Tramways Co Ltd	3		
	1881	2	Dewsbury, Batley & Birstal Tramways Co Ltd	4		
	1881	2	Dewsbury, Batley & Birstal Tramways Co Ltd	5		
101	1881	1	Calcutta Tramways	1		
102	1881	3	Calcutta Tramways	2		
103	1881	2	Rangoon Tramways	1		
104	1881	2	Rangoon Tramways	2		
105	1881	2	Rangoon Tramways	3		
106	1881	2	Rangoon Tramways	4		
107	1881	2	Rangoon Tramways	5		
108	1881	2	Rangoon Tramways	6		
109	1881		Rhineland Tramway Co	1	RSTM1 215	
110	1881		Rhineland Tramway Co	2	RSTM2 216	
111	1881		Rhineland Tramway Co	3	RSTM3 217	
112	1881		Rhineland Tramway Co	4	RSTM4 218	
113	1881	4	North Staffordshire Tramways Co Ltd	3		
114	1881	5	North Staffordshire Tramways Co Ltd	4		
115	1881		Dutch Rhenish Rly	13	NRS13	
116	1881		Dutch Rhenish Rly	14	NRS14	
117	1881		Dutch Rhenish Rly	15	NRS15	
118	1881		Dutch Rhenish Rly	16	NRS16	
119	1881		Dutch Rhenish Rly	17	NRS17	
120	1881	2	Rangoon Tramways	7		
121	1881	2	Stockton & Darlington Steam Tramways	1		
122	1881	2	Stockton & Darlington Steam Tramways	2		
123	1881	2	Stockton & Darlington Steam Tramways	3		
124	1881	2	Stockton & Darlington Steam Tramways	4		
125	1881	2	Stockton & Darlington Steam Tramways	5		
126	1881	2	Stockton & Darlington Steam Tramways	6		
127	1881		Dewsbury, Batley & Birstal Tramways Co Ltd	6		
128	1881		Dewsbury, Batley & Birstal Tramways Co Ltd	7		
129	1881		Dewsbury, Batley & Birstal Tramways Co Ltd	8		
130	1881		Dewsbury, Batley & Birstal Tramways Co Ltd	9		
131	1882		Rhineland Tramway Co Holland	5	RSTM 5 219	Std
132	1882	6	North Staffordshire Tramways Co Ltd	10		4 0
133	1882	6	North Staffordshire Tramways Co Ltd	11		4 0
134	1882	3	Barcelona à San Andrés	11		Metre
135	1882	3	Barcelona à San Andrés	12		Metre
136	1882	3	Barcelona à San Andrés	13		Metre
137	1882	3	Barcelona à San Andrés	14		Metre
	1882	3	Barcelona à San Andrés	15		Metre
138	1883	3	Stockton & Darlington Steam Tramways Co Ltd	7		4 0
139	1883	3	Stockton & Darlington Steam Tramways Co Ltd	8		4 0
140	1883	4	Alford & Sutton Tramway	2		2 6

MERRYWEATHER & SONS LTD.

Works No	Date	Type	Customer	No		Gauge
141	1883		Sydney Tramways	55		
142	1885		North London Tramways	1		Std
143	1885		North London Tramways	2		Std
144	1885		North London Tramways	3		Std
145	1885		North London Tramways	4		Std
146	1885		North London Tramways	5		Std
147	1885		North London Tramways	6		Std
148	1885		North London Tramways	7		Std
149	1885		North London Tramways	8		Std
150	1885		North London Tramways	9		Std
151	1885		North London Tramways	10		Std
152	1885		North London Tramways	11		Std
153	1885		North London Tramways	12		Std
154	1885		North London Tramways	13		Std
155	1885		North London Tramways	14		Std
156	1885	2	Rangoon Tramways	8	(i)	Std
157	1885	2	Rangoon Tramways	9		Std
158	1885	2	Rangoon Tramways	10		Std
159	1885	2	Rangoon Tramways	11		Std
160	1885	2	Rangoon Tramways	12		Std
161	1885	2	Rangoon Tramways	13		Std
162	1885	2	Rangoon Tramways	15		Std
163	1885	2	Rangoon Tramways	14		Std
164	1885	2	North London Tramways	15		Std
165						
166						
167						
168						
169						
170						
171						
172						
173	1892		Dutch Rhenish Rly	18	NRS 18	
174	1892		Dutch Rhenish Rly	19	NRS 19	

Type No	Cylinders inches	WD* ft in	Approx Weight Empty ton cwt	TUBES No	TUBES Dia inches	TUBES Lgth ft in	TUBES H/S sq ft	F'box H/S sq ft	Total H/S sq ft	Grate Area sq ft	W'base ft in
1	6 x 9		4 0								
2	6½ x 10	2 2	5 0				146·56	26·85	173·41	3·66	4 6
3	7 x 11	2 0	6 10	78	1¾	4 0	139·0	20·75	159·75	4·25	4 6
4	7½ x 12	2 4	7 10	85	1¾	4 0					4 6
5	8½ x 14		8 10								
6	10 x 14		9 10								

(a) Vertical boiler (b) For trial purposes returned to makers 1878 to Denmark? (c) No 26 in Whitcombe's list (d) Title in Full: Tranvía de Barcelona à San Andrés de Palomar (e) Paris Exhibition 1878 named EUREKA (f) Rijnlandsche Stoomstramweg? (g) To Invercargill Borough Council N Z (h) May have been works nos 94–99 duplicated (i) Numbers had prefix J W D

J W D 15 recorded as 1887

* Variable.

METROPOLITAN RAILWAY

Neasden

Three *E* class 0-4-4Ts were built at Neasden in 1896. Their running numbers were 1, 77, and 78. They had inside cylinders 17½″ x 26″, 5′ 6″ driving wheels and were fitted with condensing gear. Another four of the same class were built by R. & W. Hawthorn Leslie: Nos. 79 and 80 in 1900, and Nos. 81 and 82 in 1901. The condensing gear was subsequently removed.

The locomotives were initially used for the main line trains between Aylesbury and Verney Junction and performed very well. They were designed by Mr T. F. Clark, the Mechanical Engineer of the railway (1893–1906). The main dimensions were: Heating surface Tubes 1050 sq. ft; Firebox 95·6 sq. ft; Total 1145·6 sq. ft; Grate Area 16·7 sq. ft; Boiler Pressure 160 lb/sq. in.; Tank Capacity 1300 gallons; Coal Capacity 2 ton 4 cwt.

Weight in working order was 54 tons 10 cwt and the diameter of bogie wheels was 3′ 9⅛″.

Fig. 384 **Metropolitan Rly** 1896 0-4-4T No. 77 on train of the period

MIDLAND RAILWAY

Derby

These works were originally built by the North Midland as the main repair depot for that railway's locomotives. No new ones were built by the NMR and under the Incorporation Act of 10 May, 1844 it became part of the Midland Railway together with the Midland Counties, and the Birmingham & Derby Junc. Railways.

From this date until his death in May 1873, Matthew Kirtley was Locomotive Superintendent at Derby and responsible for all the locomotives from the three constituent companies, amounting to seventy-seven passenger and eighteen goods locomotives. The North Midland locomotives had been under the care of Thomas Kirtley, the Midland Counties J. Kearsley, and Matthew Kirtley had had charge of the Birmingham & Derby locomotives. The locomotive superintendents for the Midland Railway were: 1844–73 Matthew Kirtley; 1873–1903 Samuel Waite Johnson; 1904–9 Richard M. Deeley; 1909–23 Henry Fowler.

Early in 1842 the North Midland Railway pointed out to Robert Stephenson, who was the company's consulting engineer, that the chimneys and smokeboxes were being destroyed in a short time by the excessive heat. Experiments were carried out and it was found that a temperature of 773°F was indicated.

Stephenson decided to lengthen the tubes on certain locomotives from 9 ft to 13 or 14 ft and due to this elongation of the boiler barrel the trailing axle was placed in front of the firebox instead of behind. They were called 'long boilered'. They proved successful with a temperature of 442°F. All locomotives at this time had been built by many different locomotive builders to their own designs and when breakdowns occurred it was frequently necessary to take the locomotive out of service to wait for spare parts from the makers.

Matthew Kirtley strongly urged the directors to enlarge the repair shops at Derby to enable him to build his own locomotives and to provide a better repair service for existing stock. It was agreed that the outside contractors should build locomotives for the Midland Railway to the railway's designs.

In 1849 he designed a standard goods locomotive with double frames, inside cylinders 16″ x 24″ and six-coupled wheels 5 ft diameter on a 16′ 6″ wheelbase. Fifty were ordered from outside firms, none being built at Derby.

1851 was the year of first new construction and consisted of ten passenger 'singles' similar to the *Jenny Lind* design of E. B. Wilson. They were numbered 150–159 but strangely enough the first to appear was No. 158. Ten standard goods followed and then Kirtley designed a large 'single' for passenger traffic, 6′ 6″ diameter driving wheels had outside and inside bearings, and the leading and trailing wheels had outside bearings only. Inside cylinders were 16″ × 22″.

Derby Works was not at this time large enough to carry out the building of all new locomotives required, and some trouble was experienced with Stephensons and Sharp Stewart on their being asked to tender for Kirtley designed locomotives. As these would be a long delivery, Kirtley had to accept locomotives to the makers' designs. A strong point in the makers favour was that they were so satisfactory in running, that Kirtley with some superficial alterations built fourteen at Derby in 1854.

Building went on with 'single' 2-2-2s and 0-6-0 goods locomotives. Four 0-6-0Ts were built in 1861 for the Lickey incline with 4 ft diameter wheels and 16½″ x 24″ cylinders.

The first 2-4-0 was actually a rebuilt 2-2-2, additional adhesion being required on the heavier trains. Six new 2-4-0s were built in 1862 for the Leicester to King's Cross services, especially the special exhibition traffic for that year. The driving wheels were 6′ 2″ diameter and the boiler was the same as for the 0-6-0 goods. They were numbered 80–85. Fifteen others were built with raised firebox casings and smaller cylinders.

In 1863 more 'singles' were built with 6′ 8″ diameter driving wheels and 16½″ x 22″ cylinders. These replaced the earlier Wilson *Jenny Lind* class. They had an improved cab which gave more protection than the customary weatherboards. They were a very successful type with a good turn of speed and economical on coal.

Many 0-6-0 goods locomotives were built,

Fig. 385 Derby works early 1900s

Fig. 386 **Midland Rly** 1867 Kirtley 6′ 2″ 2-4-0 No. 9 originally No. 1

mostly by outside contractors, including two by Beyer Peacock & Co. fitted with special boilers and fireboxes designed by Mr Beattie of the London & South Western Railway, intended to burn the coal without producing smoke. Unfortunately one blew up at Nottingham and both were rebuilt with standard boilers.

Mr Kirtley introduced a new type of frame which instead of being built up and rivetted in sections was in one piece, and arched over each axlebox.

Further goods 0-6-0s were built with 17″ x 24″ cylinders and the standard 5′ 2″ diameter wheels. The boiler and firebox produced 1096 sq. ft of heating surface and the grate area was 17 sq. ft. Weight in working order was 34 tons 13 cwt, and they had six wheeled tenders holding 2000 gallons of water.

A larger 2-4-0 appeared in 1870 with the same boiler as the previous 0-6-0s. The locomotive weighed 35 tons 18 cwt in working order. The boiler pressure was the standard 140 lb/sq. in. St Pancras had now been opened and these locomotives were used on the main expresses and accelerated times were possible at the outset.

Kirtley's last main line locomotive was the *890* class 2-4-0 with 6′ 8″ diameter wheels and cylinders 17″ x 24″, the leading wheels having outside bearings. The boiler had 232 1½″ tubes giving 1020 sq. ft of heating surface and the firebox 92 sq. ft. Working pressure was 140 lb/sq. in. Weight in working order without tender was 36 tons 14 cwt. Forty-two were built at Derby and twenty by Neilson & Co.

Matthew Kirtley unfortunately died in office in May 1873 leaving behind a respectable stud of locomotives, all of which had inside cylinders, round top fireboxes, which in the case of most passenger types were raised, whereas the boilers for goods locomotives were flush topped. Kirtley favoured sandwich frames but later introduced outside plate frames. The dome was placed on the boiler combined with spring balance safety valves. The whistle was usually mounted on a bolted cover on top of the firebox. Cabs varied from a weatherboard to a roofed cab, with a very short roof.

In 1859 Kirtley had commenced experiments for converting the fireboxes to burn coal and with the addition of brick arches and deflector plates over the fire doors proved successful.

His locomotives built at Derby and by outside contractors were, on the whole, extremely long lived due to the design and the workmanship and although rebuilt from time to time many lasted eighty years and more.

Samuel Waite Johnson joined the MR in 1873. He had served his apprenticeship in the

Fig. 387
Midland Rly
1889 Johnson 7′ 4½″ 4-2-2 No. 615 – 18″ × 26″ cylinders

Fig. 388
Midland Rly
1892 Johnson 0-4-4T No. 1346 fitted with auto gear

Fig. 389
Midland Rly
1902 Johnson 6′ 9″ 4-4-0 No. 814 became No. 724

Fig. 390 (Right)
Midland Rly 1907 Deeley 0-6-4T No. 2002 5′ 7″ D.W. 18½″ × 26″ cylinders

works of E. B. Wilson & Co, under James Fenton. After leaving Wilson's he became manager of the Peterborough shops of the Great Northern Railway, then Works Manager at Gorton on the Manchester, Sheffield & Lincolnshire Railway, which he left to become Locomotive Superintendent of the Edinburgh & Glasgow Railway. He succeeded Robert Sinclair on the GER where he served from 1866 to 1873.

Mr Johnson had already produced four-coupled passenger locomotives for the Great Eastern Railway and in 1876 similar ones were built by Kitsons and Dubs, with 17½" x 26" cylinders and 6' 6" diameter driving wheels, for the Midland Railway.

Although four-coupled locomotives were firmly established and were used to work the steeper gradients on the line, for the lighter expresses 'singles' were still built and in 1887 the first five 4-2-2s were built at Derby with 18" x 26" cylinders and 7' 4" diameter driving wheels. They had double frames but with single outside bearings for the trailing wheels and inside frames on the bogie. Larger cylinders and wheels were fitted to the 1889 batch, 18½" x 26" and 7' 6" respectively, and the boiler pressure was raised to 170 lb/sq. in.

The ultimate Johnson 'singles' were built in 1900 with larger boilers at 180 lb/sq. in. One was named PRINCESS OF WALES, one of the few locomotives which were named on the Midland Railway. It won a Gold Medal at the Paris Exhibition of 1900. A 7' 6" single No. 1853 won a similar medal at the Paris Exhibition of 1899.

In the same year Mr Johnson introduced the Belpaire firebox with a new batch of two-cylinder 4-4-0s with 6" 9" diameter driving wheels, 19½" x 26" cylinders and a boiler having a total heating surface of over 1500 sq. ft.

Safety valve practice was also altered. Hitherto lock-up valves had been fitted over the firebox and a pair of Salter spring balance valves on the dome. With the Belpaire fireboxes Ramsbottom valves were used on the firebox. A new type of chimney was also introduced.

In January 1902 appeared the first 4-4-0 compound, a class, with some modifications, which was to multiply to a total of 240 during Midland and LMS days and which were to penetrate all over the larger LMS system. Its forerunner was a Worsdell NER 4-4-0 which was rebuilt, under the direction of Mr W. M. Smith from a two-cylinder compound to a three-cylinder, with one high pressure and two low pressure cylinders.

The advantage of the Smith compound was that it could be worked in three ways – as a three-cylinder simple, a semi-compound or fully compound. Much can be found describing these famous locomotives and the marvellous work they did. Five were built to Johnson's design, numbers 2631–2634.

Regarding freight locomotives, Johnson designed a simple inside frame 0-6-0 with 4' 11" diameter wheels which were subsequently increased to 5' 3" for later batches. From 1875 to 1902 no less than 865 were built. They had round topped fireboxes with brass covered safety valve and combined dome and safety valves on the centre of the boiler barrel. Although classed as goods they rapidly became mixed traffic locomotives and some were fitted with vacuum brakes.

Mr Johnson retired in 1903 and was succeeded by Mr Richard M. Deeley.

Deeley served his apprenticeship at Derby Works and after being given various posts in the works, became Works Manager in 1902 and soon after became Assistant Locomotive Superintendent. He succeeded Mr Johnson on 1 January, 1904 and Mr Cecil W. Paget, son of Sir Ernest Paget the company's chairman, became Works Manager.

Fig. 391 **Midland Rly** 1909 Deeley 4-4-0 compound as rebuilt in 1913 running as LMS No. 1040

Fig. 392 **Midland Rly** 1909 Deeley 4-4-0 No. 995 with patent valve gear

Deeley at first carried on where Johnson had left off and 4-4-0s and 0-6-0s continued to be built. Two rail motors were built for the Morecambe and Heysham line, and they followed the conventional lines of rail motors built by other companies with vertical boilers, four-coupled bogie with outside Walschaerts valve gear. The bodies were built in the adjacent carriage shops. In his second year of office he modified the compound design, the chief differences being the boiler pressure up rated to the high figure of 220 lb/sq. in., the boiler heating surface modified, grate area increased, and Smiths' reducing and regulating valves omitted. A larger cab was fitted, and a redesigned chimney which tapered outwards, with a capuchon on the rim which became standard. A new design of six-wheeled tender was introduced to replace the huge eight-wheel tenders fitted to the original five compounds. Forty of these locomotives were built and the original five were rebuilt to the same design.

In 1907 Deeley introduced a renumbering scheme and locomotives were grouped according to wheel arrangement and class. Before this no attempt had been made to issue blocks of numbers for different classes which made the records very complicated indeed although not quite as bad as the LNWR and NER, to name but two.

Deeley rebuilt a large number of the Johnson six-coupled classes with larger boilers. Three were rebuilt with 6 ft wheels and were probably tried out as forerunners of the 0-6-4Ts he designed for suburban work. These three 0-6-0s must have been a rare sight.

An interesting batch of ten 4-4-0 locomotives were designed by Mr Deeley and the first one appeared in 1907 followed by nine more in 1909. They were two-cylinder simple, with 6′ 6½″ diameter driving wheels and cylinders 19″ x 26″ and 8¾″ diameter piston valves. The most interesting feature was the special valve gear designed by Mr Deeley. The travel of the valve for lead is given by a pendulum link and the expansion link is oscillated by a rod attached to the crosshead of the opposite motion. By this means eccentrics were dispensed with and for all positions of the gear a good distribution of steam was effected. The bogie was of the swing-link type. The boiler had a Belpaire firebox with a total heating surface of 1557·4 sq. ft and a pressure of 220 lb/sq. in. Another new feature was a blast pipe with moveable cap for varying the draught. Its weight without tender was 58 tons 10 cwt in working order.

The initial locomotive No. 999 was built to compare it with the three-cylinder compounds, the boilers being almost the same, the *999* class having slightly more tube heating surface. They spent all their working life stationed at Carlisle and worked mainly on the Carlisle–Leeds expresses.

Some neat outside cylinder 0-4-0Ts were built by Deeley with side tanks the whole length of the boiler and smokebox. To assist in maintenance and running repairs, outside Walschaerts valve gear was fitted. They were employed on brewery and colliery sidings.

The 0-6-4Ts previously referred to also had full length tanks but with inside cylinders it was necessary to cut away the tanks between the leading and driving wheels for access to the motion and oiling points. They were not very successful and, due to rolling, they were involved in a number of serious derailments.

The two-cylinder 4-4-0s built by Johnson were also rebuilt with larger boilers and at a later stage some were entirely rebuilt or, more correctly, renewed. This Deeley period was noteworthy in that there was more rebuilding than new work.

Mention has been made of Cecil Paget. In 1907 he had been promoted General Superintendent, his place as Works Manager being filled by Henry Fowler. Paget had designed a revolutionary locomotive and it was built at Derby in 1908.

The engine had eight single acting cylinders 18″ x 12″ with rotary steam distribution valves. Cecil Paget had always been interested in the Willans high speed central valve engines some of which had been installed in Derby Works power house, and he designed this locomotive incorporating these principles. To provide room for the cylinders outside frames were used. The valves were placed horizontally over each cylinder. Gearing was used for driving the valves and reversing them. The cut-off was adjusted by turning the valve liners in the steam chests. The cylinders were so placed that two drove the leading coupled wheels, two each side of the centre driving wheels were connected to that axle, and two drove the trailing coupled wheels, the coupling rods extending to the rear driving a jack shaft which carried the reversing gearbox. A swinging bolster truck was fitted at each end of the coupled wheels. The boiler was large, being 6′ 8″ diameter at the front. The firebox was embodied in the boiler shell and had 55 sq. ft of heating surface. It was lined with firebricks

Fig. 393
Midland Rly
1919 Fowler
4-cylinder
0-10-0 for
Lickey Bank

Fig. 394
1924
4-4-2T No. 2117 for London Tilbury & Southend section

Fig. 395
1929
Fowler 2-6-4T No. 2337

and had a brick bridge 9′ 0″ from the firing door. The locomotive had many unique features and theoretically the design was meticulously thought out, with special attention being paid to the comfort of the crew and to maintenance.

It was built at Derby, under a private order and Paget spent all his money on it which unfortunately did not cover the whole cost and the company paid an additional sum to complete it. It was tragic that this project met with intense hostility, so much so that it is surprising that it was ever built.

Although the boiler steamed well, trouble was experienced in leakage in the glands and piston rings, and the sleeves which were made of phosphor bronze and fitted into the cast iron chest resulted in unequal expansion and the ultimate seizure of the sleeves. After numerous trial trips of varying fortunes, the locomotive gravitated to the paint shop at Derby and was eventually scrapped in 1915.

Deeley and Paget were not on the best of terms and it was probably this animosity that accelerated Deeley's departure. Henry Fowler took the post of Chief Mechanical Engineer in 1909. Rebuilding continued and superheating was more generally used.

The most interesting locomotive built during this period was undoubtedly the 0-10-0 tender banker for the Lickey incline. It was built in 1919. It had four cylinders 16¾″ x 28″ and 4′ 7½″ diameter wheels. The boiler had a Belpaire firebox and a superheater giving 445 sq. ft of heating surface out of a total of 2163·25. The cylinders were inclined at 1 in 7. The tender had a cab fitted for working in reverse down the incline. The cylinders of this mammoth locomotive have been preserved at Derby.

Besides building for its own requirements, locomotives were built from time to time for the Somerset & Dorset Railway and for the Midland & Great Northern Joint Committee.

They were all standard Midland types except a new design of 2-8-0 for the S&DR. They were built in 1914 and had outside inclined cylinders 21″ x 28″ and coupled wheels 4′ 7½″ diameter (S&DR Nos. 80–85).

In 1911 two new 0-6-0s No. 3835/6 were built and were more powerful than previous locomotives of the same type. The cylinders were 20″ x 26″, fitted between the frames. The boiler was superheated with a moderate working pressure of 160 lb/sq. in. and a total heating surface of 1,483 sq. ft. The grate area was 21·1 sq. ft.

They were subjected to exhaustive tests and it was not until 1917 that building began in earnest and 192 were built by the Midland Railway. This class was to become the most numerous in the country at the time of building. 772 were built altogether, including the original 192, up to the year 1940.

Henry Fowler was knighted for war-time services.

Midland locomotives were of average size and no larger types, such as 4-6-0s, were contemplated, the policy being that if one locomotive cannot pull the train, couple on another one. Nevertheless standardisation was an established law and was carried on to the end of Midland days and unfortunately after that, when the MR became part of the LONDON MIDLAND AND SCOTTISH RAILWAY on 1 January, 1923.

For details of locomotive building during this period 1923–47 see under London Midland and Scottish Railway and also tables showing locomotives built at Derby.

From 1 January, 1948 the LM&SR became part of the nationalised BRITISH RAILWAYS and at first work proceeded with the building of standard LMS classes.

Fig. 396
1930
Royal Scot class
6161 THE KING'S OWN–
Liverpool &
Manchester
Rly Centenary,
Wavertree

Fig. 397
1952
BR *Class 4* 2-6-4T
80006

Fig. 398 **Midland Rly** 1956 BR *Class 5* 4-6-0 No. 73127–one of 30 built with Caprotti valve gear

Derby Works

The Works were first brought into use on or about 11 May, 1840 coincident with the opening of the North Midland Railway from Derby to Rotherham, and the new passenger station at Derby. The original repair shops were in the form of a polygon shaped building with a central turntable and radiating pits, the area of ground covered by all the locomotive department being 8½ acres. The first locomotive was built in 1851. The NMR became part of the Midland Railway in 1844.

In 1873 2000 workmen were employed and in the same year the Carriage and Wagon departments were separated from the Locomotive side and new shops built. This enabled the Locomotive department to be enlarged.

A new erecting shop was built in 1892, 450 ft long with three 50 ft bays, each bay having three lines of pits with a total accommodation for 108 locomotives.

1895 figures give 4346 workmen employed and an average output of forty new and 800 repaired steam locomotives and 120 new boilers. The locomotive stock stood at a total of 2400.

At the time of the 1923 amalgamation the Midland Railway possessed the largest number of steam locomotives of any railway in the United Kingdom with a total of 3019. Nine wheel arrangements comprised this total, five tender 2-4-0, 4-2-2, 4-4-0, 0-6-0 and one 0-10-0. The tank types were 0-4-0T, 0-6-4T, and 0-6-0T, so that standardisation as far as wheel arrangement was concerned was quite an achievement but in many ways, due to the policy of double-heading trains.

It was not until 1930 that Derby built its first six coupled express passenger tender locomotive although the 2-8-0s for the Somerset and Dorset Joint Railway were a radical departure for Derby.

In the 1930s the works were reorganised to carry out new construction and repairs on the progressive system similar to Crewe but on a smaller scale.

In 1955 the area covered by offices and workshops was given as 13 acres and total staff was approximately 3750. The number of classified repairs was 1000 out of a total of 2650 locomotives for which Derby was responsible.

Matthew Kirtley started a system of order numbers in 1875 appertaining to all work carried out in the locomotive shops. Commencing at No. 1 and having reached 9999 in 1938 a new series was commenced (2nd series) and this had reached the 9000s in 1956 embracing the last order for steam locomotive building.

Derby used these order numbers for boilers, cylinders and other parts for the NCC and S&DJR. Material was also supplied to other railways from time to time including the Furness, Midland & Great Northern Joint, London Brighton & South Coast, London Tilbury & Southend, and South Eastern & Chatham railways. Other orders were issued for locomotive rebuilding and modifications, repairs to London & North Western Railway locomotives and even for the cutting up of condemned locomotives. An excellent plan for costing purposes and also for an almost complete record of what went on from 1875 at least.

The Midland Railway bought many of its locomotives from outside contractors – more so than most of the other home railways.

Date	Type	Cylinders inches	DW ft in	Heating Surface sq ft				Grate Area sq ft	WP lb/sq in	Weight WO ton cwt	Class
				Tubes	S'heat	F'box	Total				
KIRTLEY											
1856	2-2-2	15 x 22	6 6	963	—	81	1044	14·5	120	28 9	110
1866	2-4-0	16½ x 22	6 2	960	—	85	1045	14·8	140	33 5	156
1870	2-4-0	17 x 24	6 8	989	—	99	1088		140	36 3	800
1873	2-4-0	17 x 24	6 8½	1025	—	92	1117	16·0	140	36 12	890
S. W. JOHNSON											
1876	0-4-4T	17 x 24	5 7	1150	—	104	1254	16·0	140	49 7	1252
1882	4-4-0	18 x 26	6 9	1032	—	110	1142	17·5	140	41 19	1562
1885	4-4-0	18 x 26	7 0½	1151	—	110	1261	17·5	160	42 19	1740
1890	4-2-2	18½ x 26	7 6	1123	—	117	1240	19·7	160	43 3	1853
1896	4-2-2	19½ x 26	7 9	1105	—	128	1233	21·3	170	47 2	115
1900	4-2-2	19½ x 26	7 9½	1070	—	147	1217	24·5	180	50 3	2601
1900	4-4-0	19½ x 26	6 9	1374	—	145	1519	25·0	180	51 12	**700 BELP***
1903	0-4-0T	15 x 20	3 9¾	697	—	64	761	10·5	150	30 19	1139A
1901	4-4-0	19/21 x 26	7 0	1448	—	150	1598	26·0	200	59 10	2631

* 1907 renumbering

Date	Type	Cylinders inches	DW ft in	Tubes	S'heat	F'box	Total	Grate Area sq ft	WP lb/sq in	Weight WO ton cwt	Class
R. M. DEELEY											
1905	4-4-0	19/21 x 26	7 0	1305·5	—	152·8	1458·3	28·4	220	59 16	1005
1907	4-4-0	19 x 26	6 6½	1404·6	—	152·8	1557·4	28·4	220	60 5	999
1903	0-6-0	18½ x 26	5 3	1303	—	125	1428	21·1	175	43 16	3775
1907	0-6-4T	18½ x 26	5 7	1222	—	125	1347	21·1	175	75 11	2000
1911†	4-4-0	19 x 26	6 6½	1170	360	151	1681	28·4	180		999

† Date superheated built 1907

Date	Type	Cylinders inches	DW ft in	Tubes	S'heat	F'box	Total	Grate Area sq ft	WP lb/sq in	Weight WO ton cwt	Class
H. FOWLER											
1911	0-6-0	20 x 26	5 3	1045	313	125	1483	21·1	160	49 2	3835
1914	2-8-0	21 x 28	4 7½	824	347	151	1322	28·4	220	64 15	S&D 80-85
1919	0-10-0	4/16¾ x 28	4 7½	1560	445	158·25	2163·25	31·5	180	73 13	2290
*1923	4-4-2T	1 x 26	6 6						170	71 10	2110
*1924	4-4-0	19/2 x 26	6 9	1170	272	147	1589	28·4	200	61 14	1045

* LMSR

O/No	Date Built	Type	Numbers
1	1875	0-4-4T	1226 – 1235
97	1876	2-4-0	147 – 156
107	1876	2-4-0	187 – 196
179	1877	2-4-0	197 – 206
204	1878	0-6-0T	1660 – 1669
218	1878/9	0-6-0T	1670 – 1689
232	1879	2-4-0	207 – 216
239	1879	0-6-0T	1690 – 1699
240	1880	0-6-0	3040 – 3049
262	1880	0-6-0T	1700 – 1709
273	1881	2-4-0	232 – 241
275	1881	2-4-0	272 – 281
279	1880	2-4-0	222 – 231
283	1880	2-4-0	217 – 221
289	1881/2	0-4-4T	1266 – 1285
340	1882	0-6-0T	1710 – 1719
341	1883	0-4-0ST	1500 – 1504
370	1882	4-4-0	328 – 337
400	1883	4-4-0	338 – 347
414	1883	0-6-0T	1720 – 1729
415	1883	0-4-4T	1286 – 1290
430	1883	4-4-0	348 – 357
444	1884	4-4-0	483 – 492
460	1883/4	0-4-4T	1291 – 1310
496	1884	0-6-0T	1730 – 1744
499	1885	0-6-0T	1745 – 1757
530	1885	0-6-0	3130 – 3139
538	1885	0-4-4T	1311 – 1320
544	1885	0-6-0	3140 – 3149
554	1885/6	4-4-0	358 – 367
589	1886	0-4-4T	1321 – 1330
615	1886/7	4-4-0	368 – 377
617	1886	0-6-0	3150 – 3159
633	1887	0-6-0	3160 – 3169
663	1887/8	0-6-0	3170 – 3189
678	1888	4-4-0	378 – 387
713	1888	0-6-0	3190 – 3199
734	1888	4-4-0	388 – 392
763	1889	0-4-4T	1331 – 1340
816	1889/90	0-4-OST	1505 – 1507
824	1889/90	0-6-OT	1760 – 1769
854	1890	0-6-OT	1770 – 1779
869	1890	0-6-OT	1780 – 1789
872	1891	4-4-0	S & DJR 15-18
883	1890	0-6-OT	1795 – 1804
920	1891	4-4-0	393 – 402
924	1890	0-6-OT	1790 – 1794
968	1891	0-6-OT	1805 – 1814
981	1892	0-4-4T	1341 – 1350
991	1891	0-6-OT	1815 – 1824
1162	1893	0-4-OST	1508 – 1512
1235	1894	4-4-0	443 – 452
1276	1894	4-4-0	453 – 462
1353	1894/5	0-6-0	3460 – 3469
1395	1895	0-6-OT	1845 – 1854
1410	1895	4-4-0	463 – 472
1431	1896	4-4-0	S & DJR 67-68
1449	1896	0-6-0	S & DJR 62-66
1458	1896	4-4-0	423 – 427
1482	1897	4-4-0	S & DJR 14 & 45
1534	1897	0-4-OST	1513 – 1517
1552	1897	0-4-OST	1518 – 1522

O/No	Date Built	Type	Numbers
1597	1897	4-4-0	493 – 502
1602	1898	0-4-4T	1401 – 1410
1635	1898	4-4-0	523 – 532
1834	1899	4-4-0	533 – 542
1869	1900/1	4-4-0	700 – 709
2041	1901	4-4-0	543 – 552
2109	1902/3	4-4-0	1000 – 1004
2135	1902	4-4-0	710 – 719
2250	1902	4-4-0	720 – 729
2328	1903	0-6-0	3765 – 3774
2458	1903	4-4-0	730 – 739
2517	1903	0-4-0ST	1523 – 1527
2530	1903	0-6-0	3775 – 3784
2588	1903	4-4-0	S & DJR 69-71
2601	1903/4	4-4-0	740 – 749
2652	1904	0-6-0	3785 – 3794
2692	1904	0-6-0	3795 – 3804
2726	1904/5	4-4-0	750 – 759
2798	1905	4-4-0	760 – 769
2821	1906/7	0-6-0	3805 – 3814
2833	1905	4-4-0	MR 63-66 NCC
2889	1905	4-4-0	1005 – 1014
2915	1905	2-2-0RM	MR 90 & 91 NCC
2918	1905	4-4-0	770 – 779
2998	1906	4-4-0	1015 – 1034
3031	1907	0-4-0T	1528 – 1532
3139	1907	4-4-0	999
3187	1907	0-6-4T	2000 – 2019
3258	1907	0-6-4T	2020 – 2039
3306	1908	2-6-2	2299
3310	1907	4-4-0	S & DJR 77 & 78
3344	1908	0-6-0	3815 – 3834
3371	1909	4-4-0	990 – 998
3385	1908	4-4-0	MR 67 & 68 NCC
3410	1908/9	4-4-0	1035 – 1044
4000	1911	0-6-0	3835
4001	1911	0-6-0	3836
4209	1914	2-8-0	S & DJR 80-85
4369	1914	4-4-0	MR 69 & 70 NCC
4482	1919	0-10-0	2290
4991	1917	0-6-0	3837 – 3851
5064	1918	0-6-0	3852 – 3861
5127	1918	0-6-0	3862 – 3871
5168	1918/9	0-6-0	3872 – 3886
5233	1919	0-6-0	3887 – 3901
5308	1920	0-6-0	3902 – 3916
5335	1920/1	0-6-0	3917 – 3936
5469	1921	0-6-0	3987 – 4006
5528	1921/2	0-4-0T	1533 – 1537
5530	1921/2	0-6-0	4007 – 4026

O/No 444 Locos 483 – 492 when rebuilt 1896-1901 under O/Nos 1460, 1707 & 2072 were 10 additional new locos.

STEAM LOCOMOTIVES ORDER NOS
1ST SERIES (contd)

O/No	Date Built	Type	Numbers
5648	1922	4-4-0	LMSNCC14 & 15
5649	1923	0-6-0	LMS 71-73 NCC
5871	1924	4-4-2T	2110 – 2119
5938	1924	4-4-0	1045 – 1064
6066	1924	4-4-0	1065 – 1084
6213	1924/5	0-6-0	4027 – 4056
6293	1925	4-4-0	1085 – 1114
6438	1925/6	0-6-0	4207 – 4226
6460	1926	0-6-0	4227 – 4246
6473	1926	0-6-0	4247 – 4266
6486	1926	0-6-0	4267 – 4286
6632	1926/7	0-6-0	4287 – 4301
6751	1927	4-4-2T	2125 – 2134
6807	1927/8	2-6-4T	2300 – 2324
6841	1927	0-6-0	4407 – 4436
6901	1928	4-4-0	(a)
7080	1928	4-4-0	(b)
7120	1929	2-6-4T	2325 – 2334
7137	1928/9	0-6-OT	11270 – 11279
7224	1929	2-6-4T	2335 – 2354
7237	1929	2-6-4T	2355 – 2374
7403	1927/30	4-4-0	(c)
7406	1930	4-4-2T	2151 – 2160
7467	1930/1	2-6-2T	15500 – 15524
7575	1931	2-6-2T	15525 – 15549
7580	1930	4-6-0	6150 – 6169
7753	1931	2-6-2T	15550 – 15559
7854	1931/2	4-4-0	661 – 685
7860	1932/3	0-4-4T	6400 – 6409
8027	1932	2-6-4T	2375 – 2384
8052	1932	2-6-2T	15560 – 15569
8078	1932	4-4-0	935 – 939
8207	1934	2-6-0	LMS 90–93 NCC
8241	1933	2-6-4T	2385 – 2394
8338	1933/4	2-6-4T	2395 – 2424
8425	1934	2-6-4T	2500 – 2504
8503	1934	2-6-4T	2505 – 2536
8610	1934/5	4-6-0	5655 – 5564
8638	1935	2-6-2T	71 – 90
8880	1935	2-6-2T	91 – 110
8882	1935	2-6-2T	111 – 144
8884	1935	2-6-4T	2537 – 2544
9204	1936	2-6-4T	2425 – 2444
9206	1936	2-6-4T	2445 – 2464
9208	1936/7	2-6-4T	2465 – 2494
9696	1937	2-6-2T	145 – 159
9710	1937/8	2-6-2T	160 – 174

(a) Numbers 563–571, 573/4/7-9, 581–592, 601. S & DJR 44-46
(b) Numbers 572, 593–600, 602–612
(c) Numbers 575/6, 580, 613-632.
Under O/Nos 7656 and 8179 twelve Claughtons were rebuilt but not classed as new locos.

2ND SERIES

O/No	Date Built	Type	Numbers
46	1938	2-6-2T	175 – 184
112	1938	2-6-4T	2618 – 2627
114	1938	2-6-4T	2628 – 2637
116	1838/9	2-6-4T	2638 – 2652
303	1939	0-6-0	4577 – 4586
650	1939	0-6-0	4587 – 4596
653	1940/1	0-6-0	4597 – 4606
657	1940/1	2-6-4T	2653 – 2662
660	1942/3	2-6-4T	2663 – 2672
3836	1943	4-6-0	5472 – 5481
4141	1943/4	4-6-0	5482 – 5496
4888	1944	4-6-0	5497 – 5499, 4800 – 4806
8277	1945	2-6-4T	2673 – 2677
8278	1945	2-6-4T	2678 – 2687
8283	1945	4-6-0	4807 – 4825
8467	1945	2-6-4T	2688 – 2699, 2200 – 2202
8468	1945/6	2-6-4T	2203 – 2222

3RD SERIES

O/No	Date Built	Type	Numbers
669	1946	2-6-4T	LMS 5 - 8 (a) NCC
672	1946	2-6-4T	2223 – 2232
675	1946	2-6-4T	2233 – 2252
678	1946/7	2-6-4T	2253 – 2272
676	1947	2-6-4T	2273 – 2292
678(b)	1947	2-6-4T	2293-2299, 2187-2189
1674	1947	2-6-4T	LMS 1-4, 9 & 10(a) NCC

(a) For Belfast 5' 3" gauge (b) O/No 678 order completed in 1948 up to 2199.

British Railways Derby
In 1948 work proceeded with the standard *Class 4* 2-6-4Ts and forty-six were built in that year followed by thirty in 1949 and thirty in 1950.

In 1951 the first *Class 5* 4-6-0s appeared, and Derby continued to build them until 1957 when the works had completed 110 out of a total of 155.

In 1952 ten of Ivatt's standard *Class 2* 2-6-2Ts were built and in the same year the BR 2-6-4T design was commenced. Ten were built in this year, four in 1954 and one in 1955, the majority being built at Brighton.

The last steam locomotive to be built at Derby was *Class 5* 4-6-0 No. 73154 which was steamed on 13 June, 1957.

MIDLAND RAILWAY

The works were heavily engaged on building diesel locomotives up to 1967 but none were built during 1968 and the works are engaged on repair work.

Steam locomotives built for British Railways were:

46	LMS	2-6-4T	Class 4	42147–82, 42190–99	1948
30	LMS	2-6-4T	Class 4	42107–32, 42183–86	1949
30	LMS	2-6-4T	Class 4	42050–65, 42133–46	1950
29	BR	4-6-0	Class 5	73000–28	1951
1	BR	4-6-0	Class 5	73029	1952
10	BR	2-6-4T	Class 4	80000–09	1952
10	LMS	2-6-2T	Class 2	41320–29	1952
20	BR	4-6-0	Class 5	73030–49	1953
4	BR	2-6-4T	Class 4	80054–57	1954
25	BR	4-6-0	Class 5	73050–74	1954
25	BR	4-6-0	Class 5	73075–99	1955
1	BR	2-6-4T	Class 4	80058	1955
20	BR	4-6-0	Class 5	73125–44	1956
10	BR	4-6-0	Class 5	73145–54	1957
261					

The number of locomotives built at Derby from 1851 to 1957 was 2995.

O/No	Date Built	Type	NUMBERS		
			LMS Designs	BR Designs	
1678	1948	2-6-4T	42190 – 42199		
2420	1948/9	2-6-4T	42147 – 42186		
3282	1949/50	2-6-4T	42107 – 42146		
3283	1949	2-6-4T	NCC 50–53		For Belfast (5′ 3″ gauge)
4310	1950	2-6-4T	42050 – 42065		
4332	1950	2-6-4T	NCC 54 – 57		For Belfast (5′ 3″ gauge)
5122	1951/2	4-6-0		73000 – 73029	
5124	1952	2-6-4T		80000 – 80009	
5125	1952	2-6-2T	41320 – 41329		Boilers from Crewe
6230	1953	4-6-0		73030 – 73049	
6231	1954/5	2-6-4T		80054 – 80058	
6735	1954	4-6-0		73050 – 73059	
8025	1954	4-6-0		73065 – 73074	
8026	1954	4-6-0		73060 – 73064	
8241	1955	4-6-0		73075 – 73089	
8845	1955	4-6-0		73090 – 73099	
9247	1956/7	4-6-0		73125 – 73154	Fitted with Caprotti valve gear

Locomotives in 40,000 series: some of the 1948 2-6-4Ts came out with LMS numbers, then had 40,000 added. All shown with BR numbers.

SUMMARY OF LOCOMOTIVES—BUILT AT DERBY

Period	MR	NCC	S & DJC	LMS	BR	Totals
1851 – 1922	1560	10	24	–	–	1594
1923 – 1947	–	19	3	1110	–	1132
1948 – 1957	–	8	–	–	261	269
	1560	37	27	1110	261	2995

MIDLAND RAILWAY
Northern Counties Committee

See BELFAST & NORTHERN COUNTIES RAILWAY.

MIDLAND & GREAT NORTHERN JOINT RAILWAY

Melton Constable

The works came into operation in 1883 having been completed by the Eastern and Midlands Railway before the formation of the M&GNJR on 1 July, 1893. The works were compact and well equipped, the erecting shop had two roads, and there was a machine shop, boiler shop, forge, smithy and other ancillary shops. A tremendous amount of rebuilding was carried out at Melton Constable.

Two types were built. A 0-6-0 Side tank which replaced six of the ex Cornwall Minerals Railway locomotives. Eight had been purchased from Sharp Stewart & Co. after that firm had received them back from Cornwall and they were sent to Norfolk equipped with four-wheel tenders and cabs. Four were rebuilt at Melton Constable to 2-4-0 tender type.

The other class was a 4-4-2T introduced in 1904 due to the increase in passenger traffic. They were well proportioned locomotives with outside cylinders and side tanks sloping to the front.

The new 0-6-0Ts had two 16" × 20" outside cylinders, 3' 6½" diameter wheels, boiler with round top firebox with a total heating surface of 737 sq. ft. Those which had old boilers fitted operated at 140 lb/sq. in. and those with new boilers 150 lb/sq. in. Their total weight in working order was 37 tons 13 cwt. Grate area was 11·3 sq. ft.

As will be seen from the list, three lasted long enough to have British Railways numbers. The 4-4-2 tanks had two outside cylinders 17¼" × 24", 3' 0" diameter bogie wheels, 6' 0"

Built	Type	Cylinders inches	DW ft in	1st No	Reno 1907	1st LNE No	LNE Reno	MGN Class	LNE Class	W'drawn
1897	0-6-0T	16 x 20	3 6½	14A	93	093	(8483)	MR	J93	1944
1898	0-6-0T	16 x 20	3 6½	1A	94	094	8488	MR	J93	1/1948
1899	0-6-0T	16 x 20	3 6½	3A	95	095	8485	MR	J93	12/1947
1899	0-6-0T	16 x 20	3 6½	11A	96	096	8484	MR	J93	5/1948
1900	0-6-0T	16 x 20	3 6½	15	–	015	(8486)	MR	J93	11/1945
1902	0-6-0T	16 x 20	3 6½	12A	97	097	–	MR	J93	
1902	0-6-0T	16 x 20	3 6½	17A	98	098	8482	MR	J93	1/1947
1903	0-6-0T	16 x 20	3 6½	2A	99	099	(8487)	MR	J93	1945
1905	0-6-0T	16 x 20	3 6½	16	–	016	8489	MR	J93	1949
1904	4-4-2T	17¼ x 24	6 0	41	–	041	(7503)	A	A*	1944
1909	4-4-2T	17¼ x 24	6 0	20	–	020	–	A	A	1942
1910	4-4-2T	17¼ x 24	6 0	9	–	09	(7504)	A	A*	1944

M&GN Class letters are doubtful *Became LNE Class C17 in 1942

Fig. 399 **M&GNJR** 1902 0-6-0T No. 17A re-numbered 98 with enlarged coal bunker

MIDLAND & GREAT NORTHERN JOINT RAILWAY

diameter drivers and 3′ 6½″ trailing wheels. The boiler had a round top firebox with a total heating surface of 1099 sq. ft. Boiler pressure was 160 lb/sq. in. and the total weight in working order was 68 tons 9 cwt. The tank capacity was 1650 gallons of water and 2 tons of coal.

All locomotives built by the M & GN incorporated parts of those they replaced but the large amount of work entailed could justify them termed as 'new'.

Mr William Marriott was the Resident Engineer and Locomotive Superintendent during the period of new building.

The works closed in October 1930.

MIDLAND GREAT WESTERN RAILWAY

Broadstone, Dublin

It was common practice in Ireland for the contractors who built the railways to contract for working them as well. In early days when one contractor's time had expired another would outbid him and would get the contract for the next period. In this way contracting could change hands quite a number of times.

On the MGWR the first contractor was a John Dawson who owned the Phibsborough carriage works. They were bought by the railway company when Dawson's contract expired in 1851 and Broadstone works built on the site. After a number of alterations to the works, the final locomotive erecting shop had four roads each 160 ft long.

The first locomotives were not built until September 1879, all previous stock being supplied by many British builders and a few

Fig. 400
M&GNJR 1904
4-4-2T No. 41

Fig. 401
MGWR 1891
0-6-0 No. 74
LUNA

MIDLAND GREAT WESTERN RAILWAY

Fig. 402 **MGWR** 1897 *Class K* 2-4-0 No. 27 CLIFDEN

from Thomas Grendon & Co. of Drogheda. The predominant types were 2-2-2, 2-4-0, 0-6-0 and 0-4-2s and a great deal of rebuilding had been carried out at Broadstone before new locomotives were contemplated.

The 1879 locomotives were the first two of six 0-6-0s laid down. They had 18″ x 24″ cylinders and 5′ 1″ diameter wheels and were named MARQUIS, VISCOUNT, BARON, REGENT, DUKE and EARL. The practice of naming the locomotives started with the appearance of all new locomotives until 1855 when the names were all removed, but in 1868 the policy was reversed and naming recommenced.

The next batch, built in 1880, was four 0-6-0Ts for banking duties and they were mostly employed in the Dublin area. No more building took place until 1884 and from this date successive batches of 2-4-0s and 0-6-0s were put in hand.

The Locomotive Superintendent at this time was Mr Martin Atock who had taken office in 1872. Apart from the all-over cabs of some tank locomotives, the cabs for tender locomotives were of peculiar design, the roof having a turned upwards flare which gave the locomotive a rather archaic appearance. The chimneys were neat but tall with a pronounced flare at the top. The turned up cab was intended to minimise wind resistance and to keep the steam and smoke above the cab. This was satisfactory except when the locomotive was running tender first!

Some form of standardisation was introduced although many old parts were used where feasible in the building of some of the 'new' replacements.

The 0-6-0s had 18″ x 24″ cylinders, and 5′ 1″ diameter wheels. The wheels were enlarged to 5′ 3″ diameter for the 1885 batch and subsequent batches and altogether thirty-two were built at Broadstone from 1885 to 1893. They had domed boilers with round top flush fireboxes with a total heating surface of 1065 sq. ft. Others were built by Sharp Stewart & Co. (five) and Kitson & Co. (five) in 1895.

The 2-4-0s had 16″ x 22″ cylinders and 5′ 8″ diameter coupled wheels, and a domed boiler with flush round top firebox. The total heating surface was 914 sq. ft. In 1894 an enlarged edition appeared with 17″ x 24″ cylinders and 6′ 3″ diameter coupled wheels and a larger boiler giving a total heating surface of 1115 sq. ft. Six were built in 1889–90. They were intended for the fast main line passenger traffic.

In 1893 a further class of 2-4-0s was built with the same cylinders and boilers but the coupled wheel diameter was reduced to 5′ 8″. Twenty were built from 1893 to 1898.

In 1900 Mr Atock introduced the 4-4-0 passenger type on to the MGWR by drastically rebuilding six 2-4-0s which had been built by Beyer Peacock & Co. in 1880. They were not completed before Mr Atock's retirement.

He was replaced by Mr E. Cusack in 1901. Martin Atock had left behind small but efficient locomotives during his twenty-one years in

MIDLAND GREAT WESTERN RAILWAY

Fig. 403 **MGWR** 1903 *Class A* 4-4-0 No. 127 TITANIC in shop grey

office, keeping the classes to a minimum with a maximum amount of standardisation. He rebuilt many of the older locomotives and indeed carefully used many parts of withdrawn locomotives in building some of his 'new' ones.

His tenders were unique with springs placed over the axleboxes but inside the frames. They were so awkward to change that slots were cut opposite each one in the tender frame to make changing and servicing the springs an easier matter.

In 1902 Mr Cusack built his first class of six 4-4-0s. They were much larger than Atock's rebuilt 2-4-0s, indeed at the time they were the largest locomotives in Ireland. They had inside cylinders 18" x 26" and 6' 3" diameter coupled wheels.

In 1909 he introduced a further class of 4-4-0s with a smaller boiler, although they were subsequently rebuilt with a similar boiler to his 1902 class. These were the last passenger locomotives to be built by the MGWR at Broadstone. A total of nine were built.

In 1915 Mr Cusack retired and his place was taken by his chief draughtsman W. H. Morton who had more to do with the design of locomotives during Mr Cusack's term of office than Mr Cusack himself. Only one further MGWR design was built at Broadstone, it was for a larger 0-6-0 with 19" x 26" cylinders, 5' 8" diameter wheels and a boiler with a Belpaire firebox. Eighteen were built from 1921 to 1924 and were known as *Class F.* With their 5' 8" wheels, they were intended for mixed traffic work and turned out to be eminently suitable for all types of traffic. The Belpaire firebox was a new departure for new locomotives at Broadstone. Many of the earlier 0-6-0s had also been fitted.

Superheating was also introduced by Morton and one of the early Broadstone 0-6-0s (No. 5) was also fitted with a firebox superheater in 1919 devised by Mr Cusack and Mr Morton.

When the last Belpaire 0-6-0 appeared in June 1924, 120 locomotives had been built at Broadstone, and whilst many had been built by outside contractors up to Mr Cusack's accession very few were ordered in Great Britain after that. Armstrong Whitworth supplied five of the *Class F,* 0-6-0s in 1921, and the North British Locomotive Co. four 0-6-0s in 1904 with 19" x 26" cylinders and 5' 3" diameter wheels and known as *Class B.*

The works were always occupied in repairs and rebuilding and conversion of many boilers to superheating. Some older 0-6-0s had Belpaire boilers fitted, and the *Class 13* 2-4-0s of Atock design, introduced in 1893, were also superheated.

On 12 November, 1924 the MGWR, Cork Bandon & South Coast Railway, and the Great Southern & Western Railway undertakings were merged into one company and designated: THE GREAT SOUTHERN RAILWAY CO.

Just previous to this amalgamation twelve sets of parts had been purchased by the MGWR of the Woolwich 2-6-0. They were erected at Broadstone from 1925 to 1927 and were known as *Class K1.* They had two outside cylinders 19" x 28" and 5' 6" diameter coupled wheels and weighed 62 tons 4 cwt in working order, without tenders.

Little work was done apart from erecting these locomotives and strictly speaking the total of 120 locomotives built at Broadstone is correct. A total of 132 were erected.

Mr J. R. Bazin, who had charge of Inchicore

on the GS&WR, now took over the locomotive departments of the other railways in the amalgamation. Even at this time traffic was decreasing due to road transport competition and coupled with the poor financial state of the company very few new locomotives appeared after 1927 and those that did were built at Inchicore. Repairs were carried out at Broadstone until 1933 when the last repaired locomotive left the works on 30 June of that year. The works were taken over by the bus department who still use them.

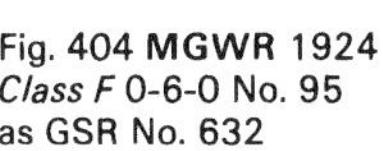

Fig. 404 **MGWR** 1924
Class F 0-6-0 No. 95
as GSR No. 632

Fig. 405 1925
Class K1 2-6-0 No. 373
erected from Woolwich
Arsenal parts

P No	Date	Type	Cylinders inches	DW ft in	First No	Name	GSR No	W' Drawn
1	9/1879	0-6-0	18 x 24	5 1	49	Marquis	(563)	1928
2	9/1879	0-6-0	18 x 24	5 1	50	Viscount	(564)	1925
3	5/1880	0-6-0	18 x 24	5 1	51	Regent	(565)	1926
4	6/1880	0-6-0	18 x 24	5 1	52	Baron	(566)	1927
5	9/1880	0-6-0	18 x 24	5 1	53	Duke	567	1950
6	9/1880	0-6-0	18 x 24	5 1	54	Earl	(568)	1925
7	1880	0-6-0T	18 x 24	4 6	100	Giantess	614	1955
8	1880	0-6-0T	18 x 24	4 6	101	Giant	615	1951
9	1880	0-6-0T	18 x 24	4 6	102	Pilot	616	1950
10	1880	0-6-0T	18 x 24	4 6	103	Pioneer	617	1959

MIDLAND GREAT WESTERN RAILWAY

P No	Date	Type	Cylinders inches	DW ft in	First No	Name	GSR No	W'drawn
11	1883	2-4-0	16 x 22	5 8	41	Regal		1915
12	1883	2-4-0	16 x 22	5 8	42	Ouzel		1921
13	1884	2-4-0	16 x 22	5 8	1	Orion		1922
14	1884	2-4-0	16 x 22	5 8	4	Venus		1910
15	1884	2-4-0	16 x 22	5 8	5	Mars		1910
16	1884	2-4-0	16 x 22	5 8	6	Vesta		1961
17	1885	0-6-0	18 x 24	5 3	55	Inny	594	1961
18	1885	0-6-0	18 x 24	5 3	56	Liffey	595	1957
19	1885	0-6-0	18 x 24	5 3	57	Lough Corrib	596	1959
20	1885	0-6-0	18 x 24	5 3	58	Lough Gill	597	1959
21	1885	0-6-0	18 x 24	5 3	59	Shannon	598	1965
22	1885	0-6-0	18 x 24	5 3	60	Lough Owel	599	1964
23	1886	0-6-0	18 x 24	5 3	85	Meath		1924
24	1886	2-4-0	16 x 22	5 8	45	Queen		1916
25	1886	2-4-0	16 x 22	5 8	47	Viceroy		1921
26	1886	0-6-0	18 x 24	5 3	104	Wren	611	1925
27	1887	2-4-0	16 x 22	5 8	43	Leinster		1916
28	1887	2-4-0	16 x 22	5 8	44	Ulster		1911
29	1887	2-4-0	16 x 22	5 8	46	Munster		1921
30	1887	2-4-0	16 x 22	5 8	48	Connaught		1922
31	1887	0-6-0	18 x 24	5 3	68	Mullingar	606	1963
32	1887	0-6-0	18 x 24	5 3	71	Galway	609	1954
33	1888	0-6-0	18 x 24	5 3	61	Lynx	600	1957
34	1888	0-6-0	18 x 24	5 3	62	Tiger	601	1959
35	1888	0-6-0	18 x 24	5 3	63	Lion	602	1959
36	1888	0-6-0	18 x 24	5 3	64	Leopard		1923
37	1888	0-6-0	18 x 24	5 3	65	Wolf	603	1965
38	1888	0-6-0	18 x 24	5 3	67	Dublin	605	1957
39	1888	0-6-0	18 x 24	5 3	72	Sligo	610	1963
40	1889	0-6-0	18 x 24	5 3	66	Elephant	604	1961
41	1889	0-6-0	18 x 24	5 3	69	Athlone	607	1962
42	1889	0-6-0	18 x 24	5 3	70	Ballinasloe	608	1959
43	1889	2-4-0	17 x 24	6 3	7	Connemara		1909
44	12/1889	2-4-0	17 x 24	6 3	10	Faugh-a-Ballagh		1910
45	3/1890	2-4-0	17 x 24	6 3	11	Erin-go-Bragh		1922
46	6/1890	2-4-0	17 x 24	6 3	9	Emerald Isle		*c* 1912
47	8/1890	2-4-0	17 x 24	6 3	8	St Patrick		1914
48	10/1890	2-4-0	17 x 24	6 3	12	Shamrock		1910
49	1891	0-6-0T	18 x 24	4 6	105	Hercules	618	1949
50	4/1891	0-6-0	18 x 24	5 3	84	Dunkellen	(572)	1925
51	6/1891	0-6-0	18 x 24	5 3	80	Dunsandle	574	1963
52	9/1891	0-6-0	18 x 24	5 3	74	Luna	576	1957
53	10/1891	0-6-0	18 x 24	5 3	75	Hector	612	1961
54	1/1892	0-6-0	18 x 24	5 3	83	Lucan	(571)	1925
55	3/1892	0-6-0	18 x 24	5 3	82	Clonbrook	583	1963
56	5/1892	0-6-0	18 x 24	5 3	76	Lightning	(569)	1925
57	8/1892	0-6-0	18 x 24	5 3	73	Comet	582	1959
58	9/1892	0-6-0	18 x 24	5 3	77	Star	589	1963
59	11/1892	0-6-0	18 x 24	5 3	79	Mayo	(578)	1927
60	2/1893	0-6-0	18 x 24	5 3	78	Planet	(570)	1925
61	3/1893	0-6-0	18 x 24	5 3	81	Clancarty	613	1963
62	6/1893	2-4-0	17 x 24	5 8	13	Rapid	659	1961
63	6/1893	2-4-0	17 x 24	5 8	14	Racer	650	1959
64	6/1893	2-4-0	17 x 24	5 8	18	Ranger	652	1954
65	1894	2-4-0	17 x 24	5 8	17	Reindeer	661	1959
66	1894	2-4-0	17 x 24	5 8	19	Spencer	653	1963
67	1894	2-4-0	17 x 24	5 8	20	Speedy		1923
68	5/1895	2-4-0	17 x 24	5 8	21	Swift	662	1955
69	7/1895	2-4-0	17 x 24	5 8	23	Sylph	664	1961
70	9/1895	2-4-0	17 x 24	5 8	22	Samson	663	1959
71	12/1895	2-4-0	17 x 24	5 8	15	Rover	660	1959
72	12/1895	2-4-0	17 x 24	5 8	16	Rob Roy	651	1959

MIDLAND GREAT WESTERN RAILWAY

P No	Date	Type	Cylinders inches	DW ft in	First No	Name	GSR No	W'drawn
73	3/1897	2-4-0	17 x 24	5 8	28	Clara	654	1962
74	8/1897	2-4-0	17 x 24	5 8	27	Clifden	666	1957
75	1897	2-4-0	17 x 24	5 8	24	Sprite	665	1959
76	1897	2-4-0	17 x 24	5 8	29	Clonsilla	655	1961
77	1/1898	2-4-0	17 x 24	5 8	32	Ariel	668	1959
78	4/1898	2-4-0	17 x 24	5 8	30	Active	656	1957
79	8/1898	2-4-0	17 x 24	5 8	34	Aurora	658	1954
80	9/1898	2-4-0	17 x 24	5 8	33	Arrow	657	1961
81	11/1898	2-4-0	17 x 24	5 8	31	Alert	667	1957
82	1900	4-4-0	16 x 22	5 8	36	Empress of Austria	530	1949
83	3/1900	4-4-0	16 x 22	5 8	2	Jupiter	534	1949
84	3/1900	4-4-0	16 x 22	5 8	26	Britannia	532	1949
85	9/1900	4-4-0	16 x 22	5 8	37	Wolf Dog	533	1953
86	3/1901	4-4-0	16 x 22	5 8	25	Cyclops	531	1945
87	5/1901	4-4-0	16 x 22	5 8	3	Juno	535	1949
88	6/1902	4-4-0	18 x 26	6 3	129	Celtic	546	1959
89	12/1902	4-4-0	18 x 26	6 3	128	Majestic	549	1931
90	9/1903	4-4-0	18 x 26	6 3	127	Titanic	545	1955
91	4/1904	4-4-0	18 x 26	6 3	126	Atlantic	548	1955
92	1905	4-4-0	18 x 26	6 3	125	Britannic	547	1954
93	1905	4-4-0	18 x 26	6 3	124	Mercuric	550	1957
94	7/1909	4-4-0	18 x 26	6 3	7	Connemara	540	1953
95	12/1909	4-4-0	18 x 26	6 3	10	Faugh-a-Ballagh	543	1959
96	5/1910	4-4-0	18 x 26	6 3	5	Croagh Patrick	539	1952
97	12/1910	4-4-0	18 x 26	6 3	4	Ballynahinch	538	1950
98	1911	4-4-0	18 x 26	6 3	6	Kylemore	542	1959
99	5/1912	4-4-0	18 x 26	6 3	9	Emerald Isle	537	1959
100	2/1913	4-4-0	18 x 26	6 3	12	Shamrock	536	1951
101	6/1913	4-4-0	18 x 26	6 3	8	St Patrick	541	1959
102	5/1915	4-4-0	18 x 26	6 3	11	Erin-go-Bragh	544	1955
103	4/1921	0-6-0	19 x 26	5 8	39		633	1957
104	1921	0-6-0	19 x 26	5 8	40		634	1959
105	1921	0-6-0	19 x 26	5 8	41		635	1957
106	1921	0-6-0	19 x 26	5 8	42		639	1963
107	3/1922	0-6-0	19 x 26	5 8	43		640	1960
108	7/1922	0-6-0	19 x 26	5 8	36		636	1959
109	7/1922	0-6-0	19 x 26	5 8	37		637	1963
110	1922	0-6-0	19 x 26	5 8	38		638	1963
111	1923	0-6-0	19 x 26	5 8	35		623	1957
112	1/1924	0-6-0	19 x 26	5 8	87		624	1965
113	1/1924	0-6-0	19 x 26	5 8	88		625	1961
114	1/1924	0-6-0	19 x 26	5 8	89		626	1961
115	2/1924	0-6-0	19 x 26	5 8	90		627	1961
116	6/1924	0-6-0	19 x 26	5 8	91		628	1954
117	6/1924	0-6-0	19 x 26	5 8	92		629	1954
118	6/1924	0-6-0	19 x 26	5 8	93		630	1959
119	6/1924	0-6-0	19 x 26	5 8	94		631	1954
120	6/1924	0-6-0	19 x 26	5 8	95		632	1959
121	4/1925	2-6-0	19 x 28	5 6	(a)			1960
122	9/1925	2-6-0	19 x 28	5 6	373			1959
123	11/1925	2-6-0	19 x 28	5 6	374			1959
124	12/1925	2-6-0	19 x 28	5 6	375			1957
125	2/1926	2-6-0	19 x 28	5 6	376			1961
126	3/1926	2-6-0	19 x 28	5 6	377			1960
127	10/1926	2-6-0	19 x 28	5 6	378			1959
128	12/1926	2-6-0	19 x 28	5 6	379			1959
129	12/1926	2-6-0	19 x 28	5 6	380			1959
130	12/1926	2-6-0	19 x 28	5 6	381			1959
131	1/1927	2-6-0	19 x 28	5 6	382			1955
132	3/1927	2-6-0	19 x 28	5 6	383			1959

(a) Built as MGWR No 49, photographed as GSR 410 then immediately changed to No 372.

MILLER & CO.
Vulcan Foundry, Coatbridge

There is no evidence that this firm built any locomotives, but they supplied at least one to the three foot gauge Tudhoe Ironworks system (a 0-4-0ST with outside cylinders). They were probably agents.

MILLER & BARNES
Glass House Field, Ratcliffe

This firm was established in 1822 by Joseph Miller and John Barnes, changing to MILLER & RAVENHILL and then to RAVENHILL & SALKELD at unknown dates. The firm quoted for supplying a locomotive for the North Midland Railway in 1838; £1450 for the engine, and £170 for the tender and in the NMR's minute No. 219 dated 6/2/1838 it was resolved to order two.

About 1840 the firm was taken over by SIMPSON & CO. and it would appear that these two locomotives tendered for were delivered after Simpson & Co. had taken over.

MONKLAND & KIRKINTILLOCH RAILWAY
Kipps, Nr. Coatbridge

The railway was incorporated in 1824 and opened in 1826. The gauge was 4′ 6″. As stated in the Ballochney Railway section, their first Greenside shops were used until new shops were completed at Moss-side, Kipps; and all M&KR work was concentrated here. These shops survived until the mid 1960s and were known latterly as "Kipps Wagon-shop'.

George Lish took charge of the new works leaving William Dodds to concentrate on Ballochney Railway work at Greenside. The first locomotive built at Moss-side was probably a 0-4-0 named ATLAS with 13″ x 20″ inside cylinders and 4′ 0″ diameter wheels, completed in 1840. Two similar 0-4-0s named ZEPHYR and SIROCCO were completed in 1841 or 1842, the latter having 14″ x 20″ cylinders and 4′ 6″ diameter wheels.

The names THETIS and BEDLAY have been mentioned as new locomotives but these were probably old locomotives which may have been rebuilt or even renamed by 1852. BEDLAY may not even have been a name but merely 'the engine at Bedlay'.

It is likely that locomotives may have been built for local industry or coal owners adjacent to the railway and this equally applies to the Greenside works of the Ballochney Railway. After 1842, new locomotives required were built by private builders especially Neilson & Mitchell.

On amalgamation with the Ballochney and Slamannan Railways in 1848 the title became the Monkland Railways Co. and the gauge converted to 4′ $8\frac{1}{2}$″.

Unfortunately the early Minute books, before 1848, no longer exist and the above notes, and those for the Ballochney Railway have been made possible due to information received from Messrs A. G. Dunbar, Jas. F. McEwan, E. Craven and D. Martin.

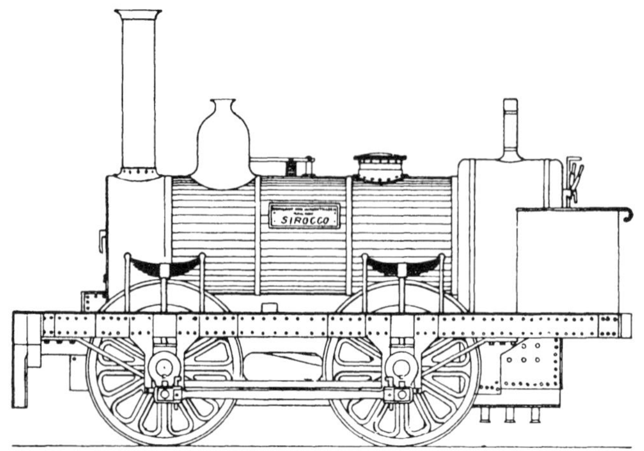

Fig. 406 **M&KR** 1841-2 0-4-0 SIROCCO

MONMOUTHSHIRE RAILWAY & CANAL CO.

Dock Street, Newport

This railway commenced operations in 1849 with locomotives built by Grylls, Neath Abbey and Stothert & Slaughter. These proved to be too heavy and rigid for the road and horses were re-employed for a period.

After relaying, W. Craig was appointed Locomotive Superintendent in September 1849 and he proceeded to order a number of locomotives from Stothert & Slaughter of varied types. He resigned in 1854 and his place was taken by Richard Laybourne who came from Crewe.

The Dock Street works were built about 1855 under his supervision and when completed all heavy repairs and rebuilding was carried out there, and the first new locomotive, a 0-6-0ST, appeared in 1867 followed by another in 1868. They had outside cylinders 16″ x 24″ and 4′ 0″ diameter wheels, the rear pair being the driven wheels.

Henry Appleby took over in 1868 and the Board decided to carry on building their own locomotives at the rate of two per annum supplementing their requirements with orders to outside builders. This policy was followed in 1870 and 1871 but only one was built in 1872, 1873 and 1875 and none in 1874.

The 0-6-0 side tanks were built from 1870 and two additional ones were obtained from the Yorkshire Engine Co. in 1871. These were the first 0-6-0T type to be built by this firm, and it would appear that Henry Appleby was responsible for the design. Although a 'standard' design, some dimensions, particularly for the boiler, firebox, and wheelbase varied from one to another. The 4-4-0 side tanks were built from 1870 also and had the same size cylinders as the 0-6-0Ts. A similar locomotive was built by the Yorkshire Engine Co. for the Neath & Brecon Railway.

NPT No	MR No	Built	Type	Cylinders outside inches	DW ft in	GWR No	W'drawn
1	37	1867	0-6-0ST	16 x 24	4 0	1338	1905
2	38	1868	0-6-0ST	16 x 24	4 0	–	1875
7	14	1870	4-4-0T	16 x 24	5 0	1304	1905
8	11	1870	0-6-0T	16 x 24	4 1	1339	1904
9	41	1871	4-4-0T	16 x 24	5 0	1305	1905
10	3	1871	0-6-0T	16 x 24	4 1	1340	1903
12	15	1872	4-4-0T	16 x 24	5 0	1306	1904
15 or 16	9	1873	0-6-0T	16 x 24	4 1	1341	1905
18	10	1875	4-4-0T	16 x 24	5 0	1307	1905
		*	0-6-0T	16 x 24	4 1	1342	1903
		†	0-6-0T	16 x 24	4 1	1383	1898

*Completed at Swindon Oct 1876. †Completed at Swindon Apr 1873.

Fig. 407 **Monmouthshire R.&C. Co.** 1875 Appleby's 4-4-0T as GWR 1307

MONMOUTHSHIRE RAILWAY & CANAL CO.

Henry Appleby also acted as a consultant to the N&BR which would account for the similarity of the 4-4-0Ts of the two railways. Two other types were standardised by Appleby, the 0-4-4T and 0-6-0ST, but all these were built by outside firms.

A series of works numbers were allocated for the new locomotives built and also for various rebuilds and conversions, but the official list has never been available and therefore the numbers shown are mostly pure assumption. Due to the limited facilities at the Dock Street works, many parts for the new locomotives were bought out including a number of boilers.

The Great Western Railway took over the working of the railway in August 1875 and although at this time two standard 0-6-0Ts were laid down, the frames, cylinders and tanks were transported to Swindon where they were leisurely completed, the boilers for both being built in the GWR works. New building ceased at Dock Street and all heavy repairs and rebuilding were carried out at Swindon and Wolverhampton.

Typical heating surfaces for the 0-6-0Ts were tubes 831·7; firebox 81·69; total 913·39; grate area 13·16, and weight in working order 38 tons 16 cwt; for the 4-4-0Ts the boiler was the same, but the grate area increased to 13·99 and weight to 41 tons 3 cwt. There were a number of variations as previously mentioned.

MORRISON, ROBERT & COMPANY

Ouseburn Engine Works, Newcastle

Built ten out of an order of twenty locomotives for the East Indian Railways, and then went bankrupt. There were a number of firms building locomotives for the East Indian Railways about 1855–6 and this would probably be the period of building for R. Morrison & Company.

MURDOCH, AITKEN & CO.

Hill Street Foundry, Glasgow

This firm were manufacturers of marine, high-pressure and stationary engines, also making pumps, machine tools and every type of millwrighting work.

They received the first order placed in Scotland for a railway locomotive, from the Monkland & Kirkintilloch Railway. Two were built, the first being completed in May 1831 and the second in the following September. George Dodds, the engineer on the Ballochney Railway was responsible for the design which took the form of a four-coupled tender locomotive with two vertical cylinders sunk into the boiler barrel, one over the centre line of each axle. The boiler had a single firetube half way down the barrel where a tube plate was fitted, supporting a quantity of small diameter tubes. The coupling rods had ball and socket joints. The first one was commenced in February and completed in May

Fig. 408 **Murdoch Aitken** 1833 Monkland & Kirkintilloch Rly 0-6-0 GARTGILL
Note method of drive

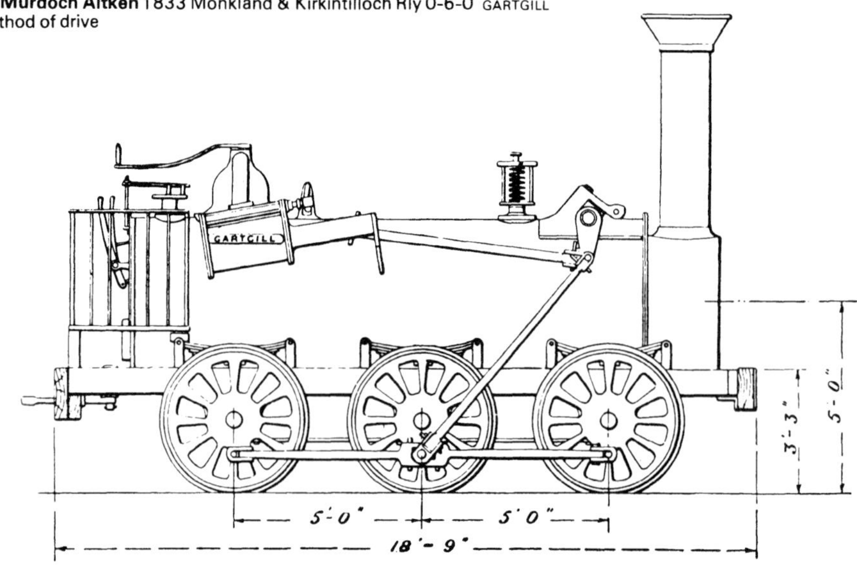

which was quite an achievement despite the fact that they were experienced in stationary and marine engines. The second may have been put in hand after the completion of the first one and took roughly the same time to build if this was so. Both did extremely well. The design was similar to George Stephenson's Killingworth locomotives which were built in collaboration with Ralph Dodds the chief viewer of the colliery – whether he was a relative of George Dodds is not known.

The next order came from the Garnkirk and Glasgow Railway for three different types, a 0-4-0, 0-6-0 and 2-2-0. The tenders were built at the company's repair shops at St Rollox (the Caledonian Railway's works were not established until 1854). Little is known about these three locomotives. The boiler pressure was 50 lb/sq. in. The 2-2-0 weighed 10 tons, the 0-4-0 8½ tons and the 0-6-0 13 tons. All were withdrawn in 1847. The 0-4-0 was probably similar to the ones built earlier.

The London & Southampton Railway ordered a 0-4-0 Ballast locomotive which was delivered in March 1836 for assisting in the construction of the line and was eventually bought by Brassey. In 1848 it was written off after a head-on collision.

Building locomotives was evidently of secondary importance to the firm and after two 'singles' for the Paisley & Renfrew Railway built in 1837, one further 0-4-0 was built in 1841 for the Slamannan Railway.

More may have been built during these long intervals but cannot be traced.

MURDOCH, AITKEN & CO. GLASGOW.

Date	Type	Cylinders inches	D.W. ft in	Customer	No	Name
5/1831	0-4-0	10½ x 24	3 9	Monkland & Kirkintilloch Rly		Monkland
9/1831	0-4-0	10½ x 24	3 9	Monkland & Kirkintilloch Rly		Kirkintilloch
6/1833	0-6-0	12½ x 21	4 4	Garnkirk & Glasgow Rly	4	Gartgill
3/1834	0-4-0	10½ x 24	3 9	Garnkirk & Glasgow Rly	3	Garnkirk
1836	2-2-0	11 x 16	4 0	Garnkirk & Glasgow Rly	6	Jenny
1836	0-4-0			London & Southampton Rly		Vulture
1837	2-2-2			Paisley & Renfrew Rly		Paisley
1837	2-2-2			Paisley & Renfrew Rly		Renfrew
1841	0-4-0			Slamannan Rly		Glenellrig

All built to 4′ 6″ gauge except VULTURE

NASMYTH GASKELL & COMPANY

Bridgewater Foundry, Patricroft

The Bridgewater Foundry was established in 1836 by James Nasmyth and was where he continued designing and making special purpose machine tools, steam engines and other engineering products. He took into partnership Holbrook Gaskell who wanted to invest his capital in the firm, and who remained with Nasmyth until 1850, when he had to retire due to ill health.

Nasmyth had worked with Henry Maudslay at the latter's Lambeth works until 1831 when he set up a small workshop of his own in London. After three years in the Bridgewater Foundry, Nasmyth applied his talents to locomotive building and produced nine in 1839, thirteen in 1840, eight in 1841 and sixteen in 1842. Some were built for Geo. Stephenson, Bury, and Robert Stephenson. No more were built until 1845 and only 158 in the following thirty-five years.

In 1850 the title of the firm was changed to: JAMES NASMYTH & COMPANY and in 1857 to: PATRICROFT IRON WORKS.

Ten years later Robert Wilson joined the works and with Henry Garnett became the principal partners. The title then became: NASMYTH WILSON & COMPANY and in 1882 NASMYTH WILSON & CO. LTD.

During this period the main products were pile drivers, steam hammers, rolling mill equipment, tools and ordnance work. Much of the work was for abroad. James Nasmyth retired from the firm in 1856.

In the 1880s locomotive production increased and it was noteworthy for the number built for abroad. Many were for India, New Zealand, Australia, Russia, South America, Japan, Spain and other countries.

From 1873 to 1938, out of a total of 1307 locomotives, 1188 went overseas, emphasising in no uncertain manner the tremendous overseas market the firm had built up and had maintained to the end of its existence.

NASMYTH GASKELL & COMPANY

Fig. 409 **Nasmyth Gaskell** 287/1885
Sante Fé Extension Rly.
4-6-0 No. 6
13″ cylinders 3′ 0″ D.W.

Fig. 410 **Nasmyth Gaskell** 501/1897
Toyokawa Rly Japan 0-6-0T
14″ × 18″ cylinders 3′ 0″ D.W.

Fig. 411 **Nasmyth Gaskell** 937/1911
Burma Rlys. No. 355 *Class M*
2-6-2T

Fig. 412 **Nasmyth Gaskell** 995/1913
Bombay Port Trust No. 1 2-6-0T
5′ 6″ gauge

NASMYTH GASKELL & COMPANY

In 1919 the firm became a public limited company with no change of name.

During the first World War the works were engaged in munition work, but the locomotive department was also busy and production included twenty 2-8-0s for the Chemin de Fer l'Etat of France, one hundred small petrol driven locomotives, thirty-two 2-8-0s of Robinson's well tried Great Central Railway design and many types for India. It was in 1921 that Nasmyth Wilson built their only Crane Tank, a six-coupled design used in their own works.

India continued to be the main outlet after the war but ten 4-4-2Ts and five 0-6-0s were built for the GNR of Ireland. The London Midland & Scottish Railway ordered five 4-4-2Ts for the London Tilbury & Southend section and ten 0-4-4Ts of Caledonian design.

Orders were harder to come by in the 1930s, only eighty-six being built in ten years. The firm was wound up in 1939, perhaps due to their laudable export policy with a consequent neglect of the home market. This period was a critical time for the industry as a whole, and other firms fell by the wayside in the dreary atmosphere of dwindling orders and fierce competition.

The firm had earned a reputation of good workmanship and materials, and from 1839 to 1939 had built 1531 steam locomotives. Taken over by the Ministry of Supply in 1939 it became Patricroft Royal Ordnance Factory.

Thanks are due to the Borough Librarian and Curator of the Libraries and Museum Dept. of the Borough of Eccles for photographs and information.

NEASHAM & WELCH
Stockton-on-Tees

This firm built one locomotive for the Stockton and Darlington Railway in 1840. It was S&DR No. 27 WITTON CASTLE and had inclined cylinders 14″ x 20″ driving a dummy crankshaft coupled to 4 ft diameter driving wheels. Later the cylinders were altered to 16″ x 18″ and the locomotive was replaced in 1875.

The second locomotive was for the Clarence Railway and numbered 5. It was a six wheel coupled with vertical cylinders. No further details can be discovered.

NEATH ABBEY IRON COMPANY

This firm was founded in 1792 and their main business was producing castings, cylinders and other parts for Cornish pumping engines, blowing engines and later rails and chairs.

Locomotive building commenced in 1829 and Henry Taylor was responsible for design. Those built in the earlier years showed an originality of design and are all worthy of study.

The first built, named SPEEDWELL, had a boiler of cast iron, and vertical cylinders midway along the boiler. The piston rod ends were connected to a stirrup above them and motion was transmitted to the coupled wheels by a series of levers. A feed pump was also driven off one of the levers and the feed water was preheated by passing through jacketted exhaust pipes. The wheels were flangeless. A second

Fig. 413 **Nasmyth Wilson** 1598/1932 2-8-2 for Tientsin-Puchow Rly. No. 295 being towed to docks by LMS 0-6-0T 16687

Fig. 414
Neath Abbey
*c*1830
0-6-0 Rack/Rail locomotive. Flangeless wheels – rack could be disengaged

locomotive named HERCULES was also built for the same customer but it is not known whether it was to the same design.

A unique locomotive was built in 1831 for the Dowlais Iron Company consisting of two four-coupled units with a jack shaft between them driven by a pair of inclined outside cylinders. On the jack shaft was a gear wheel which meshed with the two adjacent coupled axles on which were gear wheels. Two chimneys were fitted one for each cylinder and could be lowered to a horizontal position for passing through a tunnel. A similar locomotive was built in 1838 for the Rhymney Forge but with a conventional single chimney and the cylinders fitted at the opposite end.

The first six-coupled locomotive was supplied to the Gloucester and Cheltenham Railway. The boiler had a single flue, with corrugated firebox and water tubes. The cylinders were placed vertically over the centre pair of wheels and the drive was probably similar to the SPEEDWELL. The locomotive was transformed into a tram engine by completely encasing it with a wooden body. It was built in 1831 and named ROYAL WILLIAM.

The following year PERSEVERANCE was built and was a combined rack and adhesion locomotive. The drive was a jack shaft with a gear engaging on the rear wheels and driving an additional gear which could be lowered to engage into a rack rail. It was supplied to the Dowlais Iron Company, and a 0-6-0 was built for them the same year.

Two 0-6-0s were built for the Bodmin & Wadebridge Railway, one in 1834 and the other in 1836. The outside vertical cylinders were placed in the typical Neath Abbey manner above the centre pair of driving wheels, and the same motion was fitted. These two locomotives were dismantled and sent by ship to Wadebridge where they were re-erected.

The next recorded building was in 1848 when a double frame 0-4-2 was built. This was a more conventional type with inside cylinders but little is known about the general dimensions. It was sold to the Taff Vale Railway and lasted ten years.

The same year four 0-6-0s were built with inside frames and horizontal outside cylinders. The boiler had a central dome and spring balance safety valve, and a raised firebox with safety valve on top. The dome and safety valve covers were very similar to the elaborate ones produced by E. B. Wilson & Company. Some small four-coupled saddle tanks were built later until about 1870.

One or more locomotives of an unknown type were sent to South America. It is estimated that between twenty and twenty-five were built altogether. The chief interest in the early locomotives lies in the unusual position of the cylinders, the two 'articulated' locomotives, the first with this wheel arrangement, and the novel layout of rods between the stirrup gear and the crank pin.

The firm ceased manufacturing in the 1880s but the descendants of Henry Taylor have an engineering business, Taylor & Sons of Briton Ferry.

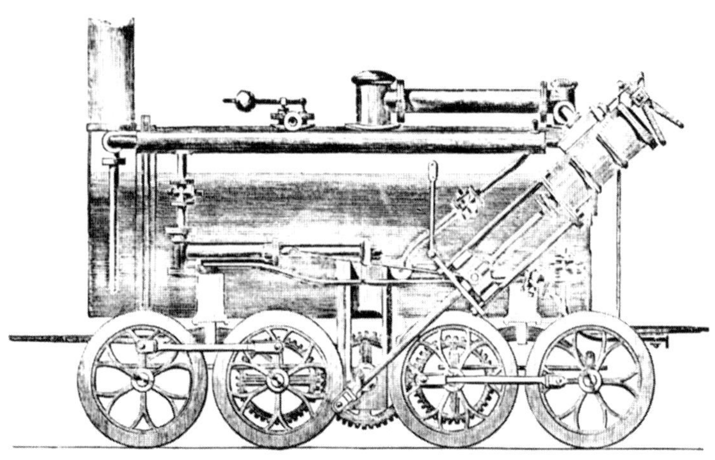

Fig. 415
Neath Abbey
*c*1830
geared locomotive

Built	Type	Cylinders inches	D.W. ft in	Customer	Name/No.	Gauge ft in
1829	0-4-0	VOC 10½ x 24		Thos Prothero –	Speedwell	4 2
1829?	0-4-0			Sirhowey Tram Road	Hercules	4 2
1830	0-4-0	10½ x 24		Harford Davies & Co, Ebbw Vale	Hawk	4 2
1831	0-4-4-0	FOC/1		Dowlais Iron Co		4 2 (a)
1831	0-6-0	VOC 10½ x 20		Gloucester & Cheltenham Rly	Royal William	3 6
1832	0-4-0GR	ROC/1 10½ x 20	3 1	Dowlais Iron Co	Perseverance	4 2 (b)
1832	0-4-0	10½ x 24		Ebbw Vale	Industry	
c. 1832					Lark	
1833	0-4-0	ROC/1 8½ x 24		Dowlais Iron Co	Mountaineer	4 2
1833	0-4-0	8½ x		Dowlais Iron Co		4 2
2/1834	0-6-0	VOC 10½ x 24	3 10	Bodmin & Wadebridge Rly	Camel	STD (c)
5/1836	0-6-0	VOC 12½ x 24	3 10	Bodmin & Wadebridge Rly	Elephant	STD (c)
1836	0-6-0	ROC/1 8½ x 20		Dowlais Iron Co	John Watt	4 2 (d)
1836	0-4-4-0			Rhymney Forge		
1837	0-4-0GR	ROC/1 8½ x 20		Dowlais Iron Co	Yn – Barod – Etto	4 2 (e)
1837/8	0-4-0			Dowlais Iron Co		
1838	0-6-0GR	FOC/1 8½ x 18	3 1	Dowlais Iron Co	Dowlais	STD (f)
1830's	0-6-0	OC 8½ x		Dowlais Iron Co	Success	4 2
1848	0-4-2	OC 15 x 24	4 6	Taff Vale Rly	Neath Abbey	STD
1848	0-6-0	OC 15 x 24	4 0	Monmouthshire Rly & Canal Co	2	STD
1848	0-6-0	OC 15 x 24	4 0	Monmouthshire Rly & Canal Co	3	STD
1848	0-6-0	OC 15 x 24	4 0	Monmouthshire Rly & Canal Co	4	STD
1848	0-6-0	OC 15 x 24	4 0	Monmouthshire Rly & Canal Co	5	STD
1855	0-6-0	OC 12 x 20	3 6	Vivian & Sons, Swansea	Caesar	4 4
1858	0-6-0			Vivian & Sons, Swansea		4 4
1864	0-4-0	8 x 16		Plymouth Iron Co Merthyr Tydfil		Std
1867	2-4-0T	OC		Rhymney Iron Co		3 0
1871	0-4-0	8½ x 16		Cyfarthfa Ironworks		3 0
Date not known						
	0-4-0ST	OC 8 x 15	2 4	Neath Abbey Iron Co		3 0
	0-4-0ST	OC 8 x 16		Aberdare Iron Co	2	
	0-4-0ST			Aberdare Iron Co	Success	
	0-4-0ST			Aberdare Iron Co		
	0-4-0ST			Pascoe Grenfell & Sons Swansea		3 2
	?			South America		

(a) Articulated, geared drive, doubtful whether this loco ever worked on Dowlais Iron Co's system. (b) Rebuilt 1840 by Neath Abbey and converted to 0-6-0. (c) Shipped in parts. (d) Rack fitted at Dowlais in 1837 by Neath Abbey. (e) 'Ready again'. (f) Ordered for 4' 2" gauge but altered to standard gauge before delivery also rack fitted. FOC/1 = outside inclined cylinders at front. ROC/1 = outside inclined cylinders at rear. VOC = outside vertical cylinders. GR = geared drive for rack.

NEILSON & MITCHELL

Hyde Park Street, Glasgow

Walter Neilson and James Mitchell founded the firm about 1836–7 and were first engaged in the manufacture of stationary and marine engines with premises in McAlpine Street.

In 1837 the firm was known as KERR, MITCHELL & NEILSON with a new works established in Hyde Park Street. The firm at first had a variety of titles – in 1840 it was KERR, NEILSON & CO, but Mr Kerr did not stay long and in 1845 the firm was known as NEILSON & MITCHELL.

The first locomotives were built in the Hyde Park works in 1843 and were three 0-4-0s for the 4′ 6″ gauge Glasgow, Garnkirk and Coatbridge Railway. For the first five years all the output went to local railways including the Monkland, Wishaw & Coltness, Glasgow and Ayr, and Edinburgh & Glasgow railways. In 1850 some 2-2-2 well tanks were built, two for the Edinburgh & Glasgow Railway named WEE SCOTLAND and LITTLE ENGLAND. They had outside cylinders 10″ x 15″ and 5′ 0″ diameter driving wheels.

Until 1855 stationary and marine engines continued to be built , and a notable stationary engine was for rope haulage of the Cowlairs incline built in 1842. This side of the business was then dropped, and in the same year the firm became NEILSON & CO. Many notable locomotive men were associated in various capacities in Walter Neilson's works: Benjamin Connor was Works Manager, Henry Dubs who subsequently started up on his own, Patrick Stirling, and James Reid who had come from Caird & Co. left for a time to work at Sharp Stewart & Co. and later returned.

Works numbers ran from 1 to 95 and then a gap occurs and the next number was 313. The missing numbers may have been for other manufactured machinery.

During this period many four-coupled tank locomotives were built for various industries at home and abroad. In most cases tenders were given separate works numbers. Their close neighbours at Cowlairs and St Rollox were excellent customers and many 2-4-0 and 0-4-2 tender types were supplied.

Large orders came from India – Punjaub, Scinde, Great Indian Peninsula, Bombay, Baroda & Central Indian; and East Indian railways all received locomotives in the 1860s and Neilson's continued to supply them all through its separate existence.

The small works could not keep pace with orders and a new foundry and boiler shop was built in Finnieston, but within a few years this additional capacity was insufficient and land was bought between the Caledonian Railway's works at St Rollox, and the North British Railway's Cowlairs works in the Springburn district. Work started in the new premises in 1861 and they were destined to become the largest private locomotive building firm in the British Isles. They were known as 'Hyde Park Works' to commemorate their previous premises. Henry Dubs left in 1864 to set up his own works at Polmadie known as 'Queens Park Works'. Unfortunately for Neilson he took a number of key men with him including the chief draughtsman Mr Goodall-Copestake. James Reid however returned from Sharp Stewart & Co. and became a partner.

Fig. 416 1772/1873 New Zealand Govt. 0-4-0ST for Castlecliff Rly.

Fig. 417 **Neilson & Mitchell** 3982/1889 Italian Govt. Rlys. RA4569 21" 0-8-0 4' 2" D.W.

Fig. 418 4426/1892 Piraeus & Latissa Rly. No. 301 4-4-0T

0-4-4Ts were popular in the 1870s and Neilson supplied the London, Chatham & Dover with nine 5' 3" diameter coupled wheels, thirty to the Midland with 5' 6" diameter coupled wheels, and twenty-five 4' 10" diameter for the Great Eastern Railway, all built in 1875–6.

There was a large order for fifty 0-4-2s, all built in 1877, with outside cylinders 11½" x 17" and 3' 3" diameter coupled wheels. They were built to the metre gauge of the Indian States Railways.

A tremendous variety of types were built, taking a random example in 1878:

Wks No. 2369	0-4-2	Tasmania
Wks No. 2370	0-4-4T	Lancashire & Yorkshire
Wks No. 2380	0-4-0WT	Beckton Gas Works
Wks No. 2381	2-4-0T	Iquique
Wks No. 2383	0-8-0ST	Spain
Wks No. 2384	4-4-0	North British Railway
Wks No. 2393	2-6-0	Great Eastern Railway

During the 1860s, 765 locomotives were built of many types and gauges. The practice of giving works numbers to tenders ceased in 1863.

The first eight coupled locomotives built by Neilson were ten 0-8-0STs in 1872 with 18" x 24" cylinders and 4' diameter wheels for the 5' 6" gauge Great Indian Peninsula Railway.

In 1873 ten 2-4-0s of Fletcher's *901* class were built for the North Eastern Railway with 7 ft coupled wheels and 17" x 24" cylinders. They were followed by ten 5' 0-6-0s and twelve 0-4-4WTs for the same railway.

2-4-0s and 0-4-2s were still in demand by the Caledonian Railway and large numbers of 0-4-2s went to the Glasgow & South Western Railway.

The GER 2-6-0 was one of fifteen designed by William Adams and was the first 2-6-0 to run in Great Britain. One bore the name MOGUL and from this all locomotives of this wheel arrangement became known as 'Moguls'.

Neilson built few 'Fairlie' type locomotives but in 1883 three were built for the Mexican railways each with two six-coupled units and double boilers.

South Africa was now ordering locomotives from Hyde Park Works and twenty 4-6-0s were built in 1883–4 and two larger ones went to the Central Argentine Railway in 1884. India was still the best customer and hundreds were built in ever increasing batches.

In 1886 the company built a 4-2-2 to exhibit at the London Exhibition of that year. It was of Caledonian lines with fittings etc. of that company who bought it after the exhibition. It was numbered 123 and has fortunately been preserved.

Walter Neilson had left the company in 1884 and like many before him had built his own locomotive works at Springburn and called them the 'Clyde Locomotive Works' hoping that some of his previous customers would transfer their orders to his new works – unfortunately for Neilson they did not do so.

James Reid was now sole proprietor of Neilson & Co. but it was not until 1898 that the firm's name was changed to: NEILSON, REID & CO.

At the end of the century there was considerable industrial strife resulting in widespread strikes. The relationship between management and the workmen had always been good and where neighbouring works went on strike for better conditions and more money, Neilson men did not – there was no need to.

The strikes had unfortunate repercussions on all locomotive contractors, they could not guarantee delivery but assumed that their regular customers would be content to wait for their locomotives. Of course they would not – locomotives are ordered because there is a need for them – and so the customers looked elsewhere. Even the home railways were in a similar situation and American 2-6-0s were ordered by the Midland, Great Northern and Great Central railways. There was a great outcry of course, and when the locomotives were delivered they were disparaged and some interesting comments were made on both sides of the Atlantic.

A lesson learned from this set-back was that to meet the increased competition particularly from America both in price and delivery, small units were uneconomic and amalgamations could be made with advantage.

Such an idea was mooted in the early 1900s in Glasgow. By the turn of the century Neilson's themselves were in a strong position, but even so methods would have to be improved to get

Fig. 419 4742/1894 Danish State Rlys. No. 501 4-4-0–with outside motion

Fig. 420 5043/1896 West Australian Rly. No. 104 *Class K* 2-8-4T 3′ 6″ gauge

the prices down to the low prices the Americans were quoting. Up to 1900 an estimated number of 5394 locomotives had been built, far more than any other manufacturer.

Small batches of 0-4-0STs were put through from time to time but main line freight and passenger locomotives were the greater part of the company's output.

Larger locomotives were now built with 4-8-0 and 4-6-4Ts for South Africa, and 4-10-2T for the Imperial Military Railway (S. Africa). The proposed amalgamation was now a fact and in 1903 Neilson Reid & Co, Dubs & Co, and Sharp Stewart & Co. combined forces to form the largest locomotive building company anywhere in the world with the exception of America. The new title was the NORTH BRITISH LOCOMOTIVE CO.

Building went on as before, but rationalisation took place and many duplications were eliminated. The last works number for Neilson Reid & Co. was 6438 and from that, the total estimated number of steam locomotives built was 5864.

At the beginning of the twentieth century Neilson Reid & Co. employed 3500 men and were turning out approximately 300 locomotives per annum. Sharp Stewart with 2000 men had an average output of 150 and Dubs & Co. with 2500 men averaged 160 locomotives per annum.

Neilson had a design staff unmatched and when Copestake went to Dubs & Co., the replacement was a Mr Edward Snowball who came from Robert Stephenson & Co. and set his own mark on Neilson's locomotives which could be seen in practically all quarters of the globe.

Perhaps the most unusual locomotive was built in 1861 for hauling trains of coaches mounted on sledges over the ice between St Petersburg and Kronstadt in Russia. No time was lost in the design work where no drawings were available and in J. Thomas's *The Springburn Story*, an example is quoted of an order received from the Nippon Railway Co. on 26 September, 1894 for twelve outside 0-6-0Ts. The first was steamed on 30 November, sixty-five days after receipt of the order and the last left the shops on the 84th day.

This would be deemed impossible today.

See NORTH BRITISH LOCOMOTIVE CO.

NEWBATTLE COLLIERY

Newton Grange, Midlothian

A locomotive is purported to have been built in the workshops of this Colliery in 1927. It had all the features of an Andrew Barclay locomotive and it is more probable that it was assembled from spare and new parts supplied by the company. It was a 0-4-0ST with outside cylinders and was No. 1 in the Newbattle list.

Apparently the Colliery were fined £200 and although the reason for this is not definitely known, it has been suggested that the drawings of Andrew Barclay Sons & Co. Ltd. may have been used without permission, being the property of that company. The locomotive was reported to be lying out of use at Lady Victoria Colliery in 1968.

(Information from The Industrial Railway Society's Bulletins 100 and 102).

NEW LOWCA ENGINEERING CO. LTD.

Whitehaven

See TULK & LEY.

NICHOLLS, WILLIAMS & COMPANY

Bedford Iron Foundry, Tavistock

This firm supplied one and possibly two locomotives around 1850, for the railway of Devon Great Consols Co. Ltd. of Tavistock. No further details known except they were probably four-coupled tank locomotives with outside cylinders.

NORTH BRITISH LOCOMOTIVE COMPANY

Glasgow

This company was formed in 1903 by the amalgamation of the three major locomotive building firms in Glasgow: Neilson Reid & Co. Hyde Park Works; Dubs & Co. Queens Park Works; and Sharp Stewart & Co. Atlas Works. So was formed the largest locomotive factory in Europe. The works numbers of each firm were added together and the first works number for the new company was 15723.

Each works had its own distinctive works plate: Hyde Park works was circular in shape; Queens Park works was diamond in shape; Atlas works was oval in shape. The company was in a position to deal with large orders and to improve delivery at the same time.

When the organisation had settled down, the production during 1906 was 582 locomotives of which 168 were built at Queens Park, 153 at Atlas and 261 at Hyde Park works. They included 4-6-0, 4-4-2, 2-8-0, 0-6-0, 4-4-0, and 0-6-2Ts for S. America; 4-6-0, 2-8-4T, 2-8-0, 4-8-0, 4-4-0, and 4-4-4s for India; 4-4-0, 0-6-0Ts for Spain; 4-6-0s for China; 4-6-2s for S. Africa; 4-4-2, 0-6-0s for North British Railway; and 4-4-0s for Highland Railway.

By 1907 the 2000th locomotive had been built by the new company, the 3000th in 1909, the 5000th in 1914. It would be tedious to list the many types and many railways these locomotives were built for, but apart from America, Russia, and parts of Europe they went all over the world.

A Turbo-Electric 0-4-4, 4-4-0 was built in 1910 but nothing seems to have developed from it. It was known as the 'Reid-Ramsay' locomotive.

During the 1914–18 war the three works were extremely busy, turning out approximately

Fig. 421 **North British Loco Co.** 16627/1905 0-4-0RM for Cape Govt. Rlys.

Fig. 422 **North British Loco Co.** 19266/1910 Reid-Ramsay 4-4-0 + 0-4-4 Turbine running on the NER.

Fig. 423 **North British Loco Co.** 19890/1912 Great Central Rly. Robinson's *Class 8K* 2-8-0 fitted with Caille-Potonie feed water heater.

1400 locomotives, including many Robinson 2-8-0s for the War Dept, 2-8-0s for the French State Railway, 2-8-2Ts for the Paris–Orleans Railway, and many types for India and other countries. Besides producing these locomotives many departments were turned over to munitions and other war work including shells, aircraft, tanks, gun-carriages and other articles.

Orders after the war were put in hand for many railways abroad – the majority for Indian Railways. Home railways were also short of locomotives and the Furness, Great Northern, North British, Great North of Scotland, Glasgow and South Western, and Caledonian railways were all supplied with various types.

To keep three large erecting shops fed with materials continuously needed a large order book and the quantity of orders being received were not up to pre-war level. It was decided to concentrate the erection of the locomotives at Queen's Park and Hyde Park works and the last order to be completed at the Atlas works in 1923 were six 4-6-0s for South Indian Railways with outside cylinders 16″ x 22″ and 4 ft diameter coupled wheels to lot No. 776. The Atlas works were advertised for sale in 1927.

Three *Mallet* 0-6-6-0s were built in 1921 for the Madras & Southern Mahratta Railway. They were compounds having cylinders 15½″ and 24¼″ x 20″ with 3′ 3″ diameter wheels.

In 1924 an order for twenty of Gresley's *Pacifics* was carried out, together with fifteen standard LMS 0-6-0Ts. Seven 5′ 3″ gauge 4-4-0s for the LMS (Northern Counties Committee) were built in 1924–25.

More orders were coming from home railways than from abroad at this period but an interesting Fairlie locomotive was built for the South African Railways in 1924. It was a 2-6-2-2-6-2 with 14″ x 23″ cylinders and 3′ 6¾″ coupled wheels.

In the same year the Reid-McLeod Turbine Condensing locomotive was built (Works No. 23.141) and exhibited at the 1924 Wembley Exhibition. It used the frames, boilers and wheels of the 'Reid-Ramsay' locomotive of 1910. The wheel arrangement in this case was 4-2-2-2-2-4 supporting one frame – it was not articulated but consisted of two turbine driving units. They were of the three-stage pressure compound impulse type and the high pressure turbine drove the trailing unit and the low pressure turbine the leading unit. Transmission was by a double helical gear and a second reduction by bevel gears on to both axles of each unit but no coupling rods were used. The boiler was at the rear end and condenser at the leading end. Of the air cooled type, the condenser was assisted by a fan, and water under pressure was connected to jet vaporisers in front of the tubes.

The boiler pressure was 180 lb/sq. in. and was superheated. The regulator was in the form of a hand wheel, turned clockwise for forward and anti-clockwise for reverse running. The reverse turbine was a single stage impulse turbine.

It was tried on the Caledonian and North British sections of the LM&SR. The total horsepower was approximately 1000 and it was found by dynamometer car tests that the tractive effort was only 15,000 lb which for such a weighty locomotive was proportionately small. After successive tests, it was stored by the

NORTH BRITISH LOCOMOTIVE COMPANY

Fig. 424
North British Loco Co.
20958/1915 South African Rlys.
No. 1661 2-6-6-2 compound *Mallet.*

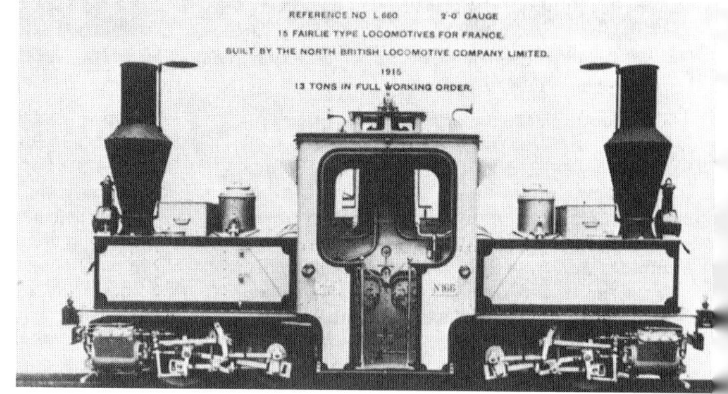

Fig. 425
North British Loco Co.
21188/1915 French War Office 'Pechot' design Fairlie 0-4-0+0-4-0

Fig. 426
North British Loco Co.
22888/1922 Glasgow & Southern Western Rly.
No. 542

makers and eventually scrapped.

In 1925 thirty *King Arthur* class were built for the Southern Railway and twenty-five LMS 4-4-0 compounds, followed by twenty-five LMS *Class 4* 0-6-0s.

Small quantity orders indicated the state of trade at this time, and many of the orders placed in 1926–8 were from the LMS, LNE and GW railways.

In 1927 the company designed and built the fifty *Royal Scot* locomotives which were so urgently needed by the LMS. Queens Park built twenty-five, and Hyde Park twenty-five.

In 1929 the controversial high pressure 4-6-0 No. 6399 FURY was built for the LMS. It was a compound with one high pressure 11½″ x 26″ and two low pressure 18″ x 26″ and had the standard 6′ 9″ diameter coupled wheels. The Schmidt boiler was in three sections having a maximum pressure of 1800 lb/sq. in., the pressure at the regulator being 900 lb/sq. in. In February 1930 it was having a trial run from Glasgow and as it reached Carstairs there was an explosion in the firebox. One man was killed – an inspector of the Superheater Co. and after this unfortunate accident it was taken back to the works. Eventually it was towed to Crewe via Derby and was rebuilt, into a conventional boilered locomotive becoming the first taper boiler *Royal Scot* renumbered 6170 and named BRITISH LEGION. It differed only in detail to the later rebuilt *Scots*.

The depression of the late 1920s reduced locomotive orders considerably. Forty outside cylinder 2-6-0s were built for the Egyptian State Railways, in 1928; ten 4-8-2s for the Mashonaland & Rhodesia Railway, twenty eight 0-6-2Ts for the Central Argentine Railway and miscellaneous small orders including one 4-10-2T for a South African colliery and a 4-8-2T for similar duty, both having outside cylinders 19″ x 27″ and 3′ 9″ diameter coupled wheels. To ease the situation orders were placed by the Great Western Railway for one hundred of their new *57XX* 0-6-0PTs, the first ten *Sandringham* 4-6-0s by the L&NER, and fifty of Stanier's *Jubilee* three-cylinder 4-6-0s, which were completed in 1934.

Fig. 427 **North British Loco. Co.** 23297/1925 South African Rlys. No. 2323 modified Fairlie

Fig. 428 **North British Loco. Co.** 24418/1938 ABT Rack System Mount Lyell Mining & Rly. Co. No. 5

NORTH BRITISH LOCOMOTIVE COMPANY

In 1933 two 4-6-2s were built for the Nizam Government State Railway's 5′ 6″ gauge, both with outside cylinders 21½″ x 28″ and 6′ 2″ diameter coupled wheels. One was fitted with booster, mechanical stoker and feed water heater, and the second with a booster only. They were modifications of the standard Indian *Class XD*.

Fifty *Class 19c*, 4-8-2s for the South African Railways were built in 1934 with outside cylinders 21″ x 26″ and 4′ 6″ diameter coupled wheels. They were all fitted with RC poppet valve gear, and twenty-five were fitted with copper fireboxes and the others with steel fireboxes. Another departure from normal practice was the fitting of special flexible stays, designed in South Africa, to the firebox water leg and the two rows of crown stays. These stays had now become standard on all SAR boilers.

Locomotives for Palestine railways, Tientsin-Pukow Railway and industrial locomotives for South Africa were built in small numbers and in addition twenty of Gresley's *K3* 2-6-0s were completed in 1935 followed by seventy-three of Stanier's two-cylinder 2-6-4Ts.

Four 2-2-2-2 tender locomotives were supplied to the Egyptian State railways in 1938. The engine units only, were supplied by the Sentinel Waggon Works Ltd. of Shrewsbury, the boilers, frames and tender being supplied by the N. B. Loco Co., and erection was carried out in the Hyde Park works (works Nos. 24413–24416). Another interesting locomotive built in the same year was a 0-4-2T for the Mount Lyell Mining and Railway Co. B.C. arranged on the ABT rack system with 12½″ x 20″ cylinders for the track and 11½″ x 15½″ for the rack drive.

At the advent of the second World War 8850 locomotives had been built since 1903 averaging approximately 250 locomotives per annum indicating a tremendous fluctuation of orders during those years.

The firm now became engaged on war work of all descriptions and from 1940 large numbers of 2-8-0s of Stanier's designs were built for the Ministry of Supply, followed by fifty-five 3′ 6″ gauge 4-6-2s for Africa.

A new 2-8-0 was then put in hand designed by R. A. Riddles and was a simplified version of Stanier's standard class with the object of facilitating production, changing materials, with a result that the number of man-hours was cut by roughly 20 per cent. This was a valuable saving – there was a serious shortage of labour due to Forces requirements and other work equally as important. A total of 545 were built, 294 at Hyde Park and 251 at Queens Park works. They were known as *Austerity* 2-8-0s and these were completed in 1944.

An *Austerity* 2-10-0 was then produced – designed by R. A. Riddles – the main difference being the larger boiler and firebox, same size cylinders 19″ x 28″ and 4′ 8½″ diameter coupled wheels. The object was to decrease the axle load from 15½ to 13½ tons to enable them to have greater route availability and to meet special conditions, including burning lower grade fuels and for this purpose the grate area was increased from 28·6 sq. ft to 40 sq. ft on the 2-10-0. One hundred and fifty were built from 1943 to 1944 all at the Hyde Park works.

At the end of the war urgent orders were fulfilled for 110 *Class XD* 2-8-2s for the Indian railways 5′ 6″ gauge and these were completed in 1946 together with forty three-cylinder metre gauge 4-6-2s for the Malayan Railway.

The L&NER ordered 250 of Thompson's new standard *B1* two-cylinder 4-6-0 class and these were completed in 1948. A number of 4-8-4T and 4-8-2T were built for various mines in South Africa and six 2-8-4Ts for the Sao Paulo Railway.

A large order was received from the South African Railways for 100 *Class 15F* 4-8-2s and fifty *Class 19D* of the same wheel arrangement.

The *Class 15F* was outstanding in size and power for the 3′ 6″ gauge of which the South African Railway operated 14,000 route miles which was the largest 3′ 6″ gauge railway in the world. Forty-four had already been delivered before the war and the post war order included various modifications, due to past running experience, and the availability of materials as an aftermath to the war. Carbon-steel plates were used instead of nickel steel and various

Fig. 429 **North British Loco. Co.** 27312/1951 South African Rlys. No. 3536 Condensing locomotive

NORTH BRITISH LOCOMOTIVE COMPANY

Fig. 430 **North British Loco. Co.** Hyde Park Works in 1948 – batch of South African *Class 19D* 4-8-2s

Fig. 431 **North British Loco. Co.** Queens Park Works

details were altered or added. The *15F* represented the maximum power possible for this wheel arrangement within the SAR loading gauge. The two outside cylinders were 24″ x 28″, coupled wheels 5 ft diameter with an axleload of 18½ tons. The large boiler had a total heating surface of 4060 sq. ft. including 665 sq. ft of 1½″ superheater tubes. The boiler pressure was 210 lb/sq. in. and with tender weighed 179 tons 10 cwt in working order. The grate area was 62·5 sq. ft and 6050 gallons of water was carried on the eight-wheeled tender, with 14 tons of coal. Provision was made in the design for fitting mechanical stokers at a later date. The main frames were of the bar type of heavy section each being machined from a slab weighing 11 tons.

Twelve inch diameter pistons were fitted, with a 7⅜″ valve travel actuated by Walschaerts valve gear.

In 1948 the L&NER again sent large orders for thirty-five *Class L1* 2-6-4Ts and seventy *Class K1* (new) 2-6-0s and further *B1* 4-6-0s.

Between 1903 and 1950 11,000 locomotives had been built and in 1950 orders were received from West and South Australian Railways, Nyasaland and Egypt. Orders for steam locomotives decreased as diesel locomotive requirements increased.

In 1951 one hundred 4-6-2s were ordered by the Indian State Railways for their metre gauge, with outside cylinders 15¼″ x 24″ and 4′ 6″ coupled wheels. These were followed by one hundred 4-8-4s for the South African Railways and were fitted with condensers. Many smaller orders were fulfilled.

In 1953 the company celebrated its 50th anniversary and at the same time the 120th anniversary of the completion of the first locomotive built by the oldest incorporated firm – Sharp Roberts & Co. when they were in Manchester.

An order for twelve 4-8-2-2-8-4 *Garratt* locomotives was received via Beyer Peacock for the South African Railways in 1957 but production was now predominantly diesel. A repeat order for twelve more *Garratts* was received and then the last order for steam locomotives – for four 2-8-2s for Nyasaland (Works Nos. 27779 to 27782) – was completed the last being steamed in the first half of February 1958.

Competition in diesel production particularly from America cut down the orders received from abroad and having concentrated on hydraulic transmission a parallel may be drawn with Beyer Peacock & Co. whose policy was the same. A much larger amount of capital was tied up in the manufacture of diesels and in 1962 the great company went into liquidation. Messrs Andrew Barclay Sons & Co. Ltd. of Kilmarnock acquired the firm's goodwill, drawings, patterns, flanging blocks, and records.

The liquidators were granted authority to sell the central area of the Queens Park works at Polmadie by the Board of Trade and a new company, the Voith Engineering Co. occupies the area. The Glasgow education committee announced in 1962 they were to buy the main offices in Flemington Street, Springburn to convert them into a central college of engineering.

The estimated number of steam locomotives built since the amalgamation to the end was 11,318 which, added to the previous output (15,437) of the three constituent companies, amounted to 26,755.

NORTH BRITISH RAILWAY
Cowlairs, Glasgow

See EDINBURGH & GLASGOW RAILWAY.

NORTH BRITISH RAILWAY
St Margarets, Edinburgh

In 1844 the North British Railway set up extensive workshops at Meadowbank for the repair of locomotives and rolling stock. Until the amalgamation with the Edinburgh and Glasgow Railway, in 1865, these works were the headquarters of the locomotive department of the NBR.

Locomotive Superintendents until the amalgamation were: 1844–51 Robert Thornton; 1852–4 William Smith; 1854 Hon. Edmund Petre; and 1855–67 William Hurst.

All the first requirements of the NBR were supplied by R. & W. Hawthorn from 1844 to 1850 with a total of seventy locomotives out of seventy-one. Considerable trouble was experienced with broken crank axles on the double framed locomotives and the problem was not solved by Mr Thornton but actually his ideas aggravated the position so that his services were dispensed with, and it was not until Mr Hurst arrived on the scene that order was restored. Locomotive building commenced in 1856 when two 2-2-2Ts were laid down. They had inside frames and inside cylinders 12″ x 18″ and 5′ 6″ diameter driving wheels. The boiler was domed. They were built for working local trains in the Edinburgh area and were numbered 31 and 32. From 1857 to 1864 Hurst built a series of 0-4-2Ts with 12″ x 18″ cylinders, 4′ 9″ diameter coupled wheels and domeless boilers for similar duties. In 1860 two 0-6-0s were built followed by two more the following year with larger cylinders. In addition, in 1862 and 1863, six more were built with 15½ x 22″ cylinders with the help of R. & W. Hawthorn, and Robert Stephenson & Co. who supplied most of the major parts. They were numbered 80 to 85.

Three of Hawthorn's 2-2-2s of 1846 were rebuilt but the rebuilding was so extensive that they were new locomotives when they re-emerged from St Margarets. In 1867 No. 36 was rebuilt with 15″ x 18″ cylinders and a new boiler. No. 37 was rebuilt in the following year as a 2-4-0 with 16″ x 21″ cylinders and in 1869 No. 38 was dealt with in a similar fashion to No. 37 but with 16″ x 24″ cylinders.

Many other early Hawthorns were rebuilt at this works but none as thoroughly as the three just mentioned.

From 1867 to 1869 a number of 0-6-0s were built with 16½″ x 24″ cylinders and in some cases parts were used off condemned locomotives. After these 0-6-0s no more new locomotives were built, the works however carried on repairing locomotives from 1869 until final closure in November 1925 under LNER ownership, when all work was transferred to Cowlairs. W. Hurst had left in 1867 to return to the Lancashire & Yorkshire Railway as Locomotive Superintendent at Miles Platting.

NBR ST MARGARETS

Date	Type	Cylinders inches	DW ft in	NBR No		W'drawn
1856	2-2-2WT	12 x 18	5 0	31	31A/$_{74}$	1877
1856	2-2-2WT	12 x 18	5 0	32	32A/$_{74}$	1875
1857	0-4-2WT	12 x 18	4 9	20		1876
1857	0-4-2WT	12 x 18	4 9	22	22A/$_{77}$	1885
1860	0-4-2WT	12 x 18	4 9	29	29A/$_{78}$	1882
1860	0-4-2WT	12 x 18	4 9	49		1878
1860	0-6-0	15 x 24	5 0	76	76A/$_{84}$ 850/$_{95}$	1897
1860	0-6-0	15 x 24	5 0	77	77A/$_{84}$ 851/$_{95}$	
1861	0-6-0	15½ x 24	5 0	78	78A/$_{84}$ 852/$_{95}$ 1052/$_{01}$	
1861	0-6-0	15½ x 24	5 0	79	79A/$_{84}$ 853/$_{95}$ 1053/$_{01}$	1904
1861	0-4-2WT	12 x 18	4 9	96		1878
1862	0-4-2WT	12 x 18	4 9	97	97A/$_{78}$	1882
1862	0-4-2WT	12 x 18	4 9	98	98A/$_{81}$	1882
1862	0-4-2WT	12 x 18	4 9	99	99A/$_{82}$	1885
1863	0-4-2WT	12 x 18	4 9	103	103A/$_{81}$	1884
1863	0-4-2WT	12 x 18	4 9	104	104A/$_{81}$	1882
1864	0-4-2WT	12 x 18	4 9	105	105A/$_{81}$	1882
1864	0-4-2WT	12 x 18	4 9	106		1878
1864	0-4-2WT	12 x 18	4 9	107	107A/$_{78}$	1886
1864	0-4-2WT	12 x 18	4 9	108		1878
1865	0-4-0	OC13 x 20	4 6	109	109A/$_{80}$	1893
1865	0-4-0	OC13 x 20	4 9	110	110A/$_{80}$	1891
1866	0-4-0	OC13 x 20	4 6	184	184A/$_{82}$	1891
1867	0-6-0	16½ x 24	5 0	17		1914
1867	0-6-0	16½ x 24	5 0	56		1913
1868	0-6-0	16½ x 24	5 0	58		1913
1868-9	0-6-0	16½ x 24	5 0	59		1913
1868-9	0-6-0	16½ x 24	5 0	131		1914
1868-9	0-6-0	16½ x 24	5 0	134		1914
1868-9	0-6-0	16½ x 24	5 0	135		1913
1868-9	0-6-0	16½ x 24	5 0	154		1912
1868-9	0-6-0	16½ x 24	5 0	155		1914
1869	2-4-0	16 x 24	6 0	38		1912

0-6-0s NBR Nos 80-85 may also have been erected from parts supplied by R & W Hawthorn & Co and R. Stephenson & Co 1862-3. Locos 76—79 had parts supplied by R & W Hawthorn & Co.

NORTH EASTERN RAILWAY

Darlington

The locomotive works were actually completed by the Stockton and Darlington Railway in 1863 but before any new locomotives were built by the S&DR the railway became part of the North Eastern Railway in July 1863. However, for some years the locomotives built were to S&DR designs originating from Shildon, and from 1864 to 1875 the predominant class built was the *1001* class, of which seventy-one were built, and two sets of parts which were sent to Shildon for erection in 1866.

The *1001* class was a general term for the Stephenson type of 'long-boilered' six-coupled mineral locomotive and originated from 1852 when Mr Bouch designed them, and ordered two from Gilkes Wilson & Co. They had 17" x 18" inside cylinders and 4' 2½" diameter wheels. No. 56 TOW LAW and No. 57 SHOTLEY – later ones of this class, built from 1860 – had 17" x 24" cylinders and 5' 0" diameter wheels. Others had 17" x 26" cylinders and one 17" x 28". Some had Bouch's feed water heater fitted which was in the form of a jacket around the chimney, the feed water passing through it from the tender before entering the boiler.

The first locomotive completed at Darlington was No. 175 CONTRACTOR in October 1864, and in 1868 No. 1001 itself was built and many older locomotives of the 'class' were rebuilt to conform.

All S&D section locomotives had 1000 added to their numbers in 1872 to avoid confusion with the others bearing the same numbers on the North Eastern.

For working passenger trains between Tebay and Darlington, where heavy gradients were encountered, Bouch put in hand in 1871 four 4-4-0s Nos. 238 to 241. They had outside cylinders 17" x 30" and 7 ft diameter coupled wheels. The flush top boiler had a total heating surface of 1217 sq. ft and a working pressure of 140 lb/sq. in. The 30" stroke was unusual, and not to be used again until the Great Western Railway used it for Churchward's two-cylinder locomotives. Another curious feature was the use of 13" diameter solid brass piston valves, the valve chests being between the frame – great trouble, as would be expected, was encountered with these piston valves and when they were rebuilt at Gateshead as 2-4-0s in 1879–82, slide valves were fitted and the stroke reduced. Six more were built in 1874 to the original design, but all gave trouble. They were known as 'Ginxs' Babies'.

In 1873 Hackworth's No. 71 (1851) a 2-4-0 was rebuilt but little of the original locomotive was re-used.

William Bouch retired in 1875 and a works manager was put in charge, Mr J. Kitching carrying on at Darlington in this capacity, governed by Edward Fletcher who had his headquarters at Gateshead and up to this time had allowed each Works Superintendent an almost free hand – at Darlington, Leeds and York.

From 1876 Fletcher held a tighter rein over the other workshops but no serious attempt was made to standardise either locomotives or fittings and many locomotives still bore traces of Bouch ancestry including a screw reversing gear he had introduced, and a steam retarder.

The 0-6-0s now built were no longer of the 'long-boiler' type and the first actual 'Fletcher' design to be built at Darlington was the class *BTP* 0-4-4WT, although others to the same design had been built at Gateshead previously. A total of twenty-five were built and even in this batch and cylinders were 16" x 22" or 17" x 22" and the coupled wheels 5' 0" diameter to 5' 8" diameter. Some were subsequently rebuilt as 0-6-0Ts. *BTP* was a contraction of 'Bogie Tank Passenger'. They were successful and popular with the footplate crews. The round top boiler had a heating surface of 1074 sq. ft and a grate area of 12·75 sq. ft, the working pressure being 140 lb/sq. in.

Fig. 432 **NER Darlington** 1/1864 – first locomotive completed at North Road Works No. 175 CONTRACTOR

Fig. 433 **NER Darlington** 282/1885 McDonnell's *59 class* 0-6-0 No. 422

The *398* class 0-6-0s were commenced in 1878 and forty-two were built up to 1883 as well as six *603* class which only differed very slightly from the *398s*. All had inside cylinders 17″ x 24″, inside frames and 5′ diameter wheels.

A number of 2-4-0s had been built including four *11* class in 1877 and four *40* class in 1882. The latter had inside cylinders 17″ x 24″ and the leading wheels had outside bearings. One, No. 1099, was fitted with Younghusband's patent valve gear. The coupled wheels were 6′ 6″ diameter.

The other remaining class that Fletcher built at Darlington was the *124* class 0-6-0WTs of which twelve were built. Gateshead at this time was the locomotive headquarters of the NER and many other important designs were built there. (See YORK, NEWCASTLE and BERWICK RAILWAY – GATESHEAD).

Edward Fletcher retired in 1883 and in April of the same year Alexander McDonnell took his place. He had previously been in charge of the Great Southern and Western Railway's locomotives at Inchicore, Dublin where he had been successful in re-organising the works and did a good deal of work towards standardising the locomotive stock there.

With so many classes and variations in the NER stock, McDonnell set about changing this state of affairs. That he was right in his overall policy was indisputable, but his tactless and autocratic manner in putting his ideas into practice condemned him from the start. He carried out some alterations to Fletcher's 2-4-0s, transferring the driving position from right hand to left hand and ruthlessly put on the scrap heap the old traditions of Bouch and Fletcher. This did not endear him with the works staff or running staff and his first express design was a 4-4-0 – they had been quite happy with 2-4-0s, so the new design was condemned out of hand before there was water in the boiler.

They had inside cylinders 17″ x 24″ and 6′ 8″ diameter coupled wheels, with a boiler working at 140 lb/sq. in., heating surface 1030 sq. ft and a grate area of 16·8 sq. ft. The chimney was cast iron, tapering outwards to the top. The bogie was the swing-link type – the first in this country. The weight in working order was 39 tons 10 cwt. According to J. S. Maclean, twelve were built by Hawthorn Leslie & Co., fifteen at Gateshead and one at Darlington, but from Mr K. Hoole's list none were built at Darlington. It is worth recording that they were not so successful as the current 2-4-0s, which made matters worse.

A more successful type introduced by McDonnell was the *59* class 0-6-0s with 17″ x 26″ cylinders and 5′ 0″ diameter wheels (except two which had 5′ 6″ diameter). Thirty-two were built at Darlington. The running plate was raised over the leading and driving wheels, and the numbers were first painted on the cab sides, but were fitted later with cast iron plates.

So much opposition was encountered by McDonnell that he resigned in 1884.

Passenger locomotives of increased power were urgently needed and a locomotive committee was set up immediately after McDonnell's resignation and a 2-4-0 design was quickly drawn up, Mr Wilson Worsdell playing a major part in its inception. Care was taken not to incorporate any details which would rouse any further controversy, and so the *Tennant* class was evolved.

In 1885 Thomas William, the elder brother of Wilson Worsdell, was appointed as the new Locomotive Superintendent. He had extensive experience in locomotive design and building having trained at Crewe, worked in America, and returned to Crewe. In 1882 he became Locomotive Superintendent on the Great Eastern Railway at Stratford until 1885.

The *Tennants* were an immediate success with their larger cylinders 18″ x 24″, 7 ft diameter coupled wheels and a boiler with a heating surface of 1215 sq. ft and grate area 17 sq. ft. The boiler pressure was stepped up to 160 lb/sq. in. They took part in the 1888 'Race to the North' and gave very creditable performances. Ten were built at Darlington in 1885 (Nos. 1463–1472) others at Gateshead.

A locomotive classification was introduced in 1886, all newly built from this time were allotted a class letter, and all those previously built, retained their class numbers.

In 1886 two new classes were built, the *B1* 0-6-2T (a new wheel arrangement for the NER) and the *E* 0-6-0T. The *B1s* were used for mixed traffic and had 18″ x 24″ cylinders and 5′ 1¼″ diameter coupled wheels, whilst the *E* class were small side tanks with 16″ x 22″ cylinders and 4′ 6″ diameter wheels.

A further class of 2-4-0s were put in hand in 1887. They were *Class G* and twenty were built 1887–8. Cylinders were 17″ x 24″, and coupled wheels 6 ft diameter. They were intended for mixed traffic work. Like all T. W. Worsdell's locomotives they were fitted with Joy's valve gear, and had a commodious cab, with two windows each side which made them far more comfortable. They were known as 'Waterburys' a well-known timepiece of the period due to their excellent time keeping.

T. W. Worsdell carried out a lot of work on compounding but this work was done at Gateshead until 1888, when five *Class B* 0-6-2Ts were built with one high pressure cylinder 18″ x 24″ and low pressure 26″ x 24″. They were on the Worsdell-von Borries system and a total of forty were built at Darlington up to 1890. Other dimensions were the same as the simple *Class B1s.*

A tender version of the 0-6-2Ts was built in 1890, having boiler, wheels, cylinders and motion interchangeable. They were *Class C* 0-6-0s and were two cylinder compounds.

Due to ill-health T. W. Worsdell retired in 1890. He had introduced many changes in design but unlike his predecessor they were brought into being more tactfully and skilfully and were accepted.

His brother, Wilson Worsdell succeeded to the post in 1890. He had also been trained at Crewe, joined his brother in America, and returned to Crewe until 1883 when he became assistant Locomotive Superintendent under McDonnell. During his term of office, great strides were made in locomotive design. He was ably assisted by Walter M. Smith chief draughtsman at Gateshead and Vincent Raven who was made assistant Mechanical Engineer in 1895.

Two major changes in policy took place – the abandonment of compounding as a general

Fig. 434 **NER Darlington** 289/1885. Designed by Locomotive Committee W. Worsdell *Class 1463* 2-4-0 No. 1468 7′ 1¼″ D.W.

Fig. 435 **NER Darlington** 591/1896 W. Worsdell *O class* 0-4-4T No. 1884

practice and the substitution of Stephenson's link motion in peference to Joy's valve gear.

The *Class C* 0-6-0s and *Class E* 0-6-0Ts continued to be built, and in 1893 a new design of 0-6-2T was built, an order for twenty being completed in 1894. They were *Class N* and had larger cylinders (17″ x 26″) than the previous *Class B1*.

A larger 0-4-4T was then put in hand with 5 ft diameter coupled wheels and 18″ x 24″ cylinders. The first one appeared in June 1894 and a total of 110 were built from 1894 to 1901, all at Darlington. They were used on branch lines all over the system and had a good turn of speed.

In 1897 the need for a powerful six-coupled goods locomotive became pressing, to replace many ageing 0-6-0s. The *Class P* was built in 1897–8 followed by the *P1* class with larger cylinders in 1899 and the *P2* class with a larger boiler in 1904 and yet a further class – *P3* in 1906, which was similar to the *P2* class except for a shallower firebox.

Twenty *Class Ps* were built with cylinders 18″ x 24″ and 4′ 7$\frac{1}{4}$″ diameter wheels. The boiler gave a heating surface of 1097 sq. ft and a grate area of 15·3 sq. ft. The pressure was 160 lb/sq. in. The *P1s* had 18$\frac{1}{4}$″ x 26″ cylinders, which was the only difference, and forty were built in 1899–1900.

The *P2s* principal difference was the boiler, which was 5′ 6″ diameter as compared with the 4′ 3″ diameter for the *P* and *P1s*, and a total heating surface of 1658 sq. ft was obtained, with a grate area of 20 sq. ft. The round topped firebox had four safety valves fitted, surrounded by a large brass casing. The original boiler pressure was 200 lb/sq. in., later reduced to 180 lb/sq. in. They were the largest 0-6-0s to be built until the LNER *J38* and *J39* classes appeared in 1926.

Thirty *P2s* were built in 1904–5 and thirty *P3s* 1906–8.

A modified version of the *Class E* 0-6-0T appeared in 1898, the difference being the reduction of wheel diameter to 4′ 1$\frac{1}{4}$″. These light tanks were so useful that further batches were built by the LNER and again in 1948 British Railways ordered twenty-eight more. They were *Class E1*.

Up to 1907 Gateshead had built most of the heavier locomotives including many 0-8-0s and in this year Darlington built twenty *Class T1* 0-8-0s in 1907–8. These had slide-valves whereas the Gateshead-built *Class T* had piston valves and a series of tests were carried out in 1906 – Gateshead had also built some *T1s* – to compare the efficiency of both types of valves. There seemed little to chose between them although the *T1s* ran more freely when coasting.

In 1908–9 ten 4-4-0s were built to *Class R1* and were the largest 4-4-0s in the country at the time. They had inside cylinders 19″ x 26″

and 6′ 10″ diameter coupled wheels. A large 5′ 6″ diameter boiler with a total heating surface of 1737 sq. ft and 27 sq. ft grate area was fitted and worked at the high pressure of 225 lb/sq. in. which was reduced when rebuilt with superheaters.

The final class was the modified *Class V* 4-4-2 which had been built at Gateshead from 1903. The class was known as *V/09* and had two outside cylinders 19½″ x 28″, whereas the Gateshead *Vs* had 20″ diameter cylinders. Ten were completed in 1910, appearing after Wilson Worsdell had retired in May 1910.

Worsdell left behind him some very fine locomotives and although Darlington were not responsible for building many of his larger types, when the new erecting shop was built in 1903 the works gradually took over the larger types, and when Gateshead built its last new locomotive in 1910 Darlington became the No. 1 works on the North Eastern Railway.

Vincent Litchfield Raven succeeded as Chief Mechanical Engineer – the title had replaced Locomotive Superintendent in 1902. He had spent his whole working life on the NER. Up to the retirement of Wilson Worsdell 850 locomotives had been built at the North Road works from 1864. In 1910 the CME's staff moved from Gateshead to Darlington and all new work was carried out here.

The first design by Raven was the *Class Y* 4-6-2T. They were intended for coal trains from the collieries to the ports and had three cylinders 16½″ x 26″, 4′ 7¼″ diameter coupled wheels. The 5′ 6″ diameter boiler gave a heating surface of 1648 sq. ft and 23 sq. ft of grate area. They were subsequently fitted with superheaters. Twenty were built 1910–11.

The first 4-6-0s to be built at Darlington were laid down in 1911 and differed from the *Class S* built at Gateshead in having larger boilers. They were known as *Class S2* with 6′ 1¼″ diameter coupled wheels, two outside cylinders 20″ x 26″. The first seven were built with saturated boilers and the next thirteen with superheaters. The last of the batch, No. 825 completed in 1913, was fitted with Stumpf cylinders patented by Professor Stumpf of Berlin. Where steam is expanded up to the end of the piston stroke, it is impossible to carry out the expansion to the limit. Professor Stumpf's idea was to utilise a proportion of this lost energy. Normally the steam is exhausted through the same ports as the inlet steam but on Stumpf's Uniflow arrangement the steam is exhausted through special ports in the centre of cylinders which were lengthened. No. 825 was converted to a standard *S2* in 1924.

The first of a new class of 0-8-0 appeared in 1913. The boiler 5′ 6″ diameter was larger than those on Worsdell's *T* and *T1* classes. Known as the *T2* class it was probably the most successful of Raven's designs. The boiler was standard with the *S2* 4-6-0s. Two outside cylinders 20″ x 26″, 4′ 7¼″ diameter wheels and a superheated boiler with a total heating surface of 1915 sq. ft, including 544·8 sq. ft of superheater elements with a grate area of 23 sq. ft formed a design which was powerful and efficient. Seventy were built at Darlington 1913–19 and fifty more came from outside contractors.

A further new type was built in 1913, the *Class D* 4-4-4T designed to replace the four-coupled tender types used on the lengthy branch lines. They had three cylinders 16½″ x 26″ and 5′ 9″ diameter coupled wheels. The boiler was superheated with a boiler providing 1332 sq. ft and a working pressure of 160 lb/sq. in. The weight in working order was 84 tons 15 cwt. The wheel arrangement was chosen to provide steady riding forward or bunker first, but unfortunately at speed the reverse was the case and the whole class of forty-five, built 1913 to 1921, were converted to 4-6-2Ts from 1931 to 1936 and then proved excellent riders.

The *Class Z* three-cylinder *Atlantic* had been built at Gateshead from 1911 and a total of twenty had been turned out. In 1914 a start was made on an order for thirty *Class Z* 4-4-2s with superheated boilers and these were built steadily through the 1914–18 War, together with forty *Class T2* 0-8-0s.

From 1915 Vincent L. Raven was made Chief Superintendent of the Royal Ordnance Factories, Woolwich; and Mr A. C. Stamer became acting CME for the war period.

Although new building was considerably reduced at this time, rebuilding continued and a large amount of superheating conversion took place. After the war Raven returned and had a knighthood conferred.

A very successful mixed traffic three-cylinder 4-6-0 with 5′ 8″ diameter coupled wheels, Class S3 the first of thirty-eight built, came out in 1919 and after amalgamation a further thirty-two were built, all at Darlington. They were built in conjunction with five three-cylinder *Class T3* 0-8-0s and both classes were fitted with the same 5′ 6″ diameter boiler, which had 2094 sq. ft of heating surface and a working pressure of 180 lb/sq. in.

Both types were successful and besides the 4-6-0s the *T3s* were also built after the amalgamation.

In the last year of independence two *Pacifics* were authorised and built in the same year. They had three cylinders 19″ x 26″, 8¾″ diameter piston valves with three separate sets of Walschaerts valve gear. Coupled wheels were 6′ 8″ diameter and the large boiler had a

Fig. 436
NER Darlington
1135/1921 3-cyl. loco rebuilt subsequently as 4-6-2T Raven *Class D* 4-4-4T No. 1527

Fig. 437
NER Darlington
1170/1922 Raven's *Pacific* No. 2400
CITY OF NEWCASTLE

Fig. 438
Darlington
1225/1924 Raven 3-cyl. 0-8-0 *Class Q7* No. 628

Fig. 439 **Darlington** 1291/1925 *Class K3* Gresley 3-cyl. 2-6-0 No. 227 fitted with electrical foam detector

Fig. 440 **Darlington** 1420/1928 Gresley *Class J39* 5′ 2″ 0-6-0 No. 2706

Fig. 441 **Darlington** 1547/1932 Gresley 3-cyl. *Hunt Class D49* No. 211 YORK & AINSTY fitted with Lentz rotary cam poppet valve gear

Fig. 442 **Darlington** 1997/1947 Thompson *Class B1* 4-6-0 No. 1017 BUSHBUCK

total heating surface of 2874·6 sq. ft including 200 sq. ft for the firebox and 510 sq. ft of superheater elements. The grate area was 41·5 sq. ft and boiler pressure 200 lb/sq. in. The trailing truck on these first two had inside bearings under a wide firebox but the three built after amalgamation had outside bearings.

The Gresley *Pacifics* were to be the standard on the LNER and although one of Raven's *Pacifics* was rebuilt with a Gresley boiler, they were not as successful as had been hoped and they had all been scrapped by 1937.

So ended a period of Darlington works history, with the total number of steam locomotives built 1163. From 1 January, 1923 the NER became part of the LONDON & NORTH EASTERN RAILWAY.

At the amalgamation Sir Vincent Raven was appointed Consultant Engineer but he resigned in 1924.

For details of locomotive building during this period 1923–47 see under London and North Eastern Railway and also the table showing locomotives built at Darlington.

From 1 January, 1948 the L&NER became part of the nationalised BRITISH RAILWAYS, and building continued with standard LNER types until 1951. The most interesting were twenty-eight *J72* 0-6-0Ts of Worsdell's design of 1898. The *L1* 2-6-4Ts of Thompson's design of 1945 were built to a total of 100 before being replaced by the BR *Class 4* 2-6-4Ts. Darlington built twenty-nine, the remainder being built by outside contractors.

Fig. 443
Darlington
2066/1949
Class A1 Pacific
No. 60147
NORTH EASTERN

Fig. 444
Darlington
2263/1954
Class 2 BR
2-6-0 No. 78016

The LMS *Class 2* and *4* 2-6-0s were built for use on the Eastern and North Eastern regions and were followed by the similar *Class 2* BR design and ten BR *Class 2* 2-6-2Ts. The last steam locomotive built at Darlington was 2-6-2T No. 84029 steamed on 11 June, 1957 – Derby Works turned out their last steam locomotive in the same month.

Darlington built 2269 steam locomotives from 1863 to 1957. Locomotive repairs were carried on after new building ceased, although the building of diesel locomotives continued into 1964. These historic works were finally closed on 2 April, 1966 and were offered for sale.

Darlington Works
Works numbers were allotted to new locomotives and certain rebuilt locomotives from 1864 to 1885. A second series of numbers was introduced by T. W. Worsdell in 1866 and continued until Wilson Worsdell's retirement in 1909. After that works numbers were not allocated by Vincent Raven and it was not until 1943 that calculations were made as to how many locomotives had been built at Darlington. This was assessed as 1902 up to the end of 1942. This would include the eleven electric locomotives built.

T. W. Worsdell introduced the lettered classification for locomotives. Before this the class was known by one of the running numbers of the class being built.

After the naming of the Stockton and Darlington locomotives it seems curious that the North Eastern only named one of their locomotives, AEROLITE No. 66, which had many rebuildings and changes.

Wilson Worsdell was a personal friend of G. J. Churchward of the Great Western Railway, and in T. W. Worsdell's time very good relations existed between the NER and the Midland Railway.

At Darlington there was room for expansion and in 1884 additional pits were provided and in 1903 a new Erecting Shop was built with three bays each having 24 pits. The shop was 500 ft long and 200 ft wide.

In 1910 the number of overhauls required were increasing and more additional shops were built including a boiler and tender shop, and a new paint shop.

During the NER period no boilers were built with Belpaire fireboxes, the majority having round flush topped fireboxes.

The locomotives were painted and finished second to none, and copper topped chimneys and brass safety valve covers were highly polished.
See BRITISH RAILWAYS.

T. W. WORSDELL

Class NE	Class LNE	Date	Type	Cylinders inches	DW ft in	Heating Surface sq ft Tubes	Heating Surface sq ft F'box	Heating Surface sq ft Total	Grate Area sq ft	WP lb/sq in	Weight WO ton cwt	No Built	
A	F8	1886-92	2-4-2T	18 x 24	5 6	994	98	1092	15·2	160	52 0	60	
B	N8	1888-90	0-6-2T	18/26 x 24	5 0	1016	110	1126	17·25	175	57 0	51	(a)
C	J21	1887-92	0-6-0	18/26 x 24	5 0	1026	110	1136	17·25	160	41 5	171	(a)
D	—	1886-88	2-4-0	18/26 x 24	6 8	1211·3	112	1323·3	17·25	170	43 5	2	(a)
E	J71	1886-95	0-6-0T	16 x 22	4 7¼	658	73	731	11·25		37 10	120	
F	D22	1891	4-4-0	18/26 x 24	6 8	1211·3	112	1323·3	17·25	170	46 10	25	(a)
G	—	1887-8	2-4-0	17 x 24	6 0	994	98	1092	15·2	160	40 5	20	(d)
H	Y7	1888-97	0-4-0T	13 x 20	3 5	448	57	505	11·2		21 15	19	
I	—	1888-90	4-2-2	18/26 x 24	7 0	1016	110	1126	17·25	175	43 15	10	(a)
J	—	1889-90	4-2-2	20/28 x 24	7 7	1029	110	1139	20·7	175	46 10	10	(a)
K	Y8	1890	0-4-0T	11 x 15	3 0	148	36	184	6·5		13 15	5	

W. WORSDELL

E1	J72	1898	0-6-0T	17 x 24	4 0	658	73	731	11·25		36 5	20	
H2	J79	1897	0-6-0T	14 x 20	3 5	448	57	505	11·25		25 0	2	
L	J73	1891-2	0-6-0T	19 x 24	4 7¼	999	98	1097	15·2		46 15	10	
M	–	1893	4-4-0	$\frac{19}{28}$ x 26	7 1	1220	121	1341	19·25	200	51 0	1	(a)
M1	D17/1	1892-4	4-4-0	19 x 26	7 1	1220	121	1341	19·5	180	50 15	20	
O	G5	1894-01	0-4-4T	18 x 24	5 0	999	98	1097	15·2	160	51 10	110	
P	J24	1894-98	0-6-0	18 x 24	4 7¼	999	98	1097	15·2	160	38 10	70	
P1	J25	1898-03	0-6-0	18¼ x 26	4 7¼	1026	110	1136	17·2	160	40 0	140	(b)
P2	J26	1904-05	0-6-0	18½ x 26	4 7¼	1531	127	1658	20·0	200	47 15	50	(c)
P3	J27	1906-22	0-6-0	18½ x 26	4 7¼	1453	142	1595	20·0	180	47 14	115	
Q	D17/2	1896-97	4-4-0	19½ x 26	7 1	1089	123	1212	19·75	180	49 15	30	
Q1	D18	1896	4-4-0	20 x 26	7 7	1089	127	1216	20·75	180	50 15	2	
R	D20	1899-08	4-4-0	19 x 26	6 10	1383	144	1527	20·0	200	51 15	60	
R1	D21	1909	4-4-0	19 x 26	6 10	1579	158	1737	27·0	225	59 10	10	
S	B13	1899-09	4-6-0	20 x 26	6 1	1638	130	1768	23·0	200	63 0	40	
S1	B14	1899-09	4-6-0	20 x 26	6 8	1639	130	1769	23·0	200	67 0	5	
T	Q5	1901-02	0-8-0	20 x 26	4 7¼	1550	125	1675	21·5	175	58 5	70	
U	N10	1902-03	0-6-2T	18½ x 26	4 7¼	1026	110	1136	17·0	160	56 15	20	
V	C6	1903	4-4-2	20 x 28	6 10	2275·8	180	2455·8	27·0	200	72 15	10	
4cc	C8	1905	4-4-2	$\frac{14½}{22}$ x 26	7 1	1782	180	1962	29·0	225	73 10	2	
W	–	1908	4-6-0T	19 x 26	5 1	1169	141	1310	23·0	170	69 0	10	(e)
X	T1	1909	4-8-0T	3/18 x 26	4 7¼	1169	141	1310	23·0	175	84 15	10	

(a) 2 cylinder compounds later converted to simples. (b) 18½" diameter cylinders for last 40 (c) WP later reduced to 180 lb/sq in
(d) Later converted to 4-4-0 (LNE D23) (e) Later converted to 4-6-2T (LNE A6)

RAVEN

Class		Date	Type	No	Cylinders inches	DW ft in	Heating Surface sq ft				Grate Area sq ft	WP lb/sq in	Weight WO ton cwt
NE	LNE						Tubes	S'heat	F'box	Total			
V1	C6	1910	4-4-2	10	19½ x 28	6 10	2275·8	–	180	2455·8	27·0	180	76 5
Y	A7	1910-11	4-6-2T	20	3/16½ x 26	4 7	1508	–	140	1648	23·0	175	87 8
Z	C7	1911-18	4-4-2	40	3/15½ x 26	6 10				2340		180	
Z1	C7	1911	4-4-2	10	3/16½ x 26	6 10	1295·8	530	180	2005·8	27·0	160	77 2
S2	B15	1911-13	4-6-0	20	20 x 26	6 1	1226	545	144	1915	23·0	160	70 15
(a)D	H1	1913-22	4-4-4T	45	3/16½ x 26	5 9	934·84	194·13	124	1252·97	23·0	160	84 15
T2	Q6	1913-21	0-8-0	120	20 x 26	4 7¼	1226	545	144	1915	23·0	160	
T3	Q7	1919-24	0-8-0	15	3/18½ x 26	4 7¼	1397·9	530·1	166	2094	27·0	180	71 12
S3	B16	1919-24	4-6-0	70	3/18½ x 26	5 8	1397·9	530·1	166	2094	27·0	180	77 14
–	A2	1922-4	4-6-2	5	3/19 x 26	6 8	2164·7	509·9	200	2874·6	41·5	200	97 0

(a) Rebuilt as 4-6-2T LNE A 8.

NORTH EASTERN RAILWAY

DARLINGTON

Date	4-6-2	4-6-0	2-6-2	2-6-0	0-6-0
1923		B 16 1437-66			J 27 5885-94
1924	A2 2402-04	B 16 1437-8		K3 1810-36	
1925				K3 1837-69	
1926					J 38 5900-34
					J 39 4700-16
1927					J 39 4717-43
1928					J 39 4744-73
1929	10000 *				J 39 4774-813
1930		817 1610-21		K3 1890-98	J 39 4814-22
1931		B17 1622-36			J 39 4823-31
1932					J 39 4832-45
1933		B17 1637-42			J 39 4846-47
1934					J 39 4848-59
1935		B17 1643-47			J 39 4860-98
1936		B17 1648-61		K3 1969-84	J 39 4899-904
1937			V2 805-24	K3 1985-92 K4 1993	J 39 4933
1938			V2 825-43	K4 1994-96	J 39 4934-70
1939			V2 844-71 V2 882-91	K4 1997-98	
1940			V2 892-917		
1941			V2 918-927		J 39 4971-88
1942		B1 1000	V2 938-962		
1943		B17 1001-04	V2 963-978		
1944	A 2 507-09	B17 1005-09	V2 979-83		
1945	A 2 510				
1946		B17 1010-14			
1947		B17 1015-39			

Date	4-8-0T	2-8-0	0-8-0	4-4-0	0-4-0T
1923					984† Y7 8086-9
1924			Q7 3465-74		
1925	TI 9918-22				
1926					
1927				D49 2700-06	
1928				D49 2707-25	
1929				D49 2726-35	
1930					
1931					
1932				D49 2736-45	
1933				D49 2746-50	
1934				D49 2751-70	
1935				D49 2771-75	
1936					
1937					
1938					
1939					
1940					
1941					
1942					
1943		LMS 8500-09			
1944		LMS 8540-42			
1945		LMS 8543-59 LMS 3125-34			
1946		LMS 3135-47			

All loco running numbers are as the 1946 renumbering scheme except the 1924 A 2s

† withdrawn before renumbering * Type 4-6-4.

NORTH EASTERN RAILWAY

A summary of steam locomotives built during British Railways' management:

Year	No.		Class	Type	Numbers
1948	29	LNE	Class L 1	2-6-4T	69001 - 15
					67717 - 30
1948	12	LNE	Class A 1	4-6-2	60130 - 41
1949	11	LNE	Class A 1	4-6-2	60142 - 52
1949	10	LNE	Class B 1	4-6-0	61350 - 59
1949	15	LNE	Class J 72	0-6-0T	69001 - 15
1950	5	LNE	Class J 72	0-6-0T	69016 - 20
1950	10	LNE	Class B 1	4-6-0	61400 - 09
1950	27	LMS	Class 4	2-6-0	43070 - 96
1951	10	LMS	Class 4	2-6-0	43097 - 4106
1951	30	LMS	Class 2	2-6-0	46465 - 94
1951	8	LNE	Class J 72	0-6-0T	69021 - 28
1952	8	LMS	Class 2	2-6-0	46495 - 4502
1952	4	BR	Class 2	2-6-0	78000 - 03
1953	8	BR	Class 2	2-6-0	78004 - 11
1954	33	BR	Class 2	2-6-0	78012 - 44
1955	10	BR	Class 2	2-6-0	78045 - 54
1956	10	BR	Class 2	2-6-0	78055 - 64
1957	10	BR	Class 2	2-6-2T	84020 - 29
	250	Total			

NORTH EASTERN RAILWAY
Gateshead

See YORK, NEWCASTLE and BERWICK RAILWAY.

NORTH EASTERN RAILWAY
York

These locomotive shops were built by the York and North Midland Railway to carry out repairs and rebuilding of their locomotive stock, which at the amalgamation of 1854 amounted to 114. No new building was attempted by the Y&NMR but in North Eastern Railway ownership a few new locomotives were built between 1854 and 1884.

At the amalgamation Edward Fletcher was made Locomotive Superintendent and he made his headquarters at Gateshead which had been the locomotive shops of the York, Newcastle and Berwick Railway. No standardisation took place for some time and York, Leeds and Darlington pursued their own policies to a large extent.

Spasmodic building took place from 1854 to 1861 giving an average of one locomotive per annum. After a lapse of years three *398* class 0-6-0s were built in September 1884 and two *BTP* class 0-4-4WTs in December of the same year. No more were built after this date, but repairs went on until the closure of the works in April 1905.

The two *BTPs* were converted to *Class 290* 0-6-0s in 1899 and 1900 respectively, due to the *BTPs* being replaced by more modern *Class O* 0-4-4Ts. Another thirty-eight of the same class were converted at York between 1899 and 1904 although these locomotives had not been built originally at York.

Locomotives repaired and built at York could be distinguished from those coming from the Leeds, Darlington and Gateshead works by the painting, especially the lining out which was black bands and white lines, whereas Gateshead and Darlington lined out with dark green and black bands with white and vermilion lines, and Leeds used just yellow lines. Number plates from York had a green background, the other works vermilion. Cabs and domes were also different and there were other small details.

The works were eventually converted into the Railway Museum where the 'large' exhibits, including historic locomotives, have been on show since the 1925 Railway Centenary, when it was realised that a permanent home must be found for all these treasures. The museum has now been extended to house further exhibits from Clapham and elsewhere.

NER YORK

Date	Type	Cylinders inches	D.W. ft in	Frames	NER No	
1854	0-4-0	IC 15 x 20	5 6	IF	278	
1856	2-4-0	IC 16 x 22	6 1	OF	271	
1857	2-4-0	IC 16 x 22	6 1	OF	293	
1857	0-6-0	IC 16 x 24	5 0	OF	305	
9/1857	2-4-0	IC 16 x 22	6 6	OF	254	
10/1857	2-4-0	IC 16 x 22	6 0	(a)	260	
1859	0-4-0	IC 15 x 20	5 0	IF	263	
1861	0-4-0	IC 15 x 20	5 0	IF	272	
9/1884	0-6-0	IC 17 x 24	5 0	IF	52	398 class (d)
9/1884	0-6-0	IC 17 x 24	5 0	IF	134	398 class (d)
9/1884	0-6-0	IC 17 x 24	5 0	IF	266	398 class (d)
12/1884	0-4-4T	IC 17 x 22	5 6	IF	290	BTP class (b)
12/1884	0-4-4T	IC 17 x 22	5 6	IF	305	BTP class (c)

(a) Outside frames – leading wheels only. (b) Rebuilt 1899 to 0-6-0T Class 290. (c) Rebuilt 1900 to 0-6-0T Class 290. (d) Boilers, motion & other parts supplied by Gateshead.

NORTH EASTERN STEEL CO. LTD.

Acklam Works, Middlesbrough

This firm became part of the Dorman Long group in 1923.

Referring to the section of Dorman Long locomotives, Acklam also replaced some locomotives, building them on similar lines to those they replaced. They were as follows:

1919 ACKLAM No. 16. replaced former unknown locomotive (possibly JOICEY?) purchased second hand from Cargo Fleet Iron Co. Ltd. in 1917. Subsequently scrapped before the 1943 renumbering.

1921 ACKLAM No. 7. replaced former Hudswell Clarke locomotive, and became No. 40 in 1943.

1922 ACKLAM No. 1. replaced former Black Hawthorn locomotive and became No. 41 in 1943.

1923 ACKLAM No. 10. replaced former Hudswell Clarke locomotive and became No. 38 in 1943.

1926 ACKLAM No. 14. replaced former Hudswell Clarke locomotive. Later renumbered ACKLAM No. 11 and became No. 37 in 1943. This locomotive was rebuilt after the firm became part of the Dorman Long Group.

NORTH LONDON RAILWAY

Bow

Locomotive Superintendents: William Adams 1853–73; J. C. Park 1873–93; and Henry J. Pryce 1893–1903. The NLR was worked by L&NWR from December 1908.

The first locomotive was built in 1863 and from this year all new locomotives required were built in Bow workshops which had been established in 1853. The works stood on thirty-one acres and included carriage and wagon shops. A new locomotive erecting shop was built in 1882 on the site of the old running shed. It had two bays 230 ft long, 45 ft wide and provided six lines of pits.

William Adams was the first locomotive superintendent and his first design was a 4-4-0T with inside cylinders 16″ x 24″ and 5′ 3″ diameter driving wheels. The boiler had a total heating surface of 1141·72 sq. ft and working pressure of 160 lb/sq. in. which at that time was considered high. These were designed for the passenger traffic which was very intensive and a total of eight were built followed by twenty-four more of similar dimensions, except the cylinders were 17″ diameter and the coupled wheels 5′ 9″ diameter.

Previous to these 4-4-0Ts the stock of the NLR comprised a very varied selection of locomotives including 2-4-0WTs (from Crewe), 0-6-0s, and one 2-4-0 and some 4-4-0Ts from Robert Stephenson & Co. which formed the basis of Adams' inside cylinder design.

In 1868 it was decided to standardise on two types, an outside cylindered 4-4-0T for the passenger traffic, and a 0-6-0T, also with outside cylinders.

The outside cylinder 4-4-0Ts were started in 1868 and incorporated a special bogie with side play and centre pivot. The cylinders were inclined 1 in 20 and the connecting rod big ends were between the coupling rod bearings and the wheels, presumably to keep the cylinder centres within gauge. The crossheads were between four slide bars. There was no foot-

Original Loco	Original Type	Maker	Wks No	Built	Rebuild Date	Rebuilt Type	1943 Dorman Long No	
Old Acklam No 16	0-4-0ST	Joicey?			1919	0-4-0ST	–	
Old Acklam No 7	0-4-0ST	HC			1921	0-4-0ST	40	
Old Acklam No 1	0-4-0ST	BH	614	1882	1922	0-4-0ST	41	
Old Acklam No 10	0-4-0ST	HC	323	1888	1923	0-4-0ST	38	
Old Acklam No 14 (a)	0-4-0ST	HC	324	1889	1926	0-4-0ST	37	(b)

(a) Later renumbered Old Acklam No 11. (b) Rebuilt after firm became part of Dorman Long Group in 1923.

Fig. 445 **NLR** Bow Works

plating from the end of the sandboxes to the front buffer beam, but the smokebox was nicely curved to meet the sheeting around the cylinders.

The boiler was domed with a pair of Ramsbottom safety valves over the round top firebox. The cab was comparatively roomy with square windows in the front plate and rear weatherboard. The chimney was distinctive, tapering inwards towards the top. The valve gear was Stephenson's link motion, the valve chests being inside the frames.

Most were rebuilt with $17\frac{1}{2}''$ cylinders and the coupled wheel diameter varied from 5′ 3″ to 5′ 5″. A total of eighty-six were built from 1868 to 1906.

Fig. 446 **NLR** 109/1865 rebuilt 1886 4-4-0T No. 109

NORTH LONDON RAILWAY

Works No	Date	Type	Cylinders inches	DW ft in	First No	Re-No.	Rebuilt	LNW No	LMS No		BR No		W'drawn
									1st	2nd			
	1863	4-4-0T	IC 16 x 24	5 3	43	43A/$_{77}$ 101/$_{77}$	—						1888
	1863	4-4-0T	IC 16 x 24	5 3	44	44A/$_{77}$ 102/$_{77}$	—						1889
	1863	4-4-0T	IC 16 x 24	5 3	45	103/$_{83}$	—						1889
	1864	4-4-0T	IC 16 x 24	5 3	46	104/$_{83}$	—						1889
	1864	4-4-0T	IC 16 x 24	5 3	47	105/$_{83}$	1889						1915
	1864	4-4-0T	IC 16 x 24	5 3	48	106/$_{83}$	1889						1915
	1865	4-4-0T	IC 16 x 24	5 3	49	107/$_{84}$	1883						1904
	1865	4-4-0T	IC 16 x 24	5 3	50	108/$_{84}$	1887						1918
	1865	4-4-0T	IC17 x 24	5 9	51	109/$_{85}$	1886	2874	(6435)			(a)	1925
	1865	4-4-0T	IC17 x 24	5 9	52	112/$_{85}$	1888						1911
	1866	4-4-0T	IC17 x 24	5 9	53	111/$_{85}$	1883						1894
	1866	4-4-0T	IC17 x 24	5 9	54	110/$_{85}$	1888						1910
	1866	4-4-0T	IC17 x 24	5 9	55	113/$_{85}$	—						1889
	1866	4-4-0T	IC17 x 24	5 9	56	114/$_{85}$	1885 1902	2648	(6436)			(b)	1925
	1866	4-4-0T	IC17 x 24	5 9	11	115/$_{86}$	1883						1891
	1867	4-4-0T	IC17 x 24	5 9	12	116/$_{86}$	—						1891
	1867	4-4-0T	IC17 x 24	5 9	13	117/$_{87}$	1886	2649	(6437)			(b)	1925
	1867	4-4-0T	IC17 x 24	5 9	14	118/$_{87}$	1886						1911
	1867	4-4-0T	IC17 x 24	5 9	57	119/$_{87}$	—						1892
	1867	4-4-0T	IC17 x 24	5 9	58	120/$_{87}$	1886						1909
	1867	4-4-0T	IC17 x 24	5 9	59	121/$_{88}$	—						1892
	1867	4-4-0T	IC17 x 24	5 9	60	122/$_{88}$	1883						1895
	1867	4-4-0T	IC17 x 24	5 9	61		—						1887
	1867	4-4-0T	IC17 x 24	5 9	62		—						1887
	1867	4-4-0T	IC17 x 24	5 9	63		—						1887
	1868	4-4-0T	IC17 x 24	5 9	64	101/$_{89}$	1890						1913
	1868	4-4-0T	IC17 x 24	5 9	65		—						1888
	1868	4-4-0T	IC17 x 24	5 9	66		—						1889
	1869	4-4-0T	IC17 x 24	5 9	15	102/$_{89}$	1890						1913
	1869	4-4-0T	IC17 x 24	5 9	16	113/$_{89}$	1890						1910
	1869	4-4-0T	IC17 x 24	5 9	17	103/$_{89}$	1885	2647	(6438)			(c)	1924
	1869	4-4-0T	IC17 x 24	5 9	18	104/$_{89}$	1883						1894
	1868	4-4-0T	OC17 x 24	5 6	2	124/$_{96}$	1883						1897
	1868	4-4-0T	OC17 x 24	5 6	8		1887 1902	2807	(6439)				1925
	1868	4-4-0T	OC17 x 24	5 6	9		1883 1899	2808	6440				1928
	1868	4-4-0T	OC17 x 24	5 6	22		1884 1900	2816	6441				1925
	1869	4-4-0T	OC17 x 24	5 6	25		1883 1896	2819	6442				1926

	1869	4-4-0T	OC17 x 24	5 6	26		—	—	—				1890
	1869	4-4-0T	OC17 x 24	5 6	27		—	—	—				1890
	1870	4-4-0T	OC17 x 24	5 6	5		—	—	—				1890
	1870	4-4-0T	OC17 x 24	5 6	6		—	—	—				1894
	1870	4-4-0T	OC17 x 24	5 6	10		1883 1899	(2809)	6446				1928
	1870	4-4-0T	OC17 x 24	5 6	21		1886 1902	2815	6451				1928
	1870	4-4-0T	OC17 x 24	5 6	1	125/06	1882 1896	2872	6443				1925
	1871	4-4-0T	OC17 x 24	5 6	3	127/07	1883 1899	(2873)	(6448)				1923
	1871	4-4-0T	OC17 x 24	5 6	4	124/07	1885 1898	(2871)	(6450)				1926
	1871	4-4-0T	OC17 x 24	5 6	7		—	—	—				1890
	1871	4-4-0T	OC17 x 24	5 6	19		1882 1902	—	—				1922
	1871	4-4-0T	OC17 x 24	5 6	20		1888 1903	2814	6447				1929
	1871	4-4-0T	OC17 x 24	5 6	24		—	—	—				1890
	1872	4-4-0T	OC17 x 24	5 6	23		—	—	—				1890
	1872	4-4-0T	OC17 x 24	5 6	28		1888 1907	(2822)	(6452)				1925
	1872	4-4-0T	OC17 x 24	5 6	29		1887 1901	(2823)	(6453)				1927
	1873	4-4-0T	OC17 x 24	5 6	39		—	—	—				1892
	1873	4-4-0T	OC17 x 24	5 6	40		1892 1904	2833	(6455)				1926
	1873	4-4-0T	OC17 x 24	5 6	41		1884 1900	2834	(6454)				1925
160	1874	4-4-0T	OC17 x 24	5 6	38	123/91	—	—	—				1894
161	1874	4-4-0T	OC17 x 24	5 6	42		—	—	—				1893
162	1874	4-4-0T	OC17 x 24	5 6	34		1885 1901	2827	6457				1928
163	1874	4-4-0T	OC17 x 24	5 6	35		1891 1905	2828	6458				1929
164	1875	4-4-0T	OC17 x 24	5 6	30		1892	2824	(6459)				1926
165	1875	4-4-0T	OC17 x 24	5 6	31		1885 1899	2825	6460				1929
166	1875	4-4-0T	OC17 x 24	5 6	32		1888	—	—				1914
167	1875	4-4-0T	OC17 x 24	5 6	33		1892 1906	2826	6497				1928
168	1876	4-4-0T	OC17 x 24	5 6	67		—	—	—				1893
169	1876	4-4-0T	OC17 x 24	5 6	68		—	—	—				1895
170	1876	4-4-0T	OC17 x 24	5 6	69		—	—	—				1895
171	1876	4-4-0T	OC 17 x 24	5 6	70		1891	2856	(6464)				1926
							1904						
172	1876	4-4-0T	OC17 x 24	5 6	36		—	—	—				1893
173	1876	4-4-0T	OC17 x 24	5 6	37		—	—	—				1893
174	1877	4-4-0T	OC17 x 24	5 6	43		1896	2836	(6467)				1926
175	1877	4-4-0T	OC17 x 24	5 6	44		—	—	—				1894
176	1878	4-4-0T	OC17 x 24	5 6	71	124/94	—	—	—				1895
177	1878	4-4-0T	OC17 x 24	5 6	72	125/94	—	—	—				1897
178	1878	4-4-0T	OC17 x 24	5 6	73		1893	(2859)	(6471)				1924
179	1878	4-4-0T	OC17 x 24	5 6	74		{1894 1909	2860	6472				1928
180	1879	0-6-0T	OC17 x 24	4 4	75	115/91	1896	2634	7503			ϕ	1933
181	1880	0-6-0T	OC17 x 24	4 4	76	116/91	1897	2650	7505	27505	58850	ϕ	1960

(a) Rebuilt Crewe 1909 (b) Sold LNWR 1909 (c) Sold LNWR 1907 ϕ Sold to LNWR in 1909

Works No	Date	Type	Cylinders inches	DW ft in	First No	Re-No.	Rebuilt	LNW No	LMS No 1st	LMS No 2nd	BR No		W'drawn
182	1881	0-6-0T	OC17 × 24	4 4	77	119/92	1895	2635	7504			φ	1930
183	1881	0-6-0T	OC17 × 24	4 4	78	121/92	1897	2651	7506	27506		φ	1937
184	1882	0-6-0T	OC17 × 24	4 4	79	104/94	1897	2631	7507	–		φ	1932
185	1882	0-6-0T	OC17 × 24	4 4	80	111/94	1896 1925	2633	7508	27508		φ	1936
186	1883	4-4-0T	OC17 × 24	5 6	45		1898	2838	(6473)				1926
187	1883	4-4-0T	OC17 × 24	5 6	46		1897	2839	6474				1928
188	1883	4-4-0T	OC17 × 24	5 6	47		1898	2840	6475				1928
189	1883	4-4-0T	OC17 × 24	5 6	48		1899	(2841)	(6476)				1926
190	1884	4-4-0T	OC17 × 24	5 6	49		1899	2842	6477				1928
191	1884	4-4-0T	OC17 × 24	5 6	50		1898	(2843)	(6478)				1923
192	1885	4-4-0T	OC17 × 24	5 6	51		1901	2844	6479				1928
193	1885	4-4-0T	OC17 × 24	5 6	52		1897	—	—				1922
194	1885	4-4-0T	OC17 × 24	5 6	53		1897	2845	6480				1929
195	1885	4-4-0T	OC17 × 24	5 6	54		1900	2846	6481				1928
196	1885	4-4-0T	OC17 × 24	5 6	55		1900	2847	(6482)				1926
197	1885	4-4-0T	OC17 × 24	5 6	56		1900	2848	(6483)				1925
198	1886	4-4-0T	OC17 × 24	5 6	11		1898	2810	6484				1929
199	1886	4-4-0T	OC17 × 24	5 6	12		1904	2811	(6485)				1927
200	1887	4-4-0T	OC17 × 24	5 6	13		1903	2812	(6486)				1926
201	1887	4-4-0T	OC17 × 24	5 6	14		1903	2813	(6487)				1926
202	1887	4-4-0T	OC17 × 24	5 6	57		1902	2849	(6488)				1926
203	1887	4-4-0T	OC17 × 24	5 6	58		1902	(2850)	(6489)				1923
204	1888	4-4-0T	OC17 × 24	5 6	59		1903	2851	(6490)				1925
205	1888	4-4-0T	OC17 × 24	5 6	60		1904	2852	6491				1929
206	1887	0-6-0T	OC17 × 24	4 4	61	122/95	1903	2652	7509	27509	58851	φ	1955
207	1887	0-6-0T	OC17 × 24	4 4	62	123/95	1904	2636	7510	27510	58852	φ	1955
208	1887	0-6-0T	OC17 × 24	4 4	63		1901	2881	7511	27511			1937
209	1888	0-6-0T	OC17 × 24	4 4	64		1904	2882	7512	27512	58853		1954
210	1888	0-6-0T	OC17 × 24	4 4	65		1905	2883	7513	27513	58854		1956
213	1889	0-6-0T	OC17 × 24	4 4	66	107/05	1904	2632	7531				1932
214	1889	0-6-0T	OC17 × 24	4 4	15		1905	2875	7514	27514	58855		1956
215	1889	0-6-0T	OC17 × 24	4 4	16		1907	2876	7515	27515	58856		1957
216	1889	0-6-0T	OC17 × 24	4 4	17		1906	2877	7516				1932
217	1889	0-6-0T	OC17 × 24	4 4	18		1903	2878	7517	27517	58857		1958
221	1890	4-4-0T	OC17 × 24	5 6	5		1906	2804	6444				1928
222	1890	4-4-0T	OC17 × 24	5 6	7			2806	6449				1929
223	1890	4-4-0T	OC17 × 24	5 6	23		1902	2817	6492				1929
224	1890	4-4-0T	OC17 × 24	5 6	24			(2818)	6493				1928
225	1890	4-4-0T	OC17 × 24	5 6	26		1905	(2820)	(6494)				1924
226	1890	4-4-0T	OC17 × 24	5 6	27		1908	2821	6495				1929

228	1891	0-6-0T	OC17 x 24	4 4	76		1907	2886	7519				1930
229	1892	0-6-0T	OC17 x 24	4 4	77		1908	2887	7520	27520	58858		1953
230	1892	0-6-0T	OC17 x 24	4 4	78		1910	2888	7521	27521			1937
231	1891	4-4-0T	OC17 x 24	5 6	38		1905	2831	(6496)				1926
235	1892	4-4-0T	OC17 x 24	5 6	39		1908	2832	(6498)				1927
236	1893	4-4-0T	OC17 x 24	5 6	42		1907	2835	6456				1928
238	1893	4-4-0T	OC17 x 24	5 6	37		1908	2830	6466				1927
239	1893	4-4-0T	OC17 x 24	5 6	67			2853	6461				1929
240	1894	4-4-0T	OC17 x 24	5 6	72			2885	6470				1928
241	1894	4-4-0T	OC17 x 24	5 6	36		1905	2829	6465				1928
242	1894	0-6-0T	OC 17 x 24	4 4	79			2889	7522	27522	58859		1957
243	1894	0-6-0T	OC17 x 24	4 4	80		1909	2890	7523	27523			1937
244	1895	0-6-0T	OC17 x 24	4 4	61			2879	7524	27524			1936
245	1895	0-6-0T	OC17 x 24	4 4	62			2880	7525	27525			1948
246	1894	4-4-0T	OC17 x 24	5 6	44		1906	2837	6468				1929
247	1894	4-4-0T	OC17 x 24	5 6	71		1910	2857	6469				1929
248	1894	4-4-0T	OC 17 x 24	5 6	6		1909	(2805)	6445			(d)	1929
249	1895	4-4-0T	OC 17 x 24	5 6	1	125/06		2872	(6443)				1925
250	1895	4-4-0T	OC 17 x 24	5 6	68			(2854)	6462				1925
251	1895	4-4-0T	OC 17 x 24	5 6	69			2855	(6463)				1926
252	1896	4-4-0T	OC 17 x 24	5 6	2	126/06		—	—				1909
253	1896	4-4-0T	OC 17 x 24	5 6	25			2819	(6442)				1926
254	1896	4-4-0T	OC 17 x 24	5 6	81			(2861)	6499				1929
255	1896	4-4-0T	OC 17 x 24	5 6	82			(2862)	(6500)				1923
256	1896	4-4-0T	OC 17 x 24	5 6	83			(2863)	(6501)				1923
257	1896	4-4-0T	OC 17 x 24	5 6	84			2864	(6502)				1926
258	1897	4-4-0T	OC 17 x 24	5 6	85			2865	6503				1928
259	1897	4-4-0T	OC 17 x 24	5 6	86			(2866)	(6504)				1924
260	1897	4-4-0T	OC 17 x 24	5 6	87			2867	6505				1928
261	1898	4-4-0T	OC 17 x 24	5 6	88			2868	6506				1928
262	1898	4-4-0T	OC 17 x 24	5 6	89			2869	(6507)				1926
263	1898	4-4-0T	OC 17 x 24	5 6	90			2870	6508				1929
283	1900	0-6-0T	OC 17 x 24	4 4	91			(2891)	7526				1935
284	1900	0-6-0T	OC 17 x 24	4 4	92			2892	7527	27527	58860		1957
285	1901	0-6-0T	OC 17 x 24	4 4	93			2893	7528	27528	58861		1953
286	1901	0-6-0T	OC 17 x 24	4 4	94			2894	7529	27529			1937
287	1901	0-6-0T	OC 17 x 24	4 4	95			2895	7530	27530	58862		1956
314	1905	0-6-0T	OC 17 x 24	4 4	66			2884	7532	27532	58863		1952
322	1906	4-4-0T	OC 17 x 24	5 6	1			2800	6509				1929
323	1906	4-4-0T	OC 17 x 24	5 6	2			2801	6510				1929
324	1906	4-4-0T	OC 17 x 24	5 6	3			2802	6511				1926
325	1906	4-4-0T	OC 17 x 24	5 6	4			2803	(6512)				1926

(d) Preserved until 1932 when it was cut up. ϕ Sold to LNWR in 1909.
Works Numbers introduced by J. C. Park the first being No 160. The first loco. built was allocated an assumed No 101 and 101 to 159 included the two rebuilt 0-4-4STs and the rebuilt 0-4-2CT. Works Nos were also allocated to major rebuilds which are not shown on the above list.

GENERAL DIMENSIONS

Type	Series	Cylinders inches	D.W. ft in	Heating Surface			Grate Area sq ft	WP lb/sq in	Weight WO ton cwt	Water Galls	Coal ton cwt	
				Tubes	F'box	Total						
4-4-0T	1863-65	IC 16 x 24	5 3	815.5	80.0	895.9	13.37	160	44 5	850		
4-4-0T	Rebuilt	IC 16½ x 24	5 3	815.5	80.0	895.9	13.37	160		850		
4-4-0T	1865-69	IC 17 x 24	5 9	846	112	958	16.72	160	42 0	1200	1 15	
4-4-0T	Rebuilt	IC 17½ x 24	5 11	1050.74	91	1141.74	16.62	160	49 13	1200	1 15	
4-4-0T	1868-74	OC 17 x 24	5 6	924	91	1015	16.72	160	43 12	844	1 5	
4-4-0T	Rebuilt	OC 17½ x 24	5 8	924	91	1015	16.72	160	41 12	844	1 10	(a)
4-4-0T	1875-	OC 17 x 24	5 5	860	91	951	16.62	160	46 0	850	1 10	
4-4-0T	Rebuilt	OC 17½ x 24	5 5	860	91	951	16.62	180	46 0	850	1 10	(b)
4-4-0T	1896	OC 17 x 24	5 3	860	91	951	16.62	160	46 0	850	1 10	
0-6-0T	1879 to No 213	OC 17 x 24	4 4	894.6	81	975.6	16.33	160	43 18¼	879	1 5	
0-6-0T	214 onwards	OC 17 x 24	4 4	875.85	81	956.85	16.33	160	43 18½	956	1 5	

(a) Some with 5' 5'' diameter coupled wheels. (b) Pressure reduced to 160 lb/sq in later

Fig. 447 **NLR** 254/1896 4-4-0T No. 81 became LMS No. 6499

Fig. 448 **NLR** 228/1891 rebuilt 1907 0-6-0T No. 76 as L&NWR No. 2886

NORTH LONDON RAILWAY

No new locomotives were built after this date. Four inside cylinder 4-4-0Ts were taken into LMS stock, the last one being scrapped in 1925. In addition seventy-four outside cylinder 4-4-0Ts were absorbed and all had disappeared by 1929. Their duties were taken over by new standard LMS 0-6-0Ts.

The 0-6-0Ts were built between 1879 and 1901 and totalled thirty locomotives. The initial design was carried out by J. C. Park who had succeeded William Adams. They were sturdy machines; the cylinders were low slung and horizontal, the crossheads running on single slide bars. The wheels were cast iron H section with only four spokes, the rest of the space taken up with massive integral balance weights, opposite which were blocks to accommodate the crank pins.

The coupling rods had no joint but were two separate rods on each side, one coupling the leading to the driving wheels, and the other coupling the driving and trailing wheels so that the driving crank pin had three rod bearings to accommodate.

The boiler arrangement was similar to the 4-4-0Ts. The first twelve locomotives weighed 43 tons 18¼ cwt and the remainder 43 tons 18½ cwt.

Nine of the first batch were renumbered into the duplicate list between 1891 and 1905 and when the London & North Western Railway entered into a working agreement with the NLR in 1909 these nine tanks received LNWR numbers and were transferred to Birkenhead.

Another interesting transfer took place much later, in LMS days, of some of these tanks to the Cromford & High Peak section and they acquitted themselves with distinction on the Hopton incline.

It speaks well for their construction that fourteen passed into British Railways ownership, the last one disappearing in 1958 at the age of seventy.

Up to 1883 all NLR locomotives were painted green with black bands edged with red and white with all brass and copper work polished. The livery was then changed to black with yellow, red and pale blue lining.

Two Beyer Peacock 0-4-2ST locomotives were converted to 0-4-4 Saddle tanks (Nos. 38 and 41) and a 0-4-0ST built by Sharp Stewart & Co. was rebuilt as a 0-4-2ST with a 3 ton steam crane mounted at the rear.

Works numbers officially ran from 101 of 1863 to 339 of 1910. They were allocated for new work, replacements and rebuilding.

The locomotives placed on the duplicate list had numbers allocated between 101 and 127 and a few locomotives in the 1870s had the letter A added to their running number.

NORTH STAFFORDSHIRE RAILWAY

Stoke-on-Trent

The initial locomotive stock was purchased from outside contractors from the formation of the railway in 1848 until 1868. The maintenance of locomotives and responsibility for ordering new ones was at first in the hands of the railway's Engineer of the time, who were 1848–65 J. C. Forsyth and 1865–70 J. Johnson.

Locomotive Superintendents: 1870–4 T. W. Dodds; 1874–5 R. N. Angus; and 1876–82 C. Clare.

Loco. Carriage & Wagon Superintendents: 1882–1902 L. Longbottom; 1902–15 J. H. Adams; and 1915–23 J. A. Hookham.

Although all repairs and heavy rebuilding took place in Stoke Works, the necessary facilities for new building were not completed until 1868 when the first three locomotives were built. They were 0-6-0STs very similar to a Hudswell Clarke product supplied in 1866. They were small and weighed just under 21 tons in working order with inside cylinders 13″ x 18″ and 3 ft diameter wheels.

The earlier locomotives of the NSR were at this period in very poor condition and steps were taken to replace the older ones and most of these were supplied by outside contractors but as the workshops received more equipment a few new locomotives were built up to 1874. The principal work carried out was rebuilding and converting a number into tank engines.

T. W. Dodds claim to fame was his patent 'wedge motion' which dispensed with links and one eccentric served one valve and reversing was achieved by sliding wedges which altered the position of the eccentric in relation to the appropriate crank.

The original patent for this was brought out by his father Isaac Dodds and a Mr Owen, and this was left to his son who carried out modifications to the wedge motion to improve the expansive working as well as effect the unusual method of reversing. This motion was fitted to a number of locomotives during the superintendency of T. W. Dodds but when he retired in 1874, and Mr Angus took over, the motion was not viewed with favour and was taken off the locomotives. The main trouble was bad starting and these engines were equipped with crow-bars to enable the locomotive to be inched forwards or back. To alleviate the shortage of motive power Angus put in hand three 2-4-0 tender and two tank versions of the same wheel arrangement and these were built in 1874, together with two with larger driving wheels. These 2-4-0s were really the first effort to provide a type suitable for express passenger working, and – what was unusual for this railway – outside cylinders.

Mr C. Clare took over in 1875 and efforts

were made to provide a certain amount of standardisation of wheel arrangements and boilers and during his term of office 2-4-0, 2-4-0T, and 0-6-0 were built. The 2-4-0Ts were under Mr Clare's classification scheme *Class A* and a peculiarity in design was that the Giffard type injectors were fitted in such a position that the overflow from them, was returned back to the side tanks, by means of a funnel instead of discharging on to the track. These locomotives had coupled wheels 4′ 6″ diameter and were soon found to be too small for local passenger work, and a modified version was built with coupled wheels a foot larger and these were more satisfactory. They were known as *Class C.* Two well known 2-4-0s were built in 1882 Nos. 54 and 55. These carried the names JOHN BRAMLEY MOORE and COLIN MINTON CAMPBELL. These were the only named locomotives on the NSR apart from the original No. 1 built by Sharp Bros. & Co. and named DRAGON.

Mr L. Longbottom became Loco. Superintendent in 1882. He had served his apprenticeship with Fenton, Murray & Jackson and with E. B. Wilson, after which he joined Kitson Thompson & Hewitson. In 1855 he was appointed Engineer to the Kendal & Windermere Railway which was taken over by the London & North Western Railway in 1860. He remained with the L&NWR for twenty-two years. His *Class D* 0-6-0T of which forty-nine were built from 1883–99 were a sound design and were all built at Stoke.

The inside cylinders were 16¾″ x 24″, but the last thirteen had the bore increased to 17″. The boiler was almost the same as that fitted to the *Class A* 2-4-0Ts. Up to 1882 the standard colour for locomotives and tenders was a bright green with a wide black lining with white lines each side. Mr Luke Longbottom altered the scheme to a gamboge brown with wide red lining with yellow lines each side. The 'Staffordshire knot' was the railway company's emblem and was painted on tender and tank sides.

Standardisation was intensified and boilers were made interchangeable with rebuilt as well as new locomotives.

In 1899 a batch of 0-6-2Ts was introduced with 4′ 6″ coupled wheels and 18″ x 24″ inside cylinders. They had a similar boiler to the *100* class 0-6-0s but their most noticeable feature was the very large bunker which carried 4½ tons of coal and 650 gallons of water. The side tanks contained 1000 gallons. The weight in working order was 55 tons 5 cwt and they were very strong locomotives.

An interesting feature of the earlier 2-4-0Ts was the 'clerestory' cab roof very similar to Stroudley's design on his tank locomotives. From 1890 this type of roof was discontinued, a plain sheet roof being substituted.

From 1902 Mr J. H. Adams was in office and intensive rebuilding and reboilering was continued. In 1907 a new design was put through, taking the form of a *Class M* 0-4-4T. This was a new type for the NSR but five only were built in 1907–8. They marked the last of the small tank designs. Adams brought out a modernised version of Longbottom's *100* class 0-6-0, known as the *New 100* class. Although dimensionally the same as the earlier type the boiler pressure was increased from 150 to 160 lb/sq. in. Only

Fig. 449
NSR 1874
outside cylinder
2-4-0 No. 71
5′ 6″ D.W.

two were built but a more powerful *Class H* 0-6-0 appeared in 1909 with inside cylinders 18½" x 26" and 5 ft diameter wheels.

Although standardisation had been the target for some years, it seems strange that so many varied types appeared.

In 1910 the first 4-4-0 passenger tender locomotive appeared mainly due to the increase in weight of the summer holiday traffic which was proving too much for the older 2-4-0s. Four were built at Stoke in 1910 with 6 ft diameter driving wheels and cylinders 18½" x 26". The boiler had a working pressure of 175 lb/sq. in. and the total heating surface was 1225 sq. ft. They were known as *Class G.* A tank version was built in 1911–12 (*Class K*) but these were fitted with Schmidt Superheaters. Other basic measurements were the same as the tender version. It was intended to build eight but the eighth was modified as a tender locomotive and was slightly larger and heavier than the *Class Gs,* in fact this sole representative of *Class Kt* was the largest locomotive the NSR possessed.

The other Adam's design of note was a more powerful version of the *Class L* 0-6-2Ts and took the form of a 0-6-4T with 20" x 26" cylinders, 5 ft diameter coupled wheels and a boiler with a Robinson's superheater. Originally this batch of tanks was equipped with condensers for pre-heating the boiler feed water, and a hot water injector and hot water feed pump. The condensers were subsequently removed presumably having given some trouble.

As with the 4' 6" 2-4-0Ts, the driving wheels had been undersized and for passenger work they were not successful.

After the sudden demise of Mr Adams in December 1915, a Mr J. A. Hookham replaced him and in 1916 brought out a further series of 0-6-4Ts (*Class F*) with larger driving wheels of 5' 6" diameter. Eight were built from 1916 to 1919. They were practically identical to the 5' 0" version and were equipped with condensers etc. as before. They were employed on expresses on the Stoke to Derby, and the Manchester main lines and were excellent machines.

The most interesting design for which Mr Hookham was responsible was for a four-cylinder 0-6-0 side tank.

The cylinders were 14" x 24", wheels 4' 6" and the cranks were set in two pairs at 90° and at 135° in relation to one another, thus giving eight exhaust beats per revolution of the wheels. The boiler was superheated and the total heating surface was 1051·7 sq. ft. The weight in working order was 56 tons 13 cwt.

This locomotive was intended to work local passenger traffic on the steep gradients around Stoke but attempts at fast running resulted in shortage of steam and, for any distance, coal and water. It was decided to convert it into a tender engine to provide more coal and water storage and the maximum cut off was altered from 50 per cent to 60 per cent. In this form it was used for goods traffic and was more successful on this work.

On 1 January, 1923 the North Staffordshire Railway lost its identity and became part of the LONDON MIDLAND & SCOTTISH RAILWAY.

Construction of locomotives at Stoke reached its finale under its new title, by building four 0-6-2Ts of the *New L.* Class. They differed

Fig. 450
NSR 1882
2-4-0 No. 55
COLIN MINTON CAMPBELL

NORTH STAFFORDSHIRE RAILWAY

Fig. 451 **NSR** 1891 2-4-0T No. 1 *Class B* at Stoke

Fig. 452 **NSR** 1915 0-6-4T No. 4 *Class C*

First Built	Class	Type	Cylinders inches	D.W. ft in	Heating Surface sq ft				Grate Area sq ft	WP lb/sq in	Weight WO ton cwt
					Tubes	Super	F'box	Total			
1878	A	2-4-0T	16½ x 24	4 6	731	–	92	823	15.5	140	43 2
1881	B	2-4-0T	16½ x 24	5 6	731	–	92	823	15.5	140	43 18
1882	C	2-4-0	16½ x 24	6 0	726	–	96	822	15.5	140	45 0
	D	0-6-0T	16½ x 24	4 6	726		92	818	15.5	150	45 0
1896	100	0-6-0	17 x 24	4 6	815	–	95	910	15.5	150	37 9
1899	DX	0-6-2T	18 x 24	4 6	815	–	106	921	17.1		55 5
1871	E	0-6-0	16½ x 24	5 0				1061			31 5
1916	F	0-6-4T	20 x 26	5 6	887	258	129	1274	21.1		77 7
1910	G	4-4-0	18½ x 26	6 0	1092	–	133	1225	21.1	175	47 11
1909	H	0-6-0	18½ x 26	5 0	1011.7	–	108.3	1120	17.8	175	42 17
1910	H[1]	0-6-0	18½ x 26	5 0	1084.5	–	113	1197.5	17.8	175	43 12
1911	K	4-4-2T	18½ x 26	6 0					21.1	160	70 10
1908	New L	2-4-2T	18½ x 26	5 0	1011.7	–	108.3	1120	17.8		59 15
1923	L Sup	0-6-2T	18½ x 26	5 0	831	261	113	1205			64 19
1907	M	0-4-4T	18½ x 26	5 6	1011.7	–	108.3	1120	17.5	175	56 0
1914	C	0-6-4T	20 x 26	5 6	887	258	129	1274	21.1	160	74 15
1922	Reb. D	0-6-0T	4/14 x 24	4 6	750	195	107	1052			56 13

from previous ones of the same class by reverting to slide valves with superheated boilers. As one would expect, the slide valves were troublesome but the class as a whole was extremely sound.

These four locomotives had been allocated NSR Nos. 1, 2, 10 and 48 and were at first painted in the North Stafford livery. They were allocated LMS numbers 2270 to 2273. These were the last to be built at Stoke and as far as can be ascertained a total of 197 steam locomotives were built from 1868 and 1923. As in most cases it is very difficult to discover what entirely new locomotives were built and what category many of the 'renewals' should be placed where varying quantities of material off the withdrawn locomotives has been used in 'building' the replacements. Stoke carried on repairing locomotives of the old NSR and a few ex L&NWR. The last North Stafford locomotive to be cut up was LMS No. 1436 which was withdrawn in April 1939 and was cut up at the beginning of 1940. Stoke Works were officially closed on 31 December, 1926 but this was not completely carried out until July 1927. All repair work was transferred to Crewe.

OLIVER & CO. LTD.
Broad Oaks Works, Chesterfield

This firm designed a 0-4-0ST locomotive specifically for ironstone quarry work and two were ordered in 1888 by the Cranford Ironstone Company. These were completed in 1889 presumably just before the firm was bought up on 5 October, 1899 by MARKHAM & CO. LTD.

The maker's plate on the second locomotive is for Oliver & Co. Ltd. so presumably the first one bore a similar plate. They were metre gauge and named WILLIAM and THE BARONET, works numbers 101 and 102 respectively. Works numbers for locomotives commenced at 101.

Markham & Co. Ltd. had an interest in various collieries and quarries, mainly the Staveley Coal & Iron Company, and a number of locomotives were built from 1889 to 1914 for metre and standard gauge and were all 0-4-0STs except one which was six-coupled, a total of nineteen all told.

The three metre gauge locomotives had outside cylinders 8″ x 10″ and 2 ft diameter coupled wheels. The small saddle tank was mounted over the boiler only. The firebox was of the raised round top type with the safety valves mounted above it with connections to the whistle, injector and blower. The brakes were hand operated on the rear wheels only. No cab was fitted and the chimney was a plain stove pipe with small rim around the top.

The standard gauge locomotives had outside cylinders 13″ x 20″ and 3′ 6″ coupled wheels later ones had 14″ x 20″ cylinders.

Fig. 453 **Oliver & Co.** 102/1889 Cranford Ironstone Co. THE BARONET Metre gauge

Fig. 454 **Oliver & Co.** 107/1892 Staveley Coal & Iron Co. later named G. BOND

Fig. 455 **Oliver & Co.** 110/1895 diagram of the only 0-6-0ST built

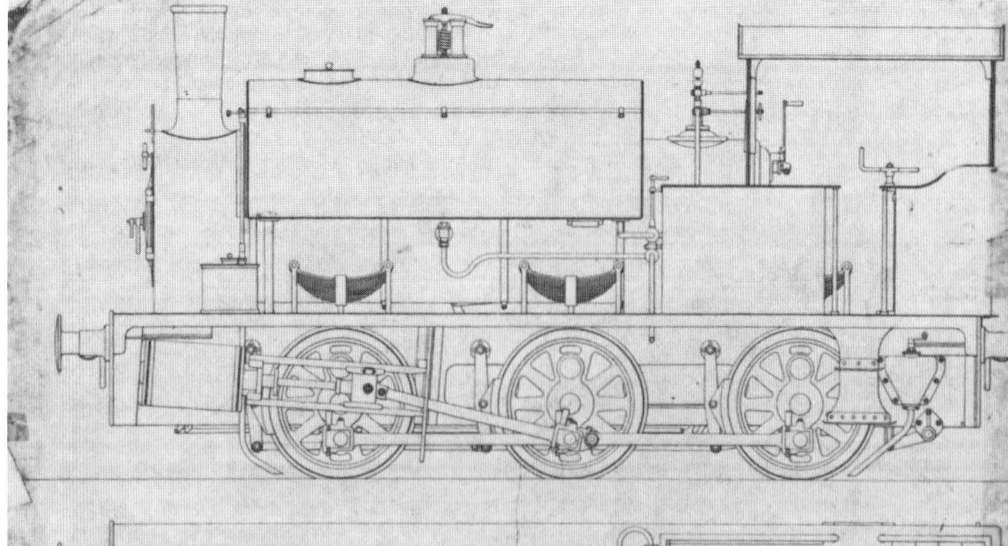

OLIVER & CO. LTD.

The only one supplied new, away from the area, was one of the 14″ 0-4-0STs which went to the Dalmellington Iron Company in 1909, and was their No. 15.

The 1909 locomotives had saddle tanks over the boiler only, and the firebox was flush with the barrel and round topped. Brakes were on all four wheels and hand operated only. The safety valves were mounted on the dome and protruded through the saddle tank. A steam fountain was mounted on top of the firebox with connections for the whistle, injector and blower. A small cab was fitted but no proper coal bunker. A stove-pipe chimney with a small rim was 11′ 0″ from rail to top of chimney. The wheels had cast iron centres and 3′ 6″ diameter on a wheelbase of 5′ 10″. The sandboxes were placed mid-way between the coupled wheels with a straight pipe down to the rails.

These locomotives were solidly built and, in most cases, were engaged in hard work and most lasted well into the 1950s.

Thanks are due to Messrs Markham & Co. Ltd. for kindly providing drawings, photographs and information.

ORMEROD, GRIERSON & COMPANY

Manchester

The *English Mechanic* for 1881 makes reference to experiments at Liverpool with steam locomotives and also with a combined steam car constructed on Apsley's patent, by this firm.

They were established as an iron and brass founder in Minshull Street in 1811, and by 1861 were known as Ormerod, Richard & Son, St George's Ironworks. In 1863 they became Ormerod, Grierson & Company, Hulme Hall Road and Chester Road, Hulme. In 1892 they became St George's Ironworks Ltd. No definite details can be found of any locomotives that may have been built for these experiments.

OLIVER & CO

Works No	Order No	Built	Type	Cylinders inches	D.W. ft in	Gauge	No/Name	Customer	W'drawn
101	1358	1889	0-4-0ST	8 x 10	2 0	Metre	William	Cranford Iron Co	
102	1358	1889	0-4-0ST	8 x 10	2 0	Metre	The Baronet	Cranford Iron Co	1960

MARKHAM & CO

Works No	Order No	Built	Type	Cylinders inches	D.W. ft in	Gauge	No/Name	Customer	W'drawn
103	1588	1891	0-4-0ST	13 x 20	3 6	Std	Duston	Staveley Coal & Iron Co	
104	1588	1891	0-4-0ST	13 x 20	3 6	Std	St Albans	Bestwood Coal & Iron Co	1956
105	1588	1891	0-4-0ST	13 x 20	3 6	Std	Staveley	Staveley Coal & Iron Co	1965
106	1588	1891	0-4-0ST	13 x 20	3 6	Std		Hickleton Main Coll	S/S
107	1588	1893	0-4-0ST	13 x 20	3 6	Std	G. Bond	Staveley Coal & Iron Co	1965
108	1855	1893	0-4-0ST	8 x 10	2 0	Metre	Dowie	Clay Cross Co	
109	1863	1894	0-4-0ST	14 x 21	3 6	Std	Gladys	Staveley Coal & Iron Co(a)	
110	1909	1895	0-6-0ST	14 x 20	3 6	Std	Violet	Staveley Coal & Iron Co	1953
111	2113	1897	0-4-0ST		3 6	Std	2	W. Cooke & Co Sheffield	
	4288	1909	0-4-0ST	14 x 21	3 6	Std	Bull	Bullcroft Main Coll	
	4288	1909	0-4-0ST	13 x	3 6	Std	Charles	Markham Main Coll	
	4288	1909	0-4-0ST	14 x 21	3 6	Std	Lily	Staveley Coal & Iron Co	1960
	4288	1909	0-4-0ST	14 x 21	3 6	Std	15	Dalmellington Iron Co	1956
	4361	1909	0-4-0ST	14 x 21	3 6	Std	13	Vickers Ltd. Sheffield	1968
	4361	1909	0-4-0ST	14 x 21	3 6	Std	10	Parkgate Iron & Steel Co	1961
	4697	1913	0-4-0ST	14 x 21	3 6	Std	11	Parkgate Iron & Steel Co	
	4697	1914	0-4-0ST	14 x 21	3 6	Std	12	Parkgate Iron & Steel Co	

(a) Preserved by Midland Railway Project Group.

OXFORD, WORCESTER & WOLVERHAMPTON RAILWAY

Worcester

The workshops were opened at Worcester in 1854 for the repairing and building of their locomotives. It was some time before it was sufficiently equipped to commence actual building and the first locomotives were completed in 1860–1, and were replacements of Hawthorn built locomotives. Some of this material was almost certainly used on the three locomotives involved. They were all outside frame 0-6-0s.

No. 7 became GWR No. 240; No. 13 became GWR No. 244; and No. 19 became GWR No. 246.

They were fitted with inside cylinders 16″ x 24″ and 5 ft diameter wheels. The frames were parallel plates with horn plates rivetted on and stayed across the base of the horn plates.

By this time the OW&WR had been amalgamated with the Newport, Abergavenny & Hereford Railway and the Worcester & Hereford Railway to form the WEST MIDLAND RAILWAY (1 July, 1860).

In 1862 four further 0-6-0s were built at the Worcester Shops, and were very similar in appearance to the first three. They were: WMR No. 38 which became GWR No. 260; WMR No. 112 which became GWR No. 261; WMR No. 113 which became GWR No. 262; and WMR No. 114 which became GWR No. 263.

The frames were solid type and they had the folding type of smokebox door similar to the Derby locomotives supplied to the OW&WR.

No further locomotives were built although extensive repairs and rebuilding took place. The majority of the work was taken on by the Stafford Road Works at Wolverhampton after the West Midland became part of the Great Western Railway.

PARFITT & JENKINS

Cardiff

The firms' premises were opposite the Cardiff Railway's Tyndall Street works. They were a general engineering firm and ship builders, and the close proximity to the Tyndall Street locomotive repair shops no doubt had some bearing on their receiving orders for locomotives.

Between 1869 and 1881 thirteen 0-6-0STs were built for the Marquis of Bute's railway which later became the Cardiff Railway. They were built as follows. The actual dates have never been able to be verified, but are usually accepted:

Marquis of Bute's Railway:
No. 9 1869; No. 10 1870; No. 11 1871; No. 12 1871, became GWR 694; No. 14 1872; No. 15 1873, became GWR 695; No. 16 1873; No. 18 1875, became GWR 696; No. 19 1875, became GWR 697; No. 20 1877; No. 21 1879; No. 22 1880; No. 23 1881.

Several of those which did not pass into GWR hands were sold to Mr H. Flint of Wigan about 1897–8 and the five which were withdrawn during that period (Nos. 9–11, 21 & 23) were probably the ones that went to Wigan, No. 14 and No. 22 were withdrawn in 1908 and 16 and 20 in 1919 and 1917 respectively.

These locomotives had inside frames, and inside cylinders with tanks over smokebox, boiler and firebox. The dome had two pop safety valves on top, and a substantial cab was fitted. The cylinders were $15\frac{1}{2}$″ x 22″, with 4′ diameter wheels, 1000 gallon tank and weight in working order of 35 tons.

No other locomotives were built by this firm.

Fig. 456 **Parfitt & Jenkins** 1875 Cardiff Rly No. 19 as GWR 697 with Kitson chimney

PARRY

No information can be found of any locomotives being built by a firm of this name. The firm itself may have been David Parry, Leeds Iron Works, Leeds which operated in the 1860s and 1870s.

PARTINGTON STEEL & IRON COMPANY
Irlam Works

This firm did not build any completely new steam locomotives, but many were supplied by Pearson & Knowles Coal & Iron Company. They did buy one second-hand Peckett 0-6-0ST, built in 1904, Peckett No. 1036, about 1937 when they gave it a major overhaul including fitting a new boiler supplied by Peckett & Sons. Many of their own locomotives were overhauled and rebuilt at the Irlam works.

PARTRICROFT IRONWORKS

See NASMYTH GASKELL & COMPANY.

PEARSON & KNOWLES COAL & IRON CO. LTD.
Dallam Forge, Warrington, Lancs.

Established about 1870 this firm developed a thriving structural and general engineering business and in addition had considerable interests in coal mining, including the pits at Moss Head, Low Hall, Maypole, Platt Bridge and Chisnall Hall collieries.

Locomotives were built by the company from 1899 for their own use and the first three were 0-4-0Ts similar to those supplied to Dallam Forge by Fletcher Jennings & Co. of Whitehaven, including the patented design of valve layout (See TULK & LEY).

Two inside cylinder 0-6-0STs were built in 1910–11 for colliery use followed by five 0-4-0STs which were very similar to the Hudswell Clarke tanks supplied. It may well be that either the drawings or the major parts were supplied in the case of the four-coupled tanks of Hudswell Clarke designs.

At least two 0-6-0STs were built 1916–17 with inside cylinders and saddle tanks fitted the length of the boiler barrel only.

It has been suggested that six 0-6-0STs were built during 1916–17 for the Inland Waterways and Docks but this cannot be confirmed. HENRY may have been one of these locomotives as it was bought by J. F. Wake from W. D. Richborough, repaired and sold to Wharncliffe Woodmoor Colliery Co. Ltd., West Riding, and given No. 2.

A total of twelve confirmed locomotives were built, but could have been sixteen to eighteen. Locomotives were also extensively rebuilt and repaired at Dallam Forge until the early 1930s.

In 1930 due to financial difficulties the coal interests held at that time were disposed of to the Wigan Coal Corporation Ltd. and the fabricating side taken over by the newly formed Lancashire Steel Corporation and this section

Fig. 457 **Pearson Knowles** 0-6-0ST JOHN 1911 at Walkden Yard

PEARSON & KNOWLES COAL & IRON CO. LTD.

became THE PEARSON & KNOWLES ENGINEERING CO. LTD. with a view to expanding further and intensively developing the structural and general engineering activities of the firm.

It was incorporated into the British Steel Corporation on nationalisation of its parent company, the Lancashire Steel Corporation.

In 1972 under further reorganisation the Warrington Branch of Redpath Dorman Long Ltd. was formed which included the fabrication shops of Pearson & Knowles.

Thanks are due to Mr H. T. Dodd of Redpath Dorman Long Ltd., Warrington for some of the information contained in this section.

PECKETT & SONS LTD.

See FOX WALKER & COMPANY.

PEEL, WILLIAMS & PEEL

Soho Works, Ancoats

The original firm was founded in the early 1800s in Millar's Street, Manchester and was known as Peel, Williams & Company. In 1811 new works were taken in Swan Street known as the Phoenix Foundry. They were ironfounders, builders of stationary engines, and textile machinery. It is reported that in 1805 one or more high pressure engines were built for Richard Trevithick. A second foundry was acquired in Ancoats known as the Soho Works. They were, up to about 1830, the biggest engineering business in Manchester and on expansion, commenced building boilers and extended their range of products. About 1825 the firm became Peel, Williams and Peel and in 1839 entered the locomotive field, but did not develop this side of business to any extent although their name was coupled with the larger locomotive builders in the area including Sharp Roberts & Company, and Nasmyth Gaskell & Company, as being of considerable importance in this field.

This was not borne out and, as far as can be discovered, they built one, probably two locomotives in 1839 and then ceased although later catalogues and letter headings included a picture of a locomotive among beam engines, hydraulic presses, looms and gearing as late as 1850. Indeed one was shown in 1835 four years before building took place. It was in the form of a 2-2-0 *Planet* type.

The SOHO, a 2-2-2, was completed in September 1839 and was tried out on the Liverpool & Manchester Railway. It had 5′ 6″ diameter driving wheels and inside cylinders 11″ x 18″. A notable feature was the lack of eccentrics, the valves being operated by gear wheels off the crank axle and a combination of wheels. There is no record of this locomotive being sold to any railway. The MANCHESTER appeared in December 1839 and ran trials on

DALLAM FORGE

Built	Type	Cylinders inches	DW ft in	Location	Name/No
1899	0-4-0T	OC 12 x 20	3 6	Dallam Forge	9
1900	0-4-0T	OC 12 x 20	3 6	Dallam Forge	10
1901	0-4-0T	OC 12 x 20	3 6	Dallam Forge	1
1910	0-6-0ST	IC 16 x 22	4 0	Moss Colly Platt Bridge	Daisy
1911	0-6-0ST	IC 16 x 22	4 0	Chisnall Hall Colly	John
1914	0-4-0ST	OC		Partington Steel & Iron Co Irlam	4
1914	0-4-0ST			Partington Steel & Iron Co Irlam	5
1915	0-4-0ST			Partington Steel & Iron Co Irlam	6
1916	0-4-0ST			Partington Steel & Iron Co Irlam	7
1916	0-4-0ST			Partington Steel & Iron Co Irlam	10
1916	0-6-0ST	IC 16 x 22	4 0	Partington Steel & Iron Co Irlam	11 (a)
1917	0-6-0ST	IC 16 x 22	4 0	?	(a)

(a) May have been built for and delivered to Inland Waterways and Docks originally.

the Manchester & Leeds Railway and although the cylinder bores are given as 13″, two inches larger than SOHO, it is possible that they were one and the same locomotive.

In the following year they put in a tender for two locomotives for the Manchester & Leeds Railway but were unsuccessful.

In 1847 an engine ran trials on the Manchester & Birmingham Railway but no success was apparent – it was probably the SOHO/MANCHESTER locomotive the firm brought out of store in an unsuccessful bid to get it off their hands.

The firm however was a noted builder of other machinery and they prospered until 1887 when the works ceased trading.

Interesting notes concerning this firm are to be found in the following: Transactions, Lancashire & Cheshire Antiquarian Soc. 1959 Vol. 69 'Peel, Williams & Co., Ironfounders and Engineers of Manchester c. 1800–1887'. A communication in Industrial Archaeology by A. E. Musson, M.A. Business History Vol 3 No. 1 December 1960. Liverpool University Press Stephenson Locomotive Society Journal No. 376 October, 1956.

PEMBERTON COLLIERY COMPANY

Lancs.

See BLUNDELL, JOHNATHAN & SON.

PEN-Y-DAREN IRONWORKS

See TREVITHICK, RICHARD.

PHILLIPS, CHARLES D.

Emlyn Engineering Works, Newport

Very little information is available on this firm. Locomotives for contractors were overhauled and probably rebuilt but no record is available of any new construction being carried out. At some period the firm was known as Watkins & Phillips.

PHOENIX FOUNDRY

Stoke-on-Trent

One locomotive was supplied to the Lilleshall Company *c* 1859. It was a 0-6-0T, had inside cylinders, was named PHOENIX No. 3 and was withdrawn in 1914. No further information is available. The firm may have been agents.

PICKERING R. Y. & COMPANY

Wishaw, Lanarkshire

This firm was established in 1864 and engaged in the manufacture of railway carriages, rail and road motor coaches and wagons of all descriptions. Commencing early in this century some steam rail cars were manufactured but unfortunately all the firm's records prior to 1960 have been destroyed and no information is available. An experimental rail motor coach was built in 1905 for the Kent & East Sussex Railway (No. 16). It was a four wheeled coach with a multitubular vertical boiler at one end feeding two $5\frac{1}{2}$″ x 8″ diameter cylinders, with a chain drive from the crankshaft to the wheel axle.

RANSOMES & RAPIER

Waterside Works, Ipswich

This firm was formed as a branch of Ransomes, Sims & Head in 1868 to concentrate on the railway plant side of the business. Chairs, rails, points and many other articles for railway use were manufactured and in 1876 three small locomotives were built and exported to China. The first one named PIONEER was used by the contractors building the Shanghai and Woosung Railway to 2′ 6″ gauge. It was a 0-4-0ST with a service truck coupled to it on which the driver sat. It had outside cylinders 5″ x 6″ and coupled wheels 1′ 6″ diameter. It weighed 1 ton 10 cwt.

The other two were outside cylindered 0-4-2Ts, 8″ x 10″ cylinders with 2′ 3″ diameter coupled wheels. The water tanks were a combination of side and saddle and completely enveloped the boiler but left the smokebox clear. They weighed 9 tons and were named CELESTIAL EMPIRE and FLOWERY LAND.

RANSOMES & RAPIER

These three locomotives were the first to be exported to China from this country. They were followed by a larger locomotive named VICEROY. Subsequent locomotives took the form of three classes although no record of building any of these classes is available. (1) *Calcutta* class 9″ cylinders 0-6-0T; (2) *Lima* class 0-4-2T; and (3) *Ipswich* class 6½″ cylinders 0-4-2T. A few inspection locomotives were also built with vertical boilers, 6″ outside cylinders driving a single pair of wheels with a pair of carrying wheels at the rear (0-2-2).

The locomotive side of the business was only a part of the whole, the main work being for steam travelling cranes, breakdown cranes, and portable and stationary engines.

One further 0-6-0T known to have been built in 1881 was for the Perak State Railway (PSR No. 1) which became Federated Malay States Railway No. 1. It had outside cylinders 9″ x 12″ and 2′ 3″ diameter wheels and bore a works number 150. Evidently there was no separate series of works numbers for locomotives.

RAYNE & BURN
Newcastle

This firm were agents only and sub-contracted any locomotive orders received, their main line of business being wagon building. They supplied three six-coupled locomotives to the General Mining Association Railways, Nova Scotia. One went to the Acadia Coal Company, Stellaton in 1853 and was named ALBION. It had inclined cylinders fixed just behind the smokebox and fastened to the boiler barrel. The frames were inside the wheels and were generally similar to three locomotives supplied by Timothy Hackworth in 1838 to the Albion Coal Mining Co. of Nova Scotia. Two other locomotives were supplied to Nova Scotia via Rayne & Burn, in 1854 named HALIFAX and PICTOU.

Fig. 458
Ransome & Rapier
1876. Drawing of 0-4-0 PIONEER with braked tender

Fig. 459
Ransome & Rapier
locomotives FLOWERY LAND and CELESTIAL EMPIRE on turntable at Shanghai on Shanghai–Woosung Rly

In 1848 a locomotive was supplied to the Great Grimsby and Sheffield Junction Railway. It was a 2-4-0 and later became No. 60 NEMESIS on the Manchester, Sheffield and Lincolnshire Railway. It was similar to those built by Fossick & Hackworth for the same railway and may have been built by them. The 0-6-0s, although built as late as 1853, had boilers with single flues, and inclined outside cylinders placed high on the barrel. Seats were provided for driver and fireman.

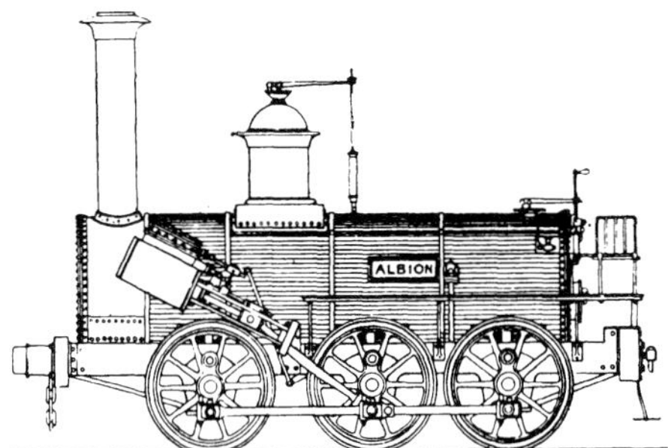

Fig. 460 1853 almost an obsolete design similar to Hackworth's 1838 type

REDRUTH & CHASEWATER RAILWAY

Devoran

Although the 0-6-0ST MINER bore a plate 'manufactured at Devoran Works 1869' this locomotive was a heavy rebuild of a 0-4-0ST supplied by Neilson (No. 81) in 1854. It had already been converted to a 0-4-2ST at Devoran about 1856. The locomotive in its final form indicates that a considerable amount of work was carried out during the conversion, entailing a lengthening of the frames. The tank was probably new and what appears to be a haystack firebox of large proportions was fitted which indicates the probability of a new boiler as well. The boiler would have been bought out, so that, by and large, a drastic rebuild instead of a new engine must be the verdict.

RENNIE, G. & J.

Holland Street, Blackfriars, London

This engineering works was established in 1824 by George and John Rennie in Stamford Street, and was moved in 1833 to Holland Street. George, the elder brother, was in charge of the mechanical side of the business and John of the civil engineering, becoming famous subsequently, as Sir John Rennie. Realising the great demand for railway locomotives they commenced building in 1838, and five single drivers were turned out for the London and Southampton Railway. They had outside frames and inside cylinders 13″ x 18″ and driving wheels 5′ 6″ diameter. The boilers had a combined dome and safety valve close to the chimney and the round top firebox, raised, had spring balance safety valves mounted on top. The valve gear was J. & C. Carmichael's patent, with one fixed eccentric.

The next order was for two 0-4-2s for the London & Croydon Railway and were similar in appearance to Alexander Allan's later design used on the Caledonian and Scottish Central railways. They had outside cylinders and the driving wheels were flangeless. The valve gear was Carmichael's and the valve chests were inside the frames.

Four further six-wheeled single driver locomotives were built; three for the London & Brighton Railway and one exported to Germany. In 1841–2 two more were delivered to the Joint railway committee.

The only 7′ Broad gauge locomotives built were two for the Great Western Railway to Gooch's designs with 7′ diameter driving wheels, slotted sandwich frames, and Gothic firebox. It resembled the original *Star* class built by Robert Stephenson & Company.

Locomotives built by G. & J. Rennie were not, in general, finished to the standards of other contemporary builders. Trouble was experienced with the five built for the London & Southampton Railway and all of them were rebuilt by W. Fairbairn & Son in 1841 and were practically new locomotives.

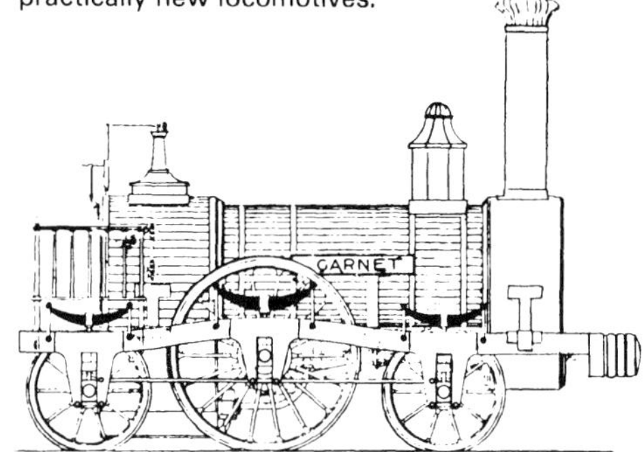

Fig. 461 **G. & J. Rennie** 1838 London & South Western Rly. GARNET

RENNIE, G. & J.

Those built for the London & Brighton Railway also gave trouble a short time after delivery. The two GWR singles fared better; both ran for twenty-nine years, perhaps the chief factors being the detailed specifications furnished by Gooch and the locomotives being to his design. For various reasons the firm decided to abandon locomotive building in favour of marine engines and in this venture certainly earned a more respected reputation and manufacture was continued into the 1890s.

RENNOLDSON, GEORGE
South Shields

Possibly built one or two locomotives for the Stanhope and Tyne Railway. The boiler of one exploded in the yard whilst undergoing steam trials on 20 November, 1837 killing two people.

RHYMNEY IRONWORKS
Tredegar

One locomotive was built in the workshops about 1866. It was designed by Mr Moyle who was the company's engineer at that time. The internal railway system was laid with L-shaped tram plates to a gauge of 3′ 0″, the angle being turned outwards in the usual fashion. Consequently all wheels were flangeless. The locomotive was a 2-4-0T, and the leading truck had adequate side play, bearing in mind that the wheels were guided by the vertical member of the tram plates engaging on the inner faces of the wheel rims. In addition the leading truck was fitted with two radial arms pivotted off a cross member of the inside frames. Even so the nature of the track made a long rigid wheelbase impossible and the distance between coupled wheels which were 2′ 8″ diameter was only 3′ 6″. The leading wheels were 2′ 0″ diameter. The outside cylinders, 8″ x 16″ were inclined and the valve chests were on top, at a steeper angle. The eccentrics mounted on return cranks drove the Stephenson's link motion – all being outside the wheels. The boiler had a dome mounted on the firebox with spring balance safety valve on top. The heating surface was 183·2 sq. ft including 19·7 sq. ft for the firebox, and the grate area was 5·32 sq. ft. Boiler pressure was 60 lb/sq. in.

As the firebox projected over the rear axle the laminated springs were fixed upside down the two ends bearing down on the axleboxes and brackets fitted to the framing behind the axle. Adjustment of tension was by means of bolts through brackets bolted on the frame and the

G. & J. RENNIE

Date	Type	Cylinders inches	D.W. ft in	Customer	Name	No		W'drawn
6/1838	2-2-2	13 x 18	5 6	London & Southampton Rly	Deer			1841
6/1838	2-2-2	13 x 18	5 6	London & Southampton Rly	Garnet			1841
8/1838	0-4-2	13 x 18	5 6	London & Croydon Rly	Croydon	2	(a)	1845
10/1838	2-2-2	13 x 18	5 6	London & Southampton Rly	London			1841
11/1838	2-2-2	13 x 18	5 6	London & Southampton Rly	Victoria			1841
1838	2-2-2	13 x 18	5 6	London & Southampton Rly	Reed			1841
5/1839	0-4-2	13 x 18	5 6	London & Croydon Rly	Archimedes	6	(a)	1845
1839	2-2-2	14 x 17	6 0	Kaiser Ferdinand Nord Bahn	Nord Stern			
4/1840	2-2-2	14 x 18	5 6	London & Brighton Rly	Eagle	6	(b)	1855
5/1840	2-2-2	14 x 18	5 6	London & Brighton Rly	Vulture	7	(b)	1853
1840	2-2-2	14 x 18	5 6	To Germany				
3/1841	2-2-2	15 x 18	7 0	Great Western Rly	Mazeppa		(c)	1870
5/1841	2-2-2	15 x 18	7 0	Great Western Rly	Arab		(c)	1870
12/1841	2-2-2	15 x 18	5 6	London & Brighton Rly	Satellite	26	(d)	1855
12/1842	2-2-2	15 x 18	5 6	Joint Committee	Man of Kent		(e)	1861
2/1843	2-2-2	15 x 18	5 6	Joint Committee	Kentish Man	28	(d)	1855

(a) One of the two 0-4-2s sold to Thomas Powell who sold it to the Taff Vale Rly who named it LLANTWIT. The other was sold to an unknown buyer. (b) Rebuilt as 2-2-2WT. (c) 7ft gauge. (d) Sold. (e) Joint locomotive committee of the London & Croydon, and South Eastern Rlys.

weight centred through this assembly. Brakes engaged the rear wheels only and were steam operated. No cab was provided.

This extremely interesting locomotive had a number of novel features and was apparently operated quite successfully on curves with a mean radius of 17′ 6″. It is described in *The Locomotive* dated 15 August, 1940 and a line diagram given. This was the only locomotive built by the company.

RICHARDSON, THOMAS & SONS

Castle Eden Foundry, Hartlepool, Co. Durham

The original works at Castle Eden were founded in 1838 on a site which is still called 'The Foundry' and consisted of a foundry, blacksmiths' shop, machine shop, pattern shop and offices. The main products were locomotives and various types of mining equipment, Castle Eden being close to some of the earliest Durham pits.

The first connection of the Richardson family with Hartlepool appears to have been in 1837 when Mr Thomas Richardson entered into partnership with a Mr Joseph Parkin to build a ship called the CASTLE EDEN on land near the Hartlepool ferry, now covered with buildings.

In 1838, 5000 yards of land was purchased from a Mr William Walker and this was followed by the establishment of the firm, controlled by T. Richardson and two of his brothers as RICHARDSON BROS. Later in 1845 the firm's title became: THOMAS RICHARDSON & SONS and continued thus until 1900.

The firm's main activities were confined to shipbuilding, but engineering of a similar character to that carried out in the Castle Eden works soon became predominant.

The first locomotives built by the firm were completed in 1840 for the Hartlepool Railway, being four six-coupled goods locomotives and probably one four-coupled. These were followed by two for the Stockton & Hartlepool Railway. In the same period many portable engines, winding and blast engines were built for neighbouring collieries and ironworks.

Although over fifty locomotives were built by Richardson from 1840 to 1857 very little information has been found in relation to the majority of them. The York, Newcastle and Berwick Railway was their best customer; at least eighteen were built and a number repaired.

It is interesting to observe that eight were built for Robert Stephenson in the 1840s, presumably acting as sub-contractors.

At this time Stephensons were building a good number of locomotives for Northern railways such as Great North of England, Newcastle & Darlington, York & North Midland, and Leeds & Bradford, and the eight built by T. Richardson & Sons might well have been for

THOMAS RICHARDSON & SONS

Works No	Date	Type	Cylinders inches	D.W. ft in	Name/Number	Customer
104	1840	0-6-0	13½ x 20	4 0		Hartlepool Rly (NER 114)
105	1840	0-6-0	12½ x 20	4 0		Hartlepool Rly (NER 115)
106	1840	0-6-0	14 x 18	4 0		Hartlepool Rly (NER 121)
107	1840	?0-4-0	14½ x 18	4 0		Hartlepool Rly (NER 128)
108	1840	0-6-0				Hartlepool Rly (?)
119	1845	0-6-0				Stockton & Hartlepool Rly
120						Stockton & Hartlepool Rly
139					Rebuild of 'Albert'	YN & BR
140	1846		OC			For R. Stephenson & Co
141	1846		OC			For R. Stephenson & Co
142	1846		OC			For R. Stephenson & Co
143	1846		OC			For R. Stephenson & Co
146	1847		OC			For R. Stephenson & Co
147	1847		OC			For R. Stephenson & Co
148	1847		OC			For R. Stephenson & Co
149	1847		OC			For R. Stephenson & Co
155					East Hetton	YN & BR
156						YN & BR
157						YN & BR
160						YN & BR
161						YN & BR

RICHARDSON, THOMAS & SONS

Works No	Date	Type	Cylinders inches	D.W. ft in	Name/Number	Customer
162					118 Kelloe	YN & BR
163					123 Heughhall	YN & BR
164	1849	0-6-0				YN & BR
165					Repaired Loco	YN & BR
166					Repaired Loco	YN & BR
167						YN & BR
168	2/1850	0-6-0	15 x 22	5 0	11	Maryport & Carlisle Rly
171	5/1850					YN & BR
175	5/1850					YN & BR
176	4/1850				No 57 Rebuilt	YN & BR
177	9/1850				No 27 Rebuilt	YN & BR
179	1850					YN & BR
181						YN & BR
182					'Canada' Rebuilt	YN & BR
188					Rebuilt Loco	Wingate Grange Coal Co
190	7/1851				Rebuilt Loco	YN & BR
191					Rebuilt Loco	YN & BR
192					Rebuilt Loco	YN & BR
208	4/1852					Consett Iron Co
209	1852					Consett Iron Co
210	1853					Consett Iron Co
212	5/1852				Repaired Loco	Fossick & Hackworth
213	7/1853				Repaired Loco	Earl of Durham Colly
216	1852				Repaired Loco	YN & BR
222	1852		14 x			YN & BR
232	6/1853					Consett Iron Co
233	9/1853					YN & BR
236						Earl of Durham Colly
237						YN & BR
238						YN & BR
239			18 x 24			
240			18 x 24			
241	1854		15 x 24	4 6	(a)	Hartlepool Dock & Rly
247	5/1854					Mounsey & Co
251	10/1854		17 x 24			Lambton Colly
252	1854	0-6-0	15 x 22			Marley Hill Colly
253	11/1854		15 x 22			NER
254			15 x 22			
255	1855		15 x 22			East Hetton Colly
256	1855		15 x 22			East Hetton Colly
262			16 x			NER
265	1856		16 x			South Hetton Colly
269	1857					NER
270		-	-	-	Small Loco Boiler	

(a) NER No 112 YN & BR – York, Newcastle & Berwick Rly

Fig. 462 **Thomas Richardson** 252/1854 Long Boiler Type 0-6-0ST after rebuilding from 0-6-0 Bowes & Partners Ltd. No. 10

one of these. Works numbers were allocated to all machinery with no separate list for locomotives. Numbers commenced at 101 and the first three were stationary engines. Locomotive building ceased in 1857 and work was concentrated on colliery and marine engines. The first marine engine was built in 1851 and the firm was the first to build triple expansion engines on the north east coast.

In 1900 a major change in the constitution of the Company took place, with the amalgamation of Thomas Richardson & Sons with the Tees firm of Sir Christopher Furness, Westgarth & Co. Ltd. and William Allan & Company, on the Wear. The newly formed Company became RICHARDSONS, WESTGARTH & CO. LTD., a firm which has developed into one of the foremost producers of turbo-alternators and ship machinery with a works area of some 15 acres and 2000 workpeople.

RIDLEY, T. D.

became Ridley Shaw & Co. Ltd., Middlesborough

Founded by Thomas Ridley, its original activities were exclusively on public works contracting but gradually changed to light engineering and the repair and rebuilding of industrial locomotives. According to P. W. B. Semmens, in an article in the *Railway Magazine* (No. 663 July 1956), during the earlier years of the 20th century a few tank locomotives were constructed. Building was found to be uneconomic due to lack of space, and since 1925 no new construction took place but rebuilding and general repairs were undertaken from time to time. At least six 0-4-0ST with outside cylinders were built between 1899 and 1913, but it is probable that they were not completely new productions.

RIDLEY SHAW

Works No	Date	Type	Customer	Name
13	1899	0-4-0ST	To Scotland	Clarence
51	1911	0-4-0ST	Armstrong Whitworth Elswick	Ormesby
54	1913	0-4-0ST		Kenneth
74	1920	0-4-0ST	Eldon Colliery Co Durham	Eldon No 1
	?	0-4-0ST	Hartlepool Gas & Water Co	
	?	0-4-0ST	Ord & Maddison Ltd. Middleton-in-Teesdale	

All had outside cylinders

Fig. 463 T. D. **Ridley** 13/1899 APCM Kent Works, Stone No. 1 CLARENCE

RIGBY, J. & C.

Holyhead Harbour Works, Holyhead

This firm was the contractor for constructing and maintaining the Holyhead breakwater which was commenced in 1848. A 7′ 0″ gauge railway was used between a quarry and the breakwater for conveying the stone used in its construction. A number of locomotives were supplied by R. B. Longridge and at least two were subsequently sold, one to South America and one to the harbour board at Ponta Delgada, a harbour on the island of São Miguel in the Azores. This locomotive carried a plate: 'J. & C. Rigby, Holyhead Harbour Works, 1861' and would probably be one of the Longridge 0-4-0WTs supplied to the contractor and overhauled at their own workshops. It is extremely unlikely that the locomotive was built at Holyhead.

RILEY BROS.

Middlesborough

One vertical boilered 0-4-0T named WASP is said to have been built by the above firm in 1890 and supplied to Wilsons Pease & Company. No other details are available.

ROBERTSON & HENDERSON

Glasgow

Built a combined car in 1877 having a vertical boiler and engine at one end and a condensing arrangement under the floor at the other. On its trial trip it blew up. No further information has been found.

Fig. 464 **Robey & Co.** 0-4-0WT SANTIAGO

Fig. 465 **Robey & Co.** 0-4-0T

ROBEY & CO. LTD.
Globe Works, Lincoln

Founded by Robert Robey in 1854 the firm became Robey & Co. Ltd. in 1861 but by 1874 had become Robey & Company and in 1893 it reverted to Robey & Co. Ltd. The major products were steam wagons, steam rollers, portable steam engines and all types of vertical and horizontal stationary steam engines. A few steam railway locomotives were built, but very little information is available.

Locomotives known to have been built were:

	0-4-0T	SANTIAGO	Robey's works loco.
1876	0-4-0T	SANTIAGO	
	0-4-0CT		
1895	0-4-0ST	PERLA i.c.	Geared loco for S. America
c 1880	0-4-0WT		
1884	0-4-0TG	No. 8001	for own use

The 0-4-0TG (8001) may have been a conversion from a road wagon having the cylinders on top of boiler, flywheel and chain drive. PERLA had an interesting transmission, the cylinders being well forward driving a geared jackshaft. There were two SANTIAGOS and the illustration shows the Robey works engine with wooden buffer beams, dome on the first ring of the boiler, and round top firebox with man-lid on the top. The 1876 one was a side tank with all over cab, inside cylinders and framing similar to the works loco. The works were at one time known as Perseverance Ironworks.

Fig. 466 **Robey & Co.** *c*1880 0-4-0WT – note solid cast iron wheels

Fig. 467 **Robey & Co.** 1895 0-4-0ST PERLA – geared drive

ROBINSON, THOMAS & SON LTD.

Railway Works, Rochdale

Established in 1838, this firm, well known for the manufacture of woodworking and flour mill machinery, at one time was involved in the production of railway material and locomotives. The first known locomotive was built for their own use in 1881. Named MARY it was a neat little machine, four-coupled with vertical boiler and two-cylinder compound. Details are shown in the illustration.

In 1884 a four-coupled outside framed tender loco was supplied to New Zealand. The cylinders were inside, driving the rear wheels by gearing. Steam pipes projected through the base of the smokebox. The boiler had a large dome combined with Salter safety valve and round top firebox. A third locomotive may have been sent to Brazil.

Articles on Robinson's locomotive can be found in the *Industrial Railway Record* Nos. 32 and 39.

ROBSON & TAYLOR

East Yorkshire Ironworks, Middlesborough

This firm of engineers and ironfounders advertised in the *Engineer*, with a drawing of a 0-4-0T as shown with outside cylinders, cross heads with single slide bars (not too robust in design). It was a 'long boiler' type with firebox behind the trailing axle, and dome on the first boiler ring. The double buffers are interesting. No records are available of any type of locomotive being built. Their chief occupation was the manufacture of semi-portable mining engines, brick-making machines, boilers and girder work.

Fig. 469 **Robson & Taylor** advertisement in the *Engineer*

Fig. 468 **Robinson** 1881 MARY built for own use

ROTHWELL, HICK & ROTHWELL
Union Foundry, Bolton

This firm was established about 1830 as general engineers. Benjamin Hick was one of the partners, but in 1832 he left to set up his own firm of Benjamin Hick & Sons in the same town and in the same business.

The firm then became ROTHWELL & COMPANY. It soon saw the possibilities in the rapid expansion of railways and the requirements for locomotives after Benjamin Hick had witnessed the Rainhill trials.

Orders were quickly secured for the American market and other foreign countries and up to 1840, out of fifty-six locomotives built, twenty-eight went abroad. The first locomotives sent out were similar to the 2-2-0 supplied to the Bolton & Leigh Railway. During 1834–5 there is no record of any locomotives being built, though much colliery work was carried out in the form of pumping plant and stationary engines, but it seems, from 1836, locomotive construction became their main line of business and 2-2-0, 0-4-0, and 2-2-2 types went to America, France and Germany. In 1836 a 4-2-0 was built for the South Carolina R.R. of 5 ft gauge to the requirements of Horatio Allen, one of America's leading locomotive engineers. The bogie was of the swivelling type and inclined outside cylinders 9½″ x 18″ were fitted. The single drivers were 4′ 6″ diameter. It was reputed to have worked for thirty-five years and carried the name TENNESSEE.

Five locomotives built for the Leipzig and Dresden Railway, comprised four 0-4-0s and one 2-2-2. They were very small and weighed between eight and nine tons. 2-2-2s were built, with slight dimensional modifications, as a standard type for the Paris & Versailles, Grand Junction, Stockton & Hartlepool, London & Southampton and Manchester & Leeds railways.

Some orders were sub-contracted by Edward Bury, and a number of 2-2-0s and 0-4-0s were built to his designs for the London & Birmingham, Glasgow, Paisley & Greenock and Midland Counties railways. From 1841 the firm received a number of orders for 7 ft gauge locomotives from the Great Western, Bristol & Exeter, South Devon and Carmarthen & Cardigan railways.

There is no doubt that the colossal 4-2-4T locomotives, of which eight were built for the Bristol and Exeter Railway to Mr Pearson's design, were the most startling and interesting of any that were built by Rothwells. For someone used to the minute 'Bury' types on the London & Birmingham Railway, to be confronted by one of these monsters at say Bristol or Exeter would have been a shattering experience. They had 16½″ x 24″ inside cylinders which protruded at the front end beneath the smokebox. The driving wheels were 9 ft diameter and were flangeless with inside and outside bearings. The two bogies had 4 ft diameter wheels and were equipped with ball and socket swivels. Brakes were fitted between the trailing bogie wheels, and large sandboxes were fitted in front of the driving wheels.

The domeless boiler was 4′ 0½″ diameter and 10′ 9″ long, the safety valves being housed above the flush top firebox. Coke was the fuel used, and the weight in working order of the locomotive was 42 tons. The water tank was fixed between the frames.

At the time (1854) they were the fastest in service and a recorded speed of 82 mph was made by one of them, running down Wellington Bank, the scene of other high speed trips.

The largest order was from the London & South Western Railway and finally amounted to twenty-eight 2-2-2s. They had inside cylinders 15″ x 22″ and 6′ diameter driving wheels and were the firm's own design and had their characteristic dome and safety valve cover, close to the chimney. The frames were outside

Fig. 470
Rothwell Hick & Rothwell
1853 Pearson's Bristol & Exeter Rly. 4-2-4WTs 9′ D.W.

Ref No	Date	Type	Cylinders inches	D.W. ft in	Customer	No/Name		W'drawn
1	1831	2-2-0	9 x 18	5 0	Bolton & Leigh Rly	Union	Spiral Flue (j)	
2	1832	2-2-0	10 x 18		Pontchartrain RR	Pontchartrain		
3	1832	0-4-0		4 6				
4	11/1832	2-2-0	9 x 18	4 6	Bangor & Piscataguis RR	Pioneer		
5	1833	2-2-0	9 x 18	4 6	Greensville & Roanoke RR	Nottoway		
6	1836	2-2-0	11 x 16	4 6	Bangor & Piscataguis RR	Bangor		
7	1836	2-2-0			Canada			
8	1836	4-2-0	OC		S. Carolina RR	Tennessee	5' Gauge	
9	1836	0-4-0	11 x 16	4 6	Leipzig & Dresden Rly	Komet		
10	1836	0-4-0	11 x 16	5 0	Leipzig & Dresden Rly	Blity		
11	1836			5 0	For stock			
12	1837	2-2-2	12½ x 18	5 0	Grand Junction Rly	6 Stentor		
13	1837	2-2-2	12½ x 18	5 0	Grand Junction Rly	9 Alecto		
14	1837	2-2-2	12½ x 18	5 0	Grand Junction Rly	10 Dragon		
15	1837	2-2-2	12½ x 18	5 0	Grand Junction Rly	11 Zamiel		
16	1837	2-2-2	12½ x 18	5 0	Grand Junction Rly	15 Phalaris		
17	1837	2-2-2	12½ x 18	5 0	Grand Junction Rly	16 Lynx		
18	1837	0-4-0	11 x 16	4 6	Leipzig & Dresden Rly	Windsbrant		
19	1837	0-4-0	11 x 16	5 0	Leipzig & Dresden Rly	Faust		
20	1837	2-2-2	11 x 18	5 0	Leipzig & Dresden Rly	Peter Rothwell		
21	1837	2-2-2	13 x 17	5 6	Paris & Versailles Rly	Aquilon		
22	1837	2-2-2	13 x 17	5 6	Paris & Versailles Rly	Bucéphale		
23	1837	0-4-0	10 x 16	4 6	Richmond, Fredericksburg & Potomac RR	Robert Morris		
24	1837	0-4-0	10 x 16	4 6	Richmond, Fredericksburg & Potomac RR	John Hopkins		
25	1837	0-4-0	10 x 16	4 6	Richmond, Fredericksburg & Potomac RR	Oliver Evans		
26	12/1837	2-2-0	12 x 18	5 6	London & Birmingham Rly	25	(b)	
27	1838	2-2-0	12 x 18	5 6	London & Birmingham Rly	26	(b)	
28	1838	2-2-0	12 x 18	5 6	London & Birmingham Rly	27	(b)	
29	1838	2-2-0	12 x 18	5 6	London & Birmingham Rly	28	(b)	
30	1838	2-2-0	12 x 18	5 6	London & Birmingham Rly	29	(b)	
31	1838	2-2-0	12 x 18	5 6	London & Birmingham Rly	30	(b)	
32	1838	2-2-2	11 x 18	5 0	Liverpool & Manchester Rly	59 Rokeby		
33	1838	2-2-2	11 x 18	5 0	Liverpool & Manchester Rly	60 Roderic		
34	1838				Firm's General Agent			
35	1838				Firm's General Agent			
36	1838	2-2-2	13 x 18	5 6	Grand Junction Rly	44 Harlequin		
37	1838	2-2-2	13 x 18	5 6	Grand Junction Rly	49 Columbine		
38	1838	2-2-2		5 0	Leipzig & Dresden Rly	Salamander		
39	1838	2-2-2	13 x 18	5 6	Stockton & Hartlepool Rly			
40	1838	2-2-2	13 x 18	5 6	Stockton & Hartlepool Rly			
41	7/1839	2-2-2	13½ x 18	5 6	London & South Western Rly	35 Vivid		6/1854
42	7/1839	2-2-2	13½ x 18	5 6	London & South Western Rly	36 Comet		7/1851

43	9/1839	2-2-2	13½ x 18	5 6	London & South Western Rly	37 Arab		2/1853
44	9/1839	2-2-2	13½ x 18	5 6	London & South Western Rly	38 Vizier		1/1870
45	10/1839	2-2-2	13½ x 18	5 6	London & South Western Rly	39 Wizard		4/1856
46	1838	2-2-2	13 x 17	5 6	Paris & Versailles Rly	Vulcan		
47	1838	2-2-2	13 x 17	5 6	Paris & Versailles Rly	Pégase		
48	1838	2-2-0	12 x 18	5 0	Glasgow Paisley & Greenock Rly	1 Lucifer		
49	1/1839	2-2-0	12 x 18	5 0	Glasgow Paisley & Greenock Rly	3 Zamiel		
50	2/1838	2-2-0	12 x 18	5 0	Glasgow Paisley & Greenock Rly	2 Hecate?	(a)	
51	1839	2-2-2	13 x 17	5 6	Paris & Versailles Rly	Le Cyclope		
52	1839	2-2-2	13 x 17	5 6	Paris & Versailles Rly	Phébus		
53	1839	2-2-2	12 x 18	5 0	Leipzig & Dresden Rly	Magdeburgh		
54	1839	2-2-2	12 x 18	5 0	Leipzig & Dresden Rly	Simson		
55	1839	2-2-2	12 x 18	5 0	Leipzig & Dresden Rly	Altenburgh		
56	1840	2-2-2	12 x 18	5 0	Leipzig & Dresden Rly	Nordlight		
57	9/1840	2-2-2	14 x 18	5 6	Manchester & Leeds Rly	17 Stockport	(b)	
58	1840	2-2-2	14 x 18	5 6	Manchester & Leeds Rly	18 Littleborough	(b)	
59	11/1840	2-2-2	14 x 18	5 6	Manchester & Leeds Rly	24 Hebden Bridge	(b)	
60	1840	2-2-2	14 x 18	5 6	Manchester & Leeds Rly	25 Mirfield	(b)	
61	7/1840	2-2-0	12 x 18	5 6	Midland Counties Rly	13	(b)	
62	8/1840	2-2-0	12 x 18	5 6	Midland Counties Rly	14	(b)	
63	1840	2-2-0	12 x 18	5 6	Midland Counties Rly	15	(b)	
64	11/1840	0-4-0	13 x 20	5 0	Midland Counties Rly	42	(b)	
65	11/1840	0-4-0	13 x 20	5 0	Midland Counties Rly	43	(b)	
66	6/1841	2-4-0	15 x 18	5 0	Great Western Rly (k)	Aries	7' gauge	6/1871
67	7/1841	2-4-0	15 x 18	5 0	Great Western Rly	Taurus	7' gauge	12/1870
68	9/1841	2-4-0	15 x 18	5 0	Great Western Rly	Gemini	7' gauge	7/1870
69	10/1841	2-4-0	15 x 18	5 0	Great Western Rly	Cancer	7' gauge	6/1874
70	10/1841	2-4-0	15 x 18	5 0	Great Western Rly	Leo	7' gauge	12/1870
71	12/1841	2-4-0	15 x 18	5 0	Great Western Rly	Virgo	7' gauge	12/1870
72	1842	2-4-0	15 x 18	5 0	Great Western Rly	Scorpio	7' gauge	6/1871
73	1842	2-4-0	15 x 18	5 0	Great Western Rly	Libra	7' gauge	12/1872
74	1842	2-4-0	15 x 18	5 0	Great Western Rly	Capricornus	7' gauge	6/1871
75	1842	2-4-0	15 x 18	5 0	Great Western Rly	Sagittarius	7' gauge	7/1870
76	1842	2-4-0	15 x 18	5 0	Great Western Rly	Aquarius	7' gauge	7/1870
77	1842	2-4-0	15 x 18	5 0	Great Western Rly	Pisces	7' gauge	6/1874
78	1841	2-2-2			For Russia			
79								
80								
81								
82								
83	1842	2-2-2			For Russia			
84	1842	2-2-2			For Russia			
85	1845-6	2-4-0			London & North Western Rly	(c)		
86	1845-6	2-4-0			London & North Western Rly	(c)		
87	1845-6	2-4-0			London & North Western Rly	(c)		

Ref No	Date	Type	Cylinders inches	DW ft in	Customer	No/Name		W'drawn
88	1845-6	2-4-0			London & North Western Rly	(c)		
89	1845-6	2-4-0			London & North Western Rly	(c)		
90	1845-6	2-4-0			London & North Western Rly	(c)		
91	1845-6	2-4-0			London & North Western Rly	(c)		
92	1845-6	2-4-0			London & North Western Rly	(c)		
93	1845-6	2-4-0			London & North Western Rly	(c)		
94	1845-6	2-4-0			London & North Western Rly	(c)		
95	1845-6	2-4-0			London & North Western Rly	(c)		
96	1845-6	2-4-0			London & North Western Rly	(c)		
97	3/1846	0-6-0	15 x 24	4 8	Midland Rly	173	(c)	
98	1846	0-6-0	15 x 24	4 8	Midland Rly	174	(c)	
99	5/1846	0-6-0	15 x 24	4 8	Midland Rly	175	(c)	
100	8/1846	0-6-0	15 x 24	4 8	Midland Rly	176	(c)	
101	9/1846	0-6-0	15 x 24	4 8	Midland Rly	177		
102	10/1846	0-6-0	15 x 24	4 8	Midland Rly	178		
103	12/1846	2-2-2	15 x 22	6 0	London & South Western Rly	73 Fireball	(e)	3/1867
104	12/1846	2-2-2	15 x 22	6 0	London & South Western Rly	74 Firebrand		4/1867
105	2/1847	2-2-2	15 x 22	6 0	London & South Western Rly	75 Fire King		5/1866
106	1847	2-2-2	15 x 22	6 0	London & South Western Rly	76 Firefly		12/1872
107	5/1847	2-2-2	15 x 22	6 0	London & South Western Rly	77 Hecla		3/1865
108	6/1847	2-2-2	15 x 22	6 0	London & South Western Rly	78 Hecate		12/1862
109	7/1847	2-2-2	15 x 22	6 0	London & South Western Rly	79 Harpy		12/1865
110	7/1847	2-2-2	15 x 22	6 0	London & South Western Rly	80 Hornet		4/1864
111	8/1847	2-2-2	15 x 22	6 0	London & South Western Rly	81 Herod		1/1867
112	8/1847	2-2-2	15 x 22	6 0	London & South Western Rly	82 Sultana		12/1865
113	9/1847	2-2-2	15 x 22	6 0	London & South Western Rly	83 Siren		9/1866
114	9/1847	2-2-2	15 x 22	6 0	London & South Western Rly	84 Styx		12/1865
115	11/1847	2-2-2	15 x 22	6 0	London & South Western Rly	85 Saracen		4/1868
116	11/1847	2-2-2	15 x 22	6 0	London & South Western Rly	86 Shark		3/1865
117	11/1847	2-2-2	15 x 22	6 0	London & South Western Rly	87 Stentor		5/1866
118	12/1847	2-2-2	15 x 22	6 0	London & South Western Rly	88 Sirius		11/1867
119	12/1847	2-2-2	15 x 22	6 0	London & South Western Rly	89 Saturn		12/1865
120	12/1847	2-2-2	15 x 22	6 0	London & South Western Rly	90 Sybil		12/1866
121	1/1848	2-2-2	15 x 22	6 0	London & South Western Rly	91 Spitfire		1/1868
122	1/1848	2-2-2	15 x 22	6 0	London & South Western Rly	92 Charon		10/1868
123	5/1848	2-2-2	15 x 22	6 0	London & South Western Rly	93 Cyclops		12/1866
124	5/1848	2-2-2	15 x 22	6 0	London & South Western Rly	94 Camilla		5/1869
125	5/1848	2-2-2	15 x 22	6 0	London & South Western Rly	95 Centaur		8/1869
126	5/1848	2-2-2	15 x 22	6 0	London & South Western Rly	96 Castor		12/1867
127	8/1848	2-2-2	15 x 22	6 0	London & South Western Rly	97 Pegasus		9/1866
128	8/1848	2-2-2	15 x 22	6 0	London & South Western Rly	98 Plutus		10/1867
129	9/1848	2-2-2	15 x 22	6 0	London & South Western Rly	99 Phlegon		4/1869

130	9/1848	2-2-2	15 x 22	6 0	London & South Western Rly	100 Python		7/1868
131	1850	0-4-2	15 x 24	5 0	Great Southern & Western Rly	112		
132	1850	0-4-2	15 x 24	5 0	Great Southern & Western Rly	113		
133	1850	0-4-2	15 x 24	5 0	Great Southern & Western Rly	114		
134	1850	0-4-2	15 x 24	5 0	Great Southern & Western Rly	115		
135	1853	4-2-4T	16½ x 24	9 0	Bristol & Exeter Rly	39	(f) 7' gauge	1868
136	1853	4-2-4T	16½ x 24	9 0	Bristol & Exeter Rly	40	7' gauge	1873
137	1853	4-2-4T	16½ x 24	9 0	Bristol & Exeter Rly	41	7' gauge	1868
138	1854	4-2-4T	16½ x 24	9 0	Bristol & Exeter Rly	42	7' gauge	1868
139	1854	4-2-4T	16½ x 24	9 0	Bristol & Exeter Rly	43	7' gauge	
140	1854	4-2-4T	16½ x 24	9 0	Bristol & Exeter Rly	44	7' gauge	
141	1854	4-2-4T	16½ x 24	9 0	Bristol & Exeter Rly	45	7' gauge	
142	1854	4-2-4T	16½ x 24	9 0	Bristol & Exeter Rly	46	7' gauge	
143	11/1854	2-2-2	18 x 24	8 0	Great Western Rly	Alma	(g) 7' gauge	6/1872
144	12/1854	2-2-2	18 x 24	8 0	Great Western Rly	Balaklava	7' gauge	10/1871
145	3/1855	2-2-2	18 x 24	8 0	Great Western Rly	Inkermann	7' gauge	10/1877
146	4/1855	2-2-2	18 x 24	8 0	Great Western Rly	Kertch	7' gauge	12/1872
147	5/1855	2-2-2	18 x 24	8 0	Great Western Rly	Crimea	7' gauge	11/1876
148	5/1855	2-2-2	18 x 24	8 0	Great Western Rly	Eupatoria	7' gauge	10/1876
149	7/1855	2-2-2	18 x 24	8 0	Great Western Rly	Sebastopol	7' gauge	10/1880
150	1855							
151	1855							
152	10/1855	4-4-0ST	17 x 24	5 6	Bristol & Exeter Rly	47	7' gauge	12/1879
153	12/1855	4-4-0ST	17 x 24	5 6	Bristol & Exeter Rly	48	7' gauge	3/1879
154	12/1855	4-4-0ST	17 x 24	5 6	Bristol & Exeter Rly	49	7' gauge	6/1884
155	12/1855	4-4-0ST	17 x 24	5 6	Bristol & Exeter Rly	50	7' gauge	6/1884
156	12/1855	4-4-0ST	17 x 24	5 6	Bristol & Exeter Rly	51	7' gauge	12/1882
157	1/1856	4-4-0ST	17 x 24	5 6	Bristol & Exeter Rly	52	7' gauge	7/1880
158	1857	2-4-0	OC17 x 20	5 1	Lancaster & Carlisle Rly	1	(h)	
159	1857	2-4-0	OC17 x 20	5 1	Lancaster & Carlisle Rly	2		
160	1857	2-4-0	OC17 x 20	5 1	Lancaster & Carlisle Rly	3		
161	1857	2-4-0	OC17 x 20	5 1	Lancaster & Carlisle Rly	4		
162	1857	2-4-0	OC17 x 20	5 1	Lancaster & Carlisle Rly	5		
163	1857	2-4-0	OC17 x 20	5 1	Lancaster & Carlisle Rly	6		
164	1857	2-4-0	OC17 x 20	5 1	Lancaster & Carlisle Rly	7		
165	1857	2-4-0	OC17 x 20	5 1	Lancaster & Carlisle Rly	8		
166	1857	2-4-0	OC17 x 20	5 1	Lancaster & Carlisle Rly	9		
167	1857	2-4-0	OC17 x 20	5 1	Lancaster & Carlisle Rly	10		
168	1857	2-4-0	OC17 x 20	5 1	Lancaster & Carlisle Rly	11		
169	1857	2-4-0	OC17 x 20	5 1	Lancaster & Carlisle Rly	12		
170	1857	2-4-0	OC17 x 20	5 1	Lancaster & Carlisle Rly	13		
171	1857	2-4-0	OC 17 x 20	5 1	Lancaster & Carlisle Rly	14		
172	1857	2-4-0	OC 17 x 20	5 1	Lancaster & Carlisle Rly	15		
173	1857	2-4-0	OC 17 x 20	5 1	Lancaster & Carlisle Rly	16		
174	1857	2-4-0	OC 17 x 20	5 1	Lancaster & Carlisle Rly	17		

Ref No	Date	Type	Cylinders inches	D.W. ft in	Customer	No/Name		W'drawn
175	1857	2-4-0	OC 17 x 20	5 1	Lancaster & Carlisle Rly	18		
176	1857	2-4-0	OC 17 x 20	5 1	Lancaster & Carlisle Rly	19		
177	1857	2-4-0	OC 17 x 20	5 1	Lancaster & Carlisle Rly	20		
178	1857	2-2-2	16 x 20	6 1	Lancaster & Carlisle Rly	21		
179	1857	2-2-2	16 x 20	6 1	Lancaster & Carlisle Rly	22		
180	1857	2-2-2	16 x 20	6 1	Lancaster & Carlisle Rly	23		
181	1857	2-2-2	16 x 20	6 1	Lancaster & Carlisle Rly	24		
182	1857	2-2-2	16 x 20	6 1	Lancaster & Carlisle Rly	25		
183	2/1858	2-4-0	OC 18 x 22	5 1	Eastern Counties Rly	301	(I) (i)	4/1873
184	3/1858	2-4-0	OC 18 x 22	5 1	Eastern Counties Rly	302		10/1873
185	10/1858	2-4-0	OC 18 x 22	5 1	Eastern Counties Rly	303		4/1873
186	10/1858	2-4-0	OC 18 x 22	5 1	Eastern Counties Rly	304		1/1873
187	11/1858	2-4-0	OC 18 x 22	5 1	Eastern Counties Rly	305		11/1875
188	12/1858	2-4-0	OC 18 x 22	5 1	Eastern Counties Rly	306		1/1873
189	8/1859	2-2-2WT	14½ x 18	5 6	Bristol & Exeter Rly	57	7' Gauge	12/1877
190	10/1859	2-2-2WT	14½ x 18	5 6	Bristol & Exeter Rly	58	7' Gauge	12/1880
191	7/1860	0-6-0	17 x 24	5 0	Bristol & Exeter Rly	59	7' Gauge	6/1887
192	8/1860	0-6-0	17 x 24	5 0	Bristol & Exeter Rly	60	7' Gauge	12/1884
193	1864	4-4-0ST	17 x 24	5 3	Carmarthen & Cardigan Rly	Etna	7' Gauge	5/1892
194	6/1864	4-4-0ST	17 x 24	5 3	Carmarthen & Cardigan Rly	Hecla	7' Gauge	5/1892
195								
196								
197								
198								
199								
200								

(a) May have been named GLASGOW. (b) Partial construction only. (d) Stephenson long boiler type. (e) Rothwell design. (f) Pearson's design. (g) Gooch's design. (h) Allan's design. (i) Sinclair's design. (j) The actual design is in doubt. Variously described as Planet and Bury types, but more probably a vertical boiler type described by Theodore West with horizontal cylinders transmitting motion through an oscillating shaft. See sketch and description in C. F. Dendy Marshall's Book *A History of Railway Locomotives Down To The Year 1831.*

One locomotive built in 1835 for the South Carolina RR named H. SCHULTZ may have been built by Rothwell, and one in 1838 named PHOENIX for the Richmond & Petersburg RR.

and the horn plates were rivetted on separately. The firebox was raised and round topped, surmounted by another safety valve. The building was spread over a period of nearly two years and no other locomotives were built during this time. They were not supplied with tenders.

The large order was due to the satisfactory performance of the previous batch of 2-2-2s supplied in 1839 and the later singles acquitted themselves equally as well. Their average life was twenty years.

Another large order came from the Lancaster and Carlisle Railway for twenty 2-4-0 goods and four 2-2-2 passenger locomotives and was completed in 1857. They were of Alexander Allan's design, both types had outside cylinders and were similar to the Crewe built locomotives.

Another 'Allan' type with outside cylinders was turned out in 1858. They were 2-4-0 goods locomotives designed by Mr Sinclair of the Eastern Counties Railway.

The last six to be built by Rothwells were all 7′ broad gauge; four for the Bristol & Exeter Railway in 1859–60 and two 4-4-0 saddle tanks for the Carmarthen and Cardigan Railway, in 1864.

Orders had declined from 1856 due to the intense competition at this time and although the firm struggled on for a few more years the works were closed and stood empty for many years.

Part of the premises was bought by the Bolton Iron & Steel Company who were also well known in the railway business, producing tyres, axles and other material. About 1907 they were taken over by Henry Bessemer & Company. A total of approximately 200 locomotives were built.

ROWAN, JAS. M. & CO.

Atlas Works, Springburn, Glasgow

The works were situated on part of the site subsequently occupied by the Clyde Locomotive Works in 1886 which were then purchased by Messrs. Sharp Stewart & Co. in 1888 when they moved from Manchester.

Two 'singles' were built for the Slamannan Railway named BOANERGES and BOREALIS, and in 1840 three inside cylindered 0-4-0s were built with cylinders 10½″ x 24″ and 4 ft diameter coupled wheels for the Wishaw & Coltness Railway. The boiler diameter was 4′ 6½″ and length 9′ 3″. The grate area was 4·2 sq. ft and the working pressure 50 lb/sq. in. They were designed by George Dodds who was the locomotive superintendent of the Monkland & Kirkintilloch Railway. According to Jas. McEwan CLELAND was based on the Stockton and Darlington Railways's LOCOMOTION but with an orthodox boiler. An additional 0-4-0 was built in 1843 for the same railway and named METEOR. It was to the bar framed *Bury* pattern.

Three outside frame 0-4-0s were built in 1842 for the Pollok & Govan Railway with inside cylinders 12″ x 18″ and 4′ 6″ diameter wheels. They had small boilers 3 ft diameter, 8′ 3″ long with a working pressure of 40 lb/sq. in. Gab motion was fitted to these locomotives.

It is possible that two 0-4-0s were built for the Monkland Railway in 1851 but no complete details of this firm's locomotives can be obtained.

Rail motor cars were built for the Gribscov and Pontiloff Railway (Russia) in the 1880s.

J. M. ROWAN

Date	Type	Cylinders inches	D.W. ft in	Customer	Name	Became
1839	2-2-2			Slamannan Rly	Borealis	CR 93
1839	2-2-2			Slamannan Rly	Boanerges	CR 94
11/1840	0-4-0	10½ x 24	4 0	Wishaw & Coltness Rly	Wishaw	
11/1840	0-4-0	10½ x 24	4 0	Wishaw & Coltness Rly	Coltness	
12/1840	0-4-0	10½ x 24	4 0	Wishaw & Coltness Rly	Cleland	
4/1842	0-4-0	IC 12 x 18	4 6	Pollok & Govan Rly	Pollok	CR 83
1842	0-4-0	IC 12 x 18	4 6	Pollok & Govan Rly	Govan	CR 84
1842	0-4-0	IC 12 x 18	4 6	Pollok & Govan Rly	Tradeston	CR 85
1843	0-4-0	IC 14 x 20	4 6	Wishaw & Coltness Rly	Meteor	
? 1851	0-4-0	15 x		Monkland Rly		CR 116
1851	0-4-0	15 x		Monkland Rly		CR 117

CR = Caledonian Rly

RUSSELL, GEORGE & CO.

Alpha Works, Motherwell Junction

Established in 1865 , this firm's advertisement in *Engineering* for 5 January, 1866 stated that they were engineers, contractors and patentees, and makers of stationary boilers and engines, steam and hand cranes, winches, hoisting engines and light locomotives. If any locomotives were built by this firm no records can be traced, but the advertisement shows a vertical boilered locomotive with four wheels coupled and vertical cylinders driving one axle through gears fixed between the frames. Later specifications give a range of cylinders 4" diameter to 8" diameter and wheels 2' 3" to 2' 9" diameter.

RUSTON PROCTOR & CO.

Sheaf Iron Works, Waterside South, Lincoln

This company was established in 1840 as Proctor & Burton with works at Waterside South and by 1849 they were classified as 'millwrights and engineers'. Shortly afterwards they were styled Burton & Proctor and when Joseph Ruston entered into partnership in 1857 the title became Ruston Proctor & Company and in 1899 they became a limited company. They amalgamated with Richard Hornsby & Sons Ltd. in 1918 and became Ruston & Hornsby Ltd., and in 1940 entered into an association with Davey Paxmán & Co. Ltd. of Colchester.

Railway locomotives were made in small numbers from 1866, and were four and six-coupled tanks. Details were published in the *Engineer* for 17 April, 1868 of a contractors tank locomotive which was sent to the Paris Exhibition of 1867. It was a 0-4-0ST with outside cylinders 9" x 16" with the valves and valve

Fig. 471 **Ruston Proctor** 1868 diagram of 0-4-0ST

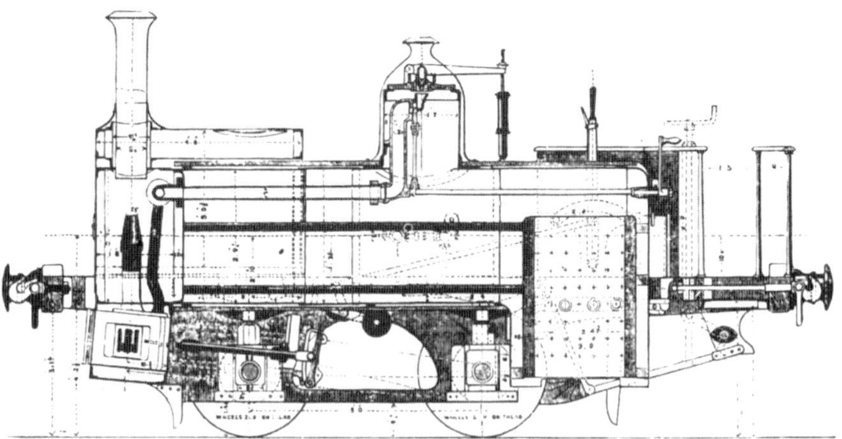

Fig. 472 **Ruston Proctor** 1868 drawing of same locomotive

gear on the inside of the frames. The saddle tank was over the smokebox and projected a short way back over the boiler barrel and came down each side to the running plate. The dome was on the back ring of the barrel with Salter spring balance safety valves bolted to the dome cover. The rear pair of wheels were fitted with hand operated screw-lever brakes. It had no cab.

In 1868 they obtained an order from the Great Eastern Railway for five 0-6-0Ts to Mr Samuel Johnson's design. Johnson was then the locomotive superintendent of the GER. They had inside cylinders inclined 1 in 10 to clear the front axle. The dome was over the round top firebox surmounted by Salter safety valves. The side tanks carried 770 gallons of water and the weight in working order was approx 35 tons.

Three were converted to crane tanks, with 3 ton jib cranes, at Stratford; one in 1891 and two in 1893. The other two were scrapped in 1889 and 1892 but of the three crane tanks one was scrapped in 1950 and the other pair lasted until 1952 at the age of eighty-four.

Very little is known of the remaining locomotives built by Ruston Proctor & Company but at least sixteen went to Argentina on the 5′ 6″ gauge systems and the T. A. Walker locomotives of 1888 were used on the con-

Fig. 473 **Ruston Proctor** 1868 Great Eastern Railway Nos. 204–208

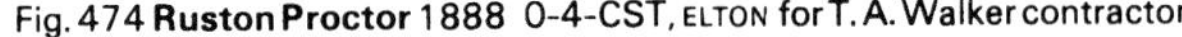

Fig. 474 **Ruston Proctor** 1888 0-4-CST, ELTON for T. A. Walker contractor

struction of the Manchester Ship Canal. They were very small saddle tanks, the tanks being more conventional than the 1867 locomotive being over the boiler barrel only, the dome remained over the firebox with two Salter safety valves on top. The boiler was fed by an injector, and a feed pump worked by an eccentric fitted to the driving axle. The weight in working order was 12 tons.

The list of locomotives is by no means complete but thanks are due to Messrs Ruston & Hornsby Ltd. and reference should be made to the Stephenson Locomotive Society's Journal of March 1960 for further details.

Works No	Date	Type	Cylinders inches	D.W. ft in	Gauge ft in	Customer	No/Name
969	5/1866	0-4-0ST		2 9			
	1867	0-4-0ST	OC 9 x 16	2 9	Std	Paris Exhibition 1867	
	5/1868	0-6-0T	IC 16 x 22	4 1	Std	Great Eastern Railway	204
	1868	0-6-0T	IC 16 x 22	4 1	Std	Great Eastern Railway	205
	1868	0-6-0T	IC 16 x 22	4 1	Std	Great Eastern Railway	206
	1868	0-6-0T	IC 16 x 22	4 1	Std	Great Eastern Railway	207
	1868	0-6-0T	IC 16 x 22	4 1	Std	Great Eastern Railway	208
	1869	0-4-0T				Traction Engine Type	
1948	7/1870	0-6-0T				T. A. Walker	
7272	1/1881	0-4-0ST	OC 9½ x 16	2 6	3 6	For Argentina	
7273	1/1881	0-4-0ST	OC 9½ x 16	2 6	3 6	For Argentina	
10866	10/1884	0-4-0ST	OC 9½ x 16	2 9	5 6	For Argentina	
10905	10/1884	0-4-0ST	OC 9½ x 16	2 9	5 6	For Argentina	
10943	10/1884	0-4-0ST	OC 9½ x 16	2 9	5 6	For Argentina	
10980	11/1884	0-4-0ST	OC 9½ x 16	2 9	5 6	For Argentina	
11024	12/1884	0-4-0ST	OC 9½ x 16	2 9	5 6	For Argentina	
11026	12/1884	0-6-0T	IC 13 x 18	3 6	5 6	For Argentina	
11063	1/1885	0-6-0T	IC 13 x 18	3 6	5 6	For Argentina	
11094	2/1885	0-6-0T	IC 13 x 18	3 6	5 6	For Argentina	(b)
11138	3/1885	0-6-0T	IC 13 x 18	3 6	5 6	For Argentina	
11735	12/1885	0-4-0ST	OC 9½ x 16	2 9	5 6	For Argentina	
11736	12/1885	0-4-0ST	OC 9½ x 16	2 9	5 6	For Argentina	
11737	1/1886	0-4-0ST	OC 9½ x 16	2 9	5 6	For Argentina	
11793	2/1886	0-4-0ST	OC 9½ x 16		2 11½	For Argentina	
11794	2/1886	0-4-0ST	OC 9½ x 16		2 11½	For Argentina	
11795	2/1886	0-4-0ST	OC 9½ x 16		2 11½	For Argentina	
11796	2/1886	0-4-0ST	OC 9½ x 16		2 11½	For Argentina	
12301	1/1887	0-4-0ST					
13000	1/1888	0-6-0T	IC 13 x 18	3 6	Std	T. A. Walker	Dunham
13003	1/1888	0-6-0T	IC 13 x 18	3 6	Std	T. A. Walker	Eccles
13044	2/1888	0-6-0T	IC 13 x 18	3 6	Std	T. A. Walker	Altrincham
13110	3/1888	0-4-0ST	OC 10 x 16	2 9	Std	T. A. Walker	Glazebrook
13111	5/1888	0-4-0ST	OC 10 x 16	2 9	Std	T. A. Walker	Widnes
13112	5/1888	0-4-0ST	OC 10 x 16	2 9	Std	T. A. Walker	Elton
21232	1/1897	0-4-0ST	OC 9½ x 16		3 0		
?	*c.* 1870	0-4-0ST	OC		Std	London & St Katherine Dock Co	18
?	By.1883	0-4-0ST	OC		Std	Ruston Proctor & Co	Meteor
?	*c.* 1888	0-4-0ST	OC		Std	Ruston Proctor & Co	Rocket (a)

(a) To Bahia Blanca & NW Rly No 33 (b) Renamed RUSTON/1900

RYLANDS & SONS

Wigan

This firm is reported to have built a number of 0-4-0STs in the 1880s. Practically no information is available except that one was built in 1881 and two in 1886. If they were actual builders and not simply rebuilders, it is possible that a few more could have been built.

ST HELENS FOUNDRY

St Helens

A locomotive is stated to have been built by this firm, in 1857 for an unknown customer. It was a 0-6-0ST with inside cylinders but nothing else is known, except that it carried the name OSWALD and was sold in 1914 to the Calder and Mersey Extract Co. Ltd., Widnes. Another locomotive reported as built at the Atlas Foundry, St Helens at an unknown date, might be by the same maker. The locomotive was a 0-4-0WT with outside cylinders named LUCY for Gaskell Deacon & Co., Widnes. The manager at this time was George Heaton Daglish.

ST HELENS & RUNCORN RAILWAY

St Helens, Lancs.

This old railway was opened in 1833 and taken over by the London and North Western Railway on 1 August, 1864 when the locomotive shops were sold to James Cross. Cross was the railway's Locomotive Engineer when a few locomotives were built in the company's shops. At least six were built between 1849 and 1860. Unfortunately very little is known about the St Helens products.

Locomotives that are known to have been built were all four-coupled types; two side tanks and four with tenders. An interesting locomotive, not strictly new however, was one on the St Helens & Runcorn Railway converted to a 2-4-2T and named WHITE RAVEN. Its main features were W. Bridge Adam's patent radial axleboxes fitted to the leading and trailing wheels, and his patent spring tyres, a steel hoop spring being fitted between the tyre and wheel centre. The latter was a patent to eliminate stress between tyre and wheel centre. This locomotive was rebuilt in this form in 1863.

Date	Type	St Helens No	Name	LNWR No 1864	Disposal
1849	-4-	7	Eden	1373	Sold 1865 Strangeways Colly (a)
1853	-4-T	23	Hero	1389	Sold 1865 (b)
1854	-4-	25	Goliath	1391	Scrapped 1865
1854	2-4-0	24	Alma	1390	Sold 1874 I. W. Boulton
1856	-4-T	4	Hercules	1370	Sold 1865 James Cross (b)
1860	-4-	27	Dee	1393	Sold 1865

(a) Thought to have been either 0-4-2 or 2-4-0. (b) Thought to have been either 0-4-2T or 2-4-0T

Fig. 475 Built in the 1880s at Rylands works

ST ROLLOX FOUNDRY COMPANY
Glasgow

Built four locomotives only, for the Garnkirk and Glasgow Railway. The three six-coupled locomotives built had vertical cylinders fastened to the boiler and connected to the coupling rods between the leading and intermediate wheels through a jackshaft. They had bar frames and an elaborate gab motion. The boiler was an adapted Cornish type with a large 3 ft diameter flue divided into four similar ones passing into the smokebox. Later the coupling rods were removed between the intermediate and trailing wheels, converting them to 0-4-2s. All three were sold in 1847 to contractors.

SANDYS CARNE & VIVIAN
Copperhouse Foundry, Hayle

This was a copper smelting works established about 1760. Ironfounding replaced the copper smelting about 1820 and engine building was established for the mining industry and in the middle 1820s the firm's name was Trevenan, Carne & Wood. Many pumping engines were built, some iron sailing ships, plant for gasworks, flour mills, saw mills, Cornish boilers, church bells, and the chains for the Clifton and Hungerford suspension bridges. In 1838 the firm is reported to have built the first locomotive in the county of Cornwall. It was named CORNUBIA and was ordered by the Hayle Railway. The history of such building by this firm is practically non-existant but it is probable that a second locomotive named CARN BREA was built for the same railway. The Foundry and Works closed down in 1869.

Built	Type	Cylinders inches	D.W. ft in	Garnkirk & Glasgow Rly		W'drawn
				No.	Name	
1835	0-4-0	OC 13 x 18	4 0	5	Frew	1848
1840	0-6-0	VC 13 x 18	3 10	7	Victoria	1847
1840	0-6-0	VC 13 x 18	3 10	8	St. Rollox	1847
1840	0-6-0	VC 13 x 18	3 10	9	Carfin	1847

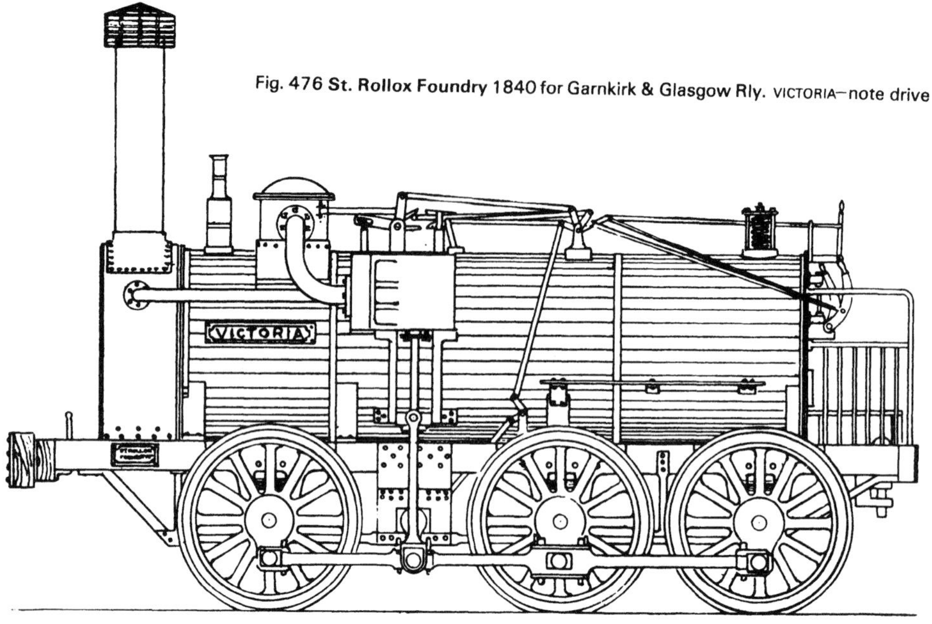

Fig. 476 **St. Rollox Foundry** 1840 for Garnkirk & Glasgow Rly. VICTORIA—note drive

SARA & COMPANY
Plymouth

One locomotive was built in January 1868 for the South Devon Railway and was used on the Sutton Harbour branch. The locomotive took the form of a flat truck on which was mounted a vertical tubeless boiler which had a vertical flue through the centre. The vertical cylinders were attached to the boiler, and drove on to a jackshaft attached to the truck framing. A gear was attached to this shaft which meshed into a gear keyed to the track wheels axle. The two axles were coupled together in the conventional manner with side rods.

A water tank was built into the main frames. The regulator valve was situated in the main steam pipe over the cylinders. The main dimensions were: cylinders 9″ x 12″; wheels 3′ 0″ diameter; boiler 2′ 8½″ diameter x 6′ 3″ high; firebox 2′ 4½″ diameter; grate area 6·13 sq. ft; water capacity 60 gallons; gauge 7 ft; wheelbase 5′ 9″; and overall height 11′ 10½″.

This locomotive worked very well and was withdrawn in 1883, but due to its good condition it was decided to use it as a stationary engine in Newton Abbot works where it performed satisfactorily until 1927. As she was the only example left of a broad gauge locomotive, she was overhauled and stood on a plinth on the down platform of Newton Abbot station.

The name TINY was carried and, when the Great Western Railway took over, the number 2180 was allocated. Although no records are available, this firm probably built one or two similar locomotives for the various local china clay mines.

SARA & BURGESS
Penrhyn, Falmouth, Cornwall

Nicholas Sara, who had been a foreman at the Perran Foundry started up in 1857 and did foundry work. At least four locomotives were built by this firm in the 1860s.

They were four coupled vertical boiler type with two vertical cylinders driving a geared shaft which engaged in another gear on one of the coupled axles. A water tank and coal box were carried on the footplating near the boiler, the tank being slung between the frames. The cylinders were 7″ x 12″ and wheels 3 ft diameter.

Fig. 477 **Sara & Co.** 1868 South Devon Rly – Sutton Harbour branch TINY

SARA & BURGESS

Three were supplied to Cox & Son of Falmouth Docks: No. 1 BLACKBIRD, No. 2 TORBAY and No. 3 BILLY. All were withdrawn about 1926.

The fourth locomotive of similar type was supplied to Port of Par Ltd., Par Harbour, Cornwall and was scrapped in 1936.

A fifth locomotive was commenced by Sara & Burgess but the parts were assembled in the workshops of Cox & Son, Falmouth.

The locomotives which were built in the 1860s were to 7 ft gauge and would have been converted to standard gauge about 1892. They were probably converted by the owners.

In 1887 the owners were E. B. Sara, Nicholas's son and John Burgess who came from the Perran Foundry. Sara & Burgess went out of business before 1918.

SAVAGES LTD.
St Nicholas Works, Kings Lynn

Although not strictly within the scope of this book, this firm built fairground engines, traction engines, steam wagons and ploughing engines. They also built a number of fairground locomotives designed to run on a circular track with divergent axles. These are fully described in Mr G. Woodcock's book *Miniature Steam Locomotives.*

SAVILE STREET FOUNDRY
Sheffield

The foundry patented a compound tramway engine with vertical boiler, horizontal cylinders and an air condenser placed under the roof. The date of construction is uncertain, but after trials at Sheffield it was transferred to Burnley in 1882 but was not adopted there. The high pressure cylinder was 8″ in diameter and the low pressure is stated to have been 21″ diameter, the stroke being 14″. The foundry also built a combined steam car of which no details are preserved.

Fig. 478 **Sara & Burgess** *c*1860 for Falmouth Docks

SCOTTISH CENTRAL RAILWAY
Perth

Locomotive Superintendents: 1848–53 Robert Sinclair; 1853–65 Alexander Allan.

This railway was also responsible for operating the Scottish Midland Junction Railway which was incorporated in 1845, the same year as the Scottish Central Railway. In addition, a few years after commencing operations it was agreed to pool the locomotives of the SCR and the Aberdeen Railway.

The Perth locomotive works was used for the repair of SMJR as well as its own locomotives from 1851 to 1854. The original stock of the 'Central' consisted of 2-2-2s which had been built to the specifications of the Caledonian Railway's locomotives of the same type, understandable as Robert Sinclair was also the Locomotive Superintendent of the latter railway too.

Building did not take place until Allan had taken office. Two 0-4-2 goods locomotives were commenced in 1856 and although the boiler was of typical Allan design with a raised firebox, dome and Salter safety valve on top and a pillar type safety valve mid-way along the boiler, the inside frames and horizontal outside cylinders were more conventional. Allan introduced his straight link motion on these locomotives. The boiler heating surface was 1049 sq. ft which included a small combustion chamber.

Date	SCR No	Type	Cylinders inches	D.W. ft in	1st CR No		W'drawn
12/1856	54	0-4-2	OC 16 x 22	4 7½	362		1888
5/1857	55	0-4-2	OC 16 x 22	4 7½	361		1885
3/1864	4	2-2-2	OC 16 x 20	6 1	306		1882
9/1864	5	2-2-2	OC 16 x 20	6 1	307		1884
11/1864	16	2-2-2	OC 16 x 20	6 1	308		1884
3/1865	81	2-2-2	OC 16 x 20	6 1	299		1888
5/1865	18	2-2-2	OC 16 x 20	6 1	309		1888
6/1865	82	2-2-2	OC 16 x 20	6 1	298		1888
10/1865	84	2-2-2	OC 16 x 20	6 1	297		1892
11/1865	83	2-2-2	OC 16 x 20	6 1	296		1883
1866	–	2-2-2	OC 16 x 20	6 1	294		1882
1866	–	2-2-2	OC 16 x 20	6 1	295		1883
Became part of the Caledonian Rly in August 1865							
1866	–	2-4-0	OC 17 x 22	7 2	472	Reno 123/76 123A/86 (a)	1888
1866	–	2-4-0	OC 17 x 22	7 2	473	Reno 124/76 124A/86 (a)	1893

(a) Parts sent from SNER Arbroath Works & completed at Perth

Fig. 479 **SCR** 1856 0-4-2 originally SCR No. 54 as Caledonian Rly No. 362

SCOTTISH CENTRAL RAILWAY

No further new building took place until 1864 when a start was made on an order for ten 2-2-2 passenger type. These had the Allan type frames and outside cylinders 16" x 20", and a slightly smaller boiler than the 0-4-2s with 1014·76 sq. ft of heating surface, the firebox being fitted with a transverse midfeather. The boiler pressure was the same at 125 lb/sq. in.

Some parts were used off condemned locomotives but the class was reasonably uniform. They were said to be good steamers. The last two, built in 1866 did not receive SCR numbers as they were now part of the Caledonian Railway. Altogether twelve new locomotives were built at Perth. Allan resigned in 1865 but his front end design was perpetuated for another thirty years. Repairs only were carried out after the amalgamation at the Perth shops and they were later used as a wagon repair shop.

SCOTTISH NORTH EASTERN RAILWAY

Arbroath

This railway was formed by the amalgamation of the Aberdeen, and Scottish Midland Junction Railways in 1856. After the SNERs amalgamation with the Caledonian Railway in 1866 work carried out at Arbroath was transferred to St Rollox and the workshops closed.

In 1859 two locomotives were built at the Arbroath works, numbered 10 and 11, but it is not certain what type they were. The facilities were meagre and it is almost certain that they would have been 'erected' from parts supplied by outside sources.

At the time of the 1866 amalgamation two further locomotives were under construction at Arbroath. They were not completed but transported to Perth where they were finally erected in 1869. They were 2-4-0s and numbered 123 and 124 in the Caledonian list. The outside cylinders were 17" x 22" and the boilers had raised fireboxes, and inside bearings for the coupled wheels and outside for the leading wheels. Thomas Yarrow was the Locomotive Superintendent of the SNER and carried out a number of experiments using coal, instead of coke, as fuel.

SCOTTISH NE RLY–ARBROATH

Built	Type	Cylinders inches	D.W. ft in	Rly Numbers SNE	Rly Numbers CAL R		W'drawn
1859	2-2-2?	16 x 20		10			1870
1859	2-2-2?	16 x 20		11			1878
1866	2-2-2	OC 15½ x 20	6 0	(12)	460		1885
4/1866	2-2-2	OC 16½ x 20	7 0½	–	461	Reno 311/77	1885
1866	2-4-0	OC 17 x 22	7 2	–	472 (a)		1888
1866	2-4-0	OC 17 x 22	7 2	–	473 (a)		1893

(a) Parts sent to Perth Shops (see Scottish Central Rly) for completion

Fig. 480 **SNER** 1866 2-4-0 No. 472 (CR) completed at Perth

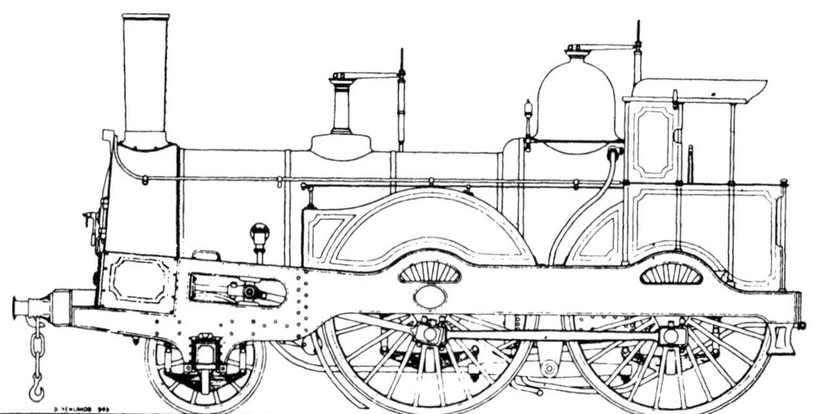

SCOTT SINCLAIR & CO.

Greenock

This well known shipbuilding firm built seventeen locomotives, three for the Caledonian Railway and fourteen for the Scottish Central Railway. The three for the Caledonian Railway were 'Allan' Crewe type 2-2-2s, identical to those built at Crewe, except for minor details, and the actual drawings were supplied by the London & North Western Railway. Robert Sinclair who was jointly superintendent of the Caledonian, and Scottish Central Railway ordered these three locomotives. His uncle was a partner in the firm and had served his apprenticeship in the locomotive shops of the Glasgow Paisley and Greenock Railway moving on to Robert Stephenson & Co. where he obtained further locomotive building experience.

The 2-2-2s had boilers 3′ 6$\frac{3}{4}$″ diameter by 9′ 9″ long with tubes giving 730·88 sq. ft of heating surface and the firebox 49·0 sq. ft. The grate area was 10 sq. ft and boiler pressure 90 lb/sq. in. Stephenson's link motion was fitted and the weight in working order was 19 tons and 28 tons, with a four-wheeled tender carrying 2 tons of coke and 800 gallons of water. The boiler feed pumps were worked off eccentrics.

Nine more of the same type were ordered for the Scottish Central Railway and were dimensionally similar being completed in 1849. At the same time five 0-4-2 goods locomotives were ordered for the same railway, but with inside frames, horizontal outside cylinders and boilers with raised fireboxes. Seventeen locomotives had been built by Scott Sinclair & Co., but building policy concentrated on ships etc., and the locomotive side was dropped.

SENTINEL WAGGON WORKS LIMITED

Shrewsbury

In 1906 the first Sentinel patent steam unit was built and by 1923 over 7000 had been constructed principally for road vehicles. In this year the unit was incorporated in the first Sentinel-Cammell rail car in association with the Metropolitan-Cammell Carriage, Wagon & Finance Co. At the same time the unit was applied to a railway locomotive design for shunting and light goods traffic.

The rail car was intended to combat the severe road competition claiming that the cars could be run at less than half the running costs of a branch line train with locomotive and carriages and lower operating costs per seat-mile than the road omnibus. They were constructed to gauges from 2′ 6″ to 5′ 6″ and the standard designs were:

AA	Single engined	100–125 HP	Single Car
B	Double engined	200–250 HP	Single Car
C	Single engined	100–125 HP	Double Car
D	Double engined	200–250 HP	Double Car
E	Single engined	175–200 HP	Single Car
F	Double engined	350–400 HP	Single Car
G	Single engined	175–200 HP	Double Car
H	Double engined	350–400 HP	Double Car

In the single car the body was carried on two bogies, and in the double car on three, the central bogie being arranged on the 'Gresley' patent system. Horse powers shown above for each type indicate the power at early cut-off, and on intermediate cut-off respectively.

Describing the standard rail cars the design of the coach varied according to the railway's requirements and were built round the engine unit. The water tank was usually placed over the driving bogie and filled through a panel at either

SCOTT SINCLAIR & CO–GREENOCK

Works No.	Date	Type	Cylinders inches	D.W. ft in	Rly No.	Customer	1st CR No.	W'drawn
1	1847	2-2-2	OC 15 x 20	6 0	28	Caledonian Rly	–	1869
2	1847	2-2-2	OC 15 x 20	6 0	40	Caledonian Rly	–	1869
3	1847	2-2-2	OC 15 x 20	6 0	41	Caledonian Rly	–	1870
4	1848	2-2-2	OC 15 x 20	6 0	27	Scottish Central Rly	336	1869
5	1848	2-2-2	OC 15 x 20	6 0	28	Scottish Central Rly	335	1870
6	1848	2-2-2	OC 15 x 20	6 0	29	Scottish Central Rly	334	1869
7	1849	2-2-2	OC 15 x 20	6 0	30	Scottish Central Rly	333	1867
8	1849	2-2-2	OC 15 x 20	6 0	31	Scottish Central Rly	332	1871
9	1849	2-2-2	OC 15 x 20	6 0	32	Scottish Central Rly	331	1872
10	1849	2-2-2	OC 15 x 20	6 0	33	Scottish Central Rly	329	1872
11	1849	2-2-2	OC 15 x 20	6 0	34	Scottish Central Rly	328	1871
12	1849	2-2-2	OC 15 x 20	6 0	35	Scottish Central Rly	327	1872
13	1848	0-4-2	OC 16 x 18	4 7	44	Scottish Central Rly	377	1870
14	1848	0-4-2	OC 16 x 18	4 7	45	Scottish Central Rly	376	1870
15	1849	0-4-2	OC 16 x 18	4 7	46	Scottish Central Rly	371	1877
16	1849	0-4-2	OC 16 x 18	4 7	47	Scottish Central Rly	370	1876
17	1849	0-4-2	OC 16 x 18	4 7	26	Scottish Central Rly	369	1868

side of the car. For single engine units the capacity was 400 gallons and 600 gallons for the double engine unit. Single engine units with articulated cars usually had a tank holding 450 gallons.

Coal bunkers were fixed at the front left hand side opposite the controls with capacities of 15 cwt and 30 cwt for single and double engined units which on an average was sufficient for 120 to 150 miles. On articulated units consisting of two cars the central bogie was arranged on the 'Gresley' patent system.

An interesting car built in 1928 for the Leopoldina Railway was used as an inspection car with a day saloon and pantry at one end, and bathroom and sleeping accommodation at the other end with the boiler in the centre compartment. The coach could be driven from either end.

The power unit consisted of a vertical six-cylinder single-acting engine with 6″ x 7″ cylinders arranged for working with steam at 300 lb/sq. in. The cylinder heads were detachable and incorporated the valve chests and ports. The valves were of the mushroom type with cast iron valve guides, camshaft operated, with case-hardened steel cam followers. The camshafts were driven by spur gearing from the crankshaft. 'Notching up' was done by sliding the camshafts which were connected to the driver's controls by a suitable mechanism. The crankshaft ran in a totally enclosed crankcase. The entire engine unit was suspended from the underframe and the drive transmitted to a gearbox mounted on the driving axle, by means of a Hardy-Spicer cardan shaft with a disc joint at the engine end and a universal joint at the gearbox end.

A spur gear on the layshaft meshed with the final drive pinion pressed on the axle. The standard gear ratios were arranged to give a speed of 30 or 38 miles per hour at 500 rpm of the engine with a maximum speed of 40 to 45 or 50 to 55 mph respectively. The controls were arranged at both ends of the car dispensing with turntable requirements.

The standard 100 HP Boiler consisted of two cylindrical shells, an outer and inner, the inner shell being provided with a series of diagonally disposed corrugations forming landings for three banks of tubes arranged spirally round a central 'chute' through which the fuel was fed into the grate below. The superheater consisted of two coils of solid drawn steel piping lying in the space round the stoking chute between the top tubes and the boiler top. One end connected with the collector steam pipe round the top of the steam space in the boiler and the other with the main stop valve.

The heating surface of the tubes, excluding the superheater, was 73 sq. ft and the grate area 5·1 sq. ft and, on good coal, would evaporate 2300 to 2400 lbs of water per hour at a pressure of 300 lb/sq. in. and a steam temperature of 700°–750°F.

Other boilers were designed for various conditions; the 'Bengal' boiler which was adapted for burning inferior coal in India, also oil fired boilers and for the larger applications for the double engined cars a 200 HP boiler of water tube type with a 27″ diameter steam drum and tubular superheater elements, coal or oil fired.

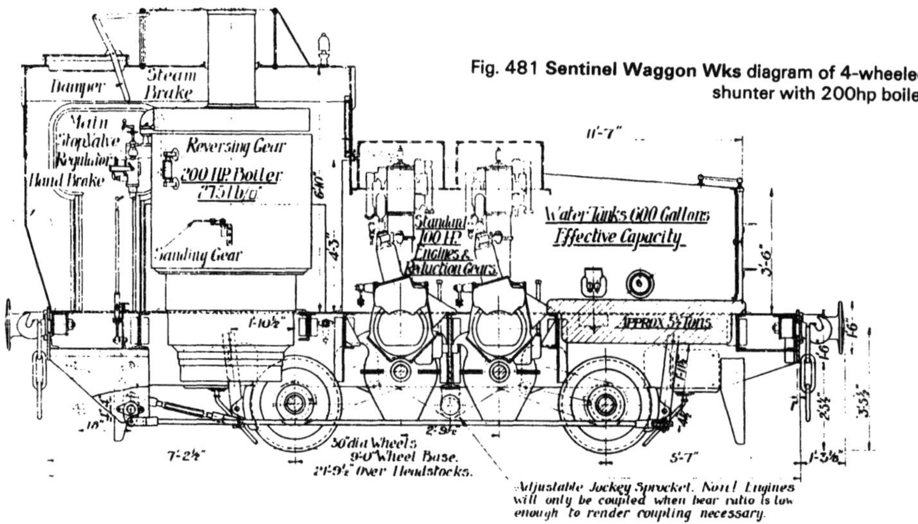

Fig. 481 **Sentinel Waggon Wks** diagram of 4-wheeled shunter with 200hp boiler

These cars had very good acceleration characteristics and were economical in fuel consumption.

The standard Sentinel shunting locomotives had four wheels, with a vertical water tube boiler, fired from the top down the centre flue, was equipped with a superheater and worked at a pressure of 275 lb/sq. in. The first locomotives had horizontally positioned cylinders 6¾" x 9" but from 1925 many were fitted with vertical cylindered engines, the cylinders being the same size and the crankshaft driven at 500 rpm. On the 200 HP class two engines were installed.

The transmission to the wheels was by means of chains, connecting sprockets keyed on to the end of the crankshaft and bolted on to the wheels. The ratio between sprockets varied according to the work to be done. For heavy slow work an additional reduction gear was provided by a spur gear between the crankshaft and a counter shaft on which the driving sprockets were keyed.

The tensioning of the drive chains was effected by radius bars which could move the axles in either direction. Another interesting feature was the braking system comprising the usual hand brake and steam brake and a counter-pressure brake which gave accurate stopping and retardation and was operated by drawing in air and compressing it in the cylinders and exhausting by means of a valve operated by a foot pedal. The fire could be dropped easily and soot and ash in the tubes expelled by a hand operated steam jet.

The majority of the earlier locomotives went abroad to India, South Africa, Egypt, Denmark, Sweden, France, Australia, Ceylon and New Zealand.

In 1925 Sentinel built two steam railcars for the L&NER and one for the LM&SR who subsequently sold it to the Jersey Eastern Railway.

Fig. 482 **Sentinel Waggon Wks** diagrams of 100hp and 200hp boilers

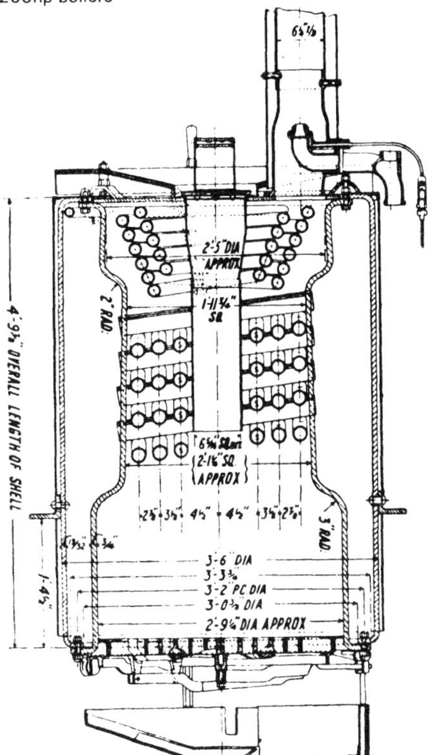

100-H.P. Boiler.

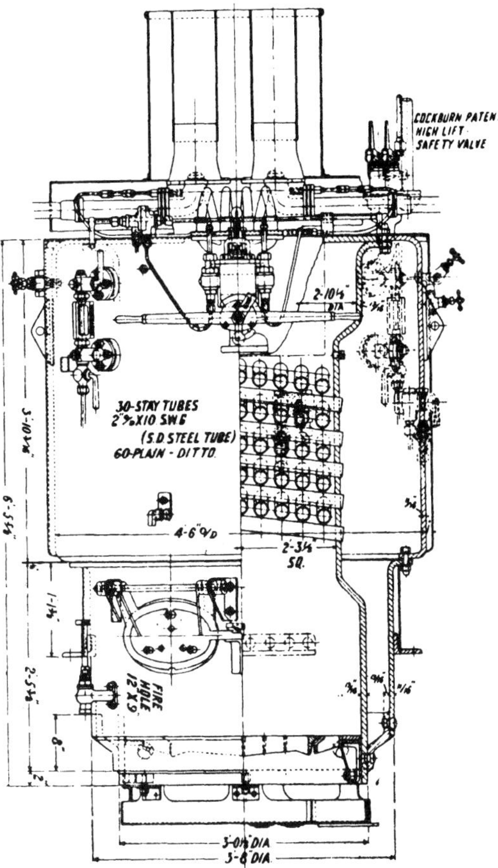

200-H.P. Boiler.

The two supplied to the L&NER were fitted with a jackshaft below, the engine being driven by a chain from the sprocket fixed to the engine crankshaft. From the jackshaft were sprocket and chain drives to the two axles.

In 1926 a unique articulated locomotive was built for the Kettering Coal & Iron Co. Ltd. being, in effect, two standard four-wheeled vehicles back to back. It was expected to effect considerable economies, but apparently it did not come up to expectations and did not do a great amount of work.

In this country the L&NER were the biggest customers having bought fifty-eight of three types: (1) 100 HP 2′ 6″ wheels 1 speed; (2) 2′ 6″ wheels with 2 speeds and larger boiler and (3) double ended 'Super' Sentinel with 3′ 2″ wheels, specially designed as tram locomotives for the Wisbech and Upwell Tramway. Many steam railcars were also supplied to this railway.

The power units were of the Sentinel standard type and the coach body at the driving end was pivoted. Twenty-four of these chain driven cars were supplied in conjunction with Messrs Cammell Laird & Co. of Nottingham, between 1925 and 1928, in which year an improved type of coach was produced with the engine unit mounted rigidly in the body but driving a flexible bogie through a cardan shaft. Only one of these improved coaches was supplied as again in the same year a six-cylinder unit was put on the market with a boiler with an increased pressure of 300 lb/sq. in. and no less than forty-nine of this improved version were supplied to the L&NER between 1928 and 1931. The six-cylinders were 6″ x 7″ and the bogie wheels were 3′ 1″ diameter.

A further type designed for heavy gradients comprised a unit of two sets of six-cylinder engines producing a total of 200 HP. Five were supplied. Finally one articulated twin car set was supplied in 1930 with separate 100 HP units.

In 1933 some compound engines were constructed, including three for railcars ordered by the Colombian National Railways, each car having two six wheel bogies and a Woolnough water tube boiler.

In the 1930s the Doble steam car was introduced. Abner Doble came to Sentinels from New Zealand in 1931 and he brought a steam road car with him for demonstration purposes. He applied his principles to railway traction but the design was extremely complicated and to drive and maintain such a steam locomotive not only required a steam engineer conversant with the design, but an electrical engineer as well, both to be on the locomotive most of the time. The major parts required were: condenser; exhaust steam turbine driving condenser fan; boiler with oil burner, blower, spark plug igni-

Fig. 483 **Sentinel Waggon Wks** 7275/1928 London & North Eastern Rlv. railcar NETTLE

SENTINEL WAGGON WORKS LIMITED

Fig. 484
Sentinel Waggon Wks
9530/1952
National
Coal Board
New Hucknall
Colly.
STIRLING CASTLE

tion, float chamber and booster turbine; by-pass solenoid and valve; electric motor, fuel pump, breaker spark coil, check valves, thermostats and switches; pressure regulators, throttle valve; magnetic steam valves, water pump, air pump; water tank, fuel oil tank, petrol tank; first feed heater, second feed heater; main engine (compound); dynamo and battery.

In 1933 a rail bus was constructed for the Southern Railway, but in this case a standard Sentinel boiler was used, without condenser, the fuel being coal. The engine which had 4½″ and 7½″ diameter cylinders and a stroke of 6″ was nose-suspended on the power bogie and power was transmitted via gearing to the track wheels on one axle.

An automatic stoker was used, operated by a subsidiary steam pump driving the ratchet pawl on the stoker coal screw. The rail bus was used on the Dyke branch and other parts of the system. It was reasonably successful and Sentinels were asked to quote for five more of improved design. The price apparently was too high (far more than the prototype) and the whole project petered out. Others were built for abroad but the cost was not competitive and the conventional two, four and six-cylinder standard engine units were used in most cases.

Whereas the Doble engine was mounted on the axle, the Sentinel had a cardan shaft drive to the axle or chain drive, with gearbox if required.

Rail cars continued to be built and were sent to France, Peru, Belgium, Paraguay, Egypt, Tasmania and many other countries.

Shunting locomotives were also produced for home and abroad, usually with two, four or six vertical cylinders: 6″ x 7″ or 6¾″ x 9″ or 7″ x 9″, and wheels ranging from 1′ 8″ to 2′ 6″ diameter. Building continued to 1958. In 1957 Rolls-Royce paid £1·5 million for the 16 acre Sentinel site and plant, inheriting about a dozen incomplete steam shunters some of which were finished before the reorganisation was completed and diesel locomotives started to roll out of the erecting shop.

In 1957 a gyro-electric locomotive was built in conjunction with the Swiss firm Oerlikon, the principle employed being the kinetic energy of a flywheel driving a generator to supply electric current to the driving motors. Charging points were necessary for starting. Apparently it was not successful, having run trials in conjunction with the National Coal Board, and it was later converted to a diesel hydraulic shunter (nothing to do with steam locomotives, but an interesting experiment).

Approximately 850 steam locomotives were built from 1923 to 1957. Up to works number 9362 all vehicles made by Sentinel were included, but from 9363 of 1945 the series was for railway steam locomotives exclusively except for Nos. 9402 to 9501 which were steam road wagons for the Argentine and 9502 (a steam dumping wagon for Penderyn Quarries, Aberdare) and finally 9503–9510 which were blank. Works numbers ended at around 9650. Many of the boilers were made by Abbott & Co. (Newark) Ltd.

SHANKS, ALEXANDER & SON LIMITED

Dens Iron Works, Arbroath, Forfarshire

Established in 1840 this firm built, principally, portable steam engines, combined engines and boilers of the semi-portable type, iron bridges, steam cranes and in the 1870s some four-coupled saddle tank locomotives.

In their catalogue, around the early 1870s this type of tank engine was put forward for use on branch lines at docks, collieries as well as for contractor's purposes. With slight modifications they could be used for colonial estates, and would prove invaluable where horse-power was insufficient. The saddle tank could be enlarged when intended for foreign service. Four- and six-wheeled locomotives could be supplied. Copper fireboxes and brass tubes were standard and the boiler capacity would admit wood or

Fig. 485 **A. Shanks & Co.** Page from firm's catalogue

Fig. 486 **A. Shanks & Co.** 1877 London & South Western Rly No. 108 COWES

other fuels being used where coal was unobtainable. The boilers were tested to 250 lb/sq. in. with a working pressure of 120 lb/sq. in. A pump and injector were fitted, and spark arrester and canopy was available if required.

Their main customers were the Southampton Dock Co. and contractors. Eleven locomotives can be traced; three were supplied new to the London & South Western Railway for working the Southampton docks. At least two were equipped with condensing apparatus which was particularly efficient. The 10¼" x 20" outside cylinders were horizontal and on the centre line of the wheels, which made them low slung. They had square sided saddle tanks holding 400 gallons and 3 ft diameter wheels which had cast iron H section spokes. The connecting rods were circular with cottered brasses.

Four others had been sent abroad to a contractor engaged on Cuxhaven harbour and on their return were bought as indicated. They were slightly smaller and had neater rounded saddle tanks. The condensing apparatus on the Southampton dock locomotives may have been fitted by the railway company.

In the firm's catalogue the drawing of their standard tank locomotive has very little in common with those known to have been built. Although not traced they built some vertical cross tube boilered locomotives in the 1870s with 3:1 spur gearing.

Works No.	Built	Type	Outside Cylinders inches	D.W. ft in	Customer	No. Name	W'drawn
	1/1872	0-4-0ST	10 x 20	3 0	Southampton Dock Co	Sir Bevis	1893
	7/1872	0-4-0ST	10 x 20	3 0	Southampton Dock Co	Ascupart	1893
	1872	0-4-0ST	10 x 20	3 0	Southampton Dock Co	Arbroath (a)	
	1872	0-4-0ST	10 x 20	3 0	London & St Katherine Dk. Co.	9 Victoria (b)	Sold 1901
	1872	0-4-0ST	10 x 20	3 0	London & St Katherine Dk. Co.	10 Albert (b)	1902
	1873	0-4-0ST	10 x 20	3 0	Millwall Dock Co	Cuxhaven (c)	Sold 1912
468	1873	0-4-0ST	10 x 20	3 0	London & South Western Rly	Ritzebuttel (d)	Sold 1915
	9/1876	0-4-0ST	10¼ x 20	3 0	London & South Western Rly	Southampton	Sold 1915
	1877	0-4-0ST	8 x		Otago (N.Z.) Rlys	A 6 Mouse	
	1877	0-4-0ST			Otago (N.Z.) Rlys	Kangaroo	
	11/1877	0-4-0ST	10¼ x 20	3 0	London & South Western Rly	Cowes	Sold 1915

(a) This locomotive not confirmed. (b) Bought second hand 12/1878. (c) Bought second hand 4/1879. (d) Bought second hand 12/1879.

The second hand locos were bought at auction at Southampton. They had been used on Cuxhaven Harbour contract.

SHARP ROBERTS & CO.

Atlas Works, Great Bridgewater Street, Manchester

Fig. 487 **Sharp Bros.** Manchester 1847 drawing of typical early 2-2-2

Originally established in 1828 by Thomas Sharp and Richard Roberts, they began by constructing all kinds of machinery used in the cotton spinning mills in the vicinity, also machine tools such as planing, punching and shearing machines, lathes and drilling machines. A few stationary steam engines were also built.

Demand for railway locomotives was realised and the first one was designed and built in 1833. This was Liverpool & Manchester Railway No. 32 EXPERIMENT and ran on four wheels as a 2-2-0 with outside bearings for the leading wheels and inside bearings for the driving wheels. The 11″ x 18″ cylinders were placed vertically over the leading wheels and the crossheads were connected to the crankpins by connecting rods and bell crank levers. There were no eccentrics and the valves were operated by a short lever on the bell crank giving the travel by means of a long rod extending to the rocking gear placed on the footplate. The valves were piston type with no lap or lead. The 11″ diameter pistons had no rings and the working surfaces were of white metal. This locomotive was unsteady at 'speed' and of course leakage past the pistons and valves was inevitable. Within a year the position of the cylinders and bell cranks was altered and a pair of trailing wheels was added. Three very similar locomotives were built for the Dublin & Kingstown Railway and these gave considerable trouble with steam leakages, axles fracturing and springs breaking. When in good mechanical order they had a good turn of speed but the track was not strong enough to withstand sustained high speeds.

After their experience with these four locomotives the vertical cylinders and bell crank levers were abandoned and a new 2-2-2 was designed with the inside cylinders under the smokebox, outside frames to all wheels with additional inside bearings for the driving axle and the inside framing was extended to the front of the firebox to which it was attached. A combined dome and safety valve was fixed on the boiler close to the chimney – a Sharp feature which was in evidence on their locomotives for many years. An additional safety valve was fixed over the firebox. The leading and driving springs had screw links for weight adjustment. Without the tender the locomotive weighed 12 tons 5 cwt empty. Ten were supplied to the Grand Junction Railway in 1837 and so began a long line of 'single' express locomotives, becoming an early standard product, and some 600 were built from 1837 to 1857 with cylinders ranging from 12½″ to 17″ in diameter and driving wheels from 5 ft to 6 ft. A well tank version was also built together with 0-4-2, 2-4-0, 0-6-0 and 0-4-0ST types, but the 2-2-2 was the principal product and on many railways was the mainstay of passenger services. Many Sharp singles were delivered to France, Belgium, Holland, Germany, Russia, Italy and Spain.

In 1843 Roberts left the firm and the title was changed to: SHARP BROS. Among the

Fig. 488 Manchester 1774/1867 0-8-0ST for Mauritius Govt. Rly. later named CYCLOPS

Fig. 489 Manchester 2062/1870 2-4-0 for Rybinsk & Bologae Rly, Russia No. A3

Fig. 490 Manchester 2987/1881 0-6-0 for Royal Sardinian Rly. UMBERTO 1°

locomotives built in this period were twenty of the famous 2-2-2 *Bloomer* class for the Southern division of the London & North Western Railway during 1851 and 1852. In 1852 the senior partner, John Sharp, retired and Charles Patrick Stewart took his place and the title was again changed, to: SHARP STEWART & COMPANY.

Thomas Sharp also retired shortly after and was succeeded by Stephen Robinson. In 1864 the firm became a limited company and Mr Stewart remained Chairman until his death in 1882. The firm had acquired the sole rights for manufacturing Giffard's patent injector in 1860. The positioning of the dome on the boiler at the front end close to the chimney was done to reduce priming, which was a common complaint in those early days. The early 'piston' valve, developed and patented by Roberts, was a failure, due principally to no rings being employed, and the idea was abandoned.

It was not until 1862 that anything larger than 0-6-0s and 2-4-0s were built, but in this year, the first of three 4-6-0ST appeared for the Great Indian Peninsula Railway designed for the Bhore and Thul Ghatt inclines by Mr J. Kershaw. It had two inside cylinders 20″ x 24″, 4′ 4″ diameter coupled wheels, a tank capacity of 1050 gallons and sledge brakes. Two similar locomotives but of 2-6-0ST formation were built the following year.

Another interesting class was six 0-8-0 tender locomotives for the Oudh & Rohilkund Railway with 16½″ x 24″ cylinders, 4′ 4″ diameter wheels, and intermediate driving shaft, built in 1865–8.

In 1872 some 4-2-4STs were built for the Lisbon Steam Tramways with 11″ x 18″ cylinders and 3′ 9″ diameter driving wheels. They were unique in that the leading and trailing wheels engaged on a mono-rail whereas the driving wheels drove off the roadway. A total of sixteen were built.

Besides locomotive building, the firm carried on the business of iron and brass foundry work, machine tools and iron merchants. A full order book made it imperative to expand and as this was not possible in their present works they decided to move to Glasgow where the Clyde Locomotive Works was for sale. The last locomotive completed at Manchester was a 0-6-0ST for the Earl of Ellesmere named WARDLEY, Works No. 3423, and the total output of the Manchester factory was 3442.

Fig. 491 **Sharp Stewart,** Glasgow – Atlas Works in 1912 – South African Rlys 4-8-2s under construction

SHARP STEWART & CO. LTD.

Atlas Works, Glasgow

Fig. 492
Sharp Stewart,
Glasgow
4300/1897
Midland & South Western Jnction Rly No. 17

Fig. 493
Sharp Stewart,
Glasgow
4462/1899
Demarara Rly (E. Div.) No. 9
ABARY

The Clyde Locomotive Works having been taken over and reorganised, it was not long before the output exceeded that attained at Manchester. The name Sharp Stewart was already famous all over the world in the locomotive field.

Early developments at Sharps had been a form of wheel centre with H section spokes used for a number of heavy goods locomotives and they were early users of cylinders with the valves and chest located underneath.

A number of compounds were built for the Argentine Central Railway, some 4-4-0s in 1889 with 15″/21½″ x 22″ cylinders and 4′ 6″ diameter driving wheels and 2-8-0s with 16″/23″ x 24″ cylinders and 3′ 6″ diameter driving wheels.

A large order from the Midland Railway was executed in 1892–3 comprising thirty-five 4-4-0s having two cylinders 20½″ x 26″ and 7′ 0½″ diameter driving wheels and forty 0-6-0s with 18″ x 26″ cylinders and 5′ 3″ diameter wheels.

In 1894 the first 4-6-0s for Great Britain were built, the famous 'Jones Goods' for the Highland Railway, with two outside cylinders 20″ x 26″ and 5′ 3″ diameter driving wheels. They were not, by any means, the first 4-6-0s built by Sharp Stewart, their first being for the Buenos Ayres, Ensenada & South Coast Railway in 1884.

The first of many orders from South African railways came in 1895 from the Cape Government Railway for twelve 4-8-0Ts for the 3′ 6″ gauge and thirteen 4-8-0 tender locomotives with two outside cylinders 17″ x 23″ and 3′ 6⅜″ diameter coupled wheels.

To list all the customers would be tedious, in fact it would be easier to list those places where Sharp Stewart locomotives could not be found. Most of the home railways were supplied and many small industrial saddle tanks were sandwiched in between 0-8-0Ts for the Bombay, Baroda & Central India Railway and 4-4-0s for the Indian North Western Railway, and 0-6-0Ts for Imperial Japanese Railway rubbing buffers with 2-6-2Ts for Spain.

In 1903 the 'big three' of Glasgow, Neilson Reid & Co., Dubs & Co., and Sharp Stewart & Co. Ltd. came to a final agreement to form one, all embracing, company to be known, from 12 February, 1903, as the NORTH BRITISH LOCOMOTIVE CO. LTD. and Sharp Stewart & Co., Ltd. would retain the name 'Atlas Works'.

From their beginning as Sharp Roberts & Co. to the time of amalgamation they had built 5088 locomotives made up as follows:

1st works numbers: Letters A to U	21
Works numbers	5083
	5104
Less 4 built by Hunslet, 4 built by Yorkshire Engine Co and 8 blank numbers (3629–3636)	16
Total	5088

See NORTH BRITISH LOCOMOTIVE CO. LTD.

SHELTON IRON STEEL & COAL CO. LTD.

Stoke-on-Trent

Two locomotives were built in the firm's workshops; the first one in 1911 No. 14 SHELLINGFORD and in 1912 No. 15 GLENALMOND. SHELLINGFORD was a 0-6-0ST with inside frames and cylinders. It had strong Peckett characteristics and some parts were probably supplied by this firm. Similarly GLENALMOND, a 0-4-0ST, had strong Scottish features and it is probable that the locomotive was erected from parts supplied by Andrew Barclay Sons & Co. Ltd. of Kilmarnock.

Fig. 494
Sharp Stewart, Glasgow
4630/1900 Cape Govt. Rlys.
No. 260 4-6-0 3′ 6″ gauge

Fig. 495 **Shelton**
1911 0-6-0ST SHELLINGFORD

SHEPHERD & TODD

Railway Foundry, Leeds

Mr Todd left Messrs Todd, Kitson and Laird and with Mr Shepherd set up the 'Railway Foundry' in 1838 to manufacture locomotives, carriages and wagons. They received the first order in 1839 for two locomotives without tenders.

In 1840–1 a number of 2-2-2s were built for the North Midland Railway and four-coupled goods for the Manchester & Leeds Railway. In the same period two of an unknown type were sent to France for the Paris-Orleans Railway.

In 1840 two 6 ft singles were built for the Hull & Selby Railways to Mr John Gray's designs. They had inside cylinders 13″ x 24″ with two domes with spring balance safety valves on top, one dome on the centre of the barrel and one on the raised firebox. The collecting pipe from the front dome was joined to the main pipe from the back dome. They had outside frames and the motion was Gray's patent expansion gear, known as the 'horse-leg' motion and were probably the first two locomotives to use expansion gear, except one Liverpool and Manchester Railway locomotive No. 46 CYCLOPS built by the Haigh Foundry in 1836 and fitted experimentally with Gray's gear. Two 0-6-0s, also of Gray's designs, were built in 1844 for the Hull & Selby Railway with 16″ x 24″ cylinders and 5′ 6″ diameter wheels. The cylinders were set low under the smokebox, inclined so that the piston rods were under the leading axle. Trouble was experienced with the cylinders working loose as a satisfactory attachment could not be made to the frames. Two 'singles' for the York & North Midland Railway – No. 9 ANTELOPE and No. 10 ARIEL, built in 1840, were also fitted with Grays' motion.

Todd left the firm in June 1844 and Shepherd was joined by E. B. Wilson in 1845 but the latter only stayed a year. The firm was taken over by James Fenton in 1846 becoming FENTON CRAVEN & COMPANY.

In this period six 2-2-2s were built for the East Lancs Railway (1846–7) with 15″ x 20″ outside cylinders and inside frames with 5′ 6″ diameter driving wheels. Four of these were rebuilt in the Bury shops of the ELR, first as 2-4-0s and then 0-6-0s. The majority of locomotives built by Fenton Craven were Stephenson's 'long boiler' type with all wheels in front of the firebox. For passenger work a number of outside cylinder 2-4-0 long boilered types were built, including ten for the York & North Midland Railway, five for the Eastern Counties Railway and a few others. For passenger work they were a failure, mainly due to the overhang at each end, particularly at the rear causing violent oscillations resulting in frequent derailments. The six-coupled goods locomotives built to the same patent were not so troublesome, the cylinders being inside and speed not being a requirement.

Further passenger locomotives were built with single drivers 6 ft diameter and outside cylinders. Some went to the London & North Western and Eastern Counties railways.

The partnership of Fenton and Craven lasted less than a year and at the end of 1846 Mr E. B. Wilson appeared once more and took over the business completely, retaining Mr Fenton as Works Manager. The title of the company now became E. B. WILSON & COMPANY. In some instances the maker's plates bore the name 'The Railway Foundry, Leeds', Wilson's name not being shown.

Plans were implemented to increase the capacity of the works in building locomotives from about ten per annum to over fifty and extensions were put in hand including a new

Fig. 496 **Shelton** 1912 0-4-0ST GLENALMOND

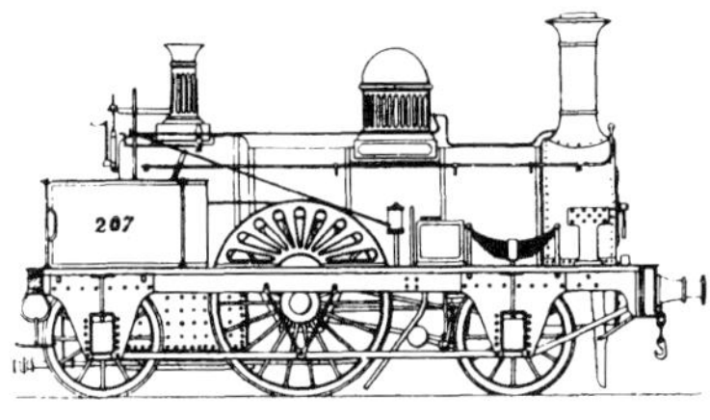

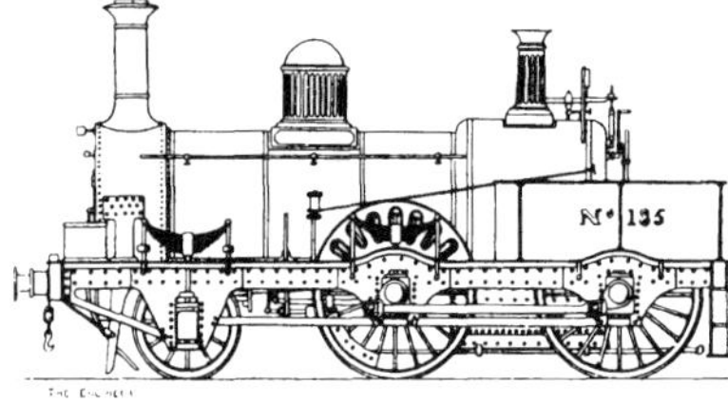

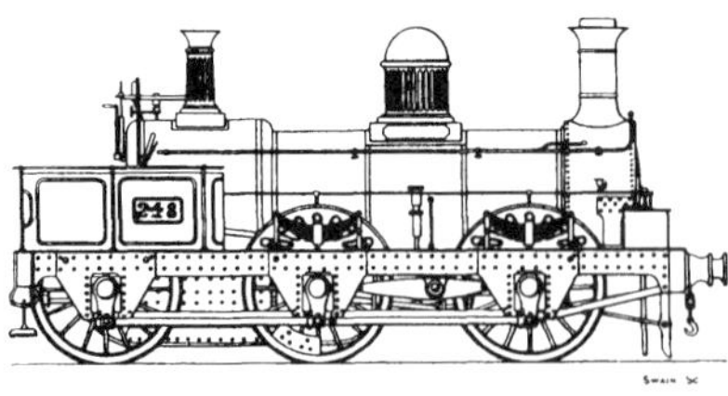

Fig. 497 **Shepherd & Todd** diagrams of typical locomotives built

erecting shop with twelve pits. Building for the home market certainly increased and, as the firm's reputation for good workmanship and reliability became known over a wider area, orders multiplied.

In 1847 the most famous design was evolved in the form of a 'single' express passenger locomotive known as the *Jenny Lind* type and was a development of John Gray's singles for the Brighton Railway with the characteristic 'mixed' framing with outside bearings to the leading and trailing wheels only. There has been much controversy as to who was actually responsible for the design. David Joy was chief draughtsman and if one refers to that interesting series of articles in the *Railway Magazine* 'Some links in the evolution of the Locomotive' the part published in June 1908 indicates that Joy claimed to have carried out all the design work and drawing. Other claims were for Wilson himself, and James Fenton, but whoever it was, the *Jenny Lind* type was extremely successful. The average dimensions were inside cylinders 15″ x 20″ with large and smooth exhaust passages, 6 ft diameter driving wheels, a boiler with a heating surface of 800 sq. ft which was increased later to 1000 sq. ft.

The boiler pressure was 120 lb/sq. in., a big increase over the normal boilers of the time and this made a large contribution to the locomotive's success. The weight in working order was 24 tons and with tender 39 tons 14 cwt. The dimensions of cylinders and boiler were increased from time to time to suit heavier traffic conditions. Over seventy were built, and no less than twenty-four were sold to the Midland Railway.

The fluted dome cover and safety valve casing became almost a trademark of Wilson's locomotives and the design of these fittings are attributed to the manager's wife, Mrs James Fenton. The locomotives were highly finished with mahogany strips around the boiler, polished and varnished. The feed pumps were driven off an eccentric fixed to the outside of the driving axle. Front and rear springs were of thick slabs of Indian rubber in circular cast iron pots. They were later changed to leaf springs as it was found that the rubber lost its elasticity very quickly.

Outside frames were provided for the leading and trailing wheels, and inside frames for the drivers. The inside frames stopped short of the firebox to enable a wide firebox to be used. The inside cylinders had the slide valves between them, operated by Stephenson's motion.

David Joy left in 1850 to take charge of the Nottingham and Grantham Railway which had just opened.

There is no doubt that the *Jenny Lind* 2-2-2 and 2-4-0s that followed became established types for English passenger locomotives in the 1850s and 1860s.

Joy was later appointed Locomotive Superintendent of the Oxford, Worcester & Wolverhampton Railway but resigned in 1856 and returned to the Railway Foundry. His valve gear was patented in 1879 and F. W. Webb of the London & North Western Railway was one of the first to use it in 1880.

The largest *Jenny Lind* was the SALOPIAN, built for the Shrewsbury and Birmingham Railway in 1849, with 15½″ x 22″ cylinders, and a boiler having 1271 sq. ft of heating surface.

Besides new construction Wilson contracted for repairs and rebuilding and many belonging to the Midland Railway were rebuilt at the Railway Foundry as Derby had not the capacity to deal with all its stock in the 1850s and 1860s.

In 1853 locomotives were supplied to the Hanoverian State and West Flanders railways followed by orders from Denmark, India, Belgium, Spain, Portugal and Ireland.

E. B. Wilson and Company were noteworthy in developing standardisation to a marked degree and between 1848 and 1858 the majority of locomotives built to 2-2-2, 2-4-0 and 0-6-0 wheel formations were to a standard pattern and any modifications required by a customer were charged for. The standard 0-6-0 goods had double frames with outside cranks, inside cylinders 16″ x 24″ and wheels 5′ to 5′ 3″ diameter. The outside frames were the sandwich pattern with oak planks between iron plates, the inside frames extending only as far as the front of the firebox. Approximately 160 of this type were built and sold to many home railways and abroad.

The standard 2-4-0 was modified when Archibald Sturrock ordered fifteen with outside bearings to all wheels in 1851. The standard sizes for this type had 16″ x 20″ cylinders and 5′ 9″ coupled wheels, or 16″ x 22″ and 6 ft wheels, or 15½″ or 16″ x 22″ and 6′ 6″ wheels.

The Railway Foundry built few tank locomotives but a number of six-coupled tanks were built with 15″ x 20″ inside cylinders, inside frames and 4′ 8″ diameter wheels. The tanks were between the frames beneath the coal bunker. Many locomotives of standard types were built for stock, which involved a large amount of capital being tied up when customers were not forthcoming.

In addition to these standards a few specials were built including a number of 'Crampton' patent locomotives, including some four-wheeled tanks with a wheelbase of 10′ 6″ and driven by a dummy crankshaft fixed mid-way, and from inside cylinders. One was supplied to the York & North Midland Railway and three to the Nottingham & Grantham Railway.

Some 0-4-0WTs were built in 1852 for the contractors building the Portland Breakwater with 10½″ x 17″ inside cylinders, inside sandwich frames and 4 ft diameter wheels. One was bought by the Torbay & Brixham Railway in 1868.

John V. Gooch designed a 2-4-0T for the South Slesvig Railway of which six were built in 1854. The largest order received was for twenty-five 0-4-2s and ten 2-2-2s for India and many more orders were received from this country. In 1857 twenty goods locomotives were built for the Madrid, Saragossa and Alicante Railway.

E. B. Wilson & Co. manufactured other engineering goods such as pumping engines, carriages, wagons, ironwork etc., and also entered into contracts to provide motive power for the smaller railway companies such as the Oxford, Worcester & Wolverhampton, and the East & West Yorkshire railways. This placed the company in financial straits and on the death of Mr Wilson in 1856 a manager was appointed – a Mr Alexander Campbell who came from the locomotive building department of Scotts of Greenock. However, due to an action in Chancery, the works were forced to close down in 1858 despite the large number of orders which were placed with the firm. As a railway locomotive works it had become one of the most up to date in the country.

Perhaps the most well-known type built was the 0-6-0 inside cylinder double frame goods and possibly over 160 were built to a standard design. The last to be built were for the Newport, Abergavenny and Hereford Railway and four more were built to the same design for the North Eastern Railway by Manning Wardle & Company who took over part of the workshops in 1862 and Hudswell Clarke & Rogers who took other parts of the factory in 1860. It was Manning Wardle & Company who took over the goodwill and patterns.

In 1844 E. B. Wilson employed some forty men and in 1845 considerable expansion took place and by 1847 the number of workmen had increased tenfold, and at that time it was expected to turn out seventy-five locomotives per annum. The majority of locomotives built were 2-2-2, 2-4-0, and 0-6-0 types and approximately 635 were built at the Railway Foundry. They bore works plates with works numbers but very few of the earlier ones have been recorded. The last two built (Nos. 634 and 635) were for the New South Wales Railways, one was a 2-2-2WT and the other was a 2-2-2 with 5′ 9″ diameter driving wheels.

SHILDON WORKS CO., THE

See STOCKTON & DARLINGTON RAILWAY.

SHOTTS IRON COMPANY
Shotts

One 0-4-0ST with outside cylinders is reported to have been built in 1900 for the company's own use but no details can be obtained. In 1968 it was reported to be out of use at Monktonhall Colliery, Millerhill, Midlothian.

SILLEY COX & CO. LTD.
Falmouth

This firm's main business is ship repairing and engineering and as a further locomotive was required in 1918 by the Falmouth Docks and Engineering Company one was built on similar lines to those already supplied by Sara & Burgess who had gone out of business by this time. The locomotive took the form of a vertical boiler mounted on a wagon type of chassis. The cylinders were mounted vertically on the boiler barrel. The four wheels were coupled and the drive was probably geared. The firm was established as Cox, Farley & Company in 1868.

Fig. 498 **Silley Cox** 1918 diagram of 0-4-0VB locomotive

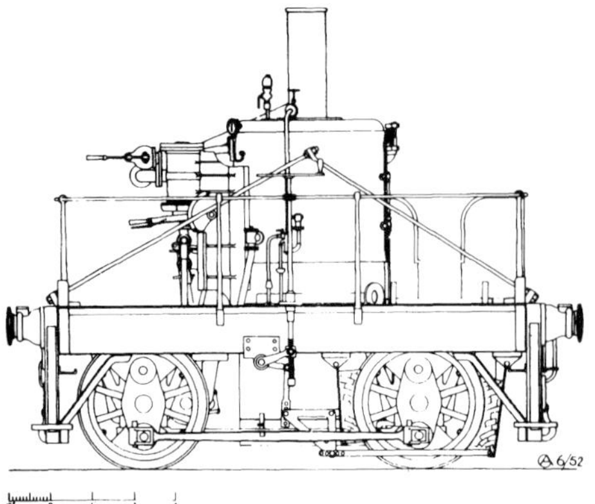

SIMPSON & COMPANY
Aberdeen

At least six 0-4-2s were built by this firm, four for the Aberdeen Railway and two for the Stirling & Dunfermline Railway. The latter were larger with 15" x 21" cylinders and double frames. The period of building was from 1845 to 1853 but the list may not be complete. There was probably a connection with this firm and Blackie & Company.

SIMPSON JAMES & COMPANY
101 Grosvenor Road, Pimlico, London

This firm of engineers specialised in pumps and pumping installations especially for water supply undertakings at home and abroad. The firm transferred to Newark-on-Trent and in 1917 a new company was formed combining the interests of the Worthington Pump Company with their own, its title becoming WORTHINGTON SIMPSON LTD.

One locomotive is known to have been supplied by the firm for the Southampton Dock Company in 1866. It was a 0-4-0ST named OSBORNE and was built by Henry Hughes of Loughborough and delivered to Southampton in September 1866. Simpson acted only as agents.

SIMPSON & CO.
Ratcliffe

See MILLER & BARNES.

SIMPSON & CO ABERDEEN

Date	Type	Cylinders inches	D.W. ft in	Railway	No	
1845	0-4-2	13 x 18	5 0	Aberdeen	6	SNE 53
1846	0-4-2	13 x 18	5 0	Aberdeen	7	SNE 54
1848	0-4-2	13 x 18	5 0	Aberdeen	20	SNE 67
1848	0-4-2	13 x 18	5 0	Aberdeen	24	SNE 71 then 40
1853	0-4-2	15 x 21	5 0	Stirling & Dunfermline (a)		E & GR 60 then NBR 247
1853	0-4-2	15 x 21	5 0	Stirling & Dunfermline		E & GR 61 then NBR 248

(a) Double frames

W. SISSON & CO. LTD.

Sisson Road, Gloucester

Steam power units were supplied by this firm in 1911 for the Cardiff Railway's two steam rail motors. Three units were supplied one for a spare. The boilers were supplied by Messrs Abbott & Co. Ltd. of Newark and the carriage bodies by the Gloucester Railway Carriage and Wagon Co. Ltd. The units had two outside cylinders 12″ x 16″ and 4′ 0″ diameter coupled wheels, and the boiler's heating surface was 660 sq. ft with a grate area of 11·5 sq. ft. The Sisson's works numbers were 971 to 973.

SKINNINGROVE IRON CO. LTD.

Carlin How

One locomotive named MARY was built in the company's workshops in 1913, and was a 0-4-0ST. The frames and buffer beams were bought in plate form, cylinders supplied by Pecketts, and the general lines were similar to Peckett's standard designs. A new feature was introduced in the form of a steel casting instead of the usual fabricated box member between the frames and to which the outside cylinders were fixed, by means of fitted bolts. All the link motion, side rods, pistons and piston rods, valves and spindles were forged and machined at Skinningrove. The boiler was bought out but the cab, smokebox, footplate etc. were fabricated in the workshops.

As a matter of interest, soon after its trial trip, the engine and a slag ladle went off the end of the slag tip. No lives were lost but the engine needed extensive repairs, sooner than anticipated. The weight of the engine was $24\frac{1}{2}$ tons, boiler pressure 140 lb/sq. in. and cylinders 14″ x 20″. The locomotive was derelict in 1954 and was broken up in 1955.

Fig. 499 W. Sissons & Co. 1911 engine unit for Cardiff Rly railmotor

SLAUGHTER & COMPANY
Slaughter Gruning & Company, Bristol

See STOTHERT SLAUGHTER & COMPANY.

SLEE & COMPANY
Earlestown

In *Mineral Railways* by R. W. Kidner (Oakwood Press) a 0-4-0VB is shown built by this firm for the 2′ 3″ gauge Plynlimmon and Hafan Railway in 1897 and named VICTORIA. This is the only indication of locomotive building by the firm, and they were probably agents.

SMITH A. & W. & COMPANY
Eglinton Engine Works, Glasgow

In the 1870s this firm advertised shunting tank locomotives as shown. No further information is available.

SMITH, JOHN
Thornton Road, Bradford, Yorks.

As far back as 1837, Smith & Richardson are shown as ironfounders in Thornton Road, and by 1842 the name of the firm was John Smith and in 1845 the firm was listed under Engineers & Millwrights. One locomotive was supplied in 1840 to the Grand Junction Railway (GJR No. 60 TANTALUS) but in the Haigh Foundry building list for 1840 a locomotive was built for Mr J. Smith for testing some improvements and was then sold to the Grand Junction Railway as above. It was a 2-2-2 with 13″ x 20″ cylinders, 5′ 6″ diameter driving wheels and a boiler heating surface of 522·42 sq. ft. From this it seems that John Smith was acting as an agent, or financed the building of a locomotive incorporating 'improvements' – but what the improvements were is not known.

SMITH, JOHN
Village Foundry, Coven

They were general engineers, building engines for road traction, portable types, winding engines, boilers, power presses and small locomotives, four- and six-coupled tanks for narrow gauge lines in local collieries and ironworks. About 1861 Smith was joined by Mr J. B. Higgs and in 1862 the agricultural side of

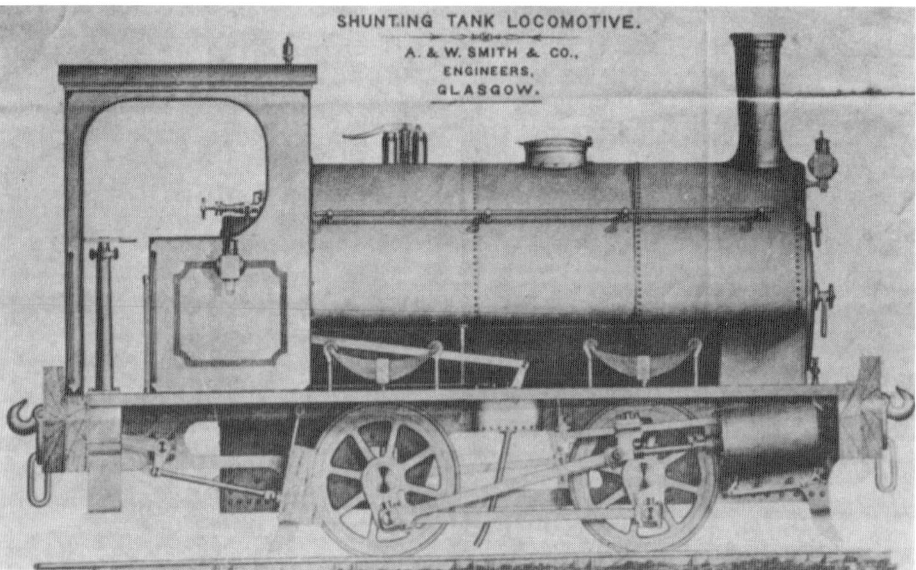

Fig. 500 **A. & W. Smith** sketch of 0-4-0ST

the business was largely dropped in favour of the industrial plant manufacture. In the same year the firm was approached by Fletcher, Solly and Urwick of Willenhall Furnaces and Smith obtained an order for a suitable locomotive for their work.

Between twenty and thirty locomotives were built, but demand became negligible by 1874 and the works were sold by auction on 30 March, 1874. Among the contents were 'locomotive and other engines'. The foundry buildings were retained and the firm as Smith & Higgs continued as threshing and ploughing contractors.

SMITH, JOHN

Village Foundry, Coven

No specific details can be obtained regarding the majority of locomotives built. One of a batch of six geared locomotives is described in *The Chronicles of Boulton's Sidings* by A. R. Bennett and also diagrams of the LION geared engine and the 0-4-0T No. 122. John Smith allocated works numbers from 101 and approximately twenty-five were built.

SMITH & HIGGS

See SMITH, JOHN – COVEN.

Fig. 501
John Smith
1860s North
British Iron Co.
PORTOBELLO

Wks No	Date	Type	Cylinders inches	D.W. ft in	Gauge ft in	Name	Customer
101	1/1863	0-4-0ST	OC 6½ x 12	2 6	2 6 (c)		Fletcher, Solly & Urwick
102	7/1863	0-4-0T	OC 10 x 14	2 6	2 6		Fletcher, Solly & Urwick
103	9/1863	0-4-0T					Chillington Iron Co
		0-4-0T					Chillington Iron Co
		0-4-0T					Chillington Iron Co
		0-4-0T					Chillington Iron Co
					2 3½		Earl of Dudley
(a)		0-4-0WTG	7 x 12	2 6	2 3½	Lion	Earl of Dudley
		0-6-0T	16 x 22	3 6	Std	Pelsall	Pelsall Coal & Iron Co Ltd
		0-6-0T	16 x 22	3 6	Std	Victor	Pelsall Coal & Iron Co Ltd
		0-4-0T	IC		3 2¼	Portobello	New British Iron Co Ltd
	c. 1869	0-6-0T			Std	Success	W. Harrison Ltd. Brownhills
	c. 1874	0-4-0T	OC		*c.* 2 6	President	Ashmore Park Coll Co Ltd
		0-4-0T	OC 10 x			Vulcan	G. & R. Thomas Ltd. Bloxwich
		0-4-0T	OC 10 x			Ajax	
(a)		0-4-0WTG	7 x 12	2 6			North East Colly
122		0-4-0T	OC 10 x 16	2 6			(b)

Also (un-identified): One hired to LNWR Bushbury. One sent to Stafford. One sent to an exhibition.

(a) It is understood that at least six geared locomotives were built gearing ratio 5:1. LION was bought second hand by I. W. Boulton. (b) Bought second hand by I. W. Boulton. (c) Wheelbase 4' boiler 2' 4" diameter pressure 100 lb/sq in grate area 4.28 sq ft.

SOMERSET & DORSET JOINT RAILWAY

Highbridge, Somerset

The locomotive building and repair policy after the formation of the Joint railway was that the Midland Railway would deal with heavy repairs, rebuilding and renewals, and the Highbridge works would undertake running repairs and overhauls. Derby's Locomotive Superintendents and later the Chief Mechanical Engineers acted for the S&DJR with a resident Locomotive Superintendent at Highbridge from 1875. The men at Highbridge were: 1875–83 B. S. Fisher; 1883–9 W. H. French; 1889–1911 A. H. Whittaker; 1911–13 M. F. Ryan; and 1913–30 R. C. Archbutt.

From 1930, when the locomotives were taken into LM&SR stock, the post of Locomotive Superintendent was abolished, and the works closed in May 1930, having served the railway since 1862.

For a small railway the workshops were well equipped. The erecting shop had two parallel repair pits and locomotives were lifted and positioned in a curious fashion. Having only one travelling crane, the leading end of the locomotive was lifted on to a small carriage running on rails, braced by girders and inclined. The trailing end was then lifted and the front end was moved over to the pits, at right angles to them by means of the inclined rails and the movement of the crane at the other end.

Officially three locomotives were built at Highbridge, but it is probable that many parts were used from other locomotives which they replaced. All three were intended for working the colliery branches. The first appeared in 1885 as a 0-4-2ST, numbered 25A, with outside cylinders and a saddle tank.

Mr W. H. French unfortunately died as a result of an accident at Highbridge and Mr Alfred H. Whittaker took over the position. As with Mr French, he was a 'Derby' man having been a pupil of Matthew Kirtley. He was responsible for the re-organisation of Highbridge works. The 4-4-0 was introduced, and he developed a table exchange apparatus, for single line working, which was fitted on the locomotive tender. The same apparatus was used on the Midland and Great Northern Joint Railway and also the Great Western Railway.

Two 0-4-0STs were built during his term of office, in 1895, numbered 26A and 45A. They were officially classed as rebuilds, a number of parts being used off older locomotives. No. 45A replaced an old tank locomotive named BRISTOL which had been bought second-hand in 1883 from an unknown source. Both had outside cylinders 10″ × 15″, 3 ft diameter wheels, and saddle tanks over the smokebox and boiler barrel. The boiler was domeless with a round top firebox on which was fixed a shapely dome with spring balance safety valves. The cylinders were slightly inclined and single bar crossheads were fitted. All three locomotives were scrapped in 1929. The 0-4-2ST of 1885 was *Class L* and the two 0-4-0STs *Class M.*

Although closed in 1930 the lifting of locomotives for wheel removal was carried out for some time until a wheel-drop was installed at the Bath running sheds.

Fig. 502 **Somerset & Dorset Rly.** 1895 0-4-0ST No. 45A

SOUTH DEVON RAILWAY

Newton, later Newton Abbot

When the amalgamation took place with the GWR three broad gauge locomotives were under construction at Newton in 1875. Most of the parts, including boilers, had been supplied by the Ince Forge Co. of Wigan and were to have been 2-4-0STs. The work was transferred to Swindon and they were built as 2-4-0 Side Ts to standard gauge. They were numbered 1298 to 1300. On the South Devon Railway they were to have carried names SATURN, JUPITER and MERCURY but when they reappeared in 1878 from Swindon they only bore numbers.

SOUTH EASTERN RAILWAY

Ashford, Kent

The first locomotive completed at Ashford in 1849 was the 0-4-0 tank which had been sent from Bricklayers Arms and is referred to in the South Eastern Railway – Bricklayers Arms section.

James I'Anson Cudworth had been appointed Locomotive Superintendent in 1845 and it was not until 1853 that Ashford had completed its first new passenger locomotive. This was the first of ten 2-4-0s of the *Hastings* class. Even then the boilers were supplied by an outside locomotive building firm as the works were not as yet geared up to produce the necessary parts quickly enough for quantity production. Teething troubles occurred of course although it must be realised that the policy at this time favoured obtaining the majority of new locomotives from contractors. These first 2-4-0s,

Fig. 503 **SDR**
1875 2-4-0T
Domeless Boiler as
GWR No. 1300
at Swindon

Fig. 504 **SER**
1875 2-4-0
Class 118
No. 57

Nos. 157–166, had inside cylinders 15″ x 20″, outside frames, 5′ 6″ diameter coupled wheels and a firebox fitted with a midfeather.

Cudworth introduced compensating levers in the springing of many locomotives and these 2-4-0s had a single spring between the coupled wheels and compensating levers. Four of his standard 0-6-0 goods class were then put in hand with double frames, inside cylinders 16″ x 24″ and 4′ 9¾″ diameter wheels. There were no compensating beams on the springing of the locomotive but the two pairs of leading wheels on the six-wheeled tenders had such beams between them. The boiler had a dome halfway along the barrel with spring balance safety valves attached. The raised round topped firebox had an additional safety valve in a distinctive cover. These were built in 1856, but with modifications to the boiler and firebox and modified wheelbase, an additional forty-nine were built from 1859 to 1876.

Cudworth also introduced a double frame 2-4-0 known as the *118* class which was a handsome design, with a boiler the same diameter (3′ 10½″) as the standard goods but 9″ shorter. The cylinders were the same size and the coupled wheels 6′ 0½″ diameter. The patent Cudworth 'coal burning' firebox was fitted with the midfeather which necessitated two firing doors.

One hundred and twenty-four of these were built and were practically identical apart from modifications to the heating surface. In this case no equalising beams were fitted. They were used on all types of passenger trains and at one time represented half of the South Eastern's total locomotive stock. The *118* class was built from 1860 to 1875 at Ashford and others were built by the Vulcan Foundry, Dubs & Co. and George England & Co. Ashford built sixty-eight out of a total of 110.

Some 2-2-2s were built from 1861 to 1866 known as the *Mail* class; two in 1861 were slightly larger than the six built in 1865–6. They had double frames with cylinders 16″ x 22″ and 7′ diameter driving wheels and were used on the continental mail trains.

The only tank locomotives built at Ashford whilst Cudworth was in office were six 0-4-2WTs. They were built in 1867–9 and were intended for local passenger trains on the north Kent line. They had outside frames and inside cylinders 16″ x 22″ and 5′ 8″ diameter coupled wheels which had springs fitted with equalising beams between. The combined dome and safety valve was surrounded by a square casing at the base, housing sandboxes each side of the boiler. Front and rear spectacle plates were fitted, but no roof. The chimney was of the stove pipe variety.

Cudworth had done a tremendous job in standardisation, and as frequently happens, he was loth to depart from these standard designs despite the fact that heavier, faster and more powerful locomotives were badly needed in the 1870s. Cudworth would not however budge from his policy and matters came to a head when the Board invited John Ramsbottom to make a report on the state of locomotive affairs on the SER. It was agreed that twenty of Ramsbottom's *Newton* class should be built. It was soon after this impasse that Cudworth tendered his resignation and he finally departed in October 1876. Alfred Watkin, who happened to be the chairman's son was appointed Locomotive Superintendent and Richard Mansell was made Locomotive Engineer at Ashford. Watkin's appointment was not very popular, and being elected MP for Great Grimsby in 1877 he was asked to resign in the autumn. All that Watkin had actually done in the locomotive department was to bestow names on a few locomotives of various classes.

Richard Mansell was placed in charge as a temporary measure, which lasted until March 1878. During this period of upheaval the twenty Ramsbottom *Newtons* had been delivered, ten from Sharp Stewart and ten from Avonside Engine Co. Three 0-6-0Ts had been designed by Cudworth as replacements for the old *Bulldog* class, which had been built by R. Stephenson with four-coupled wheels, with a dummy crankshaft in between driven off the inside cylinders. The new 0-6-0Ts were completed in 1877 by Mansell and went to work the Folkestone Harbour branch. They had inside frames, which was a new departure for Ashford, side tanks, and dome without safety valves, the twin valves being on top of the round top firebox. A neat, all over cab and short bunker was fitted. An unusual feature was that each cylinder had two piston rods passing above and below the leading axle, as the cylinder centre line coincided with the leading axle centre line. They had 17″ x 24″ cylinders and 4′ 6″ diameter wheels.

Mansell then brought out a 0-4-4T, with outside frames to the coupled wheels. These had strong Cudworth features with combined dome and spring balance safety valves. Nine were completed in 1878 and when new went to London for the increasing suburban traffic and were known as 'Gunboats'. They too had 17″ x 24″ inside cylinders and the coupled wheels were 5′ 7½″ diameter. All the SER locomotives, during Cudworth's time, had an excellent finish with brass dome and safety valve covers and all other copper and brasswork was highly polished.

Three 0-6-0 goods locomotives were also built by Mansell in 1879 with 17″ x 24″ cylinders and 5′ diameter wheels, otherwise they were very similar to the previous standard goods class. Approx 174 locomotives had been built at Ashford up to the time Mansell relinquished his temporary appointment and resumed his position as Works Manager at Ashford.

In March 1878 James Stirling, brother of Patrick Stirling, was appointed Locomotive Superintendent. He came from Kilmarnock where he had been Locomotive Superintendent of the Glasgow & South Western Railway. There was still a pressing need for larger locomotives and although it was intended that ten 0-6-0s were to be built to Mansell's designs only three were built as related above, the other seven being cancelled by Stirling.

To expedite delivery of the larger locomotives Stirling based his first designs on those he had recently built for the G & SWR. Domeless boilers were now the standard, and the first 4-4-0s to be built at Ashford were twelve inside framed, inside cylinder locomotives with the typical Stirling safety valves mounted on the centre of the boiler. They were similar to his northern products and had 18″ x 26″ cylinders, 6′ 0½″ diameter coupled wheels and a moderate heating surface of 948·5 sq. ft, compared with 1108·9 sq. ft for the earlier *118* class 2-4-0s. Boiler pressure remained at 140 lb/sq. in. Other features were steam reversing gear and the rounded pattern cabs. James Stirling did more than any other engineer to develop the inside frame and inside cylinder 4-4-0, beginning with his well-known 'bogies' of 1874.

His next class of 4-4-0 were the *F* class from 1883–98. The cylinders were increased to 19″ x 26″ and coupled wheels 7′ diameter. The boiler was larger giving a total of 1020·5 sq. ft of heating surface. The *F1* class were *F* class rebuilds with 18″ cylinders but a larger heating surface of 1123·99 sq. ft. The *B* class and *B1* class were the final 4-4-0 designs of Stirling and ten were built in 1898–9 at Ashford and twenty by Neilson Reid. The same pattern of rebuilding occurred as with the *F* and *F1* classes. These 7 ft, 4-4-0s formed the mainstay of express passenger workings replacing Cudworth's 2-2-2s and 2-4-0s.

One of the *F* class, No. 240 built in 1889, was sent to the Paris Exhibition and named ONWARD for the occasion. A gold medal was obtained and it ran trials for a period on the PLM railway of France where she acquitted herself well in the high speed trials. The last *F1* was withdrawn in 1949 (No. 231) and the last *B1* in 1951 (No. 443) although the last Ashford built *B1* (No. 217) went on to 1950. Both the *F1* and *B1* rebuilds had domed boilers, and the last 4-4-0s, the class *B*, had a new cab design of more orthodox shape than the original wrap-around variety.

The most numerous class of Stirling's design was the *O* class 0-6-0, with 18″ x 26″ cylinders and 5′ 1″ diameter wheels. 122 were built, with various modified boilers, and Ashford were responsible for buiding fifty-seven out of this total between 1882 and 1899.

The initial batch were a disappointment, mainly due to indifferent steaming and weak

Fig. 505 **SER** 1886 0-6-0 *Class O* No. 332

Fig. 506 **SER** 1889 4-4-0 *Class F* No. 194

frames. These were subsequently rebuilt. This class remained the standard goods throughout Stirling's term of office.

He designed two types of tank locomotives. The *Q* class 0-4-4T, which was a modified version of his Kilmarnock 0-4-4T, was to be the standard suburban tank and was the second largest class with a total of 118 of which Ashford built forty-eight. There were a few variations in the period of building (1881–97) but all had 18″ x 26″ cylinders and 5′ 6″ diameter coupled wheels. There were a number of different bogies fitted, and others had condensing gear and short chimneys. Domeless boilers were fitted, but the rebuilds known as *Q1* class had domed boilers. Stirling's tanks had a pronounced flare on the top edge, matched by a flare on the bunker.

The second tank class was a 0-6-0T known as the *R* class, for shunting use. Very few genuine shunting tanks had been built and to cure this deficiency twenty-five were built 1888–98, all at Ashford. They had domeless boilers, and when rebuilt as *R1* they received domed boilers. The cylinders were 18″ x 26″ and wheels 5′ 2″ diameter.

The boilers for the *Class A* 4-4-0s, *Class Q* 0-4-4T and the *Class O* 0-6-0 were dimensionally similar although the tube heating surface varied.

James Stirling retired at the end of 1898, which also marked the end of the South Eastern Railway. During his stay at Ashford, Stirling had created another series of standard locomotives to a total of 394, and even so the demand for more powerful locomotives was not met to the

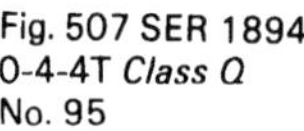

Fig. 507 SER 1894 0-4-4T *Class Q* No. 95

full. Indeed, the running department were perpetually in trouble, the loading capacity of the passenger locomotives not keeping pace with current design.

On 1 January, 1899 the railway married its old rival the London Chatham & Dover Railway and became the SOUTH EASTERN & CHATHAM RAILWAY.

The Locomotive Superintendent of the newly formed SE&CR was Harry S. Wainwright the former Carriage & Wagon Superintendent. William Kirtley who had managed locomotive affairs on the old London, Chatham & Dover Railway retired at the same time as James Stirling. Up to the end of 1898 approximately 409 locomotives had been built at Ashford.

It was decided that Ashford would be the chief locomotive workshops and a modernisation programme, with extensions to the works, was carried out. Longhedge continued to build a few new locomotives until 1904 when all new work was allocated to Ashford, and Longhedge continued as a repair shop for some time. During 1899 some Stirling *Class B* 4-4-0s were built and five 0-6-0 *Class O*.

Wainwright's chief draughtsman was Robert Surtees who was transferred from Longhedge and was a powerful influence 'behind the throne'. The first class to emerge was the *Class C* 0-6-0, the original design having been prepared at Longhedge. A total of 109 were built, Ashford building fifty-five and Longhedge nine. They had 18½″ x 26″ cylinders and 5′ 2″ diameter wheels, a higher boiler pressure of 160 lb/sq. in. and a total heating surface of 1200 sq. ft. They were a simple, strongly built, typical 0-6-0 with domed boiler, round topped firebox and Ramsbottom safety valves. A drastic alteration in finish was apparent, with polished brass dome cover, safety valve cover, number plates and brass edging, all highly finished and instead of black they were painted Brunswick green with the frames in reddish brown. They were used on main line goods, semi-fast passenger and excursion trains and were successful from the start. Building took place at Ashford from 1900–8. Once more it became a standard class.

The well-known *Class D* 4-4-0 appeared in 1901 with 19¼″ x 26″ cylinders, 19″ on later batches. The driving wheels were 6′ 8″ diameter. Similarly the initial boiler pressure was 180 lb/sq. in. reduced in later boilers to 175 lb/sq. in. A domed boiler gave 1505 sq. ft of heating surface – domeless boilers were now a thing of the past – and yet a lot of trouble was caused by priming.

Steam reversing was used as before, and steam sanding. The finish was superb and besides the similar finish to the *Class C* 0-6-0s, the chimney had a polished copper cap and the engine numbers were of brass, separately rivetted on the splashers.

A development of the *D* class was the *E* class, the principal difference being the diameter of the driving wheels being reduced to 6′ 6″ to give more power on the banks and a boiler with a Belpaire firebox increasing the heating surface to 1532 sq. ft. Minor differences were four windows instead of two on the cab front, completely filled in splashers, flush rivetted smokebox and fluted coupling rods. The finish was of the same high standard. They were the first locomotives to be fitted with Belpaire fireboxes at Ashford.

Further suburban tanks were required for the increasingly heavier trains and the *Class H* 0-4-4T appeared in 1904 and altogether sixty-six were built up to 1915. Cylinders were 18″ x 26″ with the *C* class motion, the coupled wheels being 5′ 6″. The boiler had a round top firebox with a total heating surface of 1104·7 sq. ft and in working order they weighed 54 tons 8 cwt. A new design of cab had the cab roof projecting over the side panels. The tank sides were straight but the bunkers were still flared out at the top and rear.

In 1909–10 eight small 0-6-0Ts were built, and were known as the *P* class. They replaced some rail motors supplied by Kitsons in 1905–6 on some services. The 0-6-0Ts had inside cylinders 12″ x 18″ and wheels 3′ 9″ diameter. They were very small and had only a slight edge over the rail motors. Wainwright's last design was the *Class J* 0-6-4Ts, and five were built in 1913. They were intended for the business trains, to Tonbridge and the coast, but were not able to keep the tight timings on those services and were put on the Redhill route. They had 19½″ x 26″ inside cylinders and coupled wheels 5′ 6″ diameter. The boilers were fitted with Schmidt superheaters and Belpaire fireboxes. The total heating surface was 1233 sq. ft, the superheater contributing 234 sq. ft to that total. The weight in working order was 70 tons 14 cwt. The large bunker carried 3½ tons of coal. Piston valves were fitted and tail rods projected on to the buffer beams. They were very similar to the Metropolitan Railway's *Class G* 0-6-4Ts built in 1915. The tail rods were subsequently removed from the SE&CR 0-6-4Ts.

Harry Wainwright resigned through ill health on 30 November, 1913. He had improved the locomotive stock in some degree and being a carriage and wagon man, his carriages were comparable with most of the other railways. As previously mentioned, his chief draughtsman was Robert Surtees and it is acknowledged that the designs of Wainwright's period were strongly influenced by Surtees. Wainwright's successor was R. E. L. Maunsell who had come from the Inchicore works of the Great Southern & Western Railway of Ireland where he had been Works Manager.

Maunsell found the locomotive position still in a parlous state and to relieve the shortage some Great Northern Railway 2-4-0s of Patrick Stirling's design, and 0-6-0s from the Hull & Barnsley Railway of Matthew Stirling's design were borrowed. Could there be any other instance when locomotives designed by three brothers were running on the same railway?

Some larger 4-4-0s known as the *L* class had already been designed and ordered by Wainwright, twelve from Beyer Peacock, and, because of guaranteed delivery, ten from A. Borsig of Berlin. The latter were put together at Ashford by Borsig's fitters and the last one ran trial only one month before war was declared on Germany!

Maunsell's first design for express passenger work was a 2-6-4T *K* or *River* class and due to the War this locomotive did not appear until June 1917. Almost simultaneously, a tender version was built in the form of a 2-6-0 and was completed in August the same year.

A comparison of dimensions is interesting.

	2-6-4 T (K) No. 790	Common Dimensions	*2-6-0* (N) No. 810
Cylinders/2 outside	19" x 26"		19" x 28"
Coupled Wheels dia.	6'–0"		5'–6"
Leading Wheels dia.		3' 1"	
Boiler dia.		4' 7¾ x 5' 3"	
Boiler Length		12'–6"	
Tubes Htg. surface sq ft		1390·6	
Superheater surface sq ft		203·0	
Firebox surface sq ft		135·0	
Total surface sq ft		1728·6	
Boiler pressure lb/sq in		200	
Grate area sq ft		25	
Weight in W.O.	82 tons 12 cwt		59 tons 8 cwt
Tender in W.O.	–		39 tons 5 cwt
Coal Tons	2½		5
Water Gallons	2000		3500

Fig. 508 **SER Ashford Erecting Shop** 1907

Fig. 509 1922 2-6-0 *Class N* two-cylinder No. A820

As will be seen both types had the same boiler, tapering up to a Belpaire firebox, and very akin to Great Western Railway practice with smokebox regulator, long lap piston valves, and top feed. This was probably the influence of Maunsell's assistant, G. H. Pearson from Swindon. There were Midland characteristics too which James Clayton, the new chief draughtsman who came from Derby, introduced such as the cab and tender on the 2-6-0, the cab and chimney and other details on the 2-6-4T. The outside Walschaerts valve gear and single slide bar were new.

Only one tank was built but eleven more *N* class 2-6-0s were built in 1920–2. They were classed as mixed traffic locomotives, and undoubtedly extremely successful and free running after minor modifications were carried out. Features that certainly assisted in the good running of this class were the 10″ piston valves and the valve travel of 6½″. A three-cylinder version was completed in December 1922 and was the last locomotive built at Ashford under SE&CR auspices. The cylinders were 16″ x 28″ and were cast in two parts, the LH cylinder as one casting, and the inside cylinder and the RH outside as the other. The saddle was cast separately and all bolted together. The two outside steam chests were served by two sets of Walschaerts valve gear and the centre piston valves were operated by Holcroft's conjugate gear by means of a long extension rod fastened to an arm on the pedulum lever and attached to a lever in front of the cylinders behind a deep buffer beam, the lever having a 2·33 to 1 ratio and floating lever 1·33 to 1 ratio. The *N1* three-cylinder class were found to be more economical, freer steaming and running and with full regulator working, shorter cut-offs could be used.

Besides new building Maunsell did some extensive rebuilding, and in the case of one of the *E* class 4-4-0s, No. 179, this was fitted with new cylinders, long travel piston valves and superheated boiler. Wainwright had already tried superheating but the additional weight barred them from the Chatham lines. Maunsell's rebuild was more carefully designed to keep the weight down. This *E* class was so successful that a total of ten were sent to Beyer Peacock for rebuilding. Ten *Class D* were similarly dealt with and others of the same class rebuilt at Ashford. They were then reclassified *E1* and *D1* respectively.

On 1 January, 1923, the SE&CR lost its identity to become the *Eastern section* of the newly formed SOUTHERN RAILWAY. Up to this date as far as can be estimated 639 locomotives had been built at Ashford. When the two railways amalgamated the South Eastern Railway locomotive running numbers were not altered but the LC&DR stock had 459 added to their numbers. In about 1912 the majority of brass and copper work was painted over and the fine finish previously employed was discontinued. In 1915 the running numbers appeared in large white figures on tank and tender sides and during the 1914–18 war all locomotives were painted battleship grey, and this continued until the 1923 amalgamation.

Class	First Built	No.	Cylinders inches	D.W. ft in	Heating Surface sq. ft Tubes	F'box	Super	Total	Grate Area sq ft	B.P. lb/sq in	Weight WO ton cwt	*	Type
A	1879	12	18 x 26	6 0½	858·3	90·2	–	948·5	15·75	140	37 19	S	4-4-0
B	1898	29	19 x 26	7 0	976·0	111·5	–	1087·5	16·5	160	46 1	S	4-4-0
B1	1910	27	18 x 26	7 0	1011·19	112·8	–	1123·99	17·5	170	45 2	W	4-4-0
D	1901		19 x 26	6 8	1381·0	124	–	1505·0	20·3	175	50 0	W	4-4-0
D1	1921	21	19 x 26	6 8	1149·85	127·1	251	1527·95	24·0	180	51 5	M	4-4-0
E	1906	26	19 x 26	6 6	1396·0	136	–	1532·0	21·15	180	52 5	W	4-4-0
E1	1920	11	19 x 26	6 6	1149·85	127·1	251	1527·95	24·0	180	52 10	M	4-4-0
F	1883	88	19 x 26	7 0	917·0	103·5	–	1020·5	16·78	150	41 10	S	4-4-0
F1	1903	76	18 x 26	7 0	1011·19	112·8	–	1123·99	17·5	170	45 2	W	4-4-0
G	1900	5	18 x 26	6 1	1093·5	113·5	–	1207·0	18·0	165	44 12	W	4-4-0
L	1914	22	20½ x 26	6 8	1252·0	160·0	319	1731·0	22·5	160	57 9	M	4-4-0
L1	1926	15	19½ x 26	6 8	1252·5	154·5	235	1642·0	22·5	180	57 16	M	4-4-0
O	1875	122	18 x 26	5 1	1040·0	89·0	–	1129·0	15·25	140	35 9	S	0-6-0
C	1900	109	18½ x 26	5 2	1089·33	110·66	–	1200·0	17·0	160	43 16	W	0-6-0
R	1888	25	18 x 26	5 2	833·35	90·25	–	923·6	15·75	140	42 10	S	0-6-0T
P	1909	8	12 x 18	3 3	387·3	51·7	–	439·0	9·1	180	28 10	W	0-6-0T
J	1913	5	19½ x 26	5 6	887·0	112·0	234	1233·0	17·6	160	70 14	W	0-6-4T
Q	1881	118	18 x 26	5 6	858·3	90·2	–	948·5	15·75	140	46 12	S	0-4-4T
R	1900	15	17½ x 24	5 6	971·7	99·3	–	1071·0	16·25	155	51 9	W	0-4-4T
H	1904	66	18 x 26	5 6	1002·5	102·2	–	1104·7	16·66	160	54 8	W	0-4-4T
W	1932	15	3/16½ x 28	5 6	1390·6	135·0	203	1728·6	25·0	200	90 14	M	2-6-4T
N	1917	80	19 x 28	5 6	1390·6	135·0	203	1728·6	25·0	200	59 8	M	2-6-0
N1	1923	3	3/16 x 28	5 6	1390·6	135·0	203	1728·6	25·0	200	62 15	M	2-6-0
U	1928	50	19 x 26	6 0	1390·6	135·0	203	1728·6	25·0	200	62 6	M	2-6-0
U1	1928	7	3/16 x 28	6 0	1390·6	135·0	203	1728·6	25·0	200	65 6	M	2-6-0

* S = James Stirling W = H.S. Wainwright M = R.E.L. Maunsell

Boiler dimension, weights etc are for initial locomotives and modifications took place in some cases on later building

The following steam locomotives were built and rebuilt at Ashford during this period.

1923	3	2-6-0	Class N	823, 824, A825
1924	20	2-6-0	Class N	A826–A845 Woolwich parts
1925	30	2-6-0	Class N	A846–A875
1925	1	2-6-4T	Class K1	A890
1928	2	2-6-0	Class U	A620 & A621
1929	8	2-6-0	Class U	A622–A629
1930	5	2-6-0	Class N1	A876–A880
1931	10	2-6-0	Class U	A630–A639
1932	6	2-6-0	Class N	1400–1405
1933	7	2-6-0	Class N	1406–1412
1934	2	2-6-0	Class N	1413 & 1414
1935	7	2-6-4T	Class W	1916–1922
1936	3	2-6-4T	Class W	1923–1925
1942	20	0-6-0	Class Q1	C17–C36
1943	11	2-8-0	Class LMS	8610–8612, 8618–8624, 8671
1944	3	2-8-0	Class LMS	8672–8674
REBUILDING				
1926	1	4-4-0	Class D to D1	A470
1927	8	4-4-0	Class D to D1	A492, A505, A509, A736, A739, A741, A743, A745,
1928	7	2-6-0	Class U	A797–A802, A805
1928	1	2-6-0	Class U1	A890

Class U and *U1* from 2-6-4T *Class K* and *K1* respectively.

One of the earliest *Class N* 2-6-0s, No. A819, built in 1922, was fitted with a Worthington feed pump in 1924 and ran with it until 1927. The most interesting modification was to A816 when a patent exhaust steam condenser marketed by 'The Steam Heat Conservation Co.' was fitted in 1931 as an experiment to investigate the claims of its inventor Mr H. P. H. Anderson that a saving of 30 per cent in fuel could be achieved. The condenser is fully described in Mr H. Holcroft's book *Locomotive Adventure* Vol. 1. After many teething troubles and trial runs the main trouble was with the back pressure set up, and condenser tubes. Various modifications were carried out but money was short and the locomotive went back into traffic in 1935.

No. 1850 (previously A850) was fitted with Marshall's valve gear in 1933 but it was soon replaced by the standard Walschaerts valve gear the following year. Efficient at slow speeds, the Marshall gear was not suitable for high speed work.

As previously stated, the last steam locomotive built at Ashford of LM&SR design, was completed in March 1944. The majority of Ashford's locomotives, especially during Maunsell's reign, were strongly built and of simple and often handsome lines. In the 2-6-0 designs could be seen strong Great Western and Midland characteristics influenced by G. H. Pearson, H. Holcroft and others from Swindon and J. Clayton from Derby. Nevertheless the locomotives had many original features and there were few classes which did not acquit themselves in an efficient manner. Although Ashford ceased steam locomotive building in 1944, it was kept busy on repair work and a certain amount of rebuilding. The most important modifications were to the 2-6-0 classes, a large number of which were rebuilt from 1954 with new front ends, with cylinders with outside steam pipes, some receiving new frames and others BR blast pipes and chimneys.

Fig. 510 1935 2-6-4T *Class W* No. 1916

Gradually locomotive repairs to various classes which Ashford had allocated, were moved to Eastleigh and by July 1962 repairs were being carried out on steam cranes, and about this time the locomotive section of the works ceased to operate.
See SOUTHERN RAILWAY.

SOUTH EASTERN RAILWAY
Bricklayers Arms, London

In 1848 W. Fernihough was instructed to build a small locomotive combined with a carriage for use as a Directors' Saloon. The locomotive was put in hand but for some reason the work on it proceeded slowly, and about twenty-one months after the building began James I'Anson Cudworth, the Locomotive Superintendent, was told to complete the locomotive with all speed. In July 1850 the parts were packed off to the new works at Ashford and quickly completed.

It had a vertical boiler, the cylinders were 5¾" x 9" and the coupled wheels 3' 6" diameter. It was numbered 126 and went by the name of COFFEE POT. Most of the work was done at the Bricklayers Arms works, but neither works could really lay claim to completely building it.

SOUTH EASTERN & CHATHAM RAILWAY
Ashford & Longhedge

This railway was the result of the amalgamation sanctioned on 5 August, 1899 of the London, Chatham & Dover Railway and the South Eastern Railway but joint working had commenced on 1 January, 1899.
See LONDON CHATHAM & DOVER RAILWAY – LONGHEDGE; SOUTH EASTERN RAILWAY – ASHFORD.

At the formation of the Southern Railway on 1 January, 1923, the London & South Western, London Brighton & South Coast and the South Eastern & Chatham railways were the main constituents of the new company, with locomotive works at Eastleigh, Brighton and Ashford respectively. Mr R. E. L. Maunsell of the SE&CR was appointed Chief Mechanical Engineer.

The locomotives of each section retained their existing numbers, but the following prefixes were added: A (Ashford) for SE&CR, B (Brighton) for LB&SCR, E (Eastleigh) for L&SWR, and the Isle of Wight railways had the prefix W. From 1931 the prefix was omitted and all E section retained their numbers, the A section had 1000 added to their numbers and B section numbers were increased by 2000. The only exceptions were the eight 0-8-0Ts bearing numbers A950–A957, built after grouping, which became 950 to 957.

The L&SWR duplicate list had the number 0 as a prefix and these locomotives had the figure 3 as a substitute for the 0. No new designs were built in the first three years except the *K1* three-cylinder 2-6-4T RIVER FROME at Ashford in 1925. It was the only one built and was a tank version of the *N1* class 2-6-0 built in 1923. The tank held 2000 gallons. It was converted into a tender locomotive in 1928.

Brighton had built the two-cylinder prototype 2-6-4T in 1917 and added ten more in 1926 whilst Armstrong Whitworth erected nine more from Woolwich parts and boilers made by the North British Locomotive Co.

In 1925 names were added to the Urie *N15* 4-6-0s which had been so successful that Maunsell had fifty-four more built in 1925–7, Eastleigh building twenty-four and the North British Locomotive Co. thirty, all of which were built in 1925 at their Hyde Park works. They had modifications, including long travel valves, outside steam pipes and improved smokebox. The chimney was Maunsell's handsome design with a deep flare at the top.

The names chosen were all associated with King Arthur and the West Country and were then known as *King Arthur* class. Smoke deflectors were fitted to clear the exhaust from the cab windows and all other large types were so fitted.

In 1927, due to the increased loads on the freight side, Maunsell modified the *S15* class of 4-6-0s. They had cylinders 20½" x 28", the same as the later *King Arthur* class, 11" piston valves and the same boiler giving total heating surface of 2215 sq. ft, including 337 sq. ft of superheating surface. Twenty-five were built from 1927 to 1928 at Eastleigh.

An important step to modernise the old

Wainwright *Class D* and *E* 4-4-0s had taken place in 1920 when eleven *Class Es* were entirely rebuilt, ten by Beyer Peacock and one at Ashford. The old frames were used and other parts but the boiler and front end incorporated superheating and 10″ piston valves. Their appearance was very similar to the Midland *Class 2* rebuilds even to the cab, chimney and running plate. In the same way the *Class D* was rebuilt, eleven at Ashford 1922–7 and ten by Beyer Peacock & Co. In their rebuilt state as classes *E1* and *D1* they did good work on many expresses from Cannon Street.

In 1926 appeared, from Eastleigh, Maunsell's first four cylinder 4-6-0. Required for hauling 500 ton trains at an average speed of 55 mph, Maunsell built the prototype and it was not until 1928 that further locomotives of the same class were built, Maunsell wishing to thoroughly test the new design before any quantity were built.

No. 850 LORD NELSON had many distinctive features. The inside cylinders were slightly forward of the outside cylinders, and drove the leading coupled wheels. The outside cylinders drove the middle pair. Four sets of Walschaerts valve gear were fitted and the angles of cranks were set to give eight exhaust beats per revolution of the driving wheels. The boiler barrel was 5′ 9″ diameter and had a Belpaire firebox 10′ 6″ long. The total heating surface was 2365 sq. ft including 376 sq. ft of superheater surface. The grate area was 33 sq. ft. Cylinders were 16½″ x 26″ and the coupled wheels 6′ 7″ diameter. The boiler pressure was 220 lb/sq. in. A large eight-wheeled tender was coupled on with 5000 gallons of water and 5 tons of coal. Maunsell's own design of superheater header was used incorporating air relief valves, which protruded through the smokebox. A soot blower, exhaust injector and sight feed lubricator were provided. The weight in working order was 83 tons 10 cwt, and 140 tons 4 cwt with tender. The coupled axle loads were all 20 tons 13 cwt.

Due to the shortage of main line express locomotives in 1925 more *King Arthur* class were ordered to bridge the gap, which delayed the appearance of LORD NELSON. In addition its stated duty had to be derated so that the *King Arthurs* could take on the same train load if required.

Fourteen more *Lord Nelson* class were built in 1928–9. They were, for a time, Britain's most powerful express passenger locomotive, and no opportunity was lost by the publicity department to make this known to the public. They were not the most successful of Maunsell's designs and one difficulty was maintaining steam pressure. A number of modifications were carried out on some of the class including the fitting of 6′ 3″ diameter coupled wheels to No. 859, the alteration of cranks to 90° settings on No. 865, and fitting a Kylchap blastpipe and double chimney to No. 862.

Two had boiler alterations, No. 860 had a longer boiler barrel and No. 857 had a round topped firebox and a combustion chamber fitted. Some of these modifications were carried out by Mr Bulleid who took Mr Maunsell's place and when the front end had been redesigned there was indeed a transformation in their performance.

Ashford at this time was pre-occupied in turning out 2-6-0s of various classes including the erection of a number of Woolwich built 2-6-0s. In addition the 2-6-4Ts were converted to 2-6-0 tender locomotives in 1928–9, all three works participating. They were then known as the *U* class and a three-cylinder version known as *U1* class was built.

All were evolved from Maunsell's *N* class design for the SE&CR, the first one being built in 1917 at Ashford (No. 810) and the class eventually grew to a total of 80. They had two outside cylinders 19″ x 28″, 5′ 6″ diameter coupled wheels, and a taper boiler with Belpaire firebox with a total heating surface of 1728·6 sq. ft including a Maunsell superheater of 203 sq. ft. Their outside Walschaerts valve gear was introduced for ease of maintenance, and the cross-heads worked on single side bars. The 6½″ valve travel was comparative with Great Western Railway practice and the design in parts was similar to the *4301* class built at Swindon.

The *N1* class was a three-cylinder version of *Class N*, the valve events of the inside cylinder being effected by Holcroft's conjugated motion. An interesting feature was that the left-hand and centre cylinders were cast in one piece. Six were built at Ashford, one in 1923 and five in 1930.

The third class of 2-6-0s the *U* class was the result of an order for twenty more *River* class 2-6-4Ts but due to the Sevenoaks accident with A800 RIVER CRAY, the company decided to build them as tender locomotives. Ten were built at Brighton in 1928 and ten at Ashford in 1929 followed by ten more at Ashford in 1931. In 1928 the ill-fated 2-6-4Ts were also converted, thus forming a class of fifty. The main difference between the *Class N* and *Class U* was that the latter had 6′ 0″ diameter coupled wheels.

The solitary three-cylinder 2-6-4T No. A890 RIVER FROME was converted to a tender locomotive at Ashford in 1928 and was the prototype of the *U1* class, twenty of which were built at Eastleigh in 1931. The three cylinders were 16″ × 28″ and the same boiler was fitted to all four classes of 2-6-0s.

In 1930 appeared the most powerful 4-4-0 running in this country and indeed Europe. They had three cylinders 16½″ x 26″ and 6′ 7″ diameter coupled wheels. The boiler was a shortened version of the *King Arthur* class with pressure increased to 220 lb/sq. in. Very few teething troubles ocurred, the most serious being the bogie frames which had to be strengthened. They were an outstanding success and, many think, Maunsell's best design. Their design duty was 400 ton trains at average speed of 55 mph. They were named after public schools in the Southern area and a total of forty were built between 1930 and 1935, all at Eastleigh. Officially *Class V,* they were known as the *Schools* class.

They were later fitted with smoke deflector plates and later still about half the class had Lemaitre multiple jet blast pipes. The Eastern section had the first batches which were allocated to Dover, Ramsgate and St Leonards, others going to Bournemouth and Bricklayers Arms.

A modern 0-6-0 goods locomotive was needed to replace the older Drummond and Wainwright designs and Maunsell designed a modern locomotive with long travel piston valves, superheated boiler with Belpaire firebox. The cylinders were 19″ x 26″ wheels 5′ 1″ diameter, and heating surface 1247 sq. ft with an additional 122 sq. ft of superheater elements. They were known as *Q* class and were not actually built until after Mr Maunsell had retired in October, 1937, the first *Q*, No. 530, appearing in 1938 the first of twenty built at Eastleigh in 1938–9.

His successor was Mr O. V. S. Bulleid who filled the position from 1 November, 1937. He had come from the London & North Eastern Railway where he had served under Sir Nigel Gresley, as his principal assistant.

No new designs were forthcoming until after the War started. In 1941 his first three-cylinder *Pacific* appeared and it was evident from its design and appearance that there was no intention to follow previous designs or develop them – here was something strikingly new, and the Southern Railway had its first 4-6-2. Six were built at Eastleigh bearing the numbers 21 C1 to 21 C6. Interesting features were: a chain driven valve gear which together with the inside connecting rod was enclosed in an oil tight casing, all three sets of valve gear being inside the frames. 11″ diameter piston valves with outside admission were fitted. All cylinders drove one axle, the inside cylinder having an inclination of 1 in 7½. The boiler was tapered, the second ring tapered downwards towards the firebox, the upper surface remaining horizontal (the opposite to the Great Western taper boiler, and the London & North Eastern tapered all round). Two Nicholson thermic syphons were built into the firebox which combined with a combustion chamber to give 275 sq. ft of heating surface. Tubes and flues gave 2176 sq. ft and the superheater the large figure of 822 sq. ft. The firebox was steel and gave a grate area of 48·5 sq. ft. The safety valves were fitted to the front ring, and the working pressure was 280 lb/sq. in. Exhaust passed through a multiple jet blast pipe.

Steam reversing gear and steam operated fire doors were introduced. The wheels were cast steel 'Boxpok' double disc pattern, and the 6′ 2″ diameter coupled wheels had clasp brakes. Streamlining took the form of a casing carried on the frames and not the boiler, and the large cab was a continuation of the casing. At the front end the casing was brought forward of the smokebox and the top formed a catchment duct for air to pass through for exhaust clearance.

The tender was all-welded with the filler at the cab end instead of the rear. A generator was fitted for electric light. The total weight in working order, without tender, was 92 tons 10 cwt of which 63 tons were available for adhesion.

Apart from the oil tight casing, which did not remain oil tight for very long, and the chain driven valve gear, the class gave an excellent account of itself and was put to work on the heavy war-time trains principally on the London–Exeter line. They were known as the *Merchant Navy* class and were named after the main shipping lines, each name bearing the appropriate house flag. Thirty were built from 1941–9, all at Eastleigh.

For working west of Exeter, and other main lines on the Eastern and Central section, a lighter version was built with a maximum axle load of 18 tons 15 cwt as compared with 21 tons on the *Merchant Navy* class. The cylinders were 16⅜″ diameter and piston valves 10″, otherwise they had the same general lines, and they weighed 86 tons in working order. The boiler was smaller with a heating surface of 2122 sq. ft including the firebox, and the superheater was reduced to 545 sq. ft. These were known as the *West Country* class, the majority being built at Brighton.

Between these two classes of 4-6-2s appeared Bulleid's 0-6-0 *Class Q1*. Strictly utilitarian, these locomotives probably caused a greater stir in locomotive circles than the *Pacifics.* A powerful locomotive was needed which could work over most of the Southern Railway, and to meet these conditions the major problems of weight and loading gauge clearance were overcome by ruthlessly cutting away all superfluous detail.

The boiler was tapered with a full cone and the same flanging blocks were used for the firebox backplate as for the *Lord Nelson* class. The heating surface was 1472 sq. ft, plus 170 sq. ft of firebox and 218 sq. ft for the superheater. The grate area was 27 sq. ft and the working pressure 230 lb/sq. in. Like the *Pacifics*, the boiler cleading was fastened to the frames. The cylinders were 19″ x 26″ and wheels 5′ 1″ diameter. Long travel, long lap piston valves were fitted and a five jet nozzle blast pipe. Reversing was by steam and Boxpok wheels were again used. An interesting flashback in history was the use of one forging for piston rod and crosshead – used by Stroudley. The maximum axle load was 18 tons 15 cwt and the weight in working order 51 tons 5 cwt with a tender weighing 38 tons. The maximum travel of the valves was $6\frac{7}{8}''$ and Stephenson's link motion was used. The appearance was austere and ugly and came in for much criticism, but not the performance. The *Q1* class was the most powerful 0-6-0, by a large margin, in this country, and to achieve this with a weight of just over 50 tons was a remarkable feat in design and engineering. They were numbered C1 to C40 and were built at Brighton and Ashford in 1942.

In 1943 and 1944 all three works built the standard *8F* Stanier 2-8-0, twenty-three at Eastleigh, fourteen at Ashford and no less than ninety-three at Brighton which had certainly come back to the forefront in locomotive construction.

Ten more *Merchant Navy* class were built at Eastleigh in 1944–5 and then no further steam locomotives were built there for the Southern Railway. Brighton continued building *West Country* class *Pacifics* in 1945–7 and a total of seventy were turned out in these three years.

Ashford, after the completion of fourteen *8F* 2-8-0s, also ceased building steam locomotives for the SR, in fact in March 1944, with the steaming of LMS No. 8674 no further steam locomotives were built at all and by then the total number which had been built from 1853 to 1944 was 790 or more accurately, if the 50 Woolwich 2-6-0s are deducted, a total of 740. On 1 January, 1948 the Southern Railway became the Southern Region of BRITISH RAILWAYS.

See BRITISH RAILWAYS; LONDON BRIGHTON & SOUTH COAST RAILWAY; LONDON & SOUTH WESTERN RAILWAY; SOUTH EASTERN & CHATHAM RAILWAY.

R. E. L. MAUNSELL

Type	Class	First Built	Total	Cylinders inches	D.W. ft in	Boiler H/S Tubes & Flues	Boiler H/S Super	Boiler H/S F'box	H/S Total sq ft	Grate Area sq ft	Press lb/sq in	Weight WO ton cwt
0-6-0	Q	1938	20	19 x 26	5 1	1125	185	122	1432	21.9	200	49 10
0-8-0T	Z	1929	8	3/16 x 28	4 8	1173	–	106	1279	18.6	180	71 12
4-4-0	V	1930	40	3/16½ x 26	6 7	1604	283	162	2049	28.3	220	67 2
4-6-0	S.15	1927	25	20½ x 28	5 7	1716	337	162	2215	28.0	200	80 14
4-6-0	KA*	1925	54	20½ x 28	6 7	1716	337	162	2215	30.0	200	80 19
4-6-0	LN	1926	16	4/16½ x 26	6 7	1903	376	194	2473	33.0	220	83 10

* Improved N.15 Class

O. V. S. BULLEID

Type	Class	First Built	Total	Cylinders inches	D.W. ft in	Boiler H/S Tubes & Flues	Boiler H/S Super	Boiler H/S F'box	H/S Total sq ft	Grate Area sq ft	Press lb/sq in	Weight WO ton cwt
0-6-0	Q.1	1942	40	19 x 26	5 1	1302	218	170	1690	27.0	230	51 5
4-6-2	MN	1941	30	3/18 x 24	6 2	2175.9	822	275	3272.9	48.5	280	94 15
4-6-2	MN REB	1956	30	3/18 x 24	6 2	2175.9	612	275	3062.9	48.5	250	97 18
4-6-2	WC	1945	110	3/16⅜ x 24	6 2	1869	545	253	2667	38.25	280	86 0
4-6-2	WC REB	1957	60	3/16⅜ x 24	6 2	1869	488	253	2610	38.25	250	90 1

Totals include those built after 1947

SOUTH YORKSHIRE RAILWAY
Mexborough

The shops were brought into use in 1855 for repairs and, in 1861–2, two 0-6-0 locomotives were built or 'erected'. No. 20 in 1861 became MSLR No. 171 and No. 22 in 1862 became MSLR No. 173. They both had inside cylinders 16" x 24" with 5' wheels. No. 171 was replaced in 1893 and No. 173 in 1889. The SYR was leased to the MSLR in 1864 and repairs were continued for some time after this.

SPENCE, WILLIAM
Cork Street Foundry and Engineering Works, Dublin

The entire output of steam locomotives built by this firm went to the Guinness Brewery. Mr Samuel Geoghehan, the Brewery company's engineer, had encountered considerable trouble with locomotives already supplied for working the company's 1' 10" gauge internal railway. The principal difficulty was that the valve motion was too near the ground and maintenance consequently heavy. Mr Geoghehan decided to

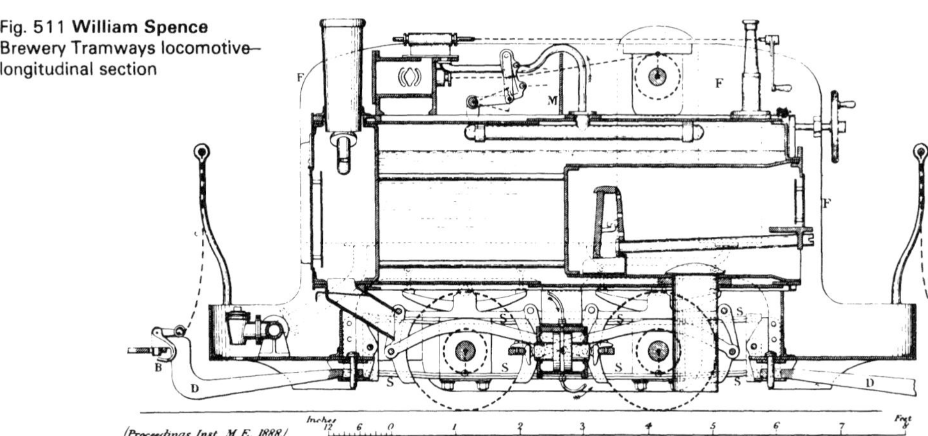

Fig. 511 **William Spence** Brewery Tramways locomotive—longitudinal section

Fig. 512 **William Spence** 1902 Brewery Locomotive No. 18

design a more suitable locomotive and he did this by placing the cylinders and crankshaft horizontally above the boiler. To avoid bolting the cylinders to the boiler the frames were carried to the full height to allow for stays above and below the boiler and the two cylinders were bolted between these frames. The cylinders drove a horizontal crankshaft from which vertical connecting rods drove the rear pair of wheels which were coupled to the front pair. The crankshaft was housed in horn blocks but the axleboxes on the track wheels had no hornplates and were fixed between horizontal spring frames with a vertical movement only.

The coupled wheelbase was 3 ft, so that the sharp radius curves could be easily negotiated. Steam brakes were fitted. The main dimensions were: cylinders 7″ x 8½″; wheel diameter 1′ 10″; boiler diameter 2′ 5″; tubes 64; tubes diameter 1½″; tubes length 2′ 10⅜″; Heating surface Tubes 72·61 sq. ft; Heating surface Firebox 13·75 sq. ft; Heating surface Total 86·36 sq. ft; grate area 3·24 sq. ft; pressure 180 lb/sq. in.; weight 7 tons 8 cwt; side tanks 80 gallons; coal capacity 3½ cwt.

The boiler was of the Ramsbottom type with circular firebox. Thus the locomotives successfully met the special requirements.

To provide motive power on the 5′ 3″ gauge lines, special haulage wagons were built on which the narrow gauge locomotives were mounted. When they had been lowered into the haulage wagon frames, the loco wheels engaged on rollers turned to the mating profile of the flanges. On the roller shafts were fixed gear wheels which meshed with gears mounted on the track axles of the haulage truck, which had outside sprung frames on two axles. These were also made by Spence, two in 1888, one in 1893 and one in 1903. The gear ratio was 3:1.

These locomotives and trucks were unique and no similar application is known. Details of manufacture of locomotives: 1887 Nos. 7, 8, 9; 1891 Nos. 10, 11, 12; 1895 Nos. 13, 14, 15; 1902 Nos. 16, 17, 18 & 19; 1905 Nos. 20, 21; 1912 Nos. 22; 1921 Nos. 23, 24. Total Built: 18.

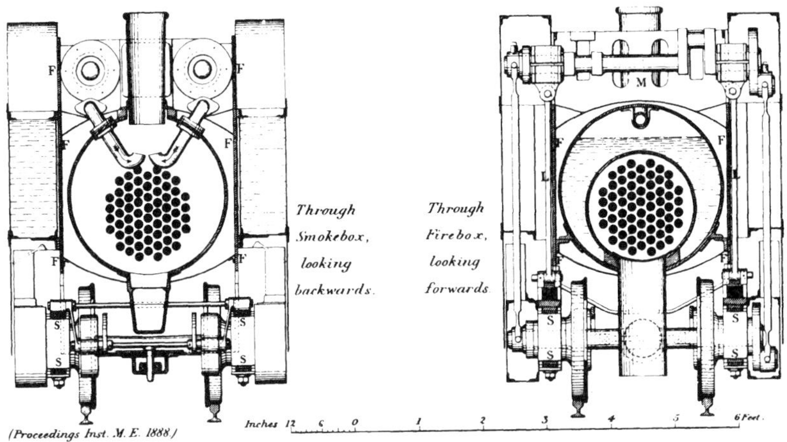

(Proceedings Inst. M. E. 1888.)

Fig. 513 **William Spence** Brewery Tramways locomotive – Cross sections

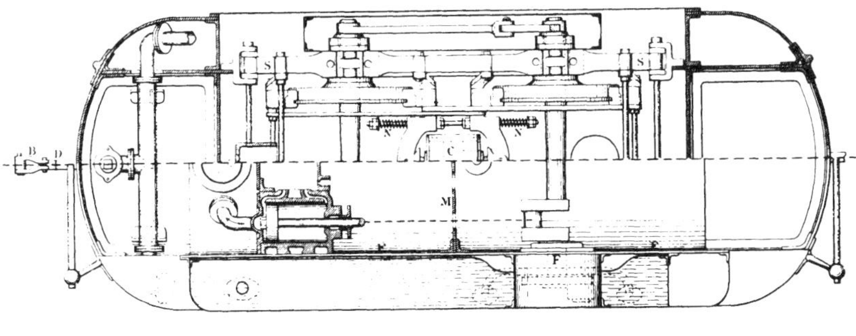

(Proceedings Inst. M. E. 1888.)

Fig. 514 **William Spence** Brewery Tramways locomotive-plan

Inevitably they were superseded by diesel traction and were withdrawn as follows: 1948 Nos. 7, 8; 1949 Nos. 9, 10, 11; 1951 Nos. 14, 16, 18, 19; 1954 Nos. 12; 1956 Nos. 13(a), 20(b); 1957 Nos. 15(c), 22; 1959 Nos. 21 (stored); 1962 Nos. 17(d); 1965 Nos. 23(e), 24.

Fortunately a number of these locomotives have been preserved:

(a) No. 13 Tal-y-lyn Railway Museum at Towyn, 1956.

(b) No. 20 Belfast Transport Museum, presented in 1956.

(c) No. 15 donated to the Irish Steam Preservation Society in 1966.

(d) No. 17 preserved at the Brewery.

(e) No. 23 and haulage truck No. 4 to Brockham Museum 1966.

SPITTLE, THOMAS

Cambrian Iron Foundry, Newport, Mon.

Six neat 0-4-0STs were built by this engineering firm over a period of about twenty-five years. Whether they were all intended for the firm's own use when new is not clear, but two went to the Devon Great Consols Ltd. at Tavistock and were returned in 1903. They may have been on hire, but as the Tavistock quarries closed down about 1901, it is possible they were sold back to Thomas Spittle.

Fig. 515 **William Spence** 1921 locomotive No. 23 on haulage wagon

T. SPITTLE

Works No	Built	Type	Outside Cylinders inches	D.W. ft in	Customer	Name	
1		0-4-0ST	10 x 15	2 8	T. Spittle	Gert	
2	1877	0-4-0ST	10 x 15	2 8	T. Spittle	Tom	
3	1882	0-4-0ST	10 x 15	2 8	Devon Great Consols Co Ltd	Hugo	(a)
4	1894	0-4-0ST	10 x 15	2 8	T. Spittle	Stanley	(b)
5	*c* 1896	0-4-0ST	10 x 15	2 8	Devon Great Consols Co Ltd	Ada	(a) (c)
6	1899	0-4-0ST	10 x 15	2 8	T. Spittle	Trevor	(d)

(a) Returned to T. Spittle in 1903. (b) Sold to Tredegar Estates Rly. (c) Sold to Belgian Government in 1918. (d) Sold to A. R. Adams & Son Newport in 1927.

STANHOPE & TYNE RAILWAY

South Shields

At least one locomotive was built by the railway. It was an 0-6-0 goods tender engine with outside frames and inside cylinders 14½" x 18"' and 4 ft diameter wheels. It was named PROJECTOR and was built in 1839. The railway failed in 1841 and most of the locomotives were sold to the Pontop and South Shields Railway. It has been reported that two or three further new locomotives were built by the new owners but this is doubtful as the stock of the P&SSR was taken over by the York & Newcastle Railway in 1847 and numbered 82 to 94. They consisted of ten by Robert Stephenson & Company two by Longridge and one by R. & W. Hawthorn. If any more had been built they would surely have survived until 1847.

STARK & FULTON

Glasgow

This firm built a few locomotives for the Glasgow, Paisley, Kilmarnock and Ayr Railway and three for the Midland Counties Railway. They were all 2-2-0 'Bury' type locomotives and may well have been subcontracted by Edward Bury. Two 2-2-2s were built in 1849 for the contractors of the Caledonian and Dumbartonshire Junction Railway which became part of the Edinburgh & Glasgow Railway in 1862. No further details are available.

STARK & FULTON

Date	Type	Cylinders inches	DW ft in	Railway	No/Name	
1839	2-2-0			GPK & A	Stuart	Bury Type
1839	2-2-0			GPK & A	Bute	Bury Type
1839	2-2-0	12 x 18	5 6	Midland Counties	Hawk	Bury Type
1839	2-2-0	12 x 18	5 6	Midland Counties	Vulture	Bury Type
1839	2-2-0	12 x 18	5 6	Midland Counties	Eagle	Bury Type
1840	2-2-0			GPK & A	No 1 Mercury	Bury Type
1840	2-2-0			GPK & A	No 2 Mazeppa	Bury Type
1840-1	2-2-0			GPK & A	Wallace	
1840-1	2-2-0			GPK & A	Queen	
1849	2-2-2			Contractors of the Caledonian & Dumbarton Rly		

GPK & A Glasgow, Paisley, Kilmarnock & Ayr Rly DB Stark appointed Loco superintendent to GPK & A in 1840 at Cook St. but retired after four months service.

Fig. 516 **Spittle** 3/1882 Devon Great Consols Ltd. HUGO

STAVELEY COAL & IRON CO. LTD.
Staveley

Although the above firm have no records of having built locomotives, it is almost certain that they built two and probably four. The two generally accepted were JOHN GREEN built in 1880, and BELVOIR built in 1885. They were 3′ 0″ gauge with vertical cylinders, and saddle tanks, and were supplied to the Eastwell Iron Ore Co. Ltd. which were managed by the Staveley Coal & Iron Co. Ltd.

The other two, which were supplied to the Waltham Iron Ore Co. Ltd., also managed by the same firm, were: GEORGE BOND and RUTLAND. They were the first locomotives to work at Waltham, so their building dates would be approximately the same as for the first two. The railway was metre gauge.

The photograph of RUTLAND shows it to be a very interesting and novel machine with two outside vertical cylinders fastened to a large cast iron frame surrounding the smokebox. The outside frames housed a jack shaft with outside cranks which presumably drove the first coupled axle through gearing. All the locomotives were four coupled. Whether the cabs were fitted to the locomotives initially is doubtful, such voluminous cabs were indeed rare.

Ref: *The Ironstone Railways & Tramways of the Midlands* by E. S. Tonks. Correspondence with Mr Davenport, who kindly gave permission to reproduce the photograph of RUTLAND.

Industrial Railway Record No. 32 – In an article on the above locomotives, OLIVER & CO. (later Markham & Co.) are put forward as an alternative maker.

STEPHENSON, GEORGE

It was in 1812, whilst working at Killingworth colliery, that due to his outstanding skill as a fitter and knowledge of colliery engines and pumping equipment which he had repaired and improved from time to time, that he was made engine wright. In a very short time he was appointed Chief Engineer to that group of collieries including Killingworth and Long Benton in Northumberland, and Mount Moor, Derwent Moor and South Moor in Durham, which were leased by the 'Grand Allies'.

Having now to travel in his own area, he was able to visit other collieries and it was at Wylam and Coxlodge that he was able to see the locomotives built by Hedley and Blenkinsop. He was so impressed that he convinced the owners that this form of transport would result in cheaper movement of coal from colliery to staithes and he was given permission to build a 'travelling engine' forthwith. Due to his preoccupation with his routine work and many trials and errors his first locomotive took ten months to build, and it was eventually completed in July, 1814.

It was a four wheeled, geared locomotive of very modest dimensions and it was named BLUCHER. Due to it running on edge rails, insufficient adhesion resulted in a poor performance, even after various modifications were made. In collaboration with Ralph Dodds, chief viewer at Killingworth, a locomotive was built which used a crank pin drive and was finished in March, 1815. A similar locomotive to BLUCHER may have been built before this one, as a statement by Nicholas Wood mentioned that there were two used up to the end of 1814 and two more since. A total of four locomotives

Fig. 517
Staveley Built in the 1880s
RUTLAND

were built at Killingworth for the use of the colliery from 1814 to 1821.

Meanwhile with the assistance of his friend William Losh, owner of the Walker Iron Works a design was completed for a much larger machine with six wheels, chain driven, with two cylinders let into each end of the boiler vertically and each driving the front and rear axles respectively, by means of crank pins. Steam suspension was used. The intermediate axle was connected to the front and rear axles by chains and sprocket wheels. This locomotive was probably built at the Walker Iron Works or at Kilmarnock as it would be highly unlikely that permission would have been granted for this locomotive to be built at Killingworth. It had been ordered by the Kilmarnock and Troon Railway and was the first locomotive to work in Scotland. It was named THE DUKE. Unfortuntely, as in so many instances, the rails were not strong enough to support its weight and it had to be put aside. The locomotive was completed in 1816.

In 1819 he was asked to rebuild the waggonways connecting Hetton colliery with the staithes on the river Wear at Sunderland, making them suitable for steam locomotive working. Three and a half miles of track was altered but the remaining four and a half miles had a number of self-acting inclines and two more worked by stationary engines.

Locomotives for this railway were built at Hetton Colliery under Stephenson's supervision and five were built from 1820 to 1822. Four carried the names HETTON, DART, TALLYHO and STAR. The official opening took place on 18 November, 1822. (See HETTON COAL COMPANY.)

They were similar to THE DUKE except that the wheels were 3′ 9″ and two axles were used instead of three. Each cylinder drove one pair of wheels through connecting rods and crank pins. The two axles were connected with spur gears and chains.

The steam springs were later removed, due principally to steam leakage. After initial teething troubles they performed their duties satisfactorily and, following on these successes, George Stephenson decided to establish a

Fig. 518 **Geo. Stephenson** 1814 Killingworth locomotive from original drawing

locomotive building factory of his own, and with the assistance and backing of Edward Pease and Thomas Richardson an establishment in Forth Street, Newcastle was founded under the title of Robert Stephenson & Company in June, 1823. Robert was his son, and gradually took over the responsibility of the works. His father had other interests and during this period was Chief Engineer to the Liverpool and Manchester Railway which opened in September, 1830. In 1840 he bought Tapton Hall near Chesterfield and slowly withdrew from his engineering activities. He died on 12 August, 1848 at the age of 67.

A locomotive working at Heaton colliery was ascribed to George Stephenson. In a letter from William Losh to Edward Pease dated November 1821 a large 'travelling engine' is mentioned which may have been Chapman's second one transferred from Lambton.

Another enigma is the quotation from a speech made by George Stephenson at Darlington in 1844 'Lord Ravensworth & Company were the first parties that entrusted me with money to make a locomotive, which engine was made thirty-two years ago and was called MY LORD. It would carry about 30 tons at the rate of 4 mph'. If this was an accurate statement, this engine must have preceded BLUCHER by two years which was the year he was appointed engine wright. That the steam locomotive did not emanate from George Stephenson's brain is well known, and to call him the 'Father of Railways' is not giving credit to the earlier pioneers, but he did more than anyone else to foster the advantages of the steam locomotive and to instil confidence, especially at a time when the future of locomotion was on a knife edge. Some did go the other way and reverted to horse traction but this momentous period, which started a revolution in transport which was to change our social and industrial worlds, could not be held back, and George Stephenson was the prime mover for which, in all reverence and humility, we should thank him for the 'Steam Age' – an age which lasted over 150 years.

Date	Type	Cylinders inches	DW ft in	Where Built	Name	Notes
7/1814	4WG	8 x 24	3 0	Killingworth Colliery	Blucher	Geared Drive from Jackshafts
? 1814	4WG	8		Killingworth Colliery		Geared Drive from Jackshafts
3/1815	0-4-0	8	3 0	Killingworth Colliery		with Dodds Crankpins on Wheels*
1816	0-6-0		3 2	Walker Iron Works (?)	The Duke	Axles connected by chains & sprockets
				for Kilmarnock & Troon Rly		Losh & Stephenson's patent *
1820	0-4-0	9 x 24	3 9	Hetton Colliery	Hetton?	Axles connected by chains & sprockets *
1821	0-4-0	9 x 24	3 9	Hetton Colliery	Dart?	Axles connected by chains & sprockets *
1821	0-4-0	9 x 24	3 9	Hetton Colliery	Tally Ho?	Axles connected by chains & sprockets *
1821	0-4-0	9 x 24	3 9	Hetton Colliery	Star?	Axles connected by chains & sprockets *
1821	0-4-0	8 x 24	3 0	Killingworth Colliery		Axles connected by chains & sprockets *
1822	0-4-0					
1822	0-4-0	9 x 24	3 9	Hetton Colliery		Axles connected by chains & sprockets *

In 1825 Geo. Stephenson said he had built about fifty-five engines – sixteen being locomotives * Steam springs

All vertical cylinders set into top of boiler, one at each end. Boiler pressure 50lb/sq. in.

STEPHENSON, ROBERT & COMPANY

Forth Street, Newcastle-on-Tyne

Fig. 519 **Geo. Stephenson** Newcastle1(1st series)/1825 Stockton & Darlington Rly. LOCOMOTION NO. 1

The first locomotive works in the world were established in 1823 by George Stephenson with his son Robert, Edward Pease, and Michael Longridge as partners. Locomotives had been built elsewhere, of course, before these works were built but in most cases they were erected at collieries with elementary tools and by millwrights who had no experience of locomotive construction but were guided by the designers who were practical men as well.

Much of the work fell on Robert Stephenson's shoulders, his father being away much of the time superintending railway construction and other work.

The original factory occupied 8 acres extending from Orchard Street on the east to Regent Street on the west and from Forth Street on the north to South Street. Much of the machinery was designed and built by George Stephenson.

The first locomotive built was LOCOMOTION NO. 1 for the Stockton & Darlington Railway followed by three more named HOPE, BLACK DIAMOND and DILIGENCE. Two six-coupled, with inclined cylinders, were also supplied to the same railway in 1827–8 named VICTORY and EXPERIMENT. The twelfth to be built, a four-coupled locomotive with two 9″ x 24″ outside cylinders and 4 ft diameter wheels went to the Delaware & Hudson R.R. and a six-coupled name WHISTLER was ordered by the Boston & Providence R.R. and was reported lost at sea. It is interesting to note that between 1814 and 1825 George and Robert Stephenson were the only builders of locomotives in this country and only two, both of which were failures, were built anywhere else. These were the first two to be built in Germany, in 1816 and 1818.

The Rainhill trials of 1829 resulted in the triumph of ROCKET designed and built by Robert Stephenson & Company. The cylinders were 8″ x 17″, inclined at an angle of 35° but subsequently lowered to an almost horizontal position. Henry Booth of the Liverpool and Manchester Railway laid claim that the multi-tubular boiler was his idea, although Marc Seguin in France had built a boiler of this kind eighteen months before the ROCKET but had not fitted it to a locomotive until after the ROCKET was built. Henry Booth was also the inventor of the screw coupling.

In addition, an outside firebox was fitted which was an important step in boiler construction. The driving wheels were 4′ 8″ diameter and the trailing wheels 2′ 6″ and the long wheelbase was 7′ 2″.

The twentieth locomotive was a 0-2-2 for the Canterbury & Whitstable Railway, built in 1830 and named INVICTA. This was the first to have the cylinders at the front, or chimney, end. They were still inclined. Another point about INVICTA was that it was the first passenger train locomotive.

The Rainhill ROCKET (there had been a ROCKET built in the same year for the Stockton & Darlington Railway) was sold to the Liverpool & Manchester Railway and Stephensons built seven more very similar ones, L&MR Nos. 2–8 in 1830. In the same year appeared the *Planet* class 2-2-0s with inside cylinders under the smoke box and outside frames, and this was adopted as a standard on the Liverpool &

STEPHENSON, ROBERT & COMPANY

Fig. 520
Robt. Stephenson
Newcastle 42(2nd series)/1833 Saratoga & Schenectady R.R. DAVY CROCKETT

Manchester Railway. A further development was the PATENTEE which had a pair of trailing wheels fitted behind the firebox making it a 2-2-2.

Orders for these two types were so numerous that many had to be passed on to other locomotive builders, such as Tayleur & Company and others.

By 1831 at least one hundred locomotives had been built in the world as follows: 93 British (probably a few more); 2 French; 2 American; and 2 German.

It should be noted that in 1831 the old series of works numbers up to approx No. 40 was discontinued for some unknown reason and a fresh series started from No. 1 again. At least five locomotives had works numbers in both series: Liverpool & Manchester Railway 0-4-0s:

SAMSON old No. 34 New wks No. 14; JUPITER old No. 35 New wks No. 15; GOLIATH old No. 36 New wks No. 16; SATURN old No. 37 New wks No. 18; SUN old No. 39 New wks No. 19.

Robert Stephenson & Company supplied many to America up to 1840. The European market continued to be very demanding for many years up to the 1870s and, to a lesser degree, in the early 1880s. The two main types built were 2-2-2s and 0-4-2s and it is right to say that where there were railways very few had no locomotives from this builder.

In 1841 another important development took place with the appearance of the 'long boiler' type. All axles were in front of the firebox. The advantage claimed for this type of boiler was that instead of increasing the diameter of the boiler to obtain a greater heating surface, the boiler could be lenthened instead at the same time obtaining a better ratio of the tube length to tube diameter, and the wheelbase of the locomotive would not have to be increased. Within certain limits, such a boiler would transfer more heat from the flue gases to the water, the gases having to travel a greater distance with a consequent greater drop of temperature at the smokebox tubeplate. At the same time more ash was deposited in the tubes instead of being thrown out of the chimney which necessitated more frequent tube cleaning. The 'long boiler' was also applied to six-coupled locomotives. Many were built and were especially popular in France. In this country the North Eastern Railway had many which were built in this fashion up to 1866.

Various valve gears were fitted to early locomotives. Four fixed eccentrics were fitted as early as 1836 but they were by no means the first to employ them. Gab motions were improved upon but the major step forward was Stephenson's link motion of 1842, usually ascribed to William Howe, a pattern maker in the employ of Robert Stephenson & Company. Another claimant was William Williams, a Stephenson draughtsman.

In 1839 two 2-2-2s were supplied for the 7 ft gauge of the Great Western Railway, and they were the first to appear on this railway. They had originally been ordered by the New Orleans Railway to 5′ 6″ gauge but due to financial complications they were not delivered and

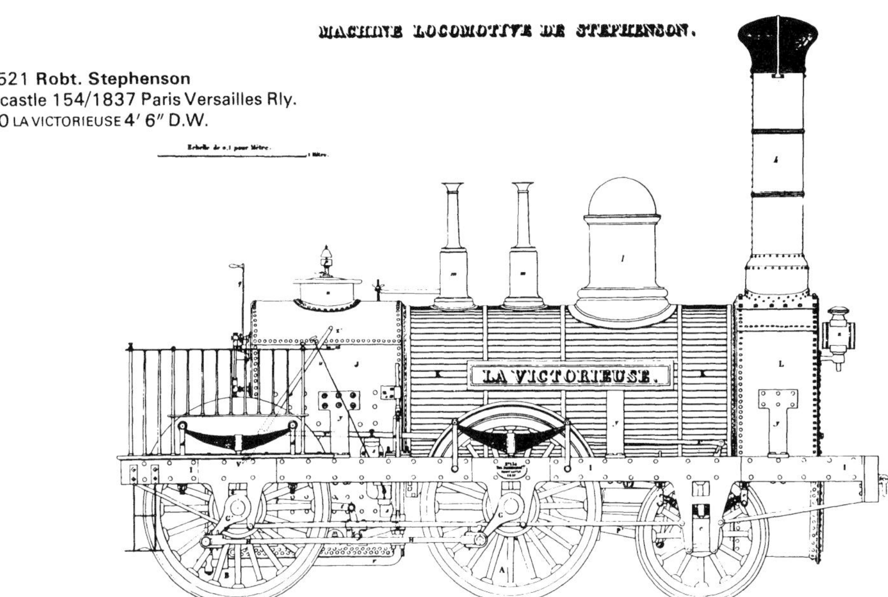

Fig. 521 **Robt. Stephenson** Newcastle 154/1837 Paris Versailles Rly. 2-4-0 LA VICTORIEUSE 4′ 6″ D.W.

were converted to Brunel's 7 ft gauge. The first was named NORTH STAR and was delivered in 1837, and it was this locomotive that compelled Brunel to state ' – we have a splendid engine of Stephenson's, it would be a beautiful ornament in the most elegant drawing room – ' The other locomotive MORNING STAR did not arrive until January, 1839. Ten further 2-2-2s, very similar to the previous two, were delivered from 1839 to 1841. They were without a doubt the only reliable locomotives running on the GWR for some time and they formed a basis for Gooch's later standard designs.

The works were extremely busy despite intense competition from the rapidly growing number of locomotive builders and many 2-4-0, 2-2-2, and 0-6-0s were delivered at home and abroad.

In 1846 a long boiler 4-2-0 was built for the gauge trials, standard 4′ 8½″ gauge versus Brunel's 7 ft gauge. This locomotive was put on the York & North Midland Railway and was known as 'A'. It had outside cylinders, a low set boiler, and Bury type firebox. The cylinders were 15″ x 24″.

The 7ft gauge locomotive chosen for the trials was a 2-2-2 named IXION, with 15″ x 18″ cylinders, built by Fenton, Murray & Jackson of Leeds in 1841. IXION attained a higher speed with a heavier load than 'A' did, and although the broad gauge lost their battle the Commission considered the 7 ft gauge superior in every way, but the standard guage was bound to have its way on account of mileage alone – which was a pity.

Two, three-cylinder 4-2-0s were built for the

Fig. 522 **Robt. Stephenson** Newcastle 1747/1867 North Eastern Rly 'Long Boiler' 0-6-0 No. 658

York, Newcastle & Berwick Railway in 1846 with 'long boilers', two small outside cylinders $10\frac{1}{2}$″ x 22″ and one inside cylinder $16\frac{3}{4}$″ x 18″, the centre cylinder having the same volume as the two outside.

A number of 'Cramptons' were built, including two 4-2-2s for the London & North Western Railway with 7 ft diameter driving wheels, and two outside cylinders 18″ x 24″. The boiler had a total heating surface of 1348·67 sq. ft and a grate area of 16·125 sq. ft. They were completed in 1848, and were a development of the 'A' 4-2-0.

By 1855 over 1000 locomotives had been built and orders were varied and widely dispersed. Some 0-4-0s were supplied to Norway in 1857 and fifty 'singles' to the Lombardo-Venetian Railway. Locomotives were sent to Australia, Egypt, Turkey, Ceylon, Holland, Belgium, Luxembourg and India.

In 1882 the first order was obtained from China and large six-coupled locomotives, 4-6-0s, were supplied to the Natal Government Railway, 2-6-0s to the Central Uruguay Railway and four 4-6-2Ts to the Bengal Nagpur Railway and by 1900, 3000 locomotives had been built.

Home railways were also well provided for particularly the North Eastern, Midland, London & South Western, Welsh, and the Great Northern Railways.

A private company was formed in 1886, which was voluntarily wound up in 1899 and a new public limited company formed as: ROBERT STEPHENSON & CO. LTD.

Some 3 ft gauge 4-4-0Ts were supplied to the Cavan & Leitrim Railway and ten 0-6-0s, with 17″ x 24″ cylinders and 5′ 3″ diameter wheels, to the Midland Great Western Railway of Ireland.

Fig. 523 **Robt. Stephenson** Newcastle 2309/1879 maker's plate

It was decided at the turn of the century to find more convenient accommodation. The Forth Street works could expand no more and the layout had become very uneconomic. New works were built at Darlington and production commenced in 1901–2. Part of the old works was taken over by their neighbour, R. & W. Hawthorn Leslie.

STEPHENSON, ROBERT & CO. LIMITED
Springfield, Darlington

Fifty-four acres of land had been purchased by the company and the new works were completed in 1902 when all machinery and equipment was moved from Newcastle. New construction commenced and the first locomotive was steamed in October, 1902.

Some 4-4-0s for the Oudh and Rohilkund Railway were built in 1903–4, followed by five for the SE&CR, and five for the Cambrian Railway. Some larger types were put in hand, some 2-8-0s for the Bengal Nagpur Railway and various 4-6-0s for the Bengal Nagpur, Oudh & Rohilkund, Central Argentine and Burma railways. Up to 1914 the majority of orders came from abroad and most of those from India, Burma and South America, to the designs of the individual companies or the Crown Agents. Nevertheless in certain details one can see the Stephenson characteristics even though the general design was not Stephensons' responsibility.

In other cases the overall dimensions, maximum axle weight and duty would be specified and a design built around the specification. The 2-8-0s built for the Bengal–Nagpur Railway in 1903, with two outside cylinders 21″ x 26″, coupled wheels 4′ 8″ diameter, had a heating surface of about 2000 sq. ft and a grate area of 32 sq. ft, the working pressure being 180 lb/sq. in. A Belpaire firebox was incorporated. These locomotives performed so well that the Standards Committee for the Indian State Railways adopted it as a standard type, with slight modifications. The first 2-10-0 to be built in Britain was laid down in 1905 for the Argentine Great Western Railway's 5′ 6″ gauge. It had two outside cylinders $19\frac{1}{2}$″ x 28″, 4′ 3″ diameter coupled wheels, and a total heating surface of 2440 sq. ft. It weighed 79 tons 10 cwt without the tender.

Fig. 524 **Robt. Stephenson** Darlington 3926/1926 Kenya Uganda Rly 2-8-2 No. 6 metre gauge

Fig. 525 **Robt. Stephenson** Darlington 3982/1928 Central Argentine Rly. No. 863 *Class CS8A* compound cylinders 21″ & 31½″ × 26″ D.W. 5′ 2″ 5′ 6″ gauge

Fig. 526 **Robt. Stephenson** Darlington 4028/1930 Buenos Ayres Pacific Rly. No. 2367 4-6-4T D.W. 5′ 7″

Fig. 527 **Robt. Stephenson** Darlington 4085/1934 Appleby-Frodingham Steel Co. Ltd. No. 50

STEPHENSON, ROBERT & CO. LIMITED

Some interesting tank locomotives were built during this period including 2-8-2Ts for the Bengal–Nagpur, and 0-8-2Ts for the Ottoman Railway. At home some powerful 0-6-2Ts were ordered by the Brecon & Merthyr, and Rhymney railways.

For the South African 3′ 6″ gauge some powerful 4-8-2s were built, and for the Sudan Government Railways some *Mikado* 2-8-2s with bar frames, in 1923. Stephenson's valve gear was fitted as standard to the majority of locomotives built, but Walschaerts gear was fitted to many locomotives of the larger types when specified.

Few locomotives were built in 1915 and none in 1916, the company being engaged in munition work, gun mountings and parts, and Admiralty work. However in 1917 the War Ministry urgently required locomotives for operating the railways taken over on the Continent and eighty-two Robinson type 2-8-0s were built in 1917–20 and thirty 0-6-0Ts for the Metre gauge with outside frames and cylinders 11″ x $15\frac{3}{4}$″ and wheels 2′ 10″ diameter. These locomotives had sheeting around and were roofed over similar to a tram locomotive.

The 1920s became a period of unrest, with labour troubles and as a consequence many overseas orders were cancelled. The largest order in 1921 was for thirty-five standard 2-6-0 mixed traffic locomotives for the Great Western Railway, which up to this time had built the majority of its requirements at Swindon.

Small orders started to come through from India, Africa and Argentina supplemented by home orders from the L&NER for thirty 0-6-2Ts, and Somerset & Dorset for five 2-8-0s.

In 1928 ten compounds were built as 2-8-2s for the Central Argentine Railway with cylinders $31\frac{1}{2}$″ x 26″ and 21″ x 26″, the coupled wheels being 5′ 2″ diameter. A few industrial locomotives were built from time to time, including two 0-6-0STs with 18″ outside cylinders and outside Stephensons' valve gear, for the Moss Bay Steel Company in 1934.

Orders were indeed scarce around this time and in 1936 and 1937 the total built was forty-six. Among these were eleven 4-6-0s of the *Sandringham* class for the L&NER and seventeen six-coupled industrial tanks.

Stephenson's works, although modern in their equipment, were laid out for small batch orders, but all their locomotives were carefully built, quality coming before quantity.

On 1 January, 1937 Robert Stephenson & Co. Ltd. combined with R. & W. Hawthorn Leslie & Co. Ltd. and became: ROBERT STEPHENSON & HAWTHORNS LTD.

The locomotives built by R. & W. Hawthorn Leslie totalled 2783, and the last works number by Robert Stephenson & Co. Ltd. was 4155. These were added together making a total of 6938 and the first new combined works No. of RS&H was 6939. This, of course, is not the total built, and it will be recalled in the Robert Stephenson & Company Newcastle section that a second series of works numbers commenced in 1831 at No. 1, the first series from 1825 to 1831 reaching No. 40 at least. A few were duplicated in the second series, so that approximately thirty-five early locomotives should be added to the RS&H totals.

STEPHENSON ROBERT & HAWTHORNS LTD.

Darlington & Newcastle-on-Tyne

On 1 January, 1937 Robert Stephenson & Co. Ltd., purchased the locomotive department of R. & W. Hawthorn Leslie Co. Ltd., (excluding the boiler department) and the goodwill and business for the sum of £40,260 to be satisfied by allotment of 161,040 five shilling (25p) shares. R. & W. Hawthorn, Leslie & Co. Ltd. were also to subscribe in cash, at par, 113,960 shares of Robert Stephenson & Co. Ltd.

The total number of engines built by R. & W. Hawthorn Leslie & Co. Ltd. were 2783 and those by Robert Stephenson & Co. Ltd. 4155. The numbers were added together making 6938. The first locomotive built by the amalgamated firms was given the works number 6939.

The policy was to concentrate the building of main line locomotives at Darlington and industrial lcomotives at the Forth Banks works of the former Hawthorn Leslie company. Some industrial locomotives were built at Darlington from time to time depending upon what type of orders were received at any particular time.

In 1938 RS&H acquired the goodwill of the locomotive business of Kitson & Co. Ltd. of Leeds together with Manning Wardle & Co. Ltd.; and spares for locomotives built by these two firms were supplied as and when required.

Standard 0-4-0ST and 0-6-0ST continued to be built in large numbers from 1938 for collieries, steelworks, chemical plants, the Air Ministry, Ministry of Supply, dockyards and many other industries. Very few orders were obtained from abroad but war time requirements were such that the works were fully extended. In 1943 an order for ninety 0-6-0STs of the Ministry of Supply's *Austerity* class was put in hand and completed in 1945.

Fig. 528 **R. Stephenson & Hawthorn** 6982/1941 Iraq Rly for heavy expresses – Kotchek to Baghdad (330 miles) oil fired trough system

Fig. 529 **R. Stephenson & Hawthorn** 7754/1954 Richard Thomas & Baldwin Ltd. No. 81 standard 0-6-0ST $18\frac{1}{2} \times 26$ cylinders 4′ 3″ D.W.

Fig. 530 **R. Stephenson & Hawthorn** 7813/1954 Standard 0-4-0ST APCM Holborough Works TUMULUS

STEPHENSON, ROBERT, & HAWTHORNS LTD.

In 1944 the Vulcan Foundry Ltd. acquired a substantial stock-holding in the Company and as a result there has been close association with the two firms. They became members of the English Electric Co. Ltd. and the works at Newton-le-Willows, Darlington, Newcastle, Bradford and Preston formed the main units of the railway traction division. The Vulcan Foundry was founded in 1830 by Robert Stephenson and Charles Tayleur and so the wheel turns a full circle.

In the same year six 4-6-0s were built for the Ceylon Government Railways with outside cylinders 18½" x 26" and 4' 5½" diameter wheels. Six more were built in 1947.

In 1948–9 ten 4-6-2s *Class YB* were built for the Indian Government Railways with outside cylinders 16" x 24" and 4' 9" diameter coupled wheels and orders from Tasmania, Jordan, India, Tanganyika, New Zealand, and Australia were received and executed.

Two orders were received from British Railways, the first for thirty-five Eastern region *Class L1* 2-6-4Ts which were built in 1949–50, and then for eighty 0-6-0PTs for the Western region which were built from February 1950 to January 1953. In 1951 five powerful 0-6-0STs were supplied to Dorman Long's Lackenby works. They had 18" x 24" outside cylinders and weighed 53 tons in working order, similar locomotives were supplied to various units of the National Coal Board. The largest 0-6-0STs were two supplied to Messrs Richard Thomas & Baldwins Ltd. for their Ebbw Vale Steelworks. They had 18½" cylinders and weighed 69 tons 15 cwt.

The British Electricity Authority preferred six-coupled side tanks and a number were supplied in the 1950s, closely following in design a similar locomotive supplied to Birmingham Power Station at Ham's Hall in 1936 by R. & W. Hawthorn Leslie.

Fireless locomotives continued to be built, and the last steam locomotive built by RS&H was a six coupled fireless locomotive ordered by the National Coal Board for their Glasshoughton Coking Plant and delivered in January 1959, works No. 8082. The last conventional steam locomotive built was an inside cylinder 0-6-0ST for Messrs Stewarts & Lloyd's Hungarford Quarry completed in October 1958 (works No. 8051).

The Forth Banks works were closed in 1960 and all diesel and electrical locomotive building was concentrated at Darlington. Forth Banks works, established in 1823 by Robert Stephenson & Company, adjoined the works of R. & W. Hawthorn Ltd. and so a very important link with the past has gone. It was virtually the birthplace of the locomotive. Approximately 1000 steam locomotives were built by the joint company.

As from 1 January, 1962 the works became: ENGLISH ELECTRIC CO. LTD., Stephenson Works, Harrowgate Hill, Darlington. The name of the works happily perpetuates the long record of locomotive building of George and Robert Stephenson.

STEWART, WILLIAM
Newport, Monmouth

An unknown design of locomotive was built for the Park End Colliery Co., Forest of Dean in 1814 and was tried on the Severn & Wye tramroad. Apparently it was successful, but was not accepted by the Park End Colliey Co., and was taken back to Newport. No further details are known.

STIRLING, JAMES & CO.
East Foundry & Victoria Foundry, Dundee

Under the name of the Dundee Foundry Co. the firm was established at the end of the eighteenth century, principally for the production of iron castings. Later steam engines and ships' machinery were manufactured and in the 1830s the firm became James Stirling, Dundee Foundry Co. James's brother Robert was also a partner. It was here that Robert's son Patrick served his apprenticeship, later becoming Locomotive Superintendent of the Glasgow & South Western Railway (1853) and obtaining a similar position on the Great Northern Railway in 1866 where he gained his great reputation as a locomotive engineer.

It was decided to enter the field of locomotive construction and in 1834 TROTTER was built for the Dundee and Newtyle Railway. It was stated to be similar to the two built by J. & C. Carmichael, but smaller. The Carmichael pair were 0-2-4s with vertical cylinders and the Stirling locomotive was of the same wheel arrangement. The cylinders were 11" x 18", the driving wheels 4' 6" diameter, and the weight 6 tons. The gauge was 4' 6½". It was withdrawn in 1849.

In November 1837 a tender was accepted for three locomotives for the Arbroath and Forfar Railway at a price of £1145 each, including tenders. The railway was ready for opening, in part, before the delivery of the first locomotive and horses had to be employed for hauling the first passenger trains. It was not until January 1839 that the first appeared, named VICTORIA. It was a 2-2-2 with inclined outside cylinders adjacent to the smokebox 13″ x 18″. The driving wheels were 5 ft diameter with short connecting rods implemented by extended piston rods. The slide valves were fitted horizontally above the cylinders. The boiler was domed with a raised firebox surmounted by a man-hole cover complete with whistle. Wheel guards were fitted in a continuous strip over all wheels.

According to Wishaw, the boiler was 3′ 9″ diameter with 105 2″ OD tubes, 8′ 5″ long. The firebox was 4 ft long x 2′ 6″ wide x 6 ft high.

The leading and trailing wheels were 3′ 6″ diameter and all wheels had inside bearings. The four-wheeled tender held 540 gallons of water and 18 cwt of coke.

Fig. 531 **James Stirling** 1839 Arbroath & Forfar Rly. VICTORIA

Date	Type	Cylinders inches	D.W. ft in	Railway	Name	Gauge ft in	
3/1834	0-2-4	VC 11 x 18	4 6	Dundee & Newtyle	Trotter	4 6	
1837				Newtyle & Glamis	Victoria		
1838	0-4-2?	OC 12 x 18	4 6	Arbroath & Forfar	Princess	5 6	
1839	2-2-2	OC 13 x 18	5 6	Arbroath & Forfar	Victoria	5 6	
1839	2-2-2	OC 13 x 18	5 6	Arbroath & Forfar	Britannia	5 6	
1840	2-2-2	OC 13 x 18	5 6	Arbroath & Forfar	Caledonia	5 6	
1840	2-2-2	14½ x 18	4 6	Arbroath & Forfar	Albert	5 6	

GOURLAY MUDIE

Date	Type	Cylinders inches	D.W. ft in	Railway	Name	Gauge ft in	
1847	0-4-0	14 x 18	4 6?	D. P & A Junc	18 Caledonia	Std	
1848	0-4-0	14 x 18	4 6	D. P & A Junc	19 Gowrie	Std	
1848	0-4-0?			Aberdeen	21	Std	SNE 68
1849	0-4-2	IC 14 x 20	4 6	D. P & A Junc	20	Std	
1849	0-4-2	IC 14 x 20	4 6	D. P & A Junc	21	Std	

c 1843 firm taken over by Gourlay, Mudie & Co at some time known as Gourlay Bros.
D. P. & A. Junc: Dundee, Perth & Aberdeen Junction Rly.

STIRLING, JAMES & CO.

Two other similar locomotives were turned out – the BRITANNIA in 1839 and CALEDONIA in 1840. The gauge of the railway at this time was 5′ 6″. Other locomotives which may have been built by Stirling were one 0-4-2 PRINCESS, one 2-2-2 ALBERT, and one 0-4-0 VICTORIA but their location and further particulars have not, as yet been verified.

About 1843 the firm was taken over and renamed: GOURLAY, MUDIE & CO. who undertook to alter two of Stirling's locomotives in 1846. Their principal business being shipbuilding and marine engineering, very few locomotives were built but a few four-coupled tender types were turned out. It is doubtful whether any were built after 1850 although no records can be traced. The total number built would be fifteen to twenty. Daniel Gooch was a draughtsman here from 1836 to 1837.

STOCKTON & DARLINGTON RAILWAY

New Shildon

These workshops were established in 1826 and consisted of the bare essentials for carrying out repairs to the locomotive stock and conditions for the workman must have been extremely hard and primitive but gradually the hand tools were supplemented by machine tools and standards improved.

The original stock of the S&DR consited of four locomotives built by Robert Stephenson & Co., two in 1825 and two in 1826 and although new locomotives they were very troublesome in maintaining steam and in general maintenance. In fact, at one period the railway practically decided to revert to horse haulage.

The locomotive engineer was Timothy Hackworth and he proposed to build a locomotive which would fulfil all the requirements needed. He was given a free hand and in 1827 he produced the famous ROYAL GEORGE. This was not entirely new as the boiler of CHITTAPRAT, which had been built by Robert Wilson and tried out on the S&DR without success was used. The general design of the ROYAL GEORGE was quite original. It was the first to have six-coupled wheels. An interesting feature was that the leading and intermediate wheels were mounted on a pair of inverted leaf springs but the rear driving wheels were not sprung as the vertical cylinders were mounted

Fig. 532 **S&DR** 1827 Hackworth's ROYAL GEORGE parts used from CHITTAPRAT

above them with a direct drive to them. A shaft bearing the two loose eccentrics was mounted above the boiler in front of the cylinders, and the slide valves and reversing gear were operated by them.

ROYAL GEORGE proved the superiority of the steam locomotive over the horse and in 1829 the VICTORY was put in hand. It was similar to the ROYAL GEORGE except for the motion, the crosshead having slide bars and the position of the eccentrics altered to the rear axle. The cylinders were slightly larger. Although the next type that Hackworth designed (for passenger traffic), was not built at Shildon but at Newcastle by Robert Stephenson & Co., it had some special features which are worth noting. It was much smaller than the previous 0-6-0s, the boiler having a single flue with radial tubes passing through the flue tube and may have been the fore runner of the well-known 'Galloway' type of stationary boiler. Surmounting the boiler was a large copper dome which acted as a steam drier. Other new features were inside cylinders driving a cranked axle, and the valve gear being reversed by a single lever.

One wonders, if Hackworth had built this locomotive at New Shildon, whether there would have been any doubt as to it being the first inside cylinder locomotive with a cranked axle. As it turned out both Stephenson and Bury laid claim to this innovation as well as Hackworth.

The next type was the 0-6-0s of the *Majestic* class, of which six were built at New Shildon although it is possible that R. &. W. Hawthorn supplied parts for three and R. Stephenson & Co. for the other three; the locomotives being erected by Hackworth in the S&DR works. These were another radical departure as far as the drive was concerned. The vertical cylinders were fixed to an extended platform in front of the chimney and drove a 'jackshaft' – a crankshaft without wheels – and the cranks were coupled to the leading wheels which were also coupled to the wheels on the other two axles. It has been stated that counter weights were fixed at the back end to counteract the overhung weight and the tendency of this class to 'shoulder'.

Hackworth resigned from the service of the Stockton and Darlington railway in 1840 and set up a works of his own in the vicinity. The post was taken over by William Bouch who was a brother of the designer of the first Tay Bridge – Thomas Bouch.

The six locomotives of the *Shildon* class, Nos. 29 to 34, had cylinders 15" x 24" with the 'standard' diameter of 4 ft for the wheels of mineral locomotives of the time. The characteristic long boiler was fitted to this class with a length of 13 ft and a total heating surface of 1363 sq. ft and a small, disproportionate grate area of 10 sq. ft. No. 35 COMMERCE was slightly different with horizontal cylinders coupled to the centre pair of wheels. Two outside cylinder 2-4-0s were built with inside frames and 6 ft diameter driving wheels.

In 1849 an agreement was made of a most unusual nature. William Bouch with Oswald Gilkes, who was the Secretary of the Stockton and Darlington Railway, negotiated with the company to provide motive power for the railway and build new locomotives as required in the same workshops at New Shildon.

From 1854 to 1858 four such new locomotives were built by the newly formed company which was known as: THE SHILDON WORKS COMPANY. In July 1863 the Stockton and Darlington Railway became part of the NORTH EASTERN RAILWAY. Before this amalgamation it was planned to build a new works at Darlington which would be more central and would enable the workshops to be planned more efficiently.

Building however proceeded at New Shildon until 1867 with a new series of 0-6-0 mineral locomotives. William Bouch was also busy planning the new works at Darlington which was opened in 1863.

After 1863, besides the six new locomotives built at New Shildon, four more were erected with parts supplied by Darlington. The works continued to repair locomotives until 1871 when the works were closed and work was transferred to Darlington and so ended an important phase of locomotive development, stemming from the ROYAL GEORGE of 1827.
See NORTH EASTERN RAILWAY – DARLINGTON.

STOCKTON & DARLINGTON RAILWAY

BUILT BY T. HACKWORTH
SHILDON

Built	Type	Cylinders inches	D.W. ft in	S & D No	S & D Name	
11/1827	0-6-0	11½ x 20	4 0	5	Royal George	(a) Sold 1840
9/1829	0-6-0	12 x 22	4 0	8	Victory	(a)
1831	0-6-0	14½ x 16	4 0	12	Majestic	(b)
1831	0-6-0	14½ x 16	4 0	13	Coronation	(b)
1831	0-6-0	13 x 16	4 0	14	William the Fourth	(b)
1831	0-6-0	13 x 16	4 0	15	Northumbrian	(b)
1832	0-6-0	14½ x 16	4 0	17	Lord Brougham	(b)
1832	0-6-0	14½ x 16	4 0	18	Shildon	(b) Sold 1838

BUILT BY W. BOUCH

Built	Type	Cylinders inches	D.W. ft in	S & D No	S & D Name	
3/1842	0-6-0	15 x 18	4 0	7	Prince	
1843	2-2-2	12 x 18	5 2	50	Meteor	
6/1845	0-6-0	15 x 24	4 0	29	Miner	
6/1845	0-6-0	15 x 24	4 0	30	Wear	
9/1845	0-6-0	15 x 24	4 0	31	Redcar	
2/1846	0-6-0	15 x 24	4 0	32	Eldon	
2/1846	0-6-0	15 x 24	4 0	33	Shildon	
3/1846	0-6-0	14 x 24	4 0	34	Driver	
1/1847	0-6-0	16 x 24	4 0	35	Commerce	
5/1847	0-6-0	16 x 24	4 0	36	Guisboro	
6/1847	0-6-0	16 x 24	4 0	37	Gem	
6/1847	2-4-0	14½ x 24	6 0	38	Rokeby	
6/1847	2-4-0	14½ x 24	6 0	39	Ruby	
2/1849	0-6-0	15 x 24	4 6	63	Birkbeck	
2/1849	0-6-0	15 x 24	4 6	64	Larchfield	

BUILT BY SHILDON WORKS CO (W. BOUCH & O. GILKES)

Built	Type	Cylinders inches	D.W. ft in	S & D No	S & D Name	
5/1854	0-6-0	18 x 18	4 0	80	Duke	
7/1855	2-4-0	16 x 20	5 4	99	Ayton	
7/1855	0-6-0	16 x 20	5 4	100	Stobart	
11/1858	0-6-0	18 x 18	4 0	135	Eden	

BUILT UNDER OWNERSHIP OF NORTH EASTERN RAILWAY

Built	Type	Cylinders inches	D.W. ft in	S & D No	S & D Name	
12/1863	0-6-0	17 x 24	5 0	171	Gladstone	
12/1863	0-6-0	17 x 24	5 0	172	Barrow	
6/1864	0-6-0	17 x 24	5 0	173	London	
6/1864	0-6-0	17 x 24	5 0	174	John Dixon	
6/1867	0-6-0	17 x 24	5 0	202	Ireland	
6/1867	0-6-0	17 x 24	5 0	203	England	

FOLLOWING PROBABLY ERECTED AT SHILDON FROM PARTS SUPPLIED BY DARLINGTON NORTH ROAD WORKS

Built	Type	Cylinders inches	D.W. ft in	S & D No	S & D Name	
9/1865	0-6-0	17 x 24	5 0	189	Spring	Long Boiler Type
9/1865	0-6-0	17 x 24	5 0	190	Summer	Long Boiler Type
10/1866	0-6-0	17 x 24	5 0	194	Alice	Long Boiler Type
10/1866	0-6-0	17 x 24	5 0	195	Helena	Long Boiler Type

(a) Vertical cylinders at rear, driving the rear coupled wheels. (b) Vertical cylinders at chimney end driving jackshaft.

STOTHERT, SLAUGHTER & COMPANY

Avon Street, Bristol

This firm was founded by Henry Stothert in 1837 as Henry Stothert & Company, and in 1841 Edward Slaughter became a partner forming the firm of Stothert Slaughter & Company. Locomotive building commenced in that year with an order for two 7 ft gauge 2-2-2s for the Great Western Railway to Sir Daniel Gooch's designs. They had sandwich frames, inside cylinders 15″ x 18″ and 7 ft diameter driving wheels. The boilers were domeless with large domed fireboxes with safety valves on top. The tenders had four wheels and no brakes were fitted. They were part of an order for sixty-two *Firefly* class and were actually built by seven locomotive builders. They were named ARROW and DART. Eight *Sun* class followed for the same railway. They were smaller with 6 ft diameter driving wheels but were similar in shape. The firm anticipated further orders from the GWR but more came from the South Devon and Bristol and Exeter Railways, also with 7 ft gauge track. An order for two 5 ft gauge 2-4-0s was then dealt with for the Great Eastern Railway.

Unfortunately there are large gaps in the records up to 1859 and some after that date, so that the complete range of locomotives built during that period cannot be assessed.

Locomotives built were for: Great Western Railway of Canada; London & Southampton; East Indian; Minho & Douro – Portugal; East Lancashire and other railways.

In 1844 the works were named Avonside Ironworks and in 1851 a shipbuilding yard was acquired at Hotwells, Bristol and Henry Stothert took charge of this work, many ships being built up to the 1870s. A Mr G. K. Stothert was building ships from 1844 and this may have been the same shipyard.

Further broad gauge locomotives were built for the Birmingham and Gloucester Railway, the order being for six 2-2-2s two 0-6-0s and three 2-4-0s. They were completed in 1844 but when the conversion from 7 ft to standard gauge took place in 1854 none of these locomotives were converted. The firm supplied the locomotives and at the same time contracted to be responsible for the management of the locomotive department, supplying all the necessary labour and stores, for a period of two years.

When Stothert went over to the ship building side, Edward Slaughter remained at the locomotive works and in 1856 Mr Gruning became a partner and the firm became: SLAUGHTER, GRUNING & COMPANY.

Up to this time approximately 340 locomotives had been built. For the next ten years many railways abroad were supplied including: East Indian Railway; Tarragona Barcelona & France Railway; Victorian Government, Railways; Northern Railway of Spain; S. Australian Railways; Canterbury, (N.Z.); Buenos Ayres Northern Railway; Queensland Railways; Sweden; Norway; Chile; and the Great Indian Peninsula Railway.

In 1864 two interesting 0-8-0Ts appeared for the Vale of Neath Railway with outside cylinders $18\frac{1}{2}$″ x 24″, 4′ 6″ diameter wheels, Gooch's stationary link motion and side tanks holding 1300 gallons. They were the country's first 0-8-0Ts to work on a main line and these two worked the banks around Glyn Neath. Unfortunately the amalgamation with the Great Western Railway took place in the following year and they evidently did not come under the approval of that railway's locomotive department and both disappeared in 1871. Two

Fig. 533 **Stothert Slaughter** 1846 London Brighton & South Coast Rly. 0-6-0 No. 112

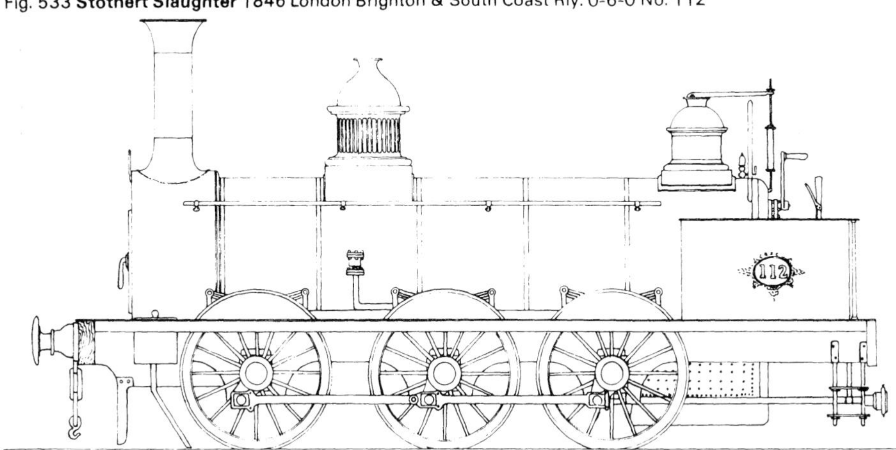

Fig. 534
1392/1898 North Mount Lyell Co. *Class TA* 4-6-0 3′ 6″ gauge

further 0-8-0Ts were built for the Great Northern Railway in 1866. They were larger, and were fitted with condensing gear. All four were fitted with Messrs Slaughter & Caillet patent control springs which were fitted to the leading and trailing wheels to provide side play.

A further order for 7 ft gauge GWR locomotives, amounting to twenty 2-4-0s of the *Hawthorn* class, were built in 1865–6 with inside frames and domeless boilers. One of them was named SLAUGHTER but, for obvious reasons, for the peace of mind of would-be passengers the name was tactfully changed to AVONSIDE. Another of the class HEDLEY, although altered from time to time, finishing up as a stationary engine and was not cut up until 1929.

At the beginning of 1866 the firm's name was again changed to; AVONSIDE ENGINE COMPANY and by this time 600 locomotives had been built.

Archibald Sturrock of the Great Northern Railway ordered a large number of steam tenders and five were built by Avonside. These were in effect the forerunners of the 'booster' tender introduced by Gresley on the LNER. From the same railway orders were received for five 0-4-2WTs for the Metropolitan trains and ten small 2-4-0s, all completed by 1868.

The largest orders for this period were:

1866	20	2-4-0s	for Great Indian Peninsula Rly
1867	30	2-4-0s	for East Indian Rly
1867	20	0-6-0s	for GIPR
1870	15	0-6-0s	for Oudh & Rohilkund Rly

In the 1870s Avonside built a large number of locomotives to Fairlie's patent and a few to Fell's patent. The Fairlies, up to 1870, had been built by Geo. England & Co., James Cross and by Fairlie himself when he took over the Hatcham works of Geo. England. The works were sold in 1872 but orders for this type of locomotive were numerous and a number of locomotive firms built them. Avonsides built more than any other firm and they went to Australia, Canada, India, Africa, Burma, Cuba, Peru, Brazil, Mexico and other countries. The Fairlie locomotives built were 'double' and 'single' types. The 'double' had two boilers with

Fig. 535
1402/1900 Stanton Ironworks STANTON NO. 2

a common firebox and two engine units. The 'single' types had one boiler and engine unit.

In most cases the 'double' Fairlie locomotives were allocated two works numbers so that the final works number of Avonside does not indicate the total number built.

The 'Fell' locomotives were four 0-4-2Ts built for the New Zealand Government Railways and were designed for the Rimutaka incline on North Island. A separate single rail was laid between the two running rails but laid horizontally instead of vertically. When operating on the incline two horizontal wheels on each side of the centre rail were engaged by powerful springs. On the level these locomotives had two outside cylinders 14" x 16" driving in the normal way the 2' 8" diameter coupled wheels. Two inside cylinders 12" x 14" drove the friction wheels which could be engaged or disengaged as required. They were built in 1875.

The last double Fairlies built were for the Indian States and Burma railways in 1880, followed by seven single 0-6-4T. Fairlies for the New Zealand Government Railways with 13" x 16" cylinders and 3' 0½" diameter wheels, were the completion of a total order for twenty-one of this type, the last of which was steamed in 1881. After twenty-five 0-6-0s were built for the East Indian Railway in 1880–1 and twenty 2-6-0s for Indian State Railways in 1881 orders were few and far between and from 1882 to 1900 only eighty-five were built, an average of less than nine per annum.

The capacity of the works was such that good deliveries for large orders of main line locomotives were impossible to attain. The firm were in the doldrums, Edward Slaughter had died which led to financial difficulties. Edwin Walker, who came from the neighbouring firm of Fox Walker & Company set about re-organising the works and the policy was changed to the production of industrial tank locomotives, Walker having had considerable experience in this field. Unfortunately he met with little success and the firm went into liquidation in the early 1880s. Despite this set-back Walker formed a new company but bearing the same name, with the intention of still building locomotives but on a smaller scale in physical dimensions and quantities.

So two firms in the same city had the same policy of building industrial locomotives – the new Avonside Engine Company, and the successors to Fox Walker & Company, Peckett & Sons. Both firms were to become famous names in this type of locomotive manufacture.

From 1889 Avonside built a standard design of four-coupled and six-coupled saddle tanks. In addition three interesting 2-6-4Ts, for the Junin Railway of Chile, which were three-cylinder compounds, were built, with one high pressure cylinder 15" x 16" inside the frames and two low pressure cylinders 15" x 20". They were to 2' 6" gauge with outside frames, and as the HP cylinder drove the front coupled axle, the frames were considerably extended. These three locomotives were the only compounds built by Avonside. Up to 1905 the industrial saddle tanks predominated and the total number of locomotives built from 1841 was approximately 1475.

In this year the firm moved to new premises at Fishponds, Bristol and the building of industrial types continued. A few larger locomotives were built from time to time including some 0-8-2Ts with 20" x 26" cylinders for the

Fig. 536
1865/1922
Croydon Gas Co.
ELIZABETH

E. Greta Coal Mining Company. In the same period a number of 0-4-0Ts were built for narrow gauge lines. During the 1914–18 war, sixteen 1′ 6″ gauge 0-4-0Ts were built for Woolwich Arsenal with outside cylinders 8½″ x 12″ and 2′ 1″ diameter wheels. Many standard tanks were built for Steelworks, collieries and docks.

In 1926 the Great Western Railway ordered six 0-4-0Ts for use in their docks at Swansea. They had outside cylinders 16″ x 24″ and 3′ 9½″ diameter wheels and were unusual in having outside Walschaerts valve gear which was a reversal of usual GWR practice of having the valve gear between the frames.

At the time of the 1930–1 trade depression, Avonside fared no better than most firms and orders were very hard to come by. Although petrol and diesel locomotives were being built (the first petrol driven locomotive was produced as far back as 1913), from 1932 to 1935 only eleven steam locomotives were built and in November 1934 the firm again went into voluntary liquidation and in July 1935 the Hunslet Engine Company bought the goodwill of the firm together with the drawings, patterns and stock. So ended a renowned firm of builders, stemming from Broad Gauge days, producing industrial locomotives in the main, bearing their own well known characteristics.

The last works number was 2078 which included tenders (as separate orders) petrol and diesel locomotives, a few cancelled orders and some Fairlie locomotives which had two works numbers. There were also at least six 92 mm Klatz-Wilson safety valves which were allocated works numbers. If it is assumed that the blanks in the early locomotive lists were steam locomotives then the approximate number of steam locomotives built was 1960.

SUMMERS GROVES & DAY

Millbrook Foundry, Southampton

This engineering firm was founded in 1834 and did a lot of repair work on the London & South Western Railway's locomotives into the 1850s. A few locomotives were also built there although they may have acted as agents as well. FLY, SOUTHAMPTON and the London & Greenwich Railway No. 7 are all reported as being inspected during construction at Southampton, confirming that they were actually builders.

The first to be built was JEFFERSON in 1837 which was sent to America. The remainder were all built in 1839. No. 7 was rebuilt as a 2-2-2 later in 1839 and ran satisfactorily in this form. FLY also was converted to a 2-2-2 in 1841 and SOUTHAMPTON was sent to Fairbairns in 1841 and entirely rebuilt, everything being new except the boiler.

The Bourne Bartley pair were both withdrawn before 1846.

This record is incomplete and other new locomotives may have been built during this period. By 1847 the firm had become SUMMERS DAY & BALDOCK, Mill Place Iron Works, Foundry Lane, Millbrook, subsequently moving to Northam, becoming DAY SUMMERS & COMPANY.

SWAINSON, TAYLOR

See WHITEHAVEN COLLIERIES.

SUMMERS, GROVES & DAY.

Date	Type	Cylinders inches	D.W. ft in	Customer	Name/No
1837	2-2-0			Richmond, Fredericksburg & Potomac R. R.	Jefferson (a) (b)
3/1839	2-2-0	11 x 18	5 0	London & Greenwich Rly	No 7 (a)
1839	2-2-2	12 x 18	5 0	Bourne Bartley & Co	St. George (c)
1839	2-2-2	12 x 18	5 0	Bourne Bartley & Co	St. David (d)
6/1839	2-2-0			London & Southampton Rly	No. 40 Fly (a)
8/1839	2-2-2		5 6	London & Southampton Rly	Southampton

(a) 'Planet' type. (b) Alternative builder: Braithwaite, Milner & Co. (c) Sold to North Union Rly No. 13. (d) Sold to Bolton & Leigh Rly.

TAFF VALE RAILWAY
West Yard, Cardiff Docks

This was the only Welsh railway to build locomotives of any quantity, and was the oldest company in Wales. West Yard works date back to the mid-1840s and at first it was of modest size, coping with the maintenance of little more than a dozen locomotives. This state of affairs rapidly altered and at the end of 1863 the stock had risen to fifty-one many of the older locomotives having been scrapped by this time, and an endeavour was made to bring some order into locomotive affairs.

Up to this time no less than nine locomotive builders had contributed five different types, with very few alike. It is no wonder there had been a succession of locomotive superintendents up to 1846 when Henry Clements took over.

Locomotive Superintendents: George Bush 1840–1; Edward Bage 1841–December 1842; William Brunton December 1842– ; Richard Gregory; William Craig; Alexander Colville; Henry Clements December 1846–January 1858; Joseph Tomlinson January 1858 – July 1869; B. S. Fisher July 1869 – October 1873; Tom Hurry Riches October 1873 – September 1911; and J. Cameron September 1911–22.

The first locomotive built at West Yard was really a rebuild of a 0-4-2 named CARDIFF built by Sharp, Roberts in 1841. It was converted to a double frame 0-6-0 with cylinders and wheels the same size as the original, namely 14″ x 18″ and 4′ 6″ diameter, respectively. A few years after its rebuilding it was recorded as a new locomotive and as such has always been acknowledged as No. 1 of 1856.

Works plates were fixed to each locomotive but there is no evidence forthcoming that any works numbers were shown on any locomotive built at West Yard. Three series of works numbers are known but, to add to the confusion, rebuilds were also included. At first only names were carried and no running numbers, but this was altered in 1863 when each locomotive was given a number and the names were gradually removed.

The first genuine new locomotive was built in 1857 and was a double frame 2-4-0 named VENUS with $14\frac{1}{2}$″ x 20″ cylinders and 4′ 6″ diameter driving wheels. In 1859 West Yard commenced building a standard double frame 0-6-0 goods, designed by Tomlinson. More of this type were built than any other. They had 16″ x 24″ cylinders and 4′ 6″ wheels and forty-four were built up to 1872, numerous additions coming from Kitson, Hawthorn and Slaughter Gruning.

By 1864 two 2-4-0 double frame passenger locomotives had been completed, with 14″ x 20″ cylinders and 5′ 0″ diameter driving wheels. They had modest boilers and a small grate area of 11 sq. ft and four wheeled double framed tenders.

Tom Hurry Riches was appointed locomotive superintendent in October 1873 and designed an inside frame 0-6-0 and completed the last 2-4-0 in 1879. It is interesting to note that the 2-4-0 had equalising beams between the springs of the leading wheels and the front coupled wheels, in common with many previous locomotives of the same type, as indeed had many of the old 0-6-0s between the leading and intermediate coupled wheels. This assisted weight distribution and better riding qualities over some of the rougher sections of the track. The double framed locomotives gave a lot of trouble with broken axles despite the ad-

Fig. 537 **Taff Vale Rly** 1884 T. Hurry Riches 4-4-0T *Class I* No. 67

Fig. 538 **TVR** 305/1895 0-6-2T *Class O* sold to Lilleshall Co. No. 1 in 1932

ditional axleboxes. This was a common fault, not only on the Taff Vale Railway, and it was this that mainly influenced Mr Riches when he considered his 0-6-0 goods design. Eighty-five were built, mainly by Kitson but West Yard produced twelve from 1874 to 1889. They were basically of Kitson design.

The Taff Vale Railway was predominantly a mineral railway, with coal traffic as its mainstay. It was extremely intense, one train following another down the Aberdare, Merthyr and Rhondda valleys with another stream of traffic consisting of returned empties to the multitude of collieries.

These 0-6-0s were used for this work up to the 1920s, supplemented by the 0-6-2Ts which started to appear in 1885 and eventually became the best known 'Welsh' wheel arrangement on most of the railways of South Wales.

Between 1876 and 1883 four of the older double frame 0-6-0s were converted to 0-4-4 side tanks for working various branch lines. Cylinders were 16″ x 24″ with 4′ 8″ diameter coupled wheels. These were quite successful and despite the adaption of the double frames, and inside frame trailing bogies they looked very neat and were finished in the customary style of the TVR at that period, with polished brass dome cover, copper capped chimneys, brass beading, and fully lined out.

In 1884–5 three 4-4-0Ts were built, very similar in appearance to those running on the Monmouthshire Railway & Canal Co., at that time. (See illustration.) They had outside cylinders 16″ x 24″ with 2′ 9″ diameter bogie wheels and 5′ 3″ diameter driving wheels. The boiler had 956·14 sq. ft of heating surface, a grate area of 16 sq. ft and a pressure of 140 lb/sq. in. Their weight in working order was 45 tons 8 cwt. They were used on the branch lines and were eventually equipped for auto train working.

1891 heralded the first 0-6-2T built at West Yard, but it was not the first on TVR metals, thirty-eight having been built by Kitsons from 1885. Three were built at Cardiff in 1891–2. Known as *Class M,* they had 17½″ x 26″ cylinders, 4′ 6″ diameter coupled wheels, 1061 sq. ft of heating surface, 18 sq. ft of grate area, side tanks holding 1500 gallons of water and weighed 49 tons 6 cwt in working order. These were mixed traffic locomotives and later, when replaced by more powerful 0-6-2Ts, were used on push-and-pull services. Their numbers were 4, 5, and 54.

In 1894 six *Class O* 0-6-2Ts were started. They had the same cylinders and wheels as the *M* class but larger boilers, giving 1148·5 sq. ft of heating surface with a grate area of 19·14 sq. ft. They weighed 56 tons 6 cwt in working order.

These were followed by six *Class Ol* with the same wheel arrangement, and the only dimensional difference from the *Class O* was a greater overhang at the front end, increasing the weight to 56 tons 8 cwt. These were the last locomotives built at West Yard except for a rail motor unit built in 1903. They were allocated works numbers 300 to 311. The rail motor unit had outside cylinders 9″ x 14″ and 2′ 10″ diameter wheels. The horizontal boiler had a heating surface of 338·5 sq. ft, grate area 8 sq. ft, and a pressure of 160 lb/sq. in. It was the forerunner of seventeen more, making the Taff Vale the most extensive user of this mode of passenger service in Wales.

TAFF VALE RAILWAY

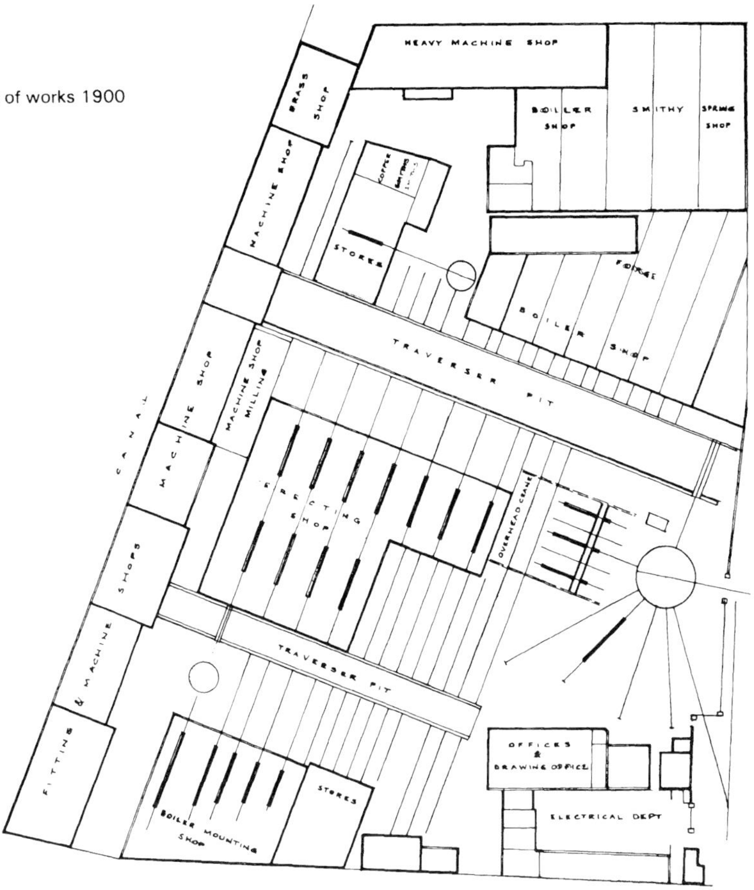

Fig. 539 **TVR** Plan of works 1900

Works No	Date	Type	Name	Cylinders inches	D.W. ft in	1st TV No	GWR No	TVR Class	W'drawn
1	1856	0-6-0	Cardiff	14 x 18	4 6	37			1874
2	1857	2-4-0	Venus	14½ x 20	4 6	41			1878
3	1859	0-6-0	Aberaman	16 x 24	4 6	17			1882
	1860	0-6-0	Stuart	16 x 24	4 6	18			1884
	1861	0-6-0	Dinas	16 x 24	4 6	4		R 0-4-4T	1893
	1861	0-6-0	Llancaiach	16 x 24	4 6	5		R 0-4-4T	1893
	1861	0-6-0	Merthyr	16 x 24	4 6	6			1887
	1861	0-6-0	Cambrian	16 x 24	4 6	10			1890
	1861	0-6-0	Llandaff	16 x 24	4 6	11			1888
	1862	0-6-0	Plymouth	16 x 24	4 6	3			1885
	1862	0-6-0	Newbridge	16 x 24	4 6	12			1891
	1862	0-6-0	Treforest	16 x 24	4 6	21			1895
	1863	0-6-0	Aberdare	16 x 24	4 6	7			1886
	1863	0-6-0	Cymmer	16 x 24	4 6	23			1894
	1863	0-6-0		16 x 24	4 6	51			1894
	1863	2-4-0		14 x 20	5 0	22		-	1886
	1864	2-4-0		14 x 20	5 0	24		-	1887
	1864	0-6-0		16 x 24	4 6	13		A	1884
	1864	0-6-0		16 x 24	4 6	14		A	1893
	1864	0-6-0		16 x 24	4 6	52		A	1887
	1864	0-6-0		16 x 24	4 6	53		A	1894
	1864	0-6-0		16 x 24	4 6	54		B	1898
	1865	0-6-0		16 x 24	4 6	15		B	1904

TAFF VALE RAILWAY

Works No	Date	Type		Cylinders inches	D.W. ft in	1st TV No	GWR No	TVR Class	W'drawn
	1865	0-6-0		16 x 24	4 6	16		B	1904
	1865	0-6-0		16 x 24	4 6	55		B	1906
	1865	0-6-0		16 x 24	4 6	56		A	1893
	1866	0-6-0		16 x 24	4 6	19		A	1885
	1866	0-6-0		16 x 24	4 6	20		A	1889
	1866	0-6-0		16 x 24	4 6	64		B	1904
	1866	0-6-0		16 x 24	4 6	65		A	1895
	1866	0-6-0		16 x 24	4 6	70		A	1898
	1866	0-6-0		16 x 24	4 6	71		A	1889
	1866	0-6-0		16 x 24	4 6	72		A	1895
	1866	0-6-0		16 x 24	4 6	73		A	1899
	1867	0-6-0		16 x 24	4 6	25		A	1895
	1867	0-6-0		16 x 24	4 6	26		A	1900
	1867	0-6-0		16 x 24	4 6	74		A	1893
	1867	0-6-0		16 x 24	4 6	75		B	1906
	1868	0-6-0		16 x 24	4 6	27		A	1896
	1868	0-6-0		16 x 24	4 6	37		A	1899
	1868	0-6-0		16 x 24	4 6	76		A	1895
	1868	0-6-0		16 x 24	4 6	77		A	1896
	1869	0-6-0		16 x 24	4 6	78		A	1897
	1869	0-6-0		16 x 24	4 6	79		A	1903
	1870	0-6-0		16 x 24	4 6	28		A	1899
	1871	0-6-0		16 x 24	4 6	29		A	1896
	1872	0-6-0		16 x 24	4 6	30		A	1902
	1872	0-6-0		16 x 24	4 6	40		A	1898
150	1874	0-6-0		17¼ x 26	4 6	38	(930)	L	1923
151	1875	0-6-0		17¼ x 26	4 6	39	(926)	L	1923
	1876	0-4-4T		16 x 24	4 8	4	–	J	1893
	1878	0-4-4T		16 x 24	4 8	5	–	J	1893
	1879	2-4-0		16 x 24	5 3	41	–	J	1895
162	1877	0-6-0		17¼ x 26	4 6	57	–	L	1911
163	1877	0-6-0		17¼ x 26	4 6	58	–	L	1910
164	1880	0-6-0		17¼ x 26	4 6	1	–	L	1914
165	1880	0-6-0		17¼ x 26	4 6	2	–	L	1914
166	1881	0-4-4T		16 x 24	4 8	59	–	J	1906
167	1882	0-6-0		17¼ x 26	4 6	6	–	L	1919
168	1882	0-6-0		17¼ x 26	4 6	18	913	L	1925
169	1884	0-6-0		17¼ x 26	4 6	19	912	L	1926
170	1884	0-6-0		17¼ x 26	4 6	20	(923)	L	1923
172	1883	0-4-4T		16 x 24	4 8	66	–	J	1902
	1884	4-4-0T		16 x 24	5 3	67	1133	I	1925
	1885	4-4-0T		16 x 24	5 3	68	1184	I	1925
	1885	4-4-0T		16 x 24	5 3	69	999	I	1925
	1889	0-6-0		17½ x 26	4 6	55	–	L	1915
	1889	0-6-0		17½ x 26	4 6	56	–	L	1915
	1891	0-6-2T		17½ x 26	4 6	4	442	M	1925
	1892	0-6-2T		17½ x 26	4 6	5	443	M	1926
	1892	0-6-2T		17½ x 26	4 6	54	466	M	1928
300	1894	0-6-2T		17½ x 26	4 6	21	446	O	1929
301	1894	0-6-2T		17½ x 26	4 6	25	447	O	1925
302	1894	0-6-2T		17½ x 26	4 6	26	448	O S	1930
303	1894	0-6-2T		17½ x 26	4 6	33	452	O	1925
304	1895	0-6-2T		17½ x 26	4 6	34	453	O	1926
305	1895	0-6-2T		17½ x 26	4 6	190	581	O S	1930
306	1897	0-6-2T		17½ x 26	4 6	28	450	O.1 S	1926
307	1897	0-6-2T		17½ x 26	4 6	60	471	O.1	1927
308	1897	0-6-2T		17½ x 26	4 6	61	472	O.1	1925
309	1897	0-6-2T		17½ x 26	4 6	62	473	O.1	1928
310	1897	0-6-2T		17½ x 26	4 6	63	474	O.1	1927
311	1897	0-6-2T		17½ x 26	4 6	64	475	O.1 S	1930
	1903	0-2-2		9 x 14	2 10	RM1	–		1919

S = Sold

TAFF VALE RAILWAY

The works were situated between Bute Street and the Glamorganshire canal, with Bute West Dock on the other side of Bute Street. The works access was by means of a track across Bute Street from a turntable near the Docks Station and all locomotives and wagons had to cross this busy street from and to the works. Horses were originally used for hauling everything across and these were supplemented by various 0-4-0STs from time to time.

When the TVR became a constituent company to the Great Western Railway in 1922, 274 locomotives were being maintained at West Yard. It could not expand in any direction although additional shops were badly needed and a plan to build a new factory at Radyr was in active consideration just prior to amalgamation.

With all the additional locomotives added to GWR stock, a central workshop was considered of paramount importance and eventually it was decided to enlarge and modernise the Caerphilly works of the old Rhymney Railway. Between 1924 and 1926 the works were gradually run down and work transferred to Caerphilly, and West Yard ceased repairs in the Autumn of 1926. It is difficult to assess the number of new locomotives built but as far as can be estimated eighty-four were built including the major rebuilds. The original locomotives of the company were painted green with red underframes. Mr Riches changed the colour scheme to reddish-brown, with the majority of goods locomotives in black. Later passenger locomotives were painted red. All were lined out. The number plate was an oval shaped brass casting. Tenders were numbered with the same number as its locomotive.

Locomotives were given duplicate or surplus stock numbers which were in the 200 and 300 series. Original numbers only are given in the list.

TAIT JAS. JNR. & PARTNERS
Middlesborough

This firm carried out extensive rebuilding and repairing of locomotives. Works numbers were issued for such work. No. 69 of 1920 was a 0-4-0ST with outside cylinders supplied to Stanley Bros., Stockingford Colliery, Nuneaton, and was supposed to be a new locomotive. This is extremely doubtful and it is more likely to have been assembled from parts from various locomotives.

TAYLEUR, CHARLES & COMPANY
Vulcan Foundry, Near Warrington

These works were established in 1830. After the opening of the Liverpool & Manchester Railway it was realised that a locomotive works in Lancashire would obviate the cost of transport from Newcastle and provide a better service for the railways in that county and beyond.

Chas. Tayleur was a local merchant and also a director of the Liverpool & Manchester Railway, and he entered into partnership with Robert Stephenson in 1832. The Foundry was built as a general engineering works intended to be capable of supplying girders, bridges, lifts, winding engines, cranes, points and crossings turntables and all the necessary equipment required for running a railway. Stephenson's colleagues, over at Newcastle, were not very keen on his interest being divided and as far as is known the partnership did not last more than a year or two.

Fig. 540 Jas. Tait Jun. 69/1920 Stanley Bros. Stockingford WESTWOOD

The first locomotive was completed in 1833 and was the first of an order for two 0-4-0s placed by Mr Hargreaves for the North Union Railway. It was similar to Stephenson's *Planet* design 'A' and had two inside cylinders 11″ x 16″ and 4′ 8″ diameter wheels on a wheelbase of 4′ 11″. It was named TAYLEUR and the second one STEPHENSON. In the same year three 2-2-0s were built for the Warrington & Newton Railway and were similar to Stephenson's *Planet* design 'B', with two inside cylinders 11″ x 16″, leading wheels 3 ft diameter and driving wheels 5 ft diameter.

Two 4-2-0s were built and sent to America for the Philadelphia & Colombia R.R. followed by three more of the same type in 1835 for the South Carolina R.R. They were the first 'bogie' locomotives to be built in Britain but unfortunately no details are available as to how the bogie was mounted or pivotted. Eight 2-2-2s were built for the Paris & St Germain Railway (1835–6), one 0-4-2 for the Belgian State Railway (1835), two 0-4-0s for the Raleigh & Gaston R.R. (1836), one 2-2-2 to Russia (1837) and two 0-4-2s to Vienna in the same year. These exports were the beginning of an excellent export trade which the firm built up and maintained throughout. The Stephenson influence was strong and in the 1840s many Stephenson 'long boilered' types were built.

In 1847 the firm's name was changed to THE VULCAN FOUNDRY COMPANY and in 1864 became a limited liability company: THE VULCAN FOUNDRY CO. LTD., Newton-le-Willows.

In 1852 the first locomotives to run in India were built. They were to the 5′ 6″ gauge for the Great Indian Peninsula Railway and comprised eight 2-4-0s, with inside cylinders 13″ x 20″ and 5 ft diameter coupled wheels. This initial order for India was to be the first of many spanning over a century, when average output over this period was one locomotive every fortnight.

The first Fairlie locomotives to be built by Vulcan Foundry were two for the Dunedin & Port Chalmers Railway (3′ 6″ gauge) and two for the Peruvian Government Railway (2′ 6″ gauge). All were 0-4-4-0Ts built in 1872. The firm built a number of Fairlie locomotives including the well known single Fairlie TALIESIN for the Festiniog Railway (1876), and the large 0-6-6-0Ts in 1910 for Mexican Railways, with 19″ x 25″ cylinders and 4 ft diameter wheels.

In 1870 the first locomotive to run in Japan was built. It was a 2-4-0T, with outside cylinders 12″ x 18″, 4′ 3″ diameter driving wheels for 3′ 6″ gauge and weighed 14 tons 15 cwt. An unusual 0-4-0ST was supplied in 1873 to the Tredegar Iron Company with 3′ 0″ diameter flangeless wheels, to run on 4′ 4″ angle iron rails to a gauge of 2′ 11$\frac{1}{4}$″. From 1870 many of the well known Kirtley 0-6-0 double frame goods tender locomotives were supplied to the Midland Railway.

On 1 January, 1898 the firm's title became VULCAN FOUNDRY LTD.

A steady flow of orders came from the home railways, and from Ireland, Russia, Peru, Uruquay, Chile, Germany, New Zealand, Australia, Spain etc., but the dominating orders

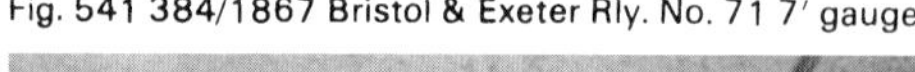
Fig. 541 384/1867 Bristol & Exeter Rly. No. 71 7′ gauge

were from India and the Argentine. To indicate how strong the Indian and Ceylon market was, out of 1032 locomotives built from 1907 to 1918, 977 were exported to those countries including 4-6-0, 0-6-0, 2-8-0, 0-6-4T, 4-4-0, 0-4-2, crane tanks and 4-4-2 types.

After the 1914–18 war the pattern continued and, in addition, thirty eight 4-8-0s were supplied to the Kenya & Uganda Railway's metre gauge (1922–4). Twenty standard LMS 0-6-0Ts were built in 1924 (LMS Nos. 7100–7119) followed in 1926–7 by one hundred more (LMS Nos. 16460–16509 & 16550–16599). Pursuing the 'small engine' policy of the Midland Railway, an order was placed by the LMS for sixty-five compound 4-4-0s with 19"/21" x 26" cylinders and 6' 9" diameter driving wheels (LMS Nos. 1160–1184, built 1925 and 1185–1199, 900–924 in 1927).

Orders in the early 1930s were still predominantly from India, with smaller orders from Tanganyika and the Argentine. The trade recession was felt in 1932–3 and in 1934 a large order from the LMS for Stanier designed two cylinder 4-6-0s and 2-8-0s helped considerably in maintaining work in the shops. Twenty-four 4-8-0s were built in 1935 for the Chinese National Railway, with outside cylinders $20\frac{1}{8}$" x $29\frac{1}{2}$" and 5' 9" diameter driving wheels.

Few locomotives were built in the early years of the war but in 1943 work was concentrated on building 2-8-0 to Ministry of Supply design of which 396 were built in 1943–5 after which a batch of fifty 0-6-0ST of standard design were built for the MOS. At the end of the war one hundred and twenty 2-8-0 *Liberation* class were built for UNRRA for service in Europe. The two outside cylinders were $21\frac{5}{8}$" x 28", coupled wheels 4' $9\frac{1}{8}$" diameter, boiler heating surface

Fig. 542 1773/1901 Burma Rlys. Fairlie 0-6-0+0-6-0 metre gauge

Fig. 543 2330/1908 East Indian Rly No. 1300 4-4-2 5' 6" gauge cylinders 19" × 26" D.W. 6' $6\frac{1}{2}$"

Fig. 544 4674/1936 Chinese National Rly No. 607 4-8-4 std. gauge

Fig. 545 5367/1946 UNRRA *Liberation* class cylinders $21\frac{5}{8}''$ × 28″ D.W. 4′ $9\frac{1}{8}''$

Fig. 546 5974/1951 Iranian State Rlys 2-10-2 cylinders 24″ × 26″ D.W. 4′ 3″

Fig. 547 6273/1956 East African Rlys *Class 31* No. 3108

A 1933 Advertisement

total 2933 sq. ft, grate area 44 sq. ft and a pressure of 227 lb/sq. in. The maximum axleload was $18\frac{1}{2}$ tons and the weight of locomotive and tender was $142\frac{1}{2}$ tons. The tender was mounted on two bogies and carried 5500 gallons of water and 10 tons of coal. For their wheel arrangement they were gigantic machines.

There was also a grave shortage of locomotives in India. Replacements had been held in abeyance due to the war and when India became independent in 1947 the new government produced a five year plan of standardisation and replacement.

Class HPS 4-6-0s, with $20\frac{1}{2}$" x 26" cylinders and 6′ 2″ diameter coupled wheels were built in 1949 and 1950 and a newly designed 4-6-2 *Class WP* with bar frames, 5′ 7″ coupled wheels and $20\frac{1}{4}$" x 28" cylinders built. Due to intense competition none were built in this country but came from America, Canada, Poland, Austria – all purchased under the Colombo Plan Aid programme. Nevertheless Vulcan Foundry and Robert Stephenson & Hawthorn, who had been acquired by Vulcan Foundry in 1944, received orders for ten *WL Pacifics*, sixty *Class WM* 2-6-4Ts, six *XD* 2-8-2s and ten new design *WP* 2-8-2s, subcontracted from the North British Locomotive Co. The large orders previously won were now a thing of the past, for two reasons. India was now equipped to build a large quantity of locomotives per annum herself and those that were not built in India were obtained under fierce competition by Germany, Japan, Italy, Austria and America and the delivery required was very exacting.

So the re-equipping of the Vulcan Foundry started with diesel and electric locomotives replacing the steam locomotive. In March 1955 Vulcan Foundry became a member of the English Electric group together with Robert Stephenson & Hawthorn. So ended a truly magnificent record of steam locomotive building with a total of 6210 units.

TAYLOR, H. E.
Chester

Henry Enfield Taylor was a mining engineer with a small works at 15 Newgate Street in Chester. Between 1877 and *c.* 1885 he built at least five 3′ 6″ gauge 0-4-0Ts, with outside cylinders $9\frac{1}{2}$" x 18″ and 2′ 6″ diameter wheels, for John Bazley White & Bros. Ltd., Swanscombe Works, as follows: 1877 CHESTER; 1879 MILLBANK; 1882 IRON HORSE; 1882 DEAD HORSE; and LIVERPOOL (date unknown). The first four were scrapped in 1930.

Fig. 548 **H. E. Taylor** 1882 John Bazley White & Bros Swanscombe IRON HORSE 3′ 6″ gauge

TAYLOR J. & A.

Smith Street, Ayr

This firm were builders of mining equipment including haulage and winding engines. A locomotive was built during the period 1869 to 1871 for the Dalmellington Iron Co. It was a 0-4-0ST, becoming the Iron Co's No. 8, with outside cylinders 13½″ x 21″ and wheels 3′ 8″ on a 6 ft wheelbase. The slide bars which had shields over them were practically identical to those fitted to a locomotive supplied to the same company in 1866 by the Lilleshall Co. (Dalmellington Iron Co. No. 6). No. 8's dome and safety valves were over the firebox and the saddle tank did not cover the smokebox or firebox. It was scrapped in 1912. An interesting account of the Dalmellington Iron Company's locomotives may be found in David L. Smith's book. No further locomotives were built.

TEESIDE IRON & ENGINE COMPANY

See GILKES, WILSON & COMPANY.

TENNANT, T. M. & COMPANY

Bowershall Iron & Engine Works, Leith

This firm is recorded in the Edinburgh directory 1854–6 as screw bolt manufacturers and machine makers, at 74 Clerk Street. The initials of the firm are shown as S. M. but this was probably a misprint. In 1856–7 they were described as engineers, in addition to the previous description, and their address was now Newington Works, East Sciennes Street. In 1860–1 an additional address is given at St Leonards, which is a district on the south side of the town. In 1862–3 it was given as above and also at Bowershall Iron Works, Leith. By 1863–4 the sole address was Bowershall Iron and Engine Works, Leith, and in March 1866 they became a limited company, as: T. M. TENNANT & CO. LTD.

From records, the company was registered on 27 March, 1866 with the object of 'manufacturing . . . rolling stock, railway plant of all descriptions'. They had a share capital of £50,000 in 500 shares of £100 each. The registered offices were at Bowershall Engine Works, Leith.

Information regarding their locomotive building activities is practically nil. That they did build some vertical boilered four-coupled locomotives in this period is possible but how many and what sort of transmission they had is not known. The seal of the company shows a typical portable agricultural engine carried on four wheels, with large flywheels and cylinder over the firebox.

Fig. 549 J. & A. Taylor *c*1870 Dalmellington Iron Co. No. 8

TENNANT, T. M. & COMPANY

A resolution was passed on 9 June, 1871 that the company was to be wound up voluntarily but this was not finally accomplished until 28 February, 1878 so that it would appear from this that the assets were disposed of piecemeal over that period. The premises are at present occupied by 'Bonnington Castings Ltd.' An interesting point is that the streets in the vicinity of the works are Jane Street, Waddell Street and Tennant Street, and at some time the entrance to the old works could have been from the last named street.

THAMES BANK IRONWORKS

Canning Town, London

These works were established in 1835, by Joseph Ditchburn and Charles Mare, primarily for shipbuilding and civil engineering contracts. The works were extended to the west bank of the River Lea at Blackwall and, at their busiest, employed about 3000 men. They were notable in building, in 1860, Britain's first ironclad battleship the WARRIOR in addition to many gunboats, and passenger ships for home and abroad. It has been stated that one or more locomotives were built but no records of any such activity has been forthcoming.

THOMAS, ALFRED R.

Cardiff

This firm was asked to tender for supplying 3′ 8″ gauge locomotive for the Severn & Wye Railway in 1864 and although £600 quoted was the lowest, Fletcher Jennings & Co. got the order. No locomotives are known to have been built by this firm.

THOMPSON BROS.

Wylam

Thompson Bros were colliery owners and were probably engaged in colliery engineering at Wylam. Six locomotives were built for the Newcastle and Carlisle Railway from 1839 to 1841. It has also been stated that they built one or more for the Great North of England Railway, but particulars are not known. The N&CR locomotives were four 0-6-0s and two 2-4-0s. Whether there was any connection between Hawks & Thompson and Thompson Bros cannot be confirmed, but Thompson Bros carried on building locomotives for the N&CR the year after Hawks & Thompson's last locomotive. Another detail, which may have a bearing, is that the boiler heating surface of 530 sq. ft for Thompson Bros first three 0-6-0s was the same as for VICTORIA built in 1838 by Hawks & Thompson. The boiler barrel dimensions were also identical. Some details of these locomotives may be found in J. S. MacLean's book *The Newcastle & Carlisle Railway*.

THOMPSON & COLE

Hope Foundry, Little Bolton

Very little information is available regarding this firm, and the locomotives that were built. Mr Thompson was possibly the same gentleman who became one of the partners of Kitson, Thompson & Hewitson when this firm was reformed in 1842. Only five locomotives have been traced, all 0-4-2s with the same size cylinders and coupled wheels. The two locomotives for the Birmingham and Derby Junction Railway had outside frames and inside cylinders. According to the *Locomotive Magazine* No. 628 the firm 'was believed to have been located in Carlisle' but Little Bolton is correct.

THOMPSON BROS WYLAM

Date	Type	Cylinders inches	D.W. ft in	Heating Surface sq ft	Pressure lb/sq in	Newcastle & Carlisle Rly No	Newcastle & Carlisle Rly Name	NER No
11/1839	0-6-0	14 x 18	4 0	530	60	21	Matthew Plummer	469
4/1840	0-6-0	14 x 18	4 0	530	60	22	Adelaide	470
9/1840	0-6-0	14 x 18	4 0	530	80	23	Mars	471
11/1840	2-4-0	14 x 18	4 9	591	90	24	Jupiter	472
1/1841	2-4-0	14 x 18	4 9	563	70	25	Venus	473
4/1841	0-6-0	14 x 18	4 6	594	80	26	Saturn	474

THOMPSON, JAMES & SONS

Kirkhouse, Nr Brampton

In 1837 this firm became the lessees of the Earl of Carlisle's collieries in Cumberland and the railway connected with them, known as the Brampton Railway. The workshops were at the Kirkhouse colliery and the work carried out there was mainly the manufacture and maintenance of pumping and other machinery used at the collieries. In addition all the locomotives of the Brampton Railway were repaired, and extensive rebuilding took place. The 'new' locomotives which were built at Kirkhouse were almost certainly reconstructions and not entirely new.

The following are recorded as being built there: 1848 LOCH 0-6-0 oc 13″ x 22″ 4′ DW Long boiler type. S1885; 1866 GARIBALDI 0-6-0 oc 14″ x 22″ 4′ DW Long boiler type S1900.

James Thompson was a director of the Newcastle and Carlisle Railway. The lease of the Brampton Railway and associated collieries terminated in 1908. It was Mr Thompson who bought the famous ROCKET from the Liverpool and Manchester Railway in 1837.

THOMPSON, R. & J. A.

Pot House Bridge Works, Bilston, Staffs.

Built one well tank locomotive for the Parkfield Iron Co., Wolverhampton 2′ 5½″ gauge in 1863.

THORNEWILL & WARHAM LTD.

Burton-on-Trent

This was a brewery engineering business managed by John R. Warham and Robert Thornewill founded in the 1840s. Locomotive building commenced in 1861 as an adjunct to the normal brewery machinery building and repairs they caried out. All those built were for Bass, Ratcliffe and Gretton Ltd.

THOMPSON & COLE

Date	Type	Cylinders inches	D.W. ft in	Railway	Name	
7/1840	0-4-2	14 x 18	5 0	North Midland		
1840	0-4-2	14 x 18	5 0	North Midland		
9/1840	0-4-2	14 x 18	5 0	North Midland		
c. 1841	0-4-2	14 x 18	5 0	Birmingham & Derby Junc	Kingsbury	(a)
c. 1841	0-4-2	14 x 18	5 0	Birmingham & Derby Junc	Willington	(b)

(a) Sold 1851. (b) Sold 1852

Fig. 550 **James Thompson** LOCH

As will be seen from the list most of them worked their entire life around the breweries. Building continued as and when the brewers required a new locomotive either to replace an older one or for increased traffic requirements, and ceased in 1890. Unfortunately no details can be discovered regarding these interesting locomotives apart from the fact that some of them are reported as having superheaters and piston valves which would probably be before they were introduced on main line locomotives. The valve gear also had some special features. This is a case where the firm's successors, with new policies, destroyed all records and drawings concerning these locomotives.

The first and last two were saddle tanks but the remainder were well-tanks, but at least four of these were converted to saddle tanks at the turn of the century by Hunslet Engine Company. The saddle tank locomotives had tanks the full length covering smokebox, boiler and firebox. A combined dome and spring balance safety valves were centrally mounted. A very narrow chimney was fitted with a considerable top flare. The coupling and connecting rods had cottered brasses. Steam brakes were fitted and as far as is known they all had outside cylinders and inside motion. Thirteen new locomotives have been traced. The reference or works numbers seem to have included other pieces of equipment built by the firm.

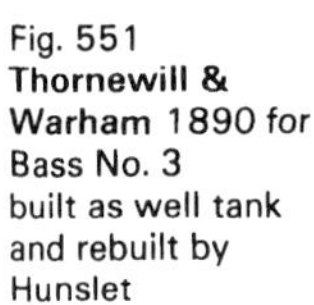

Fig. 551 **Thornewill & Warham** 1890 for Bass No. 3 built as well tank and rebuilt by Hunslet

THORNEWILL & WARHAM LTD

Works No	Built	Type	Cylinders inches	DW ft in	BR & G No (a)	Rebuilt Saddle Tank	
	c. 1850	2-4-0WT					Napoleon (b)
	1861	0-4-0ST					
224	1863	0-4-0WT					
	1864	0-4-0WT			1		
	1864	0-4-0WT			2		
303	1869	0-4-0WT			3		
	1872	0-4-0WT			4		
373	1873	0-4-0WT	14 x 20	4 0	5	1897	
	1874	0-4-0WT	14 x 20	4 0	6	1900	
400	1875	0-4-0WT	14 x 20	4 0	7	1899	
420	1876	0-4-0WT	14 x 20	4 0	8	1898	
	1877	0-4-0WT	14 x 20	4 0	9		
	1880	0-4-0ST	14 x 21	3 6	2		
	1890	0-4-0ST	14 x 21	3 6	3		

(a) Bass, Ratcliffe & Gretton. (b) To Babbington Coal Co Notts in 1855 from a contractor. Rebuilt to 0-6-0ST by T & W but may not have been built by them.

THWAITES & CARBUTT

Vulcan Works, Thornton Road, Bradford, Yorks

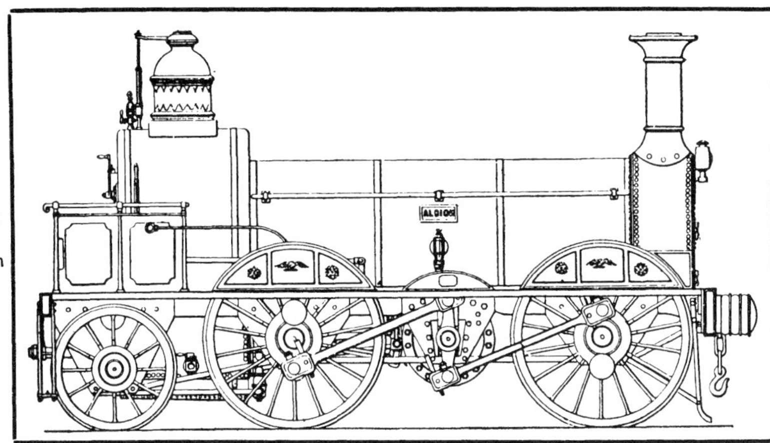

Fig. 552 **Thwaites & Carbutt** 1848 constructed on 'Cambrian' system of drive

This firm was founded by Mr Robinson Thwaites and subsequently became known under the title of Thwaites and Carbutt. In the 1880s the name became Thwaites Bros and, a short time after this, it was converted into a limited liability company trading as Thwaites Bros Ltd. with Lieut-Colonel Thwaites as Managing Director. These works in their day were the largest of their kind in Bradford.

The principal products of the firm were the Root's patent blower of various sizes, the 'Vulcan' forges, smelting plant, cranes and hoists, pumps, gas engines, steam hammers, steam engines, compressors and rolling mill engines, and many other types of machinery. To this great variety of products they added one more, for in 1848 they took on the manufacture of a railway locomotive.

Apparently E. B. Wilson & Company had been asked to build it, but had refused to do so (probably very wisely). It was a patent of a Mr John Jones of Bristol, and after the drawing office at Wilson's had spent a long time puzzling out the design, the project was taken over by Thwaites and Carbutt. The design incorporated a transverse segmental cylinder fixed between the frames between two pairs of driving wheels.

The object of the design was to present a balanced locomotive in which the reciprocating parts mutually balanced each other. Probably the correct term to use is oscillating weights, as the cylinder, instead of being placed parallel to the longitudinal centre plane of the engine, was placed transversely and provided an oscillating vane piston fixed on a shaft passing through the centre of the cylinder, the shaft being provided with double arm rocking levers at the outer ends. To the ends of these levers, connecting rods were attached to operate the driving wheels. It will be seen that the moving parts really balance each other longitudinally but the upward and downward inclination of the connecting rods produced a certain amount of 'pitching' due to steam reactions. This on its own would not be sufficient to condemn the arrangement but the weakness lay in the construction of the cylinder and piston with their attendant difficulties in the way of ensuring steam tightness between the rocking shafts and the cylinder abutments and between the ends of the piston and the ends of the cylinder. Two pairs of driving wheels were used and their crank pins were set at 180° to each other. There was a pair of trailing wheels at the rear of the firebox.

Despite the attendant troubles the locomotive was run on the South Yorkshire Railway as their No. 5 ALBION and was replaced in 1870, so it would appear that some work was done during this time. The arrangement was known as the Cambrian system. The driving wheels were 5′ 6″ diameter and the trailing wheels 3′ 9″. ALBION became Manchester, Sheffield and Lincolnshire Railway No. 156 when the SYR was absorbed in 1864.

Three 0-6-0s with inside cylinders were also built. They were bought by Boulton at Cardiff in 1866 but their building dates were unknown. They had domeless boilers with haycock fireboxes with the safety valves mounted above the fireboxes. According to Mr A. R. Bennett in his *Chronicles of Boulton's Sidings* they had cast iron wheels with wrought iron tyres. They were all subsequently rebuilt by Boulton as saddle tanks.

Sharp Stewart & Company built a locomotive of unknown type, Works No. 2478, in 1875 for Thwaites & Carbutt, who were probably acting as agents.

TODD, CHARLES

Sun Foundry, Dewsbury Road, Leeds

Charles Todd left the firm of Shepherd & Todd in 1844 and set up his own works. He built his first locomotives in 1844 which were two six-coupled goods with inside cylinders 16″ x 24″ and 5′ 2″ diameter driving wheels. He built approximately twenty locomotives, including further 0-6-0 types for the York & North Midland and the Belgian Eastern Railways. He also supplied two 0-6-0WTs and three 2-4-0s to the Grand Central Belge Railway. The works were closed in 1858 and were taken over by Messrs Carrett Marshall & Company.

TODD KITSON & LAIRD

Airedale Foundry, Leeds

The works were established in 1835 by James Kitson and were known as the Airedale Foundry. A partnership was formed with Charles Todd and in 1838 they were joined by David Laird. Charles Todd had been apprenticed to Matthew Murray at the Round Foundry and had some knowledge of locomotive building whereas David Laird was a wealthy farmer who invested capital in the firm and, expecting quick returns for his money got impatient when they were not forthcoming. He left in 1842. On the appearance of Laird, Charles Todd left and started up a similar business with Mr Shepherd in adjacent premises known as the Railway Foundry. Hunslet became a centre for locomotive building – the canals and roads making transport convenient.

In the early days the firm's title changed frequently and these are often given in different order of names. Up to 1838 it was probably JAMES KITSON, from 1838 it was TODD KITSON AND LAIRD, and in the same year it became KITSON and LAIRD. Another version was LAIRD & KITSON. By 1842 Isaac Thompson and William Hewitson had joined and the title changed to KITSON, THOMPSON & HEWITSON.

Thompson left in 1858 but Hewitson remained with the firm until his death in 1863. Some time in this period the firm became KITSON & HEWITSON. About 1863–4 it became KITSON & COMPANY.

Returning to the beginning, only locomotive parts for other builders were made at first, but in 1838 the first complete locomotive appeared and before it could be delivered, a wall of the factory had to be pulled down. It is not certain whether this was the LION for the Liverpool and Manchester railway or one for the North Midland Railway. LION was a 0-4-2 with 5′ coupled wheels, inside cylinders 11″ x 20″ and outside frames. The boiler had a haycock firebox and a pressure of 50 lb/sq. in. Fortunately it has gained a permanent place in locomotive history, for after working on the railway for twenty-one years it was sold to the Mersey Docks & Harbour Board and was used as a stationary

CHAS. TODD

Date	Type	Cylinders inches	D.W. ft in	Railway	Rly No	
1844	0-6-0	16 x 24	5 6	Hull & Selby		
1844	0-6-0	16 x 24	5 6	Hull & Selby		
9/1846	0-6-0	15 x 24	4 6	York & North Midland	78	
11/1846	0-6-0	15 x 24	4 6	York & North Midland	77	
12/1846	0-6-0	15 x 24	4 6	York & North Midland	76	Swallow
12/1846	0-6-0	15 x 24	4 9	Newcastle & Darlington		
3/1847	0-6-0	15 x 24	4 8	Newcastle & Darlington		
11/1847	0-6-0	16 x 24	5 0	York & North Midland	96	
12/1847	0-6-0	16 x 24	5 0	York & North Midland	99	
1/1848	0-6-0	16 x 24	5 0	York & North Midland	101	
2/1848	0-6-0	16 x 24	5 0	York & North Midland	104	
3/1848	0-6-0	16 x 24	5 0	York & North Midland	109	
5/1848	0-6-0	16 x 24	5 0	York & North Midland	111	
12/1848	0-6-0	16 x 24	5 0	York & North Midland	121	
1855	0-6-0WT	15 x 20	4 6$\frac{1}{8}$	Grand Central Belge	94	
1855	0-6-0WT	15 x 20	4 6$\frac{1}{8}$	Grand Central Belge	95	
1855	2-4-0	16 x 22	5 11½	Grand Central Belge	23	
1855	0-6-0	16$\frac{1}{16}$ x 24	4 11¼	Belgian Eastern	62	
1855	0-6-0	16$\frac{1}{16}$ x 24	4 11¼	Belgian Eastern	63	
1856	2-4-0	16¼ x 22	5 5	Grand Central Belge	24	
1856	2-4-0	16¼ x 22	5 5	Grand Central Belge	25	

engine. It was jacked up off its wheels and coupling rods were removed and connected to pumps for pumping water out of Prince's Dock Liverpool. It was 'discovered' almost in its original condition and it was decided to restore it, this being done at Crewe. It took part as a live exhibit at the centenary of the Liverpool & Manchester Railway at Wavertree, Liverpool in October 1930 and since then has been used in films and exhibitions. Its original cost was £1100. The complete order from the Liverpool & Manchester railway was:

2 Luggage Engines

L&M	No. 57	LION	0-4-2	11" x 20"	1838	£1100
L&M	No. 58	TIGER	0-4-2	11" x 20"	1838	£1100

2 Coaching Engines

L&M	No. 62	LEOPARD	2-2-2	11½" x 18"	1839	£1060
L&M	No. 64	PANTHER	2-2-2	11½" x 18"	1839	£1060

2 Banking Engines

L&M	No. 65	ELEPHANT	0-4-2	13" x 20"	1839	£1130
L&M	No. 67	BUFFALO	0-4-2	13" x 20"	1839	£1130

All had 5ft driving wheels

Fig. 553
1508/1868 Thos. Nelson contractor RESOLUTE

Other customers in the early days were: North Midland, Manchester & Leeds, York & North Midland, Midland, Lancashire & Yorkshire, Altona & Kiel, Orleans & Bordeaux, South Eastern, Eastern Counties and Leeds & Thirsk railways.

From 1850 Kitsons gave two works numbers to any tender locomotive built, one for the locomotive and one for the tender. The first example was for six 2-4-0s for the North Staffordshire Railway Nos. 48 to 53, which had works numbers 226 to 231 and then 232 to 237 for the tenders. This practice was discontinued in 1875. There were also many blanks in the early series so that the last works number, believed to be 5487 of 1938 bears no relationship to the actual number of locomotives built.

In 1855–6 began a long association with Indian railways when thirty-five 0-4-2s were built and two small 2-2-2WTs with 6 ft driving wheels. In the same period orders were carried out for the Madras, Indian Midland, Great Indian Peninsula, Scinde Punjaub and Delhi railways, together with substantial orders from Madrid, Saragossa & Alicante and Copiabo & Caldera railways.

Kitsons did not build large numbers of industrial tank locomotives. The first was probably a 0-6-0T for Guest Keen & Nettlefolds in 1859 named SAMSON, works No. 701. Some interesting 0-8-0STs were built for the Great Indian Peninsula Railway in 1865–6, with inside cylinders 18″ x 24″ and 4 ft diameter wheels. They were used for banking, and twenty were built.

Kitsons had a large share of Kirtley's Midland Railway double frame goods, the first order being completed in 1866. Besides India, Kitsons had important contracts with Russia. Fifteen 2-4-0s were built in 1869–71, with outside cylinders 16″ x 24″ and 5′ 6″ driving wheels, for the Tamboff and Saratoff Railway followed by five slightly smaller ones of the same type for the Novotorjock Railway, then in 1870 some 0-6-0s were built, eight for the Moscow-Riazan Railway, fifteen for the Grande Société de Russe, and twelve for the Brest Litovsk and Smolensk.

Other orders were: 1871–2 ten 0-4-0ST Voronezh-Rostov; 1872 eighteen 2-4-0 Odessa; and 1872 six 0-6-0 Yaroslav & Vologna.

Orders from abroad were keeping the works extremely busy during this period and, besides the above, locomotives were being sent to Germany, Ceylon, Argentine, Denmark, Australia, Japan, Spain, S. Africa, Trinidad, India and Mauritius. At the same time many orders from home railways were sandwiched in between.

Kitsons although not the first to build steam tramway locomotives entered the field in 1876 and built some combined steam cars to W. R. Rowan's design. In 1878 they built three to their own design with vertical boilers, four coupled wheels and inclined outside cylinders. Motion was by means of a modified version of Walschaerts valve gear. All was enclosed in bodywork and the wheels and motions were surrounded by protective plates. The condensing system, which was an essential part of this type of locomotive, was placed on the roof and consisted of a series of copper tubes through which the exhaust steam passed, the surrounding air cooling the steam and the condensate returning to the feed water tank. After many trials it was decided to replace the vertical boiler by a horizontal type and this was standardised for future steam trams. Various types of condensers were tried and the final type was a series of arched transverse tubes which were a great improvement.

More than 300 units were built and besides supplying many to the tramway systems of the British Isles, others were sent to New Zealand, Australia and the continent. The last one built was in 1901 for the Portstewart Tramway (Works No. T302). Works numbers for this type of locomotive were kept separate and bore a prefix T.

An interesting order from the Great Eastern Railway was for ten 4-2-2s in 1881–2. All wheels had inside frames, the outside cylinders were 18″ x 24″ and the driving wheels 7′ 6″ diameter which were the largest wheels ever used on the GER. The locomotives were designed by Mr Massey Bromley. An unusual feature was the bogie which had large wheels of 4 ft diameter. The splashers were slotted and the crosshead was single slide bar type. They were short-lived being broken up in the 1890s.

An order from the Western Railway of France in 1882–3 was for twenty 2-4-0Ts with 16½″ x 22″ cylinders and 5′ 5″ diameter coupled wheels. A large order for 4-6-0s was carried out in 1889–91 for the Cordoba Railway in the Argentine. They had 16″ x 22″ cylinders and 4 ft diameter coupled wheels. Five were lost at sea and five replacements had to be built.

In 1894 three Meyer 0-6-6-0 locomotives were built for the Anglo-Chilean Nitrate and Railway Company, with four outside cylinders 14″ x 18″ and 2′ 10¾″ diameter wheels. The Meyer system of articulated locomotives was for a single loco type boiler and the wheels arranged in two independent groups. In the two bogie type, the front bogie carried the load on a central pivot and the rear one carried the load on hemi-spherical side bearings.

A Kitson-Meyer articulated type was designed similar to the Meyer, but with the cylinders each facing the centre and the tanks carried on the framing instead of on the bogies. Two were built in 1903 for the Rhodesian Railways. They were 0-6-6-0, with 15″ x 23″ cylinders and 3′ 6¾″ diameter wheels. In 1904 three were built for the Jamaica Government Railways, with the same wheel arrangement. The cylinders were 14″ x 18″, wheels 3′ 6″, heating surface 1458 sq. ft and they carried 2500 gallons of water and 4 tons of coal. Their weight in working order was 80 tons 15 cwt.

Over fifty Kitson-Meyer locomotives were built including two for the GS of Spain Railway which were 2-8-8-0 with 4 ft wheels, 14¾″ x 24″ cylinders, a heating surface of 1901 sq. ft and weighing 101 tons. It was estimated that either of these two locos could haul 449 tons at 10 mph up a gradient of 1 in 50. Most of this type of locomotive were built from 1903 to 1913 and the last two were sent to the Girardot Railway in 1935, these being 2-8-8-2s.

Some special Kitson-Meyer 0-8-6-0s were built for the Argentine Trans-Andine Railway Metre gauge, for rack and adhesion purposes. There were four cylinders for the normal adhesion wheels 16″ x 19″ and two 18½″ x 19″ for

Fig. 554
3580/1894
Lambton, Hetton
& Joicey Collieries
No. 56

Fig. 555
4580/1908
Kitson-Meyer
2-8-0 + 0-8-0
Gt. Southern Rly of
Spain No. 50

rack working. Others of the same wheel arrangement for the same railway had two 13″ x 14″ cylinders for the small rack, two 18″ x 19″ for the large rack and 16½″ x 19″ for adhesion.

In common with other builders, Kitsons built some rail motor units, among which were three for the Belfast & County Down Railway, one for S. Australian Government Railway, six for the South Eastern & Chatham Railway, two for Madras and one for the Central South African Railway.

During the 1914–18 war a large number of 2-8-0s were built for the Indian railways and in 1918 thirty-two Robinson Great Central Railway type 2-8-0s were built for the Government. Orders were few and far between in the early 1920s and the most interesting locomotive which appeared in this period was the Kitson-Still 2-6-2T experimental locomotive built in 1924 (Works No. 5374). It had a circular boiler and firebox and steam was initially raised by means of an oil burner. The engine had eight cylinders horizontally placed, four in front and four in the rear, with a common crankshaft in the centre. Oil was applied by airless injection to the outer ends of the cylinders and steam to the inner ends. It was a four stroke cycle.

The pressure in the boiler was maintained during running by waste heat recovery of the ic engine as the water jacketted cylinders were in direct contact with the boiler, and the exhaust gases were led through tubes into the boiler water space. The crankshaft was connected to a jackshaft by means of a pair of double helical gears and power transmitted to the wheels by coupling rods off the crankpins of the jackshaft. The admission of steam to the cylinders was effected by two gears only, operating on the Hackworth principle. The locomotive was tried out on the North Eastern section of the London & North Eastern Railway pulling freight trains between York and Hull. The locomotive was not a success, the mechanism was complicated, prone to leakage and the experiment was abandoned.

The last large order Kitsons received was from the London & North Eastern Railway for twelve 'Improved' *Director* 4-4-0s for service in Scotland, principally to supplement the hard-worked *Scott* class. They had inside cylinders 20″ x 26″ and 6′ 9″ driving wheels and were completed in 1924.

Small orders came through from 1925 to 1938 but the total built in these years was less than 100, far less than they were producing per annum in their busy years. A receiver had been appointed in 1934 and efforts had been made to rejuvenate the firm, but this was not to be, and a single 0-8-0T for the Jamaican Government was completed in 1938.

Patterns, drawings and the goodwill were taken over by Messrs Robert Stephenson & Hawthorn. In 1960 all drawings etc. relating to industrial locomotives were acquired by the Hunslet Engine Company together with those of Manning Wardle & Company.

Kitson locomotives had very marked characteristics and where industrial locomotives were supplied to their own designs, they could easily be recognised. The taper chimney had a bold rim at the top. The bunker side had a small flare towards the top and the rear cab side met it with a large radius. Wherever practicable the works plate was put on the leading sand box. Cast iron wheels in general were used up to 1885 when cast steel was introduced. Kitsons were responsible for the use of the Naylor safety valve in 1866. It was patented by W. Naylor of the Great Indian Peninsula Railways. As the valve opened under blowing off pressure it lifted one end of a balance, the other end of which was attached to a spring.

It is possible that the building of steam tenders for the Great Northern Railway, to Archibald Sturrock's design, in 1865 started the idea of developing the articulated locomotive. The advantages were appreciated by Kitson and in collaboration with J. J. Meyer of Mulhouse an improved Meyer was evolved and, as stated, many were built.

Gab motion was used on the earlier Kitson locomotives, but the slotted link motion gradually replaced it. They also used a modified Walschaerts valve gear, particularly for industrial tank locomotives. In this form the drive for the reversing link was taken from the coupling rod to a horizontal arm on that link.

It is difficult to assess the total number of locomotives built as it was the firm's custom to give a tender locomotive two works numbers until about 1870 so that, although the works numbers extended to 6322, allowing for cancellations and tender numbers the approximate number of locomotives built was 5405.

Fig. 556 5374/1924 The Kitson-Still 2-6-2T

TODD, LEONARD J.

Leith

Todd constructed a locomotive for the Tramvia de Santander in 1871. It had a locomotive type boiler and a pair of cylinders 6½″ x 9″ placed on top of the firebox and acting on a crankshaft whence motion was communicated through a pair of spur wheels to a single pair of driving wheels 5′ 6″ diameter, formed of a disc of wood and steel tyred.

The firebox end, which included the driver's cab, was supported on a bogie truck with 1′ 9″ diameter wheels at 3′ centres. In 1875 he designed a fireless steam car with reservoirs and machinery placed under the floor.

He contested the claim of Professor Stumpf of Berlin regarding the 'Uniflow' system and the straight-flow steam engine and its thermal principles. In a letter to the Editor of *The Mechanical Engineer* for 24 February, 1911 he wrote 'I would beg to make the following observation – this thermal gradient principle of action with its consequent economic advantages was first explained by myself at great length in Patent Specification No. 7301 of the year 1885. Subsequently I made an experimental engine which was kept in operation for several years in London, and with which the steam was weighed from hundreds of single, compound, triple and quadruple experiments'.

Ref: *Journal of Inst. of Loco Engineers* Vol. XXVII No. 137 *History of the Steam Tram* by Dr H. A. Whitcombe.

TREDEGAR IRON WORKS

These works were owned by Samuel Homfray and were erected in 1800. He also owned the Penydarren Iron Works where Trevithick's locomotive was built and taken to the rails in 1804. Thomas Ellis was appointed engineer in 1828. He had already spent most of his time with steam haulage engines, blast engines, and pumps. His father had been at Penydarren during the erection of Trevithick's locomotive and evidently Thomas' interest in steam locomotion stemmed from his father's description of the construction of this unique machine. One of Robert Stephenson's locomotives went to Tredegar. It was six wheeled, named BRITANNIA, and built in 1829. A steam locomotive was put in hand by Thomas Ellis at the Tredegar Iron Works in 1830 and took two years to complete. He used the BRITANNIA as a guide. The main difference was that in Ellis' locomotive the chimney was fixed to the top of the front end of the boiler. It began work in 1832 and was named ST DAVID. It was very successful, and was rebuilt in 1848, withdrawn in the 1880s and remained intact until 1909 when it was, unfortunately, broken up.

Thomas Ellis built nine locomotives between 1832 and 1854. They were all built to a gauge of 4′ 2″ and it was not until 1865 that the gauge of the Sirhowy Tram Road on which the Tredegar locomotives were operated was converted to standard gauge, thus confining Ellis' locomotives to the tracks of the Iron Works. Daniel Gooch, who was to become Chief Locomotive Superintendent of the Great Western Railway received his training under Mr Ellis.

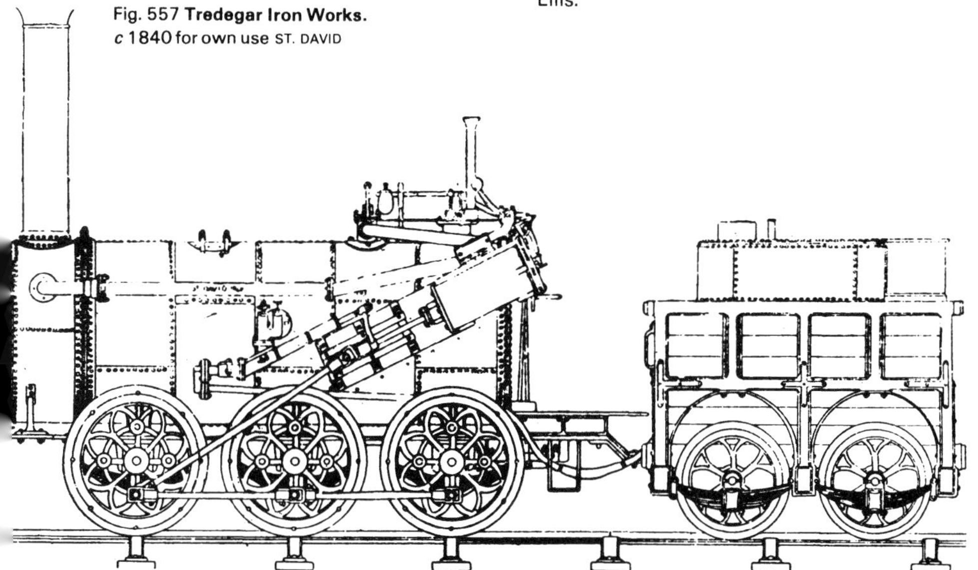

Fig. 557 **Tredegar Iron Works.**
c 1840 for own use ST. DAVID

TREDEGAR IRON WORKS

All the locomotives built in this period were six-coupled with tenders.

1832: ST DAVID
? TREDEGAR
? JANE
? LORD RODNEY
? LADY SALE
? PRINCE ALBERT
? FANNY
1853: BEDWELLTY
1854: CHARLOTTE

Four were built between 1846 and 1848. Two other names quoted were LAURA and DISPATCH, the latter being a four coupled shunter. Also two locomotives were built for the 3′ gauge.

TREVITHICK, RICHARD 1771–1833

He was the son of a Cornish mining engineer and became an eminent engineer in his own right. It was not long before he clashed with Boulton and Watt who had a virtual monopoly for supplying pumping engines and other equipment in this area. Trevithick differed from Boulton and Watt in that he was keen to develop the high pressure boiler whereas the latter would not budge from the low-pressure boiler and engine with condenser.

In 1801 he went into partnership with Andrew Vivian and by the end of that year they had built and tested a steam road carriage at Camborne, with rather disastrous results. Undetered they built a second in 1802 and took it to London where it was driven round the streets, but

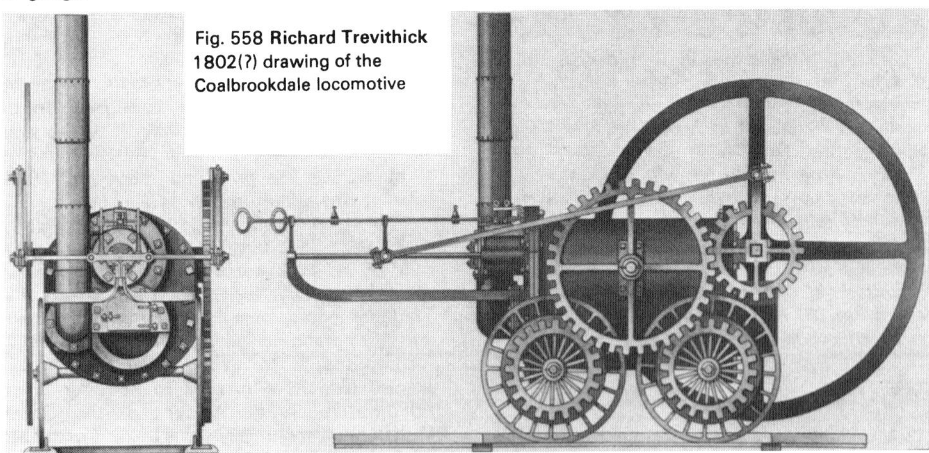

Fig. 558 **Richard Trevithick** 1802(?) drawing of the Coalbrookdale locomotive

Fig. 559 **Richard Tevithick** 1804 model of the Pen-y-Darren locomotive

no one was sufficently interested to order one. These were preliminaries to his better known work on steam propulsion and the building of locomotives.

In August 1802 he was at the Coalbrookdale Company's works and there he built an engine to demonstrate the possibilities of high pressure steam. It was put to work to force water to a specific height to measure the work done. The boiler was 4 ft in diameter and the steam cylinder 7″ diameter x 36″ stroke, with a water cylinder 10″ diameter. It pumped water to a height of 35½ feet, and during the tests the boiler pressure was allowed to reach a pressure of 145 lb/sq. in. with the engine working at forty strokes per minute.

As a result of these tests, according to various accounts, the Coalbrookdale Company, built for Trevithick a railway locomotive. This is strengthened by a letter from Trevithick to D. Giddy,a bosom friend of his, dated 22 August, 1802, in which he writes 'The Dale Company have begun a carriage at their own cost for the railroads and are forcing it with all expedition'. The history of this locomotive is controversial. Examining the dates of this letter and the trials of the pumping engine, it is evident that the locomotive was not built with the same boiler and engine as the high pressure trial engine. Coalbrookdale's locomotive if it was completed was the first to be used on a railroad.

Trevithick next visited the Pen-y-daren Ironworks owned by Samuel Homfray and came to an arrangement with him to construct a railway locomotive or tram-waggon as it was termed, to haul iron from the works to a basin on the Glamorganshire Canal at Abercynon, a distance of nine miles. Unfortunately the weight of the locomotive was too much for the tram plates which were frequently fractured and it was used as a stationary engine at the works. The boiler was of wrought iron, 4′ 3″ diameter x 6′ long and had a return tube, the chimney being adjacent to the fire door. It was mounted on four wheels. The cylinder was horizontal fixed into the boiler barrel. The piston rod extended beyond the rear with its support. The crankshaft was mounted between the front of the boiler and the chimney and a large flywheel was fitted to the crankshaft which drove a chain of gears. A gear was fixed to the 3′ 7″ diameter wheels on one side only. Dimensions have been given, for the diameter of the cylinder, as 8¼″ and 8″ and the stroke 54″. The weight was 5 tons approx. The locomotive was completed in 1804, the first trial being on 11 February.

The steam distribution to the ends of the cylinder was by means of a four-way cock operated by a projection on the crosshead striking two collars fixed to a bar. The wheels were flangeless and loose on their axles. The exhaust steam was projected into the chimney and so was the initial form of a blast pipe.

The next locomotive was built to the order of Christopher Blackett, owner of Wylam Colliery, and was actually built at John Whinfield's foundry at Pipewellgate, Gateshead in 1805 by John Steel who had assisted Richard Trevithick at Pen-y-daren in the construction of that locomotive.

It was similar to the Pen-y-daren locomotive in principle, the cylinder being 7″ x 36″ and wheels 3′ 2″ diameter except that the cylinder was at the opposite end to the flue entry and exit. Unfortunately Mr Blackett did not accept the locomotive, probably due to its weight and the weak track and it was used as a stationary engine in the foundry.

In 1808 appeared another locomotive, in what was to be Trevithick's last endeavour to foster the interest of the public in this method of propulsion. This was a four-wheeled vehicle with a cylinder sunk vertically into the boiler with the chimney and fire door at the other end. The cylinder drove the wheels adjacent and the other pair was not coupled. It was built at Bridgnorth by Hazeldine and Rastrick and was named CATCH ME WHO CAN.

A circular track was laid near Euston Square in London.surrounded by palisades and a charge of five shillings (25p) was made in July 1808 which was reduced to two shillings by September. Despite publicity the expected revenue did not materialise and the income from carrying passengers did not defray expenses and the venture was closed down.

After these attempts Trevithick concentrated on stationary engines, pumps, ship propulsion, and patented a tubular superheated boiler in 1832. This was after spending many years abroad in Peru, Colombia and Holland. He succumbed to pneumonia on 23 April, 1833 and was buried in a pauper's grave at Dartford.

Richard Trevithick had been an extremely able and inventive engineer, and if he had stayed in England and persisted in the development of his ideas, steam technology would have made more rapid strides, but lack of finance, over optimism and poor manufacturing facilities severely hampered his progress. This does not alter the fact that he was one of the foremost engineers of his time and truly the 'father of railways'.

He was also father of four sons, and one, Francis, became Locomotive Superintendent of the Northern division of the London & North Western Railway.

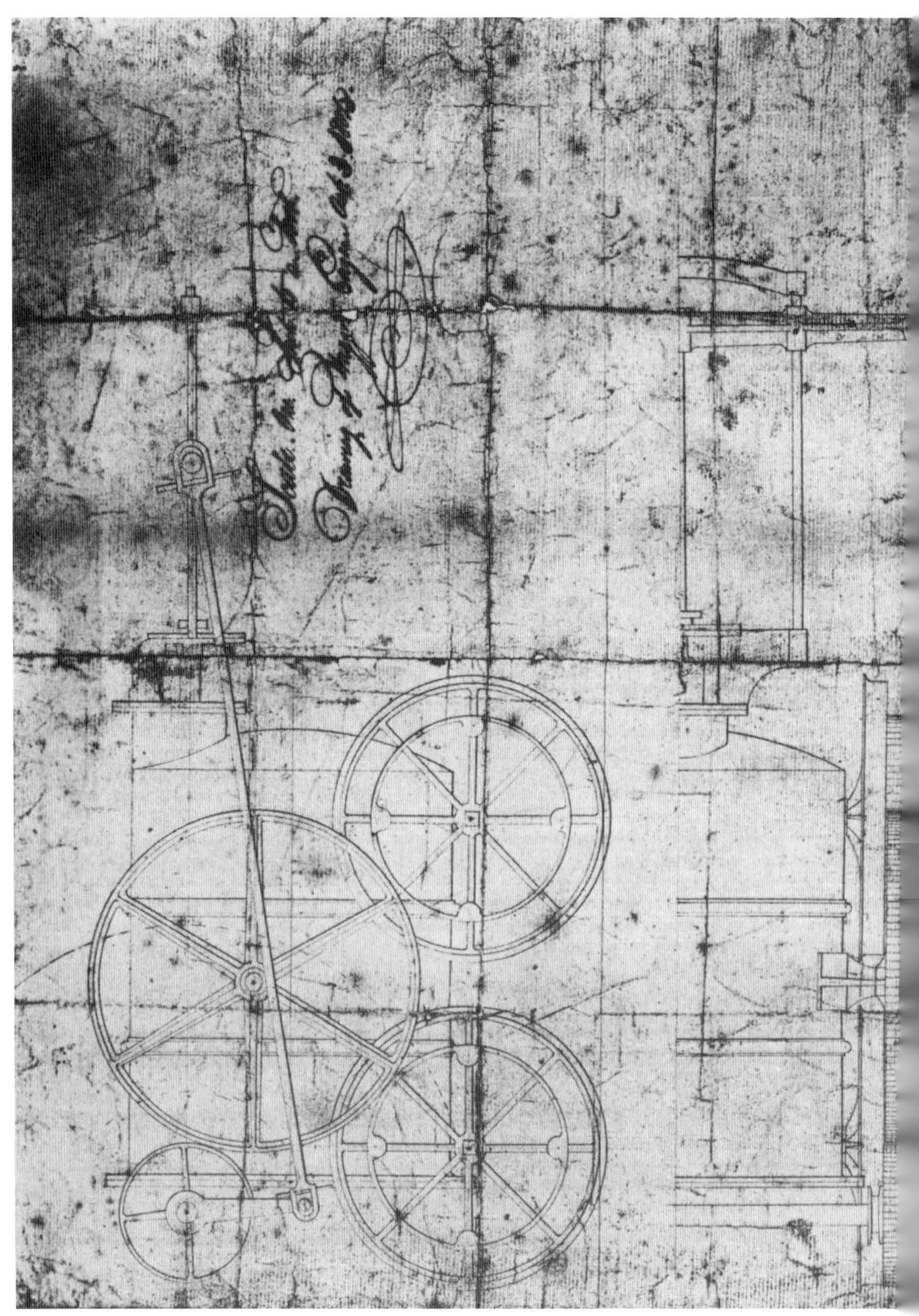

TULK & LEY
Lowca Works, Whitehaven

Founded in 1763 by Thomas Heslop who manufactured hardware, the firm became Millward & Co. in 1808 and in 1830 Tulk & Ley. Under this name the building of locomotives commenced in 1840, the firm having secured orders from the Maryport & Carlisle Railway which was opened in part on 15 July, 1840. Two locomotives were supplied initially, a 2-2-2 with 5 ft driving wheels and a six-coupled goods. A further 2-2-2 was built in 1843. The 0-4-2 was a popular type with Scottish influences, and Tulk & Ley built at least five and probably six for various Cumberland railways.

Their chief claim to fame however were the eight locomotives built under T. R. Crampton's patent. These were six 4-2-0s and two 2-2-2s. The first two 4-2-0s were ordered for the Namur and Liege Railway with outside cylinders 16″ x 20″ and 7 ft driving wheels. It is doubtful whether either of these Cramptons ever reached Belgium as the railway was behind schedule and when the locomotives were ready for delivery there was no railway available. The first, named NAMUR, ran trials on the LNWR who had also ordered a Crampton and eventually both NAMUR and the second one, named LIEGE, were sold to the South Eastern Railway. In addition a third 4-2-0 intended for Belgium was completed in 1847 and this too was purchased by the SER. A curious feature of design was that the horn blocks were inverted for the driving wheels, similar to the Bristol &

Fig. 560 (left) **Richard Trevithick** 1805 drawing of the Newcastle locomotive (see also fig. 571)

Fig. 561 (right) 172/1880 (Fletcher Jennings) Dorking Greystone No. 4 TOWNSEND HOOK

Fig. 562 (below) 234/1898 For Chas. Cammell & Co. Ltd. No. 28 CYCLOPS at Workington Iron & Steel Co.

Fig. 563 245/1906 for Astley & Tyldesley Colliery GEORGE PEACE

Exeter Railway's 4-2-4Ts built by Rothwell & Company. Another peculiarity was that a separate regulator was fitted for each cylinder. They were not successful on the SER, difficulties with steaming and rough riding making them unpopular and they were withdrawn in the late 1860s.

Cramptons never fared well in this country, but much better performances could be got out of those that went on the Continent. Similar Cramptons were built for the Dundee, Perth & Aberdeen Junction Railway named KINNAIRD, and one for the Maryport & Carlisle Railway.

The LNWR Crampton was larger, with 18" x 20" cylinders and 8 ft driving wheels. A large boiler gave a total heating surface of 1529 sq. ft with a grate area of 16 sq. ft and a working pressure of 100 lb/sq. in. With a light load this locomotive was reported to have reached a speed of 74 mph. All the above mentioned Cramptons had slightly oval boilers. Little is known about the two 2-2-2 Cramptons built

Fig. 564 250/1912 (New Lowca) Millom Ironworks No. 3

for the Sheffield, Ashton-under-lyne and Manchester Railway. It is possible that the 2-2-2 wheel arrangement, generally stated, is an error.

An interesting problem of delivery is mentioned in Mr J. Simmons' book on the Maryport & Carlisle Railway. Due to the bad and steep roads in the vicinity the M&CR locomotives built by Tulk & Ley were sent to Maryport by raft.

Only seventeen locomotives had been built in fifteen years and twenty in eighteen years. The last locomotive built (Works No. 20) is a mystery and it is possible that parts were supplied to the M&CR to erect at their Maryport workshops in 1857. In this year the firm was taken over by FLETCHER JENNINGS LTD. and from 1857 to 1884 the works were owned by Mr H. A. Fletcher. During this period, and indeed right up to the closing of the works no tender locomotives were built. The policy was to concentrate on the production of four-coupled and six-coupled industrial tanks, the majority of which comprised four coupled tanks of the saddle, side, and well tank variety.

Fletcher patented a locomotive design in which the rear axle was brought as close as possible to the front of the firebox. Being so close no eccentrics could be fixed to this axle so they were put on the front axle and the eccentric rods faced to the rear to drive the Allan linkage. Fletcher did not favour outside valve gear, hence this arrangement to enable the wheelbase to be brought to a maximum. With hindsight this idea does not seem logical, especially when many locomotives of this type were used in works abounding with sharp curves where a short wheelbase would have been an advantage, with less wear on tyres, axleboxes, and frame racking. However they were substantially built and the majority put in many years of hard work. A dome, with safety valves mounted on top, was invariably used and early types had solid cast iron wheels with cored holes or shaped integral spokes and balance weights incorporated.

Fletcher Jennings, by 1884, had built 171 locomotives and rebuilt two from other makers. Most went to industry in England and Wales and at least twenty-four went abroad. In 1884 the business was made into a limited company and its title changed to: LOWCA ENGINEERING CO. LTD. The firm's policy continued on the same lines but for standard gauge 0-4-0ST and 0-6-0T types more substantial designs were brought out and class letters were allocated.

In 1905 the firm's name was again changed to: NEW LOWCA ENGINEERING CO. LTD. presumably due to a rearrangement of the Board. Unfortunately, in 1912 there was a disastrous fire which gutted most of the factory and as orders had fallen off considerably, only nine had been built from 1903 to 1912, it was decided to close down completely. Altogether the factory had produced 245 locomotives including the twenty by Tulk & Ley. The works finally wound up in 1927.

Wks No	Date	Type	Cylinders inches	D.W. ft in	Customer	No	Name	Gauge	
1	1840	2-2-2	12 x 18	5 0	Maryport & Carlisle Rly	1	Ellen	Std	
2	1840	0-6-0	14 x 18	4 6	Maryport & Carlisle Rly	2	Brayton	Std	
3	1843	2-2-2	13 x 18	5 6	Maryport & Carlisle Rly	4	Harrison	Std	
4	1845	0-4-2	14 x 21	4 9	Maryport & Carlisle Rly	7	Lowca	Std	
5	1847	0-4-2	14 x 21	4 9	Maryport & Carlisle Rly	9		Std	
6	1847	0-4-2	14 x 18	4 6	Whitehaven & Furness Jc Rly	1	Lowther	Std	
7	1847	0-4-2	14 x 18	4 6	Whitehaven & Furness Jc Rly	2	Whitehaven	Std	
8	1846	?	14 x		Cockermouth & Workington Rly		Derwent	Std	
9	1845	0-4-2	14 x 21	4 9	Maryport & Carlisle Rly	8	Harris	Std	
10	1847	4-2-0	OC 16 x 20	7 0	South Eastern Rly	81		Std	(a)
11	1847	4-2-0	OC 16 x 20	7 0	South Eastern Rly	82		Std	
12	1847	4-2-0	OC 18 x 20	8 0	London & North Western Rly	200	London	Std	
13	1847	4-2-0	OC 16 x 20	7 0	South Eastern Rly	85		Std	
14	1847	4-2-0	OC 16 x 20	7 0	Dundee, Perth & Aberdeen Jt Rly	17	Kinnaird	Std	
15	1847	4-2-0	OC 17 x 20	6 0	SA & M Rly	35	Pegasus	Std	(b)
16	1848	4-2-0	OC 17 x 20	6 0	SA & M Rly	36	Phlegon	Std	
17	1850	4-2-0	OC 18 x 20	7 0	Maryport & Carlisle Rly	12		Std	(c)
18	1855	0-6-0	16½ x 22	4 9	Whitehaven & Furness Jc Rly	12	Big Ben	Std	
19	1855	0-4-0T	OC 12 x 16		Crutwell Levick & Co Blaina			Std	
20	?				Maryport & Carlisle Rly	?		Std	(d

(a) Works Nos 10, 11 and 13 ordered by G. Rennie for Namur & Liège Rly, but not delivered and were sold to the SER in December 1849. (b) Works Nos 15 and 16 may have been 2-2-2s. (c) Sold to the Maryport & Carlisle Rly in 1854. (d) Probably a rebuilt locomotive.

TURNER & OGDEN

Leeds

This firm were probably agents only, getting the locomotives built by nearby locomotive builders. Locomotives purported to have been built by Turner and Ogden were: York and North Midland Railway – One four coupled locomotive (no further details have been discovered); Great North of England Railway – One four coupled locomotive with 13″ x 18″ cylinders and 5 ft diameter coupled wheels. The latter became North Eastern Railway No. 11 and was replaced in 1848. Both were built about 1840.

TWELLS & COMPANY

Birmingham

Supplied a locomotive in 1838 to the London & Greenwich Railway, No. 8 *Thames,* which was sold to the Clarence Railway. It is probable that this firm acted as agents only.

ULSTER RAILWAY

Great Victoria Street, Belfast

This railway was partially opened in 1839 from Belfast to Lisburn and built to the unusual gauge of 6′ 2″ but was converted to 5′ 3″ gauge in 1849. Workshops were built in Belfast and dealt with repairs and rebuilding. In 1867 two Sharp 2-2-2s were drastically rebuilt as 2-4-0s with new frames and boilers, followed by two more in 1869 which had practically the same treatment but were classified as 'new engines'. The first pair were officially rebuilds costing £2700 but as the 'new engines' cost £2500 all four could be regarded as new building. Other Sharp singles which were dealt with did not involve so much work and from a practical point of view were rebuilds.

Four 0-4-2s were built in 1871–4 and these were completely new with inside cylinders 16″ x 22″, and 5 ft diameter coupled wheels. The boiler was 3′ 11″ diameter and 10′ 4″ long, the tubes giving 860 sq. ft of heating surface and the firebox 96 sq. ft, giving a total of 956 sq. ft. The weight in working order was 30 tons.

Two 0-6-0s built in 1872–3 were also definitely new, with 17″ x 24″ inside cylinders and 5 ft diameter wheels, with boilers similar to those fitted to the previous 0-4-2s but longer.

The next major conversions to take place were two Beyer Peacock 2-2-2s which were transformed into 2-4-0s, one in 1874 and the other in 1876. It is probable that the old leading and driving wheels were used, together with parts of the motion. The Beyer singles were inside framed locomotives and the frames were not used as the 2-4-0s were double framed.

Fig. 565 **Ulster Rly.** 1872 No. 6 TORNADO shown as GNR (I) No. 106

From 1 April, 1876 the Ulster Railway became part of the newly formed GREAT NORTHERN OF IRELAND RAILWAY. It was known as the Northern division and Mr John Eaton, who had been Locomotive Superintendent on the Ulster Railway, carried on at Belfast and until J. C. Park took over both Belfast and Dundalk workshops ten or twelve years after the amalgamation, Great Victoria Street carried on much as before with virtual independence.

Ulster Railway locomotives were renumbered into the GNR(I) stock by having 100 added to their numbers. In 1876 two 0-6-0s were built, and two more, one in 1877 and 1878, using a few parts from four old Beyer Peacock 0-4-2s. The new boilers had a tube heating surface of 893 sq. ft and firebox 90 sq. ft, totalling 983 sq. ft and a grate area of 14·5 sq. ft. Inside frames and cylinders were fitted.

Finally in 1880–2 three 0-4-2s were built to replace two Sharp 0-4-2s whose wheels were probably re-used in the case of numbers 4 (1880) and 2 (1881). The third replaced a Fairbairn single and it is doubtful whether any more material than the trailing wheels were used. The boilers for all three were the same, having 158 2″ tubes, the barrel being 3′ 11″ diameter and 10′ 1″ long. The working pressure was 140 lbs/sq. in. All new boilers had flush round top fireboxes.

The number of new locomotives built, counting the major rebuilds mentioned, was nineteen of which eight were entirely new but the remaining eleven in most cases using very few second hand parts could be placed in the 'new' locomotive category. The works were closed about 1881 and work transferred to the Dundalk works established in that year.

ULSTER TRANSPORT AUTHORITY

Belfast

See BELFAST & NORTHERN COUNTIES RAILWAY.

ULSTER RLY BELFAST

Date	Type	Cylinders inches	D.W. ft in	UR No	Name	1st GNR No	Notes	W'drawn
1867	2-4-0	15 x 20	5 6	8	Lucifer	108	Reno 113	1906
1867	2-4-0	15 x 20	5 6	12	Vulcan	112	Reno 122 then 113	1905
1869	2-4-0	15 x 20	5 6	10	Jupiter	110	Reno 122 then 112	1905
11/1869	2-4-0	15 x 20	5 6	13	Spitfire	113		1888
1871	0-4-2	16 x 22	5 0	7	Cyclone	107	(a)	1908
1872	0-4-2	16 x 22	5 0	6	Tornado	106	(a)	1906
1872	0-6-0	17 x 24	5 0	37	Stromboli	137	(a)	1939
1873	0-6-0	17 x 24	5 0	38	Volcano	138	(a)	1948
1874	0-4-2	16 x 22	5 0	5	Typhoon	105	(a)	1911
1874	0-4-2	16 x 22	5 0	39	Tempest	139	Reno 104 (a)	1908
1874	2-4-0	16 x 22	5 6	23	Tyrone	123		1903
GREAT NORTHERN RLY (I)								
4/1876	0-6-0	17 x 24	4 7	14	Vesuvius	114	Reno 144 (a)	1924
6/1876	0-6-0	17 x 24	4 7	15	Hecla	115	Reno 147 (a)	1924
8/1876	2-4-0	16 x 22	5 6	1	Lagan	101	Reno 114	1913
12/1877	0-6-0	17 x 24	4 7	3	Etna	103	Reno 139	1925
6/1878	0-6-0	17 x 24	4 7	16	Teneriffe	116	Reno 148	1925
1880	0-4-2	16 x 22	4 6	4	Owenreagh	104		1899
1881	0-4-2	16 x 22	4 6	2	Blackwater	102		1901
1882	0-4-2	16 x 22	4 6	9	Pluto	109		1907

(a) Entirely new

USKSIDE IRON COMPANY
Newport, Mon.

Established in 1827, this firm carried out repairs on steam locomotives and although the present staff is extremely doubtful whether any were completely built there, two locomotives are recorded as being built for the Blaenavon Co. Ltd.: *c* 1849 0-4-0ST 3′ 3″ gauge, and *c* 1860 0-6-0T GAN-YR-ERW. Both had outside cylinders 12″ x 18″ and 3′ 6″ diameter wheels.

VIVIAN & SONS
Hafod Foundry, Swansea

One locomotive was built by this firm, which was one of the largest in the copper industry, established in 1810 for smelting copper ore obtained from Cornwall. The locomotive was a 0-4-0ST built in 1877 for their Pentre Colliery, with outside cylinders 10½″ x 15″ and 3′ 3″ diameter wheels. A photograph appeared in the *Locomotive Magazine* Vol. XXV dated 15 September, 1919.

VERNON, THOMAS & COMPANY
Regent Street Foundry, Liverpool

In the 1840s this firm was engaged in building ships and according to Mr D. L. Bradley, in *The Locomotives of the South Eastern Railway* (RCTS), when discussing the subject of where Chanter's locomotives were built suggests that this firm might have built some of the earlier ones. No further information is available. See CHANTER, JOHN.

VULCAN FOUNDRY COMPANY
Newton-le-Willows

See TAYLEUR, CHARLES & CO.

Fig. 566 **Walker Bros**, Wigan – Photograph of 0-4-0ST from catalogue

WALKER, J. SCARISBRICK & BROS.

Pagefield Ironworks, Wigan

This firm was principally engaged in iron founding and general engineering and was founded in the 1870s. John S. Walker was managing the firm in 1875. In addition to such general work a number of steam locomotives were built at intervals. They were all tank locomotives, four- and six-coupled and mostly saddle tanks, the six-coupled having inside cylinders and the four-coupled outside cylinders.

The photograph shows a very neat four-coupled saddle tank. The coupling and connecting rods were fitted with split brasses, straps and cotters. An interesting feature is the crosshead which appears to have adjustment for slipper wear. The rear wheels only had brakes, with wooden brake blocks. A domeless boiler was fitted with a pair of Ramsbottom safety valves above the firebox. This would have been built in the 1870s. BURNLEY was the final phase of building with raised firebox with Ramsbottom safety valves on top, a domeless boiler and all over cab.

About 1880 the firm's name was changed to WALKER BROS. (WIGAN) LTD. Customers supplied were local collieries and contractors. Building continued until about 1888, but this is not certain as no records are available. About twenty steam locomotives were built. The firm subsequently manufactured diesel passenger railcars.

WALKER, RICHARD & BROTHER

Bury

Richard Walker, the head of this firm, was also a director of the East Lancashire Railway. In 1846 the firm offered the ELR one locomotive, a 2-2-2, for the sum of £1200 which was an exceedingly cheap price for a locomotive of this size. It was purchased together with two more of a similar type becoming ELR Nos. 1, 3 and 18.

During the next six years the Walker brothers built twenty more locomotives all of which were ordered from the same railway and as far as can be learned they did not supply locomotives to any other railway or works. Three types were built 2-2-2, 2-4-0 and 0-6-0, all of which had outside frames and inside cylinders.

In 1847 a large order was placed for forty locomotives with tenders. Some were to be singles and some four-coupled and delivery was to extend to 1849. Evidently some members of the railway board took exception to this, Richard Walker being in the happy position of being in both camps, but finally it was resolved to reduce the order to twenty.

The 2-2-2s were to be a copy of Sharp Bros. locomotive No. 14 AURORA which had been delivered to the ELR in March 1847. This was done but it would be interesting to know how they did the copying without a lawsuit. No orders were placed with any other firm for the other twenty locomotives which were

PAGEFIELD IRONWORKS

Works No	Built	Type	Cylinders inches	D.W. ft in	Customer	Name/No
334	1872	0-4-0ST	OC 10		G. B. Deakins Winsford	
412	*c* 1873	0-4-0ST	IC		Bridgewater Colls.	Graham
	?	0-4-0ST	10 x 16	2 9	Widnes Alkali Co	(a)
440	1873	0-4-0ST			United Alkali Co	Breidden (b)
444	1873	0-4-0ST			Swan Lane Brick & Coal Co	(c)
	1873	0-4-0ST	OC		For own use	Harry
	1874	0-4-0ST	OC		Brinsop Hall Coal Co Ltd	
	1874	0-4-0ST				Jesse
1367	1877	0-6-0ST	IC		Garswood Hall Colls Co Ltd	Ashton
1544	1877	0-4-0T			Taylor, Greenall & Kidd	
	1878	0-4-0ST	OC		Winstanley Colly Co Ltd	Eleanor (4 ft gauge)
1938	1878	0-4-0ST	OC		Winstanley Colly Co Ltd	Walter
1939	1879	0-6-0ST	IC		Abram Coll Co	No 3
1364	1880	0-6-0ST	IC		Abram Coll Co	No 1 (d)
	1881	0-4-0ST	OC		Salt Union Co	Bostock
2262	1881	0-4-0ST			Braddock & Matthews	Edleston
	1888	0-6-0T	IC		T. A. Walker Ltd	Burnley
		4WTGVB			Shap Granite Co	Wasdale
		0-4-0ST	OC		Winstanley Colly	Walter
		0-4-0ST	OC		Douglas Bank Colly	Nellie
		0-4-0ST	OC		Salt Union Ltd	George Deakins (e)
		0-6-0ST	IC		Park Lane Colly Co	No. 4

(a) Sold to L & SWR P. W. Dept No 8 Mina in 1874. (b) Rebuild of Hawthorns (Leith) Loco.
(c) Rebuild. (d) Date built may be incorrect if works No in sequence. (e) May have been works no 334.

Date	Type	Cylinders inches	D.W. ft in	East Lancs No.	Name	W'drawn
1838	2-2-2	13 x 18	5 6	52*	Diomede	1851
1838	2-2-2	13 x 18	5 0	54*	Medusa	1846
4/1846	2-2-2	15 x 20	5 2	1	Medusa	1866
5/1846	2-2-2	15 x 20	5 2	3	Hecate	1866
5/1847	2-2-2	15 x 20	5 6	18	Lynx	1867
5/1847	2-4-0	15 x 20	5 6	25	Venus	1873
6/1848	2-4-0	15 x 20	5 6	26	Lightning	1881
8/1848	2-4-0	15 x 20	5 6	27	Camilla	1881
9/1848	2-4-0	15 x 20	5 6	28	Lucifer	1878
10/1848	2-4-0	15 x 20	5 6	29	Ariel	1880
11/1848	2-4-0	15 x 20	5 6	30	Phaeton	1880
12/1848	2-4-0	15 x 20	5 6	31	Orion	1892
2/1849	2-4-0	15 x 20	5 6	33	Mazeppa	1878
5/1849	2-4-0	15 x 20	5 6	40	Fire King	1882
5/1849	2-4-0	15 x 20	5 6	42	Vampire	1879
5/1849	2-4-0	16 x 20	5 0	44	John Bull	1880
5/1849	2-4-0	16 x 20	5 0	45	Caliban	1880
4/1850	2-2-2	15 x 20	5 6	49	Gazelle	1880
4/1850	2-2-2	15 x 20	5 6	50	Banshee	1873
10/1850	2-2-2	15 x 20	5 6	53	Vivid	1880
11/1850	2-2-2	15 x 20	5 6	54	Reindeer	1873
11/1852	0-6-0	15 x 24	4 9	55	Rossendale	1883
1/1853	0-6-0	15 x 24	4 9	56	Agamemnon	1881
7/1853	0-6-0	15 x 24	4 9	57	Hannibal	1882
2/1854	0-6-0	15 x 24	4 9	58	Dugdale	1880

All had outside frames and inside cylinders. 0-6-0s were Long Boiler Type. 600 added to numbers by L & Y Rly .
* Built for Grand Junction Rly

cancelled, and only seven or eight new locomotives were built by other firms before the amalgamation of the ELR with the Lancashire and Yorkshire Railways in 1859.

The 2-4-0s built had dimensions similar to the Sharp 2-4-0s. The boiler heating surface was 975 sq. ft, grate area 14·7 sq. ft, and weight in working order was 24 tons 16 cwt.

The 0-6-0s had dimensions similar to those built by Fenton Craven & Company. The heating surface of the boiler was 1042·5 sq. ft, grate area 15·0 sq. ft and weight in working order 29 tons 19 cwt.

The 2-2-2s were built as a copy of Sharp's singles and had boilers with 750 sq. ft of heating surface, 12·2 sq. ft of grate area and

Fig. 567 **R. Walker & Bros.** 1850 East Lancs Rly No. 54 running as L&YR No. 654

they weighed 26 tons 8 cwt in working order. All three types had a boiler pressure of 120 lb/sq. in.

No new locomotives were built after 1854 although repairs were carried out for the East Lancashire Railway. In the Grand Junction Railway list two 2-2-2s are ascribed to Richard Walker & Brother. They were No. 52 DIOMEDE and No. 54 MEDUSA and were built in 1838. There seems a large gap between 1838 and 1846 when the first was built for the East Lancashire Railway. Others may have been built in this period. A study of the ELR locomotives may be found in the Stephenson Locomotive Society's Journal No. 444.

WARRINGTON, HENRY & SON

Berry Hill Colliery, Fenton

A locomotive is claimed to have been built at this colliery in January, 1898. This is a classical example of a locomotive being erected but not built.

Henry Warrington & Son at this time owned the colliery and also adjacent ironworks and foundry. The locomotive was officially classed as 0-4-0ST *Peckett* type although it bears a stronger relationship with Avonside Engine Co's products. Parts purchased from outside sources included frame plates, axle and buffer springs, tubes, safety valve and cover, boiler mountings and wheel tyres. It is more than likely that parts were used from another locomotive and could possibly be a renewal or rebuild of such a locomotive, such as Avonside Works No. 1343 of 1890 which belonged to the firm.

Dimensions of the 'new' locomotive, which was named BERRY HILL NO. 1, were: outside cylinders 15" x 21" and wheels 3' 4" diameter without tyres. It had a copper firebox and brass tubes. It was rebuilt in 1931 when a new boiler and firebox were fitted and in this instance steel tubes were used.

The ironworks and foundry closed down in 1906 and the colliery in 1961 and the locomotive was scrapped in the same year. Correspondence concerning the history of this locomotive can be found in the *Industrial Railway Record* Nos. 10, 12 and 19.

Fig. 568 **Henry Warrington** Built or rebuilt – the controversial BERRY HILL NO. 1

WATERFORD & LIMERICK RAILWAY

Limerick

Six locomotives were built at Limerick and, although all of them included a number of parts from withdrawn 0-4-2s, this was probably not more than two pairs of wheels, with or without axles, and parts of the motion. The first was a 0-6-0 No. 7 WASP, the work being carried out in 1888. At this time Mr J. G. Robinson was the locomotive engineer destined to become the Locomotive Superintendent for the Great Central Railway in 1900. Three 0-6-0s were built, one 0-4-2T, and one 0-4-4T. The practice of naming the locomotives commenced about 1886, only numbers being allocated before this. Robinson turned out his locomotives in very smart livery; some had brass domes, copper cap chimneys and brass number plates very much in the Great Western Railway, Swindon tradition where J. G. Robinson had received his training. On 1 January, 1896 the company's name was changed to the WATERFORD, LIMERICK & WESTERN RAILWAY. Under this new title one additional 0-4-4T No. 27 THOMOND was built and no further building occurred after 1899.

WATERFORD & LIMERICK RLY

Date	Type	Cylinders inches	D.W. ft in	No	Name	G.S.W. No	W'drawn
1888	0-6-0	IC 16 x 24	4 7	7	Wasp	226	1905
1890	0-6-0	IC 16 x 24	4 7	6	Ant	225	1907
1892	0-4-2T			3	Zetland	260	1912
1893	0-6-0	IC 16 x 24	4 7	5	Bee	224	1909
1894	0-4-4T	IC 16 x 24	4 7	15	Roxborough	268	1912-14
WATERFORD, LIMERICK & WESTERN RLY							
1899	0-4-4T	IC 16 x 24	5 4	27	Thomond	279	1953

Fig. 569 **Waterford & Limerick Rly.** 1894 0-4-4T No. 15 ROXBOROUGH

WATERS, THOMAS

Gateshead

After the non-acceptance by Christopher Blackett of John Whinfield's locomotive, it was realised that the biggest obstacle to steam locomotive utilisation was the strength of the tracks which in the main were wooden wagon ways. Blackett immediately set about replacing the wooden track at Wylam by cast iron angle pattern plate rails for a distance of about five miles, being completed in 1808. Trevithick was approached by Blackett to build him a locomotive, but Trevithick had already abandoned locomotive manufacture, and his search did not produce anyone willing to provide a locomotive until in 1811 Thomas Waters offered his services. He was a neighbour of John Whinfield in Gateshead and constructed steam engines 'of the Trevithick type' and was no doubt well aware of Whinfield's 1804 locomotive and its construction.

There is some doubt as to whether Waters constructed his locomotive at his own works or at Wylam, but it is more probable that he used his own works for the purpose, as it was reported as being conveyed from Gateshead to Wylam on completion early in 1813, although according to Robert Young it was completed early in 1811. The locomotive bore a striking resemblance to Trevithick's Pen-y-Darren, and Whinfield's locomotives. A number of modifications and alterations were carried out over a period of about twelve months, after which it was used at the bottom of one of the pit shafts at Wylam for haulage purposes.

Timothy Hackworth was known to have assisted in the construction of Water's locomotive. The engine had one cylinder 6" diameter and a flywheel. The boiler was cast iron with a single flue or fire tube with the chimney at the far end, so differing from the Trevithick locomotives which had a return flue.

WATKINS & PHILLIPS

See PHILLIPS, CHARLES D.

WATSON & DAGLISH

See DAGLISH, ROBERT.

WEARDALE COAL & IRON CO. LTD.

Tudhoe, Spennymoor

One 0-6-0ST with outside cylinders is reported to have been built here in 1873 for the company's own use – their No. 15. Although no confirmation can be obtained, two other 0-6-0STs, Weardale No. 6 and No. 20 may have been built at Tow Law Ironworks in the 1860s.

WEST, WILLIAM & SONS

St Blazey Foundry, St Blazey

Founded in 1848 by William West, the firm contracted to work the Newquay and Cornwall Junction Railway which was opened in 1869 with two broad gauge locomotives which belonged to them. One of these was named NEWQUAY and was hired from the Glyncorrwg Coal Company. The second was named PHOENIX and although in some quarters attributed to Brotherhoods, it is possible that it was built at St Blazey. Two other locomotives CORNWALL and ROEBUCK were used by this firm on the N&CJR and the origin of these is doubtful and again may have been built at St Blazey. PHOENIX was a six-wheeled four-coupled locomotive and the other two were six-coupled tanks. The firm's principle work was the manufacture of pumping engines for the mining industry and after West's death in 1879 the business was carried on by his two sons, William and Charles and eventually closed down in 1891.

WEST CORNWALL RAILWAY

Carn Brea, Cornwall

Built	Type	Cylinders inches	D.W. ft in	Name	W'drawn
4/1851	2-4-0T(?)			Penzance	*c* 1860
4/1852	2-4-0T	16 x 24	5 0	Camborne	*c* 1865
1853	2-4-0T(?)			Hayle	1866
1865	0-6-0	17¼ x 24	4 9	Redruth (a)	6/1887

(a) Rebuilt Newton (Devon) 12/1871 as broad gauge 0-6-0ST. Became GWR 2156

In 1846 the standard gauge Hayle Railway was absorbed by the West Cornwall Railway and the Carn Brea workshops were used for repairing locomotives. In 1851 the PENZANCE was built under the supervision of Mr Slater. The major parts were supplied by Messrs Stothert Slaughter & Company of Bristol and it is probable that all locomotives built or erected at Carn Brea had parts supplied as for PENZANCE.

WEST CORNWALL RAILWAY

The second locomotive built, CAMBORNE, was a 2-4-0T and at least one other of the first three built was the same type. HAYLE followed in 1853 and in 1865 a six-coupled outside frame goods locomotive was built which subsequently became a broad gauge saddle tank and when taken over by the Great Western Railway was numbered 2156.

WEST HARTLEPOOL HARBOUR & RAILWAY CO.

Stockton

The Hartlepool West Harbour and Dock Co. was formed in 1847 and amalgamated with the Stockton and Hartlepool Railway in 1853 and also took over the lease of the Clarence Railway. The new title West Hartlepool Harbour & Railway Co. was used until 1865 when it became part of the North Eastern Railway. As far as can be ascertained, nineteen locomotives were built in the railway's workshops comprising: four 0-4-0Ts, two 2-2-2s and thirteen 0-6-0s. No information appears to be available about these locomotives. The Locomotive Superintendent was Mr J. I. Carson.

Date	Type	Cylinders inches	D.W. ft in	WHH No	NER No
1856	0-6-0	15 x 22	4 6	54	622
1857	0-6-0	15½ x 18	4 6	61	629
1858	0-6-0	15½ x 18	4 6	62	630
1858	2-2-2	15 x 20	5 6	7	(588)
1859	2-2-2	15 x 20	5 7	64	632
1860	0-6-0	16½ x 24	4 7	65	633
1860	0-6-0	15½ x 18	4 6	63	631
1861	0-6-0	16½ x 24	4 7	66	634
1862	0-6-0	16½ x 24	4 7	67	635
1862	0-6-0	16½ x 24	4 7	68	636
1862	0-6-0	15 x 22	4 6	70	638
1862	0-6-0	16½ x 24	4 10	19	596
1863	0-6-0	16½ x 24	4 10	12	(591)
1863	0-6-0	15 x 22	4 7	13	592
1863	0-6-0	16½ x 24	4 10	14	593
1863	0-6-0	16½ x 24	4 10	15	594
1863	0-6-0	16½ x 24	4 10	16	617
1863	0-6-0	15½ x 24	4 10	47	619
1864	0-6-0	16½ x 24	4 10	21	597
1864	0-6-0	16½ x 24	4 10	25	601
1864	0-6-0	16½ x 24	4 10	26	602
1865	0-6-0	16½ x 24	4 10	9	(590)
1865	0-6-0	16½ x 24	4 10	27	603
1865	0-6-0	16½ x 24	4 10	33	608
1865	0-6-0	16½ x 24	4 10	34	609
1865	0-4-0T	14 x 20	3 6	39	611
1865	0-4-0T	14 x 20	3 6	41	613
1865	0-4-0T	14 x 20	3 6	49	619
1865	0-4-0T	14 x 20	3 6	52	620

WEST MIDLAND RAILWAY

Worcester

See OXFORD, WORCESTER & WOLVERHAMPTON RAILWAY.

WHEATLEY, THOMAS

Grimsby

This firm was contemporaneous with Isaac Boulton of Ashton. One locomotive is on record as being built in 1858, a 0-4-0ST named PERSEVERANCE but it is more than probable that this was a rebuild of an older locomotive. All available details are to be found in *Chronicles of Boulton's Sidings* by A. R. Bennett.

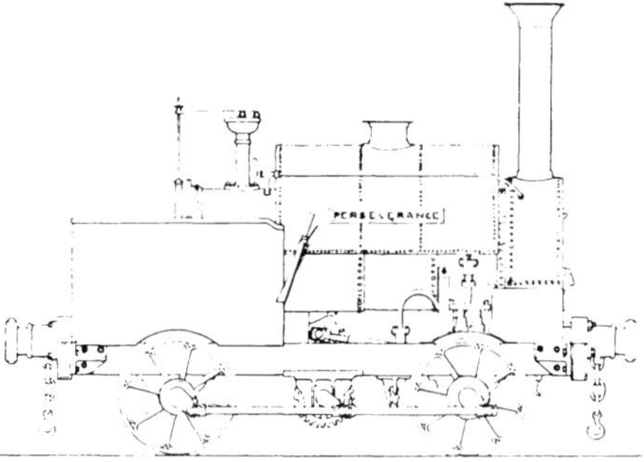

Fig. 570 **Thomas Wheatley** 1858 0-4-0ST gears and jackshaft PERSEVERANCE

WHINFIELD, JOHN

Pipewellgate, Gateshead

John Whinfield was an ironfounder and Mr Blackett, the proprietor of Wylam Colliery, learned that Whinfield was in a position to build a locomotive. He was told to go ahead with the proviso that, after completion and trial, Blackett should have the option of its purchase if the locomotive was to his requirements. Whinfield had an engineer in the person of John Steel who had worked on the Pen-y-darren locomotive and this man was responsible for the construction of Whinfield's locomotive, and it is not surprising that the design followed closely the Pen-y-darren engine, having identical gearing with large flywheel and horizontal cylinder set in the boiler. It was constructed in May 1805 and was tried on the Wylam wagon way but proved too heavy for the track. It weighed over five tons. Blackett of course refused to buy the locomotive. Dimensions known are: cylinder 7″ x 36″ and wheels 3′ 2″ diameter.

Fig. 571 **John Whinfield** 1805 Trevithick's Newcastle locomotive

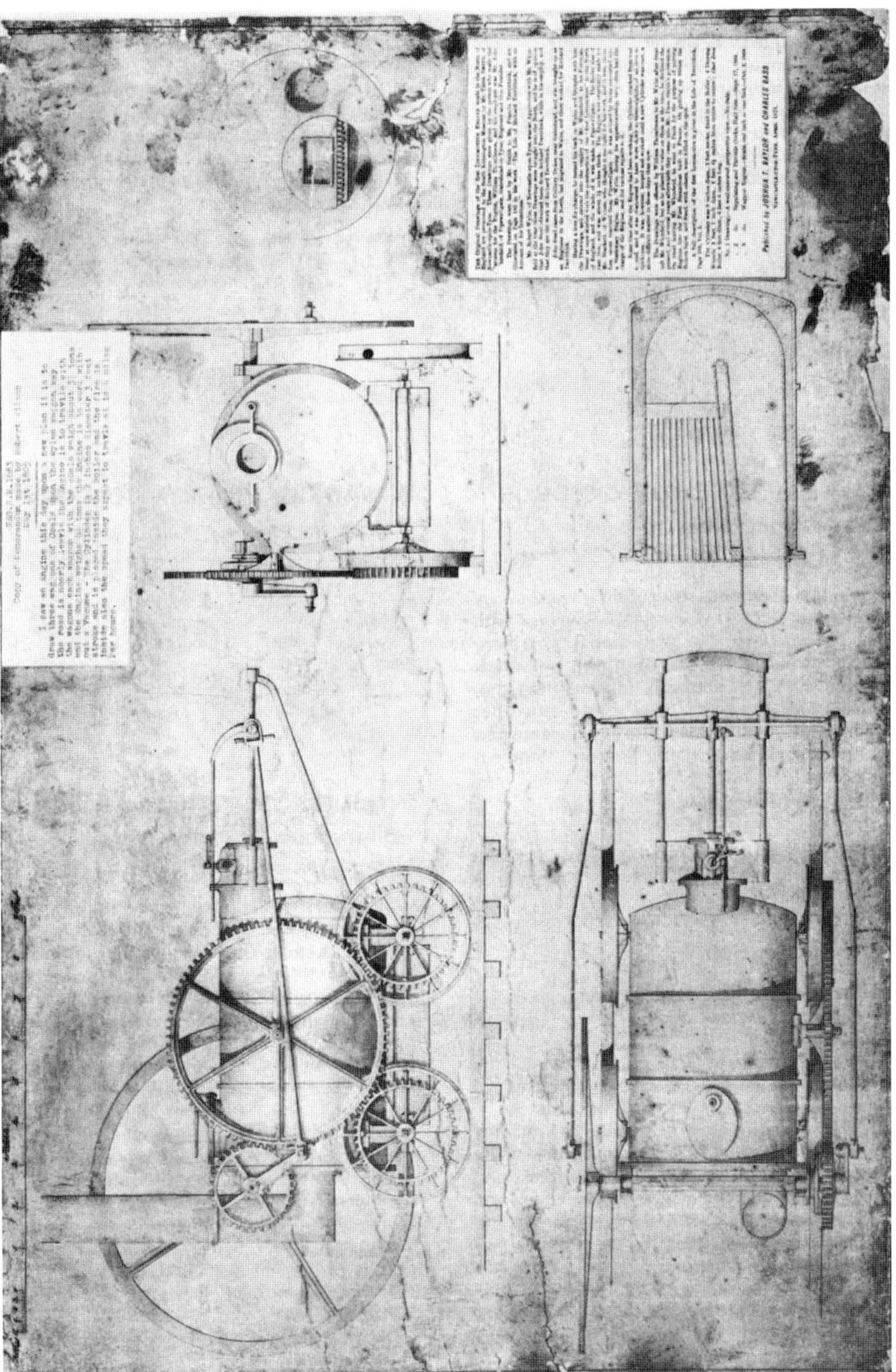

WHINFIELD, JOHN

According to J. F. Layson, in his life story of the Stephensons, 'a locomotive was constructed in October, 1804 from plans furnished by Trevithick, but the machine never left Whinfield's foundry because it was too light for the work it had to do; it is alleged that another engine proved likewise a failure; the story being that after it had been removed from Gateshead, when finished by Thomas Waters of that town and tried at Wylam, it went to pieces to the danger and alarm of the bystanders'.
See TREVITHICK, RICHARD and WATERS, THOMAS.

WHITE, JOHN BAZLEY, & BROS
Swanscombe, Kent

A locomotive named GRAVESEND was built by this firm in 1898. It was a 0-4-0 well tank with outside cylinders. The workshops did a considerable amount of rebuilding of their locomotive stock and it is fairly certain that GRAVESEND would have parts of other locomotives incorporated. The firm became part of the Associated Portland Cement Manufacturers Ltd.

WHITEHAVEN COLLIERIES
Earl of Lonsdale

Mr Taylor Swainson was the engineer to the Whitehaven Collieries, owned by the Earl Of Lonsdale, when he built a locomotive in 1812 known as the IRON HORSE. Its life was very short, due to the usual trouble encountered by these pioneers of weakness of the track, and although the wooden wagon-way had cast iron fish bellied rails, substituted a few years before the locomotive was tried, its weight was sufficient to cause enough damage for the locomotive to be removed. Mr Dendy Marshall in *History of Railway Locomotives down to 1831* deals with this subject, and it is very probable that the locomotive was very similar to that of Trevithick's at Pen-y-darren.

WHITEHAVEN & FURNESS JUNCTION RAILWAY

According to the Furness Railway locomotive list, FR No. 43 was a 0-6-0 built by the Whitehaven and Furness Junction Railway in 1866. It was replaced in 1871 by a Sharp Stewart locomotive of the same wheel arrangement. No further details are available and it is likely that the locomotive engineer received outside help in materials at least.

WIGAN COAL & IRON COMPANY.

See KIRKLESS HALL COAL & IRON COMPANY.

WILKINSON, WILLIAM
Holmeshouse Foundry, Wigan

W. Wilkinson was probably a general engineer employing a few craftsmen. About 1880 he designed a steam tram locomotive with vertical boiler and cylinders and the motive power was transmitted by means of gearing on the crankshaft engaging through one or more gears on to one of the two axles. The two pairs of track wheels were coupled and, although no condenser was fitted, a special superheater was housed in the firebox through which the exhaust steam was passed before exhausting through the chimney. The idea was to replace the horse drawn trams common on all systems, and he approached the Wigan & District Tramways Co. Ltd. to try out his first tram locomotive which he had built in 1881. It was so successful that a number of other Lancashire and Midland systems ordered locomotives of the same design.

Due to his small workshops, he could not cope with all the orders obtained and he reached agreements with three prominent locomotive builders to manufacture his tram locomotives under his patents. The firms were Beyer Peacock & Co. Ltd.; Black Hawthorn & Company and Thomas Green & Son.

Nevertheless, from 1881 to 1886 at least sixty-one units were built in Wilkinson's workshops and by then competition, particularly from Kitson & Company who had perfected a reliable condensing type of tram locomotive, grew steadily stronger and the last two were built in 1896 for the Giants Causeway Tramway, Wilkinson having already supplied two previously in 1883.

Altogether 207 locomotives were built to Wilkinson's patent: Wm Wilkinson 63; Beyer Peacock 73; Black Hawthorn 32; and Thomas Green 39.

Many of the earlier locomotives built emitted too much smoke but it was not until 1886 that Wilkinson designed and patented an air condenser. It was placed on the roof and the exhaust was taken through a box in which were fitted a number of tubes open at each end to the atmosphere and were positioned horizontally but at right angles to the track.

Consequently to induce cooling air through them deflectors were fitted. Why the tubes were not fitted in the fore and aft position to enable the air to pass freely through them seems strange. The condenser was not a success.

The most successful tram locomotives built under Wilkinson's patent were by Beyer Peacock & Company in 1886, who employed the more conventional air condensers similar to those developed by Kitson & Company. By this time the locomotive building companies had developed their own versions of tram locomotives, and due to this, orders fell short as previously mentioned.

Locomotives that were built had cylinders from 6″ to 9″ diameter and from 7″ to 11″ stroke. All boilers fitted were of the vertical type with 'Field' tubes. The boiler shell was 3′ 6″ diameter. The valve gear was Stephenson's reversing link motion and most were fitted with controls at each end. To conform with Board of Trade requirements the wheels and motion were enclosed and various types of governors were fitted.

Fig. 572 **W. Wilkinson** 1885 John Bazley White & Bros. Ltd. Swanscombe DEVONPORT

BUILT BY W. WILKINSON

Date	Wks Nos	Tramway	Nos	Gauge ft in	No Built
1881-6		Wigan & District Tramways Co Ltd	1-12	3 6	12
1882-3		Huddersfield Corporation Tramways	1-6	4 7¾	6
1882		Nottingham & District Tramways Co Ltd	—	4 8½	1
1883		Birmingham & Aston Tramways Co Ltd	7&8	3 6	2
1883		Dublin & Southern District Tramways Co	3&4	5 3	2
1883		Manchester, Bury, Rochdale & Oldham Steam Tramways Co	4-8	4 8½	5
1883		Manchester, Bury, Rochdale & Oldham Steam Tramways Co	35-38	4 8½	4
1883		Manchester, Bury, Rochdale & Oldham Steam Tramways Co	13-20	3 6	8
1883		South Staffordshire Tramways	1-2	3 6	2
1883		South Staffordshire Tramways	17-21	3 6	5
1883		Giants Causeway, Portrush & Bush Valley Rly & Tramway Co †	1&2	3 0	2
1884		Brighton District Tramways Co	1&2	3 6	2
1884		Plymouth, Devonport & District Tramways Co	—	3 6	2
1885		Manchester, Bury, Rochdale & Oldham Steam Tramways Co	61&62	3 6	2
1886		Manchester, Bury, Rochdale & Oldham Steam Tramways Co	89&90	3 6	2
1896		Giants Causeway, Portrush & Bush Valley Rly & Tramway Co †	3&4	3 0	2
					4
				Total	63

† Nos 1, 3, 4, named 'Wartrail', 'Dunluce Castle', and 'Brian Boroimhe' respectively.

BUILT BY BEYER PEACOCK & CO

Date	Wks Nos	Tramway	Nos	Gauge ft in	No Built
1883	2377-2382	Manchester, Bury, Rochdale & Oldham Steam Tramways Co ϕ	21-26	3 6	6
1883	2383-2392	South Staffordshire Tramways *	3-12	3 6	10
1883	2393	North Staffordshire Tramways	1	4 0	1
1883	2411-2429	North Staffordshire Tramways	2-20	4 0	19
1884	2593-2594	Coventry & District Tramways Co	1&2	3 6	2
1885	2595-2602	South Staffordshire Tramways *	22-29	3 6	8
1886	2713-2732	Manchester, Bury, Rochdale & Oldham Steam Tramways Co	63-82	3 6	20
1886	2733-2738	Manchester, Bury, Rochdale & Oldham Steam Tramways Co	83-88	4 8½	6
1886	2799	Manchester, Bury, Rochdale & Oldham Steam Tramways Co ϕ	91	4 8½	1
				Total	73

BUILT BY BLACK HAWTHORN & CO

Date	Wks No	Tramway	Nos	Gauge ft in	No Built
1883	735	Alford & Sutton Tramway †	1	2 6	1
1883		North Shields & Tynemouth District Tramways	1-3	3 0	3
1883-9		Gateshead & District Tramways	1-15	4 8½	15
1884		Manchester, Bury, Rochdale & Oldham Steam Tramways Co	42-51	3 6	10
1885-6		Huddersfield Corporation Tramways	7-9	4 7¾	3
				Total	32

BUILT BY THOS. GREEN & SON.

Date	Wks No	Tramway	Nos	Gauge ft in	No Built
1882		Leeds Tramways Co Ltd	5&6	4 8½	2
c1882		Mr. Lee Australia			1
1882		Manchester, Bury, Rochdale & Oldham Steam Tramways Co	1-3	4 8½	3
1883		Manchester, Bury, Rochdale & Oldham Steam Tramways Co	9-12	3 6	4
1883		Manchester, Bury, Rochdale & Oldham Steam Tramways Co	27-34	3 6	8
1883		North Shields & Tynemouth District Tramways	4&5	3 0	2
1883		South Staffordshire Tramways	13-16	3 6	4
1883		Bradford Tramways & Omnibus Co Ltd	7	4 0	1
1884		Bradford & Shelf Tramways Co Ltd	1-4	4 0	4
1884		South Staffordshire Tramways	30-37	3 6	8
1884		Coventry & District Tramways	3&4	3 6	2
				Total	39

ϕ Cylinders 8½" x 14" wheels 2' 8" diameter
* Cylinders 7½" x 11" wheels 2' 9" diameter
+ Cylinders 7¼" x 11" wheels 2' 3½" diameter
All tram locomotives 0-4-0T.

WILSON, E. B. & COMPANY
Railway Foundry, Leeds

See SHEPHERD & TODD.

WILSON, J. H. & CO. LTD.
Sandhills, Liverpool

An outside cylinder geared 0-4-0T was built named MOLE for the Welsh Slate Company, Blaenau Festiniog about 1875. It had a marine type boiler with return flue. The motion and gearing were similar to the machinery used on the firm's travelling cranes, (the firm were also general engineers). The locomotive is described by Mr J. I. C. Boyed in *Railways* for November 1951, and in the *Industrial Railway Record* No. 34.

WILSON, ROBERT & COMPANY
Forth Street, Newcastle

This firm built a locomotive named STOCKTON for the Stockton & Darlington Railway (their No. 5) and it was completed in March 1826. It ran on four cast iron wheels 4 ft diameter on a wheelbase of 5′ 9″. The interesting feature of the design was that each pair of wheels were driven separately by a pair of vertical cylinders, the crankpins being at right angles to each other. Steam could be fed to either or both cylinders, and the two sets of valve gear could be reversed independently. The exhaust steam was discharged into the chimney by two separate blast pipes, each pair of cylinders having a blast pipe. The cylinders were 6″ x 18″.

Very little is known about this design apart from a rough sketch made by the French engineer Marc Seguin. This locomotive was probably the first four-cylinder locomotive.

In October 1826 it was involved in a collision at Stockton and suffered serious damage. As it had not proved itself successful it was broken up, but its boiler was salvaged and after being lengthened by Mr John Wright of Lumley Forge it was used by Timothy Hackworth for his ROYAL GEORGE of 1827. Wilson wanted to build an improved version of STOCKTON for the S&DR but his offer was not accepted. The locomotive was also known as CHITTAPRAT but this was probably only a nickname emanating from the strange exhaust noises it made.

WOOLWICH ARSENAL

Towards the end of the Great War in 1918, the Government, anticipating considerable redundancy at Woolwich Arsenal, decided to construct locomotives as an interim measure. A series of meetings were held by the Association of Railway Locomotive Engineers and a 2-6-0 mixed traffic locomotive was decided upon. This closely followed the *N* class 2-6-0, designed by R. E. L. Maunsell at Ashford with various modifications, but such a time was taken in discussion and design, that in the end the Government decided to use the *N* class drawings without alteration.

The design was a radical departure from the traditional South Eastern & Chatham Railway practice and incorporated a taper boiler with top feed, smokebox regulator, outside Walschaerts valve gear and a boiler pressure of 200 lb/sq. in. The prototype *N* class No. 810 built at Ashford had proved an outstanding success, a contributory feature being the fitting of long lap valves. After the grouping of the railways in 1921, the Arsenal could not find any purchasers for their 2-6-0s but eventually the Southern Railway agreed to buy fifty of them at a bargain price.

It is presumed that these fifty were complete locomotives and according to Mr H. Holcroft in his *Locomotive Adventure* Vol II they were taken into Ashford works for 'vetting' before being accepted for traffic, and were not comparable with the Ashford built locomotives until they had passed through the shops for their first general overhaul. This was put down to the Arsenal men's inexperience in locomotive construction.

Twenty locomotives were purchased in 1924 (SR Nos. A826–845) and thirty in 1925 (SR Nos. A846–875). The remainder were in parts and the sets were disposed of as follows:

1924	6 sets to the Metropolitan Railway. These were erected by Messrs. Armstrong Whitworth & Company and took the form of 2-6-4Ts. (Metropolitan Railways Nos. 111–116)
1925	4 sets to Great Southern Railway of Ireland erected at Broadstone (GSR Nos. 372–375)
1926	6 sets to GSR Broadstone, (GSR No's 376–381).
1926	2 sets to GSR Broadstone, (GSR Nos. 382–383).
1927	3 sets GSR Inchicore, (GSR Nos. 384–386)
1928	3 sets GSR Inchicore, (GSR Nos. 387–389)
1929	2 sets GSR Inchicore, (GSR Nos. 390–391)
1930	6 sets GSR Inchicore, (GSR Nos. 393–398)

This makes eighty-two sets disposed of. The other eighteen sets, according to D. L. Bradley in his book on the locomotives of the South Eastern & Chatham Railway, were sold to the Roumanian State Railways who erected twelve locomotives from the parts in 1926.

In the *Engineer* dated 3 June, 1921, it was stated: 'The attempt which has been made with the object of averting unemployment, to convert a portion of Woolwich Arsenal into a permanent locomotive factory has not been as successful as some people had anticipated. In his report on the annual accounts of the ordnance factories for the year 1919–20, Sir H. J. Gibson the Comptroller states that the manufacture of locomotives was undertaken as part of the peace time programme and work for various railway companies amounting to £255,000 was carried out. The report adds that the Ministry of Munitions informed the Treasury when seeking sanction for capital expenditure that it had not been possible at the close of the year under review to arrange for the disposal of surplus locomotives or indeed to fix a price for them. It is understood that no sale of the locomotives has down to the present time been effected. The general question of the financial arrangements in connection with the manufacture of locomotives and other products since the ordnance factories were transferred to the War Office is under consideration and also that of putting the factories in a position to act as sub-contractors to engineering firms.

Again in the *Engineer* dated 5 January, 1923 it was stated 'there were fifty uncompleted locomotives at Woolwich which had cost £551,000. The fifty completed ones had cost £780,000. None were sold and were not to be finished, and their parts sold as they lie'. The sale of sets of parts was undertaken by Cohen Armstrong Disposal Corporation.

The dimensions of these locomotives were: cylinders (2) 19″ x 28″; wheels, leading 3′ 1″ driving 5′ 6″; boiler 4′ 7$\frac{3}{4}$″ and 5′ 3″ diameter x 12′ 6″ long; firebox length 8′ 0″; grate area 25 sq. ft; heating surface – tubes 1390·6 sq. ft; superheater 203·0 sq. ft; firebox 135·0 sq. ft; total 1728·6 sq ft.

Weight in working order: engine 59 tons 8 cwt; tender 39 tons 5 cwt; total 98 tons 13 cwt.

The tender held 5 tons of Coal and 3500 gallons of water.

WORCESTER ENGINE COMPANY
Worcester

This works was established in 1865 for the production of locomotives and seventy were built between then and 1870. Alexander Allan left the Caledonian Railway's locomotive shops at Perth (previously Scottish Central Railway) and joined the firm in 1866. The first locomotive was a six-coupled saddle tank named SALFORD, but nothing is known about the five succeeding locomotives.

Ten 0-6-0 goods were ordered by the North Stafford Railway which were built in 1866–7. They had 16$\frac{1}{2}$″ x 24″ cylinders and 5′ diameter wheels. An order followed for forty similar engines for the Great Eastern Railway with 5′ 3″ diameter wheels, these being delivered from 1867 to 1869.

Six 0-6-0 goods and two 2-4-2T passenger locomotives were supplied to the Bristol and Exeter Railway's standard gauge section. The 2-4-2Ts were interesting, having domeless boilers, inside cylinders and frames, a large

Fig. 573 Worcester Engine Co. 35/1868 Metropolitan Rly No. 34

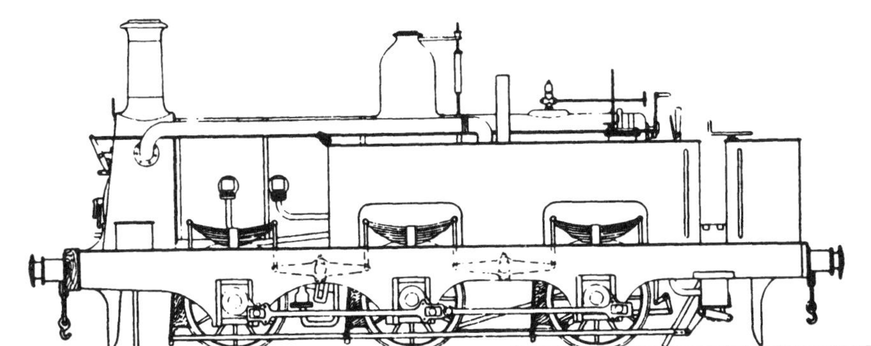

WORCESTER ENGINE COMPANY

Works No	Date	Type	Cylinders inches	D.W. ft in	Customer	No	Name
1	1865	0-6-0ST	11½ x 18	3 2			Salford
2							
3							
4							
5							
6							
7	1866	0-6-0	16½ x 24	5 0	North Staffordshire Rly	90	
8	1867	0-6-0	16½ x 24	5 0	North Staffordshire Rly	91	
9	1867	0-6-0	16½ x 24	5 0	North Staffordshire Rly	92	
10	1867	0-6-0	16½ x 24	5 0	North Staffordshire Rly	93	
11	1867	0-6-0	16½ x 24	5 0	North Staffordshire Rly	94	
12	1867	0-6-0	16½ x 24	5 0	North Staffordshire Rly	95	
13	1867	0-6-0	16½ x 24	5 0	North Staffordshire Rly	96	
14	1867	0-6-0	16½ x 24	5 0	North Staffordshire Rly	97	
15	1867	0-6-0	16½ x 24	5 0	North Staffordshire Rly	98	
16	1867	0-6-0	16½ x 24	5 0	North Staffordshire Rly	99	
17	1867	0-6-0	16½ x 24	5 3	Great Eastern Rly	437	
18	1867	0-6-0	16½ x 24	5 3	Great Eastern Rly	438	
19	1867	0-6-0	16½ x 24	5 3	Great Eastern Rly	439	
20	1868	0-6-0	16½ x 24	5 3	Great Eastern Rly	440	
21	1868	0-6-0	16½ x 24	5 3	Great Eastern Rly	441	
22	1868	0-6-0	16½ x 24	5 3	Great Eastern Rly	442	
23	1868	0-6-0	16½ x 24	5 3	Great Eastern Rly	443	
24	1868	0-6-0	16½ x 24	5 3	Great Eastern Rly	444	
25	1868	0-6-0	16½ x 24	5 3	Great Eastern Rly	445	
26	1868	0-6-0	16½ x 24	5 3	Great Eastern Rly	446	
27	1867	0-6-0	17 x 24	4 6	Bristol & Exeter Rly	77	
28	1868	0-6-0	17 x 24	4 6	Bristol & Exeter Rly	78	
29	1868	0-6-0	17 x 24	4 6	Bristol & Exeter Rly	79	
30	1868	0-6-0	17 x 24	4 6	Bristol & Exeter Rly	80	
31	1868	0-6-0	17 x 24	4 6	Bristol & Exeter Rly	81	
32	1868	0-6-0	17 x 24	4 6	Bristol & Exeter Rly	82	
33	1868	2-4-0T	15 x 24	6 4	Bristol & Exeter Rly	83	
34	1868	2-4-0T	15 x 24	6 4	Bristol & Exeter Rly	84	
35	1868	0-6-0T	20 x 24	4 0	Metropolitan Rly	34	
36	1868	0-6-0T	20 x 24	4 0	Metropolitan Rly	35	
37	1868	0-6-0T	20 x 24	4 0	Metropolitan Rly	36	
38	1868	0-6-0T	20 x 24	4 0	Metropolitan Rly	37	
39	1868	0-6-0T	20 x 24	4 0	Metropolitan Rly	38	
40	1868	0-6-0	16½ x 24	5 3	Great Eastern Rly	447	
41	1868	0-6-0	16½ x 24	5 3	Great Eastern Rly	448	
42	1868	0-6-0	16½ x 24	5 3	Great Eastern Rly	449	
43	1868	0-6-0	16½ x 24	5 3	Great Eastern Rly	450	
44	1868	0-6-0	16½ x 24	5 3	Great Eastern Rly	451	
45	1868	0-6-0	16½ x 24	5 3	Great Eastern Rly	452	
46	1868	0-6-0	16½ x 24	5 3	Great Eastern Rly	453	
47	1868	0-6-0	16½ x 24	5 3	Great Eastern Rly	454	
48	1869	0-6-0	16½ x 24	5 3	Great Eastern Rly	455	
49	1869	0-6-0	16½ x 24	5 3	Great Eastern Rly	456	
50	1869	0-6-0	16½ x 24	5 3	Great Eastern Rly	457	
51	1869	0-6-0	16½ x 24	5 3	Great Eastern Rly	458	
52	1869	0-6-0	16½ x 24	5 3	Great Eastern Rly	459	
53	1869	0-6-0	16½ x 24	5 3	Great Eastern Rly	460	
54	1869	0-6-0	16½ x 24	5 3	Great Eastern Rly	461	
55	1869	0-6-0	16½ x 24	5 3	Great Eastern Rly	462	
56	1869	0-6-0	16½ x 24	5 3	Great Eastern Rly	463	
57	1869	0-6-0	16½ x 24	5 3	Great Eastern Rly	464	
58	1869	0-6-0	16½ x 24	5 3	Great Eastern Rly	465	
59	1869	0-6-0	16½ x 24	5 3	Great Eastern Rly	466	
60	1869	0-6-0	16½ x 24	5 3	Great Eastern Rly	467	
61	1869	0-6-0	16½ x 24	5 3	Great Eastern Rly	468	
62	1869	0-6-0	16½ x 24	5 3	Great Eastern Rly	469	

Works No	Date	Type	Cylinders inches	D.W. ft in	Customer	No	
63	1869	0-6-0	16½ x 24	5 3	Great Eastern Rly	470	
64	1869	0-6-0	16½ x 24	5 3	Great Eastern Rly	471	
65	1869	0-6-0	16½ x 24	5 3	Great Eastern Rly	472	
66	1869	0-6-0	16½ x 24	5 3	Great Eastern Rly	473	
67	1869	0-6-0	16½ x 24	5 3	Great Eastern Rly	474	
68	1869	0-6-0	16½ x 24	5 3	Great Eastern Rly	475	
69	1869	0-6-0	16½ x 24	5 3	Great Eastern Rly	476	
70	1870	?	?	?	For Russia		15 ordered Apr. 1870

bunker, and being fitted with well and back tanks. The leading and trailing wheels were equipped with Adams' radial axle boxes. The driving wheels were 6′ 4″ diameter.

The most well known locomotives built by the Worcester Engine Company were five for the Metropolitan Railway. They were 0-6-0Ts designed by R. H. Burnett, the Chief Mechanical Engineer of the railway. They were powerful locomotives with 20″ x 24″ cylinders (the largest in the country at that time), double frames and 4 ft wheels. The size of the cylinders were designed to compensate for the drop in boiler pressure due to the condensing in the tunnels and absence of blast. It was found that they were too powerful for the duties required and were sold in 1873 and 1875.

An order for fifteen locomotives (type unknown) for Russia was received and the first one despatched in 1870. It is not known if the order was completed or if another builder took over the other fourteen. The business failed and was sold in September 1872 and part of the premises were occupied by Mackenzie and Holland and part by Heenan & Froude.

WORSDELL, THOMAS

Berkeley Street, Birmingham

This firm manufactured cranes, overhead traversers, winches, jacks, material testing machines, and a few steam rollers. According to records of the Potteries, Shrewsbury and North Wales Railway, Worsdell supplied one or more locomotives to this railway, but no details are available and it is probable that he was an agent in the 1860s for a locomotive manufacturing firm. Having built steam cranes, it is possible he could have built these – they may have been vertical boilered tank locomotives. At some time the firm was WORSDELL & EVANS.

WREXHAM, MOLD & CONNAH'S QUAY RAILWAY

Rhosddu, Nr Wrexham

A small locomotive repair shop was built here in 1876. Previously all heavy repairs had been carried out by the Lancashire & Yorkshire Railway at Miles Platting and the Cambrian Railway at Oswestry. Nevertheless money had not been available for proper equipment and the locomotive department led a hand-to-mouth existence.

The number of conversions and rebuilds were legion, and perhaps the most interesting was the career of an old Manchester & Birmingham Railway 0-6-0, built by Sharp Bros in 1846, becoming LNW 431, being rebuilt as a saddle tank and numbered 1829 and then purchased in that form by the WMCQ in 1876. It was given the number 6 and named QUEEN. In 1880 she became a 0-8-0ST but because of the long rigid wheelbase the trailing coupled wheels were replaced by a radial truck in 1888 so becoming a 0-6-2ST. After an accident in 1890, which almost reduced the locomotive to scrap, rebuilding took place with new frames and boiler, and was completed in 1892 as a 0-8-0ST. This was practically a new locomotive although the frames and boiler would have been bought out. A number of other locomotives were rebuilt almost as drastically as No. 6.

The other claimant as a 'new' locomotive started life on the South Staffordshire Railway as a Stephenson long-boilered 0-6-0 No. 12 PELSALL. Built in 1851, and taken over by the LNWR in 1860, she became Southern division No. 309, was renumbered 909 in 1862, was rebuilt as a 0-6-0ST in 1865 and renumbered 1188 in 1867, and again renumbered 1806 in 1871. Sold to a contractor she was eventually purchased by the WMCQ in 1876 and was numbered 7 and then 3.

WREXHAM, MOLD & CONNAH'S QUAY RAILWAY

Her metamorphosis started in 1882, when she became a 0-6-2ST, and after being condemned in 1899 it was decided to use all the parts that could be salvaged, to help in building another No. 3 which was completed in 1901 as a 2-6-0T. This was more of a new locomotive than No. 6 and a neat looking side tank was the result, with domed boiler, round top firebox, and all over cab with side windows. The leading wheels had coiled springs above the running plate. The inside cylinders were 18" x 24" and the coupled wheels 4' 8" diameter. In 1904 she became Great Central Railway No. 400B and was then scrapped in 1907. No. 3 had the distinction of being the only 2-6-0T built in Great Britain to the standard gauge with inside cylinders. The Locomotive Superintendent during this period was Frederick Willans.

WYLAM COLLIERY

See TIMOTHY HACKWORTH, STEPHENSON and TREVITHICK.

YORK, NEWCASTLE & BERWICK RAILWAY

Gateshead

The original works were completed in 1844 by the Newcastle & Darlington Junction Railway. In 1847 the York, Newcastle & Berwick Railway was formed by the amalgamation of the Newcastle & Darlington Junction Railway and the York & Newcastle Railway. Gateshead was made the locomotive headquarters of the new railway. A 0-6-0 with $14\frac{1}{2}$" x 22" cylinders and 4' 6" diameter wheels was built in 1849 but whether this was a completely new locomotive is not known.

New workshops were built in 1853–4 on a 30 acre site with a fitting and erecting shop 349 ft long and 88 ft wide. There was no iron foundry and all heavy castings, such as cylinders and wheel centres, were bought out as also were heavy forgings.

The York, Newcastle & Berwick Railway, together with the Leeds Northern and York & North Midland railways were the main constituent companies which amalgamated in 1854 to form the NORTH EASTERN RAILWAY.

Gateshead was made the headquarters of the locomotive department of the new railway and Edward Fletcher was the Locomotive Superintendent, and for some time allowed the York shops of the Y&NMR and the Leeds shops of the LNR to have almost complete freedom of action and this also applied to the Stockton & Darlington's shops at Darlington and Shildon when they too were absorbed in 1863. No further locomotives can be traced as being built before the NER took over at Gateshead.

Fig. 574 **Wrexham Mold & C.Q. Rly** 1901 No. 3 2-6-0T using parts of former No. 3

YORK, NEWCASTLE & BERWICK RAILWAY

Fig. 575 1884 McDonnell's *Class 38* No. 576

In 1859 a start was made on four *450* class 2-2-2s which were enlargements of the *220* class supplied by Robert Stephenson & Company in 1854. They had double frames and inside cylinders 16″ x 22″. The boiler had a flush top firebox on which was mounted a plain manhole cover. The combined dome and spring safety valves were mounted centrally on the boiler barrel. These were the last 'singles' to appear on the NER until 1888. The standard express passenger type was to be the 2-4-0.

In 1860 the first of the *162* class 2-4-0s was built, with outside frames, cylinders 16″ x 22″ and 6 ft diameter coupled wheels. Nine were built from 1860 to 1870 with a number of variations but at the time of building they were the most powerful passenger class run by the company.

Following these nine, a series of thirty-four smaller 2-4-0s were built from 1870 to 1875. They were presumably intended for lighter duties with 15″ x 22″ cylinders and 5′ 6″ diameter wheels. They were known as the *675* class and suffered from lack of steaming capacity with small boilers. They had inside frames. In this period no fewer than eight different classes had been built, some by outside contractors, and in 1871, Fletcher introduced a further class of 2-4-0s known as the *901* class. They were larger than previous classes and had 7 ft diameter coupled wheels, with cylinders varying from 17″ x 24″ for the initial fifteen, then 17½″ x 24″ for the next eight and 18″ for the remainder. Twenty were built by outside contractors and thirty at Gateshead from 1872 to 1882.

They had inside frames and cylinders, very square cabs with flat tops. The 4′ 3″ diameter boilers had a heating surface of 1097 sq. ft and a grate area of 15·3 sq. ft. This class, which differed in detail from maker to maker, was of robust and handsome proportions. A feature was the fitting of Fletcher's exhaust cocks which could be manually opened to divert some of the exhaust from the chimney and so soften the blast. It was the removal of these fittings that made the next Locomotive Superintendent, McDonnell, so unpopular. Another device fitted was the combined lever and screw reversing gear – evolved by R. Stephenson & Company.

For the ever increasing demands for freight working various classes of double framed 0-6-0s were supplied – the majority by outside firms. The next step was an inside framed 0-6-0, first built in 1872 and eventually totalling 324 locomotives. Known as the *398* class they were built up to 1881 and 160 had been built by five different locomotive building contractors, the rest being dealt with by Gateshead and Darlington. General dimensions were: inside cylinders 17″ × 24″, 5 ft diameter wheels and a boiler heating surface of 1138 sq. ft, with a grate area of 17·0 sq. ft. For such a large class, from diverse builders, there was bound to be variations, although an endeavour was being made at this period to introduce a certain amount of standardisation, particularly with boilers.

For branch line work Fletcher designed the *BTP* 0-4-4WT of which 130 were built from 1874 to 1884. Gateshead's share was sixty-five, built from 1876 to 1883. They were all of the same basic design with cylinders 16″ x 22″ and 5 ft diameter coupled wheels. The boiler heating surface for the majority was 1074 sq. ft and working pressure 140 lb/sq. in. The main variation was the coupled wheels diameter which had examples of 5′, 5′ 3″, 5′ 6″ and 5′ 8″. They were very successful and forty-six lasted long enough to be taken over by the L&NER, the last one not being withdrawn until 1929.

In 1876 Fletcher introduced a 6 ft version of his *901* class 2-4-0 known as the *1440* class of which fifteen were built at Gateshead in 1876–82. They were used on the more steeply graded sections. This was Edward Fletcher's last design and in 1883 he retired. He had built simple and robust designs, adhering to the 2-4-0 for main line express working. He had inherited a tremendous variety of locomotives, but no well laid down standardisation policy came from him and infinite variety was added, rather than the establishment of fewer classes.

The new man, Alexander McDonnell, coming from the Great Southern & Western Railway of Ireland, had been instrumental in bringing Inchicore works to a state of high efficiency and a certain amount of standardisation in the locomotive stock. His short stay at Gateshead was a series of clashes with the design, workshop and footplate personnel which resulted in his resignation in 1884. McDonnell's 4-4-0 and 0-6-0 classes are briefly described in the NER-Darlington section.

In 1885 Thomas William Worsdell took office, and all new classes of passenger locomotives were built as compounds on the Worsdell-von Borries principle. His first design was a 2-4-0 with one HP cylinder 18″ x 24″ and one LP 26″ x 24″ fitted with slide valves and operated by Joy's valve gear, which was standard practice. The coupled wheels were 6′ 8″. Known as *Class D* – Worsdell introduced a lettered locomotive classification – only two were built, in 1887–8, at Darlington.

From this class he developed a 4-4-0 on the same lines, with boiler interchangeable with the *Class D*. He built twenty-five, known as *Class F* and ten as simple engines (*F1*) so that a comparison could be made of their relative performances. Worsdell introduced the continuous splasher over the driving wheels which became a characteristic of most classes. The running of the compounds satisfied Mr Worsdell, and the first two 'singles' since 1863 were built in 1888, followed by three in 1889, and five in 1890. As *Class I*, they were 4-2-2s with inside bearings to all wheels. The inside cylinders were 18″/26″ x 24″ and they had 7 ft diameter driving wheels. The 4′ 3″ diameter boiler gave a heating surface of 1126 sq. ft and the weight in working order was 43 tons.

At the same time a further class of 'singles' was built at Gateshead with 7′ 7″ diameter driving wheels and cylinders 20″/28″ x 24″ and were actually intended to replace the existing 4-4-0s, on the York to Edinburgh expresses. To get both cylinders between the frames the valve chests were fitted outside the frames and were operated by rocking levers from the Joy's valve gear fitted inside the frames. The position of the valve chests gave trouble due to their exposed position but this class, known as *J*, gave a good account of itself, and ten were built in 1889–90. They were rebuilt in 1894–5 as two-cylinder simples with piston valves and Stephenson's link motion.

For local suburban traffic T. W. Worsdell introduced the 2-4-2T for the first time. They were *Class A* with 17″ x 24″ cylinders and 5′ 6″ diameter coupled wheels. Webb's radial axleboxes were used for the leading and trailing axles and they had Joy's valve gear. The boiler, which became a standard type, had a working pressure of 160 lb/sq. in. with a heating surface of 1092 sq. ft. They were handsome and well proportioned tanks, and did service for over forty years in most cases. Later batches had 18″ diameter cylinders. A total of sixty was built at Gateshead from 1886 to 1892.

For goods traffic the *Class C* compound 0-6-0 was built in large numbers. Out of a total of 171 built, Gateshead built 141 from 1886 to 1892. They had cylinders 18″/26″ x 24″ and, for comparative tests, thirty simple *Class C1* 0-6-0s were built, with 18″ x 24″ cylinders, at Gateshead in 1886–95. The compounds did excellent service and a considerable fuel saving was claimed. T. W. Worsdell's successor, however, converted most of his compounds to simples.

Two 0-6-0T crane locomotives, *Class H1* were built for use in the works in 1888. No. 590 had 13″ x 20″ cylinders and No. 995 had 14″ x 20″. Small domeless boilers were fitted and the wheels were 3′ 5″ diameter.

For dock shunting at Hull five small 0-4-0Ts were built with 11″ x 15″ inside cylinders and marine type boilers. They weighed 17 tons 15 cwt in working order and were built in 1890 as *Class K*.

For general shunting a six-coupled side tank was designed and building took place from 1886–95, by which time 120 had been completed. Inside cylinders 16″ x 22″, wheels 4′ 6″,

Fig. 576 1899 Worsdell *Class R* No. 2013

and a small boiler with 731 sq. ft of heating surface, and grate area 11·25 sq. ft were fitted.

T. W. Worsdell retired through ill-health in 1890, his place being taken by his brother Wilson.

Wilson Worsdell had been at Gateshead since 1883. Building of *Class C* 0-6-0 compounds continued but in 1894 the *Class C1* was produced with two 18″ x 24″ cylinders, but was otherwise similar to the *Class C*.

The first new design of express passenger locomotive appeared in 1892 as *Class M1* 4-4-0 with two inside cylinders 19″ x 26″ and 7′ 1″ diameter coupled wheels. As with the *Class J* 4-2-2 compounds, the steam chests were fitted outside the frames. The eccentrics were fitted on the driving axle inside the frames and the valves were operated by rocking shafts through the frames. These were large machines with boilers giving 1341 sq. ft of heating surface and a grate area of 19·5 sq. ft. The weight was 50 tons 15 cwt. Although Wilson Worsdell had officially abandoned compounding, one was built dimensionally similar to the *M1s* with a high pressure cylinder 19″ x 26″ and low pressure 28″ x 26″. This was No. 1619, *Class M* built in 1893 for comparing its performance with the simple locomotives of which twenty were built from 1892 to 1894, the last one being fitted with patent segmental piston valves.

The two-cylinder compound was rebuilt with three cylinders. The high pressure cylinder between the frames remained at 19″ x 26″ and the two outside low pressure cylinders were 20″ x 24″. All cylinders drove on the one axle. A larger 4′ 6″ diameter boiler was fitted with a pressure of 220 lb/sq. in. which included seventeen water tubes in the firebox. The total heating surface was 1328 sq. ft and the grate area 23 sq. ft. They were all excellent locomotives and took charge of the East Coast expresses.

Another class with smaller boilers, known as *Q* class, was built in 1896–7. The cylinders however were 19½″ diameter instead of 19″ and despite this anomaly they performed very efficiently. An innovation was the large cab with a clerestory roof with ventilators, and the chimney had a copper cap.

Two more 4-4-0s formed yet another class *Q1* with 7′ 7¼″ diameter coupled wheels. Built in 1896, it is said that they were specially built to break more records in the 'Race to the North', but by the time they were ready the races, by agreement with the rival companies, were over.

For mineral traffic – and the North Eastern Railway hauled the largest tonnage of coal and mineral traffic of any railway in the United Kingdom – a new series of 0-6-0s was put in hand in 1894 and were known as the *P* class. Gateshead built fifty and Darlington twenty from 1894 to 1898. A boiler, standard with the *A* 2-4-2Ts, *G* 4-4-0s and *L* 0-6-0Ts, was fitted, the diameter being 4′ 3″ and the barrel 10′ 3″ long. Two inside cylinders 18″ x 24″ with slide valves and Stephenson's valve gear were fitted, Worsdell preferring that type of motion to Joy's. The wheels were 4′ 7¾″ diameter.

A larger type of 0-6-0 was then built in 1898 as the *P1* class with a 5′ 1¼″ diameter boiler as used on his brother's *Class C* 0-6-0. The cylinders were increased to 18¼″ x 26″. Eighty were built at Gateshead and forty at Darlington from 1898 to 1902.

In 1904 a further class of 0-6-0 was designed the first batch of thirty being built at Darlington in 1904–5 and twenty at Gateshead in 1905.

These were much larger locomotives with cylinders 18½″ x 26″ diameter, and boiler with a total heating surface of 1658 sq. ft and grate area 20 sq. ft. Weight in working order was 46 tons 16 cwt. Four Ramsbottom safety valves were fitted in pairs and were surrounded by a

Fig. 577 1906 four cylinder compound *Class 4CC* 4-4-2 No. 730

large brass casing which was a distinctive feature of this class and others to follow. These were powerful machines, and with the *P* replaced many of the older six-coupled mineral classes. Worsdell's last 0-6-0 class was the *P3* but apart from minor modifications they were the same as the *P2s* – none were built at Gateshead.

A larger 4-4-0 *Class R* was introduced in 1898 with 19″ x 26″ inside cylinders, piston valves, 6′ 10″ diameter coupled wheels, and 4′ 9″ diameter boiler giving a total heating surface of 1527 sq. ft and a working pressure of 200 lb/sq. in. The outside admission piston valves were offset and under the cylinders and were 8¾″ diameter. Thirty were built from 1899 to 1902, all at Gateshead. The long elliptical splashers were fitted as for previous classes with brass beading following the contour of each coupled wheel. They were extremely reliable and useful locomotives and did well on Scottish expresses.

In the same year as the first *R* class was being constructed, the first six-coupled express passenger locomotives were being built. They were the *S* class with two outside cylinders 20″ x 26″, slide valves fitted to the first three (Nos. 2001–2003) and a 4′ 9″ diameter boiler with 1769 sq. ft total heating surface. The grate area was 23 sq. ft. The appearance of this class was marred by a shortened cab with one window either side, so that the locomotive could be turned on a 50 ft turntable. The fourth and subsequent locomotive of the class had a footplate two feet longer and had the customary double window cabs. A total of ten were built from 1899 to 1900 and numbered 2001 to 2010. No. 2006 was sent to the Paris Exhibition in 1900 and was awarded a gold medal. The last five were fitted with piston valves and all the remainder of the class were fitted with them shortly afterwards.

Five *Class S1* 4-6-0s were built in 1901 similar to the later *S* class but with 6′ 8¼″

Fig. 578 1906 Two cylinder *Class S* 4-6-0 No. 763 designed for heavy East Coast, Newcastle-Edinburgh services

diameter coupled wheels. Neither class were an unqualified success, the *S1s* proved freer running, a significant factor being the diameter of the blast pipe – 5″ for the *R* 4-4-0s and the two 4-6-0s – and in overall performance the *R* class was their superior.

Returning to the freight side, in 1901 the first eight-coupled mineral locomotive was built by Wilson Worsdell. Standard diameter wheels 4′ 7¼″, two outside cylinders 20″ x 26″ with piston valves, and a 4′ 9″ diameter boiler, with 1675 sq. ft of heating surface, were fitted. These locomotives were painted, polished and finished in the same way as the passenger types, with brass top chimneys, brass safety valve covers, company's crest and fully lined out. They were indeed handsome – and handsome is as handsome does – their haulage capacity was tremendous. They were designed to haul sixty wagon trains which they did with ease and on a trial trip from Tyne Dock one of this class hauled a rake of wagons giving a gross load of 1326 tons. They were classed *T* and forty were built at Gateshead from 1901 to 1904. Ten *Class T1* were built in 1902 exactly the same, except for slide valves being fitted instead of piston valves. They were built for comparative purposes and in tests there was little to choose. Coal consumption was slightly less for the slide valve type and maintenance costs were less.

The first *Atlantic* was built in 1903. Far larger than the 4-6-0s, the *Class V* had two outside cylinders 20″ x 28″ with piston valves, a 5′ 6″ diameter boiler giving a greatly increased heating surface of 2455·8 sq. ft and a grate area of 27 sq. ft. The weight in working order was 72 tons. The coupled wheels were 6′ 10″ diameter. To test this class, Mr Churchward lent the Great Western dynamometer car so that Mr Worsdell could get all the technical data he required for the *Atlantics* and other classes. Nine more were built in 1904.

In 1906 two *Atlantic* Smith compounds were built with two outside high pressure cylinders 14¼″ x 26″ and two LP inside 22″ x 26″, a boiler 5′ diameter – smaller than the previous *Atlantics* – with a total heating surface of 1991 sq. ft, and a grate area 29 sq. ft. A striking deviation from NER firebox design was the introduction of the Belpaire type on these two *Class 4cc* locomotives. The valve gear for HP and LP cylinders operated the HP cylinders through rocking shafts. No. 730 had Stephenson's link motion and the second, No. 731 Walschaerts valve gear. These were magnificent locomotives and in running proved superior to the *V* class. So far no locomotive had been equipped with a superheated boiler.

W. M. Smith had been the mastermind behind these and previous compounds, besides those initial 4-4-0s on the Midland, and all credit is due to him for the successful designs he had built – not without some opposition.

Wilson Worsdell had designed many classes and seemed no nearer standardisation than at the beginning, but in most cases each successive design was an improvement on the last.

In 1907 five 4-6-0Ts were built, *Class W*, and known as 'Whitby' tanks they were designed for the Middlesbrough/Scarborough line with its heavy gradients. Unlike most classes of Worsdell's they had inside cylinders 19″ x 26″ and coupled wheels 5′ 1″ with a moderate sized boiler. The bunker capacity was meagre and they were converted to 4-6-2Ts giving increased bunker capacity, by Vincent Raven.

The last design to be carried out and built at Gateshead was for ten *Class X* 4-8-0Ts for shunting and hump duties. They had three cylinders 18″ x 26″ and 4′ 7″ coupled wheels and again a moderate size boiler similar to the 4-6-0Ts. A feature of the *X* class was the monobloc casting for all three cylinders. They were built in 1909 and 1910 and so ended all new construction at Gateshead. The chief drawing office was transferred to Darlington and only rebuilding and repairs carried out until 1932, when they had a short, new lease of life during the 1939–45 war, but finally closing in 1959. For the continuation of NER locomotive building see NORTH EASTERN RAILWAY – DARLINGTON.

Gateshead works

Until 1910 Gateshead had been the locomotive headquarters of the North Eastern Railway, but like so many other factories, expansion had reached its limit. They were hemmed in by the main line and the river, and although various extensions had been made from time to time, including a new repair shop 280′ x 185′ in 1883–4, with the increased output required and the growth in size and weight of the locomotives being built the company decided to make Darlington the headquarters for all new building work and the CME's staff was gradually transferred there, leaving Gateshead to carry out overhauls only. The last new locomotive to leave the Works was *Class X*, 4-8-0T No. 1359 in 1910.

Information on the locomotives built in the 1850s and 1860s is incomplete. Over 1000 locomotives have been built at Gateshead and from existing data the approximate total is 1023 but this is more an under-estimate due to lack of knowledge of the early building.

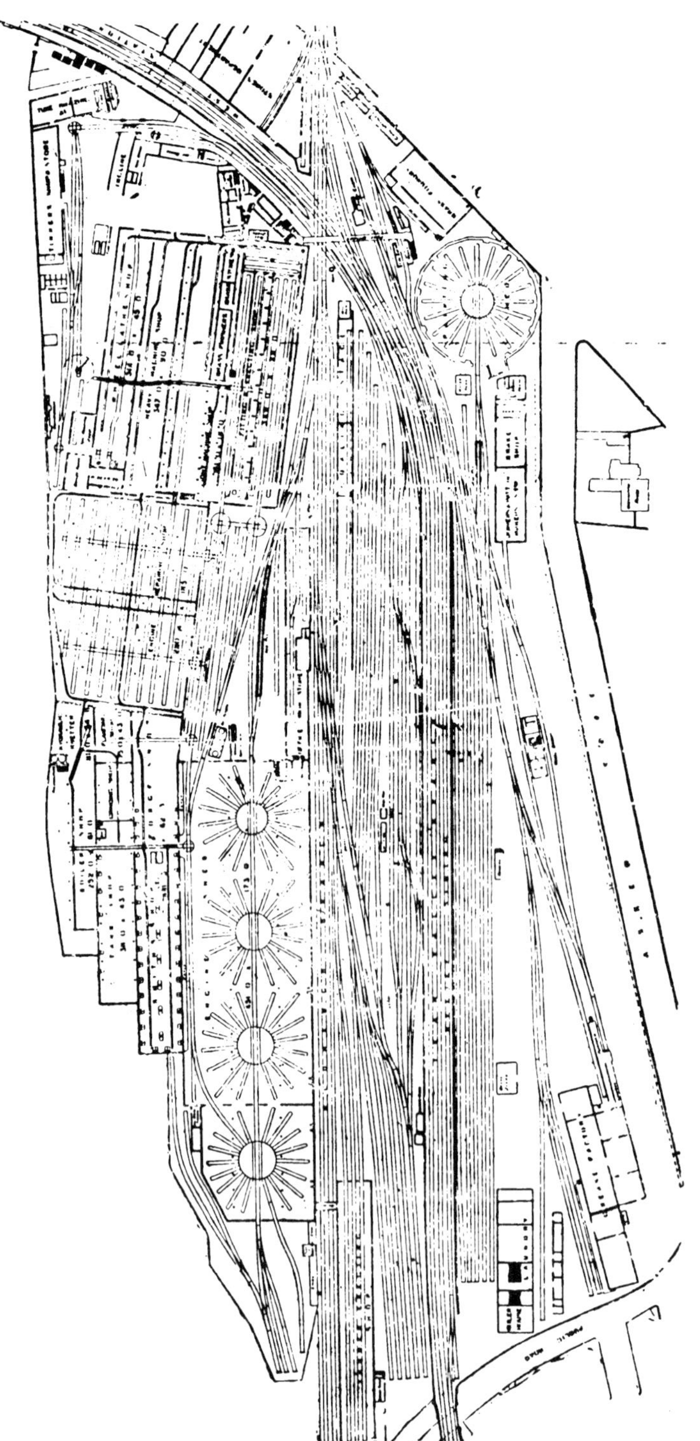

Fig. 579 Gateshead Works

YORK, NEWCASTLE & BERWICK RAILWAY

No works numbers were issued until 1885 when a system was introduced giving a works number and the last two numbers of the year, i.e. 2/86 denoted the second built in 1886. This did not run true to form as some locomotives due to be built in one particular year were completed the following year but carried the previous years digits. This lasted until 1903 when a new series of plain works numbers was introduced, ending with No. 123 in 1910.

All heavy castings such as cylinders, and wheel centres were obtained from other manufacturers as were large forgings and stampings. There were two erecting shops one used for new building and the other for repairs, the latter having accommodation for sixty locomotives.

In August 1932 the company was compelled to close the works, due to the falling off of traffic but during the 1939–45 war they were reopened for repair work and finally closing down in 1959.

An important step in safety measures was introduced in 1896, the work of Vincent Raven who at that time was Locomotive Running Superintendent. It was a form of cab signalling apparatus, consisting of a trip gear operated from the track, causing a warning whistle to operate in the cab when a signal was at danger. Many route miles on the main lines were equipped with this gear and later Raven developed a more sophisticated system, using ramps, with electrical circuits and visual signals in the cab. It was very successful but was discontinued at the time of grouping.

YORKSHIRE ENGINE COMPANY
Meadowhall Works, Sheffield

Established in 1865 the firm became well known for its haulage engines, and a lot of work was done for collieries and quarries. Locomotive building commenced in 1866 and an interesting note in *Engineering* for 1 January, 1866 quotes 'it is intended to complete no less than 400 locomotives yearly, or more than one a day is progressing towards completion'. However it was not until 1891, twenty-five years later that the four hundredth locomotive was turned out of the Meadowhall works.

The first order was for three 2-4-0s for the Great Northern Railway with inside cylinders 17″ x 24″ and 7 ft driving wheels. Ten similar machines were built in 1868 but with 6′ 7″ diameter driving wheels. Meanwhile a series of 0-6-0 tender locos were put in hand for the East Indian, and Great Indian Peninsula Railways. At this time the bulk of their orders came from abroad and locomotives were delivered to the following railways: Jamaica Government; Tamboff & Kosloff; Lemburg & Czernowitz; Poti & Tiflis; Moscow & Riazan; Buenos Ayres Great Southern; Victoria Government, Australia; Mexican; Rio de Janeiro; Compostelano; Peruvian; and Trancaucasian.

The first industrial locomotives for the home market were three 0-4-0ST for the Earl Fitzwilliam colliery in 1869.

In 1872 ten 0-6-6-0 Fairlies were built for the Mexican Railways, having four cylinders 16″ x 20″ and 3′ 9″ diameter wheels for the first five, and 3′ 6″ diameter for the second batch of five.

A considerable number of Fairlie type locomotives were built including: thirty-five 0-6-6-0s; three 0-4-4-0s and two 2-6-6-2s, all built between 1872 and 1908.

Fig. 580 **Yorkshire Engine Co.** 326/1881 Chatterley-Whitfield Colliery Ltd. POLLIE

Fig. 581 **Yorkshire Engine Co. 382/1884** for Buenos Ayres Great Southern Rly No. 25 (5′ 6″ gauge)

Fig. 582
Yorkshire Engine Co.
947/1907 Goldendale Iron Co. Ltd.
CLIFFORD 3′ 3″ dia wheels – solid centres

Fig. 583
Yorkshire Engine Co.
1945/1924 Nitrate Rlys (Chile) No. 101

From 1885, industrial tank locos were predominant with occasional batches of main line types including ten 4-6-0s for the 3′ 6″ gauge Queensland Government Railways and fifteen 0-6-0s for the Hull, Barnsley and West Riding Junction Railway and Dock Company.

In the early 1900s the orders received turned the other way and the larger types were built (comparatively speaking): 0-6-0s for Bombay Baroda & Central India Railway; 0-6-0s for Great Central Railway; 0-8-0s for Hull & Barnsley Railway; 0-6-6-0 Fairlies for the Nitrate Railways of Chile; and 4-4-2Ts for North British Railway.

From 1913 to 1927 only fifty-eight complete locomotives were built, but the works were busily engaged on building haulage gear and equipment, for the mining industry, and locomotive boilers.

In 1925 the Yorkshire Engine Company acquired the sole rights of the Poultney patent. The only locomotive built under this patent by the firm was a conversion of the Ravenglass and Eskdale Railway's RIVER ESK to a 2-8-2 – 0-8-0. The eight coupled unit was under the tender portion.

In 1930 during the industrial depression an order for twenty-five GWR 0-6-0PTs was received and during this period a few industrial tank locomotives were built. Besides complete locomotives the firm built many locomotive boilers and produced many iron and steel forgings. Locomotive repairs were also carried out.

After the Second World War the company was acquired by the United Steel companies, who in 1948 took over the management. At this time the capacity of the works was thirty new locomotives per annum and the works were re-organised and re-equipped and orders for spares and work other than locomotives was undertaken.

In 1949 an order for fifty 0-6-0PTs was received from British Railways, Western Region and this was completed in 1956. BR No. 3409 (YE Works No. 2584) was the last of the batch and the last steam locomotive built.

The first diesel electric shunter was produced in 1949 and from 1956 work was concentrated on diesel electric and diesel hydraulic units for industry. In 1965 Rolls Royce Ltd. took over the firm and work was transferred to their Sentinel works at Shrewsbury. Nearly 800 steam locomotives had been built. Works numbers were used from the start and Nos. 1 to 358 were exclusively locomotives, after which spares, boilers and other machinery were included in the same series of numbers.

STANDARD INDUSTRIAL LOCOMOTIVES 1950 STANDARD GAUGE

Dimensions	TYPE IV 0-4-0ST	TYPE V 0-4-0ST	TYPE VI 0-6-0ST	TYPE VII 0-6-0ST
Cylinders (outside) - inches	14 x 22	16 x 22	16 x 22	18 x 24
Wheel diameter - ft in	3 6	3 6	3 8	3 8
Wheelbase - ft in	5 6	6 6	11 0	11 6
Total Htg Surface sq ft	660	812	807	1031
Grate Area sq ft	10	14	14	18
Boiler Pressure lb/sq in	180	180	180	180
Weight in working order - tons	31	37	42½	55
Water capacity—gallons	850	1000	1185	1220
Fuel capacity—tons	¾	1¼	1¾	1
Tractive effort @ 75% B.P.	13860	18103	17280	23858
Ratio of Adhesion	5	4·5	5·5	5
Max Height - ft ins	10 11½	11 2⅜	11 5	11 11½
Max Width Outside Cylinders - ft ins	7 11½	8 4⅛	8 4⅛	8 6¼

YORKSHIRE PATENT STEAM WAGON COMPANY, THE

Pepper Road, Hunslet, Leeds

This firm was noted for its steam road wagons but in common with others, in the early 1930s some steam units were built for rail cars, comprising boiler and engine. In 1931 a unit was supplied to the Birmingham Railway Carriage and Wagon Co. Ltd., who were building an articulated rail car for the Entre Rios Railway.

The Yorkshire boiler had a horizontal shell with double return tubes, central vertical flue and firebox underneath. Smokebox doors were fitted at each end of the boiler barrel to give access to the tubes. The heating surface was 158 sq. ft, combustion volume 21 cub. ft. and working pressure 275 lb/sq. in. A superheater was fitted and firing was by means of an oil burner.

The engine had three vertical cylinders $5\frac{3}{4}''$ x 8″, double acting with Joy's valve gear. The totally enclosed crankshaft drove a jackshaft through two trains of gears. The coupled wheels were 3′ 6″ diameter. Similar units were supplied to the same carriage builders for articulated rail cars ordered by Egyptian State Railways, Belgian National Railways and others.

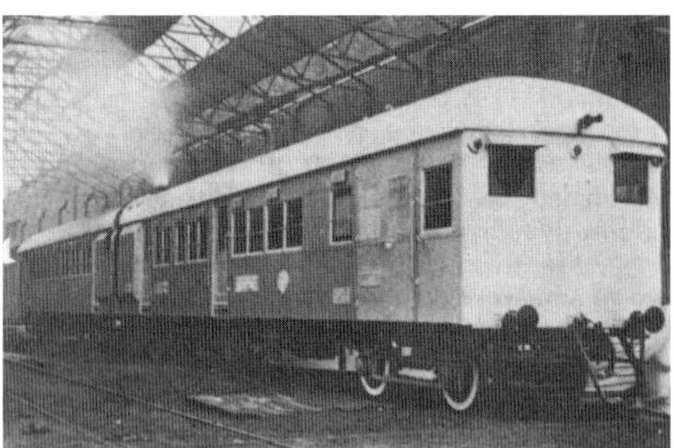

Fig. 584 **Yorkshire Patent S.W.** *c*1930 articulated rail unit

YOUNG, JOHN & THOMAS

Vulcan Foundry, Newton Green, Ayr

This firm was principally engaged in the manufacture of ships' auxiliary engines, colliery and agricultural machinery and, at one period, employed 800 men. In the early 1860s they built a small colliery locomotive and the only reference to it appeared in the *Ayr Advertiser* for 9 March, 1863 – 'we believe this to be the first locomotive engine ever to be made in Ayr – about 25 horse power; no tender; water tank above boiler, coal boxes each side of the driver'. It was built for Messrs Taylor, Ayr Colliery, which was believed to have had standard gauge track and was closed about 1870.

In the *Engineer* dated 24 February there was an advertisement for a sale of contractor's plant at Crick Station, Northants; 'Locos by Black Hawthorn & Co., Manning Wardle & Co., J. & T. Young'. Whether this was the same one that went to Messrs Taylor is not known but it can be said that at least one locomotive was built. They also were invited to tender for locomotives for the Festiniog Railway in 1862.

Fig. 585 **Yorkshire Patent S.W.** *c*1930 engine unit for railcar

BIBLIOGRAPHY
and main reference sources

Ahrons E. L. *The British Steam Locomotive from 1825 to 1925*. Locomotive Pub. Co., 1961.

Aldrich C. L. *Locomotives of the Great Eastern Railway 1862–1962.* 1969.

Bennett A. R. *The Chronicles of Boulton's Siding.* Locomotive Pub. Co., 1927.

Beyer Peacock Publications, Various.

Birmingham Loco. Club. *Industrial Locomotive Pocket Books*, 1951 onwards.

Boyd, J. I. C. *Narrow Gauge Rails to Portmadoc.* Oakwood Press, 1949.

Burtt G. F. *Locomotives of the London Brighton and South Coast Railway.* 1903.

Clark D. K. *Railway Machinery.* London, 1854.

Clark, E. K. *Kitson's of Leeds.* Locomotive Pub. Co., 1938.

Clark R. H. *Steam Engine Builders of Norfolk.* Augustine Steward Press, 1948.

Clark R. H. *Steam Engine Builders of Suffolk, Essex and Cambridgeshire.* A.S.P., 1950.

Clark R. H. *Steam Engine Builders of Lincolnshire.* Goose and Son, 1955.

Clark R. H. *Midland and Great Northern Joint Railway.* Goose and Son, 1967.

Cox E. S. *British Railways Standard Steam Locomotives.* Ian Allan, 1966.

Colburn Z. *Locomotive Engineering and the Mechanism of Railways.* Collins, 1871.

Dow G. *The Great Central, Vol. 1* 1959; *Vol. 2* 1962; *Vol. 3* 1965. Locomotive Publishing Company.

Higgins S. H. P. *The Wantage Tramway.* The Abbey Press, 1958.

Industrial Locomotive Society. *Steam Locomotives in Industry.* David & Charles, 1967.

Leleux S. A. *Brotherhoods, Engineers.* David & Charles, 1965.

Livesay H. F. F. *Locomotives of the LNWR.* Locomotive Publishing Company, 1948.

Marshall C. F. Dendy. *A History of Railway Locomotives Down to 1831.* Locomotive Publishing Company, 1953.

Marshall C. F. Dendy. *Two Essays in Early Locomotive History.* London, 1928.

Marshall C. F. Dendy. *The First Hundred Railway Engines.* Locomotive Publishing Company. 1928.

MacDermot E. T. *History of the Great Western Railway, Vol. 1* 1927; *Vol. 2* 1931. GWR.

MacLean J. S. *Newcastle and Carlisle Railway.* R. Robinson and Co. Ltd., 1948.

MacLean J. S. *Locomotives of the North Eastern Railway.* R. Robinson and Co. Ltd., 1905.

Mason E. *Lancashire and Yorkshire Railway in the Twentieth Century.* Ian Allan, 1954.

Mitchell Library, Glasgow. *Scotland's Locomotive Builders.* 1971.

Nixon J. H. R. *Brush Traction 1865–1965.* Brush, 1965.

North British Locomotive Company Limited. *A History of N.B. Loco Co.* 1953.

Oakwood Press. *Railway Histories.*

Owen J. A. *A Short History of Dowlais Iron Works.* Merthyr Tydfil, 1972.

Railway Correspondence and Travel Society. *Locomotives of the SER, LCDR & SECR.*

RCTS. *Locomotives of the LB&SCR, Vol. 1* 1969

RCTS. *Locomotives of the L&SWR Vol. 1* 1965; *Vol. 2* 1967.

RCTS. *Locomotives of the GWR* from 1951.

Rolt L. T. C. *A Hunslet Hundred.* David & Charles, 1964.

Sentinel Waggon Works. *Sentinel.* London, 1927.

Smith D. L. *The Dalmellington Iron Company.* David & Charles, 1967.

Stretton C. E. *The Locomotive and its Development 1803–1895.* Crosby Lockwood & Son, 1896.

Stephenson, Robert & Hawthorns Limited. E.E.C., 1944.

Thomas J. *The Springburn Story.* David & Charles, 1964.

Tonks E. S. *The Ironstone Railways and Tramways of the Midlands.* Locomotive Publishing Co., 1959.

Warren J. G. H. *A Century of Locomotive Building, R. Stephenson & Co. 1823–1923.* Andrew Reid and Company Limited, 1923.

Weight R. A. H. *Great Northern Locomotives 1847–1947*, 1947.

Whitcombe H. A. *History of the Steam Tram.* Oakwood Press, 1961.

Wishaw F. *Railways of Great Britain and Ireland.* London, 1842.

Young R. *Timothy Hackworth and the Locomotive.* Locomotive Publishing Company, 1923.

ILLUSTRATION CREDITS

All Illustrations are Copyright and appear by kind permission of the owners:

Abbott, R. 8, 85, 262
Allan, Ian Ltd. (from the Locomotive Magazine) 9, 66, 274, 294, 408, 460, 570
Alliez, G. (the Executors of) 6, 54, 56, 281, 318, 377, 463, 498, 553, 566
Author's Collection 1, 53, 79, 168, 225, 236, 261, 290, 309–312, 314, 362, 364, 375, 386–388, 392, 393, 456, 469, 487
Aveling–Barford 15
Baguley-Drewry 21–23
Beyer Peacock 39
Billington, M. 67, 88, 244, 297, 299, 303, 457, 523, 551, 561, 580
Birmingham Art Gallery 38
Boston, Rev. E., Collection 16, 17, 19, 26, 47–49, 57, 177, 278, 283, 298, 340, 422, 458, 459, 534, 535, 581, 582
British Railways Board 60, 152, 210, 214, 226, 232–234, 248, 265, 304, 325, 432, 477, 478, Bookjacket
Charlton, L. G. 90, 462
Clarke Chapman & Co. 84
Clark, R. H. 62, 292
Clements, R. N. 243
Colvilles Ltd. 158
Craig, A. F. & Co. 91
Davenport, W. J. T. 517
Dow, George (from Great Central Vol. 1) 100
Dow, George (from Great Central Vol. 3) 574
Eccles Public Library (Nasmyth Collection) 409–413
Engineer, The 471–474, 497
English Electric Co. Ltd. 256, 524–530, 541–547
Festiniog Railway Company 129, 136
Fleming, J. M. 252
Fodens Limited 137
Fowler, John, & Co. (Leeds) Ltd. 142, 143
Glamorgan Record Office 245, 246
Glasgow Art Gallery 4, 126, 153
Glasgow Museum of Transport 69
Good, W. Leslie 213, 218, 224, 235, 239, 365, 368–370, 391, 395, 396, 503
Green, Thomas, & Son 241
Griffith, E. 356
Hawthorn, R. & W. Catalogue 253–255
Hick Hargraves & Co. Ltd. 64, 92, 267–271
Houston, J. H. 36, 37
Hume, J. R. 102
Hunslet Engine Company Ltd. 'Frontispiece', 251, 285–289, 555
Imperial Chemical Industries 55
Irish Railway Record Society 34, 97, 103–107, 132, 193, 195, 197, 199, 201–203, 205, 209, 282, 402, 403, 565, 569
Institution of Mechanical Engineers 359, 360, 511, 513, 514
Jones, F. 3, 7, 13, 14, 18, 20, 24, 25, 27, 29, 46, 50, 52, 93, 98, 99, 145, 150, 178, 257, 262, 263, 277, 279, 280, 291, 293, 319–321, 414, 415, 475, 500, 512, 515, 536, 554, 562, 563, 568
L & G R P 76, 128, 211, 212, 407, 537
Lee, M. J. 376
Leicester Library (C. E. Stretton Collection) 78, 121, 131, 135, 138—140, 159, 295, 519, 522, 532
Leicester Musuem 284
Leleux, S. A. 61, 80, 260, 324, 540, 564
Locomotive Publishing Company 2, 51, 227, 240, 250, 276, 301, 305, 385, 445, 479, 516, 548, 567, 572
London Transport Executive 11, 300, 384
Lysaght, J. Ltd. 147, 358
Manchester Public Libraries 40–45
Markham & Co. 453–455
McEwan, Jas. F. 302, 406, 408, 476
Merryweather & Sons Ltd. 380–383
Mitchell Library, Glasgow 488–490
Mowat, J. M. 30–33
Newlands, D. 480
North British Locomotive Company 108–112, 416–421, 424, 425, 427–431, 491, 493, 494
North Thames Gas Board 149
Owen, J. A. 87
Peckett Catalogue 144, 146
Photomatic Ltd. 343, 344, 363
Railway Correspondence and Travel Society 59, 82, 296, 328, 329, 331, 357, 461, 486, 508, 533
Railway Magazine 127, 259, 327, 531, 539, 550, 552, 557, 573, 579
Real Photographs Co. Ltd. 5, 10, 12, 28, 35, 68, 70–75, 77, 86, 89, 94–96, 101, 113–115, 117–120, 122–125, 130, 133, 154–157, 160–167, 169–176, 179–192, 194, 196, 198, 200, 204, 206–208, 215–217, 219–223, 228–231, 237, 238, 266, 272, 273, 306–308, 313, 315–317, 322, 323, 330, 332–339, 341, 342, 345–355, 361, 366, 367, 371, 372, 378, 389, 390, 394, 397–401, 404, 405, 423, 426, 433–444, 446–452, 483, 484, 492, 495, 496, 502, 504–507, 509, 510, 538, 556, 575–578
Robey & Co. 464–467
Round Oak Steel Works 141, 373, 374
Science Museum, London 58, 116, 151, 247, 264, 326, 379, 470, 518, 521, 560, 571
Science Museum, London (Crown Copyright) 63, 65, 81, 134, 148, 249, 258, 520, 558, 559
Sentinel Waggon Works Catalogue 481, 482
Shanks, A. & Son Ltd. 485
Sissons, W. & Co. 499
Smith, David L. 549
Smith, F. D. 468
Weaver, C. R. 275
Yorkshire Engine Co. Catalogue 583
Yorkshire Patent Steam Waggon Co. 584, 585

INDEX

Heavy type relates to Figure numbers, not page numbers

Works

Persons

Persons (*cont.*)

Persons (*cont.*)

Railways

Railways (*cont.*)

Railways (*cont.*)

Railways (*cont.*)

Tramways

Locomotive Types

Locomotive-Types (*cont.*)

Locomotive Names

Locomotive Names (*cont.*)

Locomotive Names (*cont.*)

Locomotive Names *(cont.)*

Locomotive Names (*cont.*)

Locomotive Names (*cont.*)

Locomotive Names (*cont.*)

Locomotive Names (*cont.*)

Customers and other firms

Customers & Other Firms (*cont.*)

Customers & Other Firms (*cont.*)

Customers & Other Firms (*cont.*)

Countries

Countries (*cont.*)

General

General (*cont.*)